Predicting Outdoor Sound

Predicting Outdoor Sound

Second Edition

Keith Attenborough
and
Timothy Van Renterghem

CRC Press
Taylor & Francis Group
Boca Raton London New York

CRC Press is an imprint of the
Taylor & Francis Group, an **informa** business

Second edition published 2021
by CRC Press
6000 Broken Sound Parkway NW, Suite 300, Boca Raton, FL 33487-2742

and by CRC Press
2 Park Square, Milton Park, Abingdon, Oxon, OX14 4RN

First edition published by CRC Press 2007

CRC Press is an imprint of Taylor & Francis Group, LLC

ISBN: 978-1-4987-4007-4 (hbk)
ISBN: 978-0-4294-7080-6 (ebk)

Typeset in Sabon
by SPi Global, India

Contents

Preface

Outdoor sound propagation is of wide-ranging interest not only for predicting noise exposure but also in animal bioacoustics and in military contexts. Based on wide-ranging backgrounds in research and consultancy, the 2nd edition of 'Predicting Outdoor Sound' aims to provide a comprehensive reference on aspects of outdoor sound propagation and its prediction that should be useful to practitioners yet is respectable from the academic point of view. Despite the significant progress in theories and, particularly, in numerical methods for the various phenomena that are involved in outdoor sound propagation since the 1st edition of this text was published in 2006, current prediction schemes for outdoor sound remain largely empirical. While empirical approaches are understandable in view of the complicated source characteristics and complex propagation paths that are often of interest, numerical methods and theories have been validated extensively by comparisons with data and help with our understanding of the important effects. This text aims to bring the leading theories and data together and to provide the noise consultant or relevant practitioner with the basis for deciding between models and schemes for use in any given situation. Enough detail is presented to make the reader aware of the inherent approximations, restrictions and difficulties of the prediction methods discussed.

The text does not attempt to duplicate the comprehensive treatments of theoretical and numerical models in "Computational Atmospheric Acoustics" by Erik Salomons published in 2001 and in "Acoustics in Moving Inhomogeneous Media" by Vladimir Ostashev and Keith Wilson (2nd edition, a paperback version was published in 2019). While these texts are excellent, neither of them includes any data. In contrast, this text emphasizes data and introduces aspects of research not mentioned in these texts. Those interested in outdoor sound prediction would benefit from all three texts.

I am delighted that Timothy Van Renterghem has helped to produce this 2nd edition. He is well-known internationally for his work on many aspects of outdoor sound propagation. We worked together on an EC FP7 project[1] "Holistic and sustainable abatement of noise by optimized combinations of natural and artificial means" (HOSANNA) and it has been a pleasure to continue working with him. Some of the material in this second edition results from the HOSANNA project. The additions include comparisons between predictions and data for noise from road traffic, railways and wind turbines (Chapters 1 and 12), extended descriptions of the modelling of source characteristics, including the HARMONOISE model and its propagation modules (Chapters 3 and 12), predictions of propagation over rough seas, parallel low walls and lattices (Chapters 5, 8 and 9), descriptions of numerical methods (chapter 4), gabion and sonic crystal noise barriers performance and design (chapter 9), meteorological effects on noise barrier performance (Chapter 9), numerical designs of tree belts for road traffic noise reduction (Chapter 10) and requirements for auralization (Chapter 12). I would like to acknowledge the support of my previous co-authors, Kai Ming Li and Kirill Horoshenkov. Most of their contributions to the 1st edition are retained, except that, given the imminent appearance of a book by Maarten Hornikx and Timothy Van Renterghem on Urban Sound Propagation, the chapter devoted to this topic has been removed.

Keith Attenborough
June 2020

NOTE

1 https://cordis.europa.eu/project/id/234306

Authors' Biographies

Keith Attenborough is Professor in Acoustics at the Open University, a former Editor-in-Chief of *Applied Acoustics*, and a former Associate Editor of the *Journal of the Acoustical Society of America* and *Acta Acustica*. He is co-author with Oleksandr Zaporozhets and Vadim Tokarev for *Aircraft Noise* (CRC Press, 2017), and has co-authored several chapters in *Environmental Methods for Transport Noise Reduction* (CRC Press, 2019). He is Chair of ANSI S1 WG20 on the measurement of outdoor ground impedance.

Timothy Van Renterghem is Associate Professor in Environmental Sound at Ghent University and holds a MSc. degree in Bioengineering (Environmental Technologies) and a PhD in Applied Physical Engineering. He is Associate Editor of *Acta Acustica*, the journal of the European Acoustics Association, and Elsevier's Urban Forestry and Urban Greening. His main research interests include the impact of local meteorology on sound propagation outdoors, green noise reducing measures, and urban sound propagation with a strong focus on (detailed) numerical modelling.

Chapter 1

Introduction

1.1 EARLY OBSERVATIONS

The way in which sound travels outdoors has been of interest for several centuries. Initial experiments were concerned with the speed of sound [1]. In 1640, the Francisan (Minimite) friar, Marin Mersenne (1588–1648), timed the interval between seeing the flash and hearing the report from guns fired at a known distance and obtained a value of 450 m/s. In 1738, the French Academy of Science used the same idea with cannon fire and reported a speed of 332 m/s which is remarkably close to the currently accepted value for standard conditions of temperature (20°C) and pressure (at sea level) of 343 m/s. William Derham (1657–1735), the rector of a small church near London, was first to observe the influence of wind and temperature on sound speed and remarked on the difference between the sound of the church bells at a certain location over newly fallen snow compared with their sound at the same location without snow but with a frozen ground surface.

Many records of the strange effects of the atmosphere on the propagation of sound waves have been associated with war [2, 3]. In June 1666, Samuel Pepys wrote that the sounds of a naval engagement between the British and Dutch fleets were heard clearly at some spots but not at others a similar distance away or closer. Pepys spoke to the captain of a yacht that had been positioned between the battle and the English coast. The captain said that he had seen the fleets and run from them, '*...but from that hour to this hath not heard one gun...*'. The effects of the atmosphere on battle sounds were not studied in a scientific way until after the First World War (1914–1918). During that war, acoustic shadow zones, similar to those observed by Pepys, were observed during the battle of Antwerp. Observers also noted that battle sounds from France only reached England during the summer months and were best heard in Germany during the winter. After the war there was great interest in these observations among the scientific community. Large amounts of ammunition were detonated throughout England and the public was asked to listen for sounds of explosions.

Although there was considerable interest in atmospheric acoustics after the First World War, the advent of the submarine encouraged greater efforts in underwater acoustics research during and after the Second World War (1939–1945). Nevertheless, subsequently, the theoretical and numerical methods widely deployed in predicting sound propagation in the oceans have proved to be useful in atmospheric acoustics. A meeting organized by the University of Mississippi and held on the Mississippi Gulf Coast in 1981 was the first in which researchers in underwater acoustics met with scientists interested in atmospheric acoustics and this has stimulated the adaptation of the numerical methods used in underwater acoustics, for predicting sound propagation in the atmosphere [4].

1.2 A BRIEF SURVEY OF OUTDOOR SOUND ATTENUATION MECHANISMS

Outdoor sound is influenced by distance, by topography (including natural or artificial barriers), by interaction with the ground and with or without vegetative ground cover and by atmospheric effects including refraction and absorption. Velocity vectors of sound and wind are additive. When the source is downwind of the receiver, sound from the source propagates upwind. As height in the atmosphere increases, wind speed increases and the wind speed vector component between source and receiver is subtracted from the speed of sound increases, leading to a negative sound speed gradient. A negative sound speed gradient means that there is upward refraction of sound. If sound is considered to propagate as rays, then a negative sound speed gradient causes rays to curve upwards. When the source is near the ground during upward refraction, there is a limiting ray that leaves the source and just grazes the ground at some point before reaching a receiver at any given height. This defines the start of the shadow zone. Receivers at the same distance from the source below the limiting ray and receivers at the same height but located further away are in the shadow zone. The distance of the start of a sound shadow from the source caused by upward refraction depends on the sound speed gradient. However, ray tracing (considered in more detail in Chapter 11) ceases to be valid beyond this limiting ground-grazing ray. The shadow zone caused by upward refraction is penetrated by sound scattered by atmospheric turbulence thereby limiting the reduction of sound levels within the sound shadow. The many effects of atmospheric turbulence are considered further in Chapter 10.

A negative sound speed gradient results also when the temperature decreases with height. This is called a temperature lapse condition and is the normal condition on a dry sunny day with little wind. A combination of slightly negative temperature gradient, strong upwind propagation and air absorption has been observed, in carefully monitored experiments, to reduce sound levels, 640 m from a 6 m high source over relatively hard ground, by up to 20 dB more than would be expected only from spherical spreading [5].

The total attenuation of a sound outdoors can be expressed as the sum of the reductions due to geometric spreading, atmospheric absorption and the extra attenuations due to ground effects, scattering, visco-thermal effects in vegetation, refraction in the atmosphere and diffraction by barriers. Atmospheric absorption results from heat conduction losses, shear viscosity losses and molecular relaxation losses and increases rapidly with frequency. It acts as a low pass filter at long range.

Ground effects (for elevated source and receiver) are the result of interference between sound travelling directly from source to receiver and sound reflected from the ground. The influence of the ground depends on the source and receiver locations as well as the nature of the ground surface. If the interference is constructive, there is an enhancement compared with the free field level which tends to occur mainly at low frequencies. If the interference is destructive, there is attenuation. Porous ground surfaces allow sound waves to penetrate the pores where they are affected by viscous friction and thermal exchanges, so the reflected sound suffers a change in phase as well as amplitude. For a given source–receiver geometry, destructive interference occurs at lower frequencies than over ground which is not porous. But, irrespective of the porosity of the ground surface, if it is rough, there can be additional sound attenuation due to scattering.

The root zone created by any vegetation or ground cover tends to make the surface layer of ground more porous. Moreover, layers of partially decayed leaf matter on the floors of forests are highly porous. Propagation through bushes, hedges and crops involves ground effects, since they are planted on acoustically soft ground, scattering by stems, visco-thermal scattering by foliage and, possibly, acoustically induced leaf vibrations. Propagation through trees involves reverberant scattering by tree trunks. Ground effects, scattering phenomena and foliage attenuation are explored in more detail in Chapters 2, 5, 6 and 10.

This chapter continues by presenting data illustrating ground effects. Subsequently, data is explored that illustrates the combined effects of ground (including topography) and atmospheric refraction. There follow outlines of methods for classifying and representing the variation of sound speed with height for use in models for predicting outdoor sound propagation.

1.3 DATA ILLUSTRATING GROUND EFFECT

1.3.1 Propagation from a Fixed Jet Engine Source

Pioneering studies of the combined influences of the ground surface and meteorological conditions were carried out by Parkin and Scholes [6–10] using a fixed Rolls Royce Avon jet engine as a source at two airfields (Hatfield and Radlett). In his 1970 Rayleigh Medal Lecture, one of the investigators, the late Peter Parkin, remarked [6],

> *These horizontal propagation trials showed up the ground effect, which at first we did not believe, thinking there was something wrong with the*

> *measurements. But by listening to the jet noise at a distance, one could clearly hear the gap in the spectrum.*

These studies were among the first to quantify the change in ground effect with type of surface. The Parkin and Scholes data showed a noticeable difference between the ground effects due to two types of grass cover. The ground attenuation at Hatfield, although still a major propagation factor, was less than at Radlett and its maximum value occurred at a higher frequency. A change in weather conditions during their measurements also enabled them to remark the effects of snow cover.

> *... measurements [were] made at Site 2 [Radlett] with 6 to 9 in. of snow on the ground. The snow had fallen within the previous 24 hours and had not been disturbed. The attenuations with snow on the ground were very different from those measured under comparable wind and temperature conditions without snow ... .The maximum of the ground attenuation appears to have moved down the frequency scale by approximately 2 octaves ...*

Examples of the Parkin and Scholes data are shown in Figure 1.1. These data are of the *corrected level difference*, i.e. the difference in sound pressure

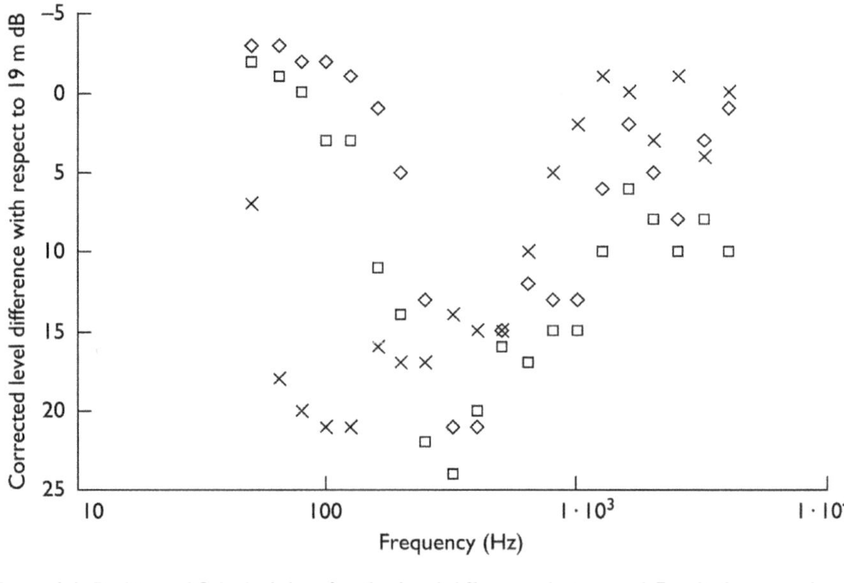

Figure 1.1 Parkin and Scholes' data for the level difference between 1.5 m high microphones at 19 m and 347 m from a fixed jet engine source (nozzle-centre height 1.82 m) corrected for wavefront spreading and air absorption. The symbols and ◇ represent data over airfields (grass-covered) at Radlett and Hatfield respectively with a positive vector wind between source and receiver of 1.27 m/s (5 ft/s). Crosses (×) represent data over approximately 0.15 m thick (6–9 in.) snow at Hatfield with a positive vector wind of 1.52 m/s (6 ft/s).

levels at 19 m (used as a reference location) and each of more distant locations corrected for the decrease expected from spherical spreading and air absorption. They provide further evidence that ground effect is sensitive to the acoustical properties of the surface which, in turn, depend on the substance of which the surface is composed. Different ground surfaces have different porosities. Soils have volume porosities of between 10% and 40%. Snow, which has a porosity of around 60%, and many fibrous materials, which have porosities of above 90%, have relatively low flow resistivities whereas a wet compacted soil surface will have a rather high flow resistivity. Also, the thickness of the surface porous layer is important and whether it has an acoustically hard substrate. The Parkin and Scholes data revealed the large attenuation at low frequencies (63 Hz and 125 Hz octave bands) in the presence of thick snow. It should be noted that, even without snow, there are significant differences in the Parkin and Scholes' Radlett data between summer and winter. Seasonal variations in ground effects are discussed further in Chapter 6, section 6.4.

1.3.2 Propagation over Discontinuous Ground

The extent to which discontinuities in surface impedance can influence the rate of attenuation is illustrated by data from measurements of noise levels during aircraft engine run-ups made at distances of up to 3 km from the source with the aim of defining noise contours in the vicinity of airports [11]. Measurements were made for a range of power settings during several summer days with near to calm weather conditions (wind speed < 5 m/s, temperature between 20 and 25°C). Between 7 and 10 measurements were made at every measurement station and the results were averaged. Example jet engine noise spectra at four angles from the direction forward of the aircraft normalized to a reference distance of 1 m are shown in Figure 1.2.

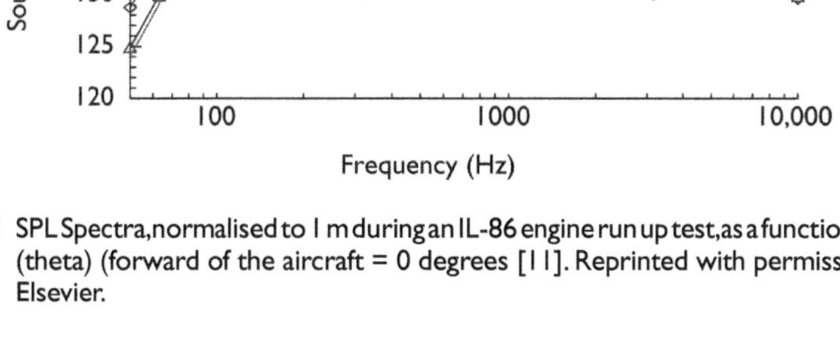

Figure 1.2 SPL Spectra, normalised to 1 m during an IL-86 engine run up test, as a function of angle (theta) (forward of the aircraft = 0 degrees [11]. Reprinted with permission from Elsevier.

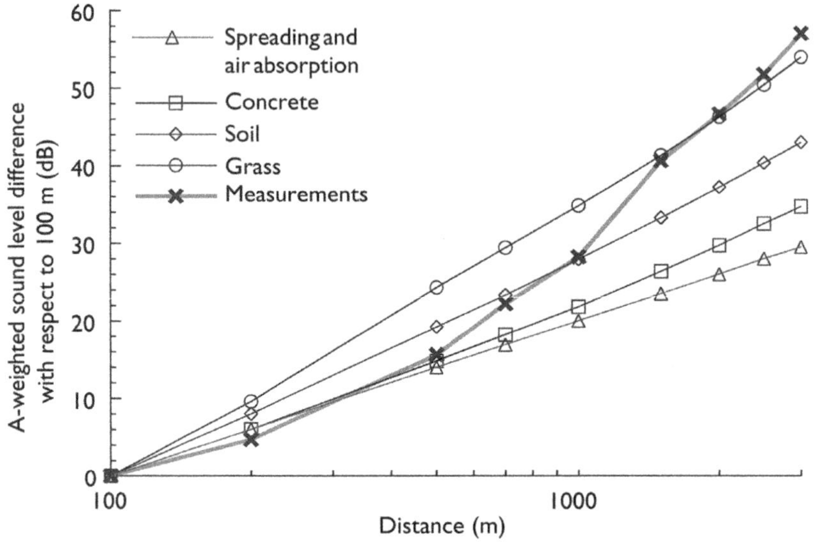

Figure 1.3 Measured differences (joined crosses) between the A-weighted sound level at 100 m and those measured at ranges up to 3 km during an Il-86 aircraft's engine test in the direction of maximum jet noise generation (~40° from exhaust axis) and predictions for levels due to a point source at the engine centre height assuming spherical spreading plus air absorption and various types of ground [11]. Reprinted with permission from Elsevier.

Example measured differences in levels between a receiver at a reference distance of 100 m and other receiver locations between 200 m and 2.7 km are shown in Figure 1.3.

Also shown in Figure 1.3 are predictions based on models detailed in Chapters 2 and 5. The predictions use equation (2.40) and a one-parameter impedance model (see Chapter 5, equations (5.1) and (5.2)), assuming effective flow resistivities of 20,000 kPa s m^{-2}, 300 kPa s m^{-2} and 2000 kPa s m^{-2}, respectively, for concrete, grass and soil. The predictions show that, up to 500 m distance, the data are consistent with propagation in acoustically neutral conditions over an acoustically hard surface but beyond 1 km are more representative of levels predicted over an acoustically soft surface. Figure 1.4 shows similar data and predictions for the attenuation from a propeller aircraft.

The averaged data in both the maximum jet noise and the maximum propeller noise directions indicate a change from one rate of attenuation to another at distances greater than 500 m which can be attributed to a variation with range in the nature of the ground surface. The run-ups took place over the concrete surface of an apron. Further away (i.e. between 500 m and 700 m from the aircraft in various directions), the ground surface was 'soil' and/or 'grass'. A good fit to the data is obtained by predictions that include an impedance discontinuity between 500 and 1000 m from the source [11].

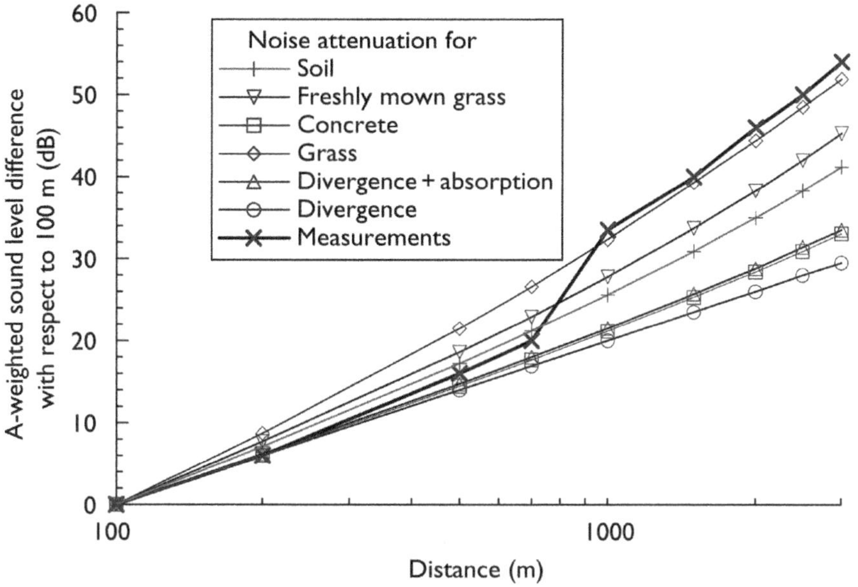

Figure 1.4 Measured differences (joined crosses) between the A-weighted sound level at 100m and those measured at ranges up to 3km from a turbo-prop engine (on an An-24 aircraft) in the direction of maximum propeller noise generation (~80° from axis of engine inlet) [11]. Reprinted with permission from Elsevier.

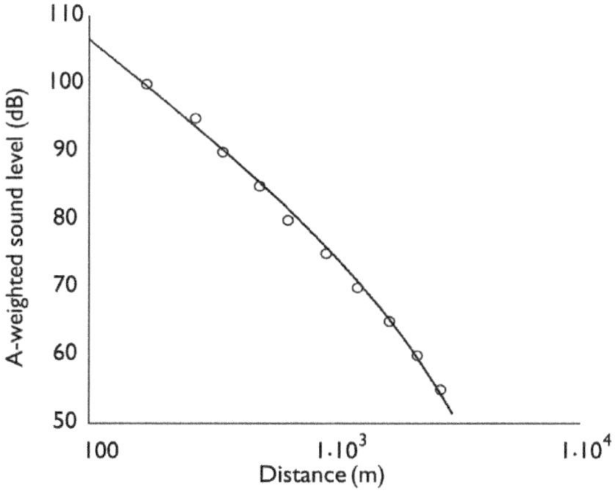

Figure 1.5 Measured sound levels in the direction of maximum propeller noise from a turboprop aircraft and the fit given by equation (1.1).

The empirical fit shown in Figure 1.5 to the data beyond 100 m in the direction of maximum propeller noise is given by

$$L_A = 162 - 27.5\log(d) - 0.005d, \tag{1.1}$$

where d (m) is the horizontal distance.

1.4 DATA ILLUSTRATING THE COMBINED EFFECTS OF GROUND AND METEOROLOGY

1.4.1 More Fixed Jet Engine Data

Although the wind speed was measured, the classical experiments by Parkin and Scholes involved relatively little meteorological monitoring. The important role of atmospheric turbulence was not appreciated at the time, so, for example, the fine-scale fluctuations in wind speed were not monitored. Similar measurements to those carried out by Parkin and Scholes using a fixed jet engine source have been made but they were augmented by more comprehensive meteorological data. The fixed jet engine was operated as a broadband noise source at a Rolls Royce test facility in a disused airfield at Hucknall, Nottingham, UK [12]. Simultaneous acoustic and meteorological measurements were made. In addition to wind and temperature gradient measurements, the fluctuation in wind velocity measurements was recorded and used as a measure of turbulence. Some of the data obtained under low wind and low turbulence conditions over continuous grassland are shown in Figure 1.6. Also shown is the third octave power spectrum of the Avon engine source between 100 Hz and 4000 Hz deduced from the measured spectrum at 152.4 m after correcting for spherical spreading and ground effect. The data obtained at the longest range is limited by background noise above 3 kHz. The significant dips in the received spectra between 100 and 500 Hz are clear evidence of ground effect. However, it is noticeable that the ground effect at Hucknall is different from that measured at either Radlett or Hatfield.

The influence of small changes in the wind speed and turbulence strength on the measured spectra at the longest range is demonstrated in Figure 1.7. The associated meteorological conditions are detailed in Table 1.1. The ground effect between 100 Hz and 400 Hz is relatively stable and significantly greater at the low microphones where it is shifted in frequency compared with ground effect at the high microphones. The data for both microphone heights show considerable variability between 400 Hz and 2 kHz which can be attributed to changes in wind velocity and turbulence.

Figures 1.8 and 1.9 show A-weighted levels deduced from average spectra in consecutive 26 s periods measured at 1.2 m height and ranges of 152.4 m, 457.6 m, 762.2 m and 1158.4 m over grassland at Hucknall. Figure 1.8 shows data for low wind speed (less than 2 m/s from source to receiver) and low turbulence conditions. Figure 1.9 shows data for moderate downwind conditions (approximately 6 m/s from source to receiver) and for higher turbulence intensities. The details of the meteorological conditions corresponding to Figures 1.8 and 1.9 are listed in Tables 1.2 and 1.3, respectively.

Figure 1.8 shows the considerable spread in the measured levels at the longer ranges resulting from the variation in wind speed and direction (up to approximately 2 m/s downwind at 6.4 m height) and turbulence levels. The data for stronger downwind conditions (up to approximately 6.5 m/s at 6.4 m height) in Figure 1.9 indicate consistently higher levels than those during the relatively low wind speed conditions. On the other hand, they have

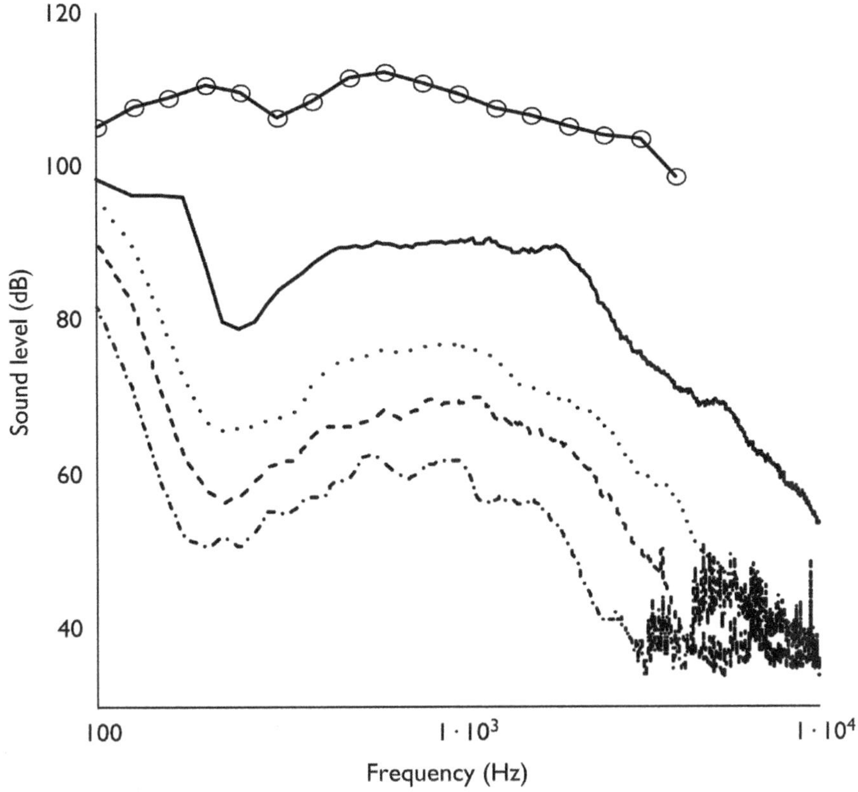

Figure 1.6 Data recorded at 1.2 m high receivers at horizontal ranges of 152.4 m (solid line), 457 m (dotted line), 762 m (dashed line) and 1158 m (dash-dot line) from a fixed Rolls Royce jet engine source with the nozzle centre 2.16 m above an airfield at Hucknall, Notts. These data represent simultaneous recordings averaged over 26 s during zero wind and low turbulence conditions (block 20 of run 454, see Figure 1.3). Also shown (connected circles) is the deduced third octave power spectrum of the Avon jet engine source after subtracting 50 dB.

a smaller spread. Although only four averages are shown in Figure 1.9, their spread is smaller than for any four averages exhibited in Figure 1.8. This is consistent with the assertion in ISO 9613-2 [13] that the variation in sound levels is less under 'moderate' downwind conditions. The average downwind level measured at Hucknall is about 10 dB higher than the levels for the lowest wind speed and turbulence conditions at 1.1 km from the source.

1.4.2 Road Traffic Noise Propagation over Flat Terrain under Strong Temperature Inversion

Figure 1.10 shows temperature profiles derived from series of air temperature measurements at different heights (up to 13 m) near Scottsdale in the Phoenix valley (Arizona, US) [13]. The area can be characterized meteorologically as one with very light synoptic winds and clear skies, giving rise to strong diurnal

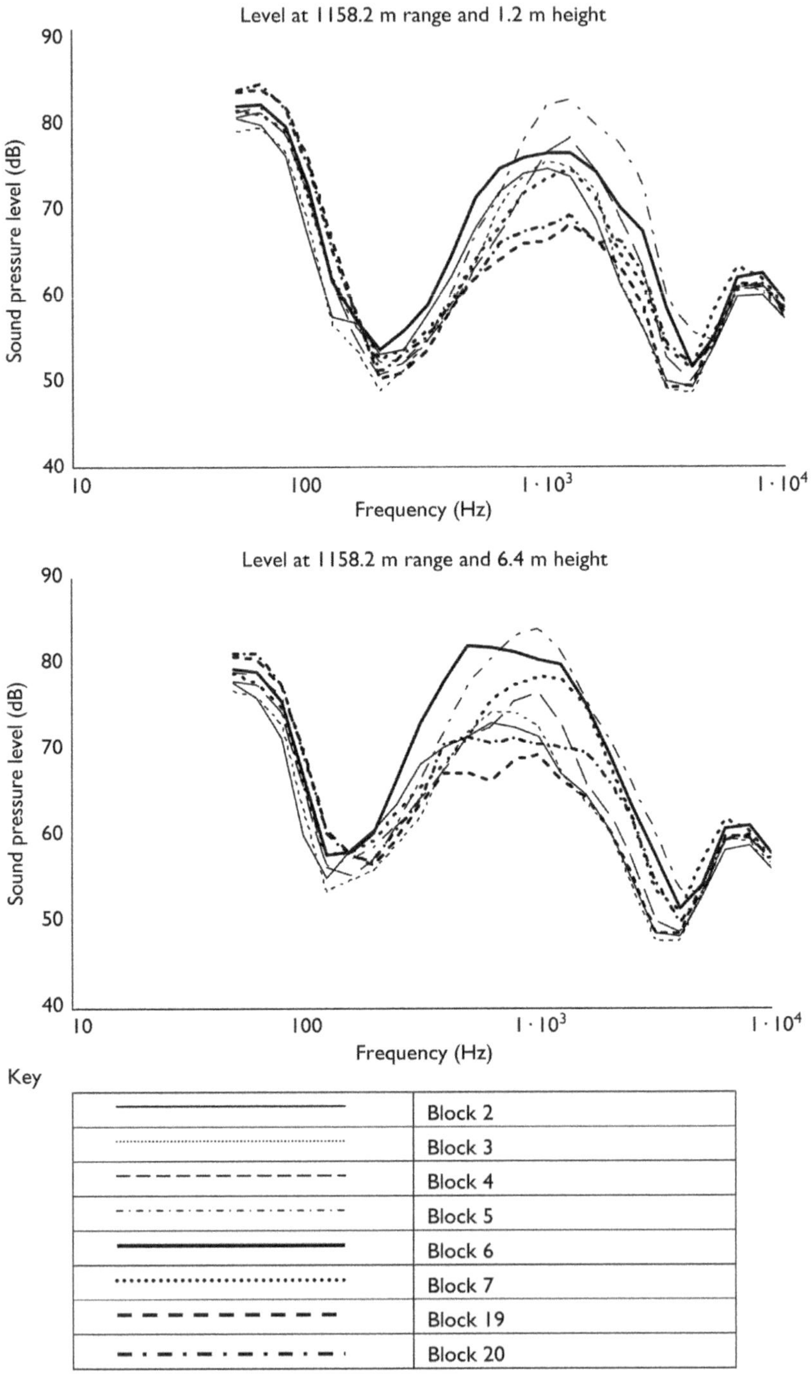

Figure 1.7 Simultaneously measured narrow band (25 Hz interval) spectra at low (1.2 m – upper graph) and high (6.4 m – lower graph) microphones between 50 Hz and 10 kHz at 1158.2 m from a fixed Avon jet engine source averaged over 26 s intervals during low wind, low turbulence conditions at Hucknall (Notts. UK). The conditions are specified in Table 1.1 and the key.

Table 1.1 Meteorological conditions corresponding to data in Figure 1.7

Run 454 Block No.	Wind speed at ground (0.025 m) (m s^{-1})	Wind speed at 6.4 m height (m s^{-1})	Direction relative to line of mics. (°)	Temperature at ground (°C)	Temperature at 6.4 m (°C)	Turbulence variable
2	1.57	1.86	23.3	10.4	9.9	0.0486
3	1.34	1.61	26.9	10.4	9.9	0.0962
4	1.27	1.96	349.0	10.5	9.8	0.0672
5	0.00	1.57	343.2	10.5	9.8	0.0873
6	0.00	1.46	346.0	10.5	9.8	0.1251
7	0.00	1.81	342.8	10.7	9.9	0.2371
19	0.00	0.00	301.6	10.2	9.8	0.0000
20	0.00	0.00	236.9	10.2	9.8	0.0000

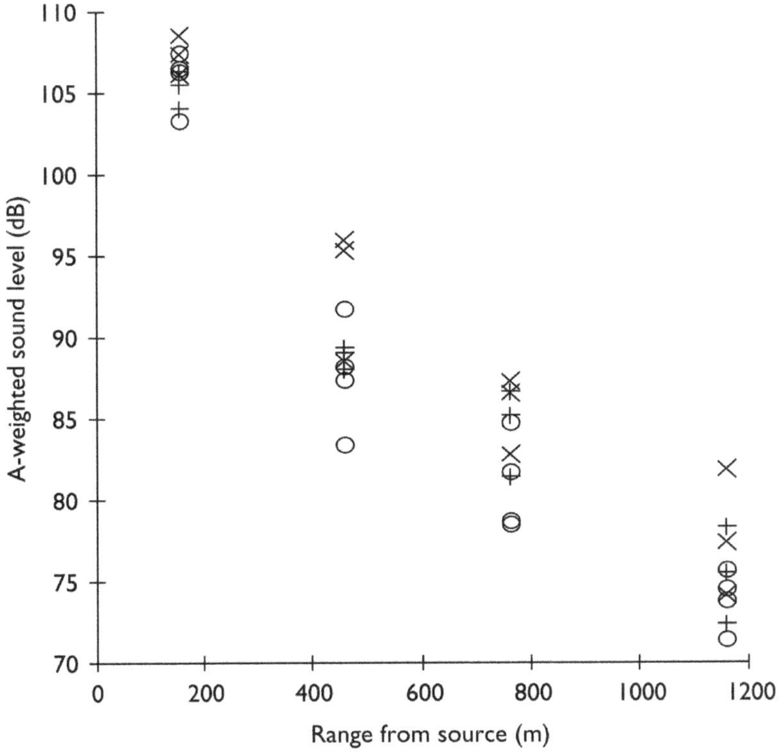

Figure 1.8 Comparison of A-weighted sound levels (26 s averages) deduced from low wind, low turbulence octave band measurements at 1.2 m height over grassland at Hucknall (Run 454 blocks 11,12,13(×); 14,15,16(+); 17,18,19,20(O)).

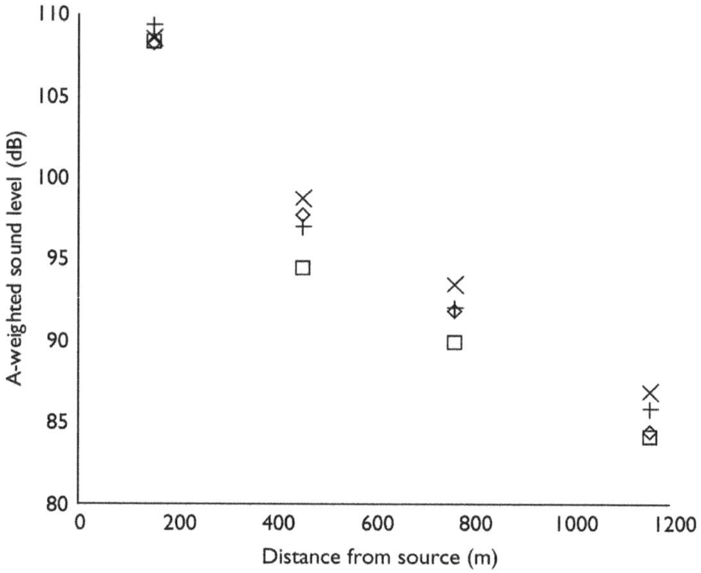

Figure 1.9 Comparison of A-weighted sound levels deduced from consecutive 26 s average downwind octave band measurements at 1.2 m height above grassland at Hucknall (Run 453 blocks 3(×); 4(+); 5() and 6(◇)).

Table 1.2 Meteorological data corresponding to sound level data shown in Figure 1.8

Run 454 Block No.	*Wind speed at ground (0.025 m) (m s^{-1})*	*Wind speed at 6.4 m height (m s^{-1})*	*Direction relative to line of mics. (°)*	*Temperature at ground (°C)*	*Temperature at 6.4 m (°C)*	*Turbulence variable*
11	0.00	1.97	348.1	10.5	9.8	0.0805
12	0.00	1.97	324.8	10.6	9.8	0.0607
13	0.00	1.09	356.9	10.6	9.8	0.0678
14	0.00	0.01	357.7	10.4	9.9	10.1489
15	1.02	1.53	50.6	10.4	9.9	0.0764
16	0.00	1.58	38.5	10.4	10.0	0.0928
17	0.00	0.92	20.9	10.2	9.9	0.1792
18	0.00	1.16	14.0	10.2	9.9	0.0424
19[a]	0.00	0.00	301.6	10.2	9.8	0.0000
20[a]	0.00	0.00	236.9	10.2	9.8	0.0000

[a] Also listed in Table 1.1.

variations in the temperature profile, either increasing or decreasing with height at low altitude, depending on the time of the day. Figure 1.10 shows the typical difference between daytime temperature lapse conditions in which temperature decreases with height and a night-time ground-based

Table 1.3 Meteorological data corresponding to sound level data in Figure 1.9

Run 453 Block No.	*Wind speed at ground (m s⁻¹)*	*Wind speed at 6.4 m height (m s⁻¹)*	*Direction relative to line of mies. (°)*	*Temperature at ground (°C)*	*Temperature at 6.4 m C(°)*	*Turbulence variable*
3	4.09	6.44	10.0	15.0	15.0	0.1202
4	4.09	6.11	20.5	14.9	15.0	0.1606
5	4.33	5.93	17.8	14.9	15.0	0.1729
6	3.89	6.07	10.8	14.9	15.0	0.1028

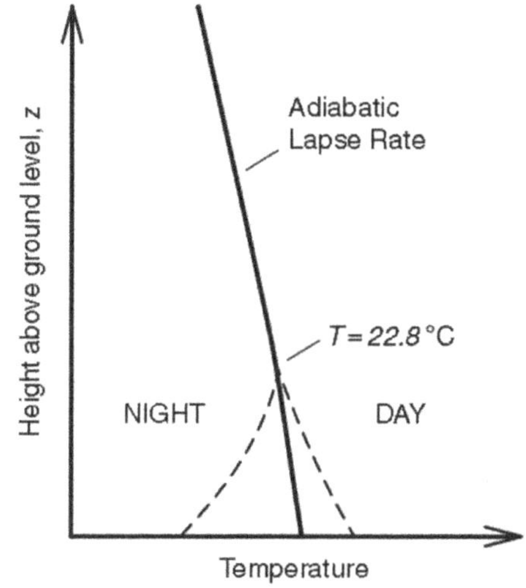

Figure 1.10 A schematic representation of the transition between a ground-based temperature inversion condition at night and a temperature lapse with height during daytime. The turning height is based on measured on-site meteorological data [13].

temperature inversion where the temperature increases for the first few metres and then decreases. As well as temperature profiles, sound level measurements due to a highway were made at several locations and show significant influences of the changing temperature profiles. Figure 1.11 shows that the ground between the highway and the measurement locations was relatively flat, consisting of either freshly ploughed agricultural land or low-density single-story buildings. Figure 1.12 shows the sound levels measured at sites 1 and 2 (the temperature profiles were measured near site 2) during 24 hours from midnight to midnight.

From about 6:00 till 10:00, the levels at site 1 were nearly constant (upper part of Figure 1.12). However, at site 2 (microphone height 1.5 m), the measured sound pressure levels show a strong variation lower part of Figure 1.12).

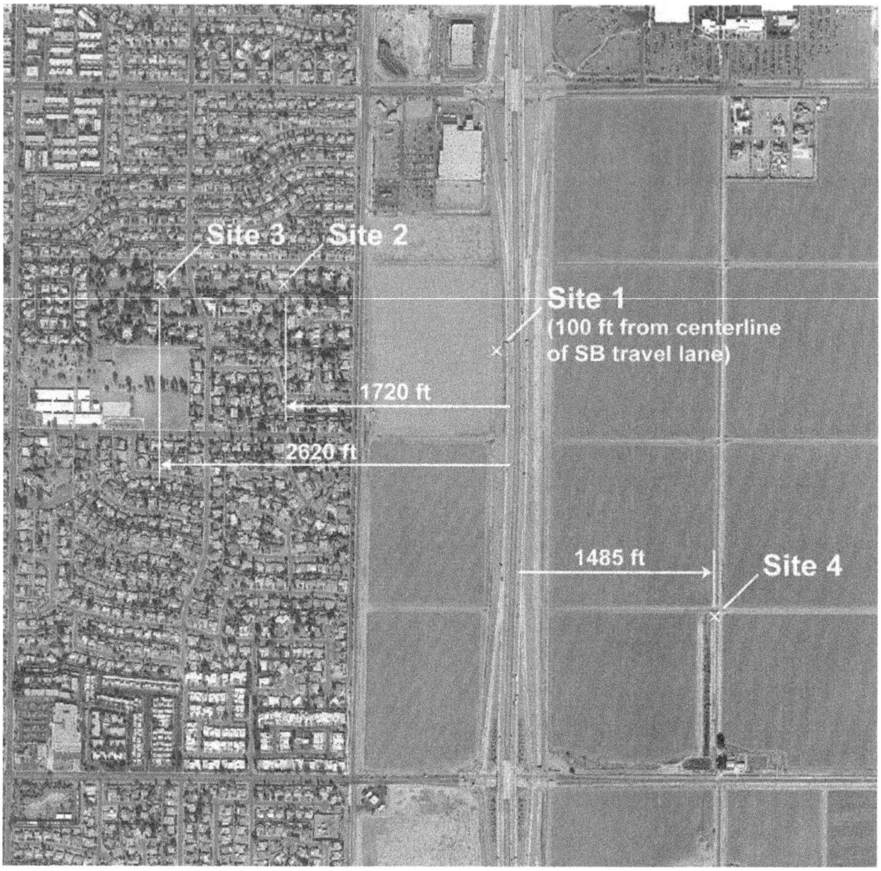

Figure 1.11 An areal photograph of a highway in Arizona, showing locations of the closest receiver (site 1, at 30 m) and a further away receiver (site 2, at 524 m) [13].

The levels at site 2 increase as the nocturnal inversion layer reached its maximum just before sunrise at about 07.00. After sunrise, the inversion layer broke up quickly and, consequently, the sound pressure levels decreased rapidly. The relatively constant sound pressure level at site 1 during this period shows that the sound level variation at site 2 is mainly a consequence of the changing temperature profiles.

1.4.3 Meteorological Effects on Railway Noise Propagation over Flat Terrain

Several 4 m high microphones were positioned up to 200 m either side of a roughly East-West railway line (at approximately 30 km from Vienna, Austria) (see Figure 1.13). Sound exposure level (SEL) measurements of individual passages of trains show the combined effect of atmospheric

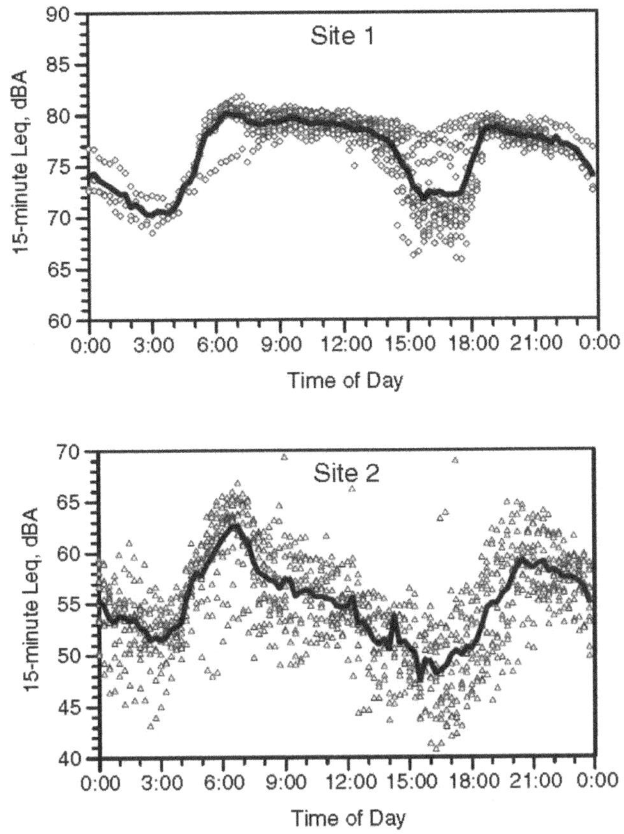

Figure 1.12 Sound pressure level variation during the selected monitoring day at the receiver (site 1) 30 m from the highway and at a receiver (site 2) 524 m from the highway [13].

stability and wind [14]. The closest microphones at 25 m from the line, either in the northerly or southerly directions, were used for reference. The surrounding flat agricultural land was covered with crops which were approximately 0.6 m high at the time of the measurements.

The monitoring period considered started at 18:30 in the evening of 2 July and continued until 8:00 the next morning (3 July). The radiation balance measured on site showed positive values up to 20:00 (including some cloudy periods), negative values during the night (more or less constant at –50 W/m^2) and positive values again the next morning from about 7:00. The positive radiation corresponds to negative Obukhov lengths and indicates an unstable atmosphere (see Section 1.6), whereas negative radiation corresponds to positive Obukhov lengths (estimated as between 15 and 30 m) and a stable atmosphere. So the measurement period involved a transition from an unstable/neutral atmosphere to a stable atmosphere after sunset

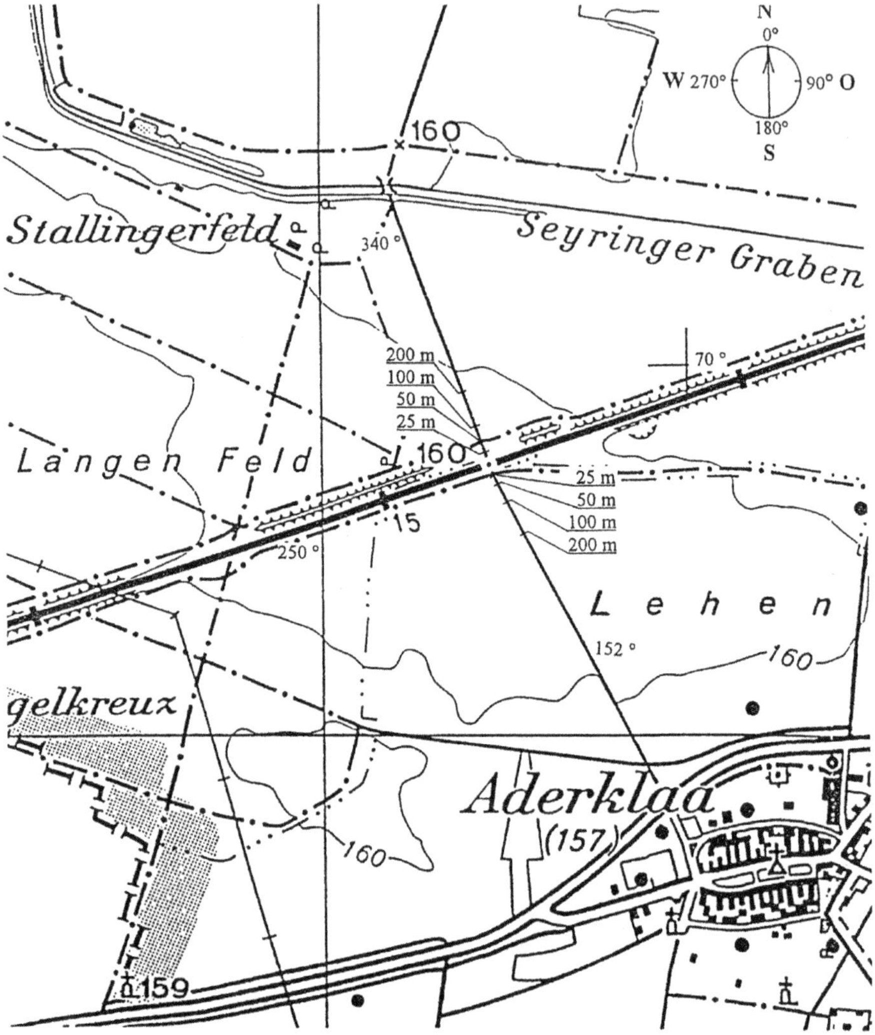

Figure 1.13 Map of a railway line 30 km from Vienna showing microphone locations [14].

and during the night, and then, again, towards a neutral or unstable atmosphere after sunrise.

A tethered balloon measured air temperature and wind speed up to a height of approximately 100 m (see Figure 1.14). Between 19:35 and 20:00, a ground-based temperature inversion was developing and was followed by a fully developed and strong temperature inversion during the night which persisted until the next morning. After sunrise, there was a transition

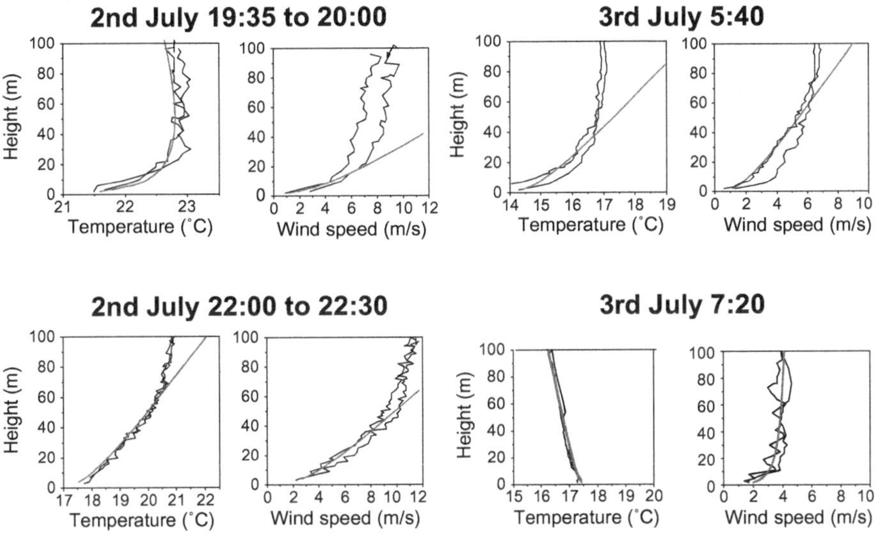

Figure 1.14 Air temperature and wind speed profiles measured using a weather balloon. The two lines indicate either up or down movement of the balloon during a measurement cycle [14].

towards a neutral and unstable atmospheric boundary layer at 7:20, indicated by the temperature decrease with height.

The wind was blowing more or less at right angles to the railway line, from a southerly direction with an average wind speed of 3.5 m/s. Sound propagation towards the north could be categorized as downwind and, therefore, downward refracting, whereas sound propagation towards the south was upwind and upward refracting. Note, however, that, since the trains were moving, various propagation directions relative to wind direction are combined in the measured Single Event Level (SEL).[1] As shown in Figure 1.14, the wind speed profiles are also influenced by the stability of the atmospheric boundary layer.

The relative sound pressure level measurements between the reference and far locations plotted according to the atmospheric stability (stable atmosphere vs neutral/unstable atmosphere) are shown in Figure 1.15.

It is reasonable to assume that, at the reference locations close to the railway, refraction effects are more or less absent. But the different refraction conditions resulted in significantly lower SELs for sound propagation in southerly direction than for propagation in the northerly direction. Exposure levels measured, while the atmospheric boundary layer was stable, were significantly higher than when it was unstable. The fact that the main differences in sound level between the atmospheric conditions were between about 250 Hz and 2000 Hz indicates that there is a strong interaction between the meteorology and the ground effect.

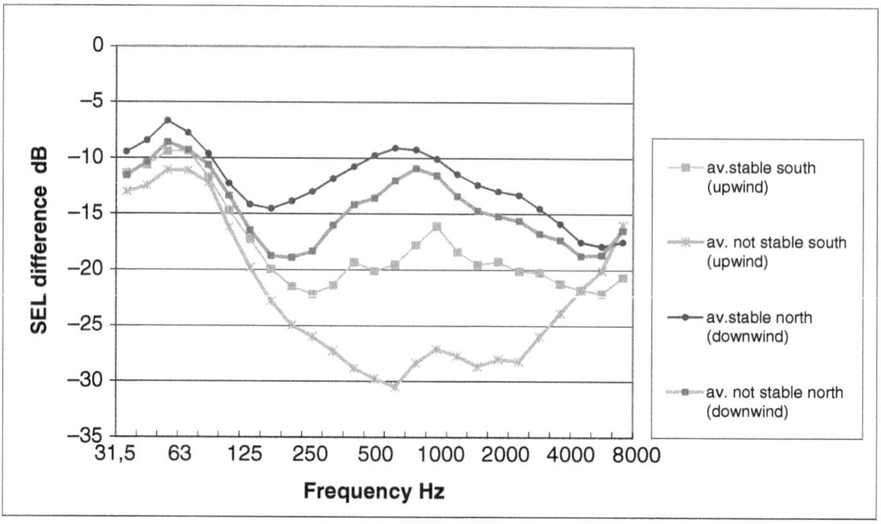

Figure 1.15 SEL differences, in 1/3 octave bands, during the selected period (2–3 July) between the close measurement point (at 25 m from the railway line) and the microphone at 200 m [14].

1.4.4 Road Traffic Noise Propagation in a Valley

Mountainous areas are a special context for outdoor sound propagation since, often, they involve extremes of terrain undulation, ground cover and meteorology. In a monitoring campaign in the Unterinntal region (near Schwaz, Austria) noise levels were measured in two cross sections of the valley and at three locations in each: one on the valley floor where a highway was located, and two on the slopes, up to 166 m above the source (see Figure 1.16) [15]. There was snow cover during the measurements. Only windless periods were selected.

Figure 1.17 shows the measured differences in total A-weighted road traffic noise between the microphone close to the road, and microphones upslope. Also shown in Figure 1.17 are the air temperature profiles measured at various heights upslope. Although positions MP3 and MP5 are much further from the road than MP2 (see cross section 2 in Figure 1.16), their higher elevations result in direct sound propagation without shielding by terrain or obstacles causing their level differences to be close to zero. Also, their elevated locations mean that the potentially very strong attenuation due to ground effect in the presence of the snow does not affect the levels at these points, in contrast to its influence on sound propagation towards MP2. In addition, levels at MP2 might be slightly increased by downward refraction in the thin temperature inversion layer that was present very close to the ground. On the other hand, sound propagation towards MP3 and MP5 will be influenced mainly by the upward-refracting part of the atmosphere.

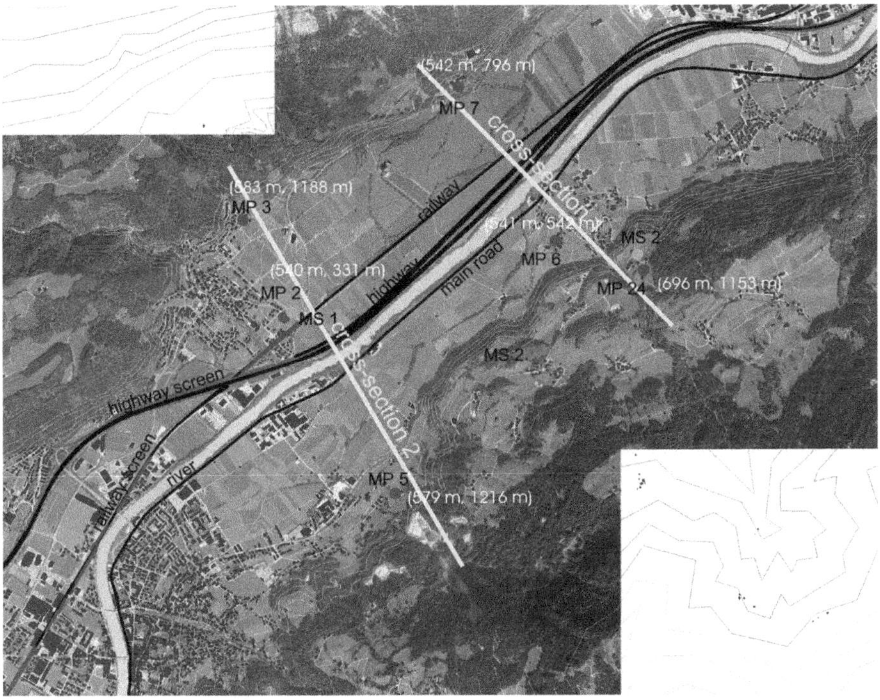

Figure 1.16 Elevation map of the highway in valley example, showing microphone positions (MP2, MP3, MP5 in section 1; MP24, MP6 and MP7 in section 2). The pairs of numbers in brackets are ground elevations at the microphone position and the orthogonal distances relative to the highway respectively. The microphones are 2 m above the ground. MS1 and MS2 indicate the locations of meteorological monitoring stations [15]. Reproduced from T. Van Renterghem, D. Botteldooren and P. Lercher, Comparison of measurements and predictions of sound propagation in a valley-slope configuration in an inhomogeneous atmosphere, *J. Acoust. Soc. Am.* **121**(5) 2522–2533 (2007) with the permission of the Acoustical Society of America.

In cross section 1, the effects of both the elevation and the inhomogeneous atmosphere tend to decrease the difference in sound pressure level between the reference point (MP6, which was on the roof of a small building rather than at the 2 m height above ground of the other microphones) and the more distant points. Quantitative analysis is made difficult here, among other reasons, by the strong variations in the terrain profile near MP24.

Full-wave calculations can be used to quantify the separate influences of the various propagation effects. Predictions made with the rotated reference frame Green's Function parabolic equation method [15], accounting for both the actual relief and the actual air temperature profile, lie close to (i.e. within 3 dB of) the measured level differences (see Figure 1.17). The ground beneath the sources was considered acoustically hard and the surface impedance of the surrounding snow-covered ground was modelled using a

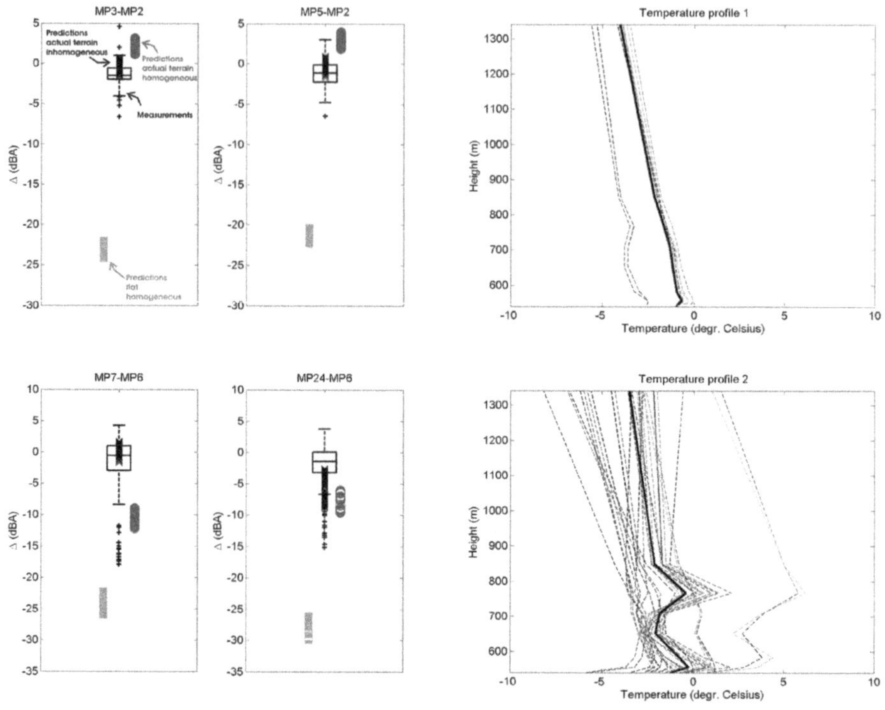

Figure 1.17 A-weighted sound level differences measured (boxplots, where the plusses denote outliers) and simulated (crosses, open circles, and squares) in each of the two sections across the valley and the highway (see Figure 1.16). In each cross section, a typical temperature profile is selected. The thick black lines show the averaged profiles within the cluster for each section [15]. Reproduced from T. Van Renterghem, D. Botteldooren and P. Lercher, Comparison of measurements and predictions of sound propagation in a valley-slope configuration in an inhomogeneous atmosphere, *J. Acoust. Soc. Am.* **121**: 2522–2533 (2007) with the permission of the Acoustical Society of America.

single-parameter model (see Chapter 5) with an effective flow resistivity of 30 kPa s m^{-2}.

Additional simulations have been performed for a flat ground and for the actual terrain profile in a homogeneous atmosphere. The valley-slope terrain is predicted to result in a large increase in the sound pressure level of up to 30 dBA at the distant elevated points compared to propagation over flat ground. The complex temperature profiles in this zone are predicted to account for level changes between −3 dBA and +10 dBA relative to a homogeneous atmosphere.

Comparisons with predictions have been made for clustered measurements having more or less similar air temperature profiles. Inevitably, as indicated by the boxplots in Figure 1.17, some variation in the measured levels will be present as well. Traffic composition and vehicle speed will result in different source spectra during the different measurement periods. Also, there will be changes in the magnitude of the atmospheric absorption. Only the average temperature profile in the cluster has been used in the predictions. The profiles belonging to a cluster are slightly different, causing changes in the refractive state of the atmospheric boundary layer and consequently in the traffic noise levels, especially at points further away from the road.

1.5 CLASSIFICATION OF METEOROLOGICAL CONDITIONS FOR OUTDOOR SOUND PREDICTION

The atmosphere is constantly in motion because of wind shear and uneven heating of the earth's surface (see Figure 1.18). Any turbulent flow of a fluid across a rough solid surface generates a boundary layer. Most interest from the point of view of outdoor noise prediction focuses on the lower part of the meteorological boundary layer called the surface layer. In the surface layer, turbulent fluxes vary by less than 10% of their magnitude but the wind speed and temperature gradients are largest. In typical daytime conditions the surface layer extends over 50–100 m. Usually, it is thinner at night.

In most common daytime conditions, the net radiative energy at the surface is converted into sensible heat. This warms up the atmosphere thereby producing negative temperature gradients as indicated in Figure 1.18. If the radiation is strong (high sun, little cloud cover), the ground is dry, and the surface wind speed is low, the temperature gradient is large. The atmosphere exhibits strong thermal stratification. If the ground is wet, most of the radiative energy is converted into latent heat of evaporation and the temperature gradients are correspondingly lower. In unstable daytime conditions the wind speed is affected by the temperature gradient and exhibits slightly less variation with height than for the isothermal case. On the other hand, 'stable' conditions prevail at night. Radiative losses from the ground surface cause positive temperature gradients.

There is a considerable body of knowledge about meteorological influences on air quality in general and the dispersion of plumes from stacks in particular. Plume behaviour depends on vertical temperature gradients and hence on the degree of mixing in the atmosphere. Vertical temperature gradients decrease

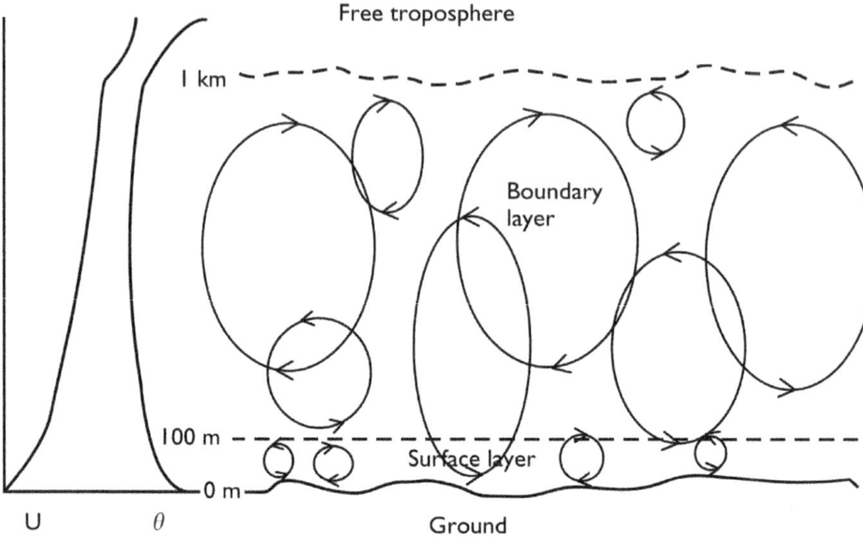

Figure 1.18 Schematic of a daytime atmospheric boundary layer and eddy structures. The sketch graph on the left shows the mean wind speed (U) and the potential temperature profiles ($\theta = T + \gamma_d z$, where γ_d = 0.098°C/km is the dry adiabatic lapse rate, T is the temperature and z is the height).

with increasing wind. The stability of the atmosphere in respect of plume dispersion is described in terms of Pasquill classes. This classification is based on incoming solar radiation, time of day and wind speed. There are six Pasquill classes (A–F) defined in Table 1.10. Data are recorded in this form by meteorological stations and so, at first sight, it is a convenient classification system for noise prediction.

Class A represents a very unstable atmosphere with strong vertical air transport, i.e. mixing. Class F represents a very stable atmosphere with weak vertical transport. Class D represents a meteorologically neutral atmosphere which has a logarithmic wind speed profile and a temperature gradient corresponding to the normal decrease with height (adiabatic lapse rate). A meteorologically neutral atmosphere occurs for high wind speeds and large values of cloud cover. Consequently, a meteorologically neutral atmosphere may be far from *acoustically* neutral.

Typically, the atmosphere is unstable by day and stable by night. This means that classes A–D might be appropriate classes by day and D–F by night. With practice, it is possible to estimate Pasquill Stability Categories in the field, for a particular time and season, from a visual estimate of the degree of cloud cover.

Table 1.4 Pasquill (meteorological) stability categories

Wind speed[a] (m s^{-1})	Daytime incoming solar radiation mW cm^{-2} >60	30–60	<30	Overcast	One hour before sunset or after sunrise	Night-time cloud cover (octas) 0–3	4–7	8
≤ 1.5	A	A–B	B	C	D	F or G[b]	F	D
2.0–2.5	A–B	B	C	C	D	F	E	D
3.0–4.5	B	B–C	C	c	D	E	D	D
5.0–6.0	C	C–D	D	D	D	D	D	D
>6.0	D	D	D	D	D	D	D	D

[a] Measured to the nearest 0.5 m s^{-1} at 11 m height.
[b] Category G is an additional category restricted to the night-time with less than 1 octa of cloud and a wind speed of less than 0.5 m s^{-1}.

The Pasquill classification of meteorological conditions has been adopted widely as the basis of a meteorological classification system for noise prediction schemes [e.g. 16]. However, it is clear from Table 1.4 that the 'meteorologically-neutral' category (C), while being fairly common in a temperate climate, includes such a wide range of wind speeds that it is not very suitable as a category for noise prediction. In the Conservation of Clean Air and Water in Europe (CONCAWE) scheme [16], this problem is addressed by defining six noise prediction categories based on Pasquill categories (representing the temperature gradient) and wind speed. There are 18 sub-categories depending on wind speed. These are defined in Table 1.5. CONCAWE

Table 1.5 CONCAWE meteorological classes for noise prediction

Meteorological category	Pasquill stability category and wind speed (m s^{-1}) (positive is towards receiver) A, B	C, D, E	F, G
1	v < –3.0	—	—
2	–3.0 < v < –0.5	v < –3.0	—
3	–0.5 < v < +0.5	–3.0 < v < –0.5	v < –3.0
4[a]	+0.5 < v < +3.0	–0.5 < v < +0.5	–3.0 < v < –0.5
5	v > +3.0	+0.5 < v < +3.0	–0.5 < v < +0,5
6	—	v > +3.0	+0.5 < v < +3.0

[a] Category with assumed zero meteorological influence.

category 4 is specified as one in which there is zero meteorological influence. So CONCAWE category 4 is equivalent to acoustically neutral conditions.

The CONCAWE scheme requires octave band analysis. Meteorological corrections in this scheme are based primarily on analysis of the Parkin and Scholes' data together with measurements made at several industrial sites. The excess attenuation in each octave band for each category tends to approach asymptotic limits with increasing distance. Values at 2 km for CONCAWE categories 1 (strong wind from receiver to source, hence upward refraction) and 6 (strong downward refraction) are listed in Table 1.6.

Wind speed and temperature gradients are not independent. For example, very large temperature and wind speed gradients cannot coexist. Strong turbulence associated with high wind speeds does not allow the development of marked thermal stratification. Table 1.7 shows a rough estimate of the probability of occurrence of various combinations of wind and temperature gradients [5].

The component of the wind vector in the direction between source and receiver is most important for sound propagation. So, the wind categories (W) must account for this. Moreover, it is possible to give more detailed but qualitative descriptions of each of the meteorological categories (W and TG, see Table 1.8).

Table 1.6 Values of the meteorological corrections for CONCAWE categories 1 and 6

Octave band centre frequency (Hz)	*63*	*125*	*250*	*500*	*1000*	*2000*	*4000*
Category 1	8.9	6.7	4.9	10.0	12.2	7.3	8.8
Category 6	−2.3	−4.2	−6.5	−7.2	−4.9	−4.3	−7.4

Table 1.7 Estimated probability of occurrence of various combinations of wind and temperature gradient

	Zero wind	*Strong wind*	*Very strong wind*
Very large negative temperature gradient	Frequent	Occasional	Rare or never
Large negative temperature gradient	Frequent	Occasional	Occasional
Zero temperature gradient	Occasional	Frequent	Frequent
Large positive temperature gradient	Frequent	Occasional	Occasional
Very large positive temperature gradient	Frequent	Occasional	Rare or never

Table 1.8 Meteorological classes for noise prediction based on qualitative descriptions

W1	Strong wind (>3–5 m s^{-1}) from receiver to source
W2	Moderate wind (≈1–3 m s^{-1})from receiver to source, or strong wind at 45°
W3	No wind, or any cross wind
W4	Moderate wind (≈1 – 3 m s~^{1}) from source to receiver, or strong wind at 45°
W5	Strong wind (>3–5 m s^{-1}) from source to receiver
TG1	Strong negative: daytime with strong radiation (high sun, little cloud cover), dry surface and little wind
TG2	Moderate negative: as T1 but one condition missing
TG3	Near isothermal: early morning or late afternoon (e.g. one hour after sunrise or before sunset)
TG5	Moderate positive: night-time with overcast sky or substantial wind
TG6	Strong positive: night-time with clear sky and little or no wind

In Table 1.9, the revised categories are identified with qualitative predictions of their effects on noise levels [5]. The classes are not symmetrical around zero meteorological influence. Typically, there are more combinations of meteorological conditions that lead to attenuation than to enhancement. Moreover, the increases in noise level (say 1–5 dB) are smaller than the decreases (say 5–20 dB).

The Harmonoise/Imagine project provided a classification of meteorological conditions, based on common observations. Five wind speed categories and five atmospheric stability categories were defined, combinations of which lead to 25 meteorological classes of practical relevance, that – at least in theory – should be considered when predicting yearly averaged exposure due to an outdoor sound source [17].

In addition, the complex-to-measure scaling parameters like friction velocity u_*, temperature scale T_* and the Monin–Obukhov length L can be estimated and used as input in theoretical sound speed profile models (see Section 1.7). The wind speed measured at 10 m height allows an estimate of the friction velocity (see Table 1.10). This classification is based on an average aerodynamic roughness length z_0 of 0.025 m.

Atmospheric stability is strongly governed by solar insolation, which can be measured with dedicated equipment. However, a useful proxy is the cloud cover fraction of the sky, which can be visually estimated, as summarized in Table 1.11. Strong solar insolation (corresponding to clear sky conditions) during the day might give rise to a pronounced temperature decrease with height (category S1 – unstable). At night, the absence of clouds is likely to give rise to a temperature inversion condition (S5 – very stable).

The temperature scale and Monin–Obukhov length will both depend on the combination of the wind category and stability category as shown in Tables 1.12 and 1.13. Clearly, such data has to be used with caution, as e.g. seasonal and topographical effects are disregarded.

Table 1.9 Qualitative estimates of impact of meteorological condition on noise levels

	W1	*W2*	*W3*	*W4*	*W5*
TG1	—	Large attenuation	Small attenuation	Small attenuation	—
TG2	Large attenuation	Small attenuation	Small attenuation	Zero meteorological influence	Small enhancement
TG3	Small attenuation	Small attenuation	Zero meteorological influence	Small enhancement	Small enhancement
TG4	Small attenuation	Zero meteorological influence	Small enhancement	Small enhancement	Large enhancement
TG5	—	Small enhancement	Small enhancement	Large enhancement	—

Table 1.10 Relationship between friction velocity and wind speed at 10 m [17]

Wind speed category	*Mean wind speed at 10 m (m/s)*	u_* *(m/s)*
W1	0 to 1	0.00
W2	1 to 3	0.13
W3	3 to 6	0.3
W4	6 to 10	0.53
W5	>10	0.87

Table 1.11 Relationship between stability category and cloud cover [17]

stability category	*time of day*	*cloud cover*
S1	day	0/8 to 2/8
S2	day	3/8 to 5/8
S3	day	6/8 to 8/8
S4	night	5/8 to 8/8
S5	night	0/8 to 4/8

Table 1.12 Temperature scale T_*(in K) dependence on wind speed category (W) and stability category (S) [17]

	S1	*S2*	*S3*	*S4*	*S5*
W1	–0.4	–0.2	0.0	+0.2	+0.5
W2	–0.2	–0.1	0.0	+0.1	+0.2
W3	–0.1	–0.05	0.0	+0.05	+0.1
W4	–0.05	0.0	0.0	0.0	+0.05
W5	0.0	0.0	0.0	0.0	0.0

Table 1.13 Dependence of Monin-Obhukhov length L^{-1} (m^{-1}) on wind speed category (W) and stability category (S) [17]

	S1	*S2*	*S3*	*S4*	*S5*
W1	–0.08	–0.05	0.0	+0.04	+0.06
W2	–0.05	–0.02	0.0	+0.02	+0.04
W3	–0.02	–0.01	0.0	+0.01	+0.02
W4	–0.01	0.0	0.0	0.0	+0.01
W5	0.0	0.0	0.0	0.0	0.0

It has been suggested [18] that noise calculation procedures should predict average levels, such as would occur under 'neutral' conditions, and that following or opposing winds or temperature inversions will cause variations of ± 10 dB about the average values. However, using the values at 500 Hz as a rough guide to the likely corrections on overall A-weighted broadband levels, it is noticeable that the CONCAWE meteorological corrections are not symmetrical around zero. The CONCAWE scheme suggests meteorological variations of between 10 dB less than the acoustically neutral level for strong upward refraction between source and receiver and 7 dB more than the acoustically neutral level for strong downward refraction between source and receiver.

Zouboff et al. [5] have carried out a series of measurements using a loudspeaker source broadcasting broadband noise with maximum energy in the 500 and 1000 Hz octave bands over a flat homogeneous area, in the South of France, covered with pebbles and sparse vegetation. Acoustical data were collected at a series of microphones positioned between 20 and 640 m from the source. Meteorological parameters (mean air temperature and wind speed at three heights, together with wind direction, solar radiation and hygrometry) were monitored on a 22-m high tower located approximately at the centre of the measurement line. One hundred and ninety-five 10-minute long samples were collected over a range of meteorological conditions and were expressed in terms of L_{Aeq}. Since the ground condition changed very little, most of the variation may be attributed to meteorological effects. Figure 1.19 shows the maximum minimum and mean total attenuation from 80 to 640 m, deduced from levels measured at 1.5-m high microphones normalized to a level of 100 dB at 20 m. These data offer further evidence for asymmetry of meteorological effects about the mean noise level. The difference between the minimum

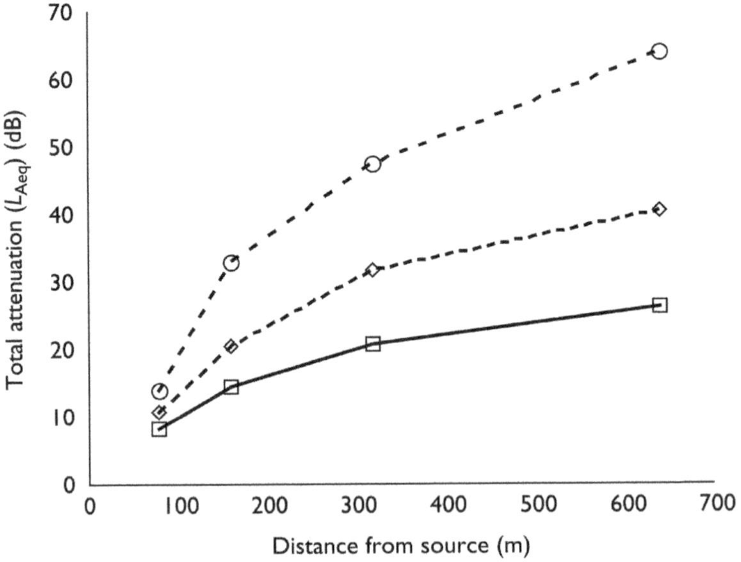

Figure 1.19 Maximum, mean and minimum total attenuation deduced from Zouboff *et al* [15] data for 10-minute L_{Aeq}.

and mean levels is considerably less than the difference between the maximum and mean levels. Smaller differences were obtained with longer averaging times. For example, the 38 dB range in 10 minute L_{Aeq} at 640 m is reduced to only 19 dB when comparing 8 hour L_{Aeq} during days differing in wind direction and cloud cover.

Longer term values of L_{Aeq} will be dominated by the highest levels, even when they are relatively infrequent. Moreover, levels observed under downward refraction conditions exhibit less variability than those measured under upward refraction conditions. For these reasons, the ISO Scheme [19] predicts for 'moderate' downwind conditions and distinguishes long-term L_{Aeq} (say seasonal or monthly) from short-term (say daily) L_{Aeq}.

1.6 TYPICAL SOUND SPEED PROFILES

Outdoor sound prediction requires information on wind speed, direction, temperature, relative humidity and barometric pressure as a function of height near to the propagation path. These gradients determine the sound speed profile. Ideally, the heights at which the meteorological data are collected should reflect the application. If this information is not available, then there are alternative procedures. It is possible, for example, to generate an approximate sound speed profile from temperature and wind speed at a given height using Monin–Obukhov similarity theory [20] and to input this directly. According to this theory, the wind speed component (m/s) in the source–receiver direction and temperature (°C) at height z are calculated from the values at ground level and other parameters as follows:

$$u(z) = \frac{u_*}{k}\left[\ln\left\{\frac{z + z_M}{z_M}\right\} + \psi_M\left(\frac{z}{L}\right)\right] \tag{1.2}$$

$$T(z) = T_0 + \frac{T_*}{k}\left[\ln\left\{\frac{z + z_H}{z_H}\right\} + \psi_H\left(\frac{z}{L}\right)\right] + \Gamma z, \tag{1.3}$$

where

u^*	Friction velocity (m/s)	(depends on surface roughness)
z_M	Momentum roughness length	(depends on surface roughness)
z_H	Heat roughness length	(depends on surface roughness)
T^*	Scaling temperature °K	The precise value of this is not important for sound propagation. A convenient value is 283 °K.
κ	Von Kármán constant	(= 0.41)
T_0	Temperature °C at zero height	Again it is convenient to use 283 °K

Γ	Adiabatic correction factor	= –0.01 °C/m for dry air Moisture affects this value but the difference is small.
L	Obukhov length (m) > 0 → stable, < 0 → unstable	$= \pm \frac{u_*^2}{kgT_*}\left(T_{av} + 273.15\right)$, the thickness of the surface or boundary layer is given by $2L$ m.
T_{av}	Average temperature °C	It is convenient to use $T_{av} = 10$ so that $(T_{av} + 273.15) = \theta_0$
ψ_M	Diabatic momentum profile correction (mixing) function	if $L < 0$ $= 5(z/L)$ if $L > 0$
ψ_H	Diabatic heat profile correction (mixing) function	$= -2\ln\left(\frac{(1+\chi_M)}{2}\right)$ if L < 0 $=5(z/L)$ if L > 0 or for $z \leq 0.5L$ [2]
χ_M	Inverse diabatic influence or momentum function	$= \left[1 - \frac{16z}{L}\right]^{0.25}$
χ_H	Inverse diabatic influence function for momentum	$= \left[1 - \frac{16z}{L}\right]^{0.5}$

For a neutral atmosphere, $1/L = 0$ and $\psi_M = \psi_H = 0$.

The associated sound speed profile, c(z), is calculated from

$$c(z) = c(0)\sqrt{\frac{T(z) + 273.15}{273.15}} + u(z) \tag{1.4}$$

Note that the resulting profiles are valid in the surface or boundary layer only but not at zero height. In fact, the profiles given by the above equations, sometimes called Businger–Dyer profiles [21], have been found to give good agreement with measured profiles up to 100 m. This height range is relevant to sound propagation over distances up to 10 km [22]. However, improved profiles are available that are valid to greater heights. For example [23],

$$\psi_M = \psi_H = -7\ ln\ (z/L) - 4.25/(z/L) + 0.5/(z/L)^2 - 0.852, \text{ for } z > 0.5L. \tag{1.5}$$

Often z_M and z_H are taken to be equal. The roughness length varies, for example, between 0.0002 (still water) and 0.1 (grass). More generally, the roughness length can be estimated from the Davenport classification [24].

Figure 1.20 shows examples of sound speed (difference) profiles, $(c(z) - c(0))$, generated from equations (1.3)–(1.5) using

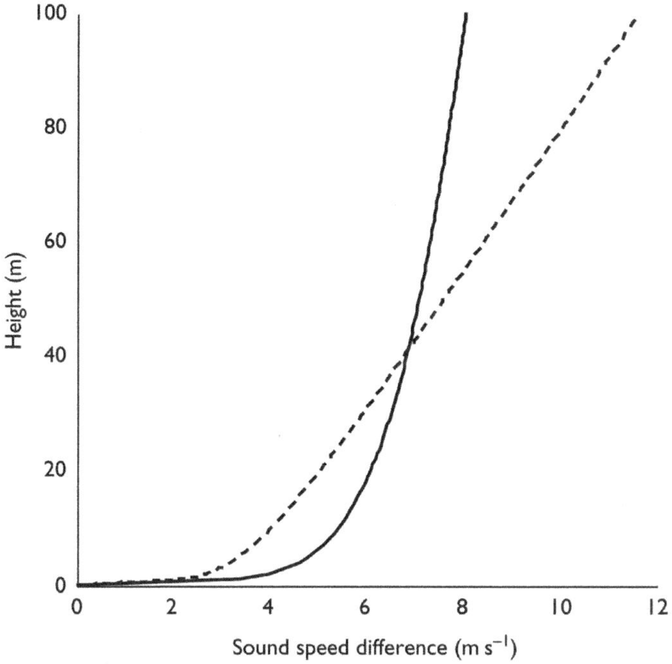

Figure 1.20 Downward refracting sound speed profiles relative to the sound speed at the ground obtained from similarity theory. The continuous curve is approximately logarithmic corresponding to a large Obukhov length and to a cloudy, windy night. The broken curve corresponds to a small Obukhov length as on a calm clear night and is predominantly linear away from the ground.

(a) $z_M = z_H = 0.02$, $u^* = 0.34$, $T^* = 0.0212$, $T_{av} = 10$, $T_0 = 6$, (giving $L = -390.64$)
(b) $z_M = z_H = 0.02$, $u^* = 0.15$, $T^* = 0.1371$, $T_{av} = 10$, $T_0 = 6$, (giving $L = -11.76$)

and $\Gamma = -0.01$. These parameters are intended to correspond to a cloudy windy night and a calm clear night, respectively [25].

Salomons et al. [26] have suggested a method for obtaining the remaining unknown parameters, u^*, T^* and L from the relationship

$$L = \frac{u_*^2}{kgT^*} \tag{1.6}$$

and the Pasquill Category (P).

From empirical meteorological tables, approximate relationships between the Pasquill class P the wind speed u_{10} at a reference height of 10 m and the fractional cloud cover N_c have been obtained. The latter determines the incoming solar radiation and therefore the heating of the ground. The former is a guide to the degree of mixing. The approximate relationship is

$$P(u_{10},N_c) = 1 + 3\left[1 + \exp(3.5 - 0.5u_{10} - 0.5N_c)\right]^{-1} \text{during the day}$$
$$6 - 2\left[1 + \exp(12 - 2u_{10} - 2N_c)\right]^{-1} \text{during the night} \tag{1.7}$$

A relationship between the Obukhov length L m as a function of P and roughness length $z_0 < 0.5$ m is

$$\frac{1}{L(P,z_0)} = B_1(P)\log(z_0) + B_2(P), \tag{1.8}$$

where

$$B_1(P) = 0.0436 - 0.0017P - 0.0023P^2 \tag{1.9}$$

and

$$\max(0,0.025P-0.125) \text{ for } 4 \le P \le 6 \tag{1.10}$$

Alternatively, values of B_1 and B_2 may be obtained from Table 1.14.

Equations (1.7) and (1.8) give

$$L = L(u_{10}, N_c, z_0) \tag{1.11}$$

Also u_{10} is given by equation (1.2) with $z = 10$ m, i.e.

$$u(z) = \frac{u^*}{k}\left[\ln\left\{\frac{10 + z_M}{z_M}\right\} + \psi_M\left(\frac{10}{L}\right)\right] \tag{1.12}$$

Equations (1.6), (1.11) and (1.12) may be solved for u^*, T^* and L. Hence it is possible to calculate ψ_M, ψ_H, $u(z)$ and $T(z)$.

Figure 1.21 shows the results of this procedure for a ground with a roughness length of 0.1 m and two upwind and downwind daytime classes defined by the parameters listed in the caption.

A consequence of atmospheric turbulence is that instantaneous profiles of temperature and wind speed show considerable variations with both time and position. These variations are eliminated considerably by averaging

Table 1.14 Values of the constants B_1 and B_2 in (1.7) for the six Pasquill classes

Pasquill class	A	B	C	D	E	F
B_1	0.04	0.03	0.02	0	−0.02	−0.05
B_2	−0.08	−0.035	0	0	0	0.025

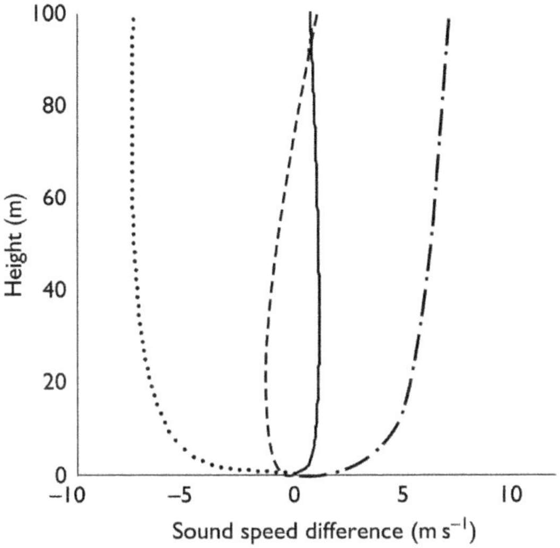

Key

Line type	Wind speed at 10 m in m s^{-1}	Cloud cover in octels	Pasquill class	Direction
———————— (solid)	1	0	A	Downwind
– – – – – – – – (dashed)	1	0	A	Upwind
— · — · — · — (dash-dot)	5	4	C	Downwind
·························· (dotted)	5	4	C	Upwind

Figure 1.21 Two daytime sound speed profiles (upwind – dashed and dotted; downwind - solid and dash-dot) determined from the parameters listed in Table 1.10.

over a period of the order of ten minutes. The Monin–Obukhov or Businger–Dyer models give good descriptions of the averaged profiles.

The Pasquill Category C profiles shown in Figure 1.21 are approximated closely by logarithmic curves of the form

$$c(z) = c(0) + b\ln\left[\frac{z}{z_0} + 1\right], \tag{1.13}$$

where the parameter b (>0 for downward refraction and <0 for upward refraction) is a measure of the strength of the atmospheric refraction. Such logarithmic sound speed profiles are realistic for open ground areas without obstacles particularly in the daytime.

A better fit to night-time profiles is obtained with power laws of the form [23]

$$c(z) = c(0) + b(z/z_0)^{\alpha}, \tag{1.14}$$

where

$$\alpha = 0.4(P-4)^{\frac{1}{4}}.$$

The temperature term in effective sound speed profile given by equation (1.4) can be approximated by truncating a Taylor expansion after the first term to give

$$c(z) = c(T_0) + \frac{1}{2}\sqrt{\frac{\kappa R}{T_0}}\left(T(z) - T_0\right) + u(z) \tag{1.15}$$

By comparing with 12 months of meteorological data obtained at a 50 m high meteorological tower in Germany, Heimann and Salomons [27] have found that (1.15) is a reasonably accurate approximation to vertical profiles of effective sound speed even in unstable conditions and in situations where Monin–Obukhov theory is not valid.

Also, the prediction of outdoor sound propagation requires information about turbulence. Specifically, it requires values of the mean square refractive index, the outer length scale of the turbulence and a parameter representing the transverse separation between adjacent rays (see also Chapter 10). The mean squared refractive index may be calculated from the measured instantaneous variation of wind speed and temperature with time at the receiver.

$$\mu^2 = \frac{\sigma_w^2 \cos^2\alpha}{C_0^2} + \frac{\sigma_T^2}{4T_0^2}, \tag{1.16}$$

where σ_w^2 is the variance of the wind velocity, σ_T^2 is the variance of the temperature fluctuations, α is the wind vector direction and C_0 and T_0 are the ambient sound speed and temperature, respectively.

In the absence of turbulence data, typical values of mean squared refractive index are between 10^{-6} for calm conditions and 10^{-4} for strong turbulence. The outer length scale may be approximated by the height of the receiver, as long as source and receiver are close to the ground. The greatest effect of turbulence is predicted when the path separation is set to zero. Chapter 10 contains a more detailed discussion of turbulence.

1.7 LINEAR-LOGARITHMIC REPRESENTATIONS OF SOUND SPEED PROFILES

Combining Equations (1.13) and (1.4) leads to a general effective sound speed profile c_{eff} with a linear-logarithmic dependency on height z above the ground:

$$c_{eff}(z) = a_0 + a_{lin}z + a_{log}\ln\left(1 + \frac{z}{z_0}\right), \tag{1.17}$$

where z_0 is the aerodynamic roughness length.

The coefficients a_{lin} and a_{log} can be deduced from measurements of air temperature $T(z)$ and wind speed $u(z)$ at various heights and wind direction; a_0 is the sound speed at ground level ($=c_0$). The effective sound speed at a specific height z can be estimated from an alternative version of equation (1.4):

$$c_{eff}(z) = \sqrt{\gamma R_{gas} T(z)} + u_{SR}(z), \quad (1.18)$$

where γ is the ratio of the specific heat capacities at constant pressure c_p and constant volume c_V ($\gamma = \frac{c_p}{c_V} = 1.4$ for a diatomic gas like air) and R_{gas} is the gas constant of dry air (287 J/(kg K)). The wind speed component along the source–receiver line is given by $u_{SR}(z)$ and has a positive sign in case of downwind sound propagation, a negative sign in case of upwind sound propagation and becomes zero for (exact) crosswind sound propagation. A meteorological convention is used whereby wind direction is defined as the direction from which the wind is blowing (with 0° for wind blowing from the North, 90° from the East, etc.). The angle of sound propagation is defined as the direction in which the sound is propagating (i.e. 0° for sound waves travelling towards the North, 90° to the East, etc.). The difference in angle between the wind direction and propagation angle, α, can be used to calculate $u_{SR}(z)$:

$$u_{SR}(z) = u(z) \cos(\pi - \alpha) \quad (1.19)$$

As an illustration, detailed meteorological tower data over a full year was processed to show the suitability of the linear-logarithmic relationship in equation (1.17). Continuous effective sound speed profiles have been fitted to the measured data (using equation (1.17)). The meteorological data consisted of 10-minute averages of wind speed (at 24 m, 48 m, 69 m, 78 m and 114 m above the ground) and air temperature (at 8 m, 24 m, 48 m, 78 m and 114 m), over a full year (1996, location: near the city of Mol, Belgium). Although wind direction was measured at three heights (24 m, 69 m and 114 m), only the lower sensor data was used for the current analysis. For this location, the aerodynamic roughness length was estimated to be 0.1 m. Frequency distributions of some meteorological observations are depicted in Figure 1.22.

Figure 1.23 illustrates the goodness of the fit for different types of effective sound speed profiles. A non-linear fitting procedure was used. Relatively simple linear-log sound speed profiles provide accurate fits to the data, as has been shown by similar exercises before [28]. Also, there is sufficient flexibility to represent more complex profiles, for example, containing both upward- and downward-refracting parts.

For each 10-minute averaged meteorological observation, a unique set of values of c_0, a_{lin} and a_{log} can be calculated. Note that refraction is governed by the combination (a_{log}, a_{lin}) and not by the absolute value of the sound speed at ground level c_0. Consequently, the latter was not considered during further analysis. The (a_{log}, a_{lin}) pairs are a compact way of representing the refractive state of the atmospheric boundary layer and their distribution

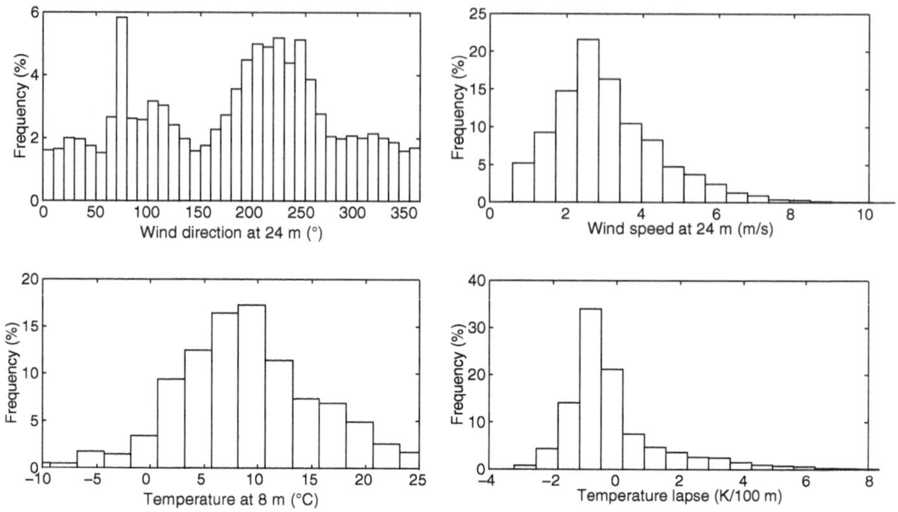

Figure 1.22 Histograms of measured wind direction at 24 m, wind speed at 24 m, air temperature at 8 m, and temperature lapse (10-minute averaged data).

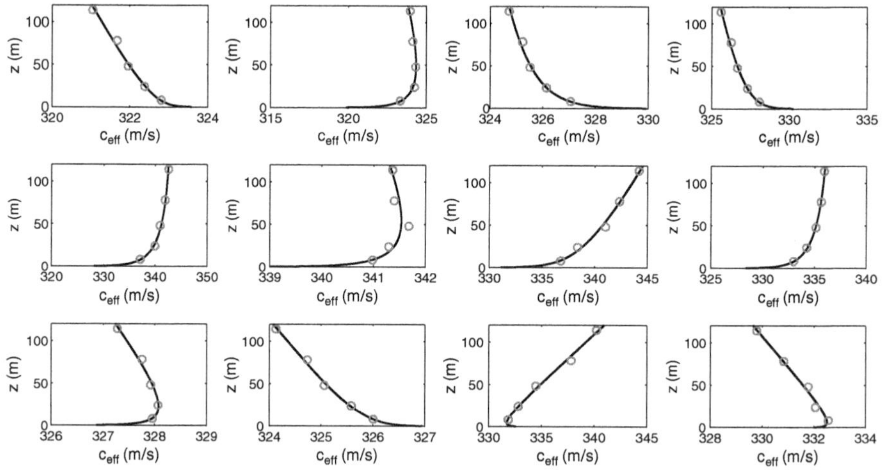

Figure 1.23 A selection of fitted effective sound speed profiles (full lines, 10-minute averaged meteorological data) using Equation (1.20a–1.20c); the open circles depict the effective sound speeds measured at the sensor heights using Equation (1.21a–1.21c). The receiver is positioned in southern direction (180°).

over the year at a specific location is of interest for long-term averaged predictions of outdoor sound propagation.

Since similar (a_{log}, a_{lin}) combinations will give rise to similar propagation conditions, they can be used in a classification scheme involving only a limited number of sound speed profiles. The class width for the a_{log} parameter was set to 0.15 m/s, and for the a_{lin} parameter to 0.015/s. Figures 1.24(a)–(d) summarize all of the meteorological profiles for four propagation directions, namely those in Northern (0°), Eastern (90°), Southern (180°) and Western (270°) directions, respectively. Note that the direction where the receiver is thought to be positioned will influence the wind velocity component experienced. The categorized effective sound speed profiles for sound propagation in Southern (180°) direction are depicted further in Figure 1.25.

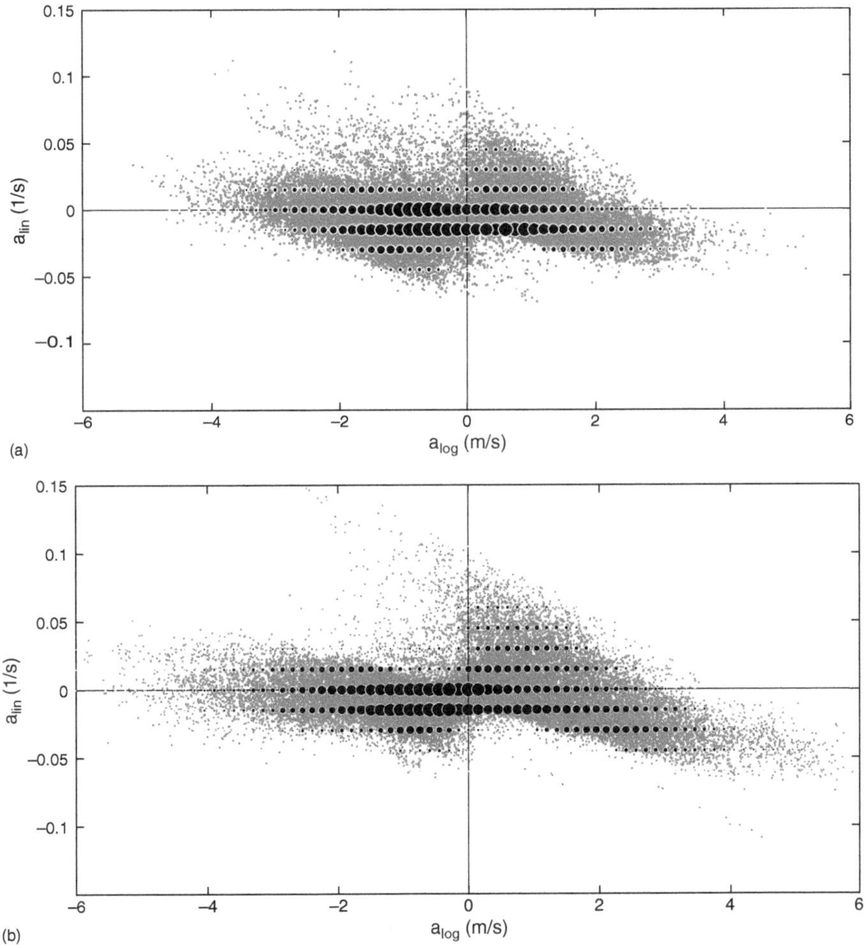

Figure 1.24 Scatter plots of best-fit (a_{log},a_{lin}) pairs over a full year of meteorological tower data. The grey dots indicate all fits (10-minute averaged data), the filled black circles the categorized (a_{log},a_{lin}) combinations, with their radii proportional to the number of occurrences in each class. The receiver is positioned (a) in Northern direction (0°), and (b) in Eastern direction (90°),

(Continued)

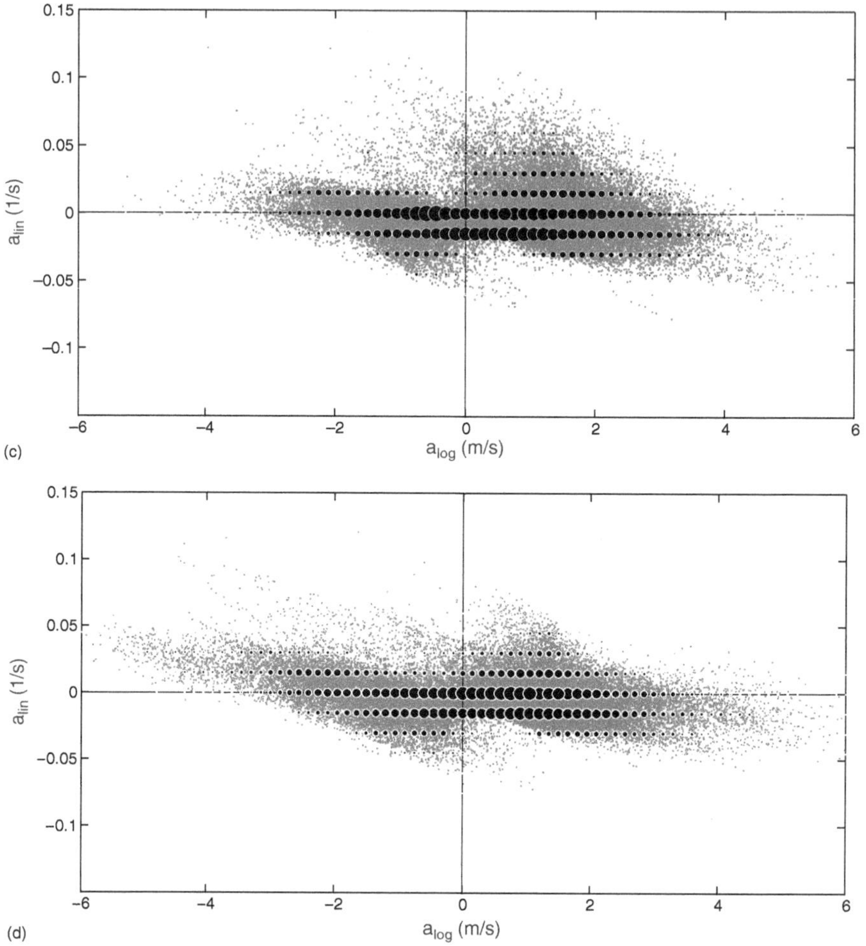

Figure 1.24 (Continued) (c) Southern direction (180°) and (d) Western direction (270°).

Note that meteorological measurements at heights below 8 m were not available. Air temperature observations closer to the ground could further improve the fit, especially under conditions of ground-based temperature inversions. On the other hand, the observation point at 114 m above the ground limits the need for extrapolation of wind speed and air temperature data. Meteorological data at this height is relevant for sound propagation up to roughly 1–2 km from a sound source.

When detailed meteorological data is not available, a_{lin} and a_{log} in equation (1.17) can be related to the physical parameters of theoretical-empirical flux-profile relationships for the case of a flat and homogeneous terrain [29–31]:

$$a_{\log} = \frac{u^* \cos(\pi - \alpha)}{\kappa} + \frac{1}{2}\frac{\gamma R_{gas}}{c_0}\frac{T^*}{\kappa} 0.74 \frac{T^*}{\kappa} \tag{1.20a}$$

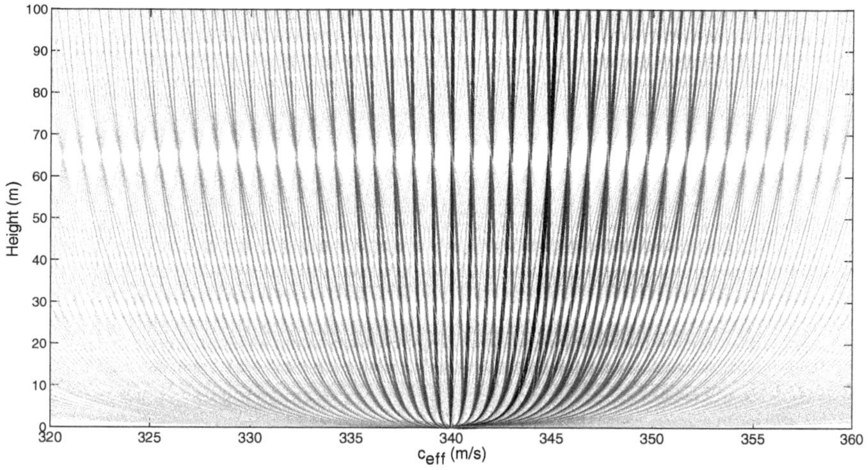

Figure 1.25 Categorized effective sound speed profiles corresponding to the data presented in Figure 1.17(c) (receiver in Southern direction, 180°), assuming a fixed sound speed of 340 m/s at ground level. The line thicknesses (and, similarly, the gray scale) are proportional to the number of occurrences in each profile. The figure presents the most frequent effective sound speed profiles over the year considered.

$$\begin{cases} \text{day} \rightarrow a_{lin} = 1.0\dfrac{u^* \cos(\pi - \alpha)}{\kappa L} + \dfrac{1}{2}\dfrac{\gamma R_{gas}}{c_0}\left(0.74\dfrac{T^*}{\kappa L} + \Gamma_d\right) \\ \text{night} \rightarrow a_{lin} = 4.7\dfrac{u^* \cos(\pi - \alpha)}{\kappa L} + \dfrac{1}{2}\dfrac{\gamma R_{gas}}{c_0}\left(4.7\dfrac{T^*}{\kappa L} + \Gamma_d\right) \end{cases}, \tag{1.20b}$$

where

$$\Gamma_d = -\frac{g}{c_p}, \tag{1.20c}$$

the dry adiabatic vertical temperature gradient, which is equal to –0.0097 K/m. The von Kármán constant κ is equal to 0.4.

Equations (1.20a–1.20c) need knowledge of the surface-scaling parameters friction velocity u_*, temperature scale T_* and the Monin–Obukhov length L. The meteorological classification scheme proposed by Harmonoise (see Section 1.6) provides averaged values for these since they are not accessible from common meteorological observations. Instead, the wind speed at 10 m height and the cloud cover fraction of the sky, together with time of the day, can be used as a rough proxy for their determination. To evaluate equations (1.20a–1.20c), wind direction is needed as well.

1.8 AIR ABSORPTION

During its passage through the air, sound is subject to two forms of dissipation of energy: classical and relaxation [32]. The classical absorption is associated with transfer of the energy of the coherent molecular motion to equivalent heat energy or random kinetic energy of translation of air molecules. The relaxation absorption mechanism is associated with redistribution of the translational or internal energy of the molecules. The relaxation mechanism may be divided into rotational and vibrational parts, the former being more significant at high values of frequency or pressure. These effects may be combined. For a plane wave, pressure p at distance x from a position where the pressure is p_0 is given by

$$p = p_0 e^{-\alpha x/2}$$

The frequency-, humidity- and temperature-dependent attenuation coefficient α for air absorption may be calculated using equations (1.21a–1.21c) [32–35].

$$\alpha = f^2\left[\left(\frac{1.84\times10^{-11}}{\left(\frac{T_0}{T}\right)^{\frac{1}{2}}\frac{p_s}{p_0}}\right)+\left(\frac{T_0}{T}\right)^{2.5}\left(\frac{0.10680e^{-3352/T}f_{r,N}}{f^2+f_{r,N}^{\;2}}+\frac{0.01278e^{-2239.1/T}f_{r,O}}{f^2+f_{r,O}^{\;2}}\right)\frac{nepers}{m\cdot atm}\right], \quad (1.21a)$$

where $f_{r,N}$ and $f_{r,O}$ are given by

$$f_{r,N} = \frac{p_s}{p_{s0}}\left(\frac{T_0}{T}\right)^{\frac{1}{2}}\left(9+280He^{-4.17\left[(T_0/T)^{1/3}-1\right]}\right), \quad (1.21b)$$

$$f_{r,O} = \frac{p_s}{p_{s0}}\left(24.0+4.04\times10^4 H\frac{0.02+H}{0.391+H}\right), \quad (1.21c)$$

where f is the frequency, T is the absolute temperature of the atmosphere in degrees Kelvin, T_0 = 293.15 K is the reference value of T (20°C), $H = \rho_{sat} r_h p_0 / p_s$ is the percentage molar concentration of water vapour in the atmosphere, r_h is the relative humidity (%), p_s is local atmospheric pressure and p_0 is the reference atmospheric pressure (1atm = 1.01325×10^5 Pa).

$$\rho_{sat} = 10^{C_{sat}},$$

where

$$C_{sat} = -6.8346 \quad (T_0 / T)^{1.261} + 4.6151$$

These formulae give estimates of the absorption of pure tones to an accuracy of ±10% for

$$0.05 < H < 5{,}253 < T < 323, p_0 < 200\,\text{kPa}$$

Outdoor air absorption varies through the day and the year [36,37].

Absolute humidity, H, is an important factor in the diurnal variation and usually peaks in the afternoon. Usually, the diurnal variations are greatest in the summer. It should be noted that use of (arithmetic) mean values of atmospheric absorption may lead to overestimates of attenuation when attempting to establish worst case exposures for the purposes of environmental noise impact assessment. Investigations of local climate statistics, say hourly means over one year, should lead to more accurate estimates of the lowest absorption values.

Figure 1.26 shows values of α versus relative humidity for air at 20°C and normal atmospheric pressure for frequencies between 2 and 12.5 kHz [38].

Table 1.15 shows numerical values for dB/km at specific frequencies.

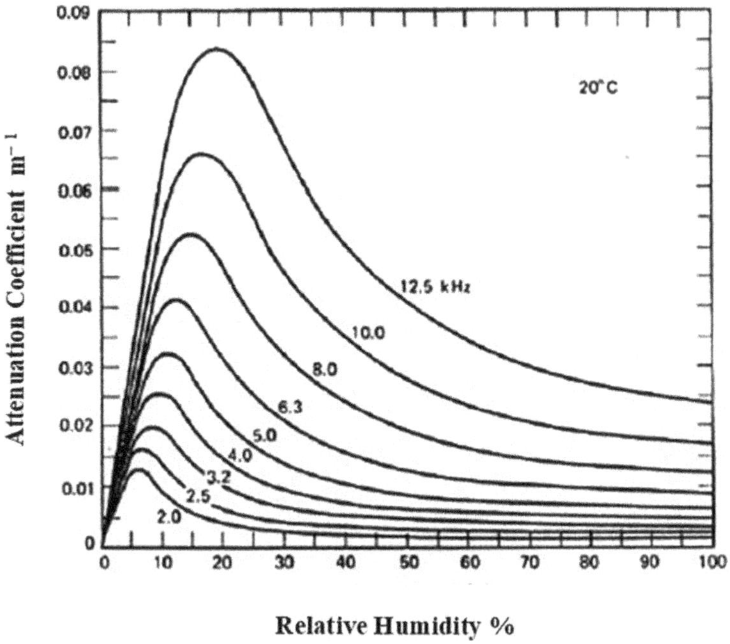

Figure 1.26 Variation of Attenuation Coefficient for air absorption with relative humidity.

Table 1.15 Total sound absorption in dB km^{-1} versus relative humidity as a function of frequency at 20°C (68°F)

Frequency (kHz)	*Relative humidity (%)*										
	0	*10*	*20*	*30*	*40*	*50*	*60*	*70*	*80*	*90*	*100*
2	4.14	38.2	17.4	10.9	8.34	7.14	6.55	6.28	6.19	6.21	6.29
4	8.84	102	62.3	38.9	28.0	22.2	18.7	16.6	15.2	14.2	13.6
6.3	14.9	154	135	90.6	65.6	51.3	42.5	36.7	32.7	29.8	27.7
10	26.3	202	261	205	155	123	102	87.3	77.0	69.3	63.5
12.5	35.8	224	338	294	232	187	156	134	1.18	106	96.6
16	52.2	250	428	423	355	294	248	214	189	170	155
20	75.4	281	511	564	508	435	374	326	289	261	238

NOTE

1 SEL is sometimes denoted by L_{AE} and is the energy equivalent continuous level over 1 s ($L_{Aeq,1s}$).

REFERENCES

[1] F. V. Hunt, *Origins in acoustics*, publ. Acoustical Society of America, AIP (1992).
[2] C. D. Ross, Outdoor sound propagation in the US Civil War, *Appl. Acoust.* 59 137–147 (2000).
[3] M. V. Naramoto, A concise history of acoustics in warfare, *Appl. Acoust.* 59 128–136 (1999).
[4] Proceedings of the first international symposium on long range sound propagation, Gulfport, MS, University of Mississippi (1981).
[5] V. Zouboff, Y. Brunet, M. Berengier and E. Sechet, A qualitative approach of atmospherical effects on long range sound propagation, in *Proc. 6th International Symposium on Long range Sound propagation*, eds. D. I. Havelock and M. Stinson, NRCC, Ottawa, publ. National research Council of Canada, 251–269 (1994).
[6] P. Parkin, Acoustical reminiscences: the Rayleigh Lecture, *Proc. Inst. Acoust.* 1 7–31 (1978).
[7] P. H. Parkin and W. E. Scholes, The horizontal propagation of sound from a jet engine close to the ground at Radlett, *J. Sound Vib.* 1 1–13 (1965).
[8] P. H. Parkin and W. E. Scholes, The horizontal propagation of sound from a jet engine close to the ground at Hatfield, *J. Sound Vib.* 2 353–374 (1965).
[9] The results of measurements of the horizontal propagation of sound at Radlett, DSIR, Building Research Station, Internal Notes B215 and B249 (unpublished).
[10] The results of measurements of the horizontal propagation of sound at Hatfield, DSIR, Building Research Station, Internal Notes IN99 (1964) (unpublished).
[11] O. Zaporozhets, V. Tokarev and K. Attenborough, Predicting Noise from Aircraft Operated on the Ground, *Appl. Acoust.* 64 941–953 (2003).
[12] K. Attenborough, K. M. Li and S. Taherzadeh, Propagation from a broadband source over grassland: comparison of models and data, *Proc. Inter-Noise 95*, Newport Beach, 1, 319 (1995).
[13] J. P. Chambers, H. Saurenman, R. Bronsdon, L. Sutherland, R. Waxler, K. Gilbert and C. Talmadge, Effects of Temperature Induced Inversion Conditions on Suburban Highway Noise Levels, *Acta Acust. united Ac.* 92 1060–1070 (2006).
[14] D. Hohenwarter and E. Mursch-Radlgruber, Nocturnal boundary layer profiles and measured frequency dependent influence on sound propagation, *Appl. Acoust.* 76 416–430 (2014).
[15] T. Van Renterghem, D. Botteldooren and P. Lercher, Comparison of measurements and predictions of sound propagation in a valley-slope configuration in an inhomogeneous atmosphere, *J. Acoust. Soc. Am.* 121 2522–2533 (2007).
[16] K. J. Marsh, The CONCAWE model for calculating the propagation of noise from open-air industrial plants, *Appl. Acoust.* 15 411–428 (1982).
[17] J. Defrance, E. Salomons, I. Noordhoek, D. Heimann, B. Plovsing, G. Watts, H. Jonasson, X. Zhang, E. Premat, I. Schmich, F. Aballea, M. Baulac and F. de Roo, Outdoor sound propagation reference model developed in the European Harmonoise project, *Acta Acust. united Ac.* 93 213–227 (2007).

[18] I. H. Flindell evidence to Heathrow T5 enquiry (unpublished).
[19] ISO 9613-2 Acoustics – Attenuation of Sound During Propagation Outdoors – Part 2: General Method of Calculation. International Standard ISO 9613-2 (1996).
[20] A. S. Monin and A. M. Yaglom, *Statistical Fluid Mechanics: mechanics of turbulence*, Vol 1, MIT press, Cambridge, Mass (1979).
[21] R. B. Stull, *An introduction to boundary layer meteorology*, Kluwer, Dordrecht, pp 347–386 (1991).
[22] A. A. M. Holtslag, Estimates of diabatic wind speed profiles from near surface weather observations, *Boundary Layer Meteorol.* 29 225–250 (1984).
[23] E. M. Salomons, Downwind propagation of sound in an atmosphere with a realistic sound speed profile: a semi-analytical ray model, *J. Acoust. Soc. Am.* 95 2425–2436 (1994).
[24] A. G. Davenport, Rationale for determining design wind velocities, *J. Am. Soc. Civ. Eng.* ST-86 39–68 (1960).
[25] W. H. T. Huisman, Sound propagation over vegetation-covered ground, Ph.D. Thesis, University of Nijmegen, The Netherlands (1990).
[26] E. M. Salomons, F. H. van den Berg and H. E. A. Brackenhoff, Long-term average sound transfer through the atmosphere based on meteorological statistics and numerical computations of sound propagation. *Proc. 6th International Symposium on Long range Sound propagation*, ed. D. I. Havelock and M. Stinson, NRCC, Ottawa, 209–228 (1994).
[27] D. Heimann and E. Salomons, Testing meteorological classifications for the prediction of long-term average sound levels, *Appl. Acoust.* 65 925–950 (2004).
[28] D. Heimann, M. Bakermans, J. Defrance and D. Kühner, Vertical sound speed profiles determined from meteorological measurements near the ground, *Acta Acust. united Ac.* 93 228–240 (2007).
[29] A. S. Monin and A. M. Obukhov, Basic regularity in turbulent mixing in the surface layer of the atmosphere, *Akad. Nauk.* SSSR 151 163–187 (1954).
[30] J. A. Businger, J. C. Wyngaard, Y. Izumi and E. F. Bradley, Flux-profile relationships in the atmospheric boundary layer, *J. Atmos. Sci.* 28 181–189 (1971).
[31] C. A. Paulson, The mathematical representation of wind speed and temperature profiles in the unstable atmospheric surface layer, *J. Appl. Meteor.* 9 857–861 (1970).
[32] L. C. Sutherland and H. E. Bass, Atmospheric absorption in the atmosphere at high altitudes, *Proc. 7th International Symposium on Long range Sound propagation*, Lyon (1996).
[33] ANSI SI.26-1995, American National Standard method for calculation of the absorption of sound by the atmosphere, Acoustical Society of America, New York (1995).
[34] ISO 9613-1: 1993, Acoustics — Attenuation of sound during propagation outdoors — Part 1: Calculation of the absorption of sound by the atmosphere, International Organization for Standardization, Geneva, Switzerland (1993).
[35] D. T. Blackstock, *Fundamentals of Physical Acoustics*, University of Texas, Austin, Texas, John Wiley & Sons, Inc (2000).
[36] C. M. Harris, Absorption of Sound in Air versus Humidity and Temperature, *J. Acoust. Soc. Am.* 40 148–159 (1966).
[37] C. Larsson, Atmospheric Absorption Conditions for Horizontal Sound Propagation, *Appl. Acoust.* 50 231–245 (1997).
[38] C. Larsson, Weather Effects on Outdoor Sound Propagation, *Int. J. Acoust. Vib.* 5 33–36 (2000).

Chapter 2

The Propagation of Sound near Ground Surfaces in a Homogeneous Medium

2.1 INTRODUCTION

Analytical methods used for predicting point-to-point propagation of sound waves above a boundary between two homogenous stationary media have been adapted from studies of the propagation of electromagnetic (em) waves, initiated by Sommerfeld [1] in 1909 and continued by Banos et al. [2]. Following these early studies of em waves, the propagation of sound above a porous ground surface has been studied, among others, by Rudnick [3] in the 1940s, by Ingard [4] and Paul [5] in 1950s, by Wenzel [6], Donato [7], Thomasson [8,9] and Chien and Soroka [10,11] in the 1970s and by Attenborough et al. [12] and Kawai et al. [13] in the 1980s. These efforts have resulted in closed form analytic approximations for the sound field from a point source in a homogeneous and still atmosphere above a flat ground which have been generalized, subsequently, by Li et al. [14–16] to allow for layered ground. Where the ground is composed of multiple layers, all of which contribute to sound reflection, the concept of effective impedance greatly simplifies the computational requirements. The analytical treatment in this chapter is based on the work of Li et al. [14–16]. In anticipation of its use in more computationally intensive numerical schemes such as FFP (see Chapter 10), the Fourier transform method is explored in considerable detail.

2.2 A POINT SOURCE ABOVE SMOOTH FLAT ACOUSTICALLY SOFT GROUND

Sound propagating from a point source over smooth flat acoustically soft ground can be idealized by the two-media problem shown in Figure 2.1. In the adopted rectangular co-ordinate system, z is the vertical axis, x and y are the horizontal axes and the plane interface is at $z = 0$. The lower medium is idealized as a homogeneous rigid-porous medium and treated as an effective fluid with complex density ρ_1 and complex propagation constant k_1. The upper medium represents the homogeneous stationary atmosphere with

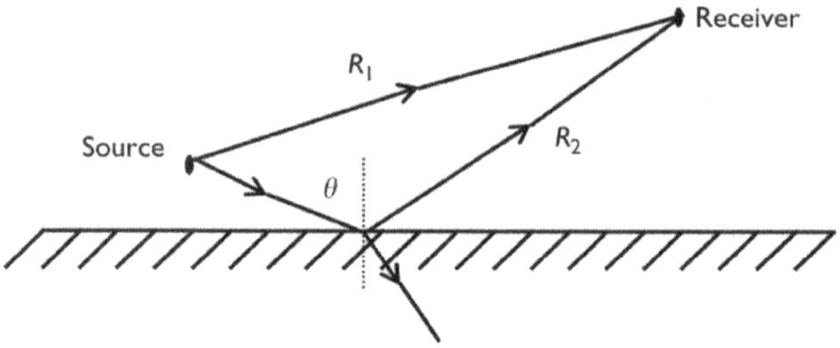

Figure 2.1 Ray-paths from a point source above an interface between two media.

constant uniform air density and speed of sound denoted by ρ and c, respectively. The acoustic pressure, p, in the upper medium is given by the inhomogeneous Helmholtz equation

$$\nabla^2 p + k^2 p = S(x_s), \tag{2.1}$$

where $k = \omega/c$ is the wave number, ω is the angular frequency of the source, time dependence factor $e^{-i\omega t}$ is understood and $S(x_s)$ is the source term at the point x_s. For convenience, but without loss of generality, the source position is assumed to be located at $(0,0,z_s)$. In the lower medium, the governing equation for the acoustic pressure, p_1, is

$$\nabla^2 p_1 + k_1^2 p_1 = 0 \tag{2.2}$$

$k_1 = \omega/c_1$ is the wave number and c_1 is the (frequency-dependent) sound speed in the porous medium. Subsequently, all variables with subscript 1 refer to the lower medium. At the plane interface, the acoustic pressure and normal particle velocity should be continuous. This implies that

$$p = p \text{ and } \frac{1}{\rho}\frac{\partial p}{\partial z} = \frac{1}{\rho_1}\frac{\partial p_1}{\partial z} \tag{2.3}$$

on $z = 0$.

The Green's function, $G_m(x|x_s)$ is to be derived for the sound field at $x \equiv (x,y,z)$ due to a point monopole source situated at x_s, where $(x_s) = -\delta(x)\delta(y)\delta(z - z_s)$.

$G_m(x|x_s)$ satisfies the inhomogeneous equation

$$\left(\nabla^2 + k^2\right)G_m(x) = -\delta(x)\delta(y)\delta(z - z_s), \tag{2.4}$$

where the second argument in the Green's function is suppressed for brevity, i.e. $G_m(x) \equiv G_m(x|x_s)$. To find the solution to this inhomogeneous equation, it is convenient to introduce a Fourier transform pair for the Green's function as follows.

$$\hat{G}_m(k_x,k_y,z) = \int_{-\infty}^{\infty}\int_{-\infty}^{\infty} G_m(x)e^{-ik_x x - ik_y y}dxdy \tag{2.5a}$$

and

$$G_m(x) = \frac{1}{4\pi^2}\int_{-\infty}^{\infty}\int_{-\infty}^{\infty}\hat{G}_m(k_x,k_y,z)e^{ik_xx+ik_yy}dk_xdk_y \tag{2.5b}$$

Then (2.4) becomes

$$\frac{d^2\hat{G}_m}{dz^2} + k_z^2\hat{G}_m = -\delta(z - z_s), \tag{2.6a}$$

where

$$k_z = +\sqrt{k^2 - \kappa^2} \tag{2.6b}$$

and

$$\kappa^2 = k_x^2 + k_y^2. \tag{2.6c}$$

This mathematical technique transforms the physical space, (x,y), into an imaginary κ-space, (k_x, k_y).

A similar Fourier transform pair for the lower medium enables (2.2) to be simplified to

$$\frac{d^2\hat{p}_1}{dz^2} + K_1^2\hat{p}_1 = 0, \tag{2.7a}$$

where

$$K_1 = +\sqrt{k_1^2 - \kappa^2} \tag{2.7b}$$

In (2.6b) and (2.7b), positive roots are chosen to ensure a finite and bounded solution for the inhomogeneous Helmoltz equation. The boundary conditions given in (2.3) do not change their form after the transformation. Hence, the boundary conditions in the transformed space, on $z = 0$, are

$$\hat{G}_m = \hat{p}_1 \text{ and } \frac{1}{\rho}\frac{d\hat{G}_m}{dz} = \frac{1}{\rho_1}\frac{d\hat{p}_1}{dz} \tag{2.8}$$

From (2.6a) and (2.7a), the acoustic field due to a point monopole source at $(0, 0, z_s)$ is

$$\hat{G}_m = A_u\ (e^{ik_zz} + \bar{U}_\infty e^{-ik_zz})\ (z \geq z_s) \tag{2.9a}$$

$$\hat{G}_m = A_d\ (e^{-ik_zz} + U_0e^{ik_zz})\ (z_s \geq z \geq 0) \tag{2.9b}$$

$$\hat{p}_1 = B_1\ (e^{-iK_zz} + U_1e^{ik_lz})\ (0 \geq z) \tag{2.9c}$$

where A_u, $\bar{U}_\infty$, A_d, U_0, B_1 and U_1 are constants to be determined from the boundary conditions. The solutions given in (2.9a)–(2.9c) contain both outgoing ($e^{ik_z z}$ for $z > z_s$, $e^{-ik_z z}$ for $z_s > z \geq 0$ and $e^{-iK_1 z}$ for $z \leq 0$) and incoming ($e^{-ik_z z}$ for $z > z_s$, $e^{ik_z z}$ for $z_s > z \geq 0$ and $e^{iK_1 z}$ for $z \leq 0$) waves. In these solutions, $\bar{U}_\infty e^{-ik_z z}$ represents a wave reflected from the 'top' of upper medium, $U_0 e^{ik_z z}$ represents that reflected from the air/ground interface and $U_1 e^{iK_1 z}$ represents that reflected from the 'bottom' of the lower medium. However, the Sommerfeld radiation condition requires that the sound field contains no incoming waves when $z \to \pm\infty$; therefore, $\bar{U}_\infty$ and U_1 must vanish for large z. By using (2.9c) with $U_1 = 0$, we can eliminate $\hat{p}_1$ from the boundary condition (2.8) to give

$$\frac{d\hat{G}_m}{dz} + ik\varsigma_1\sqrt{n_1^2 - (\kappa/k)^2}\,\hat{G}_m = 0, \tag{2.10}$$

on $z = 0$, where ς_1 is the ratio of the density in air to that in the ground and n_1 is the ratio of the sound speed in air to that in the ground, i.e. the index of refraction in the ground, given respectively by

$$\varsigma_1 = \rho/\rho_1 \tag{2.11a}$$

and

$$n_1 = k_1/k = c/c_1 \tag{2.11b}$$

In addition to the boundary condition (2.10) on $z = 0$, continuity of pressure and discontinuity of pressure gradient are required at the plane $z = z_s$, i.e.

$$\hat{G}_m\Big|_{z=z_s^+} = \hat{G}_m\Big|_{z=z_s^-} \quad \text{and} \quad \frac{d\hat{G}_m}{dz}\Big|_{z=z_s^+} - \frac{d\hat{G}_m}{dz}\Big|_{z=z_s^-} = -1 \tag{2.12}$$

The constants A_u, A_d and B_1 can be determined according to the conditions specified by (2.10) and (2.12), and, hence, the Green's function $G_m(k_x,k_y,z)$ in the transformed space can be expressed as

$$\hat{G}_m(k_x,k_y,z) = \frac{i}{2k_z}\left\{e^{ik_z|z-z_s|} + U_0 e^{ik_z(z+z_s)}\right\}, \tag{2.13}$$

where U_0 corresponds to the reflection coefficient of the air/ground interface in the transformed space and is given by

$$U_0 = \frac{(k_z/k) - \varsigma_1\sqrt{n_1^2 - (\kappa/k)^2}}{(k_z/k) + \sqrt{n_1^2 - (\kappa/k)^2}} = 1 - \frac{2\sqrt{n_1^2 - (\kappa/k)^2}}{(k_z/k) + \varsigma_1\sqrt{n_1^2 - (\kappa/k)^2}} \tag{2.14}$$

Substitution of (2.13) and (2.14) into (2.5b) leads to an integral expression for the Green's function for the sound field as follows:

$$G_m(x) = \frac{1}{4\pi^2}\int_{-\infty}^{\infty}\int_{-\infty}^{\infty}\frac{i}{2k_z}e^{ik_x x + ik_y y + ik_z|z - z_s|}dk_x dk_y$$
$$+\frac{1}{4\pi^2}\int_{-\infty}^{\infty}\int_{-\infty}^{\infty}\frac{i}{2k_z}e^{ik_x x + ik_y y + ik_z|z + z_s|}dk_x dk_y$$
$$-\frac{1}{2\pi^2}\int_{-\infty}^{\infty}\int_{-\infty}^{\infty}\frac{\zeta_1\sqrt{n_1^2 - (\kappa/k)^2}}{(k_z/k) + \zeta_1\sqrt{n_1^2 - (\kappa/k)^2}}\frac{ie^{ik_x x + ik_y y + ik_z|z + z_s|}}{2k_z}dk_x dk_y \quad (2.15)$$

The first two terms of the Green's function can be identified as Sommerfeld integrals [see, for example, Ref. 2, Chapter 2] and can be evaluated exactly as

$$\frac{1}{4\pi^2}\int_{-\infty}^{\infty}\int_{-\infty}^{\infty}\frac{i}{2k_z}e^{ik_x x + ik_y y + ik_z|z - z_s|}dk_x dk_y = \frac{e^{ikR_1}}{4\pi R_1} \quad (2.16a)$$

and

$$\frac{1}{4\pi^2}\int_{-\infty}^{\infty}\int_{-\infty}^{\infty}\frac{i}{2k_z}e^{ik_x x + ik_y y + ik_z(z + z_s)}dk_x dk_y = \frac{e^{ikR_2}}{4\pi R_2} \quad (2.16b)$$

where $R_1 = \sqrt{x^2 + y^2 + (z - z_s)^2}$ is the distance from the source located at $(0,0,z_s)$ to the receiver located at (x,y,z) and $R_2 = \sqrt{x^2 + y^2 + (z + z_s)^2}$ is the distance from the image source located at $(0,0,-z_s)$ to the same location.

The evaluation of the integral in the third term of (2.15) requires considerable effort and, in general, an exact solution is not possible. However, the integral can be estimated asymptotically by the method of steepest descents. Taking the advantage of the fact that the solution is axi-symmetric about the $z = 0$ axis, the problem may be simplified by employing a polar co-ordinate system instead of the rectangular Cartesian co-ordinates. With the transformation $(k_x, k_y) \Rightarrow (\kappa, \varepsilon)$, where κ and ε are, respectively, the magnitude and phase of the wavenumber in κ-space,

$$k_x = \kappa \cos \varepsilon \quad (2.17a)$$

$$k_y = \kappa \sin \varepsilon \quad (2.17b)$$

$$dk_x dk_y = \kappa d\kappa d\varepsilon. \quad (2.17c)$$

Similarly, (x,y) transforms to (r,ψ) for the field points in the horizontal plane of constant z, viz.

$$x = r\cos\psi \quad (2.18a)$$

and

$$y = r\sin\psi, \quad (2.18b)$$

where r is the horizontal range from source to receiver. To cover the whole integration range with k_x and k_y varying from $-\infty$ to $+\infty$, the azimuth angle ψ is required to vary from 0 to 2π and κ must vary from 0 to ∞. The unknown integral, which is denoted by I, can now be rewritten in polar form as

$$I = -\frac{1}{4\pi^2}\int_0^{\infty}\int_0^{2\pi} \frac{\varsigma_1\sqrt{n_g^2-(\kappa/k)^2}}{(k_z/k)+\varsigma_1\sqrt{n_g^2-(\kappa/k)^2}} \frac{i\kappa e^{i\kappa r\cos(\varepsilon-\psi)+ik_z(z+z_s)}}{k_z} d\kappa d\varepsilon. \tag{2.19}$$

The integral over ε can be evaluated by means of the integral expression for the Bessel function of zero order [17; Eq. 9.1.21]:

$$\int_0^{2\pi} e^{[i\kappa r\cos(\varepsilon-\psi)]} d\varepsilon = 2\pi J_0(\kappa r).$$

Hence,

$$I = -\frac{i}{2\pi}\int_0^{\infty} \frac{\varsigma_1\sqrt{n_1^2-(\kappa/k)^2}}{(k_z/k)+\varsigma_1\sqrt{n_1^2-(\kappa/k)^2}} \frac{\kappa e^{ik_z(z+z_s)}}{k_z} J_0(\kappa r)d\kappa \tag{2.20}$$

Before evaluating this integral I, two limiting cases shed light on the influence of the ground, i.e. the lower medium, on the total sound field. The first limiting case is when the ground is acoustically hard, for example a large thick metallic sheet or impermeable sealed concrete such that air is unable to penetrate the surface. In this limiting case, the density of the ground is much higher than the density of air, i.e. $\varsigma_1 = \rho/\rho_1 \ll 1$ and the contribution of the integral I to the total sound field is negligibly small. So, the Green's function is simply the sum of the two terms given in (2.16a) and (2.16b):

$$G_m(x) = \frac{e^{ikR_1}}{4\pi R_1} + \frac{e^{ikR_2}}{4\pi R_2} \tag{2.21}$$

This solution has a straightforward interpretation since the first term corresponds to the direct wave, i.e. the sound field caused by a source at $(0, 0, z)$; and the second term is the reflected wave, i.e. the sound field of a mirror image source located at $(0, 0, -z)$.

If the ground is porous (e.g. grassland or cultivated soil or a pervious asphalt road surface), air particle motion associated with incident sound penetrates the ground. In this case, ρ_1 represents the (complex frequency-dependent) density of the air in the pores (assuming a rigid solid matrix, see Chapter 5). If $|\rho_1|$ and ρ are comparable, the assumption $\varsigma_1 \ll 1$ is no longer valid and a more detailed consideration of the interaction of the sound wave with ground surfaces is required.

A consequence of visco-thermal interactions in the pores is that the sound speed within a typical outdoor ground is significantly less than that above the ground, i.e. $c > c_1$. This means that $n_1 > 1$ and, therefore, according to Snell's Law, the sound ray is refracted towards the normal as it propagates from air and penetrates the ground. For most outdoor ground surfaces including grassland, $n_1 \gg 1$, so that $\sqrt{n_1^2-(\kappa/k)^2} \approx n_1$. Sound waves at any angle of incidence will be strongly refracted towards the normal to the surface.

The specific normalized impedance, Z is

$$Z = \frac{\rho_1 c_1}{\rho c} = \frac{1}{\varsigma_1 n_1} \tag{2.22a}$$

For further developments, it is more convenient to use the specific normalized admittance, $\beta = 1/Z$. Consequently, according to (2.22a), we have

$$\beta = \varsigma_1 n_1 \tag{2.22b}$$

If $n_1 \gg 1$, the surface is called locally reacting, since the air/ground interaction is independent of the angle of incidence of the incoming waves. With this approximation, the boundary condition (2.10) becomes

$$\frac{d\hat{G}_m}{dz} + ik\beta\hat{G}_m = 0, \tag{2.23}$$

which is called the impedance boundary condition.

2.3 THE SOUND FIELD ABOVE A LOCALLY REACTING GROUND

This section will give a derivation of the classical expression for the sound field due to a monopole source above a locally reacting ground. The fields above more complicated ground surfaces will be discussed in Section 2.4.

If $n_1 \gg 1$, the integral I in (2.19) can be simplified to

$$I = -\frac{i\beta}{2\pi}\int_0^\infty \frac{\kappa}{k_z}\frac{e^{ik_z(z+z_s)}}{(k_z/k)+\beta}J_0(\kappa r)\,d\kappa \tag{2.24}$$

To facilitate the analysis, we introduce a further transformation such that

$$\kappa = k\sin\mu \tag{2.25a}$$

It is straightforward to show

$$d\kappa = k\cos\mu\, d\mu \tag{2.25b}$$

and

$$k_z = \sqrt{k^2 - \kappa^2} = k\cos\mu \tag{2.25c}$$

The integration limit for μ cannot be restricted to real values of this angle because κ is required to vary from 0 to ∞. Consequently, k_z varies from k to $i\infty$ according to (2.25c). As shown in Figure 2.2, the path of integration, Γ_1, for the new variable starts from $\mu = 0$, moves along the real axis to $\mu = \pi/2$, then makes a right-angle turn and continues parallel to the imaginary axis from $\mu = \frac{\pi}{2} + i0$ to $\mu = \frac{\pi}{2} - i\infty$.

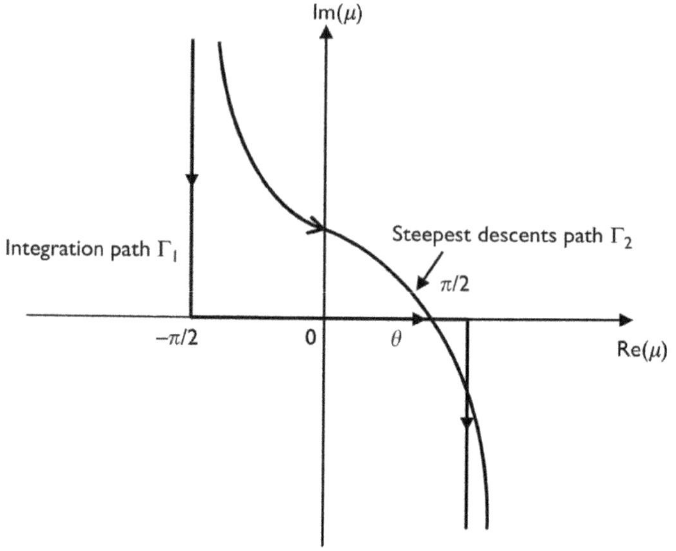

Figure 2.2 Integration paths for the integral ***I*** in spherical polar co-ordinates.

Taking the image source location as the centre, we write the separation between the source and receiver in spherical polar co-ordinates (R_2, θ, ψ):

$$z + z = R_2 \cos\theta, \quad r = R_2 \sin\theta, \tag{2.26}$$

Also, a connection between Hankel functions and the Bessel function can be used, viz.

$$H_0^{(1)}(u) = -H_0^{(2)}(-u), \quad J_0(u) = \frac{1}{2}\left[H_0^{(1)}(u) + H_0^{(2)}(u)\right] \tag{2.27}$$

Substitution of (2.25a–2.25c)–(2.27) into (2.24) leads to

$$I = -\frac{ik\beta}{4\pi}\left\{\int_0^{\pi/2-i\infty} H_0^{(1)}(kr\sin\mu)\frac{e^{ikR_2\cos\theta\cos\mu}}{\cos\mu+\beta}\sin\mu d\mu s\right.$$
$$\left.-\int_0^{\pi/2-i\infty} H_0^{(1)}(-kr\sin\mu)\frac{e^{ikR_2\cos\theta\cos\mu}}{\cos\mu+\beta}\sin\mu d\mu\right\}$$

After replacing μ with $-\mu$ in the second integral, noting the different integration limits and combining the integrals,

$$I = -\frac{ik\beta}{4\pi}\int_{-\pi/2+i\infty}^{\pi/2-i\infty}\frac{\sin\mu}{\cos\mu+\beta}\left\{H_0^{(1)}(kr\sin\mu)e^{-ikr\sin\mu}\right\}e^{ikR_2\cos(\mu-\theta)}d\mu \tag{2.28}$$

with the path of integration, Γ_2, shown also in Figure 2.2.

The integral (2.28) can be approximated by a uniform asymptotic expansion [18] that combines the steepest descent approach and the pole subtraction method [10]. Only the solution and its interpretation are outlined here.

The integral (2.28) can be evaluated asymptotically to yield

$$I = \Phi_p + \varphi_s, \tag{2.29a}$$

where

$$\Phi_p = -\frac{k\beta}{4} erfc\left(-ix_0/\sqrt{2}\right) H_0^{(1)}\left(kr\sqrt{1-\beta^2}\right) e^{-ik(z+z_s)\beta}, \tag{2.29b}$$

$$\varphi_s \approx \frac{-ik\beta}{2\pi} \frac{e^{ikR_2}}{ikR_2(\beta+\cos\theta)} \left\{1 - \frac{1}{\sqrt{2}}\left[\frac{1+\beta\cos\theta}{\sqrt{1-\beta^2}\sin\theta}+1\right]^{\frac{1}{2}} + \frac{1}{ikR_2(\beta+\cos\theta)^2}\right.$$
$$\left.\times\left([1+\beta\cos\theta] - \frac{\sqrt{1-\beta^2}\sin\theta}{8\sqrt{2}}\left[\frac{1+\beta\cos\theta}{\sqrt{1-\beta^2}\sin\theta}+3\right]\times\left[\frac{1+\beta\cos\theta}{\sqrt{1-\beta^2}\sin\theta}+1\right]^{\frac{3}{2}}\right)+\ldots\right. \tag{2.29c}$$

and erfc( ) is the complementary error function [Ref. 17, Chapter 7]. φ_s represents a correction term resulting from subtraction of the pole. The variable x_0, which fixes the location of the pole, is determined by

$$\frac{1}{2}x_0^2 = ikR_2\left\{1 + \beta\cos\theta - \sqrt{1-\beta^2}\sin\theta\right\} \tag{2.30}$$

The significance of the pole location, and hence of x_0, will be discussed at a later stage when the intriguing physical phenomenon of the associated airborne surface wave is investigated. In deriving (2.29a–2.29c), we have assumed that the following relationships hold: $kr \gg 1$, $r \gg z + z_s$ and $k(z + z_s) \gg 1$. These approximations mean that the resulting expressions will be valid only for long range and high frequency and if both the source and receiver are located close to the ground surface. The Green's function for the sound field due to a monopole source radiating sound above a locally reacting ground can be determined by summing (2.16a), (2.16b) and (2.29a) to yield

$$G(x) = \frac{e^{ikR_1}}{4\pi R_1} + \frac{e^{ikR_2}}{4\pi R_2} + \Phi_p + \varphi_s, \tag{2.31}$$

where Φ_p and φ_s are given in (2.29b) and (2.29c), respectively.

The 'exact' Green's function with (2.29b) and (2.29c) involves a considerable number of terms that may not be convenient for routine use. But additional assumptions can be made. Although there are approximations corresponding to the relatively 'soft' boundary case where $|\beta|^2 kr \gg 1$, the horizontal range where $|\beta|^2 kr \ll 1 \ll kr$ and other limiting cases, the most versatile approximation follows from the condition for a relatively hard boundary. If $|\beta|^2 \ll 1$ and $r \approx R_2$, then the factor in the curly bracket of (2.30) can be simplified to

$$1+\beta cos\theta - \sqrt{1-\beta^2}\,\sin\theta \approx \frac{1}{2}(\beta+\cos\theta)^2 \tag{2.32a}$$

The Hankel function can be replaced by the first term of its asymptotic expansion,

$$H_0^{(1)}\left\{kr\sqrt{1-\beta^2}\right\} \approx \sqrt{2/(i\pi kR_2)}e^{ikR_2} \tag{2.32b}$$

Substituting (2.32a) and (2.32b) into (2.29b), we obtain an approximation for Φ_p as

$$\Phi_p = 2i\sqrt{\pi}\left(\frac{1}{2}kR_2\right)^{1/2}\beta e^{-w^2} erfc(-ip_e)\frac{e^{ikR_2}}{4\pi R_2}, \tag{2.33a}$$

where $w\left(\equiv x_0/\sqrt{2}\right)$, the numerical distance, is approximated by

$$w = \frac{1}{2}(1+i)\sqrt{kR_2}\,(\cos\theta+\beta) \tag{2.33b}$$

Furthermore, it is possible to show that the correction term, $\varphi_s \approx 0$, so it is small when compared with Φ_p. Hence, it can be ignored in (2.31) and the total sound field above a locally reacting ground can be computed by substituting (2.33a) into (2.31) to yield an approximate Green's function:

$$G(x,y,z) = \frac{e^{ikR_1}}{4\pi R_1} + \frac{e^{ikR_2}}{4\pi R_2} + i\sqrt{k/8\pi R_2}\,e^{-w^2}\,\mathrm{erfc}(-iw)\,\beta e^{ikR_2} \tag{2.34}$$

Computation of the error function, which is complex, i.e. $w = w_r + iw_x$, can be carried out for a large range of $|w|$ [17, p. 328].

If $w_r > 3.9$ or $w_i > 3$,

$$e^{-w^2}\mathrm{erfc}(-iw) = iw\left(\frac{0.461315}{w^2-0.1901635}+\frac{0.09999216}{w^2-1.7844927}+\frac{0.002883894}{w^2-5.5253437}\right) \tag{2.35a}$$

with an absolute error of less than 2×10^{-6}.

If $w_r > 6$ or $w_i > 6$,

$$e^{-w^2}\mathrm{erfc}(-iw) = iw\left(\frac{0.5124242}{w^2-0.2752551}+\frac{0.05176536}{w^2-2.724745}\right) \tag{2.35b}$$

with an absolute error less than 1×10^{-6}.

For smaller values of w_r and w_i, another formula is available [19],

$$e^{-w^2}\mathrm{erfc}(-iw) = K_1(w_x,w_r) + iK_2(w_x,w_r) \tag{2.36a}$$

$$K_1(w_x,w_r) = \frac{hw_x}{\pi(w_x^2 + w_r^2)} + \frac{2w_x h}{\pi}\sum_{n=1}^{\infty}\frac{e^{-n^2h^2}(w_x^2 + w_r^2 + n^2h^2)}{(w_x^2 - w_r^2 + n^2h^2)^2 + 4w_x^2 w_r^2} - \frac{w_x}{\pi}E(h)$$
$$\begin{aligned} &+P && \text{if} && w_x < \pi/sh, \\ &+\frac{1}{2}P && \text{if} && w_x = \pi/h, \\ &+0 && \text{if} && w_x > \pi/h, \end{aligned} \tag{2.36b}$$

$$K_2(w_x,w_r) = \frac{hw_r}{\pi(w_x^2 + w_r^2)} + \frac{2w_r h}{\pi}\sum_{n=1}^{\infty}\frac{e^{-n^2h^2}(w_x^2 + w_r^2 - n^2h^2)}{(w_x^2 - w_r^2 + n^2h^2)^2 + 4w_r^2 w_x^2} + \frac{w_r}{\pi}E(h)$$
$$\begin{aligned} &-Q && \text{if} && w_x\ \pi/h, \\ &-\frac{1}{2}Q && \text{if} && w_x = \pi/h, \\ &+0 && \text{if} && w_x\ \pi/h, \end{aligned} \tag{2.36c}$$

where

$$\begin{cases} P = 2e^{-\left[w_r^2 + (2w_x\pi/h) - w_x^2\right]}\left[\dfrac{A_1C_1 - B_1D_1}{C_1^2 + D_1^2}\right] \\ Q = 2e^{-\left[w_r^2 + (2w_x\pi/h) - w_x^2\right]}\left[\dfrac{A_1D_1 + B_1C_1}{C_1^2 + D_1^2}\right] \end{cases} \tag{2.36d}$$

with

$$\begin{cases} A_1 = \cos(2w_r w_x) \\ B_1 = \sin(2w_r w_x) \\ C_1 = e^{-2w_x\pi/h} - \cos(2w_r\pi/h) \\ D_1 = \sin(2w_r\pi/h) \end{cases} \tag{2.36e}$$

and h is an accuracy criterion.

The error bounds $E(h)$ in (2.36b) and (2.36c) can be estimated from

$$E(h) \leq 2\sqrt{\pi}e^{-\pi^2/h^2}/\left(1 - e^{-\pi^2/h^2}\right) \tag{2.36f}$$

If $h = 1$, $|E(h)| \leq 10^{-4}$ so only three or four terms are needed of the infinite sums in $H(w_x,w_r)$ and $K(w_x,w_r)$ to meet the requirement [20]. If h is 0.8, then the magnitude of the error term becomes less than 10^{-6} and summing the series up to the fifth term will be sufficient to guarantee the required accuracy [17]. Note that a typographical error in Ref. [17] has been corrected in (2.36b).

Although (2.31) is a more accurate asymptotic solution, the approximation (2.34) is found to be sufficiently accurate for most practical purposes. Moreover, (2.34) is preferred because it can be rewritten in a form that leads to a useful interpretation of each term, thereby enhancing the physical understanding of the problem. To interpret each term in (2.34), first consider a simpler but related problem in which a plane wave impinges on an impedance boundary at an oblique angle, θ. From physical considerations and associated mathematical analysis [21], the total sound field, p_t should consist of a direct wave, p_d and a specularly reflected wave, p_r multiplied by the plane wave reflection coefficient,

$$R_p = \frac{\cos\theta - \beta}{\cos\theta + \beta} \tag{2.37}$$

Hence, $p_t = p_d + R_p p_r$.

In an analogous way, the sound field due to a point source should consist of the following:

(a) a direct contribution from the source, and
(b) a specularly reflected contribution modified by the reflection coefficient, Q, for spherical waves.

With this in mind, it is useful to rewrite the solution for the integral (2.28) in its approximate form as

$$I = 2i\sqrt{\pi}\,\frac{\beta}{\beta + \cos\theta}\,we^{-w^2}\,\mathrm{erfc}(-iw)\frac{e^{ikR_2}}{4\pi R_2} \tag{2.38}$$

So, the sound field due to a point source above a locally reacting ground becomes

$$p(x,y,z) = \frac{e^{ikR_1}}{4\pi R_1} + \frac{e^{ikR_2}}{4\pi R_2} + 2i\sqrt{\pi}\,\frac{\beta}{\beta + \cos\theta}\,we^{-w^2}\,\mathrm{erfc}(-iw)\frac{e^{ikR_2}}{4\pi R_2} \tag{2.39}$$

Regrouping the second and third terms of (2.39), the sound field can be written as

$$p(x,y,z) = \frac{e^{ikR_1}}{4\pi R_1} + \left[R_p + (1 - R_p)F(w)\right]\frac{e^{ikR_2}}{4\pi R_2}, \tag{2.40a}$$

where $F(w)$, sometimes called the boundary loss factor, is given by

$$F(w) = 1 + i\sqrt{\pi}\,w\exp\left(-w^2\right)\mathrm{erfc}(-iw) \tag{2.40b}$$

and the term in the square bracket of (2.40a) may be interpreted as the spherical wave reflection coefficient

$$Q = R_p + (1 - R_p)F(w) \tag{2.40c}$$

At grazing incidence $\theta = \pi/2$, so that in (2.37) $R_p = -1$ and (2.40a) can be simplified considerably leading to

$$p(x,y,z) = 2F(w)e^{ikr}/r, \tag{2.41a}$$

where the numerical distance, w (see 2.33b) is given by the simplified expression

$$w = \frac{1}{2}(1+i)\beta\sqrt{kr} \tag{2.41b}$$

It is important to notice that if both source and receiver are on the ground so that $R_p = -1$, then use of the plane wave reflection coefficient (2.37) instead of the spherical wave reflection coefficient (2.40c) would mean the prediction of a zero sound field which does not accord with common observation. The contribution of the second term in Q to the total field acts as a correction for the fact that the wavefronts are spherical rather than plane. This contribution has been called the *ground wave*, in analogy with the corresponding term in the theory of AM radio reception [22]. The function $F(w)$ describes the interaction of a curved wavefront with a ground of finite impedance. If the wavefront is plane, which is increasingly the case as $R_2 \to \infty$, then $|w| \to \infty$ and $F \to 0$. Alternatively, if the surface is acoustically hard, then $|\beta| \to 0$ which implies $|w| \to 0$ and $F \to 1$.

Many other accurate asymptotic and numerical solutions are available, and many numerical comparisons have been carried out between their various predictions. But, for practical geometries and typical outdoor ground surfaces, no significant numerical differences have been revealed. Equation (2.40a) is known as the Weyl-Van der Pol equation, a name borrowed from electromagnetic propagation theory [23], and is the most widely used analytical solution for predicting sound field above a locally reacting ground in a homogeneous atmosphere (without refraction or turbulence).

One of the interesting features of outdoor sound propagation over acoustically soft ground, for certain ground impedance conditions and geometries, is the existence of surface wave. This will be discussed in Section 2.5.

2.4 THE SOUND FIELD ABOVE A LAYERED EXTENDED-REACTION GROUND

While the use of (2.34) for predicting outdoor sound propagation near to the ground is satisfactory, in some outdoor situations, for example, when predicting propagation over low flow resistivity ground such as a porous road surface, gravel or a layer of newly fallen snow, modelling the ground surface as an impedance boundary may be inappropriate. The problem arises since, in such a case, the index of refraction, n_1, is not sufficiently high to assume $n_1 \gg 1$, and the refraction of sound wave depends on the angle at which sound is incident. This means that the apparent impedance depends not only on the physical properties of the ground surface but also, critically, on the angle of incidence. An improved formulation for the integral I in (2.20) is to use $\varsigma_1\sqrt{n_1^2-(\kappa/k)^2}$ instead of β in the subsequent evaluation of the integral.

Using the same transformation ((2.27)(2.29a)–(2.29c)), (2.20) can be rewritten as

$$I=-\frac{ik}{4\pi}\int_{-\pi/2+i\infty}^{\pi/2-i\infty}\frac{\varsigma_1\sqrt{n_g^2-\sin^2\mu}\,\sin\mu}{\cos\mu+\varsigma_1\sqrt{n_g^2-\sin^2\mu}}\left\{H_0^{(1)}(kr\sin\mu)e^{-ikr\sin\mu}\right\}e^{ikR_2\cos(\mu-\theta)}d\mu. \tag{2.42}$$

Attempts to evaluate the above integral asymptotically with no further approximation have resulted in forms that are rather complicated and inconvenient to compute [12]. A useful heuristic approximation is to replace $\sqrt{n_1^2-\sin^2\mu}$ by $\sqrt{n_1^2-\sin^2\theta}$.

With an 'effective' admittance, β_e defined by

$$\beta_e=\varsigma_1\sqrt{n_1^2-\sin^2\theta}, \tag{2.43}$$

(2.42) can be recast as

$$I=-\frac{ik\beta_e}{4\pi}\int_{-\pi/2+i\infty}^{\pi/2-i\infty}\frac{\sin\mu}{\cos\mu+\beta_e}\left\{H_0^{(1)}(kr\sin\mu)e^{-ikr\sin\mu}\right\}e^{ikR_2\cos(\mu-\theta)}d\mu \tag{2.44}$$

Using the analysis in Section 2.3 and the analogous forms of (2.28) and (2.44), it can be seen that both sound fields are identical [cf. (2.28) and (2.34)] except that the effective admittance (2.43) should be used instead of β. The boundary condition (2.10) can be recast in the form of an effective admittance condition, cf. (2.43), i.e.

$$\frac{d\hat{G}_m}{dz}+ik\beta_e\hat{G}_m=0 \tag{2.45}$$

The above analysis is based on the underlying assumption that the porous ground is homogeneous and semi-infinite. In many situations, for example, when considering propagation over freshly fallen snow on frozen ground or over a porous asphalt layer, there ground surface is a highly porous surface layer above a relatively non-porous substrate. So, it is useful to derive the effective admittance of the surface of ground consisting of a hard-backed layer.

Consider the reflection of plane waves on such a ground, see Figure 2.3, with the layer thickness denoted by d_1.

The technique described in Section 2.2 can be applied but supplementary boundary conditions are required to satisfy the requirement for a hard-backed

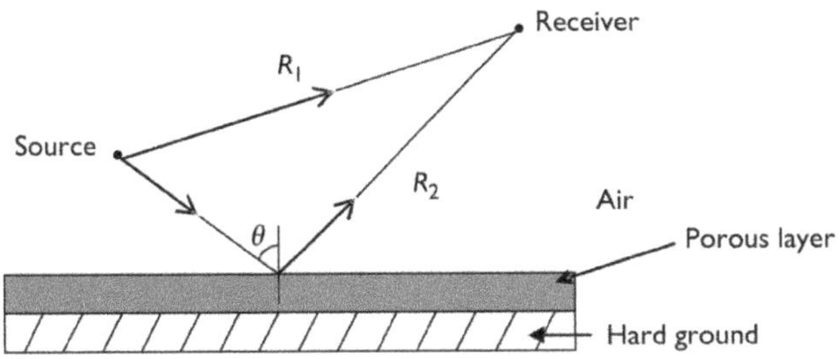

Figure 2.3 Sound propagation from a point source over a hard back layer.

layer. The boundary conditions (2.8), equations (2.9a) and (2.9b) remain unchanged but (2.9c) is replaced by

$$\hat{p}_1 = B_1(e^{-iK_1z} + U_1e^{iK_1z}) \quad (0 \geq z \geq -d_1) \tag{2.46}$$

Combining (2.8), (2.9b) and (2.46), simplifying the expressions and making U_0 the subject of the equation gives

$$U_0 = \frac{(k_z/\rho)(U_1+1)+(K_1/\rho_1)(U_1-1)}{(k_z/\rho)(U_1+1)-(K_1/\rho_1)(U_1-1)} \tag{2.47}$$

The quantity U_1 is determined by noting the condition for a hard backing, i.e. $(\partial p_1/\partial z) = 0$ at $z = -d_1$, where $d_1(>0)$ is the thickness of the porous medium. Hence, according to (2.46),

$$U_1 = e^{2iK_1d_1} \tag{2.48}$$

Then, use of the above expression in (2.47) yields

$$U_0 = \frac{(ik_z/\rho)-(K_1/\rho_1)\tan(K_1d_1)}{(ik_z/\rho)+(K_1/\rho_1)\tan(K_1d_1)} \tag{2.49}$$

Next the concept of effective admittance is extended from that for semi-infinite porous ground to that for a porous layer of finite thickness lying on a hard, impervious ground. For this case, the effective admittance can be deduced from (2.49) to give

$$\beta_e = -ik\varsigma_1\sqrt{n_1^2-\sin^2\theta}\,\tan\left(kd_1\sqrt{n_1^2-\sin^2\theta}\right) \tag{2.50}$$

At first sight, one might expect U_1 to be similar to U_0. But, if this were to be the case, then $U_1 = 1$ for a rigid ground. However, as indicated in (2.48), this is not the case. The boundary planes are located at different levels. For U_0, the reflecting plane is situated at $z = 0$ and for U_1 it is located at $z = -d_1$. As a result, a phase factor of $e^{2iK_1d_1}$ must be introduced in U_1.

To have a consistent interpretation of the interaction at each interface, a plane wave reflection coefficient at each interface is introduced such that

$$U_0 = V_0 e^{2ik_z(0)} = V_0 \tag{2.51a}$$

and

$$U_1 = V_1 e^{2iK_1 d_1} \tag{2.51b}$$

Then (2.47) becomes

$$V_0 = \frac{(k_z/\rho)\left(V_1 e^{iK_1 d_1} + e^{-iK_1 d_1}\right) + (K_1/\rho_1)\left(V_1 e^{iK_1 d_1} - e^{-iK_1 d_1}\right)}{(k_z/\rho)\left(V_1 e^{iK_1 d_1} + e^{-iK_1 d_1}\right) - (K_1/\rho_1)\left(V_1 e^{iK_1 d_1} - e^{-iK_1 d_1}\right)} \tag{2.52}$$

The plane wave coefficient of the air/ground interface, V_0, in (2.52) is a function of V_1 which, in turn, depends on the acoustical properties of subsequent layers. Typically, the reflection coefficient V_1 is 1 for a hard backing, –1 for a 'pressure-release' backing and 0 for an anechoic backing (which absorbs all incoming sound energy). Between these limiting situations, there are several cases for different types of grounds. Note that the layer thickness d_1 tends to infinity if the ground is semi-infinite. The factor, $e^{iK_1 d_1}$ is negligibly small for a large value of $d_1$because the imaginary part of K_1 is positive and non-zero. Hence, (2.52) can be approximated by

$$V_0 = \frac{(k_z/\rho) - (K_1/\rho_1)}{(k_z/\rho) + (K_1/\rho_1)} = \frac{(k_z/k) - \zeta_1\sqrt{n_1^2 - (\kappa/k)^2}}{(k_z/k) + \zeta_1\sqrt{n_1^2 - (\kappa/k)^2}} \tag{2.53}$$

which has the same form as found previously for semi-infinite porous ground, see (2.14). The minimum depth of the layer for which the ground can be treated as semi-infinite ground can be estimated. Numerical calculations reveal that if $\mathrm{Im}(K_1 d_1)$ is greater than 6 then $e^{-iK_1 d_1} \gg e^{iK_1 d_1}$. This suggests a simple condition for which the ground can be treated as a semi-infinite externally reacting ground, i.e.

$$Im\left(\sqrt{k_1^2 - \kappa^2}\, d_1\right) > 6.$$

The minimum depth, d_m to satisfy the above condition depends on the acoustical properties of the ground and the angle of incidence, but there are two limiting cases.

If $k_1 = k_r + ik_x$, then for normal incidence where $\kappa = 0$ (or $\theta = 0$), the required condition is

$$d_m > 6/k_x \tag{2.54a}$$

and for grazing incidence where $\kappa = 1$ (or $\theta = \pi/2$), the required condition is

$$d_m > 6\left[\sqrt{\frac{\left(k_r^2 - k_x^2 - 1\right)^2}{4} + k_r^2 k_x^2} - \frac{k_r^2 - k_x^2 - 1}{2}\right]^{\frac{1}{2}} \tag{2.54b}$$

Often, even a single hard-backed layer representation of the ground is inadequate. This is the case, for example, when predicting propagation through forests, where a ground surface may consist of a layer of decaying vegetation, above a humus layer which, in turn, is above relatively hard soil. This motivates extending the above analysis to allow for a multi-layered ground.

Suppose that the ground is composed of L layers of materials with different acoustical properties, see Figure 2.4.

Each layer has a different but constant depth, d_1, d_2, $d_3\cdots$ and so on. To facilitate the analysis, the location of each layer is denoted by $z_0, -z_1, -z_2, -z_3\cdots$, etc. and z_0 represents the height of the air/ground interface.

So,

$$z_0 = 0; \quad z_1 = d_1; \quad z_2 = d_1 + d_2; \quad z_3 = d_1 + d_2 + d_3\cdots$$

$$z_j = \sum_{n=1}^{j} d_n \quad \text{and} \quad d_j = z_{j+1} - z_j$$

The continuity of pressure and particle velocity is required at each layer:

$$\left.\begin{aligned} p_j &= p_{j+1} \\ \frac{1}{\rho_j}\frac{dp_j}{dz} &= \frac{1}{\rho_{j+1}}\frac{dp_{j+1}}{dz} \end{aligned}\right\} \text{at } z = -z_j \tag{2.55}$$

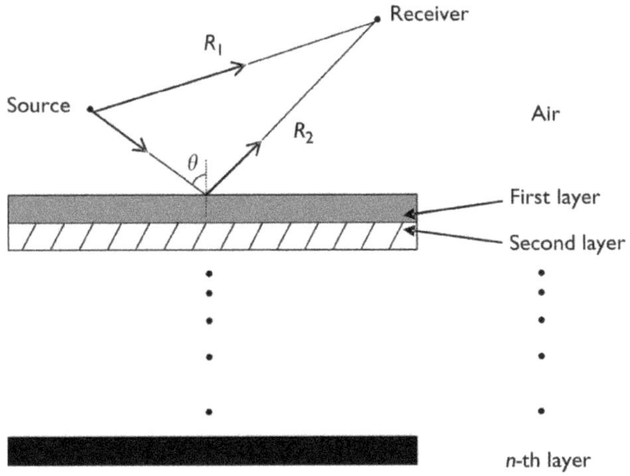

Figure 2.4 Sound propagation from a point source over a multi-layer (**n**-layer) ground.

In each layer, the pressure can be computed by summing the relevant incoming and outgoing waves in a form similar to (2.46) as follows:

$$\hat{p}_j = B_j(e^{-iK_j z} + U_j e^{iK_j z}) \text{ for } -z_j \geq z \geq -z_{j+1} \tag{2.56a}$$

where

$$K_j = +\sqrt{k_j^2 - \kappa^2} \tag{2.56b}$$

It is useful to introduce the following definitions for the density ratio and the index of refraction of each layer [see (2.11a and 2.11b)]:

$$\varsigma_j = \rho/\rho_j \tag{2.56c}$$

and

$$n_j = k_j/k = c/c_j \tag{2.56d}$$

Note that in (2.55) and (2.56a–2.56d), the variable j varies from 1, 2, ... to L.
Consider the j-th layer interface.
Use of the continuity conditions at the interface results in

$$B_j\left(e^{iK_j z_j} + U_j e^{-iK_j z_j}\right) = B_{j+1}\left(e^{iK_{j+1} z_{j+1}} + U_{j+1} e^{-iK_j z_{j+1}}\right) \tag{2.57a}$$

$$\frac{iK_j B_j}{\rho_j}\left(e^{iK_j z_j} + U_j e^{-iK_j z_j}\right) = \frac{iK_{j+1} B_{j+1}}{\rho_{j+1}}\left(-e^{-iK_{j+1} z_{j+1}} + U_{j+1} e^{-iK_j z_{j+1}}\right) \tag{2.57b}$$

Plane wave reflection coefficients V_j [see (2.51b)] for all interfaces are related to the corresponding U_j by

$$U_j = V_j e^{2iK_j z_j} \tag{2.57c}$$

Eliminating B_j and B_{j+1} by combining (2.57a) and (2.57b) and using V_j in place of U_j by means of (2.57c) gives

$$V_j = \frac{\left(K_j/\rho_j\right)\left(V_{j+1} e^{iK_{j+1} d_{j+1}} + e^{-iK_{j+1} d_{j+1}}\right) + \left(K_{j+1}/\rho_{j+1}\right)\left(V_{j+1} e^{iK_{j+1} d_{j+1}} + e^{-iK_{j+1} d_{j+1}}\right)}{\left(K_j/\rho_j\right)\left(V_{j+1} e^{iK_{j+1} d_{j+1}} + e^{-iK_{j+1} d_{j+1}}\right) - \left(K_{j+1}/\rho_{j+1}\right)\left(V_{j+1} e^{iK_{j+1} d_{j+1}} + e^{-iK_{j+1} d_{j+1}}\right)} \tag{2.58}$$

The sound field above a layered porous ground can be determined by using (2.52) and (2.58). Suppose that the ground consists of L rigid-porous layers and that the L-th interface is a rigid plane, i.e. $V_L = 1$ then, according to (2.58), V_{L-1} can be found by substituting $j = L - 1$ to give

$$V_{L-1} = \frac{\left(iK_{L-1}/\rho_{L-1}\right) - \left(K_L/\rho_L\right)\tan\left(K_L d_L\right)}{\left(iK_{L-1}/\rho_{L-1}\right) + \left(K_L/\rho_L\right)\tan\left(K_L d_L\right)} \tag{2.59a}$$

Substituting (2.58) into (2.52), putting j = 1, 2, 3 ..., $L - 2$ in turn and using (2.59) for V_{L-1}, a closed form analytic expression for the plane wave reflection coefficient V_0 can be determined. When $L = 1$, i.e. a single porous layer on a hard ground, the same expression as shown in (2.49) is obtained. With $L = 2$,

$$V_0 = \frac{(iK_0/\rho_0) - \left\{ \dfrac{\tan(K_1 d_1) + g_1 \tan(K_2 d_2)}{1 - g_1 \tan(K_1 d_1) \tan(K_2 d_2)} \right\} (K_1/\rho_1)}{(iK_0/\rho_0) + \left\{ \dfrac{\tan(K_1 d_1) + g_1 \tan(K_2 d_2)}{1 - g_1 \tan(K_1 d_1) \tan(K_2 d_2)} \right\} (K_1/\rho_1)}, \tag{2.59b}$$

where

$$g_1 = \frac{\rho_1 K_2}{\rho_2 K_1} \tag{2.59c}$$

It is interesting to consider some limiting cases of (2.59b). If d_2 becomes very small, then (2.59b) can be reduced to (2.49), which is the expression for a hardback layer. If d_1 is very large, then (2.59b) can be simplified to the expression given in (2.53), which is the expression for a semi-infinite porous medium. As before, the effective admittance of the double layer can be inferred from (c) to give

$$\beta_e = -i\varsigma_1 \sqrt{n_1^2 - \sin^2\theta} \left\{ \frac{\tan\left(kd_1\sqrt{n_1^2 - \sin^2\theta}\right) + g_1 \tan\left(kd_2\sqrt{n_1^2 - \sin^2\theta}\right)}{1 + g_1 \tan\left(kd_1\sqrt{n_1^2 - \sin^2\theta}\right) \tan\left(kd_2\sqrt{n_1^2 - \sin^2\theta}\right)} \right\}, \tag{2.60a}$$

where g_1 can be expressed in terms of ς_j, n_j and θ,

$$g_1 = \frac{\varsigma_2 \sqrt{n_2^2 - \sin^2\theta}}{\varsigma_1 \sqrt{n_1^2 - \sin^2\theta}} \tag{2.60b}$$

and ς_j and n_j are defined by (2.56c) and (2.56d).

The expression for the effective admittance for an arbitrary number of layers becomes increasingly complex for large L, although it should be simple to obtain its numerical values with the help of digital computers. On the other hand, sound waves seldom penetrate more than a few centimetres into most naturally occurring outdoor ground surfaces, so ground models consisting of more than two layers are unlikely to be required. The use of effective admittance greatly simplifies the analysis and the subsequent interpretation of the theoretical predictions. Though its introduction is somewhat heuristic, the formula gives satisfactory predictions for most practical situations. More accurate expressions have been developed, but numerical investigations have confirmed the accuracy of predictions that use the Weyl-Van der Pol formula with expressions for effective admittance [16] (see also Chapter 7).

In the following sections and most of the subsequent chapters, it is assumed that the specific normalized admittance of the ground should be treated as an effective parameter so β and β_e are used interchangeably unless otherwise stated.

2.5 SURFACE WAVES ABOVE POROUS GROUND

The Weyl-Van der Pol formula enables accurate prediction of outdoor sound propagation in neutral atmospheric conditions. Although, as discussed in Section 2.3, computation of the ground wave contribution is standard, it is instructive to consider two approximations that provide useful physical insight into its behaviour as a function of the relative size of $|w|$. In respect of the 'ground wave' term [second term of (2.40c)], it is sufficient to consider the boundary loss factor, $F(w)$. For $|w| < 1$, which implies small source/receiver separations and large impedances,

$$F(w) \approx 1 + i\sqrt{\pi} w e^{-w^2} \tag{2.61a}$$

and, for $|w| > 1$, which requires large source/receiver separations and small impedances,

$$F(w) \approx 2i\sqrt{\pi} w e^{-w^2} H[-w_i] - \frac{1}{2w^2} - \frac{1 \times 3}{(2w^2)^2} - \frac{1 \times 3 \times 5}{(2w^2)^3} - \cdots, \tag{2.61b}$$

where w_i is the imaginary part of the numerical distance, w and the Heaviside Step function is defined by

$$H[-w_i] = \begin{cases} 1 \;, & w_x < 0 \\ 0 \;, & \text{otherwise} \end{cases}.$$

The Heaviside function is a result of the definition of the complimentary error function, erfc(x), for positive and negative arguments, i.e. erfc($-x$) = 2 − erfc (x).

To satisfy the condition of $|w| > 1$ for different frequencies and ranges, we consider ranges greater than 50 m and frequencies less than 300 Hz. In this case, the Heaviside term in (2.61b) makes a significant contribution. Substituting (2.61b) into (2.40a), gives the usual form of a *surface wave*, decaying principally as the inverse root of horizontal range and exponentially with height above the ground. To allow a precise expression for the surface wave, the more accurate asymptotic expansion, i.e. (2.29a–2.29c) for (2.28) is required. The surface wave contribution is given by [14]

$$\Phi_{SW} = -H[-Im(x_0)] \frac{k\beta}{2} e^{-ik(z+z_s)} H_0^{(1)}\left\{kr\sqrt{1-\beta^2}\right\}, \tag{2.62a}$$

and is 'turned' on when $\mathrm{Im}(x_0) < 0$, i.e.

$$Im\left[-\sqrt{i}\left(1 + \beta\cos\theta - \sqrt{1-\beta^2}\sin\theta\right)\right] < 0 \tag{2.62b}$$

Use of (2.32a) and (2.32b) in (2.62a) leads to an approximate expression for the surface wave,

$$\Phi_{SW} \approx 2i\sqrt{\pi} w e^{-w^2} H[-w_x].$$

This is identical to the expression implied in (2.61b), as indeed, it should be.

By use of the condition (2.62b), we can demonstrate that the required condition can be simplified slightly to

$$\mathrm{Re}\left[\left(1+\beta\cos\theta-\sqrt{1-\beta^2}\sin\theta\right)\right]>0$$

By writing $\beta = \beta_r - i\beta_x$ (or $Z = Z_r + iZ_x$), we can represent a cut-off condition for the existence of the surface waves as

$$\frac{(\beta_r+\cos\theta)(1+\beta_r\cos\theta)}{\sqrt{1+2\beta_r\cos\theta+\beta_r^2}} > \beta_x \sin\theta. \tag{2.62c}$$

At grazing incidence, this condition is $\dfrac{1}{\beta_x^2} > \dfrac{1}{\beta_r^2}+1$.

For typical outdoor surface with large impedance, where $|\beta| \to 0$, this condition is simply that the imaginary part of the ground impedance (the reactance) is greater than the real part (the resistance). The analysis is much simpler if $-$ Im (x_0) is approximated by $-\sqrt{2}w_i$ [see (2.33b)], in which case, the cut-off for the surface wave is approximated by

$$\frac{Z_r - Z_x}{Z_r^2 + Z_x^2} + \cos\theta < 0.$$

The condition can be rearranged as the equation for a circle with centre at ($-2/\cos\theta, 2/\cos\theta$) and a radius of $1/\left(\sqrt{2}\cos\theta\right)$,

$$\left(Z_r + \frac{1}{2\cos\theta}\right)^2 + \left(Z_x - \frac{1}{2\cos\theta}\right)^2 < \left(\frac{1}{\sqrt{2}\cos\theta}\right)^2 \tag{2.63}$$

Figure 2.5 summarizes the condition (2.63) for the existence of surface waves.

As discussed in the last section, the ground wave contribution (the term in square brackets in (2.40a)), represents a contribution from the vicinity of the image of the source in the ground plane. Effectively it represents a diffusing of the image for spherical wave incidence compared with the plane wave incidence. However, the surface wave is essentially a separate contribution propagating close to and parallel to the porous ground surface associated with elliptical motion of air particles resulting from the combination of motion parallel to the surface with that normal to the surface in and out of the pores. The relative importance of each term will be shown when we discuss various ground impedance models in Chapter 5. We also note that

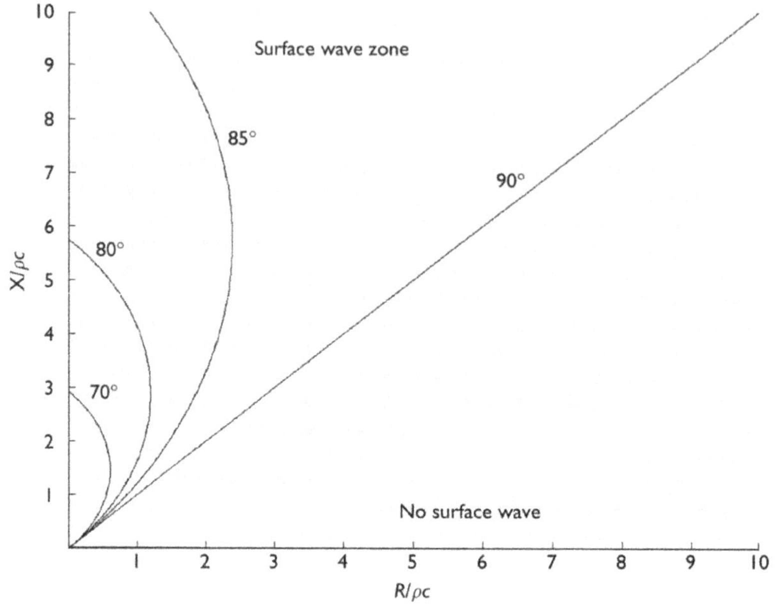

Figure 2.5 The condition for the existence of surface waves for sound propagation over a ground surface.

the surface wave speed is predicted to be less than the speed of sound in air. This forms a basis for detecting the surface wave in experiments using a pulse sound source. At a large enough range, it should arrive at the receiver separately from the main body waves.

In 1977, Piercy et al. [3] predicted the presence of airborne surface waves in outdoor sound propagation close to the ground. But, despite its predicted importance [24], its existence was the subject of considerable controversy in the early eighties from both theoretical and experimental points of view. The theoretical controversy was resolved by Raspet and Baird [25]. They analysed the propagation of spherical wave above a complex impedance plane and showed that the existence of the surface wave is independent of the body wave in air. They considered the limit where the upper half space (air) becomes incompressible and proved that the surface wave can still exist and is a true surface wave.

An analysis of the more complicated problem of a point source above a ground that is both elastic and porous by Richards and Attenborough [26] has shown that the surface wave that appears in the solution for a point source above a rigid-porous boundary is one of the two possible surface waves, the other being an air-coupled Rayleigh wave that travels below and parallel to the surface.

In a relatively low frequency range which depends on the ground impedance and source–receiver geometry, the acoustic surface wave above a porous boundary is predicted to produce total pressure levels in excess of the +6 dB that would be found over an acoustically rigid boundary.

In the mid-seventies, laboratory experiments using sources of continuous sound and artificial surfaces, designed so that the reactive part of the impedance was much greater than the resistive part thereby artificially increasing the contribution of surface waves at low frequencies, confirmed the surface wave phenomenon and showed that the surface wave term predicts both the acoustic field near-grazing incidence and its attenuation with height above the surface [27]. Further indoor experiments [28], using a pulse sound source above lighting diffuser lattices mounted on flat rigid boards, were successful in demonstrating the surface wave as a later arrival than that of a body wave thereby confirming its slightly lower speed. Other laboratory experiments on sound propagation over impedance discontinuities [29] and sound propagation over the convex impedance ground [30] have provided conclusive evidence of the existence of the surface waves (see also Chapter 6).

However, early outdoor pulse experiments in neutral atmospheric conditions at long range failed to register separate arrivals of the surface wave and direct wave [31]. The large resistive components of most outdoor ground impedances result in appreciable exponential attenuation along the surface, outweighing the potential advantage of the surface wave that it decays with the inverse square root of range (i.e. cylindrical spreading) rather than inversely (i.e. spherical spreading) with range. The ground types most likely produce measurable surface waves that may be modelled as thin hard back layers (see Chapter 5). Such is the case for propagation over a thin layer of snow above a frozen ground. By firing blank pistol shots to create impulsive sounds, Albert identified surface wave propagation during outdoor experiments over snow [32].

Surface wave arrivals have been observed also as low frequency tails in pulses recorded at a distance of 2 km from propane cannon explosions, peak energy content near 200 Hz, over relatively flat farmland both in weakly refracting and in strongly ducted downward refraction conditions. The measured surface wave contributions agreed with predictions [33,34]. Now, the presence of dispersive surface wave components during propagation above an impedance plane in a homogeneous atmosphere [35] or in meteorological conditions that lead to strongly ducted propagation [33,34] near the ground is beyond doubt.

2.6 EXPERIMENTAL DATA AND NUMERICAL PREDICTIONS

To enable the prediction of sound propagation close to the ground outdoors, the impedance of ground surface is required. Among various models for the acoustical properties of outdoor ground surfaces, a single-parameter model and a two-parameter model are used frequently. The basis for these and other ground impedance models is discussed in Chapter 5. For the moment, it is enough just to quote the relevant formulae and then to illustrate ground effects on outdoor sound through numerical computations and comparisons with experimental data.

The single-parameter model, due to Delaney and Bazley [36], describes the propagation constant, k and normalized surface impedance, Z by a single adjustable parameter known as the effective flow resistivity, σ_e which has units of Pa s m^{-2}. It is called 'effective' since in modelling outdoor ground surfaces, it rarely takes a value equal to the actual flow resistivity, i.e. a measure of the decrease in pressure per unit length in the direction of a flow of air at unit speed. The following are the expressions for propagation constant and normalized surface impedance according to the Delany and Bazley model:

$$\frac{k}{k_1} = \left[1 + 0.0978\left(f/\sigma_e\right)^{-0.700} + i0.189\left(f/\sigma_e\right)^{-0.595}\right] \tag{2.64a}$$

$$Z = \frac{\rho_1 c_1}{\rho c} = 1 + 0.0571\left(f/\sigma_e\right)^{-0.754} + i0.087\left(f/\sigma_e\right)^{-0.732}, \tag{2.64b}$$

which can be used for both locally reacting and extended reaction surfaces. For predicting outdoor sound propagating near to locally reacting ground, (2.64b) is used to calculate the impedance. If the ground is not locally reacting, (2.64a and 2.64b) is used to determine the ratios of sound speed and density of the air and ground [*c.f.* (2.11a and 2.11b)] by

$$n_1 = \frac{k_1}{k} = \frac{1}{1 + 0.0978\left(f/\sigma_e\right)^{-0.700} + i0.189\left(f/\sigma_e\right)^{-0.595}} \tag{2.65a}$$

and

$$\varsigma_1 = \frac{\rho}{\rho_1} = \frac{1}{Z n_1} = \frac{1 + 0.0978\left(f/\sigma_e\right)^{-0.700} + i0.189\left(f/\sigma_e\right)^{-0.595}}{1 + 0.0571\left(f/\sigma_e\right)^{-0.754} + i0.087\left(f/\sigma_e\right)^{-0.732}}, \tag{2.65b}$$

which allows the effective impedance to be computed.

Although this single-parameter model is physically inadmissible, for example it can predict negative real parts of complex density and impedance (see Chapter 5), it is convenient for studying the contribution of surface waves over locally reacting ground. The magnitude and phase of impedance are given, respectively, by

$$|Z| = \sqrt{\left[1 + 0.0571\left(f/\sigma_e\right)^{-0.754}\right]^2 + i7.569 \times 10^{-3}\left(f/\sigma_e\right)^{-1.464}} \tag{2.66a}$$

$$\varphi = \tan^{-1}\left\{0.087\left(f/\sigma_e\right)^{-0.732} / \left[1 + 0.0571\left(f/\sigma_e\right)^{-0.754}\right]\right\} \tag{2.66b}$$

A close inspection of (2.66b) reveals that the phase angle of impedance varies from $\pi/2$ for $(f/\sigma_e) \to 0$ to 0 for $(f/\sigma_e) \to \infty$. Also, the effective flow resistivity varies from about 15 kPa s m^{-2} (for a very soft ground such as snow-covered ground) to about 25,000 kPa s m^{-2} for a road surface made with hot-rolled asphalt.

By writing $\beta = (\cos\varphi - i \sin \varphi)/|Z|$, (2.33b) can be expanded to give

$$w = \frac{\sqrt{\frac{1}{2}kR_2\left[\left(|Z|\cos\theta + \cos\varphi\right)^2 + \sin^2\varphi\right]}}{|Z|}\exp\left[i\tan^{-1}\left(\frac{|Z|\cos\theta + \cos\left(\frac{\pi}{4}+\varphi\right)}{|Z|\cos\theta + \cos\left(\frac{\pi}{4}-\varphi\right)}\right)\right] \tag{2.67a}$$

Given the ranges of effective flow resistivity, frequencies (10 Hz–10 kHz) and source/receiver geometries of interest, the magnitude of the numerical distance varies from 0.01 to over 100. The phase of the numerical distance is limited to $\pm\pi/2$. As discussed in Section 2.3, the boundary loss factor $F(w)$ plays an important role in near-grazing propagation. Hence, it is worth showing its variation with the magnitude of the numerical distance. For grazing propagation, i.e. when the source and receiver are located on the ground surface, the numerical distance (2.67a) simplifies to

$$w = \left(\sqrt{\frac{1}{2}kR_2} / |Z|\right)\exp\left[i\tan^{-1}\left(\cos\left(\frac{\pi}{4}+\varphi\right) / \cos\left(\frac{\pi}{4}-\varphi\right)\right)\right] \tag{2.67b}$$

which is essentially the same as (2.41b) except that w is expressed in polar form.

Figure 2.6 shows a plot of 20 log $|F(w)|$ versus $|w|$ for different phase angles of impedance, ϕ. Note that, for a certain range of $|w|$ when $\varphi > \pi/4$, the boundary loss factor is greater than 1 (indicating enhancement of sound

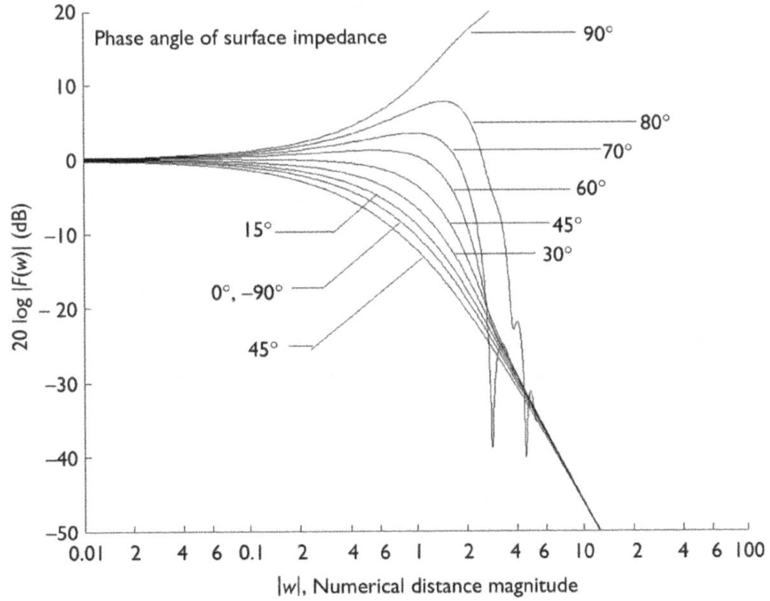

Figure 2.6 Magnitude of **F**(**w**) versus the numerical distance **w** for various phase angles of surface impedance. Source and receiver are located on the ground.

fields) due to the addition of the surface wave component. In this case, as shown in (2.67b), the numerical distance has a negative imaginary part.

Incidentally, although Figure 6 in ref. [35] shows a similar result, $|w|^2$ is plotted rather than $|w|$ as stated. The relative importance of the various contributions of the total sound field is illustrated by revisiting some of the classical Parkin and Scholes data from measurements at two disused airfields in the 1960s using a fixed jet engine source [37,38]. Figures 2.7(a) and 2.7(b) show data for the corrected difference in levels at 1.5 m height between a reference microphone and a remote microphone at different ranges (1097.3 m for Figure 2.7(a) and 35 m for Figure 2.7(b)) [39]. These figures also show theoretical predictions using a single-parameter model for the surface impedance (equation (2.64b)). The predicted contributions to the total field are divided into the direct (D) wave, the reflected (R) wave, the ground wave (G) and the surface (S) wave. The predictions that fit the data indicate that ground and surface waves are major carriers of environmental noise at low frequencies over long distances.

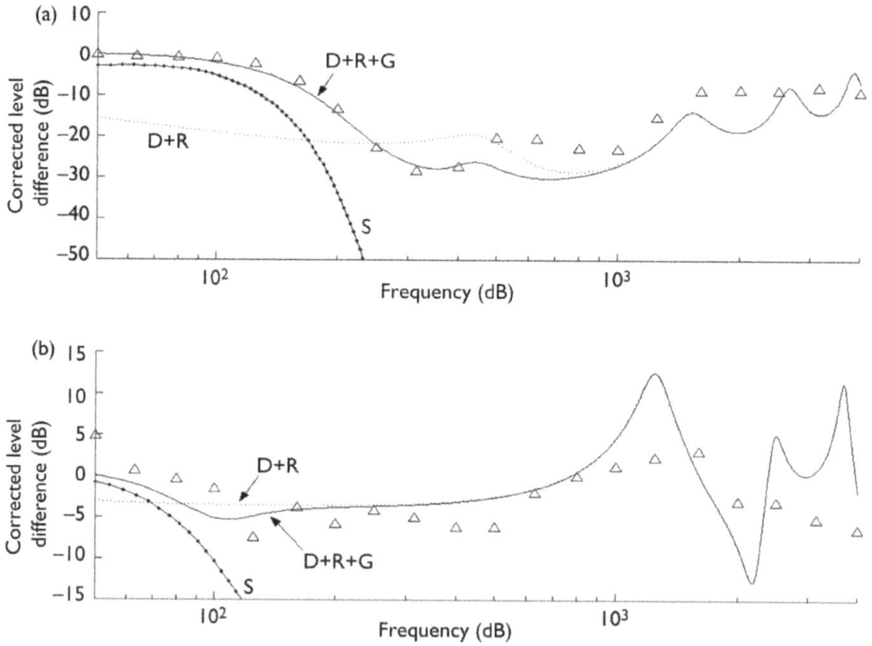

Figure 2.7 (a) Parkin and Scholes' data (Δ), reproduced from [34], for the corrected difference in levels from a fixed (1.82 m high) jet engine source between receivers at ranges of 19.5 m and 1097.3 m and at 1.5 m height over an airfield. Predictions use Delany and Bazley's impedance model with σ_e = 300 kPa s m^{-2} and show the individual contributions from the direct wave (D), the plane-wave reflected component (R), the ground wave (G) and the surface wave (S). (b) Data for sound level re free field obtained above snow (20 cm thick) at a range of 35 m (reproduced from [35]). Predictions use Delaney and Bazley's impedance model (3.1, 3.2) with σ_e = 12 kPa s m^{-2}. Again, the predicted individual components are shown.

The single-parameter model has been used to calculate the impedance of a layered outdoor ground surface. Nicolas et al. [40] have reported an extensive series of measurements over layer of snow from 5 to 50 cm thick and at propagation distance up to 15 m. Their predictions were tolerably in agreement with data, but they were unable to obtain reasonable fits of data at low frequencies by assuming either a semi-infinite ground or a hard-backed single layer ground structure. Li et al. [14] have improved the agreement between predictions and these data by introducing a double layer model. Figure 2.8 shows the measured behaviour over snow layers, nominally 6, 8 and 10 cm thick, together with various predictions.

Table 2.1 shows the values of the effective flow resistivities and layer thicknesses used for the predictions on Figure 2.8. The double layer predictions (broken lines) are in somewhat better agreement with data than those assuming a single hard-backed layer (dotted lines) or a semi-infinite layer (continuous lines). Note, however, that the introduction of the layer thickness increases the number of parameters in the impedance model and therefore reduces its convenience.

Another useful ground impedance model is the two-parameter model proposed by Attenborough [41] and revised by Raspet and Sabatier [42] (see Chapter 5, eqn. (5.38)). Although this model is suitable only for a locally

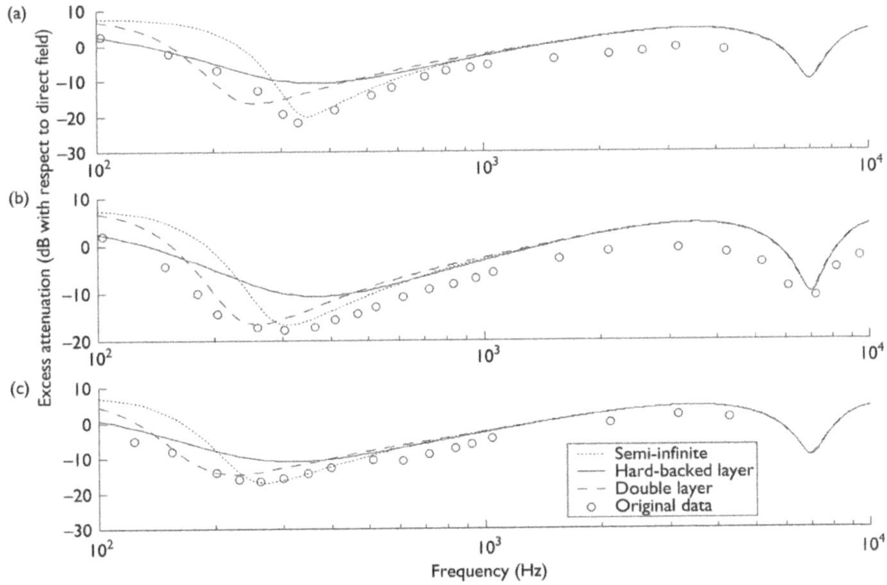

Figure 2.8 Data and predictions for sound propagation over snow. The excess attenuation data (reported in [35]) were obtained with source height 0.6 m, receiver height 0.3 m and range 7.5 m. (a) 6-cm thick layer of snow over ice. (b) 8-cm thick layer of new snow over asphalt. (c) 10-cm thick layer of freshly fallen snow above 50 cm of old hardened snow covered previously by a crust of ice.

Table 2.1 The layer thickness and flow resistivities used for the predictions shown in Figure 2.8.

Ground type	*(a)*	*(b)*	*(c)*
Hard-backed layer (dotted line)	σ_e = 15 kPa m s^{-2}	σ_e = 15 kPa m s^{-2}	σ_e = 10 kPa m s^{-2}
	d = 6 cm	d = 8 cm	d = 10 cm
Semi-infinite ground (solid line)	σ_e = 15 kPa m s^{-2}	σ_e = 15 kPa m s^{-2}	σ_e = 10 kPa m s^{-2}
Double layer (broken line)	σ_1 = 10 kPa m s^{-2}	σ_1 = 10 kPa m s^{-2}	σ_1 = 8 kPa m s^{-2}
	d_1 = = 8 cm	d_1 = 10 cm	d_1 = 12 cm
	σ_2 = 15 kPa m s^{-2}	σ_2 = 55 kPa m s^{-2}	σ_2 = 15 kPa m s^{-2}
	d_2 = 3 cm	d_2 = 1 cm	d_2 = 3 cm

reacting ground, not only is it physically admissible but also allows better agreement with data obtained over many outdoor ground surfaces than obtained with the single-parameter model (see Chapter 6). The impedance of the ground surface is given by

$$Z = 0.436(1+i)\sqrt{\frac{\sigma_e}{f}} + 19.74i\frac{\alpha_e}{f}, \tag{2.68}$$

with two adjustable parameters, the effective flow resistivity (σ_e) and the effective rate of change of porosity with depth (α_e). The effective rate of change of porosity with depth has units m^{-1}. Of course, some improvement of the agreement between predictions and data is to be expected with two 'adjustable' parameters instead of one. Impedance models using even more parameters are discussed in Chapter 5. Figure 2.9 illustrates the effects on the excess attenuation of the changes in the values of σ_e and α_e. The effect of increasing σ_e is to increase the frequency of the ground effect dip, whereas the effect of increasing α_e is to increase the magnitude of the dip.

2.7 THE SOUND FIELD DUE TO A LINE SOURCE NEAR THE GROUND

This chapter ends by considering the sound field due to a line source above a ground surface which is a two-dimensional problem often used to model highway noise. The solution can be obtained by using a similar approach to that for the three-dimensional case. In a rectangular co-ordinate system, $\mathrm{x} \equiv (x, z)$, a line source is located at $\mathrm{x}_s \equiv (0, z_s)$ and it is assumed that $x > 0$ since the fields for $x > 0$ and $x < 0$ are symmetrical. The required Green's function, $g_m(x)$, is given by

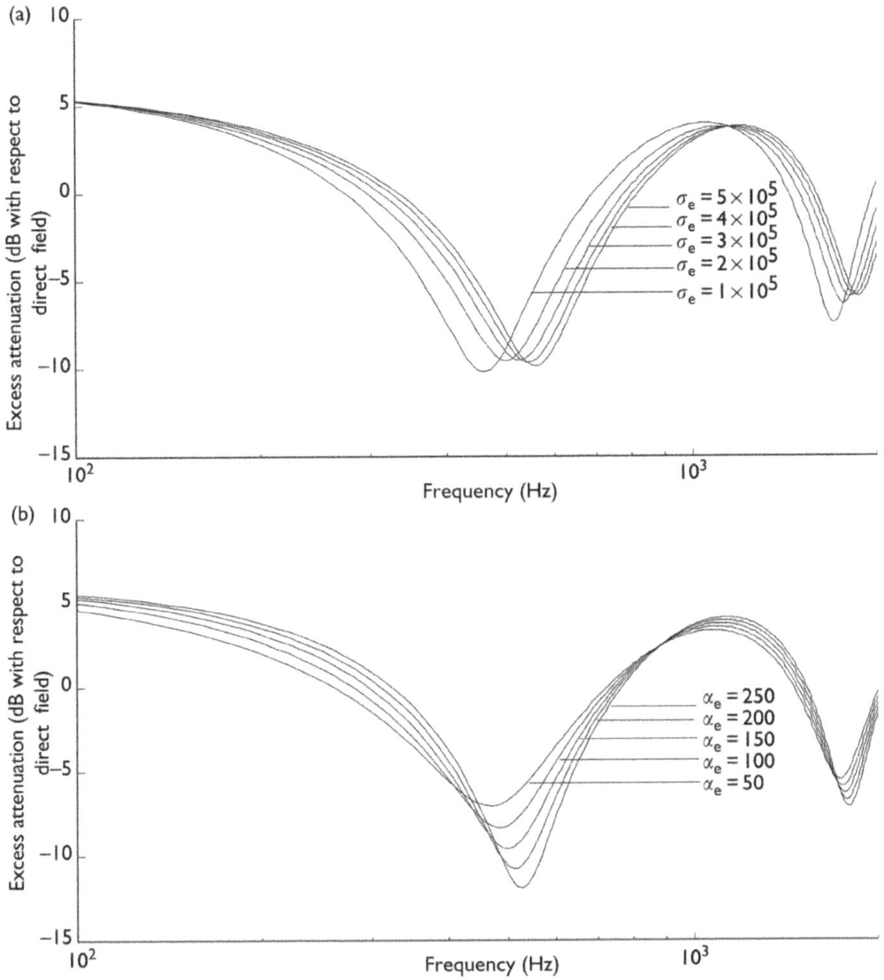

Figure 2.9 Predicted sensitivity of Excess attenuation spectra to changes in effective flow resistivity σ_e and the rate of change of porosity with depth α_e in the two-parameter model (2.68) for the effective impedance of the ground surface.

$$\left(\frac{\partial^2}{\partial x^2} + \frac{\partial^2}{\partial z^2} + k^2\right) g_m(x) = -\delta(x)\delta(z - z_s) \tag{2.69a}$$

subject to the impedance boundary condition

$$\left(\frac{\partial}{\partial z} + ik\beta\right) g_m(x) = 0 \tag{2.69b}$$

at the ground surface $z = 0$. Again, the admittance β is treated as an 'effective' parameter and used to model either a locally reacting or an extended reaction ground surface. Application of a one-fold Fourier transform on both sides of (2.69a) and (2.69b) leads to

$$\frac{d^2\hat{g}_m}{dz^2} + k_z^2\hat{g}_m = -\delta\left(z - z_s\right) \tag{2.70a}$$

$$\frac{d\hat{g}_m}{dz} = -ik\beta\hat{g}_m \text{ at z = 0,} \tag{2.70b}$$

where

$$k_z = +\sqrt{k^2 - k_x^2} \tag{2.70c}$$

and the one-dimensional Fourier transform pair is defined by

$$\hat{g}_m\left(k_x, z\right) = \int_{-\infty}^{\infty} g_m\left(x\right)e^{-ik_x x}dx \tag{2.70d}$$

and

$$g_m\left(x\right) = \frac{1}{2\pi}\int_{-\infty}^{\infty} \hat{g}_m\left(k_x, z\right)e^{ik_x x}dk_x s \tag{2.70e}$$

With the boundary condition (2.70b), (2.70a) yields

$$\hat{g}_m\left(k_x, z\right) = \frac{i}{2k_z}\left\{e^{ik_z|z-z_s|} + \frac{k_z - k\beta}{k_z + k\beta}e^{ik_z(z+z_s)}\right\} \tag{2.71}$$

Substituting (2.71) into (2.70e), the integral expression for the Green's function can be spilt into three terms as follows:

$$g_m\left(x\right) = \frac{i}{4\pi}\int_{-\infty}^{\infty}\frac{e^{ik_z|z-z_s|+ik_x x}}{k_z}dk_x + \frac{i}{4\pi}\int_{-\infty}^{\infty}\frac{e^{ik_z|z+z_s|+ik_x x}}{k_z}dk_x - \frac{ik}{2\pi}\int_{-\infty}^{\infty}\frac{\beta}{k_z+\beta}\frac{e^{ik_z|z+z_s|+ik_x x}}{k_z}dk_x \tag{2.72}$$

The evaluation of the integral can be simplified by using polar co-ordinates centred on the source for the first integral and on the image source for the other two integrals.

In the polar co-ordinate system,

$$x = R_1\cos\theta = R_2 \sin\theta,\ |z - z_s| = R_1\cos\varphi \text{ and } \quad z + z_s = R_2\cos\theta,$$

where R_1 and R_2 are the respective path lengths for the direct and reflected waves, and ϕ and θ, are their corresponding polar angles.

The required Green's function can be transformed to

$$g_m\left(x\right) = \frac{i}{4\pi}\int_{-\frac{\pi}{2}+i\infty}^{\frac{\pi}{2}-i\infty} e^{\left[ikR_1\cos(\mu-\psi)\right]}d\mu + \frac{i}{4\pi}\int_{-\frac{\pi}{2}+i\infty}^{\frac{\pi}{2}-i\infty} e^{\left[ikR_1\cos(\mu-\theta)\right]}d\mu - \frac{i\beta}{2\pi}\int_{-\frac{\pi}{2}+i\infty}^{\frac{\pi}{2}-i\infty}\frac{e^{\left[ikR_2\cos(\mu-\theta)\right]}}{\cos\mu+\beta}d\mu \tag{2.73}$$

The first and second integral can be identified as the sound field due to the source and its image source. These two integrals can be expressed in terms of Hankel functions, i.e.

$$\frac{1}{\pi}\int_{-\frac{\pi}{2}+i\infty}^{\frac{\pi}{2}-i\infty} e^{\left[ikR_1\cos(\mu-\varphi)\right]}d\mu = H_0^{(1)}\left(kR_1\right),z \tag{2.74a}$$

$$\frac{1}{\pi}\int_{-\frac{\pi}{2}+i\infty}^{\frac{\pi}{2}-i\infty} e^{\left[ikR_2\cos(\mu-\varphi)\right]}d\mu = H_0^{(1)}\left(kR_2\right) \tag{2.74b}$$

The third integral of (2.73), which we denote by I_β, cannot be evaluated exactly but an asymptotic method may be applied to evaluate the integral and yields

$$I_\beta = -\frac{\beta e^{\left[ik\left(x\sqrt{1-\beta^2}-h\beta\right)\right]}}{2\sqrt{1-\beta^2}}erfc\left(\frac{-i\varsigma_p}{2}\right) \tag{2.75}$$

Applying the same approximations as used in the three-dimensional (3-D) case, i.e. $|\beta| \ll 1$, $(z + z_s)/x \ll 1$ and $k(z + z_s)^2/x \ll 1$, and substituting (2.74a), (2.74b) and (2.75) into (2.73) gives

$$g_m(x) = \frac{i}{4}H_0^{(1)}\left(kR_1\right) + \frac{i}{4}H_0^{(1)}\left(kR_2\right) - \frac{\beta}{2}\sqrt{i\pi kR_2}\,erfc\left(-iw\right)H_0^{(1)}\left(kR_2\right). \tag{2.76}$$

The formula can be rearranged further to give the more recognizable form

$$g_m(x) = \frac{i}{4}H_0^{(1)}\left(kR_1\right)s + \frac{i}{4}\left[R_p - \left(1-R_p\right)F(w)\right]H_0^{(1)}\left(kR_2\right) \tag{2.77}$$

Hence, the two-dimensional (2-D) Green's function has the same classical form as the Weyl-Van der Pol formula for a point source.

REFERENCES

[1] N. Sommerfeld, Propagation of waves in wireless telegraphy, *Ann. Phys. (Paris)* 28 665 (1909); **81** 1069–1075 (1926).

[2] A. Banos, Jr. *Dipole Radiation in the Presence of Conducting Half-Space*, Pergamon, New York (1966), Chap. 2–4. See also A. Banos, Jr. and J. P. Wesley, The horizontal electric dipole in a conduction half-space, Univ. Calif. Electric Physical Laboratory, S10 Reference 53–33 and 54–31 (1954).

[3] I. Rudnick, Propagation of an acoustic waves along an impedance boundary, *J. Acoust. Soc. Am.* **19** 348–356 (1947).

[4] K. U. Ingard, On the reflection of a spherical wave from an infinite plane, *J. Acoust. Soc. Am.* **23** 329–335 (1951).

[5] D. I. Paul, Acoustical radiation from a point source in the presence of two media, *J. Acoust. Soc. Am.* **29** 1102–1109 (1959).

[6] A. R. Wenzel, Propagation of waves along an impedance boundary, *J. Acoust. Soc. Am.* **55** 956–963 (1974).
[7] R. J. Donato, Spherical-wave reflection from a boundary of reactive impedance using a modification of Cagniard's method, *J. Acoust. Soc. Am.* **60** 999–1002 (1976).
[8] S. I. Thomasson, Reflection of waves from a point source by an impedance boundary, *J. Acoust. Soc. Am.* **59** 780–785 (1976).
[9] S. I. Thomasson, Sound propagation above a layer with a large refraction index, *J. Acoust. Soc. Am.* **61** 659–674 (1977).
[10] F. Chien and W. W. Soroka, Sound propagation along an impedance plane, *J. Sound Vib.* **43** 9–20 (1976).
[11] F. Chien and W. W. Soroka, A note on the calculation of sound propagation along an impedance plane, *J. Sound Vib.* **69** 340–343 (1980).
[12] K. Attenborough, S. I. Hayek and J. M. Lawther, Propagation of sound above a porous half-space, *J. Acoust. Soc. Am.* **68** 1493–1501 (1980).
[13] T. Kawai, T. Hidaka and T. Nakajima, Sound propagation above an impedance boundary, *J. Sound Vib.* **83** 125–138 (1982).
[14] K. M. Li, T. Waters-Fuller and K. Attenborough, Sound propagation from a point source over extended-reaction ground, *J. Acoust. Soc. Am.* **104** 679–685 (1998).
[15] K. M. Li and S. Liu, Propagation of sound from a monopole source above an impedance-backed porous layers, *J. Acoust. Soc. Am.* **131** 4376–4388 (2012).
[16] K. M. Li and H. Tao, Heuristic approximations for sound fields produced by spherical waves incident on locally and non-locally reacting planar surfaces, *J. Acoust. Soc. Am.* **135** 58–66 (2014).
[17] M. Abramowitz and I. A. Stegun, *Handbook of mathematical functions with formulas, graphs, and mathematical tables*, Dover Publications, Inc., New York (1972).
[18] R. Wong, *Asymptotic approximations of integrals*, Academic Press, Inc., London, pp. 356–360 (1989).
[19] F. Matta and A. Reichel, Uniform computation of the error function and other related functions, *Math. Comput.* **25** 339–344 (1971).
[20] R. K. Pirinchieva, Model study of sound propagation over ground of finite impedance, *J. Acoust. Soc. Am.* **90** 2679–2682 (1991).
[21] A. P. Dowling and J. E. Ffowcs Williams, *Sound and sources of sound*, Ellis Horwood Ltd., Chichester, Ch. 4 (1983).
[22] T. F. W. Embleton, J. E. Piercy and N. Olson, Outdoor sound propagation over ground of finite impedance, *J. Acoust. Soc. Am.* **59** 267–277 (1976).
[23] L. M. Brekhovskikh, *Waves in layered media*, Academic Press, New York (1980).
[24] K. Attenborough, Predicted ground effect for highway noise, *J. Sound Vib.* **81** 413–424 (1982).
[25] R. Raspet and G. E. Baird, The acoustic surface wave above a complex impedance ground surface, *J. Acoust. Soc. Am.* **85** 638–640 (1989).
[26] T. L. Richards and K. Attenborough, Solid particle motion resulting from a point source above a poroelastic half space, *J. Acoust. Soc. Am.* **86**(3) 1085–1092 (1989).
[27] R. J. Donato, Model experiments on surface waves, *J. Acoust. Soc. Am.* **63** 700–703 (1978).

[28] H. Howorth and K. Attenborough, Model experiments on air-coupled surface waves, *J. Acoust. Soc. Am.* , **2431**(A) (1992).
[29] G. A. Daigle, M. R. Stinson and D. I. Havelock, Experiments on surface waves over a model impedance using acoustical pulses, *J. Acoust. Soc. Am.* **99** 1993–2005 (1996).
[30] Q. Wang and K. M. Li, Surface waves over a convex impedance surface, *J. Acoust. Soc. Am.* **106** 2345–2357 (1999).
[31] G. Don and A. J. Cramond, Impulse propagation in a neutral atmosphere, *J. Acoust. Soc. Am.* **81** 1341–1349 (1987).
[32] G. Albert, Observation of acoustic surface waves propagating above a snow cover, *Proc. 5th Symposium on Long range Sound Propagation*, 10–16 (1992).
[33] L. Talmadge, R. Waxler, X. Di, K. E. Gilbert and S. Kulichkov, Observation of low-frequency acoustic surface waves in the nocturnal boundary layer, *J. Acoust. Soc. Am.* **124** 1956–1962 (2008).
[34] R. Waxler, K. E. Gilbert and C. L. Talmadge, A theoretical treatment of the long-range propagation of impulsive signals under strongly ducted nocturnal conditions, *J. Acoust. Soc. Am.* **124** 2742–2754 (2008).
[35] L. C. Sutherland and G. A. Daigle, Atmospheric sound propagation, in *Handbook of Acoustics*, Edited by M. J. Crocker, Wiley, New York, pp 305–329 (1998).
[36] M. E. Delany and E. N. Bazley, Acoustical properties of fibrous absorbent materials, *Appl. Acoust.* **3** 105–116 (1970).
[37] P. H. Parkin and W. E. Scholes, The horizontal propagation of sound from a jet close to the ground at Radlett, *J. Sound Vib.* **1** 1–13 (1965).
[38] P. H. Parkin and W. E. Scholes, The horizontal propagation of sound from a jet close to the ground at Hatfield, *J. Sound Vib.* **2** 353–374 (1965).
[39] K. Attenborough, Review of ground effects on outdoor sound propagation from continuous broadband sources, *Appl. Acoust.* **24**, 289–319 (1988).
[40] J. Nicolas, J. L. Berry and G. A. Daigle, Propagation of sound above a finite layer of snow, *J. Acoust. Soc. Am.* 77 67–73 (1985).
[41] K. Attenborough, Ground parameter information for propagation modeling, *J. Acoust. Soc. Am.* **92** 418–427 (1992); see also R. Raspet and K. Attenborough, 'Erratum: Ground parameter information for propagation modeling, *J. Acoust. Soc. Am.* 92 3007 (1992).
[42] R. Raspet and J. M. Sabatier, The surface impedance of grounds with exponential porosity profiles, *J. Acoust. Soc. Am.* **99** 147–152 (1996).

Chapter 3

Predicting Effects of Source Characteristics

3.1 INTRODUCTION

The sound field due to a monopole source in a homogeneous stationary medium above an absorbing ground has no azimuthal variation. But many outdoor noise sources are inherently directional, which means that monopole source descriptions are not appropriate and descriptions in terms of multipoles may be more useful. Although a multipole source radiates sound less strongly than a monopole source of the same strength, there are many situations when the multipole contribution is larger than the contribution from monopole radiation. For example, according to Lighthill's aeroacoustic analogy, sources of a quadrupole nature may be used to model jet noise. This chapter explores the prediction of ground effect for dipole and quadrupole sources. Then the directivity and other characteristics of railway, road vehicle and wind turbine sound sources which influence the way sound propagates from them are considered.

3.2 SOUND FIELDS DUE TO DIPOLE SOURCES NEAR THE GROUND

A general expression for the sound field due to a point or monopole source above a ground surface was derived in Chapter 2. It was shown that the interaction of spherical sound waves with several types of ground surface can be modelled using the concept of effective impedance. In general, the plane wave reflection coefficient, R_p, from an extended reaction surface, i.e. a non-locally reacting ground, depends on the angle at which incident waves arrive. But as the ability for sound waves to penetrate the ground surface diminishes, the ground becomes increasingly locally reacting, the plane wave reflection coefficient becomes independent of the angle of incidence and the ground surface can be treated as an impedance boundary. This assumption greatly simplifies the analysis for sources other than a monopole.

After the monopole, the simplest source of practical importance is a dipole. Examples of sources that can be modelled successfully as dipoles include an oscillating sphere, an enclosed loudspeaker and the wheel/rail contact source associated with train noise. A dipole can be imagined to consist of two closely spaced point monopole sources with the same amplitude but opposite phase, i.e. 180° out of phase. The dipole sound field can be calculated by summing the individual contribution due to the two component monopoles and taking the limit for a vanishingly small separation. Sound radiation from a dipole has been studied but only in an unbounded medium [1,2]. This section continues with the study of the sound field due to a dipole source near to a ground surface with normalized effective admittance, β [3].

3.2.1 The Horizontal Dipole

Figures 3.1(a) and (b) portray 2-D and 3-D views, respectively, of two out-of-phase monopoles at the same height, z_s, and 2Δ apart, above a ground (the x–y plane). The line joining the two monopole sources (the dipole axis) can be aligned so that it is parallel to the x-axis. If it is not so aligned, the x–y plane can be rotated about the z-axis appropriately. The centre of the dipole is at $(0,0,z_s)$. Since the monopoles are 180° out of phase, they can be labelled as positive and negative monopoles. By convention, the direction of the dipole axis is from the negative monopole to the positive one.

The source strength of each monopole is represented by S_0. The sound field due this pair of sources can be found by solving the inhomogeneous Helmholtz equation,

$$\nabla^2 p + k^2 p = -S_0\left[\delta(x-\Delta)\delta(y)\delta(z-z_s) - \delta(x+\Delta)\delta(y)\delta(z-z_s)\right] \quad (3.1a)$$

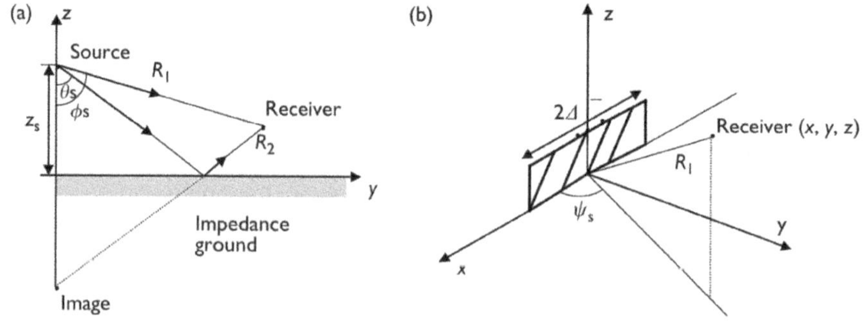

Figure 3.1 The source/receiver geometry for a dipole source above an impedance ground: (a) 2D view and (b) 3D view.

subject to the usual boundary condition at the ground, $z = 0$, of

$$\frac{dp}{dz} + ik\beta p = 0 \tag{3.1b}$$

In the limit of small Δ, the right-hand side of (3.1a) can be simplified by taking the difference of the Taylor expansions of the two terms in the square bracket to give

$$\nabla^2 p + k^2 p = S_1\left[\frac{\partial}{\partial x}\delta(x)\delta(y)\delta(z - z_s)\right] \tag{3.2}$$

where $S_1 = 2S_0\Delta$ may be regarded as the source strength of the dipole. Using the Green's function, $G_h(x)$, for the sound field due to a horizontal dipole at $x_s \equiv (0,0,z_s)$ and receivers at $x \equiv (x,y,z)$ with source strength $S_1 = 1$ together with the method of Fourier transformation (see (2.5)), the Helmholtz equation can be reduced to

$$\frac{d^2\hat{G}_h}{dz^2} + k_z^2\hat{G}_h = -ik_x\delta(z - z_s) \tag{3.3a}$$

and the boundary condition at $z = 0$ becomes

$$\frac{d\hat{G}_h}{dz} + ik\beta\hat{G}_h = 0 \tag{3.3b}$$

Equations (3.3a and 3.3b) can be solved in the manner detailed in Chapter 2. The solution is

$$\hat{G}_h(k_x,k_y,z) = \frac{k_x}{2k_z}\left\{e^{ik_z|z-z_s|} + \frac{k_z - k\beta}{k_z + k\beta}e^{ik_z(z+z_s)}\right\} \tag{3.4}$$

Hence, the sound field can be cast in an integral form, i.e.

$$\begin{aligned} G_m(x) = {} & \frac{1}{4\pi^2}\int_{-\infty}^{\infty}\int_{-\infty}^{\infty}\frac{k_x}{2k_z}e^{ik_x x+ik_y y+ik_z|z-z_s|}dk_x dk_y \\ & + \frac{1}{4\pi^2}\int_{-\infty}^{\infty}\int_{-\infty}^{\infty}\frac{k_x}{2k_z}e^{ik_x x+ik_y y+ik_z|z+z_s|}dk_x dk_y \\ & - \frac{1}{2\pi^2}\int_{-\infty}^{\infty}\int_{-\infty}^{\infty}\frac{k_x\beta}{(k_z/k)+\beta}\frac{ie^{ik_x x+ik_y y+ik_z|z+z_s|}}{2k_z}dk_x dk_y, \end{aligned} \tag{3.5}$$

where $k_z = \sqrt{k^2 - k_x^2 - k_y^2}$.

Direct evaluation of the first integral in (3.5) is challenging, if not impossible.

An alternative method for its solution is first to find the direct field, p_d in the absence of any boundary. Replacing the acoustic pressure in (3.2) by $p_d = \dfrac{\partial \phi}{\partial x}$ and reversing the order of differentiation gives

$$\frac{\partial}{\partial x}\left(\nabla^2\phi + k^2\phi\right) = \frac{\partial}{\partial x}\left[\delta(x)\delta(y)\delta(z-z_s)\right], \tag{3.6}$$

where φ is the solution for a monopole source located at $(0,0,z_s)$.
The solution of (3.6) is well known and can be evaluated exactly as $\phi = -e^{ikR_1}/4\pi R_1$. Hence, the direct wave term is simply,

$$p_d = -\frac{\partial}{\partial x}\left(\frac{e^{ikR_1}}{4\pi R_1}\right) = \frac{x}{R_1}\frac{\left(1-kR_1\right)e^{ikR_1}}{4\pi R_1^2} \tag{3.7a}$$

If the solution is expressed in the form of spherical polar co-ordinates (R_1,φ,ψ) centred on the source position, then it becomes

$$p_d = \sin\varphi\cos\psi\,\frac{\left(1-ikR_1\right)e^{ikR_1}}{4\pi R_1^2} \tag{3.7b}$$

This equation enables identification of the typical characteristics of the dipole sound field. Essentially, the solution comprises of two components: a near field dominated by the term $1/R_1^2$ and a far field term given by ik/R_1.

Similarly, the second integral of (3.5) can be identified as the sound field due to the image source situating at $(0,0,-z_s)$. By analogy, the solution is

$$p_r = -\frac{\partial}{\partial x}\left(\frac{e^{ikR_2}}{4\pi R_2}\right) = \frac{x}{R_2}\frac{\left(1-ikR_2\right)e^{ikR_2}}{4\pi R_2^2} \tag{3.8a}$$

This solution is expressed in the form of spherical polar co-ordinates (R_2,θ,ψ) centred on the image source, viz.

$$p_r = \sin\theta\cos\psi\,\frac{\left(1-kR_2\right)e^{ikR_2}}{4\pi R_2^2} \tag{3.8b}$$

To obtain the complete analytical solution for the Green's function of the horizontal dipole, it remains to evaluate third integral in (3.5), given by

$$I = \frac{-1}{2\pi^2}\int_{-\infty}^{\infty}\int_{-\infty}^{\infty}\frac{k_x\beta}{\left(k_z/k\right)+\beta}\,\frac{e^{ik_xx+ik_yy+ik_z(z+z_s)}}{2k_z}\,dk_xdk_y \tag{3.9}$$

Again, it is fruitful to transform the integral expression from (k_x,k_y) to (κ,ε) as implemented for the monopole source in Chapter 2. As a result, (3.9) can become

$$I = \frac{-1}{4\pi^2}\int_0^\infty\int_0^{2\pi}\frac{\beta}{(k_z/k)+\beta}\frac{\kappa^2\cos\varepsilon e^{i\kappa r\cos(\varepsilon-\psi)+ik_z(z+z_s)}}{k_z}d\kappa d\varepsilon, \qquad (3.10)$$

where $k_z = \sqrt{k^2-\kappa^2}$.

Using the general form of the integral expression for the Bessel function of n-th order [4, Equation (9.1.21)],

$$J_n(z) = \frac{i^{-n}}{2\pi}\int_0^{2\pi}\cos(n\varepsilon)e^{iz\cos\varepsilon}d\varepsilon \qquad (3.11a)$$

together with the identity,

$$\int_0^{2\pi}\sin(n\varepsilon)e^{iz\cos\varepsilon}d\varepsilon = 0, \qquad (3.11b)$$

I can be written in terms of the Bessel function of the first order as

$$I = \frac{-i\cos\psi}{4\pi}\int_0^\infty\frac{\beta}{(k_z/k)+\beta}\frac{\kappa^2 e^{ik_z(z+z_s)}}{k_z}J_1(\kappa r)d\kappa \qquad (3.12)$$

Making use of the following identities

$$\begin{cases} H_1^{(1)}(u) = -H_1^{(2)}(-u) \\ J_1(u) = \frac{1}{2}\left[H_1^{(1)}(u)+H_1^{(2)}(u)\right] \end{cases},$$

and substituting $\kappa = k\sin\mu$ in (3.12), it is possible derive a more familiar form:

$$I = \frac{-ik^2\beta\cos\psi}{4\pi}\int_{-\pi/2+i\infty}^{\pi/2-i\infty}\frac{\sin^2\mu}{\cos\mu+\beta}\left\{H_1^{(1)}(kr\sin\mu)e^{-ikr\sin\mu}\right\}e^{ikR_2\cos(\mu-\theta)}d\mu \qquad (3.13)$$

Compared with (2.28), (3.13) has an extra $\sin\mu$ term and the Hankel function of the first order instead of zero-th order appears since a horizontal dipole is considered here. At first sight, the different Hankel functions used for a monopole and a horizontal dipole pose a problem for the asymptotic evaluation of (3.13). Nevertheless, it is possible to generalize the method used for the monopole to obtain the field due to a ‘higher-order’ pole by using integral expressions for Hankel functions and their asymptotic forms:

$$H_n^{(1)}(z)e^{-t} = \frac{(-1)^n}{\pi z}\times\frac{4^{1-n}}{(2n-1)\ldots.5\times3\times1}\int_0^\infty y^{2n}e^{-y^2/2}\left(4iz-y^2\right)^{n-\frac{1}{2}}dy \qquad (3.14a)$$

and

$$H_n^{(1)}(z)e^{-t} = (-1)^n\left(\frac{2}{i\pi z}\right)^{1/2}\left\{1-\left(\frac{4n^2-1}{8iz}\right)+\ldots.\right\} \tag{3.14b}$$

The integral can be evaluated asymptotically in the same way as was used for the monopole and the details are described elsewhere [3]. The solution can be expressed as

$$I = -\frac{k^2\beta cos\psi}{4}\left(1-\beta^2\right)^{1/2} erfc\left(-ix_0/\sqrt{2}\right)H_1^{(1)}\left\{kr\left(1-\beta^2\right)^{1/2}\right\}e^{-ik(z+z_s)\beta}, \tag{3.15a}$$

where

$$\frac{1}{2}x_0^2 = ikR_2\left[1+\beta\cos\theta-\left(1-\beta^2\right)^{1/2}\sin\theta\right] \tag{3.15b}$$

Using (3.14b) with $n = 1$ and the approximations, $1 \gg \beta^2$ and $\theta \to \pi/2$, I can be simplified to

$$I = -ikcos\psi\left\{2i\sqrt{\pi}\left(\frac{1}{2}ikR_2\right)^{\frac{1}{2}}\beta e^{-w^2} erfc(-iw)\frac{e^{ikR_2}}{4\pi R_2}\right\} \tag{3.16}$$

Summing all contributions, (3.7b), (3.8b) and (3.16), the required Green's function for a horizontal dipole is

$$G_h(x,y,z) = \cos\psi\left\{\begin{aligned}&\sin\phi\frac{(1-ikR_1)e^{ikR_1}}{4\pi R_1^2}+\sin\theta\frac{(1-ikR_2)e^{ikR_2}}{4\pi R_2^2}\\&-ik\left[2i\sqrt{\pi}\left(\frac{1}{2}ikR_2\right)^{1/2}\beta e^{-w^2} erfc(-iw)\frac{e^{ikR_2}}{4\pi R_2}\right]\end{aligned}\right\} \tag{3.17}$$

To express the required Green's function in the form of Weyl-Van der Pol formula (see equation (2.38)), (3.16) is rewritten as

$$I = \sin\theta\cos\psi\left\{2i\sqrt{\pi}\left(\frac{1}{2}ikR_2\right)^{\frac{1}{2}}\beta e^{-w^2}\operatorname{erfc}(-iw)\frac{(1-ikR_2)e^{ikR_2}}{4\pi R_2^2}\right\}, \tag{3.18}$$

where it is assumed that $ikR_2 \gg 1$ and $\theta \to \pi/2$. Usually, these assumptions (and the earlier assumption that $1 \gg \beta^2$) are valid for outdoor sound propagation predictions. Hence, the Green's function can be cast in the classical Weyl-Van der Pol form as

$$G_h\left(x,y,z\right)=\cos\psi\left\{\sin\varphi\frac{\left(1-ikR_1\right)e^{ikR_1}}{4\pi R_1^2}+Q\sin\theta\frac{\left(1-ikR_2\right)e^{ikR_2}}{4\pi R_2^2}\right\}, \quad (3.19)$$

where Q is the spherical reflection coefficient given in (2.40c).

The Green's function can be derived also, by stating

$$G_h\left(x,y,z\right)=\frac{\partial}{\partial x}G_m\left(x,y,z\right) \quad (3.20)$$

Substituting (2.40a) into (3.20), differentiating each term with respect to x and assuming that $\partial(Q \sin \theta)/\partial x$ is negligibly small, leads to the same expression as in (3.19). Nevertheless, the fuller derivation given here extends the asymptotic analysis from that for a monopole source to that for a dipole source and gives a relatively simple 'roadmap' of the analysis required for higher-order sources.

The corresponding Green's function can be readily extended to an arbitrarily oriented horizontal dipole. Noting that the azimuthal angle is the angle measured from the dipole axis that is aligned along the x-axis. By a simple rotation of axis, the sound field can be written as

$$G_h\left(x,y,z\right)=\cos\left(\psi-\psi_l\right)\left\{\sin\varphi\frac{\left(1-ikR_1\right)e^{ikR_1}}{4\pi R_1^2}+Q\sin\theta\frac{\left(1-ikR_2\right)e^{ikR_2}}{4\pi R_2^2}\right\}, \quad (3.21)$$

where ψ_l is azimuthal angle of the axis of the arbitrary orientated dipole. A more rigorous analysis for the validity of this expression will be left to Section 3.2.3 which is concerned with the sound field due to an arbitrarily oriented dipole.

Figure 3.2a compares the excess attenuation, defined as the total sound field relative to the direct field for monopole and horizontal dipole aligned along the x-axis, i.e. $\psi_l = 0$. The source and receiver heights are 1.0 and 0.5 m, respectively, and the separation is 5.0 m. Throughout this chapter, the two-parameter ground impedance model (see Chapter 5, equation (5.38)) is used in the form:

$$\frac{1}{\beta}=0.436\left(1+i\right)\sqrt{\frac{\sigma_e}{f}}+19.74i\frac{\alpha_e}{f}$$

with default parameter values of σ_e = 38 kPa s m^{-2} and α_e = 15 m^{-1}, respectively, unless stated otherwise.

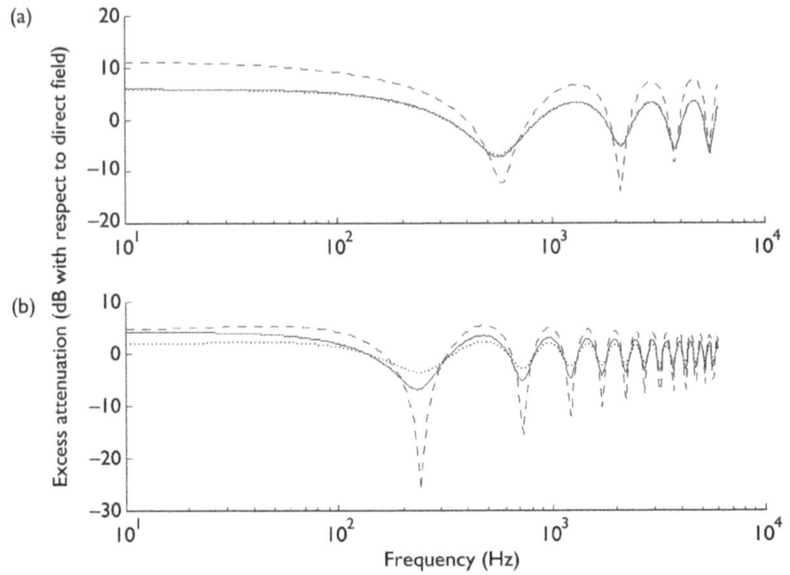

Figure 3.2 The comparison of excess attenuation spectra due to a monopole (solid line), horizontal dipole (dotted line) and a vertical dipole (dashed line). The source/receiver geometry is z_s = 1.0 m and z = 5.0 m, (a) range = 5.0 m, (b) range = 1.0 m.

For the assumed geometry, the angles for the direct and the reflected wave components, ϕ and θ respectively, are very close to each other. Hence, excess attenuation spectra for monopole and dipole sources are virtually indistinguishable, except at the interference minima. Figure 3.2b shows the same comparison but at only 1 m range. For this geometry, the angle between ϕ and θ is large and so the difference between the excess attenuation spectra predicted for monopole and horizontal dipole sources is significant. Also shown in Figures 3.2(a) and (b) are predictions of sound fields due to a vertical dipole using derivations discussed in the next section.

3.2.2 The Vertical Dipole

Suppose that the dipole is aligned so that its axis is aligned along the vertical z-axis as shown in Figure 3.3.

The governing inhomogeneous Helmholtz equation is

$$\nabla^2 p + k^2 p = \frac{\partial}{\partial z}\delta(x)\delta(y)\delta(z - z_s), \tag{3.22}$$

where a unit dipole source strength is assumed. The required Green's function

$\hat{G}_v(x,y,z)$ after the Fourier transformation is

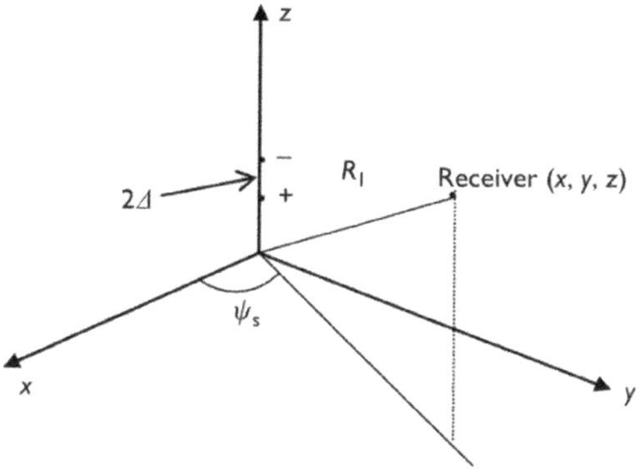

Figure 3.3 A vertical dipole above an impedance ground.

$$\frac{d^2\hat{G}_v}{dz^2} + k_z^2\hat{G}_v = \frac{\partial}{\partial z}\delta(z - z_s) \tag{3.23}$$

subject to the same boundary condition (3.3b).

Introducing a potential function such that $\hat{G}_v = d\varphi/dz$, the boundary condition for the potential function is modified to $d\varphi/dz + ik\beta\varphi = 0$ at $z = 0$. Equation (3.23) can then be simplified to

$$\frac{d^2\varphi}{dz^2} + k_z^2\varphi = \delta(z - z_s) \tag{3.24}$$

Since the original Helmholtz equation (3.22) contains a source term proportional to $\partial\delta(z - z_s)/\partial z$, there is continuity of pressure gradient but a pressure 'jump' at the plane $z = z_s$.

This condition can be stated in terms of the potential function φ (see also (2.12)) as

$$\varphi\big|_{z=z_s^+} = \varphi\big|_{z=z_s^-} \text{ and } \left.\frac{d\varphi}{dz}\right|_{z=z_s^+} - \left.\frac{d\varphi}{dz}\right|_{z=z_s^-} = -1 \tag{3.25}$$

Hence, the solution for (3.24), which satisfies the boundary conditions (3.25), is

$$\varphi = \frac{-i}{2k_z}\left\{\operatorname{sgn}(z - z_s)e^{ik_z|z-z_s|} + \frac{(k_z/k) - \beta}{(k_z/k) + \beta}e^{ik_z(z+z_s)}\right\}, \tag{3.26}$$

where, the sign function is given by

$$\operatorname{sgn}(z - z_s) = \begin{cases} 1 \text{ when } z \geq z_s \\ -1 \text{ when } z < z_s \end{cases} \tag{3.27}$$

Replacing the potential function φ by $\hat{G}_v$ and substituting it into the inverse Fourier integral, the sound field for the vertical dipole is

$$\begin{aligned}\hat{G}_v(x) = &\frac{1}{4\pi^2}\int_{-\infty}^{\infty}\int_{-\infty}^{\infty} \operatorname{sgn}(z - z_s)\frac{e^{ik_x x + ik_y y + ik_z|z-z_s|}}{2}dk_x dk_y \\ &+ \frac{1}{4\pi^2}\int_{-\infty}^{\infty}\int_{-\infty}^{\infty}\frac{e^{ik_x x + ik_y y + ik_z|z+z_s|}}{2}dk_x dk_y \\ &- \frac{1}{2\pi^2}\int_{-\infty}^{\infty}\int_{-\infty}^{\infty}\frac{k_x\beta}{(k_z/k)+\beta}\frac{ie^{ik_x x + ik_y y + ik_z|z+z_s|}}{2k_z}dk_x dk_y\end{aligned} \tag{3.28}$$

The first integral corresponds to the direct field, p_d, and the second is the sound field due to the image source, p_r. Again, the solution for a free field dipole can be used for a vertical dipole in a similar manner to that used for the horizontal dipole in the last section. The only requirement is to determine the angle between the dipole axis and the line radiating from the source to receiver, which is ϕ for the case of the direct wave. Hence,

$$p_d = -\frac{\partial}{\partial z}\left(\frac{e^{ikR_1}}{4\pi R_1}\right) = \cos\phi\frac{(1 - ikR_1)e^{ikR_1}}{4\pi R_1^2} \tag{3.29a}$$

Similarly, the angle between the dipole axis and the line radiating from the image source to receiver is θ. The corresponding sound field can be represented by

$$p_r = -\frac{\partial}{\partial z}\left(\frac{e^{ikR_2}}{4\pi R_2}\right) = \cos\theta\frac{(1 - ikR_2)e^{ikR_2}}{4\pi R_2^2} \tag{3.29b}$$

In the direct wave, the sign function has been incorporated in the cosine term by a precise definition of the direction of polar axes (dipole axis and the spherical polar co-ordinate axis). All the polar angles are measured from the direction of the negative z-axis (see Figure 3.3). When the source is above the receiver, the polar angle for the direct wave ϕ is less than $\pi/2$ which implies that $\cos\phi$ is positive. On the other hand, when the source is below the receiver, then $\pi > \phi > \pi/2$ which implies that the cosine term is negative. So, the principle of reciprocity does not hold for a vertical dipole. Exchanging the source and receiver position will not lead to the same sound field. This is illustrated numerically later after deriving the full expression for the sound field due to the vertical dipole.

The third term in (3.28) needs further investigation. Just as with the horizontal dipole, this term can be simplified to

$$I = \frac{-1}{4\pi^2}\int_0^{\infty}\int_0^{2\pi}\frac{\kappa\beta}{(k_z/k)+\beta}e^{i\kappa r\cos(\varepsilon-\psi)+ik_z(z+z_s)}d\kappa d\varepsilon \tag{3.30}$$

The integral with respect to ε can be evaluated by using the identity (3.11a) to yield

$$I = \frac{-1}{4\pi}\int_0^{\infty} \frac{\kappa\beta}{(k_z/k)+\beta} e^{ik_z(z+z_s)} J_0(\kappa r)\, d\kappa$$

which can be further reduced to

$$I = \frac{-k^2\beta}{4\pi}\int_0^{\infty} \frac{\cos\mu}{\cos\mu+\beta} \sin\mu e^{ikR_2\cos\theta\cos\mu(z+z_s)} H_0(kr\sin\mu)\, d\mu \tag{3.31a}$$

Resolving the integrand by partial fractions leads to

$$\begin{aligned} I = {} & \frac{-k^2\beta}{4\pi}\int_0^{\infty} \sin\mu e^{ikR_2\cos\theta\cos\mu(z+z_s)} H_0(kr\sin\mu)\, d\mu \\ & + \frac{k^2\beta}{4\pi}\int_0^{\infty} \frac{\cos\mu}{\cos\mu+\beta} \sin\mu e^{ikR_2\cos\theta\cos\mu(z+z_s)} H_0(kr\sin\mu)\, d\mu \end{aligned} \tag{3.31b}$$

A close examination of the above integrals reveals that the first term is akin to the direct wave term of a monopole and the second term is analogous to the boundary wave term (see equation (2.28)). Hence, no extra effort is required to obtain the solutions for these two integrals and the boundary wave term for the vertical dipole is simply

$$I = ik\beta\left\{2 + 2i\sqrt{\pi}\left(\frac{1}{2}ikR_2\right)^{\frac{1}{2}} \beta e^{-w^2}\operatorname{erfc}(-iw)\right\}\frac{e^{ikR_2}}{4\pi R_2} \tag{3.32}$$

Summing (3.29a), (3.29b) and (3.32), we get the total sound field due to a vertical pole (or the Green's function) as

$$\begin{aligned} G_v(x,y,z) = {} & \cos\phi\frac{(1-ikR_1)e^{ikR_1}}{4\pi R_1^2} + \cos\theta\frac{(1-ikR_2)e^{ikR_2}}{4\pi R_2^2} \\ & + ik\beta\left\{2 + 2i\sqrt{\pi}\left(\frac{1}{2}ikR_2\right)^{\frac{1}{2}} \beta e^{-w^2}\operatorname{erfc}(-iw)\right\}\frac{e^{ikR_2}}{4\pi R_2} \end{aligned} \tag{3.33}$$

With the extra requirement that $kR_2 \gg 1$, it is possible to write the Green's function in the classical form of the Weyl-Van der Pol formula as

$$\begin{aligned} G_v(x,y,z) = {} & \cos\phi\frac{(1-ikR_1)e^{ikR_1}}{4\pi R_1^2} + \cos\theta\frac{(1-ikR_2)e^{ikR_2}}{4\pi R_2^2} \\ & + \cos\mu_p(1-R_p)F(w)\left[\frac{(1-ikR_2)}{R_2}\right]\frac{e^{ikR_2}}{4\pi R_2}, \end{aligned} \tag{3.34}$$

where cos μ_p (μ_p being a complex angle) is determined according to

$$\cos\mu_p = -\beta \tag{3.35}$$

The term cos μ_p is used to facilitate the interpretation of the theoretical formula as it is linked to the propagation of ground wave. It can be imagined as the direction that characterizes such propagation. It is tempting, as for the horizontal dipole, to replace μ_p with θ. However, numerical experiments have suggested that it does not lead to a satisfactory approximation.

Figure 3.4(a) and (b) show predictions (solid line) of the sound field due to a vertical dipole as a function of horizontal range at 1000 and 100 Hz. The source and receiver heights are located at 2.0 and 1.0 m above the ground surface. The dotted line represents the predicted sound levels due to a monopole source at the same source/receiver geometry and frequencies as the case of vertical dipole. The source strength for the dipole is normalized to give the same sound level as the vertical dipole at 0 m range. In general, the interference pattern of the dipole is essentially the same as that of the monopole source, but the level is somewhat lower.

Predictions of the excess attenuation spectra due to a vertical dipole are also shown plotted in Figure 3.2(a) and (b) in the last section. Compared with the corresponding spectra for a monopole or a horizontal dipole, they exhibit greater interference effects. Figure 3.5 shows predictions of excess attenuation spectra with wither vertical dipole (solid line), monopole

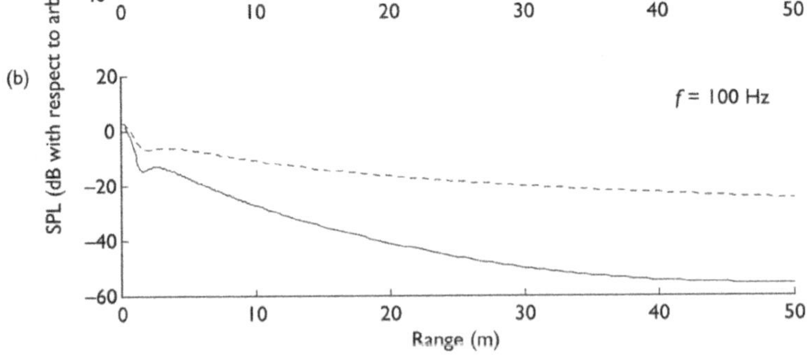

Figure 3.4 The predicted sound pressure level (SPL) from a vertical dipole (solid line) above an impedance ground. The predicted sound level due to a monopole (dotted line) is shown also. The source and receiver heights are 2.0 m and 1.0 m respectively (a) frequency, f = 1000 Hz, (b) frequency, f = 100 Hz.

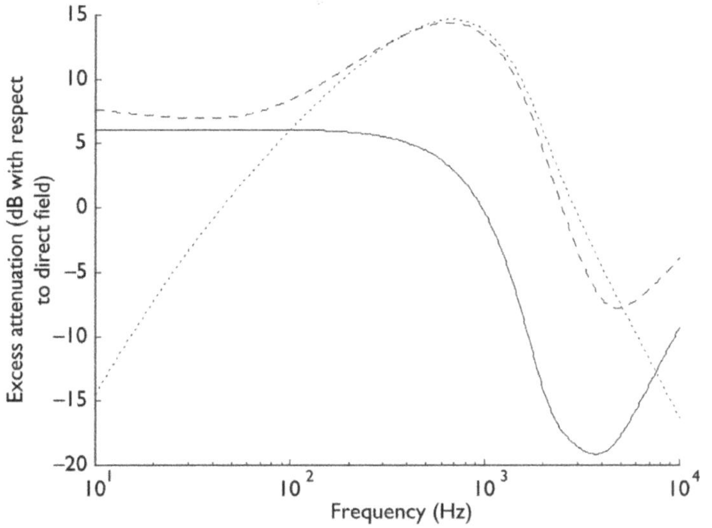

Figure 3.5 Predicted excess attenuation of sound due to a monopole (solid line) and vertical dipole (dashed line) above an impedance ground. The dotted line is the ground wave component of the sound field due to the vertical dipole. The horizontal dipole for this case is indistinguishable from that for a monopole. The assumed source and receiver heights are 0.1 m and 0.025 m respectively and the range is 2.0 m.

(dotted line) sources and receiver close to an impedance ground surface with heights of 0.1 m and 0.25 m, respectively. Propagation of near-grazing sound, the ground wave contribution due to the vertical dipole is predicted to be particularly significant. In Figure 3.5, the predicted ground wave contribution due to a vertical dipole is shown by the dotted line.

Note that the predicted sound fields are different if the positions of source and receiver are changed. This is illustrated in Figure 3.6 by excess attenuation predictions at a range of 1 m but with the source at 0.025 m and the receiver at 0.1 m over the default impedance ground surface.

The discrepancy can be explained by the fact that the polar angles joining the source directly with the receiver differ by π as shown in Figure 3.7(a) and (b). The magnitudes of the direct waves are the same for both cases, but they differ by a factor of −1. Therefore, the sound fields for both situations are not identical. Consequently, the reciprocity theorem does not hold for a vertical dipole source above the ground surface.

3.2.3 An Arbitrarily Orientated Dipole

Having gone through the analyses for the sound field due to a horizontal and vertical dipole, it is natural to expect that the Green's function for an arbitrarily orientated dipole is a combination of these two fundamental building blocks. Consider a representation of an arbitrarily orientated dipole in which

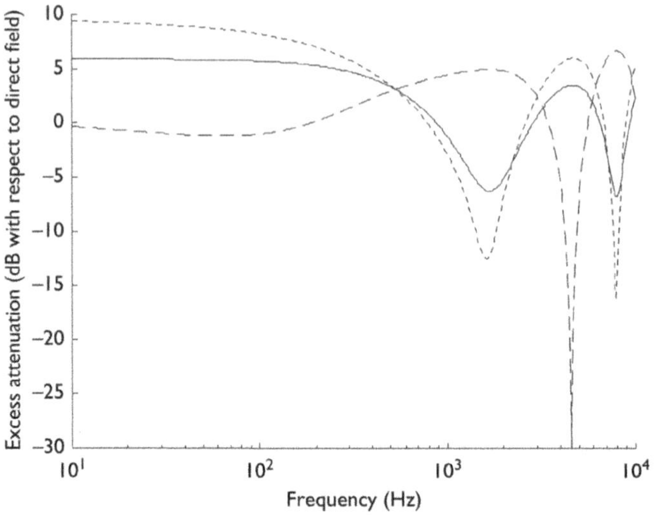

Figure 3.6 Predicted excess attenuation of sound due to a monopole (solid line) and vertical dipole (dotted line) above an impedance ground with the source and receiver heights of 0.25 m and 0.1 m respectively and range of 1.0 m. The dashed line represents the prediction for a vertical dipole with source and receiver heights of 0.1 m and 0.25 m respectively and the same range.

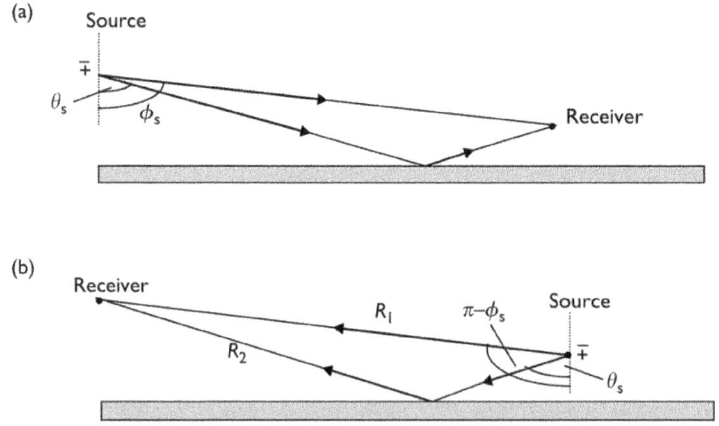

Figure 3.7 The differences in the polar angles when the position of source and receiver are interchanged.

the separation of the two equal and opposite monopoles is 2Δ. The analysis is expedited by introducing the direction cosines $l \equiv (l_x, l_y, l_z)$ of the dipole axis. Sometimes, these direction cosines are referred as the *dipole-moment amplitude* [1,2, p. 165]. As in the previous analysis, the spherical polar coordinates are frequently used. So, it is convenient to define the dipole axis as

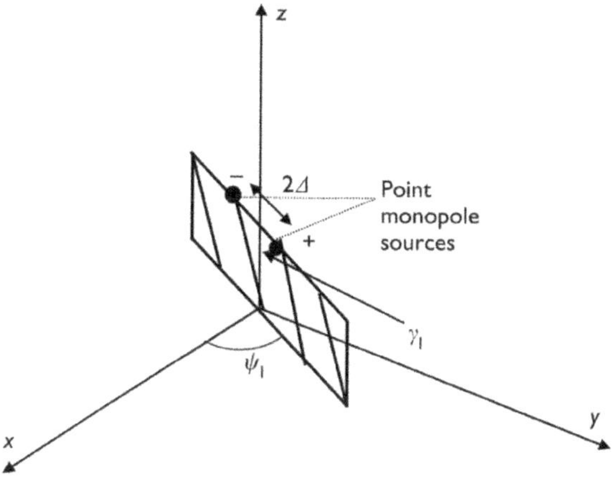

Figure 3.8 An arbitrarily oriented dipole above an impedance ground.

$$\begin{cases} l_x = \sin\gamma_l \cos\psi_l \\ l_y = \sin\gamma_l \sin\psi_l \ , \\ l_z = \cos\gamma_l \end{cases} \tag{3.36}$$

where γ_l is the polar angle measured from the negative z-axis and ψ_l is the azimuthal angle measured from the positive x-axis (see Figure 3.8). Take, e.g., the direction cosines for a horizontal dipole aligned in the x-direction is $(1,0,0)$. Hence, the polar and azimuthal angles γ_l and ψ_l are $\pi/2$ and 0, respectively. On the other hand, the direction cosines for a vertical dipole aligned in the z-direction and its corresponding polar and azimuthal angles are $(0,0,1)$, 0 and $\pi/2$, respectively.

The source on the right-hand side of (3.1a), becomes $\Gamma_1(\mathbf{r}_s)$, given by

$$\Gamma_1(\mathbf{r}_s) = -S_0\left[\delta\left(\mathbf{r} - \left[\mathbf{r}_s + \mathbf{l}\Delta\right]\right) - \delta\left(\mathbf{r} - \left[\mathbf{r}_s - \mathbf{l}\Delta\right)\right],\tag{3.37a}$$

where $\mathbf{r}_s$ and $\mathbf{r}$ are the source and receiver position. Expanding the terms in Taylor's series and grouping the resulting terms, the source term becomes

$$\Gamma_1(\mathbf{r}_s) = S_1\left[\delta'(x)\delta(y)\delta(z-z_s) + \delta(x)\delta'(y)\delta(z-z_s) + \delta(x)\delta(y)\delta'(z-z_s)\right],\tag{3.37b}$$

where the primes denote the derivatives of the delta functions with respect to their arguments. Applying the method of Fourier transformation to (3.1a) with (3.37b) as the source term, the Green's function, $G_d(x,y,z)$ for a dipole of unit strength can be reduced to

$$\frac{d^2\hat{G}_d}{dz^2} + k_z^2\hat{G}_d = -i\left[\left(k_x l_x + k_y l_y\right)\delta\left(z - z_s\right) - k_z l_z \delta'\left(z - z_s\right)\right]\tag{3.38}$$

subject to the same impedance boundary condition, $dp/dz + ik\beta p = 0$ as before.

Hence, following the same procedure as detailed in Sections 3.2.2 and 3.2.3, a solution for $G_d(k_x,k_y,z)$ is obtained as

$$\hat{G}_d(k_x,k_y,z) = \frac{1}{2k_z}\left\{\Lambda^- e^{ik_z|z-z_s|} + \Lambda^+ V e^{ik_z(z+z_s)}\right\}, \tag{3.39a}$$

where

$$\Lambda^- = k_x l_x + k_y l_y + \text{sgn}(z - z_s) k_z l_z \tag{3.39b}$$

$$\Lambda^+ = k_x l_x + k_y l_y + k_z l_z \tag{3.39c}$$

and

$$V = \frac{k_z - k\beta}{k_z + k\beta} \tag{3.39d}$$

Since each of the terms, Λ^- and Λ^+, can be decomposed into a horizontal and a vertical component, the horizontal component contains a further two terms, $k_x l_x$ and $k_y l_y$. There is only a single term, $k_z l_z$ for the vertical component. The Green's function can be considered as a sum of two components written as

$$G_d(x,y,z) = G_h(x,y,z) + G_v(x,y,z), \tag{3.40a}$$

where

$$G_h(x,y,z) = \frac{1}{4\pi^2}\int_{-\infty}^{\infty}\int_{-\infty}^{\infty}(k_x l_x + k_y l_y)\frac{e^{ik_x x + ik_y y + ik_z|z-z_s|}}{2k_z}dk_x dk_y + \frac{1}{4\pi^2}\int_{-\infty}^{\infty}\int_{-\infty}^{\infty}(k_x l_x + k_y l_y)V\frac{e^{ik_x x + ik_y y + ik_z|z+z_s|}}{2k_z}dk_x dk_y \tag{3.40b}$$

And

$$G_v(x,y,z) = \frac{1}{4\pi^2}\int_{-\infty}^{\infty}\int_{-\infty}^{\infty}\text{sgn}(z - z_s)k_z l_z\frac{e^{ik_x x + ik_y y + ik_z|z-z_s|}}{2}dk_x dk_y + \frac{1}{4\pi^2}\int_{-\infty}^{\infty}\int_{-\infty}^{\infty}k_z l_z\frac{e^{ik_x x + ik_y y + ik_z|z+z_s|}}{2}dk_x dk_y \tag{3.40c}$$

The inverse Fourier transforms for horizontal and vertical dipoles can also be written in the polar form as

$$G_h = \frac{\sin\gamma_l}{4\pi^2}\int_0^{\infty}\int_0^{2\pi}\frac{\kappa^2\cos(\varepsilon - \psi_l)}{2\sqrt{k^2 - \kappa^2}}\left[e^{ik_z|z-z_s|} + Ve^{ik_z(z+z_s)}\right]e^{i\kappa r\cos(\varepsilon-\psi)}d\kappa d\varepsilon \tag{3.41a}$$

$$G_v = \frac{\cos\gamma_l}{4\pi^2}\int_0^{\infty}\int_0^{2\pi}\frac{\kappa}{2}\left[e^{ik_z|z-z_s|} + \mathrm{sgn}(z-z_s)Ve^{ik_z(z+z_s)}\right]e^{i\kappa r\cos(\varepsilon-\psi)}d\kappa d\varepsilon, \quad (3.41b)$$

where the reflection coefficient V is transformed to

$$V = \frac{\cos\mu - \beta}{\cos\mu + \beta} \quad (3.41c)$$

The results in the preceding sections can be applied for the evaluation of (3.41a) and (3.41b) with some minor modifications.

The solutions are

$$G_v(x,y,z) = \sin\gamma_l\cos(\psi - \psi_l)\left\{\begin{array}{c} \sin\phi\dfrac{(1-ikR_1)e^{ikR_1}}{4\pi R_1^2} + \sin\theta\dfrac{(1-ikR_2)e^{ikR_2}}{4\pi R_2^2} \\ -ik\left[2i\sqrt{\pi}\left(\dfrac{1}{2}ikR_2\right)^{\frac{1}{2}}\beta e^{-w^2}\mathrm{erfc}(-iw)\dfrac{e^{ikR_2}}{4\pi R_2}\right] \end{array}\right\}$$

(3.42a)

and

$$G_v(x,y,z) = \cos\gamma_l\left\{\begin{array}{c} \cos\phi\dfrac{(1-ikR_1)e^{ikR_1}}{4\pi R_1^2} + \cos\theta\dfrac{(1-ikR_2)e^{ikR_2}}{4\pi R_2^2} \\ -ik\beta\left[2 + 2i\sqrt{\pi}\left(\dfrac{1}{2}ikR_2\right)^{\frac{1}{2}}\beta e^{-w^2}\mathrm{erfc}(-iw)\dfrac{e^{ikR_2}}{4\pi R_2}\right] \end{array}\right\}$$

(3.42b)

For an arbitrary orientated dipole, the total sound field can be calculated by substituting (3.42a and 3.42b) into (3.40a). The expression can be written in a more compact form by introducing the following unit vectors:

$$\hat{\mathbf{R}}_1 \equiv (\sin\varphi_s\cos\psi_s, \sin\varphi_s\sin\psi_s, \cos\varphi_s) \quad (3.43a)$$

$$\hat{\mathbf{R}}_2 \equiv (\sin\theta_s\cos\psi_s, \sin\theta_s\sin\psi_s, \cos\theta_s) \quad (3.43b)$$

$$\hat{\mathbf{R}}_s \equiv (\cos\psi_s, \sin\psi_s, \cos\mu_p), \quad (3.43c)$$

where $\hat{\mathbf{R}}_1$ and $\hat{\mathbf{R}}_2$ are the unit vectors pointing radially outward from the dipole centre towards the observation point for the direct and reflected wave, respectively. $\hat{\mathbf{R}}_s$ may be regarded as the unit vector that characterizes the direction of propagation of the ground wave as it involves the complex angle μ_p. Noting that l is the direction cosine of the dipole axis, it is possible to show that

$$l \cdot \hat{\mathbf{R}}_1 = \cos\varphi_s \cos\gamma_l + \sin\varphi_s \sin\gamma_l \cos(\psi_s - \psi_l) \tag{3.44a}$$

$$l \cdot \hat{\mathbf{R}}_2 = \cos\theta_s \cos\gamma_l + \sin\theta_s \sin\gamma_l \cos(\psi_s - \psi_l) \tag{3.44b}$$

and

$$l \cdot \hat{\mathbf{R}}_s = \cos\mu_p \cos\gamma_l + \sin\gamma_l \cos(\psi_s - \psi_l). \tag{3.44c}$$

For near-grazing propagation where $kR_2 \gg 1$, (3.42a and 3.42b)–(3.44a–3.44c) are substituted into (3.40a) to obtain a close form analytical solution for the sound field due to an arbitrarily orientated dipole as follows:

$$p = \frac{S_1}{4\pi}\left\{ \begin{array}{c} \left(1 \cdot \hat{\mathbf{R}}_1\right)\left[\dfrac{(1 - ikR_1)}{R_1^{\,2}}\right] e^{ikR_1} + \left(1 \cdot \hat{\mathbf{R}}_2\right) R_p \left[\dfrac{(1 - ikR_2)}{R_2^{\,2}}\right] e^{ikR_2} \\ + \left(1 \cdot \hat{\mathbf{R}}_s\right)\left(1 - R_p\right) F(w) \left[\dfrac{(1 - ikR_2)}{R_2^{\,2}}\right] e^{ikR_2} \end{array} \right\} \tag{3.45}$$

In principle, the sound field generated by a dipole can be obtained by differentiating the monopole sound field with respect to the appropriate spatial co-ordinates [see, e.g., the asymptotic expression given in Ref. 5. However, in practice the asymptotic expression for the monopole sound field is not sufficiently precise to yield satisfactory results, especially when calculating the vertical-dipole component of the sound field.

3.3 THE SOUND FIELD DUE TO AN ARBITRARILY ORIENTATED QUADRUPOLE

In the last section, the asymptotic solution of the sound field due to a dipole has been derived. It is natural to ask whether a similar asymptotic solution can be developed for an arbitrarily orientated quadrupole because it can be imagined as two closely spaced dipoles with equal but opposite dipole-moment amplitude vectors. Also, it is of interest to see whether the classical form of the Weyl-Van der Pol formula can be extended to the corresponding sound field due to the quadrupole. Study of the field due to a quadrupole source above a ground surface is relevant to modelling jet engine testing noise outdoors. In this section, we shall derive the corresponding asymptotic formula for the sound field due to the quadrupole.

As the 'construction' of a quadrupole is realizable by putting two closely spaced dipoles of equal but opposite dipole-moment amplitude vectors, the source term for the inhomogeneous Helmholtz equation is simply

$$\Gamma_2\left(\mathbf{r}_s\right) = \Gamma_1\left(\mathbf{r}_s + \mathbf{m}\Delta\right) - \Gamma_1\left(\mathbf{r}_s - \mathbf{m}\Delta\right), \tag{3.46}$$

where $\Gamma_1(\mathbf{r}_s)$ is the source term for a dipole, see (5.37) , and $\mathbf{m} \equiv (m_x, m_y, m_z)$ is the set of direction cosines that characterize the orientation of the quadrupole axis. Also, it is convenient to introduce the corresponding spherical polar co-ordinates with the polar angle of γ_m and azimuthal angle ψ_m. Using the definition for the dipole source, (3.37b) in (3.46), expanding the corresponding terms by Taylor Series and simplifying the resulting expression, the quadrupole source term can be written in a rather compact form as

$$\Gamma_2\left(\mathbf{r}_s\right) = S_2\left(\mathbf{l}\cdot\nabla\right)\left(\mathbf{m}\cdot\nabla\right)\left\{\delta\left(\mathbf{r}-\mathbf{r}_s\right)\right\}, \tag{3.47a}$$

where $\nabla \equiv (\partial/\partial x, \partial/\partial y, \partial/\partial z)$ and $S_2 = 2S_1\Delta$ is the quadrupole source strength. $\Gamma_2(\mathbf{r}_s)$ can be expanded further to yield

$$\Gamma_2\left(\mathbf{r}_s\right) = S_2\begin{bmatrix} l_x m_x \delta''\left(x-x_s\right)\delta\left(y-y_s\right)\delta\left(z-z_s\right) \\ +l_y m_y \delta\left(x-x_s\right)\delta''\left(y-y_s\right)\delta\left(z-z_s\right) \\ +l_z m_z \delta\left(x-x_s\right)\delta\left(y-y_s\right)\delta''\left(z-z_s\right) \\ +\left(l_x m_y + l_y m_x\right)\delta'\left(x-x_s\right)\delta'\left(y-y_s\right)\delta\left(z-z_s\right) \\ +\left(l_y m_z + l_z m_y\right)\delta\left(x-x_s\right)\delta'\left(y-y_s\right)\delta'\left(z-z_s\right) \\ +\left(l_z m_x + l_x m_z\right)\delta'\left(x-x_s\right)\delta\left(y-y_s\right)\delta'\left(z-z_s\right) \end{bmatrix}, \tag{3.47b}$$

where primes indicate derivatives of the delta functions with respect to their arguments.

The method for deriving the required asymptotic solution for arbitrary orientated quadrupole is similar to that for the dipole, which has been described in Section 3.2.

Although the approach is straightforward, the derivation involves tedious algebraic manipulations. So, here we simply state and discuss the solution and give some numerical predictions.

Using the method of Fourier transformation, as before, with the source term specified by (3.47b), the approximate solution for an arbitrarily orientated quadrupole of unit strength, is [6],

$$p \approx \frac{-k^2}{4\pi}\left\{\begin{aligned}&\left(\mathbf{l}\cdot\hat{\mathbf{R}}_1\right)\left(\mathbf{m}\cdot\hat{\mathbf{R}}_1\right)\frac{e^{ikR_1}}{R_1}\\&+\left[\left(\mathbf{l}\cdot\hat{\mathbf{R}}_2\right)\left(\mathbf{m}\cdot\hat{\mathbf{R}}_2\right)R_p+\left(\mathbf{l}\cdot\hat{\mathbf{R}}_s\right)\left(\mathbf{m}\cdot\hat{\mathbf{R}}_s\right)\left(1-R_p\right)F(w)\right]\frac{e^{ikR_2}}{R_2}\end{aligned}\right\}$$
$$+\frac{1}{4\pi}\times\left[\left(\mathbf{l}\cdot\mathbf{m}\right)-3\left(\mathbf{l}\cdot\hat{\mathbf{R}}_1\right)\left(\mathbf{m}\cdot\hat{\mathbf{R}}_1\right)\right]\times\frac{ikR_1-1}{R_1^2}\times\frac{e^{ikR_1}}{R_1}$$
$$+\frac{1}{4\pi}\times\left[\left(\mathbf{l}\cdot\mathbf{m}\right)-3\left(\mathbf{l}\cdot\hat{\mathbf{R}}_2\right)\left(\mathbf{m}\cdot\hat{\mathbf{R}}_2\right)\right]\times\frac{ikR_2-1}{R_2^2}\times\frac{e^{ikR_2}}{R_2} \qquad (3.48)$$

According to (3.48), the amplitude of the sound field due to a quadrupole source is a function of frequency. This characteristic is quite different from that of a monopole source in which the monopole source is proportional to $1/R$. It implies that the extra path length of the image wave (i.e. $R_1 - R_2$) becomes significant at high frequencies and the excess attenuation spectrum tends to large values. This feature, which is more prominent at smaller incident angles, is therefore not related to the effective impedance of the ground surface, but rather to the fact that the amplitude of the quadrupole field depends on frequency.

Normally two types of quadrupoles, viz. longitudinal and lateral, are identified. However, it is possible to introduce and identify the horizontal and vertical components, as was done for the dipole. In a longitudinal quadrupole, the direction cosines, **l** and **m**, are parallel but for a lateral quadrupole, they are perpendicular (see Figure 3.9).

The following plots show the numerical results of the asymptotic solution (3.48) for these two types of quadrupole. A longitudinal quadrupole aligned perpendicular to the ground surface, $\gamma_m = \gamma_l = 0$ can be described as the vertical longitudinal quadrupole. A lateral quadrupole, with $\gamma_m = \pi/2$ and $\gamma_l = 0$, is also considered.

The two-parameter impedance model with parameters σ_e and α_e of 100 kPa s m^{-2} and 50 m^{-1}, respectively, are used for the predictions in this section. Figures 3.10 and 3.11 show the predicted excess attenuation spectra for a vertical longitudinal quadrupole and a lateral quadrupole. The corresponding

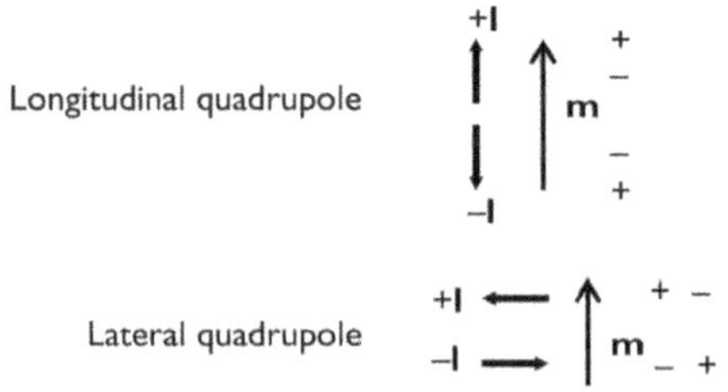

Figure 3.9 A longitudinal quadrupole and a lateral quadrupole. Arrows indicate the orientation of the dipole axis (*l*) and the quadrupole axis (*m*).

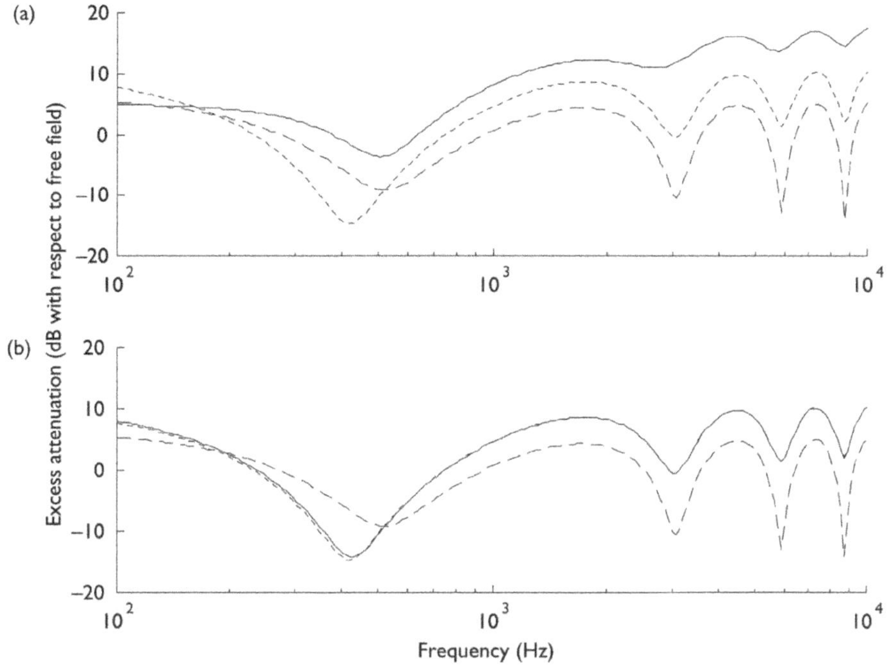

Figure 3.10 The excess attenuation spectra for a quadrupole (solid line), a vertical dipole (dotted line), and a monopole (dashed line). The source and receiver heights are 2.5 m and 1.2 m respectively and the horizontal range is 50 m. (a) A vertical longitudinal quadrupole, and (b) A lateral quadrupole with $\gamma_m = \pi/2, \gamma_l = 0$, and $\psi_m = \pi/2, \psi_l = 0$.

spectra due to a vertical dipole and monopole are also shown in the figures for comparison. The horizontal quadrupole, ($\gamma_m = \gamma_l = \pi/2$) has a similar characteristic to those for the monopole and horizontal dipole sources and will not be discussed here.

In Figure 3.10(a), the source and receiver heights are assumed to be 2.5 m and 1.2 m, respectively, and the separation is 50 m. As shown in Section 3.2.2, the predicted excess attenuation spectrum for a vertical dipole displays greater interference effects than that for a monopole [the dotted and dashed lines in 3.10(a)]. While, the frequencies of the subsequent interference dips are about the same for the monopole, vertical dipole and longitudinal quadrupole, the interferences are much less significant for the longitudinal quadrupole. Predicted spectra for a lateral quadrupole, a vertical dipole and a monopole are compared in Figure 3.10(b). For this source/receiver geometry, there is little difference in the spectra predicted for a lateral quadrupole and a vertical dipole.

Figure 3.11(a) compares predicted attenuation spectra for a vertical longitudinal quadrupole (solid line), a vertical dipole (dotted line) and a monopole (dashed line) source with the same source and receiver heights but the range is 320 m. For such near-grazing propagation, the predicted excess

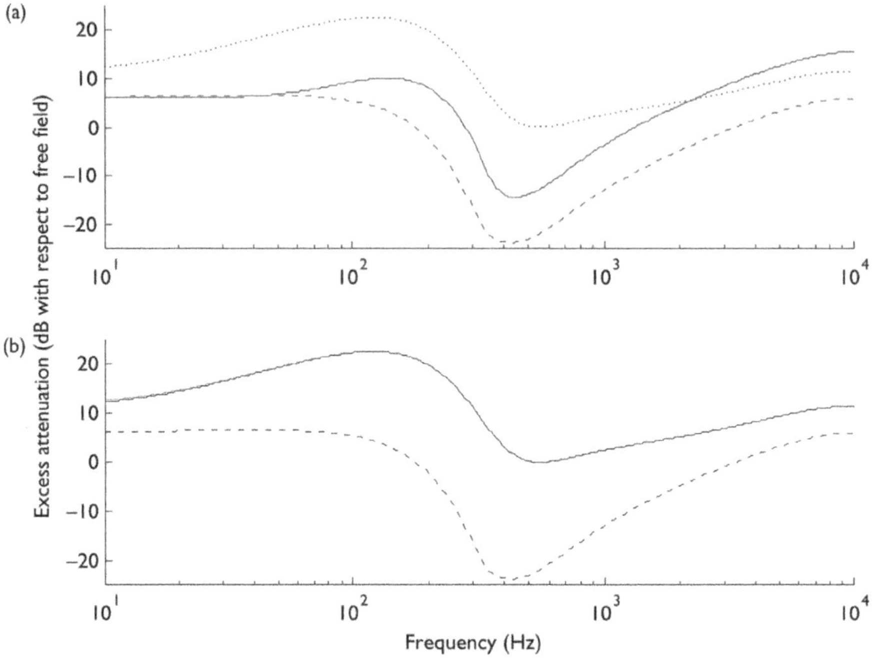

Figure 3.11 The excess attenuation spectra for a quadrupole (solid line), a vertical dipole (dotted line), and a monopole (dashed line). The source and receiver heights are 2.5 m and 1.2 m respectively and the range is 320 m. (a) A vertical longitudinal quadrupole, and (b) A lateral quadrupole with $\gamma_m = \pi/2, \gamma_l = 0$, and $\psi_m = \pi/2, \psi_l = 0$.

attenuation for the longitudinal quadrupole is lower than that from a vertical dipole, but still it is higher than predicted for a monopole. Figure 3.11(b) shows predicted attenuation spectra, assuming the same source/receiver geometry, for a lateral quadrupole (solid line), a vertical dipole (dotted line) and a monopole (dashed line). Those predicted for a vertical longitudinal quadrupole and for a vertical dipole are identical to within the width of the plotted lines. While there is not much difference in the predicted excess attenuation spectra for a lateral quadrupole and a vertical dipole, both these spectra differ from that predicted for a monopole.

A jet noise source can be idealized either as a randomly orientated longitudinal quadrupole or as a randomly orientated lateral quadrupole. The resulting formulae have been used to study the propagation of jet noise above an impedance surface by investigating the directivity pattern of a simple quadrupole [7]. This has suggested that the impedance of ground surface has a significant influence on the directivity of jet noise since there is near cancellation of the intensity field at near-grazing angles for both longitudinal and lateral quadrupoles. Furthermore, the directivity pattern is explicitly dependent on the frequency and the source height rather than the only product of kz_s as suggested by Smith and Carpenter [8], who considered the propagation of turbulent jet noise above a hard ground. The difference is a consequence of the frequency dependence of the reflected wave term.

3.4 RAILWAY NOISE DIRECTIVITY AND PREDICTION

Accurate prediction of noise from railways has become increasingly important as a result of increased rail traffic and the emphasis on high-speed rail links. There are several forms of noise generation on modern trains: traction noise, rail/wheel interaction noise, auxiliary equipment noise and aerodynamic noise. In many studies of noise from trains travelling at speeds below 200–300 km/h it is often assumed that traction and aerodynamic noise are unimportant and that the predominant noise mechanism is the rail/wheel interaction. In this respect, it has been suggested that railway noise at these speeds can be modelled by a set of sources at heights between 0 m and 1 m above either the centreline of the nearest track or the nearside rail. It has been known for many years that wheel/rail noise emitted by a train is directional and that good agreement with data is provided by an approximate representation by a line of incoherent dipole sources. Most electrically hauled trains radiate sound with dipole source characteristics. As a result, the Austrian ÖAL model [9] specifies a combination of dipole and monopole sources with the ratio of 15% monopole and 85% dipole type radiation.

An alternative numerical model based on the boundary integral equation method has been considered by Morgan [10]. According to this model the boundary integral form (8.7) is modified so that the standard 2-D Green's function for sound propagation from a monopole above an impedance ground, $G(\mathbf{r}, \mathbf{r}_s)$, is replaced by

$$G_d(\mathbf{r}, \mathbf{r}_s) = u_x \frac{\partial G(\mathbf{r}, \mathbf{r}_s)}{\partial x_s} + u_y \frac{\partial G(\mathbf{r}, \mathbf{r}_s)}{\partial y_s}, \tag{3.49}$$

where u_x, u_y are the components of, $\mathbf{u} = (u_x, u_y)$, the unit vector along the axis of the dipole, and the derivatives $\frac{\partial G(\mathbf{r}, \mathbf{r}_s)}{\partial x_s}$ and $\frac{\partial G(\mathbf{r}, \mathbf{r}_s)}{\partial y_s}$ can be calculated using a very efficient technique [11]. Usually, in railway noise calculations, the dipoles are assumed to be orientated horizontally, i.e. u = (1,0). The boundary integral equation method has been used to predict railway noise propagation in the presence of noise barriers and rigid train bodies [10]. The cross section of the track considered in this study is reproduced in Figure 3.12. Two railway tracks here are modelled separately with a pair of dipole sources at the railheads above a porous layer of ballast. The assumed spectral strength of these sources is that suggested by Hemsworth [12] for a

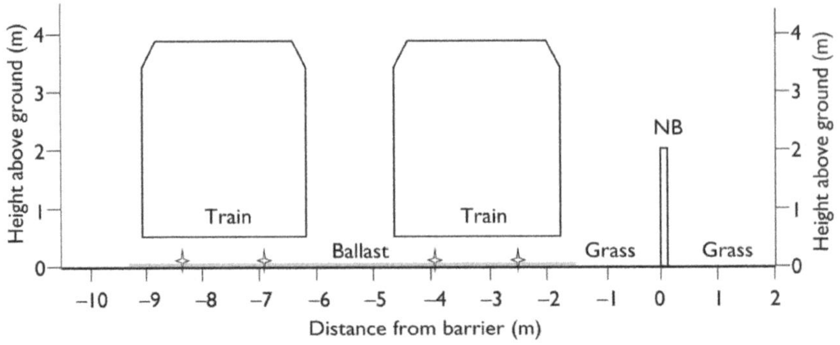

Figure 3.12 Assumed railway cross-section (adapted from [12]).

passenger coach. The ground is flat and consists of grassland on the receiver side including the vicinity of a 2.0 m high noise barrier (labelled NB). The left side of the noise barrier is absorbing.

The acoustic surface admittance of the porous ballast, grassland and absorbing barrier treatment can be modelled, e.g., using the four-parameter Attenborough model [13] and the suggested values of non-acoustic parameters listed in Table 3.1.

The receiver positions are located at distances of 20, 40 and 80 m on the right-hand side from the noise barrier and at 1.5 m above the ground. The insertion losses are calculated in 1/9-octave bands in the frequency range between 63 Hz and 3150 Hz using the boundary integral equation method (see Chapter 8) for both monopole (equation (8.10) with Green's function (8.13)) and dipole sources (equation (8.10) with Green's function (3.49)). The following formula is then used to combine the 1/9-octave band results

$$IL_B = 10\log_{10}\left(\sum_{n=1}^{N} 10^{\frac{L_0(f_n)}{10}}\right) - 10\log_{10}\left(\sum_{n=1}^{N} 10^{\frac{L_B(f_n)}{10}}\right) \tag{3.50}$$

so that the broadband insertion losses, L_B, presented in Table 3.2 [10] can be obtained. Here, $L_0(f_n)$ are the N 1/9-octave railway noise spectra predicted in the absence of the barrier and $L_B(f_n) = L_0(f_n) - IL_P(f_n)$, where $IL_P(f_n)$ is the predicted 1/9-octave barrier insertion loss.

The results demonstrate that, for the assumed railway noise spectrum, the difference between the predictions based on the monopole and dipole models is marginal and vanishes at the receiver positions more distant from the noise barrier. The small differences in predictions may be attributed mainly to the effect of acoustic scattering from the ballast, train body and from the noise barrier. A small discrepancy between these results can be observed also with an increased barrier height and this is explained by the actual difference in the emission characteristics.

Table 3.1 A summary of the non-acoustical parameters used in a four-parameter model for porous grassland, ballast and absorbing barrier treatment (adapted from [10]).

Surface type	*Flow resistivity, R, kPa s m*$^{-2}$	*Porosity,* Ω	*Tortuosity,* $q = \sqrt{T}$ *(see chapter 5)*	*Layer depth, d, m*	*Pore shape factor,* s_p
grassland	125.00	0.5	1.67	∞	0.5
ballast	9.57	0.4	1.54	0.5	0.4
absorbing barrier treatment	6.30	0.9	1.50	0.13	0.5

Table 3.2 Comparison of the broad band A-weighted insertion loss (dB) for two types of sources of railway noise (adapted from [10]).

Barrier height	*Source type*	*Nearside track*			*Farside track*		
		20m	*40m*	*80m*	*20m*	*40m*	*80m*
1.5m	monopole	12.7	9.5	6.2	6.9	3.4	0.7
	dipole	13.3	10.0	6.9	7.1	3.6	0.8
2.0m	monopole	15.6	11.7	8.0	9.6	6.3	2.6
	dipole	16.2	12.5	8.7	10.0	6.4	2.6

3.5 SOURCE CHARACTERISTICS OF ROAD TRAFFIC

This section describes the HARMONOISE/IMAGINE road traffic source power model [14], which is a refinement, based on more data, of empirical regression-type models developed originally in Northern Europe. Tyre–road interaction noise (rolling noise) and propulsion noise (engine noise) are considered separately since these contributions are excited at different heights near a car and, potentially, propagate differently. The HARMONOISE/IMAGINE model offers more spectral detail (i.e. 1/3 octave bands) in predictions of exposure to road traffic noise than the more recent CNOSSOS road traffic noise emission model [15] in which only octave bands are considered.

The HARMONOISE/IMAGINE model distinguishes between different vehicle categories. Light vehicles like passenger cars and (small) vans belong to type 1. Examples of medium–heavy vehicles (type 2) are light buses and trucks with two axles. Heavy vehicles, having more than two axles, belong to type 3. Data for other less common vehicle types are available as well [16,17], but will not be discussed here.

3.5.1 Basic Formulae and Parameters

Sound emissions from vehicles are represented by radiation from ‘effective’ point sources (see Figure 3.13), which simplifies their use in outdoor sound

Figure 3.13 Effective point sources representing sound emission from light and heavy vehicles (from [16]).

propagation models. The rolling noise point source is assumed to be 1 cm above the road surface, in the centre of the car. For light vehicles, the engine noise source height is assumed to be 30 cm above the road surface, while for heavy and medium vehicles, the assumption of a higher engine noise point source position was found to be more appropriate, namely 75 cm, similarly positioned at the centre of the vehicle.

The rolling noise source power level can be estimated from:

$$L_{W,rolling} = a_r + b_r \log_{10}\left(\frac{v}{v_{ref}}\right), \tag{3.51}$$

where v is vehicle speed, v_{ref} = 70 km/h and a_r and b_r are rolling noise coefficients (see Table 3.3) which depend on the sound frequency and vehicle category. The data given in Table 3.3 refer to a reference road surface having the acoustical characteristics of dense asphalt concrete (DAC 0/11) and stone mastic asphalt (SMA 0/11), with 'an age of 2 years or more but not at the end of its life time'. Cars are assumed to drive at constant speed ('cruising'). The air temperature equals 20 °C and the road surface is in a dry state.

The engine noise regression curve is

$$L_{W,propulsion} = a_p + b_p\left(\frac{v - v_{ref}}{v_{ref}}\right), \tag{3.52}$$

where a_p and b_p are coefficients depending on the sound frequency and vehicle category and are listed in Table 3.3 also.

Since the sound radiation of rolling and engine noise cannot be fully separated, it is advised to assign 20% of the rolling noise to the higher source height (and consequently 80% to the low source height). For the engine's acoustic energy, 80% is assigned to the higher source height (and consequently 20% to the lower source height) [17].

Figure 3.14 shows example predictions of source spectra separately for rolling and engine noise from light and heavy vehicles. In Figure 3.15, the rolling and engine noise contributions from a light vehicle are combined and A-weighted.

Table 3.3 Harmonoise/Imagine road traffic source power coefficients (in dB) to be used in Equations (3.51) and (3.52) [17].

	a_r		b_r	cat 1		cat 2		cat 3	
Frequency	*cat 1*	*cat 2 and 3**	*cat 1,2 and 3*	a_p	b_p	a_p	b_p	a_p	b_p
25	69.9	76.5	33	85.8	0	97	0	97.7	0
31.5	69.9	76.5	33	87.6	0	97.7	0	97.3	0
40	69.9	76.5	33	87.5	0	98.5	0	98.2	0
50	74.9	78.5	30	87.5	0	98.5	0	103.3	0
63	74.9	79.5	30	96.6	0	101.5	0	109.5	0
80	74.9	79.5	30	97.2	0	101.4	0	105.3	0
100	79.3	82.5	41	91.5	0	97	0	100.8	0
125	82.5	84.3	41.2	86.7	0	96.5	0	101.2	0
160	81.3	84.7	42.3	86.8	0	95.2	0	99.9	0
200	80.9	84.3	41.8	84.9	0	99.6	0	102.3	0
250	78.9	87.4	38.6	86	8.2	100.7	8.5	103.5	8.5
315	78.8	88.2	35.5	86	8.2	101	8.5	104	8.5
400	80.5	92	31.7	85.9	8.2	98.3	8.5	101.6	8.5
500	85.7	94.1	21.5	80.6	8.2	94.2	8.5	99.2	8.5
630	87.7	93.8	21.2	80.2	8.2	92.4	8.5	99.4	8.5
800	89.2	94.4	23.5	77.8	8.2	92.1	8.5	95.1	8.5
1000	90.6	92.2	29.1	78	8.2	93.8	8.5	95.8	8.5
1250	89.9	89.6	33.5	81.4	8.2	94.3	8.5	95.3	8.5
1600	89.4	88.9	34.1	82.3	8.2	95.2	8.5	93.8	8.5
2000	87.6	86.5	35.1	82.6	8.2	94.9	8.5	93.9	8.5
2500	85.6	83.1	36.4	81.5	8.2	93.3	8.5	92.7	8.5
3150	82.5	81.1	37.4	80.7	8.2	91.2	8.5	91.6	8.5
4000	79.6	79.2	38.9	78.8	8.2	89.3	8.5	90.9	8.5
5000	76.8	77.3	39.7	77	8.2	87.3	8.5	87.9	8.5
6300	74.5	77.3	39.7	76	8.2	85.3	8.5	87.9	8.5
8000	71.9	77.3	39.7	74	8.2	84.3	8.5	81.8	8.5
10000	69	77.3	39.7	72	8.2	83.3	8.5	80.2	8.5

* for category 3 vehicles, the rolling noise coefficient, a_r, should be corrected for the number of axles according to

$$a_{r,corrected} = a_r + \log_{10}\left(\frac{number\ of axles}{2}\right).$$

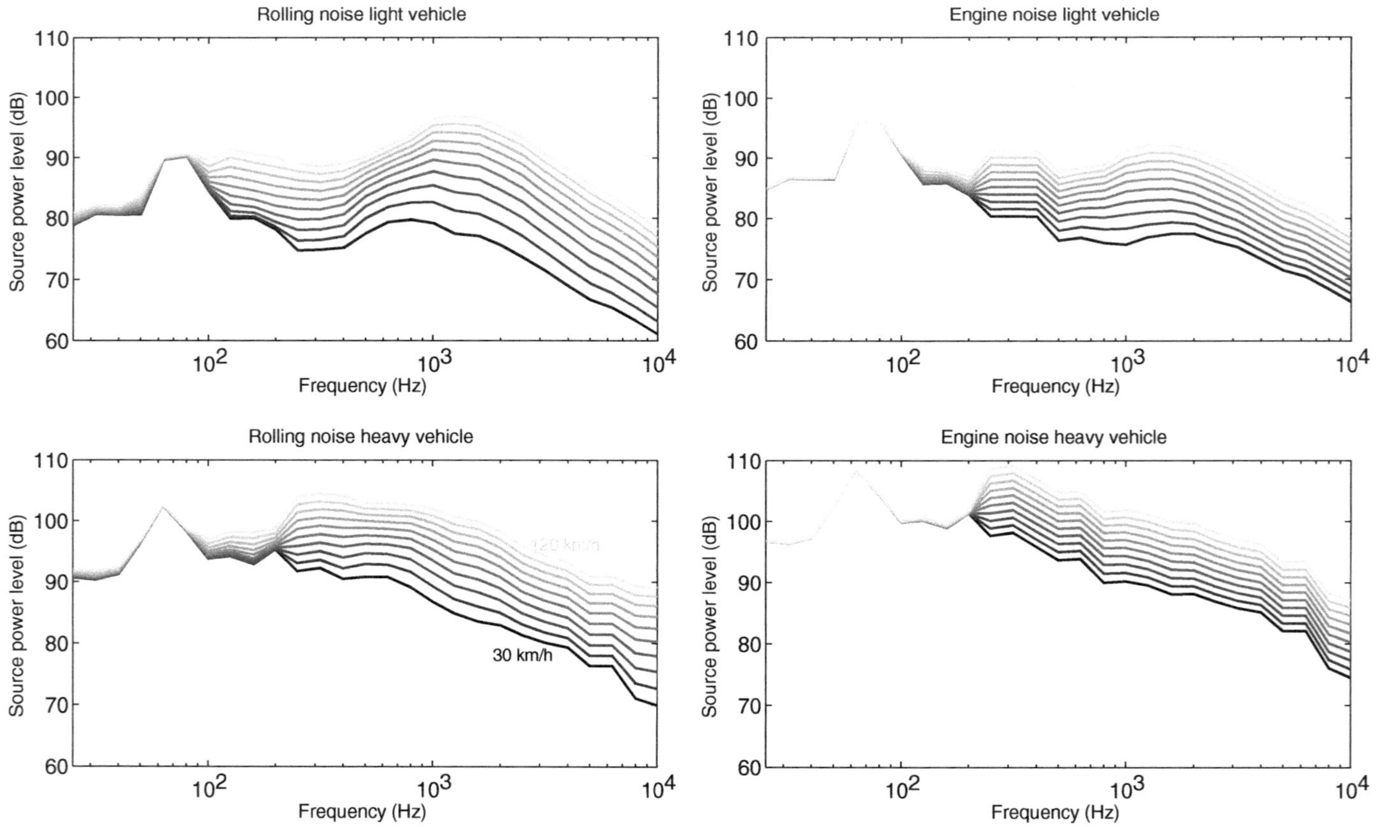

Figure 3.14 Power spectra of the effective sources of rolling noise and engine noise on light (category 1) and heavy vehicles (category 3, no correction for the number of axles, see Table 3.3) for vehicle speeds ranging from 30 km/h to 120 km/h, in steps of 10 km/h predicted by Equations (3.51) and (3.52).

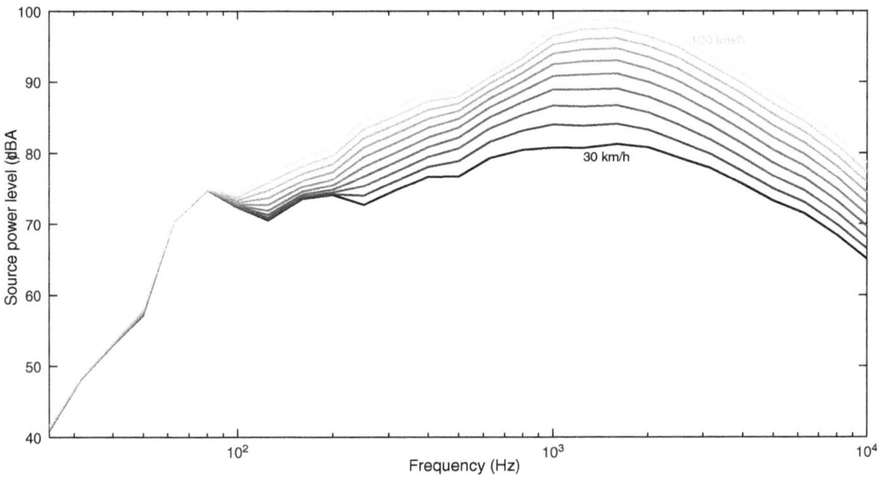

Figure 3.15 Combined A-weighted rolling and engine noise source power spectra of light (category 1) road vehicles corresponding to vehicle speeds ranging from 30 km/h to 120 km/h, in steps of 10 km/h, predicted according to Equations (3.51) and (3.52).

3.5.2 Directivity Corrections

To capture the full complexity of the sound radiation by road vehicles, directivity corrections are needed to the simplified point source radiation. Horizontal directivity results from the horn effect for rolling noise [18], and the screening of propulsion noise by the engine compartment. Vertical directivity results from screening by the bodywork of the vehicle.

The directivity correction (in dB), which depends on sound frequency, source height, horizontal φ and vertical ψ angles (in radians) (see Figure 3.16), can be calculated from:

$$\Delta L\left(f, h_s, \varphi, \psi\right) = \Delta L_H + \Delta L_V \tag{3.53}$$

For the lowest point source (rolling noise, at 1 cm), the horizontal directivity (in dB) is estimated from the following expression [19]:

$$\Delta L_H = \left[-1.5 + 2.5\left|\cos(\varphi)\right|\right]\sqrt{\cos(\psi)}, \tag{3.54}$$

except for 1/3 octave bands with centre frequencies below 1250 Hz or above 8 kHz, for which ΔL_H becomes zero [19]. No horizontal directivity correction is assigned for the engine noise source on light vehicles (height 30 cm). The following directivity function is proposed (in dB) for the heavy vehicle engine noise source (height 75 cm) [19]:

$$\Delta L_H = \left[1.546\left(\frac{\pi}{2} - \varphi\right)^3 - 1.425\left(\frac{\pi}{2} - \varphi\right)^2 + 0.22\left(\frac{\pi}{2} - \varphi\right) + 0.6\right]\sqrt{\cos(\psi)}, \tag{3.55}$$

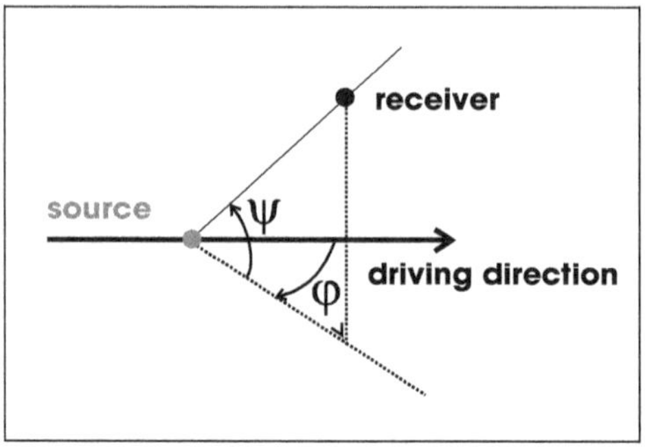

Figure 3.16 The vertical and horizontal angles, relative to the driving direction, for characterising vehicle noise emission directivity.

so that in the forward direction ($\varphi = 0$), the radiation is slightly enhanced, while in the backward direction ($\varphi = \pi$), significant screening of engine noise will be observed [19].

The dependence on vertical angle in Equations (3.54) and (3.55) ensures that, at the top of the car (a position that might appear in an automatically generated noise map), the horizontal correction becomes 0. Figure 3.17 shows plots of these horizontal directivity functions. If using either Equation (3.54) or (3.55) when ψ equals 0, then integration of the horizontal directivity factor during a complete pass-by (i.e. horizontal angles φ ranging from 0 to π) gives a contribution approaching zero. So, when predicting the

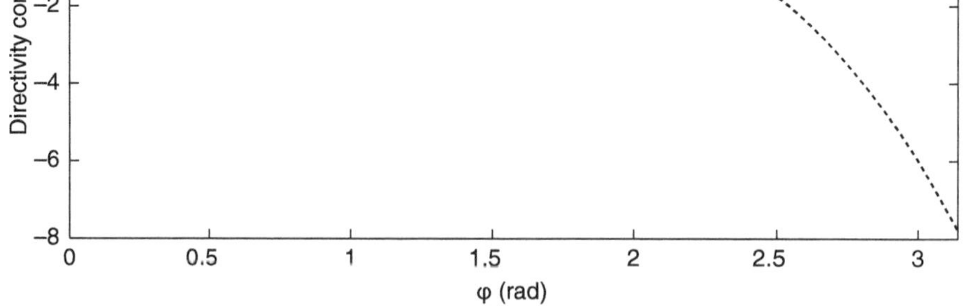

Figure 3.17 Horizontal directivity corrections for light vehicles, for the rolling (h_s = 0.01 m, Equation (3.54)) and the engine noise source at heavy vehicles (h_s = 0.75 m, Equation (3.55)). The vertical angle is fixed at $\pi/4$.

equivalent sound pressure level from a complete vehicle pass-by, horizontal directivity could be disregarded [16].

The vertical directivity functions $\Delta L_V(f, h_s, \psi)$ are summarized in Table 3.4 and illustrated in Figure 3.18 for the two point sources used to represent noise from light vehicles. Much simpler frequency-independent vertical directivity functions can be found [16]. For category 1 (light vehicles),

$$\Delta L_V(\psi) = -\frac{\psi_{\deg r}}{20}, \tag{3.56a}$$

and for category 2 or 3 (medium heavy and heavy vehicles):

$$\Delta L_V(\psi) = -\frac{\psi_{\deg r}}{30} \tag{3.56b}$$

3.5.3 Other Corrections and Limitations

The basic model can be further elaborated. Accelerating cars lead to a significant increase in the engine source power relative to cruising vehicles and estimates of this increase are available [16,17]. Other power train noise corrections allow for the fact that diesel engines are slightly noisier than petrol cars [16], or for electrically driven cars, where the engine noise contribution is much less than from internal combustion engines [20].

Road surface corrections account for pavements other than the reference DAC/SMA surface [16,21]. As is the case with the rolling noise regression Equation (3.51), this correction similarly depends on the logarithm of the vehicle speed. Second-order corrections for road temperature and road age are available as well [16,17]. While porous road surfaces reduce the tyre–road interaction noise significantly, their efficiency decreases over time [17,22,23]. Wet surfaces, especially when so wet that there is a layer of water on the road, cause increased high-frequency emission [15,17].

The possibility of specific interactions between road surface type and the type of tyre make the development of generalized source power models a challenging task. Currently quiet tyre development and road pavement optimization are separate endeavours [24], thereby partly neglecting potential interactions. Also, regional characteristics like the average age/maintenance status of the vehicle fleet and the use of studded tires could impact noise production [17]. Although the HARMONOISE/IMAGINE road traffic noise emission model produces accurate results, some inaccuracies have been observed related to the balance between rolling and engine noise for heavy vehicles which might need further tuning of the coefficients involved [25].

Table 3.4 Vertical directivity functions (in dB) for the Harmonoise/Imagine road traffic source power level [17,19].

Frequency (Hz)	$h_s = 0.01$ m			$h_s = 0.30$ m	$h_s = 0.75$ m
	cat 1	*cat 2*	*cat 3*	*cat 1*	*cat 2, 3*
50, 63, 80	0	0	0	$-2\sin(\psi)$	0
100, 125, 160	$-2\sin(\psi)$	0	0	$-4\sin(\psi)$	0
200, 250, 315	$-2[1-\cos^2(\psi)]$	$-2[1-\cos^2(\psi)]$	$-4[1-\cos^2(\psi)]$	$-5[1-\cos^2(\psi)]$	$-2[1-\cos^2(\psi)]$
400, 500, 630	$-2[1-\cos^2(\psi)]$	$-3[1-\cos^2(\psi)]$	$-5[1-\cos^2(\psi)]$	$-5[1-\cos^2(\psi)]$	$-3[1-\cos^2(\psi)]$
800, 1000, 1250	$-3[1-\cos^2(\psi)]$	$-3[1-\cos^2(\psi)]$	$-5[1-\cos^2(\psi)]$	$-6[1-\cos^2(\psi)]$	$-3[1-\cos^2(\psi)]$
1600, 2k, 2.5k	$-4[1-\cos^2(\psi)]$	$-2[1-\cos^2(\psi)]$	$-4[1-\cos^2(\psi)]$	$-6[1-\cos^2(\psi)]$	$-2[1-\cos^2(\psi)]$
3.15k, 4k, 5k	0	$-2[1-\cos(\psi)]$	$-4[1-\cos(\psi)]$	$-5[1-\cos^2(\psi)]$	$-2[1-\cos(\psi)]$
6.3k, 8k, 10k	0	$-2[1-\cos(\psi)]$	$-4[1-\cos(\psi)]$	$-8[1-\cos(\psi)]$	$-2[1-\cos(\psi)]$

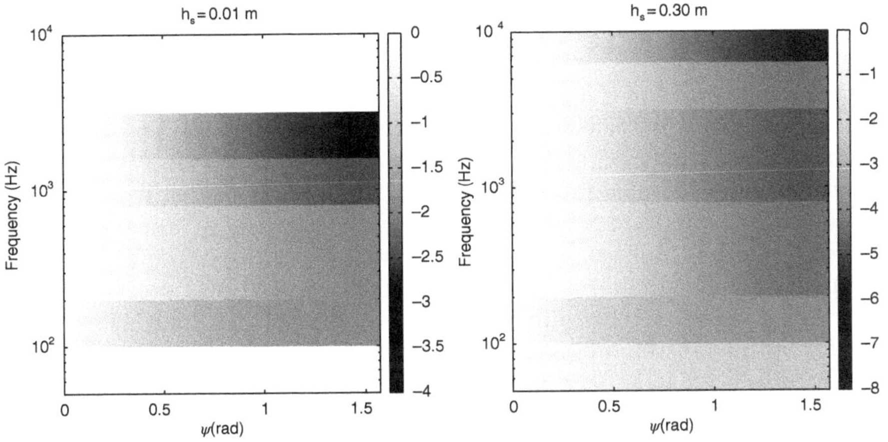

Figure 3.18 Vertical directivity corrections for light vehicle rolling (h_s = 0.01 m) and the engine noise (h_s = 0.30 m), predicted using the functions presented in Table 3.4. Simpler frequency-independent vertical directivity functions can be found in Ref. [15].

3.6 SOURCE CHARACTERISTICS OF WIND TURBINES

3.6.1 Sound-Generation Mechanisms

Wind turbines generate both mechanical and aerodynamic noise. Mechanical noise is associated with the gearbox and generator and, typically, most of the sound energy is between 20 Hz and 550 Hz [26]. While it is usual for mechanical noise to be less important than aerodynamic noise from modern large horizontal-axis wind turbines, it might be significant, particularly if there is wear or damage. Aerodynamic noise is generated by the complex interactions between turbulent flow and the turbine blades. Often blade self-noise is the main component when A-weighted sound pressure levels are of concern. Near the trailing edge of a blade, the various contributions include boundary layer-trailing edge noise, boundary layer separation-stall noise and noise from tip vortex formation [27]. Generally, the resulting sound spectra are broadband, although tonal components might be produced as well [26]. Usually, with modern blade designs, trailing edge bluntness-vortex shedding noise (which is tonal) is unimportant. Also, the interaction between the turbulent flow approaching the wind turbine and the leading edge of the blade (so-called 'inflow noise') will generate sound. Compared to the trailing edge mechanism, this process generates much lower sound frequencies, strongly dependent on the turbulent character of the wind. Potentially, the interaction between the tower and the blades is an additional noise generation mechanism, but, in the main, it is more relevant to downwind rotors and these are much less popular than upwind turbines nowadays.

3.6.2 Typical Spectra of Large Horizontal Axis Wind Turbines

In predicting sound propagation from wind turbines, the ensemble of sounds from the various physical processes is more important than their individual contributions. Many detailed outdoor sound propagation models and engineering models assume point source radiation. Accordingly, wind turbine sound emission is often represented by a single 'effective' or 'apparent' point source placed at hub height [28]. The source power spectrum can then be estimated by calculating back from a sound pressure level spectrum measured close to this virtual hub height position. In this way, the energy radiated from the noise sources distributed over the wind turbine rotor plane is assumed to be concentrated at the effective point source position.

Statistical analysis of many acoustical source power level measurements at 2MW+ wind turbines [29], following the apparent point source procedure, shows that a typical sound power spectrum can be found after normalizing for the total A-weighted level of each individual turbine (see Figure 3.19).

Typically, the sound power of modern pitch-regulated wind turbines increases with wind speed until a wind speed (measured at a height of 10 m) of roughly 8 m/s is reached, and this usually corresponds to the rated electrical power of the wind turbine [29]. Further analysis has shown that, even for wind turbines with rated powers of 200 kW, a very similar normalized spectrum is found [29]. On the other hand, in the low-frequency range, other work has shown that, with increasing rated power (and, therefore, with

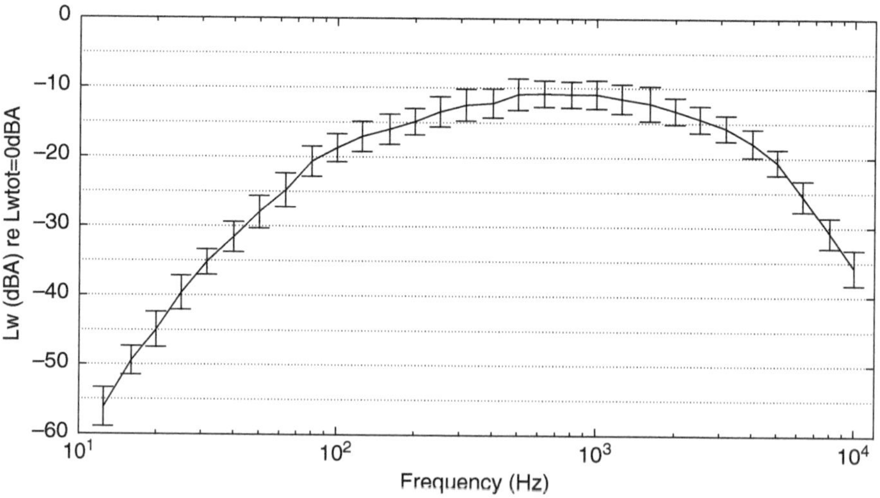

Figure 3.19 An averaged A-weighted effective point source power level spectrum for large horizontal axis wind turbines with a rated power exceeding 2MW, normalized to the total A-weighted source power level. The error bars show the 95 % confidence intervals on the means in each 1/3 octave band [29].

increasing size of the wind turbine), the amplitude at lower frequencies increases somewhat more strongly than at the higher frequencies [30]. Potentially, therefore, the simple shift of the spectrum that has been suggested [29] could lead to inaccuracies if these lower frequencies are of concern.

3.6.3 Horizontal and Vertical Directivity

The IEC 61400-11 measurement standard [28] prescribes that the sound power level by a wind turbine is to be measured in downwind direction, within a range of ±15°, following the apparent point source approach. Given that the sound-generation mechanisms of wind turbines can be directive, assuming a uniform spreading of the acoustical energy in the horizontal plane could lead to less accurate predictions.

The emission in upwind and downwind directions is rather similar and at maximum. A reduced emission is found at receivers near the ground in cross-wind directions [31].

A simplified horizontal directivity function, $\Delta L_{dir,\,\theta}$, has been proposed [31] based on measurements at receivers located about 50 m from the tower of a 1.5 MW upwind turbine rotating at more than 16 rpm:

$$\Delta L_{dir,\theta} = 10\log\left[\frac{1+a(\cos\theta)^b}{1+a}\right], \tag{3.57}$$

where θ is the angle relative to the downwind direction, and a and b are parameters deduced from curve fitting to measurements of 10-s integrated equivalent sound pressure levels. For estimating total A-weighted sound pressure levels, values of a = 1.9 and b = 0.9 are advised [31]. Fitting coefficients for a few octave bands are given in Figure 3.20. The directivity correction varies typically with 5 dB between the minimum and maximum value along the horizontal (azimuth) plane. The rpm of the turbine has a limited effect (at least, for the high rpm range considered in this data set).

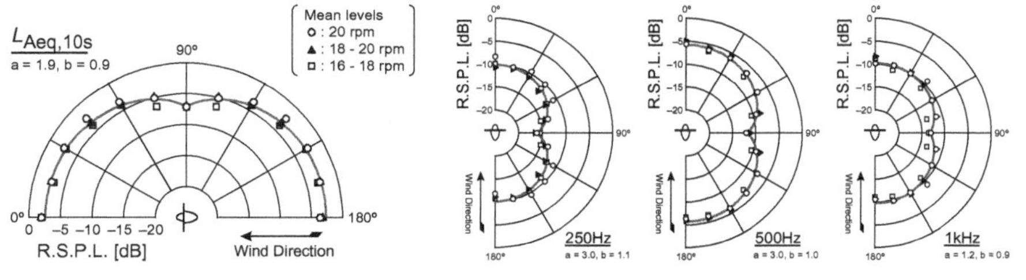

Figure 3.20 Horizontal directivity plots of wind turbine noise emissions for total A-weighted sound pressure levels and for 3 separate octave bands. The 0° direction indicates upwind sound propagation, 90° represents cross wind and 180° represents downwind [31]. Reproduced from [31] with permission from ASJ AST.

Also, wind turbine emissions have a vertical directivity pattern. Implicitly, source power estimates based on measurements relatively close to the turbine, according to the standard assessment methodology [28], include a specific vertical directivity. However, at longer range, the vertical angle might be different. A statistical analysis mentions the influence of rotor speed and possibly blade pitch regulation on vertical directivity [32].

3.6.4 Amplitude Modulation

Wind turbine sound is characterized by so-called amplitude modulation (AM), which means that the sound pressure level fluctuates periodically and rapidly. The frequency of the associated short-term variation is typically near or below 1 Hz. Often, AM is considered to be a main cause of the strong noise annoyance reaction of people living near wind turbines, even with low A-weighted sound exposure levels. This is true especially during periods with strong modulation depths.

Modulation depth D_{AM} can be estimated from the difference between the 5th ($\Delta L_{A,5}$) and 95th percentile values ($\Delta L_{A,95}$), based on the distribution of the differences between short-term (e.g. 100 ms, $L_{eq,100\text{ms},i}$) and longer term (e.g. 3 s, $L_{eq,3\text{s},i}$) equivalent sound pressure levels [33]:

$$\Delta L_{A,i} = L_{eq,100\text{ms},i} - L_{eq,3\text{s},i},\ D_{AM} = \Delta L_{A,5} - \Delta L_{A,95} \tag{3.58}$$

An alternative to $L_{eq,100\text{ms}}$ is exponential time-weighting using 'FAST' response [34]. Other methods to find the amplitude modulation depth have been reported (e.g. [35]). Figure 3.21 shows an example measurement of the

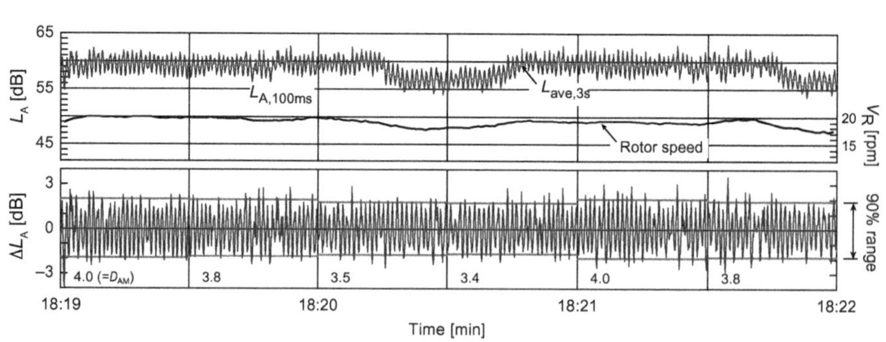

Figure 3.21 Time history of sound pressure levels measured close to a wind turbine, showing distinct amplitude modulation, and how the modulation depth can be calculated. Clearly the variations in longer-term sound pressure levels can be linked to the number of revolutions the rotor makes per unit time [33]. Reproduced from [33] with permission from ASJ AST.

temporal evolution of the exposure level near a wind turbine and shows amplitude modulation clearly. On average, at higher rpm, amplitude modulation depths are larger; although, typically, a rather wide distribution is observed [33].

An important cause of the amplitude modulation is the directivity of the trailing edge noise generation mechanisms at the blades [36]. Although a turbulent air flow, in itself, is able to generate sound, sound generation by turbulent flow becomes most efficient when the turbulent pressure fields interact with the rigid surface of the blades. This makes the associated sound sources strongly directional. Also, during a revolution, a wind turbine blade encounters quite different wind conditions. The so-called angle-of-attack varies around its optimal value for wind energy harvesting and also changes the extent to which the various aerodynamic mechanisms generate noise. This could lead to so-called transient stall [37]. Clearly, these are cyclical processes linked to the rotation rate of the wind turbine.

So far, amplitude modulation, a complex effect, is not fully understood [26]. Close to wind turbines, AM manifests as a time-varying 'swishing sound', which is most likely linked to the directivity of sound radiation from the blades. Such sounds have higher frequency content and, consequently, do not propagate very far in the environment. At longer ranges, amplitude modulation can be observed as an impulsive low-frequency sound, especially when the modulation depth is large. Such sounds are often described as 'rumbling' or 'thumping' noises, and this type of AM is sometimes referred to as 'other amplitude modulation' (in contrast to the amplitude modulation linked to the time-varying 'swishing' [35]). The impulsive amplitude modulation will be enhanced under temperature inversion conditions [38], where the wind shear is large. Although ground level wind speeds might be limited under such conditions, they can be large higher up in the atmospheric boundary layer. This means that the wind conditions encountered by the blades in the rotor plane are even more different than in neutral or unstable atmospheres. In addition to its impact on the sound-generation process, the propagation conditions are favourable. As a consequence, such sounds could be audible for several kilometres from the turbine [39].

It has been postulated that amplitude modulation at longer range could be caused by atmospheric effects during propagation towards the receiver [40–42] and/or by time-dependent interferences between direct and ground reflected sound paths from each of the rotating blades [43].

Also, amplitude modulation has non-uniform directivity characteristics [33]. Larger AM depths are observed upwind than downwind. Maximum AM depth is observed at roughly 60 degrees relative to the exact upwind direction. In contrast to equivalent sound pressure levels which decrease with distance within 200 m range from the mast, AM depths have been found not to be influenced by distance within this range [33].

REFERENCES

[1] A. D. Pierce, *Principles and applications*, 2nd Printing, Acoustical Society of America, New York (1991).

[2] P. Dowling and J. E. Ffowcs Williams, *Sound and sources of sound*, Ellis Horwood Ltd., Chichester (1983).

[3] K. M. Li, S. Taherzadeh and K. Attenborough, Sound propagation from a dipole source near an impedance plane, *J. Acoust. Soc. Am.* 101 3343–3352 (1997).

[4] M. Abramowitz and I. A. Stegun, *Handbook of mathematical functions with formulas, graphs, and mathematical tables*, Dover Publications, Inc., New York (1972). For other identities of Bessel functions, see also G.N. Watson, *A treatise on the theory of Bessel functions* Cambridge U.P., Cambridge (1944).

[5] V. Generalov, Sound field of a multipole source of order *N* near a locally reacting surface, *Sov. Phys. Acoust.* 33 492–496 (1987).

[6] K. M. Li and S. Taherzadeh, The sound field of an arbitrarily oriented quadrupole above a ground surface, *J. Acoust. Soc. Am.* 101 2050–2057 (1997).

[7] S. Taherzadeh and K. M. Li, On the turbulent jet noise near an impedance surface, *J. Sound Vib.* 208 491–496 (1997).

[8] C. Smith and P. W. Carpenter, The effect of solid surfaces on turbulent jet noise, *J. Sound Vib.* 185 397–413 (1995).

[9] ÖAL – Richtlinie, Nr. 30, 1990. *Calculation of Noise Emission from Rail Traffic* (in German).

[10] P. A. Morgan, *Boundary Element Modelling and Full-Scale Measurement of the Acoustic Performance of Outdoor Noise Barriers*, PhD Thesis, University of Bradford (November 1999).

[11] S. N. Chandler-Wilde and D. C. Hothersall, Efficient calculation of the Green's function for the acoustic propagation above a homogeneous impedance plane, *J. Sound Vib.* 180(5) 705–724 (1995).

[12] Hemsworth, Prediction of Train Noise, Ch. 15 (Figure 15.4) in P. M. Nelson (Ed.), *Transportation Noise Reference Book Butterworth*, London (1987).

[13] K. Attenborough, Acoustical impedance models for outdoor ground surfaces, *J. Sound Vib.* 99(4) 521–544 (1985).

[14] R. Nota, R. Barelds and D. Van Maercke, Technical Report HAR32TR-040922-DGMR20 Harmonoise WP 3 Engineering method for road traffic and railway noise after validation and fine-tuning Type of document: Technical Report – Deliverable 18.

[15] K. Stylianos, P. Marco and F. Anfosso-Lédée, Master report on Common Noise Assessment Methods in Europe (CNOSSOS-EU) Outcome and Resolutions of the CNOSSOS-EU Technical Committee & Working Groups (2012).

[16] B. Peeters and G. V. Blokland, The Noise Emission Model for European Road Traffic, IMA55TR- 060821-MP10, IMAGINE Deliverable 11 (2007).

[17] H. Jonasson, Acoustical Source Modelling of Road Vehicles, *Acta Acust. united Ac.* 93 173–184 (2007).

[18] U. Sandberg and J. Ejsmont, *Tyre/Road Noise Reference Book*, INFORMEX, Sweden (2002).

[19] H. Jonasson, U. Sandberg, G. van Bokland, J. Esjmont, G. Watts and M. Luminari, Source modelling of road vehicles, Deliverable 9 of the HARMONOISE project, report HAR11TR-041210-SP10, Borås (SE) (17 December 2004).
[20] J. Pang, Trend of automobile development and its challenge for noise and vibration control, *Proceedings of Internoise 2017*, Hong Kong, China (2017).
[21] Road surface corrections, Appendix 3 of the Dutch calculation and measurement methodology for environmental noise, 2012.
[22] H. Bendtsen, Q. Lu and E. Kohler, Acoustic aging of asphalt pavements, Danish Road Institute, Report 171 (2009).
[23] M. Maennel and B. Altreuther, Ageing of low noise road surfaces, *Proceedings of Internoise* (2016).
[24] L. Tan, Literature review of tire-pavement interaction noise and reduction approaches, *J. Vibro Eng.* 20(6) 2424–2452 (2018).
[25] I. Czyzewski and J. Ejsmont, Validation of harmonoise/imagine traffic noise prediction model by long term noise and traffic monitoring. *Proceedings of Joint Baltic-Nordic Acoustics Meeting 2008, 17–19 August 2008*, Reykjavik, Iceland.
[26] H. Hansen, C. J. Doolan and K. L. Hansen, *Wind farm noise: measurement, assessment and control.* John Wiley & Sons Inc, Hoboken, US (2017).
[27] T. Brooks, D. Poppe and M. Marcolini, Airfoil self-noise and prediction. *NASA Reference publication* 1218 (1989).
[28] IEC 61400-11 Wind turbines – Part 11: Acoustic noise measurement techniques. International Electrotechnical Commission (2012).
[29] Søndergaard, Low Frequency Noise from Wind Turbines: Do the Danish Regulations Have Any Impact? An Analysis of Noise Measurements, *Int. J. Aeroacoust.* 14(5–6) 909-915 (2015).
[30] H. Møller and C. Pedersen. Low-frequency noise from large wind turbines, *J. Acoust. Soc. Am.* 129(6) 3727–3744 (2011).
[31] Y. Okada, K. Yoshihisa, K. Higashi and N. Nishimura. Horizontal directivity of sound emitted from wind turbines, *Acoust. Sci. Tech.* 37 5 (2016).
[32] X. Falourd, P. Marmaroli, R. Feuz and D. Bollinger, Vertical directivity observations based on statistics of low frequency tonal components measured at downwind and upwind locations. *Proceedings of the 7th International Conference on Wind Turbine Noise*, Rotterdam, The Netherlands (2017).
[33] Y. Okada, K. Yoshihisa and S. Hyodo, Directivity of amplitude modulation sound around a wind turbine under actual meteorological conditions, *Acoust. Sci. Tech.* 40 1 (2019).
[34] Fukushima, T. Kobayashi and H. Tachibana, Practical measurement method of wind turbine noise, *Proceedings of the 6th International Conference on Wind Turbine Noise*, Glasgow, UK (2015).
[35] Institute of Acoustics, A Method for Rating Amplitude Modulation in Wind Turbine Noise, final report (2016).
[36] S. Oerlemans and J. Schepers. Prediction of wind turbine noise and validation against experiment, *Int. J. Aeroacoust.* 8(6) 555–584 (2009).

[37] H. A. Madsen, F. Bertagnolio, A. Andreas and C. Bak, Correlation of Amplitude Modulation to Inflow Characteristics, in *Internoise 2014, Conference Proceedings, (Melbourne, Australia), November* 16–19 (2014).

[38] G. van den Berg, Effects of the wind profile at night on wind turbine sound, *J. Sound Vib.* 277 955–970 (2004).

[39] K. Hansen, P. Nguyen, B. Zajamšek, P. Catcheside and C. Hansen. Prevalence of wind farm amplitude modulation at long-range residential locations, *J. Sound Vib.* 455 136–149 (2019).

[40] C. Larsson and O. Ohlund, Amplitude modulation of sound from wind turbines under various meteorological conditions, *J. Acoust. Soc. Am.* 135 67–73 (2014).

[41] R. Makarewicz and R. Gołebiewski, The Influence of a low level jet on the thumps generated by a wind turbine, *Renew. Sust. Energy Rev.* 104 337–342 (2019).

[42] G. Van den Berg, The Beat is Getting Stronger: The Effect of Atmospheric Stability on Low Frequency Modulated Sound of Wind Turbines, *J. Low Freq. Noise Vibr. Active Control* 24(1) 1–24 (2005).

[43] S. Bradley, Time-dependent interference: the mechanism causing amplitude modulation noise? *6th International Meeting on Wind Turbine Noise*, Glasgow, UK (2015.)

Chapter 4

Numerical Methods Based on Time-Domain Approaches

4.1 INTRODUCTION

Traditionally, sound propagation is simulated in the frequency domain. But analytical closed-form solutions of the governing sound propagation equations are available only for simplified cases such as described in Chapters 2, 3, 5, 7 and 11. For more arbitrary and realistic sound propagation problems, numerical techniques like the boundary element and finite element methods are widely used. Numerical approaches to solving the so-called Helmholtz equation in the frequency domain have matured during the past decades and have proven reliability. But, improving computational resources have encouraged modelling of outdoor sound propagation in the time domain.

Most environmental noise sources are broadband in nature. A major advantage of a time-domain approach is that, after exciting a short acoustic pulse at the source position, the response over a broad frequency range can be obtained with a single simulation. While analysis of a system response at a given frequency is more commonly and conveniently carried out in the frequency domain, a time-domain result can be post-processed using a Fourier transform to provide the necessary frequency-domain information. In contrast, frequency-domain techniques require a new simulation run for each sound frequency of interest. To achieve convergence, e.g. in octave or 1/3 octave bands, the number of frequencies for which calculations must be made might be rather high.

Another advantage of time-domain models is that they allow more easily for non-linear effects near high-amplitude sources. This is because a time-domain approach directly models the waveform and, therefore, the waveform distortion corresponding to the transfer of sound energy between sound frequencies can be captured. The latter is less convenient in frequency-domain techniques which calculate a single frequency at a time. Moreover, moving sources, related Doppler-shifts, and transient behaviour can be simulated directly by a time-domain technique. However, the spatial discretization employed in a time-domain simulation will limit the range of frequencies that can be resolved effectively.

Essentially, sound propagation is a time-domain process. Sound propagation is modelled by the Navier–Stokes equations describing general fluid flow. While frequency-domain analyses are dominant in acoustics, numerical techniques in the time domain are dominant in computational fluid dynamics (CFD). To incorporate the effect of a moving and inhomogeneous medium on sound propagation, it makes sense to move closer to the methods used in CFD research so time-domain approaches become attractive.

Section 4.2 describes a popular time-domain approach, the finite-difference time-domain (FDTD) method, which uses finite-differences for discretizing the spatial and temporal derivatives appearing in the governing sound propagation equations. The FDTD model allows a rather straightforward translation from continuous to discretized equations and has become adopted as a reference for modelling outdoor sound propagation. Furthermore, all complex wave-related effects influencing sound propagation from a source to a receiver outdoors can be included. These include reflection, scattering and diffraction near or in between (arbitrary) objects, and complex medium-related effects like convection, refraction and (turbulent) scattering. Perfectly matched layers are the key state-of-the-art absorbing boundary conditions in predictions of outdoor sound propagation when only a small part of the unbounded atmosphere can be described numerically. The interaction between sound waves and (natural) grounds can be incorporated efficiently by including a zone of soil in the calculation domain itself, capturing soil layering effects and non-locally reacting behaviour.

While Section 4.2 does not describe all of the possible numerical options, it aims to provide an outline of a set of basic and easily implementable (discrete) equations, capturing the major influences on outdoor sound propagation. Care is taken to include refraction effects along the sound propagation path. Although numerical efficiency is kept in mind during all steps taken in the development of this reference model, the volume-discretization technique required is not well suited to calculating sound propagation for large distances. While computer compiler optimization, programming codes on graphical processing units (GPUs) or the use of computer grids or clusters might be helpful in improving efficiency, the hybrid approaches discussed in Section 4.3 might be also needed to keep computation times within reason.

4.2 AN EFFICIENT COMPLETE FINITE-DIFFERENCE TIME-DOMAIN MODEL FOR OUTDOOR SOUND PROPAGATION

4.2.1 Sound Propagation Equations

Several authors [1–4] have proposed Linear(ized) Euler(ian) equations (LEE) to study sound propagation in the atmosphere when there is interest in not only the wave phenomena observed in a motionless medium (like reflection, diffraction and scattering by arbitrary objects) but also in flow effects like refraction, convection and (turbulent) scattering. These enable accurate predictions for many civil applications of sound propagation close to the earth's surface. The LEE are derived from the linearized equations of fluid dynamics by assuming that the atmosphere is an ideal gas, that the flow speed vector v_0 is smaller than the speed of sound c and that spatial variations in ambient air pressure and internal gravity waves can be neglected. This leads to the following closed set of coupled partial differential equations in the particle velocity vector $\boldsymbol{v}$ and acoustic pressure p:

$$\frac{\partial p}{\partial t} + v_0 \cdot \nabla p + \rho_0 c^2 \nabla \cdot v = 0, \tag{4.1}$$

$$\frac{\partial v}{\partial t} + (v_0 \cdot \nabla) v + (v \cdot \nabla) v_0 + \frac{1}{\rho_0} \nabla p = 0 \tag{4.2}$$

In these equations, t denotes time, ρ_0 is the mass density of air and there is an implicit assumption of a linear pressure–density relation.

Equations (4.1) and (4.2) only describe sound propagation. In contrast to common aero-acoustic equations, sound generation terms are absent. The flow does not generate sound but is able to deform the wavefronts. The acoustic fields do not influence the overall (macro) fluid flow. One way to describe these equations is to refer to 'sound propagation in background flow' [2,3].

Eqs. (4.1) and (4.2) do not include atmospheric absorption. Although Equation (4.2) can be easily extended with a general diffusion term (see e.g. [1]), such a direct modelling is not very efficient. Additional simulations for the wide range of air temperature and relative humidity combinations that can be observed in the atmosphere as the main drivers of the atmospheric absorption process can be avoided by appropriate filtering of the predicted time-domain signals.

In a non-moving medium, Eqs. (4.1) and (4.2) simplify to:

$$\frac{\partial p}{\partial t} + \rho_0 c^2 \nabla \cdot v = 0, \tag{4.3}$$

$$\frac{\partial v}{\partial t} + \frac{1}{\rho_0} \nabla p = 0 \tag{4.4}$$

After eliminating the particle velocities from Eqs. (4.3) and (4.4), a time-domain sound propagation equation in acoustic pressure only is obtained viz.

$$\frac{\partial^2 p}{\partial t^2} - c^2 \nabla^2 p = 0 \tag{4.5}$$

Assuming $e^{-i\omega t}$ time dependency, the well-known (frequency-domain) Helmholtz equation is obtained.

4.2.2 Numerical Discretization

An essential aspect of any numerical technique is the translation of the continuous equations to a discretized form. In contrast to frequency-domain techniques, both temporal and spatial discretization of the governing sound propagation equations are of concern, and this discretization has a strong influence on numerical accuracy, numerical efficiency and numerical stability. This section presents an efficient lowest-order discretization of Eqs. (4.3) and (4.4) in absence of flow and a discretization procedure for Eqs. (4.1) and (4.2) which is useful in the presence of low-magnitude background flow. A primary choice is whether to use only pressures to discretize the sound propagation domain (i.e. the p-FDTD approach) or to use both particle velocity components and pressures (i.e. the p-v FDTD approach) (see Figure 4.1).

While the use of pressures alone reduces the number of unknowns and offers a potential memory-related benefit, even in its simplest form, second-order derivatives are necessary. Consequently, additional fields need to be stored in memory to prevent mixing old and new fields during time-stepping, partly mitigating the initial memory-related benefit. In room acoustics and

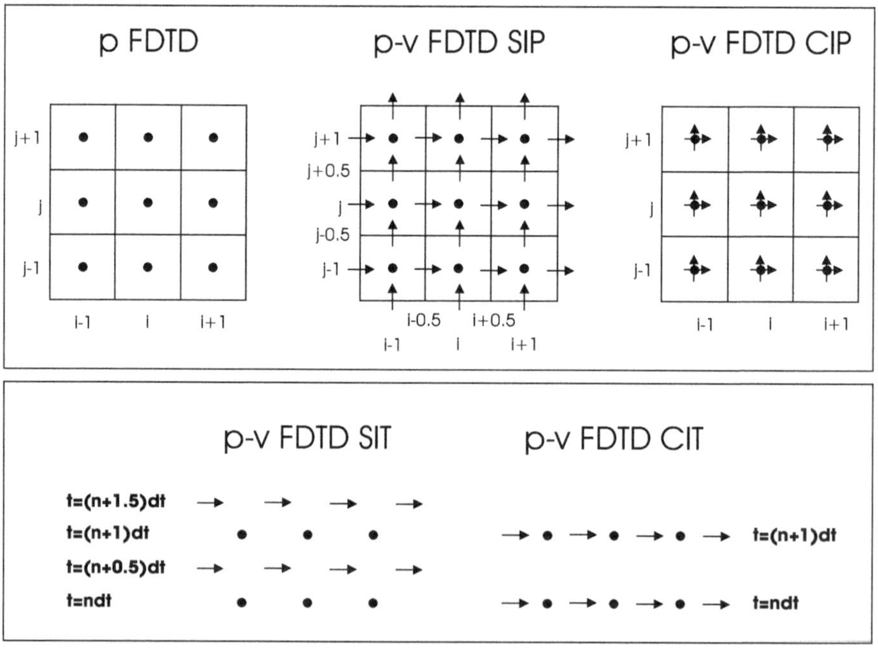

Figure 4.1 Schematics of the staggered-in-place (SIP), collocated-in-place (CIP), staggered-in-time (SIT), and collocated-in-time (CIT) discretisation setups in FDTD. Acoustic pressures are indicated with dots, the components of particle velocities are indicated by arrows.

audio applications [5–7], where flow is of limited or no concern, p-FDTD is an obvious and popular approach. However, unless simplifications are made to the flow field, it is not straightforward to derive a single equation with only the acoustic pressures as unknowns. For consistency, but also in absence of flow, a p-v FDTD discretization approach is chosen.

When using p-v FDTD, a second choice must be made between staggered and collocated spatial discretizations (see Figure 4.1). In a staggered grid, the pressures and velocities do not appear at the same physical locations in the grid. In the model presented here, the pressures are located in the computational cell centres, while the particle velocity components are located on the faces that border each cell. In a collocated grid, pressures and all components of the velocities appear at the same locations, e.g. in the centre of each computational cell.

In the remainder of this chapter, the notations used for the discretization are abbreviated by omitting the spatial discretization steps dx, dy and dz, and temporal discretization steps dt, which are constant in the model, i.e.

$$p^{ldt}_{idx,jdy,kdz} \equiv p^{l}_{i,j,k} \tag{4.6}$$

4.2.2.1 Homogeneous and Still Propagation Medium

This staggered-in-place scheme originates from the so-called Yee cell [8] developed to solve Maxwell's equations in electromagnetic applications using the FDTD technique.

At a lowest possible order and with time dependence omitted, a Taylor expansion analysis to represent the spatial derivative of the pressure based on the nearest neighbours gives

$$p_{i+1,j,k} = p_{i+0.5,j,k} + \frac{dx}{2.1!}\frac{\partial p}{\partial x}\bigg|_{i+0.5,j,k} + \frac{dx^2}{2^2.2!}\frac{\partial^2 p}{\partial x^2}\bigg|_{i+0.5,j,k} + \frac{dx^3}{2^3.3!}\frac{\partial^3 p}{\partial x^3}\bigg|_{i+0.5,j,k}, \tag{4.7}$$

$$p_{i,j,k} = p_{i+0.5,j,k} - \frac{dx}{2.1!}\frac{\partial p}{\partial x}\bigg|_{i+0.5,j,k} + \frac{dx^2}{2^2.2!}\frac{\partial^2 p}{\partial x^2}\bigg|_{i+0.5,j,k} - \frac{dx^3}{2^3.3!}\frac{\partial^3 p}{\partial x^3}\bigg|_{i+0.5,j,k} \tag{4.8}$$

Combining with an expression for the continuous gradient at spatial point $i+0.5$, where the particle velocity appears, yields

$$\frac{\partial p}{\partial x}\bigg|_{i+0.5,j,k} = \left(\frac{p_{i+1,j,k} - p_{i,j,k}}{dx}\right) - \frac{dx^2}{4.3!}\frac{\partial^3 p}{\partial x^3}\bigg|_{i+0.5,j,k} + \ldots \tag{4.9}$$

A similar expression for spatially collocating acoustic pressures and particle velocities would be

$$\frac{\partial p}{\partial x}\bigg|_{i,j,k} = \left(\frac{p_{i+1,j,k} - p_{i-1,j,k}}{2dx}\right) - \frac{dx^2}{3!}\frac{\partial^3 p}{\partial x^3}\bigg|_{i,j,k} + \ldots \tag{4.10}$$

The last term will be neglected in Eqs. (4.9) and (4.10) in this lowest-order approach. In absence of flow, the staggered representation Equation (4.9) has remarkable properties. It gives a four-fold improvement in accuracy relative to Equation (4.10). This is as if the discretization is virtually halved by taking the spatial gradients closer to the location where they are needed (namely at the particle velocity positions). Similar considerations can be made regarding spatial particle velocity gradients for use in the pressure updating (Equation (4.3)), yielding similar accuracy gains.

Also, similar benefits can be found with the staggered-temporal approach (where the pressures and velocities are not updated at the same discrete times, but in a leap-frog manner). In addition, a major advantage is the possibility for in-place computation: the old values replace the new ones in computer memory during time-integration, which is highly efficient.

Then, the discretized set of equations for p-v FDTD, in a staggered spatial and temporal grid, become

$$v_{x(i+0.5,j,k)}^{l+0.5} = v_{x(i+0.5,j,k)}^{l-0.5} - \frac{dt}{\rho_0 dx}\left(p_{i+1,j,k}^{l} - p_{i,j,k}^{l}\right), \tag{4.11a}$$

$$v_{y(i,j+0.5,k)}^{l+0.5} = v_{y(i,j+0.5,k)}^{l-0.5} - \frac{dt}{\rho_0 dy}\left(p_{i,j+1,k}^{l} - p_{i,j,k}^{l}\right), \tag{4.11b}$$

$$v_{z(i,j,k+0.5)}^{l+0.5} = v_{z(i,j,k+0.5)}^{l-0.5} - \frac{dt}{\rho_0 dz}\left(p_{i,j,k+1}^{l} - p_{i,j,k}^{l}\right), \tag{4.11c}$$

$$p_{i,j,k}^{l+1} = p_{i,j,k}^{l} - \rho_0 c_0^2 dt\left(\begin{array}{l}\dfrac{v_{x(i+0.5,j,k)}^{l+0.5} - v_{x(i-0.5,j,k)}^{l+0.5}}{dx} + \dfrac{v_{y(i,j+0.5,k)}^{l+0.5} - v_{y(i,j-0.5,k)}^{l+0.5}}{dy}\\ + \dfrac{v_{z(i,j,k+0.5)}^{l+0.5} - v_{z(i,j,k-0.5)}^{l+0.5}}{dz}\end{array}\right). \tag{4.12}$$

The Courant number (CN) is defined by

$$CN = c\,dt\sqrt{\frac{1}{dx^2} + \frac{1}{dy^2} + \frac{1}{dz^2}} \tag{4.13}$$

To ensure numerical stability of this explicit numerical scheme, the following condition must be fulfilled viz.

$$CN \leq 1 \tag{4.14}$$

Physically, this means that within a single time step (dt), the wavefront may travel a maximum of one grid cell further. Disobeying this basic stability criterion will make numerical simulations meaningless.

Numerical accuracy involves two components: the amplitude error and the phase error. Detailed analysis of the FDTD scheme in absence of flow shows that the amplitude errors are independent of the Courant number. The phase error $\Delta\varphi$ [9] depends on the propagation direction and is inherently asymmetric:

$$\Delta\varphi = 2\sin^{-1}\left(c\,dt\sqrt{\frac{\sin^2(k_x dx/2)}{dx^2} + \frac{\sin^2(k_y dy/2)}{dy^2} + \frac{\sin^2(k_z dz/2)}{dz^2}} \right), \quad (4.15)$$

where k_x, k_y and k_z are the wave number vector components in the three spatial dimensions x, y and z. According to (4.15), there are no phase errors for sound propagation along the diagonal of square cells when the Courant number is equal to one and the phase error is largest for propagation exactly along a single coordinate axis. Given the inherent phase error, a rather fine spatial discretization is needed to reduce the numerical dispersion. Usually ten computational cells per wavelength are advised in such a lowest-order discretization scheme. If more accurate phase predictions are required for a specific application, then an even finer spatial discretization should be chosen. A Courant number of exactly 1 is most interesting, not only from the viewpoint of accuracy and stability but also for numerical efficiency. In this case, a minimum number of time steps are needed for sound to reach a given distance.

Higher-order schemes enhance phase accuracy at a given spatial discretization. Higher-order spatial discretization accuracy can be increased further by applying so-called dispersion-relation preserving schemes [10], in which the Taylor coefficients are slightly adapted, depending on the specific scheme, to reduce dispersion errors further. Nevertheless, with higher-order schemes, time steps need to be reduced to keep simulations stable. So, relaxing the need for a finer spatial discretization should be weighed against the increase in calculation time. Furthermore, this could lead to a much more complicated boundary treatment than in the compact schemes used here. This is a drawback, as application of FDTD is particularly useful when there are many objects and interfaces at close distance, including arbitrary shaped objects.

Higher-order temporal discretization approaches which include more terms from the Taylor series expansion to better approach the continuous derivatives are much less interesting. For each increase in order, additional pressure and velocity fields must be kept in memory during time-stepping to avoid overwriting. This strongly increases the computational cost which is often a main bottleneck in applying the FDTD method. More advanced low-storage techniques have been developed like the one proposed by Bogey and Bailly [11] based on the Runge-Kutta approach, while still explicitly solving these equations.

4.2.2.2 Inhomogeneous Media

It is straightforward to model propagation in inhomogeneous media by using a volume-discretization technique. Each grid cell can be assigned a different temperature (and humidity), leading to a (local) sound speed. Clearly, the highest sound speed appearing in the simulation space will determine the time step, meaning that in other parts of the propagation domain the phase error cannot be minimized anymore. Using the effective sound speed approach [see Chapter 1] therefore does not complicate the sound propagation equations relative to a homogeneous medium.

4.2.2.3 *Moving Medium*

For simulating sound propagation in moving media, staggered-in-place discretization remains an appropriate choice for spatial derivatives. However, the choice of the temporal discretization method needs more care as instability and accuracy issues might appear and, moreover, this choice has an impact on the computational resources required. Methods include staggered-in-time, collocated-in-time and prediction-step staggered-in-time.

4.2.2.3.1 *Staggered-in-Time*

The (forward-difference) staggered-in-time (SIT) approach, as used in a motionless medium, would result in the following time-discrete stepping equation for the acoustic pressure:

$$p^l = p^{l-1} - dtc^2\rho_0\nabla\cdot\mathbf{v}^{l-0.5} - dt\,\mathbf{v}_0\cdot\nabla p^{l-1} \tag{4.16}$$

An explicit time-stepping algorithm is still obtained since the pressure at the new discrete time $l\ dt$ is calculated using values at the previous times $(l-1)dt$ and $(l-0.5)dt$. In-place computation is still possible. Compared to the equations that describe sound propagation in a medium at rest, additional memory is only needed to store background flow velocities. Albeit computationally fast, this approach turns out to be both unstable and inaccurate (see Section 4.2.2.4). Nevertheless, this method has been employed successfully in specific applications [1,12]. However, its use should be discouraged certainly in case of higher flow speeds and long-distance sound propagation.

4.2.2.3.2 *Collocated-in-Time*

The theoretically correct and numerically stable approach for temporal derivatives in flow is the collocated-in-time (CIT) scheme [4], involving centred finite-differences. Pressures and velocities are now updated at the same discrete times. Taking the pressure equation as an example again, the time-discrete equation now reads

$$p^l = p^{l-2} - 2dtc^2\rho_0\nabla\cdot\mathbf{v}^{l-1} - 2dt\,\mathbf{v}_0\cdot\nabla p^{l-1} \tag{4.17}$$

So, an additional particle velocity field and acoustic pressure field must be kept in memory to perform time stepping. As a result, the memory cost is doubled, while time steps are halved which means that the increase in computational cost, relative to sound propagation in a still medium, is large.

However, this discretization scheme was shown to be accurate for Mach numbers up to 1 [4].

4.2.2.3.3 Prediction-Step Staggered-In-Time

To reconcile the accuracy and stability of the CIT scheme and the calculation efficiency of SIT, the so-called prediction step staggered-in-time (PSIT) scheme has been proposed [2,3]. This procedure is based on the SIT temporal grid. The term containing the background flow velocity is now taken as half the value of the previous time $(l-1)dt$, and half the value at time ldt:

$$p^{l} = p^{l-1} - dtc^{2}\rho_{0}\nabla\cdot\mathbf{v}^{l-0.5} - 0.5dt\mathbf{v}_{0}\cdot\nabla p^{l} - 0.5dt\mathbf{v}_{0}\cdot\nabla p^{l-1} \tag{4.18}$$

Manipulating the unknown quantities to the left-hand side gives

$$p^{l} + 0.5dt\mathbf{v}_{0}\cdot\nabla p^{l} = p^{l-1} - dtc^{2}\rho_{0}\nabla\cdot\mathbf{v}^{l-0.5} - 0.5dt\mathbf{v}_{0}\cdot\nabla p^{l-1} \tag{4.19}$$

For the discretization of the spatial pressure derivatives at time $l\,dt$, neighboring values of p are involved and would divert the numerical scheme away from an explicit numerical scheme. If the gradient of the pressure at time $l\,dt$ is approximated by neglecting the background flow, then

$$p^{l}_{appr.} \approx p^{l-1} - dtc^{2}\rho_{0}\nabla\cdot\mathbf{v}^{l-0.5} \tag{4.20}$$

This step comes down to neglecting terms that are second order in Mach number.

Using equation (4.20) in (4.19) gives

$$p^{l} = p^{l-1} - dtc^{2}\rho_{0}\nabla\cdot\mathbf{v}^{l-0.5} - dt\mathbf{v}_{0}\cdot\nabla\left(p^{l-1} - 0.5dtc^{2}\rho_{0}\nabla\cdot\mathbf{v}^{l-0.5}\right) \tag{4.21}$$

The bracketed part on the right-hand side of equation (4.12) can be identified as the sound propagation equation for the pressure at time $(l-0.5)\,dt$ in a medium at rest.

$$p^{l} = p^{l-1} - dtc^{2}\rho_{0}\nabla\cdot\mathbf{v}^{l-0.5} - dt\mathbf{v}_{0}\cdot\nabla p^{l-0.5}_{noflow} \tag{4.22}$$

During the time-stepping algorithm, a 'prediction' is made of the sound field as if there were no flow. This is done after finishing the calculation of the velocity field at time $(l-0.5)dt$ and just before the new pressure field is calculated at time $l\,dt$. When updating the value of p, at time ldt, only fields at previous times are involved. There is a similarity with the CIT approach. In the latter, however, the low-Mach number approximation is not made. The advantage of the PSIT scheme is the decrease in use of computational resources relative to the CIT approach.

Similarly, the time stepping equation for the particle velocity, using this prediction-step staggered-in-time approach, reads

$$\mathbf{v}^{l+0.5} = \mathbf{v}^{l-0.5} - dt\frac{1}{\rho_{0}}\nabla p^{l} - dt\left(\mathbf{v}_{0}\cdot\nabla\right)\mathbf{v}^{l}_{noflow} - dt\left(\mathbf{v}^{l}_{noflow}\cdot\nabla\right)\mathbf{v}_{0} \tag{4.23}$$

4.2.2.4 Numerical Accuracy and Stability

The impacts of the different temporal discretization approaches on numerical accuracy and numerical stability are illustrated in Figures 4.2 and 4.3, respectively. For these calculations, the case of a steady uniform background flow of 10 m/s, directed along one axis of the grid (more precisely the zero-degree direction) is considered. While this is a simplification of the wind flow, it shows clearly the significant impact of the different numerical approaches. Note that the 'effective sound speed' Courant number (CN′) has been used, which means that the uniform flow velocity is added to the sound speed in Equation (4.13).

In absence of flow the SIT scheme is amplitude-error free, whereas, in flow, large errors are observed. The PSIT scheme results in a strong reduction in amplitude error relative to the SIT approach; however, in contrast to the CIT scheme, small amplitude errors are still present. For propagation diagonally along the grid, and since the flow direction is along one of the coordinate axes, the phase error is not zero anymore in that direction, but

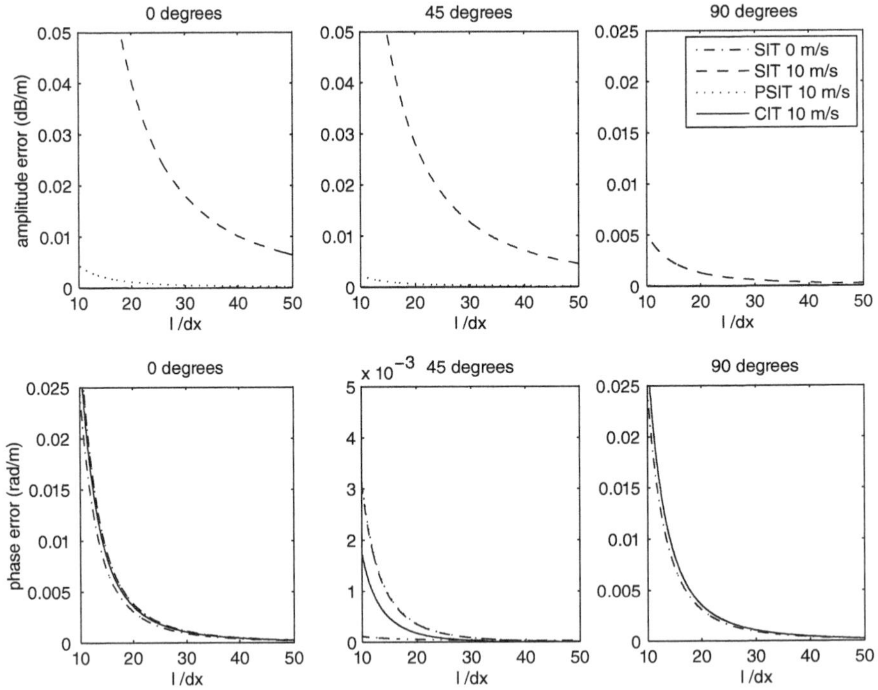

Figure 4.2 Amplitude error (upper figures) and phase error (lower figures) with increasing number of cells per wavelength, for three different propagation directions. SIT (CN' = 1), PSIT (CN' = 1) and CIT (CN' = 0.5) approaches with uniform background flow (directed along the 0° axis) at a velocity of 10 m/s are considered. For comparison, the numerical accuracy of the SIT approach in a non-moving medium (CN = 1) is shown also.

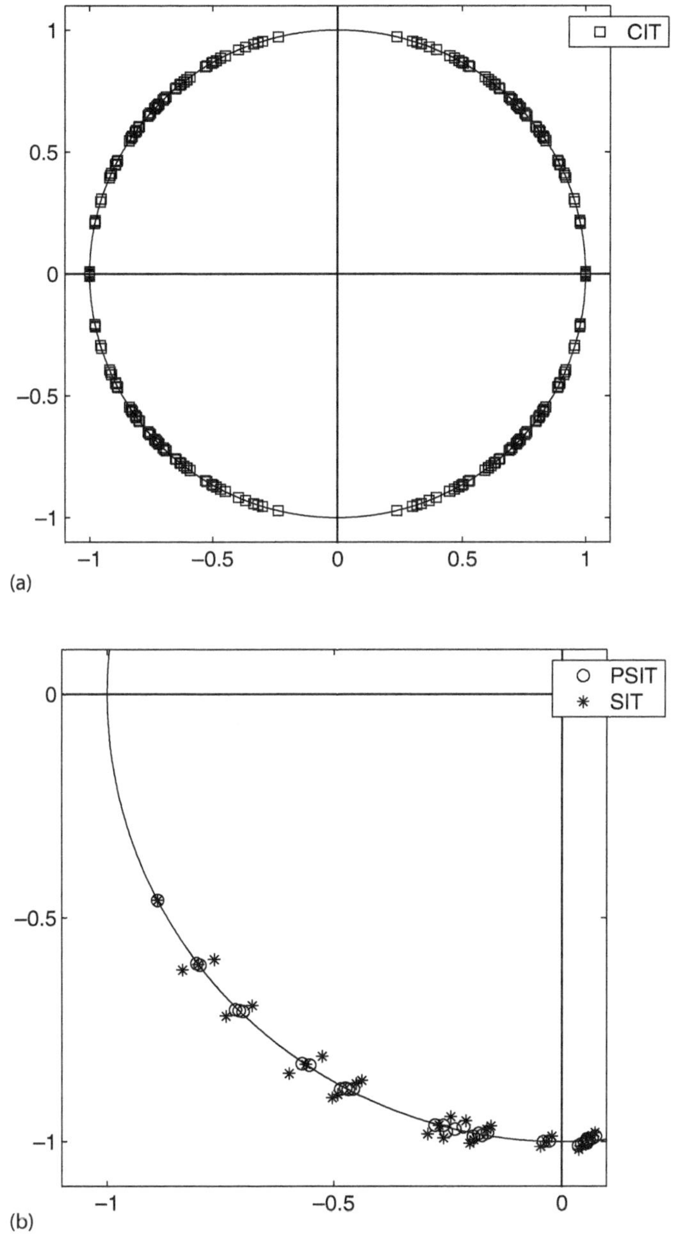

Figure 4.3 Locations of the eigenvalues in the complex plane when representing the update equations as a discrete time-delay system, for a uniform background flow velocity of 10 m/s, and an effective sound speed Courant number equal to 0.5 (CIT) and 1 (SIT and PSIT).

still very small. The flow does not influence the phase error along the coordinate axes (0 degrees and 90 degrees) in any of the schemes.

A related numerical experiment has shown that the introduction of gradients in the flow does not significantly decrease accuracy for the PSIT scheme [13]. Mainly, accuracy is determined by the maximum flow velocity appearing in the computational grid (which was kept to 10 m/s as in the uniform flow case). At a propagation distance of 100 m, the SIT approach did not allow an accuracy assessment due to instabilities.

The update equations can be written as a time-delay system, where X contains the acoustic variables:

$$X^{l+1} = AX^{l} \tag{4.24}$$

System theory shows that such a discrete system is stable only if all eigenvalues (or poles) of the matrix A have a modulus smaller than or equal to 1 or lie within a circle with radius 1 in the complex plane. The position of the eigenvalues relative to this circle gives an indication of the severity of the instability. Eigenvalues inside the unit circle indicate (numerical) damping of the system. For the CIT scheme, all poles lie exactly on the unit circle indicating stability and the absence of amplitude errors. The SIT scheme has poles both outside and inside the unit circle, indicating instability and the presence of amplitude errors. A close inspection that the PSIT poles appear only slightly outside and inside the unit circle indicate that instability and amplitude errors are much reduced compared to the SIT approach. This stability analysis has assumed cyclic boundary conditions. The presence of absorbing boundary conditions might have a stabilizing effect on the numerical schemes, even when they are inherently unstable.

This outline confirms that choice of temporal numerical discretization scheme has a strong influence on numerical stability, numerical accuracy and numerical efficiency. Given that typical FDTD calculations make high computational demands, choices should be well thought out. Although collocated temporal grids are most accurate and numerically stable, they significantly increase the computational burden. The prediction-step staggered-in-time (PSIT) approach is a compromise between accuracy, stability and computational efficiency, and is well suited to typical outdoor sound propagation applications involving wind.

4.2.3 Modelling Propagation in a Moving Unbounded Atmosphere

Perfectly absorbing boundary conditions are essential in outdoor sound propagation applications since the (unbounded) atmosphere must be truncated to a finite calculation domain. However, implementing such boundary conditions is challenging in a full-wave numerical technique. A naïve implementation of an impedance plane boundary with the same impedance as the propagation medium only leads to a limited performance. For normally incident sound waves with the SIT-SIP p-v FDTD approach in absence of background flow, typical reductions in reflection are limited to roughly –40 dB relative to the incident sound energy [3]. For obliquely incident sound waves, even larger reflections come from the borders of the computational domain, significantly affecting the propagation problem under study.

Much better performing absorbing boundary conditions can be found in 'zonal techniques' like the perfectly matched layer (PML) theory, originally proposed by Berenger for electromagnetic FDTD applications [14]. This approach is state-of-the-art in acoustic FDTD applications as well and its implementation is discussed in the remainder of this section.

A basic idea is to border the simulation region of interest with a layer of cells in which there is a gradual increase in damping. This ensures a sufficient loss in acoustic energy when the sound waves reach the borders of the computational domain, while at the same time, the change in impedance near the interface between the bulk grid and the absorbing layers is as low as possible. A second essential step is to split the acoustic pressure into orthogonal ($p_\perp$) and parallel components ($p_{//}$) relative to the interface. This split-field approach introduces an additional degree of freedom and enables the full transmission of plane waves into the absorbing layer at all angles of incidence. Clearly this approach is a purely numerical one, since, physically speaking, acoustic pressure is a scalar.

Artificial (general) damping terms (κ) are added to both the velocity and pressure PML equations. The model presented here [2,3] assumes uniform background flow in the absorbing layer. The flow is directed normal (= direction α) to the interface between the PML and the (useful part of the) computational grid:

$$\frac{\partial p_\perp}{\partial t} + c^2 \rho_0 \frac{\partial v_\alpha}{\partial \alpha} + v_{0\alpha} \frac{\partial p_\perp}{\partial \alpha} + \kappa_{1,\perp} p_\perp = 0, \tag{4.25}$$

$$\frac{\partial p_{//}}{\partial t} + c^2 \rho_0 \sum_{\gamma \neq \alpha} \frac{\partial v_\gamma}{\partial \gamma} + v_{0\alpha} \frac{\partial p_{//}}{\partial \alpha} + \kappa_{1,//} p_{//} = 0, \tag{4.26}$$

$$p_\perp + p_{//} = p, \tag{4.27}$$

$$\rho_0 \left[\frac{\partial v_\alpha}{\partial t} + v_{0\alpha} \frac{\partial v_\alpha}{\partial \alpha} \right] + \frac{\partial p}{\partial \alpha} + \kappa_{2,\perp} v_\alpha = 0, \tag{4.28}$$

$$\rho_0 \left[\frac{\partial v_\beta}{\partial t} + v_{0\alpha} \frac{\partial v_\beta}{\partial \alpha} \right] + \frac{\partial p}{\partial \beta} + \kappa_{2,//} v_\beta = 0, \quad \beta \neq \alpha \tag{4.29}$$

where β represents the coordinate direction, except for the direction of the uniform background flow, and, unless exceptions are stated, γ runs over the three coordinate directions, and $v_{0\alpha}$ is the uniform background flow velocity along coordinate axis α.

Next the so-called matching conditions for these damping terms need to be derived. The imposition of perfect absorption at all angles of incidence and at all sound frequencies leads to the following surprisingly simple 'matching' conditions [3]:

$$\kappa_{2,//} = \kappa_{1,//} = 0, \tag{4.30a}$$

$$\kappa_{1,\perp} = \frac{\kappa_{2,\perp}}{\rho_0}. \tag{4.30b}$$

A gradual increase in the absorption in the layer is achieved by scaling the damping parameters depending on the position x inside the PML:

$$\kappa_{1,\perp}(x) = \kappa_{1,\perp,MAX}\left(\frac{x}{d_{PML}}\right)^m. \tag{4.31}$$

It was found that the parameter m is preferably chosen between 3 and 5 for p-v SIP (P)SIT [2,3]. The maximum damping parameter $\kappa_{1,\perp,MAX}$ is reached at $x = d_{PML}$. The values of the damping parameters have no direct physical meaning and need to be numerically optimized, depending on the FDTD grid parameters and the number of cells constituting the PML.

Note that the PML equations are presented here for the case of a uniform background flow. This simplification of the LEE leads to the same matching conditions as in absence of flow but such simplification of the flow conditions does not significantly deteriorate its performance in practical applications [3], where significant gradients are close to the ground surface. Simulations have shown that reductions in reflection of more than 100 dB are easily obtained after tuning PMLs of reasonable thickness (e.g. 20 computational cells) [3]. Such a performance is adequate for outdoor sound propagation applications. However, perfectly matched layers have been developed for more arbitrary moving media as well [15–17].

4.2.4 Modelling Finite Impedance Boundary Conditions

4.2.4.1 Impedance Plane Approach

In the SIP approach, only particle velocities normal to the grid borders appear. A rigid boundary is therefore easily modelled by setting the particle velocity component at location i+0.5 (see Figure 4.4, assuming a right boundary) to zero:

$$v_{x(i+0.5,j,k)}^{l+0.5} = 0 \tag{4.32}$$

For finite absorbing boundary conditions, the spatial pressure gradient is treated non-symmetrically, and i+0.5 is used instead of i+1 (that would obviously fall outside the grid). Consequently, the distance over which the gradient is calculated reduces to $0.5dx$:

$$v_{x(i+0.5,j,k)}^{l+0.5} = v_{x(i+0.5,j,k)}^{l-0.5} - \frac{2dt}{\rho_0 dx}\left(p_{(i+0.5,j,k)}^{l} - p_{(i,j,k)}^{l}\right) \tag{4.33}$$

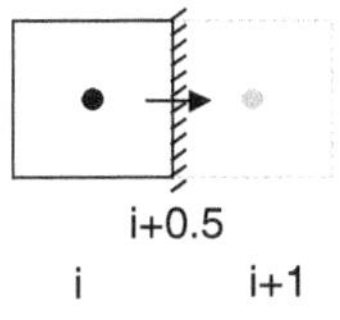

Figure 4.4 Schematic of a grid border cell for the velocity updating approach.

Note that at spatial index i+0.5, acoustic pressures are not defined. However, at this location, the definition of acoustic surface impedance can be used:

$$p_{(i+0.5,j,k)}^{l} = Z(t) * \left(\frac{v_{(i+0.5,j,k)}^{l+0.5} + v_{(i+0.5,j,k)}^{l-0.5}}{2} \right) \quad (4.34)$$

Particle velocities are not defined at time l, so the average over the times $l - 0.5$ and $l + 0.5$ is used. In this way, an equation is obtained in unknowns at correct spatial and temporal discretization points in the SIP-SIT scheme. Note that the approach symbolized in Equation (4.34) is generally applicable but involves convolution operators ($*$) which are very computationally costly at each boundary point.

The updating equation for the particle velocity on the impedance plane is greatly simplified if a constant and thus frequency-independent real impedance is assumed and becomes

$$v_{x(i+0.5,j,k)}^{l+0.5} = \left[1 + \frac{dtc_0 Z_0}{dx}\right]^{-1} \left[v_{x(i+0.5,j,k)}^{l-0.5} \left[1 - \frac{dtc_0 Z_0}{dx}\right] - \frac{2dt}{\rho_0 dx} p_{(i+1,j,k)}^{l} \right] \quad (4.35)$$

where Z_0 is the normalized impedance (relative to the impedance of air). However, the likelihood of such a simplified material behaviour is rather small.

Time-domain boundary conditions (TDBC) are an active field of research, and various approaches can be found [9,18–24]. To be attractive, frequency-impedance relationships must be approached with terms that allow an easy implementation in a time-domain equation such as linear interpolations, exponentially decaying or recursive summations and derivatives. In addition, the conditions of causality (meaning that only previous values are needed to perform time-updating), passivity (meaning that the real part of the impedance is larger than or equal to zero for all sound frequencies) and reality (the particle velocity and pressure in time domain are real values; consequently, also the time-domain representation of the frequency-impedance model must be real) need to be fulfilled [23]. This makes such boundary conditions 'physically admissible' [25]. In general, TDBC should be used with care, since for many, numerical instabilities appear, for which stability criteria are not easily derived.

4.2.4.2 *Ground Interaction Modelling by Including a Layer of Soil*

So far, the time-domain impedance plane boundary conditions discussed have assumed locally reacting materials. Non-locally reacting media can be modelled by including (part of) the ground inside the simulation domain, in contrast to only modelling the interface between two media. Sound propagation inside the material is included explicitly and requires that the sound propagation equations in the material should be known. This approach can be interesting even when only the reflected waves at the interface are of interest since, when such information is available, it enables varying the material properties with depth in the ground. The effect of shallow soil layers can be studied as well.

4.2.4.2.1 Poro-Rigid Frame Model

A popular model to account for soil reflections in FDTD simulations is the phenomenological model by Zwikker and Kosten [26]. This model assumes that the constituting part of the material, i.e. the frame, does not vibrate with the incident sound wave which is a reasonable assumption when the density of the material matrix and its stiffness are significantly higher than those in air. This condition is fulfilled by many outdoor soils [see Chapter 5, Section 5.2]. The model employs three material parameters namely the flow resistivity σ, the porosity φ and the so-called structure constant k_s, which is linked to the tortuosity:

$$\nabla \cdot p + \rho' \frac{\partial v}{\partial t} + \sigma v = 0 \tag{4.36}$$

$$\frac{\partial p}{\partial t} + \rho' c'^2 \nabla \cdot v = 0 \tag{4.37}$$

$$\rho' = \frac{\rho_0 k_s}{\varphi} \tag{4.38}$$

$$c' = \frac{c}{\sqrt{k_s}} \tag{4.39}$$

The main reason for its popularity is the ease of implementation. A general damping term is introduced in the velocity equation (4.36), which is proportional to the flow resistivity of the soil. The (scalar) modified mass density ρ' and speed of sound c' in the ground (equations (4.38) and (4.39)) include factors of the porosity and structure constant, which have constant real values, so introduce no additional numerical issues.

The discretized equations to describe sound propagation in the ground medium then become

$$v_{x(i+0.5,j,k)}^{l+0.5} = \left[1 + \frac{0.5\sigma dt\varphi}{\rho_0 k_s}\right]^{-1} \left[\begin{array}{l} v_{x(i+0.5,j,k)}^{l-0.5}\left[1 - \dfrac{0.5\sigma dt\varphi}{\rho_0 k_s}\right] \\ \quad - \dfrac{\varphi dt}{k_s \rho_0 dx}\left(p_{i+1,j,k}^{l} - p_{i,j,k}^{l}\right) \end{array}\right], \tag{4.40a}$$

$$v_{y(i,j+0.5,k)}^{l+0.5} = \left[1 + \frac{0.5\sigma dt\varphi}{\rho_0 k_s}\right]^{-1} \left[\begin{array}{l} v_{y(i,j+0.5,k)}^{l-0.5} \left[1 - \frac{0.5\sigma dt\varphi}{\rho_0 k_s}\right] \\ - \frac{\varphi dt}{k_s \rho_0 dy} \left(p_{i,j+1,k}^{l} - p_{i,j,k}^{l}\right) \end{array} \right], \tag{4.40b}$$

$$v_{z(i,j,k+0.5)}^{l+0.5} = \left[1 + \frac{0.5\sigma dt\varphi}{\rho_0 k_s}\right]^{-1} \left[\begin{array}{l} v_{z(i,j,k+0.5)}^{l-0.5} \left[1 - \frac{0.5\sigma dt\varphi}{\rho_0 k_s}\right] \\ - \frac{\varphi dt}{k_s \rho_0 dz} \left(p_{i,j,k+1}^{l} - p_{i,j,k}^{l}\right) \end{array} \right], \tag{4.40c}$$

$$p_{i,j,k}^{l+1} = p_{i,j,k}^{l} - \frac{\rho_0 c_0^2 dt}{\varphi} \left(\begin{array}{l} \frac{v_{x(i+0.5,j,k)}^{l+0.5} - v_{x(i-0.5,j,k)}^{l+0.5}}{dx} + \frac{v_{y(i,j+0.5,k)}^{l+0.5} - v_{y(i,j-0.5,k)}^{l+0.5}}{dy} \\ + \frac{v_{z(i,j,k+0.5)}^{l+0.5} - v_{z(i,j,k-0.5)}^{l+0.5}}{dz} \end{array} \right). \tag{4.41}$$

Some care is needed when using this model for high flow resistivities/low porosities. To ensure accuracy, a finer grid than needed in the bulk air should be used to accurately capture the rapid decrease in sound pressure near the soil interface. Both a sudden grid refinement with a factor 4 [12] and a gradual grid refinement, reducing spurious reflections in the refinement area near the air–ground interface [3], have been proposed. In any case, additional phase errors due to cell size changes are inevitable in the FDTD context. Depending on the available computational resources and the frequency range of interest, an overall refined grid is another option. To ensure numerical stability at higher flow resistivities, time steps should be slightly decreased. Overall, for commonly encountered parameter sets of natural soils, the additional constraints on the numerical parameters are rather limited [27].

This simple approach does not capture all relevant physics of sound waves interacting with outdoor soils, especially for propagation through soils at high frequencies and low flow resistivities [28] and if the elasticity of the ground is important (see Chapter 5.5). However, in its basic form, sound reflection is rather accurately modeled which is usually the main concern. Making the soil parameters weakly frequency-dependent extends its range of applicability [28]; however, this requires departing from the straightforward implementation discussed (see Chapter 5, Section 5.3).

4.2.4.2.2 Poro-Elastic Frame Models

When using a poro-elastic frame model, the coupled movement of the frame and air inside the porous medium needs to be accounted for. Two approaches, inspired by Biot's theory [29,30], have been reported for time-domain modelling of outdoor sound propagation. Dong et al. [31] choose to only discretize the pore air particle velocity and frame velocity inside the porous medium, while above the soil only acoustic pressures are used. Ding et al. [32] consistently follow the SIT SIP p-v FDTD approach inside the poro-elastic medium as in unbounded air (above ground). This means that inside the soil, not only the pore air pressure and particle velocity components as in a poro-rigid frame model need to be resolved, but also the frame pressure and frame velocity. These additional unknowns describing the frame behaviour are positioned at the same locations as their counterparts inside the poro-elastic medium. The corresponding continuous equations are as follows:

$$\nabla \cdot p_a + \rho_a \frac{\partial v_a}{\partial t} + \sigma\left(v_a - v_f\right) + \rho_a\left(\frac{m_t}{\varphi_a^{\;2}} - 1\right)\frac{\partial}{\partial t}\left(v_a - v_f\right) = 0, \quad (4.42)$$

$$\nabla \cdot p_f + \rho_f \frac{\partial v_f}{\partial t} + R_f v_f + \sigma\left(v_a - v_f\right) - \rho_a\left(\frac{m_t}{\varphi_a^{\;2}} - 1\right)\frac{\partial}{\partial t}\left(v_a - v_f\right) = 0, \quad (4.43)$$

$$\frac{\partial p_a}{\partial t} + K_a \phi_a \nabla \cdot v_a + \left(K_a - P_0\right)\phi_f \nabla \cdot v_f = 0, \quad (4.44)$$

$$\frac{\partial p_f}{\partial t} + K_f \nabla \cdot v_f - \frac{\phi_f}{\phi_a}\frac{\partial p_a}{\partial t} = 0, \quad (4.45)$$

where v_a and p_a are the air particle velocity and air acoustic pressure inside the material, v_f and p_f are the frame velocity and the pressure on the frame, K_a and K_f are the air and frame bulk modulus, ρ_a and ρ_f are the air and frame density, P_0 is the ambient static pressure, m_t is the tortuosity of the porous medium and φ_a and φ_f (=1 − φ_a) are the air and frame porosity.

To model frame damping in Biot's frequency-domain model, some medium parameters were assigned an imaginary part which cannot be used directly in a time-domain method. So, an additional frame damping coefficient has been used instead (the term $R_f v_f$ in Equation (4.43)).

4.3 LONG-DISTANCE SOUND PROPAGATION PREDICTION BASED ON FDTD

4.3.1 Moving Frame FDTD

After exciting a short, acoustic pulse at the source position, the moving frame FDTD performs calculations at a minimum computational cost. Instead of updating the fields throughout the full spatial grid, calculations are only made in a limited zone, centred around the propagating acoustic pulse (see Figure 4.5). This calculation frame then moves with the speed of sound in the desired direction. A moving frame approach avoids having to update acoustic variables in zones where hardly any acoustic energy is present. However, it is important that the calculation frame includes the full vertical extent of the grid. The use of a fixed calculation frame to model

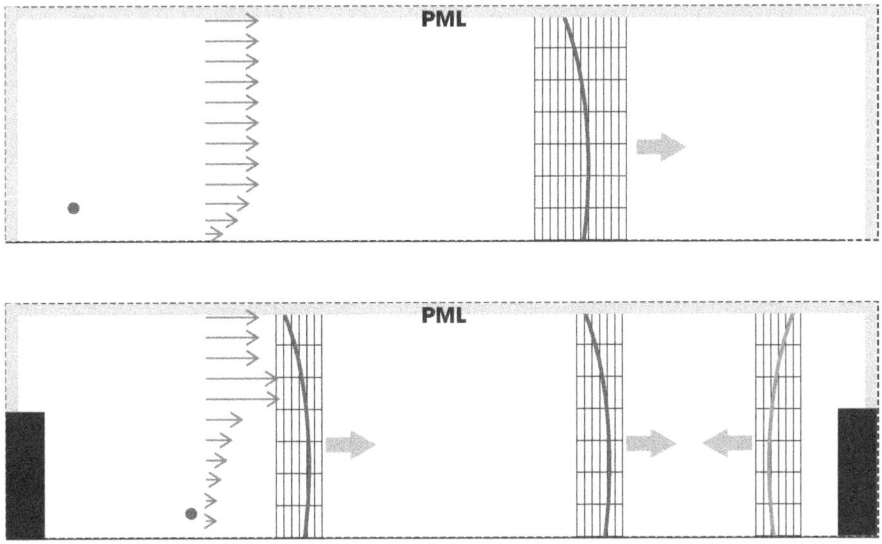

Figure 4.5 Schematic representations of static (upper) and dynamic (lower) moving frame FDTD.

sound propagation over a rigid ground plane in a refractive atmosphere has been shown not to affect the numerical accuracy [12].

When using computer coding languages allowing dynamic memory allocation, even more efficient implementations are possible. To follow distinct reflections, as depicted in the lower panel of Figure 4.5, multiple narrow frames are used through threshold-based memory allocation [33]. This enables simulation of the common case of propagation from a road source between parallel noise walls. Clearly, if the sound propagation problem becomes too reverberant, there is no advantage in the use of a moving frame.

4.3.2 Hybrid Modelling: Combining FDTD with GFPE

Many outdoor situations involve sound propagation up to a large distance, but, nevertheless, need highly detailed modelling in a source zone. Various spatial domain decomposition methods have been proposed to cope with this. Examples include coupling between the boundary element method (BEM) and the parabolic equation (PE) method [34], or between ray-tracing and analytical diffraction formulae [35]. Combinations of the BEM and ray-tracing [36] and of FDTD with a diffusion propagation model [37] have been explored. Although these approaches can be very efficient in specific applications, detailed modelling of the effect of refraction by wind, e.g., is possible in only one of the spatial domains.

The combination of FDTD with PE [38] is particularly useful for simulating outdoor acoustics. Both techniques account for the effect of (background) flow in detail. During long-distance sound propagation, open-field refraction plays a significant role along the propagation path and this is modelled efficiently by PE. However, if detailed modelling is required in the source region, e.g. to account for the object-induced refraction observed in downwind sound propagation over a noise wall (see Chapter 9, Section 9.9), the FDTD model can be used and then coupled to the PE to model subsequent propagation.

4.3.2.1 Advantages of the GFPE Method

The PE method enables efficient numerical solution of the so-called 'one-way' wave equation. The PE method uses a closely spaced vertical array of complex acoustic pressures, extrapolated (or propagated) to a nearby vertical array in a direction away from the source. Specific starting functions have been derived to model sound propagation from a point source or to assign a specific directivity [39,40]. However, any accurately calculated vertical sound pressure field can be used to initiate or to continue sound propagation up to the desired distance. As with any PE method, reflections back to the source are not modelled; although, a second PE simulation could be initiated upon reflection. The Green's Function Parabolic Equation (GFPE) method is a particularly powerful member of the PE solutions family. An overview and details of this method are given elsewhere [41,42]. The GFPE method is useful particularly for simulating long-distance sound propagation in that large step sizes are allowed in the horizontal direction. The limiting factor in practice will be the inhomogeneity of the atmosphere rather than the requirement for resolving the wavelength. The maximum acceptable range step dr varies roughly between 5 to 50 times the wavelength in case of a refracting atmosphere [41,43]. However, a much finer discretization dz is needed in the vertical direction, lower than 1/10th of the wavelength. Moreover, some care is needed to respect the minimum dr/dz ratio [43]. An absorbing layer consisting of a few tenths to even hundreds of wavelengths [43] is needed to prevent spurious reflections from the top of the PE spatial grid.

The computational cost of GFPE is strongly related to the efficiency of the Fourier Transform algorithm on which the solution scheme is based. However, very fast algorithms like the Fast Fourier Transform (FFT) are available. GFPE uses the effective sound speed approach, meaning that the (wind) flow is assumed to be parallel to the ground surface. This is a reasonable approximation in practice.

Even though, e.g., upwardly deflected flows are not captured in more complex cases, the effective sound speed approach proves to be still quite accurate [44,45]. Note that effective sound speed profiles may contain both upward- and downward-refracting zones. Locally reacting, range-dependent impedance planes can be used to model reflection from the ground. Procedures are available to include turbulent scattering [41,46–48] and terrain undulation [45,49–52] in PE models. So, it can be concluded that GFPE is a rather complete model for outdoor sound propagation applications, except for its neglect of reflections (back to the source) on vertical objects and in case of complicated flows.

4.3.2.2 Complex Source Region, Simplified Receiver Region

Many cases of outdoor sound propagation are characterized by a complex source region, with large reflecting objects like central reservations, noise walls or houses, leading to multiple reflections and complex diffraction patterns. Also, these objects influence the wind flow leading to complicated flow velocity fields. Receivers, on the other hand, are usually positioned at a longer distance. Outside this source region, reflecting objects typically become scarce and the wind flow tends to be parallel with the ground surface. In this region, sound propagation is predominantly one-way. These conditions fit the effective sound speed approach. So, it makes sense to use FDTD in the source region and GFPE in the receiver region.

4.3.2.3 Procedure for One-way Coupling from FDTD to GFPE

One-way coupling between 2D-FDTD and 2D-GFPE means that the FDTD model provides information to the GFPE model and that the sound waves, once in the PE domain, are not allowed to reenter the FDTD source region [38], consistent with the PE one-way sound propagation approach.

The interface where the spatial domains meet is considered to consist of a vertical array of points at high spatial resolution. In both FDTD and GFPE, a fine spatial discretization is needed along the vertical dimension of the computational grids. Note that when modelling broadband sources in FDTD, the highest sound frequency of interest will define the required spatial discretization. Beneficially, this involves interpolation to a less coarse grid when making the transition to PE.

At this receiver array, the full-time history has to be recorded after pulse excitation in the FDTD domain. Especially in case of highly reverberant conditions, the time response must be captured until full decay has occurred. Next, at each receiver point, a transition has to be made to the frequency domain. In this way, a set of starting functions in PE can be constructed for each sound frequency of interest. Although multiple PE calculations must be performed, a single FDTD simulation is sufficient.

The interface should be chosen where significant reflections on vertical objects are not present anymore, and at a location where the (wind) flow has become mainly horizontal. These conditions enable the fulfilling of the PE assumptions. If the interface is behind a tall vertical object, then there must be sufficient distance to that object to prevent obliquely diffracted sound waves to fall outside the PE's cone of accuracy.

A less straightforward problem is that the GFPE grid is larger than the one used in FDTD. So, there will be a need to extrapolate the pressure field up to larger heights. One way to solve this mismatch in grid height is to rely on the pressure decay present in the PML near the top of the FDTD grid. For the pressure magnitude, this occurs in a more or less optimized way and thus prevents spurious reflections in the PE domain as well. Extrapolation of the phase is best performed linearly, excluding phase jumps [38].

In the procedure explained here, continuity of the acoustic pressure is enforced at the interface between both models. In p-v FDTD, both acoustic pressures and particle velocities are used to perform simulations, while PE purely relies on acoustic pressures. Therefore, from a theoretical point of view, it is not obvious that this approach gives accurate results. Nevertheless, the numerical example described in Section 4.3.2.4 shows that the high accuracy one can find in both models is preserved.

The distance between the interface and the receiver will indicate up to what height along the interface calculations need to be accurate. The receiver distance D will thus define the necessary height H in the FDTD domain to be resolved explicitly without relying on field extrapolation towards larger heights. Typically, a ratio of D/H of 10 is a safe option. Note that this choice depends on the receiver height and the refractive state of the atmosphere. If, in downward-refraction conditions, significant sound paths only reach the ground at a large distance, then the turning point of the corresponding sound ray should appear below H.

4.3.2.4 Numerical Example

Sound propagation from an infinitely long coherent line source, positioned in between two infinitely long rigid T-shaped noise barriers, is considered (see Figure 4.6). The 2-D modelling approach assumes that the barrier cross-section is constant. The height of the barriers is 4 m and the horizontal part of the T-barrier is 2 m wide. The depiction of the sound propagation problem in Figure 4.6 shows a 1 m wide and 0.5 m high raised central reservation. Receivers are placed at a height of 2 m and 4 m. To focus on the accuracy of the proposed hybrid model itself, all surfaces are assumed to be perfectly reflecting. Possible loss of accuracy by differences in ground interaction modeling is eliminated in this way. The chosen frequencies (namely 100 Hz, 300 Hz and 500 Hz) are low enough to enable comparisons with reference full FDTD simulations to a sufficiently large distance.

Simulations were performed both in the absence and presence of wind (all in a homogeneous atmosphere). The flow fields near the screens and in the zone downwind from the source were predicted with a commercial CFD software. The k-ε turbulence closure model was used. A purely logarithmic wind speed profile is imposed as inflow boundary condition. The flow is directed normal to the barriers, and the inflow profile is characterized by a friction velocity of 0.4 m/s and a roughness length of 0.01 m. Contours of the horizontal component of the wind velocity are shown in Figure 4.7. Near the top of the screens, complex flow fields appear with strong vertical

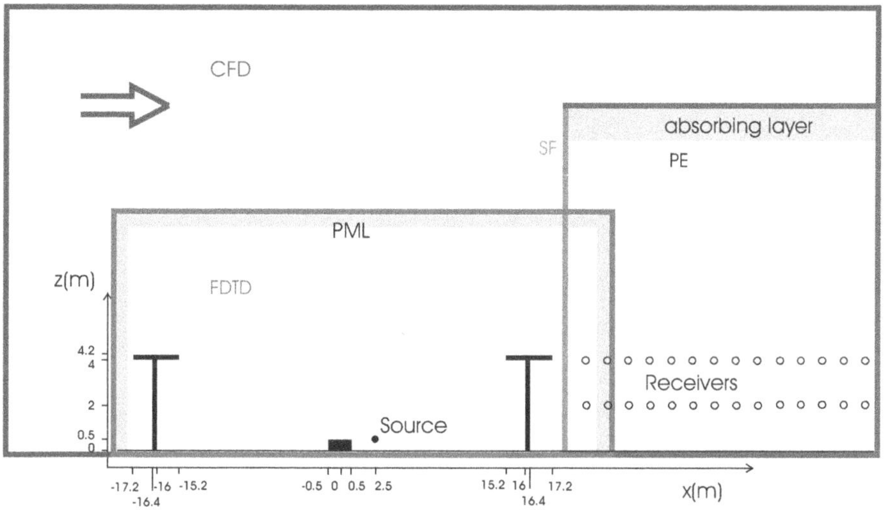

Figure 4.6 Schematics of grids for the CFD, FDTD and PE spatial domains. SF indicates the position of the starting function, forming the interface between FDTD and PE. The CFD region includes the FDTD and PE regions and their boundaries are sufficiently distant from the barrier zone to prevent coupling with the flow in that zone. Axes are not true to scale.

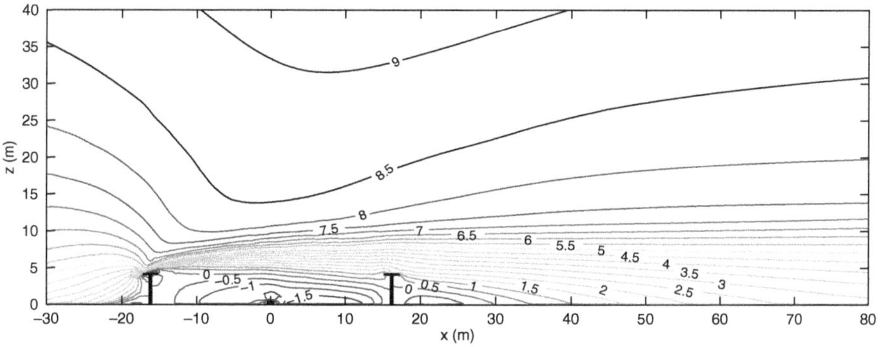

Figure 4.7 Simulated horizontal components of the wind velocity fields near the T-shaped noise barriers (in m/s). Negative values indicate the presence of recirculation zones in the wakes of the screens.

gradients in the horizontal component of the flow velocity. These lead to screen-induced downward refraction of sound, on top of the refraction due to the (undisturbed) logarithmic wind velocity profile that would appear in an open field. This means that wind flow is important in both the source and receiver region, making a hybrid FDTD-PE model an appropriate choice.

The interface between FDTD and PE was positioned at 4 m from the downwind noise barrier. The starting functions at this location are depicted in Figure 4.8. Distances up to 150 m are considered for FDTD, and up to

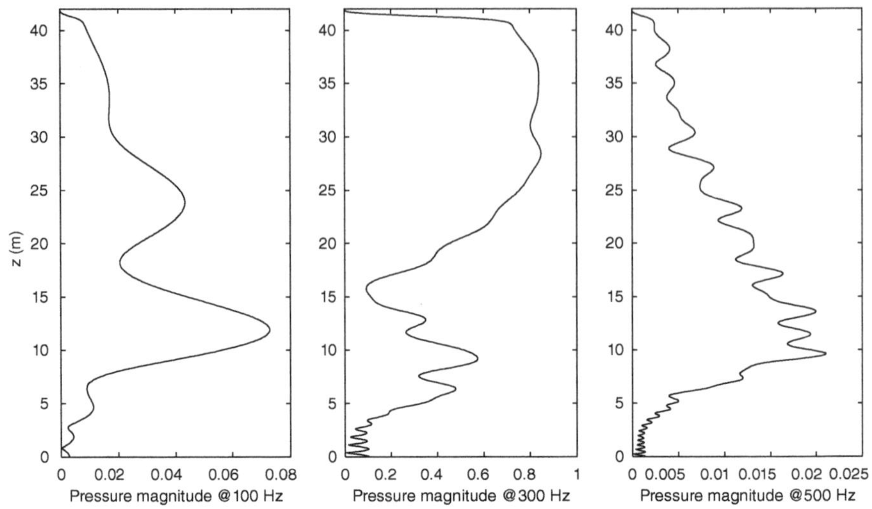

Figure 4.8 Starting fields at 3 sound frequencies, including wind flow, predicted from the FDTD simulations and subsequent Fourier Analysis (at x = 20 m). The PML starts at a height of z = 40 m in the FDTD domain. The magnitudes of the pressures depend on the properties of the pulse excited at the source position in the FDTD grid. Since final results are referred to free field sound propagation, the absolute values on the pressure scale are unimportant.

500 m for FDTD+PE. Sound pressure levels are expressed relative to free field sound propagation. Atmospheric absorption is neglected in all simulations.

Figure 4.9 shows that in absence of wind, the FDTD+PE model follows very closely the reference FDTD model, for all frequencies considered here. In a moving atmosphere, some small differences are observed especially at larger distance from the source, and near interference minima. This is caused by the (slightly) different accuracies in FDTD and PE, especially in a moving atmosphere. Possible causes are the effective sound speed approach in PE and the amplitude errors appearing in the PSIT-FDTD method. Overall, the hybrid model described does not lead to obvious accuracy errors, even when looking in detail at a single sound frequency. Some very small oscillations in the FDTD-PE results revealed by close inspection have no practical significance.

4.3.2.5 Computational Cost Reduction

There is a very large gain in computational efficiency when using the coupled FDTD-PE model, relative to a full application of FDTD. Note that in FDTD implementations (when not using the moving frame approach), each acoustic unknown appearing in the simulation domain needs to be allocated in computer memory to ease the constant memory access during the time-updating process. In contrast, since only a single vertical array of pressures

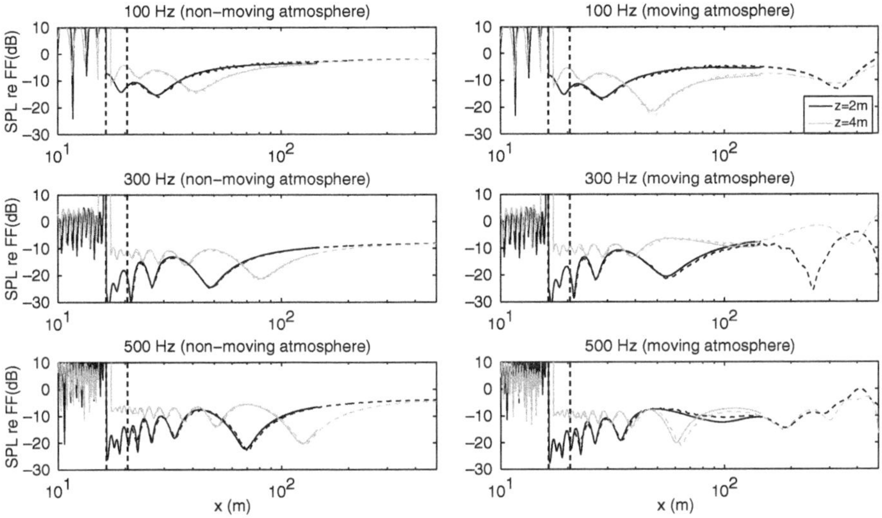

Figure 4.9 Comparison between the results of FDTD (full lines, up to 150 m) and FDTD+PE calculations (dashed lines, from the position of the starting function at 20 m up to 500 m), in a non-moving and moving atmosphere (using the flow fields shown in Figure 4.7). The distance axis is logarithmic. Dashed vertical lines indicate the positions of the downwind T-barrier (left-hand line) and the position of the starting function (right-hand line).

is needed, the memory requirements for the PE model can be (virtually) neglected compared to FDTD.

Although the details of the specific software implementations of the models are important when estimating the gains in calculation times, a rough estimate is possible. PE is estimated to be a few orders of magnitude faster (100–1000) than FDTD [38]. This means that, although FDTD can provide information on a wide range of sound frequencies by performing a single simulation, only if the required number of frequencies is very large will the same calculation time as with classical FDTD be reached. In conclusion, for a single sound frequency, the need for computational resources basically comes down to the needs in the FDTD source region.

REFERENCES

[1] R. Blumrich and D. Heimann, A linearized Eulerian sound propagation model for studies of complex meteorological effects, *J. Acoust. Soc. Am.* **112** 446–455 (2002).

[2] T. Van Renterghem and D. Botteldooren, Numerical simulation of the effect of trees on downwind noise barrier performance, *Acta Acust. united Ac.* **89** 764–778 (2003).

[3] T. Van Renterghem, *The finite-difference time-domain method for sound propagation in a moving medium*, PhD Thesis, Ghent University (2003).

[4] V. Ostashev, D. Wilson, L. Liu, D. Aldridge, N. Symons and D. Marlin, Equations for finite-difference, time-domain simulation of sound propagation in moving inhomogeneous media and numerical implementation, *J. Acoust. Soc. Am.* **117** 503–517 (2005).
[5] K. Kowalczyk and M. van Walstijn, Formulation of Locally Reacting Surfaces in FDTD/K-DWM Modelling of Acoustic Spaces, *Acta Acust. united Ac.* **94** 891–906 (2008).
[6] T. Lokki, A. Southern and L. Savioja, Studies on seat dip effect with 3D FDTD modeling, *Proceedings of Forum Acusticum*, Aalborg, Denmark (2011).
[7] C. Webb and S. Bilbao, Computing room acoustics with CUDA - 3D FDTD schemes with boundary losses and viscosity, *IEEE International Conference on Acoustics, Speech, and Signal Processing*, Prague, Czech Republic, 317–320 (2011).
[8] K. Yee, Numerical solution of initial boundary value problems involving Maxwell's equations in isotropic media, *IEEE Trans. Antennas Propag.* **14** 302–307 (1966).
[9] D. Botteldooren, Finite-difference time-domain simulation of low frequency room acoustic problems, *J. Acoust. Soc. Am.* **98** 3302–3308 (1995).
[10] C. Tam and J. Webb, Dispersion-relation-preserving finite difference schemes for computational acoustics, *J. Comput. Phy.* **107** 262–281 (1993).
[11] C. Bogey and C. Bailly, A family of low dispersive and low dissipative explicit schemes for flow and noise computation, *J. Comput. Phy.* **194** 194–214 (2004).
[12] E. Salomons R. Blumrich and D. Heimann, Eulerian time-domain model for sound propagation over a finite-impedance ground surface. Comparison with frequency-domain models, *Acta Acust. united Ac.* **88** 483–492 (2002).
[13] T. Van Renterghem and D. Botteldooren, Prediction-step staggered-in-time FDTD: an efficient numerical scheme to solve the linearised equations of fluid dynamics in outdoor sound propagation, *Appl. Acoust.* **68** 201–216 (2007).
[14] J. Berenger, A perfectly matched layer for the absorption of electromagnetic waves, *J. Comput. Phy.* **114** 185–200 (1994).
[15] F. Hu, On Absorbing Boundary Conditions for Linearized Euler Equations by a Perfectly Matched Layer, *J. Comput. Phy.* **129** 201–219 (1996).
[16] F. Hu, A stable, perfectly matched layer for linearized Euler equations in unsplit physical variables, *J. Comput. Phy.* **173** 455–480 (2001).
[17] S. Abarbanel, D. Gottlieb and J. Hesthaven, Well-posed perfectly matched layers for advective acoustics, *J. Comput. Phy.* **154** 266–283 (1999).
[18] K. Heutschi, M. Horvath and J. Hofmann, Simulation of ground impedance in finite difference time domain calculations of outdoor sound propagation, *Acta Acust. united Ac.* **91** 35–40 (2005).
[19] K.-Y. Fung and H. Ju, Time-domain impedance boundary conditions for computational acoustics and aeroacoustics, *Int. J. Comput. Fluid Dynam.* **18** 503–511 (2004).
[20] D. Dragna, B. Cotté, P. Blanc-Benon and F. Poisson, Time-Domain Simulations of Outdoor Sound Propagation with Suitable Impedance Boundary Conditions, *Am. Ins. Aero. Astro. J.* **49** 1420–1428 (2011).
[21] V. Ostashev, S. Collier, D. Wilson, D. Aldridge, N. Symons and D. Marlin, Padé approximation in time-domain boundary conditions of porous surfaces, *J. Acoust. Soc. Am.* **122** 107–112 (2007).

[22] H. Jeong, I. Drumm, B. Horner and Y. Lam, The modelling of frequency dependent boundary conditions in FDTD simulation of concert hall acoustics, *Proceedings of 19th international congress on acoustics (ICA)*, Madrid, Spain (2007).
[23] S. Rienstra, Impedance models in time domain including the Helmholtz resonator model, *Proceedings of Twelfth AIAA/CEAS Aeroacoustics Conference*, Cambridge, MA, USA (2006).
[24] C. Richter, J. A. Hay, Ł. Panek, N. Schoenwald, S. Busse and F. Thiele, A review of time-domain impedance modelling and applications, *J. Sound Vib.* **330** 3859–3873 (2011).
[25] D. Dragna and P. Blanc-Benon, Physically admissible impedance models for time-domain computations of outdoor sound propagation, *Acta Acust. united Ac.* **100** 401–410 (2014).
[26] C. Zwikker and C. Kosten, *Sound Absorbing Materials*, Elsevier, New York (1949).
[27] R. Picó, B. Roig and J. Redondo, Stability analysis of the FDTD scheme in porous media, *Acta Acust. united Ac.* **96** 306–316 (2010).
[28] D. Wilson, V. Ostashev, S. Collier, D. Aldridge, N. Symons and D. Marlin, Time-domain calculations of sound interactions with outdoor ground surfaces, *Appl. Acoust.* **68** 175–200 (2007).
[29] M. Biot, Theory of propagation of elastic waves in a fluid-saturated porous solid. I. Low-frequency range, *J. Acoust. Soc. Am.* **28** 168–178 (1956).
[30] M. Biot, Theory of propagation of elastic waves in a fluid-saturated porous solid. II. High-frequency range, *J. Acoust. Soc. Am.* **28** 179–191 (1956).
[31] H. Dong, A. Kaynia, C. Madshus and J. Hovem, Sound propagation over layered poro-elastic ground using a finite-difference model, *J. Acoust. Soc. Am.* **108** 494–502 (2000).
[32] L. Ding, T. Van Renterghem, D. Botteldooren, K. Horoshenkov and A. Khan, Sound absorption of porous substrates covered by foliage: experimental results and numerical predictions, *J. Acoust. Soc. Am.* **134** 4599–4609 (2013).
[33] B. de Greve, T. Van Renterghem and D. Botteldooren, Long range FDTD over undulating terrain, Proceedings of Forum Acusticum, Budapest, *Hungary* (2005).
[34] E. Premat, J. Defrance, M. Priour and F. Aballea, Coupling BEM and GFPE for complex outdoor sound propagation, *Proceedings of Euronoise 2003*, Naples, Italy (2003).
[35] E. Salomons, Sound propagation in complex outdoor situations with a non-refracting atmosphere: model based on analytical solutions for diffraction and reflection. *Acta Acust. united Ac.* **83** 436–454 (1997).
[36] S. Hampel, S. Langer and A. Cisilino, Coupling boundary elements to a ray tracing procedure, *Int. J. Num. Methods Eng.* **73** 427–445 (2008).
[37] S. M. Pasareanu, R. A. Burdisso and M. C. Remillieux, A numerical hybrid model for outdoor sound propagation in complex urban environments, *J. Acoust. Soc. Am.* **143** EL218 (2018).
[38] T. Van Renterghem, E. Salomons and D. Botteldooren, Efficient FDTD-PE model for sound propagation in situations with complex obstacles and wind profiles, *Acta Acust. united Ac.* **91** 671–679 (2005).
[39] J. E. Rosenbaum, A. A. Atchley and V. W. Sparrow, Source directivity in the parabolic equation method using an inverse Fourier transform technique, *J. Acoust. Soc. Am.* **129** 2442 (2011).

[40] S. N. Vecherin, D. K. Wilson and V. E. Ostashev, Incorporating source directionality into outdoor sound propagation calculations, *J. Acoust. Soc. Am.* **130** 3608 (2011).
[41] E. Salomons, *Computational atmospheric acoustics*, Kluwer, Dordrecht, The Netherlands (2001).
[42] M. West, K. Gilbert and R. Sack, A Tutorial on the Parabolic Equation (PE) Model Used for Long Range Sound Propagation in the Atmosphere, *Appl. Acoust.* **37** 31–49 (1992).
[43] J. Cooper and D. Swanson, Parameter selection in the Green's function parabolic equation, *Appl. Acoust.* **68** 390–402 (2007).
[44] R. Blumrich and D. Heimann, Numerical estimation of atmospheric approximation effects in outdoor sound propagation modelling, *Acta Acust. united Ac.* **90** 24–37 (2004).
[45] T. Van Renterghem, D. Botteldooren and P. Lercher, Comparison of measurements and predictions of sound propagation in a valley-slope configuration in an inhomogeneous atmosphere, *J. Acoust. Soc. Am.* **121** 2522–2533 (2007).
[46] V. E. Ostashev, E.M. Salomons, S. F. Clifford, R. J. Lataitis, D. K. Wilson, P. Blanc-Benon and D. Juvé, Sound propagation in a turbulent atmosphere near the ground: A parabolic equation approach, *J. Acoust. Soc. Am.* **109** 1894-1908 (2001).
[47] J. Forssen, Calculation of sound reduction by a screen in a turbulent atmosphere using the parabolic equation method, *Acustica* **84** 599-606 (1998).
[48] E. Salomons, V. E. Ostashev, S. F. Clifford and R. J. Lataitis. Sound propagation in a turbulent atmosphere near the ground: An approach based on the spectral representation of refractive-index fluctuations, *J. Acoust. Soc. Am.* **109** 1881–1893 (2001).
[49] R. Sack and M. West, A parabolic equation for sound propagation in two dimensions over any smooth terrain profile: the generalised terrain parabolic equation (GT-PE), *Appl. Acoust.* **45** 113–129 (1995).
[50] M. Collins, The rotated parabolic equation and sloping ocean bottoms, *J. Acoust. Soc. Am.* **87** 1035–1037 (1990).
[51] M. Bérengier, B. Gauvreau, P. Blanc-Benon and D. Juvé, Outdoor sound propagation: a short review on analytical and numerical approaches, *Acta Acust. united Ac.* **89** 980–991 (2003).
[52] X. Di and K. Gilbert, The effect of turbulence and irregular terrain on outdoor sound propagation. *Proceedings of the 6th International Symposium on Long-Range Sound Propagation*, Ottawa, Canada, pp. 315–333. National Research Council of Canada (1994).

Chapter 5

Predicting the Acoustical Properties of Ground Surfaces

5.1 INTRODUCTION

In ISO 9613-2 [1], ground surfaces are considered to be either 'acoustically hard', which means that they are perfectly reflecting, or 'acoustically soft', which implies perfectly absorbing. Any ground surface of low porosity is considered acoustically hard and any grass-, tree- or potentially vegetation-covered ground is considered acoustically soft. This oversimplifies a considerable range of properties and their resulting influences on near-grazing sound. Even surfaces which could be classified as 'grassland' encompass a wide range of acoustical properties.

Given the many influences on outdoor sound including discontinuous ground, meteorology, diffraction by barriers and topography, it is convenient to use as straightforward a description of the acoustical properties of a ground surface as possible. A useful description is that given by the (complex) surface impedance, i.e. the ratio of incident pressure to the particle velocity normal to ground surface. The simplest models for calculating surface impedance use a single parameter and are semi-empirical. However, as discussed later, they are physically inadmissible and, even in the relatively straightforward propagation conditions of little or no wind or temperature gradients and flat ground, better prediction accuracy will result from the use of physically admissible models that involve at least one additional parameter. This is particularly important when the governing sound propagation equations are solved in the time domain.

The interaction of sound with the ground includes several phenomena known collectively as ground effect. The spectrum of the ratio of the total sound level at a receiver to the direct sound that would be present in the absence of the ground surface is called the *excess attenuation* (EA) due to the ground surface and analytical models for it are described in Chapters 2 and 3. As long as the ground may be considered to be locally reacting, i.e. the surface impedance is independent of the angle at which sound arrives on it, the EA at a given receiver may be calculated from knowledge of the surface impedance and the source–receiver geometry.

Most naturally occurring outdoor surfaces are porous. Sound waves are able to penetrate the surface and, in doing so, are subjected to a change in phase as well as having some of their energy converted into heat. The acoustical properties of porous ground depend primarily on the ease with which air can move in and out of the surface. This is indicated by the *flow resistivity* which represents the ratio of the applied pressure gradient to the induced volume flow rate per unit thickness of material. If the ground surface has a high flow resistivity, it means that it is difficult for air to flow into or through it. Porosity influences flow resistivity. Typically, flow resistivity is high if the surface porosity is low. Conversely, flow resistivity is relatively low if porosity is high. Hot-rolled asphalt has negligible porosity and extremely high flow resistivity whereas freshly fallen snow has a low flow resistivity and high porosity. Section 5.2 reviews the various models and parameters that have been used to calculate the impedance of 'smooth' ground surfaces and Section 5.3 discusses physical admissibility and the extent to which these models are physically admissible.

As well as being porous, most naturally occurring outdoor ground surfaces are rough at some scale. Even when the scale (e.g. mean roughness height) is much smaller than the incident sound wavelengths, roughness alters the effective surface impedance near grazing incidence whether the surface is porous or not. Roughness that is added deliberately to a smooth surface that is otherwise acoustically hard can be used to change its effective acoustical properties and thereby control the EA near to the surface. The influence of ground roughness is discussed in Section 5.4.

Low-frequency and high-intensity sources in air, such as explosions, cause the ground to vibrate and, in turn, this has an influence on outdoor sound propagation which is explored in Section 5.5. Methods of measuring ground impedance and comparisons between data and predictions are described in Chapter 6.

5.2 PREDICTING GROUND IMPEDANCE

5.2.1 Empirical and Phenomenological Models

Single-parameter models, for relative normal surface impedance ($Z_c = R + iX$) and propagation constant ($k = \beta + i\alpha$), have the form

$$R = 1 + \mathrm{a}\left(\frac{f}{R_s}\right)^{\mathrm{b}} \tag{5.1a}$$

$$X - \mathrm{c}\left(\frac{f}{R_s}\right)^{\mathrm{d}} \tag{5.1b}$$

$$\alpha = \frac{\omega}{c_0}\mathrm{p}\left(\frac{f}{R_s}\right)^{\mathrm{q}} \tag{5.2a}$$

$$\beta = \frac{\omega}{c_0}\left\{1 + \mathrm{r}\left(\frac{f}{R_s}\right)^{\mathrm{s}}\right\}, \tag{5.2b}$$

where a, b, c, d, p, q, r and s are constant coefficients, $\omega = 2\pi f$ represents the angular frequency (rad/s), f is the frequency (Hz), c_0 is the adiabatic sound speed in air, ρ_0 is the density of air and R_s is the flow resistivity (Pa s m^{-2}).

The coefficient values derived by Delany and Bazley [2,3] (which were used in Chapter 2) are listed in Table 5.1 and are based on best fits to a large number of (impedance tube) measurements made with many fibrous materials having porosities close to 1.

The range of validity of these relationships is suggested to be $0.01 < \frac{f\rho_0}{R_s} < 1.0$.

Even for materials with high porosity like the fibrous materials used for deriving them, many authors have found that the Delany and Bazley model, i.e. Equations (5.1a, 5.1b) with the exponent and coefficient values listed in Table 5.1, predicts the wrong frequency dependence for normal surface impedance at low frequencies and have suggested various modifications to these formulae [4–9].

The model due to Miki [6] and a modified version of it [9], with the corresponding coefficient values in Table 5.1, have been used for representing ground impedance.

Soils have porosities much less than 1 and rather higher flow resistivities than fibrous materials. A typical flow resistivity of soil is 120 kPa s m^{-2} so, at 500 Hz, $f\rho_0/R_s = 0.0005$, i.e. well below the suggested range of validity for the Delany and Bazley model. Despite this and the fact that they are physically inadmissible (see Section 5.3), models of the form of (5.1a), (5.1b), (5.2a) and (5.2b) have been used with tolerable success to characterize outdoor surface impedances as long as the flow resistivity, R_s, is considered as an adjustable parameter i.e. as an effective flow resistivity, R_{eff}.

For the high flow resistivities typical of soils and with Delany and Bazley coefficient values, Equations (5.1a) and (5.1b) predict a reactance that

Table 5.1 Coefficient and exponent values in the Delany and Bazley [2], Miki [6,7] and modified Miki [9] models

	Coefficients and exponents							
Model	*a*	*b*	*c*	*d*	*p*	*q*	*r*	*S*
Delany and Bazley [2]	0.0497	−0.754	0.0758	−0.732	0.1696	−0.595	0.0862	−0.693
Miki [6,7]	0.0700	−0.632	0.1070	−0.632	0.1600	−0.618	0.0109	−0.618
Modified Miki [9]	0.0700	−0.632	0.1070	−0.632	0.0979	−0.632	0.1503	−0.632

exceeds the resistance over an appreciable frequency range. This is typical for the impedance spectrum of a hard-backed porous layer (see Section 5.2.3) or a porous semi-infinite medium with a rough surface (see Section 5.4). Grass-covered surfaces often have a layered structure in which the root zone near the surface has a lower flow resistivity than the substrate ([10,11], also see Section 5.4). Even though microstructural models (described later) predict that the wave penetrating the pores attenuates at a rate that increases with the square root of frequency, near-surface layers may be sufficiently shallow that they influence the surface impedance. Rasmussen [10] improved agreement with short-range propagation data over a grass-covered surface by assuming a hard-backed layer structure and by using Equations (5.1a), (5.1b), (5.2a) and (5.2b) in the formula for the impedance of a hard-backed layer of thickness d,

$$Z(d) = Z_c \coth(-ikd). \tag{5.3}$$

Studies of the extent to which various impedance models can be used to predict short-range propagation data (see [11] and Chapter 6) have found that many of these data are fitted best by assuming the grounds of interest are hard-backed layers.

Zwikker and Kosten [12] suggested a phenomenological model for the acoustical behaviour of porous materials. This introduces three parameters: flow resistivity (R_s), porosity (Ω) and a structure factor (K). This model may be expressed as

$$Z_c = \frac{\sqrt{K}}{\Omega}\left(1 + \frac{i\Omega R_s}{\omega \rho_0 K}\right), \quad k = \Omega k_0 Z_c, \tag{5.4a, b}$$

where $k_0 = \omega/c_0$. As discussed in more detail in Section 5.2.3, this model assumes that the sound speed inside the pores is adiabatic at all frequencies. Thomasson [13] used a hard-backed layer version of the phenomenological model (i.e. the combination of Equations (5.4a, b) and (5.3)) to obtain good agreement with indoor and outdoor data.

Influenced by the form of the phenomenological model (Equation (5.4a, b)), Miki [7] proposed modifications of the semi-empirical single-parameter Equations (5.1a), (5.1b), (5.2a) and (5.2b) in which R and X are multiplied by $\sqrt{K/\Omega}$ and β and α are multiplied by $\sqrt{K}$. In section 5.2.3 it is explained that $K \equiv T$ where T is the tortuosity (defined in (5.8)).

The coefficient values for the resulting three-parameter model are listed in Table 5.1.

To predict the acoustical properties of porous asphalt, Hamet [14] modified Equation (5.4a, b) to allow for the frequency dependence of viscous and thermal effects (see Section 5.2.2). Hamet's model may be written as

$$Z_C = \left(\frac{1}{\Omega}\right)\left(\frac{K}{\gamma}\right)^{\frac{1}{2}}\left\{1 + \frac{\gamma - 1}{\gamma}\left(\frac{1}{F_0}\right)^{\frac{1}{2}}\right\} F_\mu^{\frac{1}{2}}, k = \gamma \Omega k_0 Z_c, \tag{5.5a, b}$$

where $F_\mu = 1 + i\omega_\mu/\omega$, $F_0 = 1 + i\omega_0/\omega$, $\omega_\mu = (R_s/\rho_0)(\Omega/K)$, $\omega_0 = \omega_\mu(K/N_{PR})$ and N_{PR} is the Prandtl number for air.

5.2.2 Microstructural Models Using Idealized Pore Shapes

The acoustical properties of the ground may be modelled as those of a rigid-porous material and characterized by a complex density, containing the influence of viscous effects, and a complex compressibility, containing the influence of thermal effects. Thermal effects are much greater in air-filled materials than in water-filled materials. Particular forms of these complex quantities may be obtained by considering ideally shaped pores such as narrow tubes or slits. This offers a more rigorous basis for ground impedance prediction than either phenomenological [12] or semi-empirical [2,3,6,7] approaches.

According to Stinson [15], if the complex density in a (single) uniform pore of arbitrary shape is written as

$$\rho(\omega) = \rho_0 / H(\lambda), \tag{5.6}$$

where λ is a dimensionless parameter, then the complex compressibility is given by

$$C(\omega) = (\gamma P_0)^{-1}\left[\gamma - (\gamma - 1)H\left(\lambda\sqrt{N_{\mathrm{Pr}}}\right)\right], \tag{5.7}$$

where $(\gamma P_0)^{-1} = (\rho_0 c_0^2)^{-1}$ is the adiabatic compressibility of air.

$H(\lambda)$ has been calculated for many ideal pore shapes [15–16] including those of a circular capillary, an infinitely long parallel-sided slit, an equilateral triangle and a rectangle. The results are listed in Table 5.2. For a parallel sided slit of width $2b$, the dimensionless parameter $\lambda_s = b\sqrt{\omega / v}$, where $v = \mu/\rho_0$. Since $\delta = \sqrt{2v/\omega}$ is the viscous boundary layer thickness for laminar flow oscillations near a flat plate, $\sqrt{2}/\lambda_s = \delta/b$, which is the frequency-dependent

Table 5.2 Complex density functions for various pore shapes

Pore shape	λ	$H(\lambda)$
slit (width $2b$)	$b\sqrt{\omega / v}$	$1 - \tanh\left(\lambda\sqrt{-i}\right) / \left(\lambda\sqrt{-i}\right)$
cylinder (radius a)	$a\sqrt{\omega / v}$	$1 - \left(2/\lambda\sqrt{i}\right)J_1\left(\lambda\sqrt{i}\right)/J_0\left(\lambda\sqrt{i}\right)$
equilateral triangle (side d)	$\left(d\sqrt{3}/4\right)\sqrt{\omega / v}$	$1 - 3\coth\left(\lambda\sqrt{-i}\right)/\left(\lambda\sqrt{-i}\right) + 3i/\lambda^2$
rectangle (sides $2a, 2b$)	$\dfrac{2ab}{\pi\sqrt{(a^2+b^2)}}\sqrt{\omega / v}$	$\dfrac{-4i\omega}{va^2b^2}\sum_{k=0}^{\infty}\sum_{l=0}^{\infty}\left\{\alpha_k^2\beta_l^2\left[\alpha_k^2 + \beta_l^2 - \left(\dfrac{i\omega}{v}\right)\right]^{-1}\right\}$ $\alpha_k = \left(k + \dfrac{1}{2}\right)\left(\dfrac{\pi}{a}\right)$, $\beta_l = \left(l + \dfrac{1}{2}\right)\left(\dfrac{\pi}{b}\right)$

fraction of the slit-pore semi-width occupied by the viscous boundary layer. A critical frequency (or 'roll over' frequency) [17] above which inertial forces within a parallel-sided slit of width $2b$ dominate over viscous forces is given by $f_c = 3\nu/(2\pi b^2)$. At this critical frequency, $\sqrt{2/3}$, i.e. 81.6%, of a slit is occupied by the viscous boundary layers. The thermal boundary thickness is $\delta/\sqrt{N_{PR}}$. Typically, this is much smaller than the viscous boundary layer.

The dimensionless parameter λ can be related to the (steady) flow resistivity (R_s) of the bulk material by using the Kozeny–Carman formula [18],

$$R_s = \frac{2\mu T s_0}{\Omega r_h^{\,2}} \tag{5.8}$$

where the hydraulic radius, $r_h = \dfrac{\text{"wetted" area}}{\text{perimeter}}$, s_0 is a steady-flow shape factor and T is tortuosity, defined as the square of the increase in path length per unit thickness of material due to deviations of the steady-flow path from a straight line.

Consequently, $\lambda^2 = \dfrac{2 s_0 \omega T \rho_0}{\Omega R_s}$, where, for the pore shapes mentioned previously, r_h and s_0 are defined in Table 5.3. Values of porosity and tortuosity for some granular materials are listed in Table 5.4.

In terms of flow resistivity, the critical frequency above which inertial forces are greater than viscous forces within a porous medium (with identical pores) is $f_c = \Omega R_s/2\pi T\rho_0$ and, at this frequency, the viscous boundary layer thickness $\delta_c = r_h\sqrt{1/s_0}$. So, for example, at the critical frequency, 75% of a cylindrical pore area is occupied by the viscous boundary layer.

Once the complex density, $\rho(\omega)$, and complex compressibility, $C(\omega)$, of the individual pores are known then the bulk propagation constant, $k(\omega)$, and relative characteristic impedance, $Z_c(\omega)$, of the bulk porous material may be calculated from

$$k(\omega) = \omega\sqrt{T\rho(\omega)C(\omega)} \tag{5.9a}$$

$$Z_c(\omega) = \left[1/(\rho_0 c_0)\right]\sqrt{(T/\Omega^2)(\rho(\omega)/C(\omega))}. \tag{5.9b}$$

Equation (5.9b) implies that the bulk compressibility $C_b(\omega) = \Omega C(\omega)$.

Table 5.3 Hydraulic radius and steady flow shape factors

Pore shape	r_h	s_0
Slit (width $2b$)	b	1.5
Cylinder (radius a)	$a/2$	1
Equilateral triangle (side d)	$d/4\sqrt{3}$	5/6
Square (side $2a$)	$a/2$	0.89

Table 5.4 Some measured and calculated values of porosity and tortuosity

Material	*Porosity* Ω	*Method of determination*	*Tortuosity T*	*Method of determination*
Lead shot, 3.8 mm diameter particles	0.385	by weighing	1.6 1.799	Estimate from $1/\sqrt{\Omega}$ (which fits acoustic data) Cell model predictions [24]
Gravel, 10.5 mm grain size	0.45	by weighing	1.55	Deduced by fitting surface admittance data
Gravel, 5–10 mm grain size	0.4	by weighing	1.46	Deduced by fitting surface admittance data
0.68 mm diameter Glass beads	0.375	N/S [22]	1.742 1.833	Measured [20] Cell model predictions [24]
Coustone	0.4	N/S [24,32]	1.664	Measured [25], fitted [32]
Clay granulate, Laterlite, 1–3 mm grain size	0.52 0.725	N/S [93] Measured [94]	1.25	Deduced by fitting surface admittance data From assumed porosity = 0.52
Olivine sand	0.444	Measured [31]	1.626	Cell model prediction [24]

N/S = not specified

There are various published methods that allow for arbitrarily shaped pores. Attenborough [16] scaled the complex density function directly between pore shapes and introduced an adjustable dynamic pore shape parameter, s_A, so that, for example, the bulk complex density function for parallel-sided slit pores is given by [16]

$$\rho_b(\lambda) = (T/\Omega)\rho_0\left[1 - \tanh\left(\lambda\sqrt{-i}\right)/\left(\lambda\sqrt{-i}\right)\right]^{-1}, \tag{5.10}$$

where $\lambda = s_A\sqrt{\left(\dfrac{3\rho_0\omega T}{\Omega R_S}\right)}$.

The pore shape parameter $s_A = 1$ for slit-like pores and $0.745 < s_A < 1$ for pore shapes varying between equilateral triangles and slits. If the pores are cylindrical, then the dimensionless parameter is written as $\lambda = s_A\sqrt{\left(\dfrac{8\rho_0\omega T}{\Omega R_S}\right)}$.

However, this formulation has explicit dependence on s_A even in the low-frequency limit. Champoux and Stinson [19] have pointed out that, to satisfy the correct limiting behaviour for the complex density, for example that $-i\omega\rho b \rightarrow RS$ as ω tends to zero, this approach requires s_A to be frequency dependent.

Their method of allowing for arbitrary pore shapes is closer to that originally suggested by Biot [18] and the resulting pore shape parameter, s_B, has the advantage of being frequency independent. Using this alternative approach, again with reference to the functions for slit-like pores, we write the complex density for the bulk material as

$$\rho_b(\lambda) = (T/\Omega)\rho_0 + (iR_s/F_s(\lambda)), \tag{5.11}$$

where

$$F_s(\lambda) = \frac{1}{3}\frac{(\lambda\sqrt{-i})\tanh(\lambda\sqrt{-i})}{\left[1-\tanh(\lambda\sqrt{-i})/(\lambda\sqrt{-i})\right]} \tag{5.12}$$

and $\lambda = s_B\sqrt{\left(\frac{3\rho_0\omega T}{\Omega R_S}\right)}$.

$1 < s_B < 1.342$ for pore shapes varying from slits to equilateral triangles. This scales the viscosity correction or dynamic viscosity function $F(\lambda)$ instead of the complex density function. $F(\lambda)$ tends to unity in the low-frequency limit so this method of defining $\rho_b(\lambda)$ does not have any explicit dependence on a dynamic pore shape factor. In the low-frequency limit, the dependence on pore shape is only that implicit in the flow resistivity through Equation (5.8).

For a given bulk flow resistivity, porosity and tortuosity, although the complex density and complex compressibility, calculated from Equations (5.6), (5.7) and Tables 5.2 and 5.3 are found to depend on pore shape, the complex bulk propagation constant, the complex characteristic impedance (given by (5.9a) and (5.9b)) and corresponding surface impedance of a hard-backed layer (given by (5.3)) are predicted to be relatively insensitive to pore shape [20]. This is true particularly for low flow resistivities and at frequencies less than a few thousand Hz.

Other microstructural factors of significance in pore-based modelling are the variation of pore cross sections along their lengths and the associated distributions of pore sizes and shapes. To allow simultaneously for arbitrary pore shapes and for pore cross sections that change along their lengths, Johnson et al. [21] interpret $1/H(\lambda)$ as a dynamic tortuosity and introduce two characteristic lengths, Λ and Λ', which represent modifications of the dimensionless parameter λ used in Equations (5.11) and (5.12). The simplest formulations for bulk material complex density and complex compressibility resulting from this approach are based upon limiting forms for small and large characteristic lengths and may be written as [22]

$$\rho_b(\omega) = T\rho_0\left[1+\frac{iR_S\Omega}{\omega\rho_0 T}G(\Lambda)\right], \tag{5.13}$$

where $G(\Lambda) = \sqrt{\left(1-\frac{4iT\eta\rho_0\omega}{R_S{}^2\Lambda^2\Omega^2}\right)}$ and $\Lambda = s_p\sqrt{\left(\frac{8T\eta}{\Omega R_S}\right)}$

and

$$C_b(\omega) = (\gamma P_0)^{-1}\left[\gamma - (\gamma - 1)\left[1 + \frac{iR_S\Omega}{\omega\rho_0 TN_{PR}} G'(\Lambda')\right]^{-1}\right] \tag{5.14}$$

where $G'(\Lambda') = \sqrt{\left(1 - \frac{4iT\eta\rho_0\omega N_{PR}}{R_S{}^2\Lambda'^2\Omega^2}\right)}$, $\Lambda' = s_C\sqrt{\left(\frac{8T\eta}{\Omega R_S}\right)}$.

In effect, the characteristic lengths introduce two 'dynamic' pore shape factors s_ρ (equivalent to s_B) and s_C into complex density and complex compressibility, respectively. The formulations of bulk complex density and compressibility in Equations (5.13) and (5.14) omit factors of $1/\Omega$ and Ω, respectively. While this gives rise to similar expressions for propagation constant to that given by (5.9a), the resulting definition of characteristic impedance differs from that given by (5.9b), which assumes that the wave impedance is identical to the impedance of a semi-infinite homogeneous rigid-porous material. It should be noted that Allard et al. [22] introduce a factor of $1/\Omega$ into their calculation of surface impedance.

The need for two shape factors may be argued from the fact that the wider parts of each pore tend to be more important for the complex compressibility while the narrower pore cross sections dominate the complex density. Theoretically, it is expected that $s_c > s_\rho$, i.e. $\Lambda' > \Lambda$, and typically this has been found to be the case. However, for arbitrary media, the viscous and thermal shape factors must be treated as independently adjustable parameters. For example, to fit acoustic data for glass beads, porous asphalt and granular rubber values of $s_c < s_\rho$ have been found to be necessary [20,23].

A version of this model, based on the viscosity correction function for slit-like pores (Equation (5.12)), which is likely to be more accurate at intermediate values of the dimensionless parameter, may be written as follows:

$$\rho_b(\omega) = T\rho_0\left[1 + \frac{iR_S\Omega}{\omega\rho_0 T} F_s(\lambda_\rho)\right] \tag{5.15}$$

and

$$C_b(\omega) = \Omega(\gamma P_0)^{-1}\left[\gamma - (\gamma - 1)\left[1 + \frac{iR_S\Omega}{\omega\rho_0 TN_{PR}} F_s\left(\lambda_C\sqrt{N_{PR}}\right)\right]^{-1}\right], \tag{5.16}$$

where $\lambda_\rho = s_\rho\sqrt{\left(\frac{3T\omega\rho_0}{\Omega R_S}\right)}$ and $\lambda_C = s_C\sqrt{\left(\frac{3T\omega\rho_0}{\Omega R_S}\right)}$.

It has proved possible to calculate the shape factors (or characteristic lengths) for certain well-defined geometries (e.g. stacked cylinders [22] and stacked spheres [24]). Methods of measuring Λ' by non-acoustic means and Λ by ultrasonic means have been published [25,26]. By fitting data for fibrous materials [22] it has been found that $s_c \approx 2s_\rho$ ($\Lambda' \approx 2\Lambda$), whereas $s_c \approx 5s_\rho$ has been found appropriate for an ideal or model material having cylindrical pores with two different diameters along their lengths [19].

The grain shape in soils is usually far from spherical. However, a theory has been developed for the acoustical characteristics of a medium consisting of identical stacked rigid spheres in terms of the radius of the spheres and the packing fraction [24,27]. This model offers a method for estimating the likely values of the characteristic lengths in a granular medium from knowledge of the mean grain size and porosity thereby making it more feasible to use multi-parameter models for predicting acoustical properties of arbitrary granular media. If identical spheres of radius R are packed to form a granular medium with porosity Ω, the thermal characteristic length may be expressed as [27]

$$\Lambda' = \frac{2\Omega}{3(1-\Omega)}R. \tag{5.17}$$

and the relationship between thermal and viscous characteristic lengths is

$$\Lambda' = \frac{3}{2T(1-\Theta)}\Lambda, \tag{5.18}$$

$$\text{where } \Theta = \frac{3}{\sqrt{2\pi}}(1-\Omega) \cong 0.675(1-\Omega). \tag{5.19}$$

To account for dynamic thermal effects in a porous medium with arbitrary microstructure, an additional parameter, thermal permeability, has been suggested [28,29]. Hence, Equation (5.14) is replaced by

$$C_b(\omega) = (\gamma P_0)^{-1}\left\{\gamma - (\gamma-1)\left[1 + \frac{i\mu\Omega}{\omega\rho_0 k_0' N_{PR}\sqrt{1 + \dfrac{\omega\rho_0 4 k_0'^2 N_{PR}}{i\mu\Lambda'^2\Omega^2}}}\right]^{-1}\right\}, \tag{5.20a}$$

where k_0' is the thermal permeability. The thermal permeability is difficult to measure or estimate. However, the stacked sphere model [27] yields

$$k_0' = \frac{3}{2(1-\Theta)}\frac{\mu}{R_{sph}}, \quad R_{sph} = \frac{9\mu(1-\Omega)}{2R^2\Omega^2}\frac{5(1-\Theta)}{5-9\sqrt[3]{\Theta}+5\Theta-\Theta^2} \tag{5.20b, c}$$

and R_{sph} represents the flow resistivity for stacked spheres.

The stacked sphere model leads also to a relationship between tortuosity and porosity

$$T = 1 + \frac{1-\Omega}{2\Omega} \tag{5.20d}$$

which is consistent with a result derived by Berryman [30].

Values for the parameters governing the acoustical properties of some granular materials are listed in Table 5.5. These show that in some cases the

Table 5.5 Parameter values for some granular materials

Material	*Bulk density* *(kg m^{-3})*	*Viscous characteristic dimension* $\Lambda(\mu m)$	*Thermal characteristic dimension* Λ' *(μm)*	*Porosity*	*Tortuosity*	*Flow resistivity* R_s *(kPa s m^{-2})*	*Ref.*
Glass beads 0.68±0.12 mm	1525	93*	133	0.385	1.8*	45.4*	
Glass beads 1.11±0.15 mm	1575	16.8*	236	0.365	1.87*	21.25*	[93]
Glass beads 1.64±0.15 mm	1582	18.8*	263	0.362	1.88*	9.86*	
Sand 1	1638#	28	80	0.37	1.7	130	[95]
Sand 2	1560#	22	60	0.4	1.65	170	[94]
Snow	N/A	50	100	0.81	1.5	7	[96]

* Calculated from the cell model [24,27] using measured Ω, T and Λ'#Calculated assuming a density of 2600 kg/m^3 for the grains

thermal characteristic length may exceed the viscous characteristic length by a large factor.

A method of allowing for a log-normal pore size distribution, while assuming pores of identical (slit-like) shape, has been developed [31–33] based on the work of Yamamoto and Turgut [34]. The viscosity correction function for a log-normal slit-pore distribution is given by

$$F_{sd} = \frac{(-i\omega\mu\rho_0)}{\Omega R_s} \frac{\int_0^\infty b^{-1} e(b) \tanh\left(\lambda_s\sqrt{-i}\right) db}{\int_0^\infty e(b)\left[1 - \frac{\tanh\left(\lambda_s\sqrt{-i}\right)}{\left(\lambda_s\sqrt{-i}\right)}\right] db} \tag{5.21}$$

where $\lambda_s = b\sqrt{\omega / v}$, $\int_0^\infty e(b)db = \int_{-\infty}^\infty f(\phi)d\phi$, $f(\phi) = \frac{1}{\sigma\sqrt{2\pi}} \exp \left[-(\phi - \phi_a)^2 / 2\sigma^2\right]$, $b = 2^{-\phi}$, $b_a = 2^{-\phi_a}$, $R_s = \left[\frac{3\mu}{\Omega b_a^2}\right] e^{-2(\sigma \ln 2)^2}$, b_a represents the mean pore semi-width and σ is the standard deviation of the pore size distribution in ϕ units.

For slit pores with a log-normal distribution of widths along their axes [35], $R_s = \left[\frac{3\mu}{\Omega b_a^2}\right] e^{-6(\sigma \ln 2)^2}$.

The bulk complex density is calculated from

$$\rho_b(\omega) = T\rho_0\left[1 + \frac{iR_S\Omega}{\omega\rho_0 T} F_{sd}\right] = (T/\Omega)\rho_{pd}(\omega) \tag{5.22}$$

so that, using Stinson's relationship (5.7),

$$C_b(\omega) = \Omega(\gamma P_0)^{-1}\left[\gamma - (\gamma - 1)\left(\rho_0 / \rho_{pd}\left(\omega\sqrt{N_{Pr}}\right)\right)\right]. \tag{5.23}$$

The influence of pore shape can be removed by assuming that all pores have an identical geometrical shape, chosen, for example, from those listed in Table 5.2. If the flow resistivity, porosity, tortuosity and standard deviation of the pore size distribution are available from non-acoustical measurements, then a log-normal pore size distribution model does not require any adjustable parameters.

The numerical integration introduced by Equation (5.21) can be avoided by means of Padé approximations [33,35]. For a medium consisting of identical tortuous pores of (constant) arbitrary shape or of non-uniform tortuous pores, Equation (5.10) may be written:

$$\rho(\omega) = (T\rho_0 / \Omega)\left[1 + F(\varepsilon)/\left(T\varepsilon^2\right)\right], \tag{5.24a}$$

where $\varepsilon = \lambda\sqrt{(-i)}$ and $F(\varepsilon)$ is the viscosity correction function.

Low- and high-frequency asymptotes for the viscosity correction function may be expressed as

$$F(\varepsilon) = 1 + \theta_1\varepsilon^2 + O(\varepsilon^4), \varepsilon \to 0 \tag{5.24b}$$

and

$$F(\varepsilon) = \theta_2\varepsilon + O(1), \varepsilon \to \infty. \tag{5.24c}$$

Hence, a Padé approximation for the viscosity correction function in a medium with a log-normal pore size distribution has been proposed in the form

$$F(\varepsilon) = \frac{1 + a_1\varepsilon + a_2\varepsilon^2}{1 + b_1\varepsilon}, \tag{5.25}$$

where $a_1 = \theta_1/\theta_2$, $a_2 = \theta_1$ and $b_1 = a_1$.

Values of these coefficients can be determined analytically for particular pore shapes (see Table 5.6). However, as discussed in Section 5.2.5, the influence of pore shape per se on bulk acoustical properties is relatively small as long as the incident wavelength is much larger than the maximum pore cross section dimension.

Consideration of low- and high-frequency asymptotic limits for a log-normal pore size distribution enables derivation of relationships between the characteristic lengths used in the Johnson–Allard–Champoux theory and the pore size distribution parameters.

For a log normal size distribution of uniform pores [35],

$$\Lambda = \Lambda' = \bar{r}e^{-1/2(\sigma \ln 2)^2} \tag{5.26a}$$

Table 5.6 Expressions for coefficients in equation (5.23) where $\xi = (\sigma \ln 2)^2$ and σ is the standard deviation of the log-normal distribution in ϕ-units (such that a pore dimension in mm = $2^{-\phi}$)

Coefficients in (5.23)	*Slit-like pores*	*Equilateral triangles*	*Circular pores*
Uniform pores			
θ_1	$\frac{6}{5}e^{4\xi} - 1$	$\frac{7}{10}e^{4\xi} - 1$	$\frac{4}{3}e^{4\xi} - 1$
θ_2	$\frac{1}{\sqrt{3}}e^{\frac{3}{2}\xi}$	$\sqrt{\frac{3}{5}}e^{\frac{3}{2}\xi}$	$\frac{1}{\sqrt{2}}e^{\frac{3}{2}\xi}$
Non-uniform pores			
θ_1	1/5	-3/10	1/3
θ_2	$\frac{1}{\sqrt{3}}e^{-\frac{1}{2\xi^2}}$	$\sqrt{\frac{3}{5}}e^{-\frac{1}{2\xi^2}}$	$\frac{1}{\sqrt{2}}e^{-\frac{1}{2\xi^2}}$

and for a log-normal size distribution of non-uniform pores [35],

$$\Lambda = \bar{r}e^{-5/2(\sigma \ln 2)^2},\ \Lambda' = \bar{r}e^{3/2(\sigma \ln 2)^2}, \tag{5.26b}$$

where $\bar{r}$ is the mean pore dimension.

This means that the characteristic lengths ratio $\frac{\Lambda'}{\Lambda} = e^{4(\sigma \ln 2)^2}$ depends only on the standard deviation of the pore size distribution.

5.2.3 Approximate Models for High Flow Resistivities

Section 5.3 will explain why it is undesirable to use the empirical one parameter models (Equations (5.1a), (5.1b), (5.2a) and (5.2b)) to represent outdoor ground impedance, despite their comparative simplicity. On the other hand, it is impractical to use some of the more sophisticated impedance models for the acoustical properties of rigid-porous media described in the previous section since they require knowledge of up to seven parameters, many of which will not be known for soils. However, given the relatively high flow resistivities of soils compared with the porous materials used for noise control or for modifying room acoustics, it is possible to derive more usable and physically admissible one-, two- or three-parameter forms. Equations (5.8)–(5.11) lead to simple approximations for characteristic impedance and propagation constant in the limit of small λ (corresponding to low frequencies and high flow resistivities). These approximations (based on approximation of cylindrical pore functions) may be written as

$$k = \sqrt{\gamma}\frac{\omega}{c_f}\sqrt{\left[\left(\frac{4}{3} - \frac{(\gamma - 1)}{\gamma}N_{PR}\right)T + i\frac{4R_{eff}}{\omega\rho_f}\right]} \tag{5.27a}$$

$$Z_c = \left[\frac{\frac{4}{3}T\rho_f + i\frac{4R_{eff}}{\omega}}{\Omega k}\right]\frac{\omega}{\rho_f c_f}, \tag{5.27b}$$

where $R_{eff} = \Omega s_p^2 R_s$ represents an effective flow resistivity. Here s_p is equivalent to s_A used earlier but determined using circular cylinder solutions as a basis rather than those for slits. The equivalent expressions using slit-pore functions as the basis are as follows:

$$k = \sqrt{\gamma}\frac{\omega}{c_f}\sqrt{\left[\left(\frac{6}{5} - \frac{(\gamma - 1)}{\gamma}N_{PR}\right)T + i\frac{4R_{eff}}{\omega\rho_f}\right]} \tag{5.28}$$

$$Z_c = \left[\frac{\frac{6}{5}T\rho_0 + i\frac{4R_{eff}}{\omega}}{\Omega k}\right]\frac{\omega}{\rho_0 c_0}. \tag{5.29}$$

Further approximation for high flow resistivity and low frequency such that $\frac{R_{eff}}{\omega} \gg 1$ produces

$$k = \frac{1}{c_0}\sqrt{\left(\frac{\gamma\pi}{\rho_0}\right)}(1+i)\sqrt{(R_{eff}f)} \tag{5.30}$$

$$Z_c = \frac{1}{\Omega\sqrt{(\pi\gamma\rho_0)}}(1+i)\sqrt{\left(\frac{R_{eff}}{f}\right)} = \frac{1}{\sqrt{(\pi\gamma\rho_0)}}(1+i)\sqrt{\left(\frac{R_e}{f}\right)}, \tag{5.31}$$

where $R_e = s_p^2 R_s / \Omega$, i.e. $R_e = R_{eff}/\Omega^2$.

Similar approximations of the Biot/Stinson/Champoux model (Equations (5.8)–(5.10) and (5.12)) are

$$k = \sqrt{\gamma}\frac{\omega}{c_0}\sqrt{T + \frac{iR_s\Omega}{\omega\rho_0}}, \tag{5.32}$$

$$Z_c = \frac{1}{\Omega\sqrt{\gamma}}\sqrt{T + \frac{iR_s\Omega}{\omega\rho_0}}. \tag{5.33}$$

These equations are independent of pore shape (except through R_s) and give identical results to Equations (5.26a), (5.26b), (5.27a) and (5.27b) if $s_p = 0.5$.

Equations (5.32) and (5.33) are equivalent to the phenomenological model (Equations (5.4a, b)) where the structure factor is replaced by tortuosity but sound speed in the pores is assumed to be isothermal rather than adiabatic. It should be noted that the implication of isothermal conditions in the pores at low frequencies is more physically consistent than the adiabatic assumption leading to Equations (5.4a, b).

After further approximation for high flow resistivity, (5.31) becomes

$$Z_c = \frac{1}{2\sqrt{(\pi\gamma\rho_0)}}(1+i)\sqrt{\left(\frac{R_s}{\Omega f}\right)}. \tag{5.34}$$

Equations (5.29) and (5.34) are identical if s_p is set equal to 0.5 in Equation (5.29) and may be regarded as single-parameter model for the surface impedance of a homogeneous semi-infinite rigid-porous medium with high flow resistivity, or at low frequency, where the single parameter is effective flow resistivity. They predict surface impedances in which real and imaginary parts are equal and decrease as the square root of frequency.

Equations (5.28), (5.29) and (5.3) may be used together with the first two terms of the power series expansion of the hyperbolic cotangent to produce a two-parameter approximation for the impedance of a hard-backed thin high flow resistivity layer [36]:

$$Z(d) = \frac{2R_e d_e}{3\rho_0 c_0} + i\frac{c_0}{\gamma\pi f d_e} \tag{5.35}$$

where $d_e = \Omega d$ is an effective depth.

This predicts a constant real part of impedance.

In forest floors, the flow resistivity of the surface porous (litter) layer is relatively low, and it lies above a porous substrate with finite impedance, so a suitable multi-layer model is required. This may be deduced, from transmission line analysis, by means of the following equation:

$$Z(d) = Z_1\left\{\left[Z_2 - iZ_1\tan(kd)\right]/\left[Z_1 - iZ_2\tan(kd)\right]\right\}, \tag{5.36}$$

where Z_1 and Z_2 are the relative characteristic impedances of the upper porous layer and porous substrate, respectively, k is the propagation constant in the upper porous layer and d is the layer thickness.

A two-parameter approximation for a thin layer with an impedance backing such that the backing impedance is much higher than that of the layer is given by [37]

$$Z(d) = \frac{1+i}{\sqrt{\pi\gamma\rho_0}}\sqrt{\frac{R_e}{f}} + \frac{ic_0}{2\pi\gamma d_e f}, \tag{5.37}$$

where, again, $d_e = \Omega d$ and d is the layer thickness. This predicts a frequency-dependent real part of the surface impedance.

The surface impedance of a high flow resistivity, rigid-porous medium in which the porosity decreases exponentially with depth has been investigated [37–38]. A two-parameter approximation, sometimes called the *variable porosity* model, for the surface impedance of such a medium is [39]

$$Z = \frac{1+i}{\sqrt{\pi\gamma\rho_0}}\sqrt{\frac{R_e}{f}} + \frac{ic_0\alpha_e}{8\pi\gamma f}, \tag{5.38}$$

where $\alpha_e = (n' + 2)\alpha/\Omega$, n' being a grain shape factor introduced through use of the Bruggeman relationship between tortuosity and porosity ($T = \Omega^{-n'}$).

An approximation for the surface impedance of a high flow resistivity porous medium with the porosity *increasing* exponentially with depth is given by (5.32) with negative α_e. If α_e is negative, (5.38) predicts a resistance that exceeds the reactance at all frequencies.

If α_e is positive, Equations (5.37) and (5.38) imply $X > R$ and may be written generically in the form

$$Z = a(1+i)\sqrt{\frac{R_{se}}{f}} + \frac{ib\alpha_e}{f}, \tag{5.39}$$

where $a = 1/\sqrt{\pi\rho_0\gamma}$ and $b = c_0/8\pi\gamma$.

Equations (5.38) and (5.39) predict a surface impedance equivalent to that predicted for a non-hard-backed thin layer (Equation (5.37)) if $\alpha_e = 4/d_e$.

An approximation that is better at high frequencies uses either Equations (5.22) and (5.23) or (5.29) instead of Equations (5.27a), (5.27b) or (5.30) and may be written [40] as

$$Z = Z_c + \frac{ic_0\alpha_e}{8\pi\gamma f}$$

$$\text{or } Z = \frac{1}{\sqrt{\pi\gamma\rho}}(1+i)\sqrt{\frac{R_e'}{f}} + \sqrt{\frac{\pi\rho}{\gamma}}(1-i)T_e\sqrt{\frac{f}{R_e'}} + \frac{ic_0}{8\pi\gamma}\frac{\alpha_e}{f}. \quad (5.40a)$$

The third parameter, T_e, introduced in this approximation, depends on tortuosity and porosity and influences the high-frequency values of the impedance.

The number of parameters required by many models can be reduced by assuming a relationship between tortuosity and porosity such as that derived for stacked spheres (Equation (5.20d)), or the more general Bruggeman relationship $T = \Omega^{-n'}$,where n' is a grain shape factor.

Taraldsen [37] has suggested a model using a connection between permeability and porosity, of a similar form to that used in geophysics, which transforms the (three parameter) phenomenological model (5.4a, b) into a one parameter form that gives rather similar results to the Delany–Bazley model (Equations (5.1a), (5.1b), (5.2a) and (5.2b)) using the relevant coefficients and exponents in Table 5.1).

Taraldesen's model for relative admittance, β (=1/Z), may be expressed as follows:

$$\beta = \sqrt{\gamma}\,\frac{\Omega}{\sqrt{T}}\left(1+\frac{f^*}{f}\right)^{-0.5}, \quad (5.41a)$$

$$f^* = \frac{1}{2\pi\rho_0}\frac{\Omega}{\sqrt{T}}R_{eff}, \quad (5.41b)$$

$$\text{where } \frac{\Omega}{\sqrt{T}} = 100\left(10^{X/10}\right) \quad (5.41c)$$

$$X = \frac{1}{19.76}\left[-Y + 404.55 + \sqrt{546.166 + (9.35 - Y)^2}\right] \quad (5.41d)$$

$$Y = 10\log\left(R_{eff}\right). \quad (5.41e)$$

Figure 5.1 shows an example comparison between predictions of the Taraldsen and the Delany–Bazley models for a relatively low effective flow resistivity R_{eff}= 12500 Pasm^{-2}.

5.2.4 Relaxation Models

By viewing the viscous and thermal diffusion in porous materials as relaxation processes, Wilson [41] has derived models for the acoustical properties

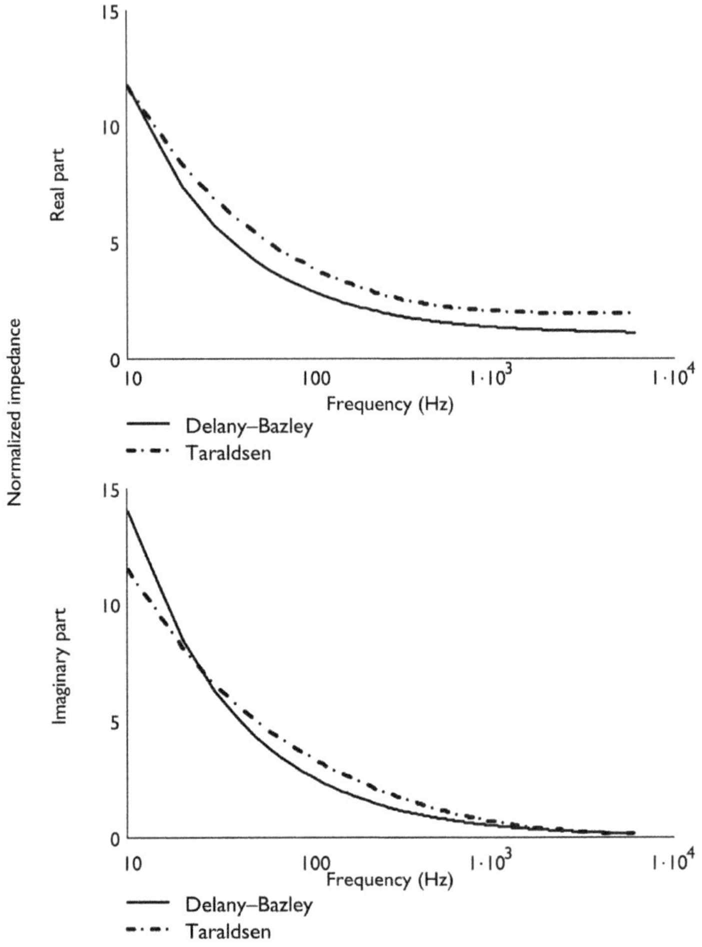

Figure 5.1 Comparison between predictions of Taraldsen's one-parameter model (equations (5.41a)–(5.41e)) and the Delany-Bazley model (equations (5.1a), (5.1b), (5.2a) and (5.2b)) for $\mathbf{R}_{eff}$ = 12500 Pa s m^{-2}.

of porous materials in simple forms that, nevertheless, enable accurate predictions over wide frequency ranges. His results may be expressed as

$$Z = \frac{\sqrt{T}}{\Omega}\left[\left(1+\frac{\gamma-1}{\sqrt{1-i\omega\tau_e}}\right)\left(1-\frac{1}{\sqrt{1-i\omega\tau_v}}\right)\right]^{-1/2} \qquad (5.42a)$$

$$k = \frac{\omega\sqrt{T}}{c_0}\left[\left(1+\frac{\gamma-1}{\sqrt{1-i\omega\tau_e}}\right)/\left(1-\frac{1}{\sqrt{1-i\omega\tau_v}}\right)\right]^{1/2}, \qquad (5.42b)$$

where, for identical uniform pores, τ_e and τ_v, the thermodynamic and aerodynamic characteristic times, respectively, are given by

$$\tau_v = 2\rho_0 T / \Omega R_s \tag{5.43a}$$

$$\text{and}\, \tau_e = N_{PR} s_B{}^2 \tau_v. \tag{5.43b}$$

Essentially this represents a single-parameter model (the parameter being either τ_e or τ_v) for a given pore shape, flow resistivity and porosity. The relaxation model is very similar in form to Hamet's modification of the phenomenological model [14]. By matching to the Delany and Bazley relationships for their range of validity, Wilson [42] has obtained the simple approximate relationships:

$$\tau_v \cong 2.1\rho_0 / R_s \tag{5.43c}$$

$$\text{and}\, \tau_e \cong 3.1\rho_0 / R_s. \tag{5.43d}$$

When substituted into Equations (5.42a) and (5.42b), these relationships provide a single-parameter model that gives similar predictions to Equations (5.1a), (5.1b), (5.2a) and (5.2b) with Delany and Bazley coefficients and exponents (Table 5.1) for $0.01 < \dfrac{f\rho_0}{R_s} < 1.0$ but is valid outside this range.

For materials in which the pore cross sections vary, Wilson has suggested a slightly more complicated form, again based on (3.42) but $\sqrt{1 - i\omega\tau_e}$ and $\sqrt{1 - i\omega\tau_v}$ are replaced by $\sqrt{1 + \dfrac{\tau_h}{\tau_l}\left(\sqrt{1 - i\omega\tau_l / \tau_h} - 1\right)}$ where τ_l and τ_h are appropriate low- and high-frequency versions of τ_e and τ_v. Wilson has shown that results equivalent to Equations (5.15) and (5.16) may be obtained with $\tau_{le} = \tau_{he} = N_{PR} s_K{}^2 \tau_v$ and $\tau_{lv} = s_\rho{}^2 \tau_{hv} = 2\rho_0 T / \Omega R_s$.

In its most general form this represents a four- or five-parameter model. Relaxation models have been used in numerical time-domain calculations [43 and Section 5.3].

5.2.5 Relative Influence of Microstructural Parameters

How are the bulk acoustical properties of a rigid-porous medium influenced by details of its microstructure? Figure 5.2 compares predictions of the surface impedance of hard-backed layers corresponding to a 'snow-like', i.e. low flow resistivity medium with identical tortuous pores of various different shapes and a microstructure of stacked spheres. The key shows the dimensions of the corresponding spheres and pores. Clearly, for the given flow resistivity and porosity, the predictions are not very dependent on the pore shape. Even smaller effects of differences in microstructure are predicted for a higher flow resistivity. The only significant difference is between the prediction of the cell model for stacked spheres and the 'uniform pore' models for the real part of the surface impedance at low frequencies. This is consistent with the finding of Pride et al. [44] who considered pores with varying cross sections.

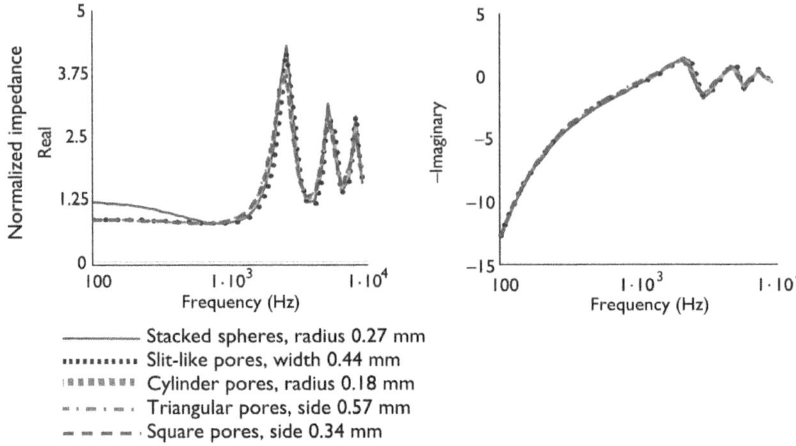

Figure 5.2 Predicted surface impedance of a 0.05 m thick layer of 'snow-like' material, with flow resistivity 10 kPa s m^{-2}, Porosity 0.6 and Tortuosity 4/3, composed of stacked identical grains or identical pores with specific shapes.

According to Pride et al. [44], an additional parameter (β) is required in the Johnson–Champoux–Allard model (Equations (5.13) and (5.14)) to account for the enhancement in the effective fluid inertia at lower frequencies caused by the cross-sectional changes in the pore size and viscous friction at the smallest apertures of the pore.

For a log-normal distribution of uniform pores it can be shown [35] that

$$\beta = 4/3e^{4(\sigma \ln 2)^2} - 1$$

and for a log-normal distribution of non-uniform pores,

$$\beta = 4/3.$$

Figures 5.3(a), (b) and 5.4(a), (b) show the predicted effects of varying the standard deviation of the assumed log-normal size distribution between 0 and 1 on the acoustical characteristics of a hard-backed layer rigid-framed porous media having values of flow resistivity, porosity, tortuosity and layer thickness intended to represent a 'grass-like' medium and a 'snow-like' medium, respectively. Increasing the standard deviation of the pore-size distribution of uniform pores is predicted to have a significantly larger effect than increasing the standard deviation of a log-normal pore size distribution of non-uniform pores. Increasing the standard deviation is predicted to increase the real part of the impedance of a porous layer with a log-normal size distribution of pores. Essentially, this is the result of lowering the frequency of the layer resonance and increasing its magnitude. The effects are predicted to be less significant if there

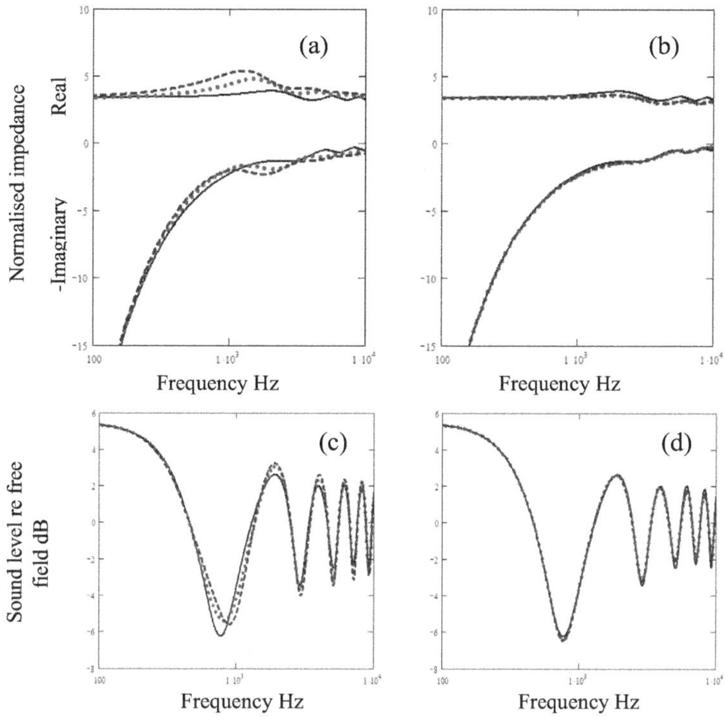

Figure 5.3 Predicted effects on surface impedance of a 'grass-like' slit-pore hard-backed rigid-framed porous medium with flow resistivity 100 kPa s m^{-2}, porosity 0.4, tortuosity $\sqrt{(5/2)}$, layer thickness 0.04 m of varying the standard deviation of the pore-size distribution (a) of uniform pores and (b) of non-uniform pores. (c) and (d) show the corresponding excess attenuation spectra predicted for source and receiver at 0.3 m height and separated by 1 m. The continuous, dotted and broken lines correspond to standard deviations of 0, 0.5 and 1 respectively.

is a size distribution of non-uniform pores. The imaginary parts of surface impedance are predicted to be relatively unaffected. Corresponding predictions of short-range EA spectra are shown in Figures 5.3 and 5.4(c), (d). Changing the standard deviation of the log-normal distribution of uniform pores is predicted to result in small, but potentially measurable, changes in the frequency and magnitude of the first destructive interference and small changes in EA spectra at higher frequencies. Much smaller but similar effects are predicted as a result of changing the standard deviation in a log-normal distribution of *non-uniform* pores.

In Section 5.4, it is shown that the presence of small-scale surface roughness is predicted to have a more significant influence on the surface impedance of an outdoor ground than changing the pore size distribution.

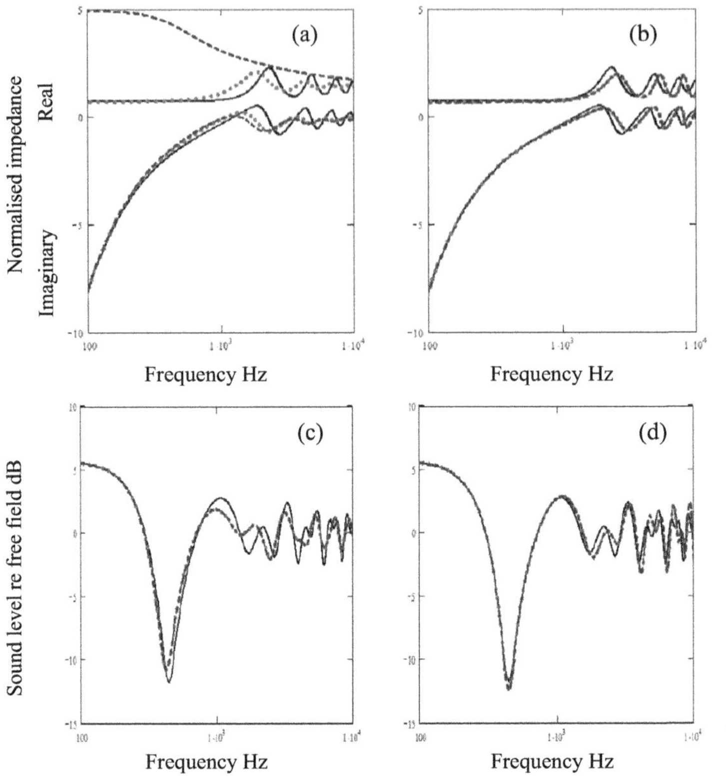

Figure 5.4 Predicted effects on surface impedance of varying the standard deviation of a log-normal distribution of (a) uniform pores and (b) non-uniform pores in a 'snow-like' slit-pore hard-backed rigid-framed porous layer with flow resistivity 10 kPa s m^{-2}, porosity 0.8, tortuosity $\sqrt{(5/4)}$ and thickness 0.06 m. The continuous, dotted and broken lines correspond to standard deviations of 0, 0.5 and 1 respectively. Figures 5.4 (c) and (d) show the corresponding predictions of excess attenuation spectra above the porous layer with source and receiver at 0.3 m height and separated by 1 m.

5.3 PHYSICAL INADMISSIBILITY OF SEMI-EMPIRICAL MODELS

While the Delany and Bazley semi-empirical model [2,3] has the advantage of using only a single parameter (effective flow resistivity) to predict the acoustical properties of ground surfaces, the model has the disadvantages that (a) it predicts negative values for the real part of the surface impedance of a layer when extrapolated to low frequencies, and (b), when an independently measured flow resistivity of a porous material is used, it overpredicts the imaginary part of the propagation constant and thereby overestimates the attenuation within a porous material [22]. To better reproduce Delany and Bazley's original measured data at low frequency and, at the same time, to avoid non-physical predictions such as a negative real part of the surface impedance of a layer of fibrous material both Miki [6,7] and Komatsu [8] have modified Delany and Bazley's relationships. Miki [6] changed the regression coefficients (see Table 5.1) but retained the structure of Equations (5.1a), (5.1b), (5.2a) and (5.2b). Komatsu [8] kept the basic parameters the same but changed the fundamental structure also.

The physical admissibility of ground impedance models is important for time-domain calculations in particular. The three conditions of reality, passivity and causality are necessary [9]. The reality condition is that $Z(\omega) = Z(-\omega)$. The passivity condition checks that $\mathrm{Re}(Z(\omega)) \geq 0$. The causality condition is equivalent to satisfying the Kramers–Kronig rule relating the real and imaginary parts of the complex impedance and ensures the absence of an impulse response for $t < 0$. Modifications to the coefficients and exponents in the Miki model [6] (identified as the 'modified Miki' model in Table 5.1) have been proposed [9] to make it physically admissible for use in time-domain calculations of outdoor propagation.

It has been mentioned that the 'passivity' condition requires that values of the real part of surface impedance should be positive or zero. But Kirby [45] has pointed out that, since impedance models of the form of Equations (5.1a), (5.1b), (5.2a) and (5.2b) represent equivalent fluid models, a more fundamental check is whether they can predict real parts of effective (complex) density and (complex) sound speed that are negative. The complex effective density ratio for an equivalent fluid impedance model can be calculated from

$$\rho / \rho_0 = \left(Z / \rho_0 c_0\right)\left(k / \left(\omega / c_0\right)\right). \tag{5.44}$$

Kirby [46] has investigated predictions of $\mathrm{Re}(\rho/\rho_0)$ for small values of f/R_s and has found that single-parameter impedance models (including that by Komatsu [8] which is not considered further here) predict negative values, whereas two other impedance models, including the model of Kirby and Cummings [47], which requires material-specific fitting parameters, and that due to Wilson [42] do not. The latter gives predictions that are more or less identical to those of models for a rigid-porous medium that assume idealized microstructures.

Kirby [46] does not specify the numerical ranges used for either frequency or flow resistivity. But checks on physical admissibility that are more particularly relevant to outdoor sound prediction consider the extremes of ranges of flow resistivity and porosity appropriate to outdoor ground surfaces.

Figures 5.5(a) and (b) compare predictions of the real parts of complex density ratio and surface (hard-backed layer) impedance using (i) a two-parameter version of the slit-pore impedance model (Equations (5.6), (5.7) and (5.10) with $s_A = 1$ and $T = \Omega^{-0.5}$), (ii) the original Delany and Bazley model [2], (iii) the three-parameter form of the Miki model [6] and (iv) the modified Miki model due to Dragna and Blanc-Benon [9] as a function of f/R_s, using parameter values consistent with those of a low flow resistivity ground such as snow or a forest floor (flow resistivity 10 kPa s m^{-2}, porosity 0.7, thickness 0.1 m [11]). For this low

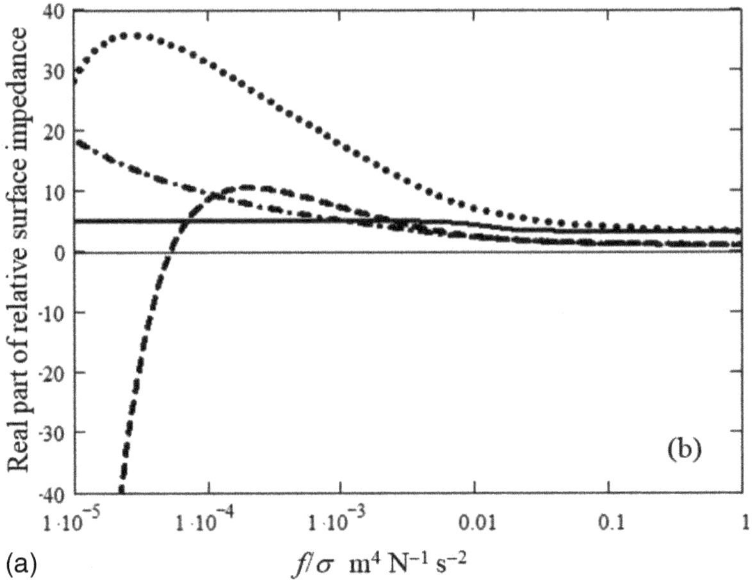

(a)

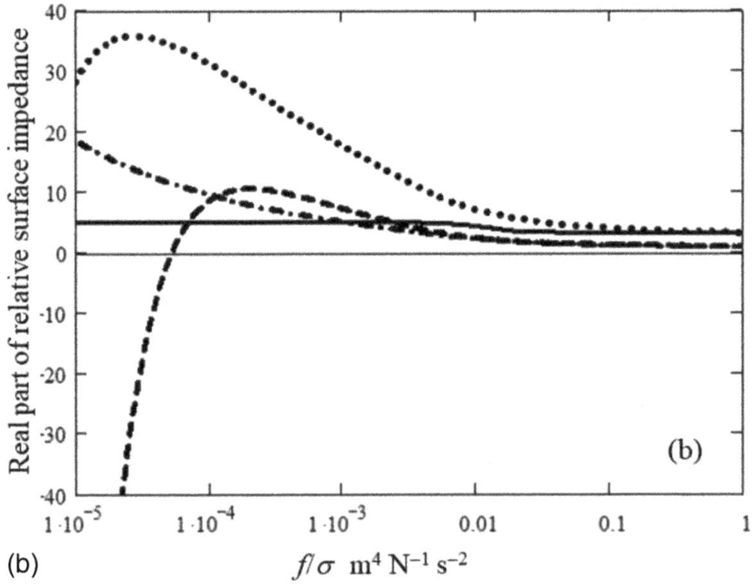

(b)

Figure 5.5 Predictions of real parts of (a) normalised complex density and (b) relative surface impedance as a function of frequency divided by flow resistivity for a 'grass-covered' ground (flow resistivity 200 kPa s m^{-2}, porosity 0.4, (hard-backed) layer thickness 0.03 m) according to slit pore (solid lines) Delany and Bazley (broken lines), three-parameter Miki (dotted lines) and modified Miki [6] (dash-dot lines) models.

flow resistivity of 10 kPa s m^{-2}, the Delany and Bazley, three-parameter Miki and modified Miki models predict that the real part of normalized effective density becomes negative below 100 Hz, whereas the slit-pore model does not. The modified Miki model predicts negative real parts of complex density ratio despite the fact that, with the same parameter values, the model predicts that the real part of surface impedance of a hard-backed layer is greater than 0 (see Figure 5.5(b)). Moreover, the modified Miki model predicts a significantly different surface impedance spectrum to that predicted by the slit-pore model. For parameter values representative of snow or forest floor, the hard-backed layer versions (using Eqn. (5)) of the Delany and Bazley and three-parameter Miki models predict negative values of Re[$Z(d)$] below 40 Hz and 5 Hz, respectively.

Figures 5.6(a) and (b) compare the predictions of these models as a function of f/σ but using parameter values more representative of a grass-covered ground (flow resistivity 200 kPa s m^{-2}, porosity 0.4, layer thickness 0.03 m [11]). For such 'grass-covered ground' parameter values, the Delany and Bazley and two Miki-based models lead to negative values of real normalized complex density below 2 kHz. Also the Delany and Bazley layer model predicts negative values of the real part of normalized surface impedance below 10 Hz. The predicted low-frequency behaviour of the dotted curve shown in Figure 5.6(b) suggests that, for 'grassland' parameters, the three-parameter Miki model will predict a negative real part of impedance also but at an even lower frequency. Again, while the modified Miki (layer) model does not predict a negative real part of surface impedance, it predicts values of the real part of the normalized surface impedance much larger than those predicted by the slit-pore model below 1 kHz.

Table 5.7 lists the physical admissibility checks and the results of these checks for several models that have been or could be used to represent ground impedance [48].

Any physically admissible impedance model can be transformed by using Padé approximations to enable implementation in the time domain [49], but, since it has a straightforward time-domain implementation, it has been common to use the Zwikker and Kosten phenomenological model for time-domain computations. As has been discussed, the Zwikker and Kosten model is, essentially, a high flow resistivity adiabatic approximation of identical pore models. While it is possible to implement the Wilson model in the time domain [43,50], and it is more accurate, particularly for low flow resistivity surfaces, it is computationally much more expensive since it requires memory-intensive evaluations of convolution integrals.

5.4 PREDICTING EFFECTS OF SURFACE ROUGHNESS

5.4.1 Boss and Stochastic Models

As well as being porous, many outdoor surfaces are rough. Surface roughness scatters the sound both coherently and incoherently. The relative strengths of the coherent and incoherent parts of the scattered energy depend on the mean

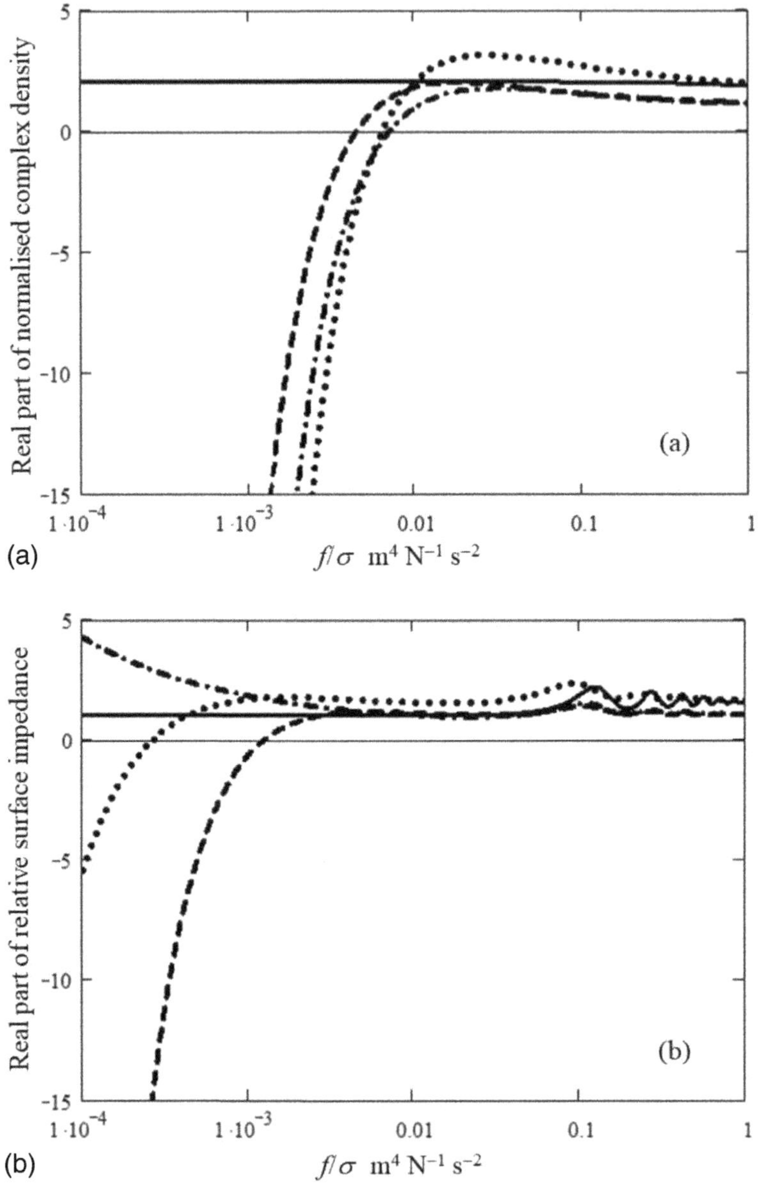

Figure 5.6 Predictions as a function of frequency divided by flow resistivity of real parts of (a) normalised complex density and (b) surface impedance for a 'snow-like' ground (flow resistivity 10 kPa s m^{-2}, porosity 0.7, (hard-backed) layer thickness 0.1 m) according to the slit pore model (solid lines), Delany and Bazley model (broken lines), three-parameter Miki model (dotted lines) and modified Miki model [9] (dash-dot lines).

Table 5.7 Results of checks on physical admissibility

	Semi-infinite				Hard-backed layer		
Ground type	*Reality*	*Passivity*		*Causality*	*Reality*	*Passivity*	*Causality*
Model		Re $\rho > 0$	Re $Z > 0$			Re $Z(d) > 0$	
Delany and Bazley	Yes	No	No	No	Yes	No	No
Miki	Yes	No	No	No	Yes	Yes	No
Modified Miki	Yes	No	Yes	Yes	Yes	Yes	Yes
Taraldsen	Yes		Yes	Yes	Yes	Yes	Yes
Slit pore	Yes	Yes	Yes	Yes	Yes	Yes	Yes
Variable porosity	Yes	N/A	Yes	Yes	N/A	N/A	Yes
Phenomenological	Yes	Yes	Yes	Yes	Yes	Yes	Yes
Hamet	Yes	Yes	Yes	Yes	Yes	Yes	Yes
Wilson	Yes	Yes	Yes	Yes	Yes	Yes	Yes

size of the roughness compared with the incident wavelength. In the frequency range of interest for applications related to surface transport noise (100 Hz–2000 Hz), the typical roughness of disked or compact-vegetation-covered soil is small compared to the wavelengths of sound. Such small-scale roughness means that a high proportion of the incident sound energy is scattered coherently in the direction of specular reflection. However, once the roughness size approaches the wavelengths of interest, as might be the case in a recently ploughed field, non-specular (incoherent) scattering dominates and interference effects associated with reflection from the ground are destroyed. The propagation of underwater sound over (or under) rough boundaries has been a topic of continuing interest for at least 50 years, but little attention has been given to equivalent situations in atmospheric acoustics until relatively recently. An important conclusion of early work [51,52], with regard to the coherent part of the scattering from the surface, is that the *impedance or admittance of the boundary is modified by the presence of small-scale roughness*. Clearly this can be considered to have an influence on the (spherical-wave) reflection coefficient and hence the ground effect, even above a ground surface that would have been considered acoustically hard if smooth.

The effects of surface roughness can be predicted using numerical methods such as the boundary element method (BEM), the Finite-Difference Time-Domain (FDTD) method or the Pseudospectral Time-Domain (PSTD) method [53]. However, there are analytical approaches also. Tolstoy [52] has distinguished between two such approaches, for predicting the *coherent* field resulting from boundary roughness where the typical roughness height is small compared to a wavelength. These are the *stochastic* and the *boss* models. Both of these reduce the rough surface scattering problem to the use of a suitable boundary condition at a *smoothed* boundary. According to Tolstoy [52], the *boss* method, originally derived by Biot [54] and Twersky [55], has the advantages that (i) it is more accurate to first order, (ii) it may be used even in conditions where the roughness shapes introduce steep slopes and (iii) it is reasonably accurate even when the roughness size approaches a wavelength.

For randomly rough surfaces, a roughness spectrum model of effective impedance derived from a stochastic small perturbation model (SPM) of electromagnetic wave propagation has been found to enable better agreement with data obtained over randomly and regularly rough surfaces in the laboratory than analytical results from the boss model [56], albeit requiring significantly more computation (see Section 5.4.2.5). On the other hand, the 'boss' approach for modelling roughness effects in outdoor sound calculations is useful when predicting propagation over artificially created 'boss-like' surfaces both for experimental validations of theory and for designing rough hard surfaces for noise reduction through destructive interference due to ground effect [57–60].

If numerical approaches are used for predicting outdoor sound, the representation of rough ground by an effective impedance can offer a significant saving in computational time and resources. Analytical and empirical approximations for the effective impedance of rough surfaces are described in the next sub-section and compared with measured roughness effects in Chapter 6.

5.4.2 Impedance Models Including Rough Surface Effects

5.4.2.1 *Hard Rough Surfaces*

Twersky developed a boss model [55,61–65] to describe coherent reflection from a hard surface containing sparse and closely packed semi-cylindrical roughnesses in which the contributions of the scatterers are summed to obtain the total scattered field and interaction between neighbouring scatterers has been included. His results lead to a real part of the effective admittance of the rough hard surface which may be attributed to incoherent scattering. Consider a plane wave incident at angle α to the normal and with an azimuthal angle φ between the wave vector and the (horizontal) axes of an array of parallel semi-cylinders of radius a and mean centre-to-centre spacing b on an otherwise plane hard boundary (Figure 5.7).

Twersky's results for the effective relative admittance $\beta_{RH,\,cyl}$ of such a rough *acoustically hard* surface containing non-periodically spaced circular semi-cylinders are

$$\beta_{RH,cyl} = \eta(\alpha,\phi) - i\xi(\alpha,\phi), \tag{5.45a}$$

$$\xi(\alpha,\phi) \approx kV\left[-1 + \left(\delta\cos^2(\phi) + \sin^2(\phi)\right)\sin^2(\alpha)\right] + O\left(k^3\right), \tag{5.45b}$$

$$\eta(\alpha,\phi) \approx \frac{nk^3\pi^2a^4}{8}\left(1-W^2\right)\left\{\begin{aligned}&\left(1-\sin^2\alpha\sin^2\phi\right)\\&\left[1+\left(\frac{\delta^2}{2}\cos^2\phi - \sin^2\phi\right)\sin^2\alpha\right]\end{aligned}\right\} + O\left(k^5\right). \tag{5.45c}$$

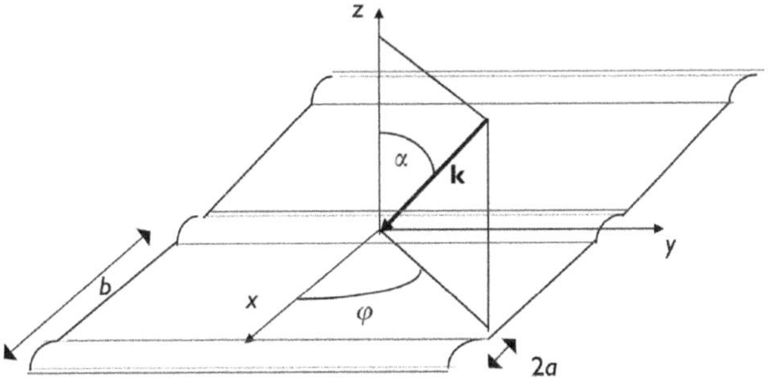

Figure 5.7 3-D representation of a plane wave incident in the direction of vector **k** on a surface containing a regularly spaced grating of semi-cylinders of radius ***a***.

$V = n\pi a^2/2$ is the raised cross-sectional area per unit length, n is the number of semi-cylinders per unit length (= $1/b$), $\delta = \dfrac{2}{1+I}$ is a measure of the dipole coupling between the semi-cylinders and $I = \dfrac{a^2}{b^2} I_2$, where

$$I_2 \cong 2W\left(1+0.307W+0.137W^2\right) \qquad \text{for } W < 0.8,$$

$$I_2 \cong \frac{\pi^2}{3}\left[1-\frac{2(1-W)}{W}\right]+6\frac{(1-W)^2}{W^2}\left[\frac{\pi^2}{6}+1.202\right] \qquad \text{for } W < 0.8$$

$$I \cong \frac{(\pi a)^2}{3b^2} \qquad \text{for } W = 1(\text{periodic}), \tag{5.45d}$$

$(1 - W)$ is a packing factor introduced to account for the degree of randomness of the distribution of 'bosses', $W = nb^* = b^*/b$, b^* being the minimum (centre-to-centre) separation between two cylinders. $W = 1$ for periodic spacing and $W = 0$ for random spacing and k is the wave number (in air).

Equations (5.45a)–(5.45d) predict that the effective admittance depends on angles of incidence (α and φ) in the planes normal to the scatterer axes as illustrated by the predictions shown in Figures 5.8(a) and (b). As might be expected, the effective impedance is predicted to increase as either α or φ increases. However, the angle dependencies are not significant near-grazing incidence and near-normal to the scatterer axes. Corresponding approximations are discussed later.

According to Equation (5.45c), the real part η of the admittance (which represents the incoherent or non-specular scattering loss term) is zero for periodic distributions of bosses. The real part of the effective impedance is predicted to increase as the degree of randomness increases. Nevertheless, small deviations from regular spacing are predicted to have a significant effect on the effective impedance. These points are illustrated for near-grazing incidence (α = 84.3°) normal to 5 cm radius semi-cylinder axes (φ = 0) and 20 cm spacing by the predictions shown in Figure 5.8(c). A deviation of only 5% (1 cm) in the locations of otherwise periodically spaced 5 cm high roughness elements with a mean centre-to-centre spacing of 20 cm is predicted to result in a real part of the effective admittance that is clearly non-zero between 500 and 5000 Hz.

At grazing incidence normal to the cylinder axes, $\alpha = \pi/2$ and the azimuthal angle $\varphi = 0$ so

$$\beta_{RH,cyl}\left(\frac{\pi}{2},0\right) = \frac{na^2k}{2}\left(\left(1-W^2\right)\frac{\left(1+\delta^2/2\right)k^2\pi^2a^2}{4} - i\pi(\delta-1)\right). \tag{5.46a}$$

With randomly distributed semi-cylindrical roughness, this simplifies further to

$$\beta_{RH,cyl,ran}\left(\frac{\pi}{2},0\right) = \left(\frac{3V^2k^3b}{2}\right)\left(1+\frac{\delta^2}{2}\right) - iVk(\delta-1). \tag{5.46b}$$

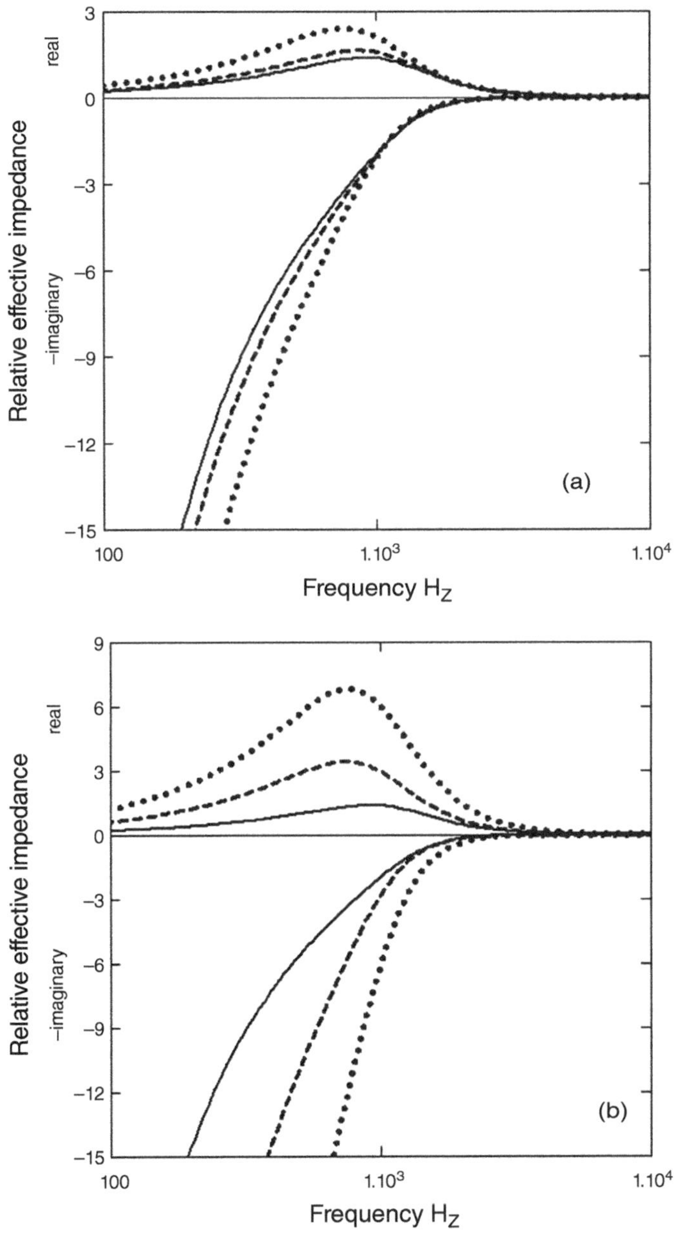

Figure 5.8 Predicted influences on the real and imaginary parts of the effective impedance of an array of 5 cm radius parallel acoustically hard semi-cylinders with mean centre-to-centre spacing 0.2 m on an acoustically hard plane of (a) the angle of incidence in the vertical plane perpendicular (azimuthal angle $\varphi = 0°$) to the semi-cylinder axes ($\alpha = 84.3°$ continuous lines; $\alpha = 75.5°$ broken lines; $\alpha = 66.4°$ black lines; $\boldsymbol{W} = 0.1$, (b) the azimuthal angle ($\varphi = 0°$ continuous lines; $\varphi = 45°$ broken lines; $\varphi = 60°$ dotted lines; $\alpha = 84.3°$; $\boldsymbol{W} = 0.1$)

(Continued)

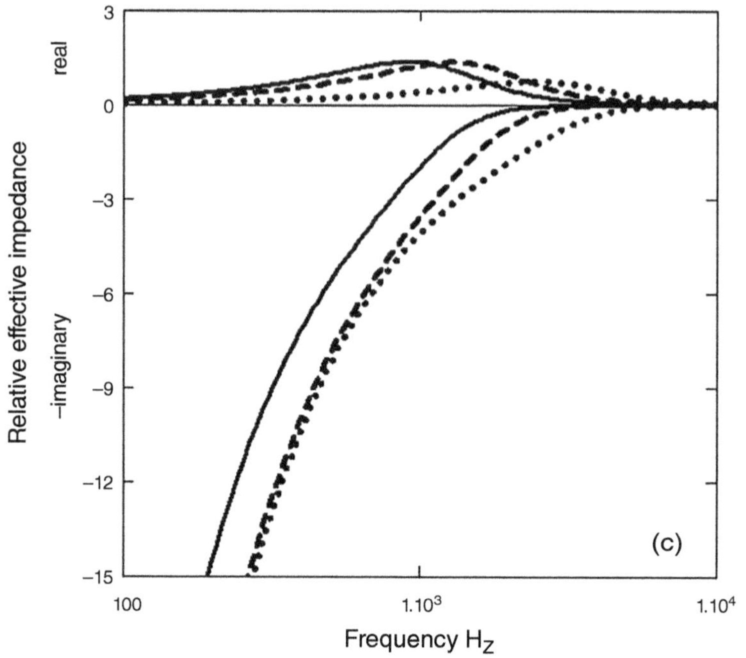

Figure 5.8 (Continued) (c) degree of randomness or regularity (W = 0.1 continuous lines; W = 0.8 broken lines; W = 0.95 dotted lines; α = 84.3°; φ = 0°).

According to Lucas and Twersky [59], for semi-elliptical cylinders with eccentricity e, so that $V = n\pi a^2 K/2$,

$$\delta = \frac{1+e}{1+I\dfrac{e(1+e)}{2}}. \tag{5.46c}$$

Twersky's semi-cylindrical boss theory (Equations (5.44)–(5.46a–5.46c)) may be generalized to scatterers of arbitrary shape as a result of comparing with the equivalent results from Tolstoy's work [52,65] for the effective admittance of a surface containing 2-D roughnesses of arbitrary shape. After correcting for a missing coefficient σ, Tolstoy's expression [65] for the effective admittance of a 2D-rough hard surface may be written as

$$\beta_{RH,2D} = -ik\varepsilon\left(\cos^2\phi - \sigma\cos^2\alpha\right), \tag{5.47a}$$

where

$$\varepsilon = V\left(\frac{2s_2}{v_2} - 1\right),\ \sigma = \left(\frac{2s_2}{v_2} - 1\right)^{-1}, \tag{5.47b, c}$$

Table 5.8 Hydrodynamic factors for use in effective admittance models

Shape of 2-D roughness	*K*
Semi-cylindrical	1
Semi-elliptical, height a', semi-base b'	a'/b'
Isosceles triangle, side u, height h	$1.05(h/u) + 0.14(h/u)^2$
Thin rectangle (slat)	1

$$\nu_2 = 1 + \frac{2\pi V s_2}{3b}, \quad s_2 = \frac{1}{2}(1+K). \tag{5.47d, e}$$

Here, ν_2 and s_2 are scatterer interaction and shape factors, respectively, K is a hydrodynamic factor depending on steady fluid flow around a scatterer, b is the mean scatterer spacing and V is the cross-sectional scatterer area above the plane per unit length. Values of K are known for various ideal shapes [66–68] and are listed in Table 5.8.

For semi-cylinders the expressions for ε and σ can be simplified to obtain

$$\varepsilon = \frac{\pi a^2}{2b}\left(\frac{2}{1+\frac{\pi^2}{3}\left(\frac{a}{b}\right)^2} - 1\right) \tag{5.48a}$$

and

$$\sigma = \frac{1}{\left(2/\left[1+\frac{\pi^2}{3}\left(\frac{a}{b}\right)^2\right] - 1\right)}. \tag{5.48b}$$

Tolstoy's result (Equation (5.47a)), that the rough hard surface admittance is purely imaginary, contrasts with Twersky's result (Equation (5.45a)) which includes a real part resulting from incoherent scattering.

On the other hand, Equation (5.45b) can be rewritten as

$$\xi(\alpha,\phi) \approx kV\left[(\delta-1)\cos^2(\phi) - \cos^2(\alpha)\left(1+(\delta-1)\cos^2(\phi)\right)\right]. \tag{5.45b$'$}$$

Hence, Twersky's and Tolstoy's expressions for the imaginary part of the effective admittance are equivalent for circular semi-cylinders if δ is replaced by $\frac{\varepsilon}{V} + 1 = \frac{2s_2}{\nu_2}$.

Twersky's result, Equations (5.46a)–(5.46c), for elliptical semi-cylinders shows the same dependence on K, the eccentricity, as obtained by Tolstoy [65], i.e. Equation (5.46c). This more general expression for δ depends on the hydrodynamic factor K (through ε) and thus on the shape of the scatterer. Consequently, forms for the real and imaginary parts of effective admittance due to roughness with any of the cross sections listed in Table 5.7 may be obtained from Equations (5.47a–5.47d, e), and the following form:

$$\eta(\alpha,\varphi) = \frac{k^3bV^2}{2}\left(1 - W^2\right)\left\{\frac{1-\sin^2\alpha\sin^2\varphi}{\left[1+\left(\frac{\delta^2}{2}\cos^2\varphi - \sin^2\varphi\right)\sin^2\alpha\right]} + O\left(k^5\right)\right\} \quad (5.49)$$

where V represents the scatterer volume per unit area (raised area per unit length in 2-D). These results allow predictions of propagation over bosses of the various shapes for which K is known. In addition to the known values for semi-cylinders, semi-ellipsoids and triangular wedges, K for thin rectangular strips is deduced by assuming that each strip affects the fluid flow as if it were a lamina [66]. The expression for the hydrodynamic virtual mass of a lamina of width $2a$ is identical to the one for a cylinder of radius a i.e. $K = 1$.

However, it should be noted that Twersky's results (Equations (5.45a, 5.45b, 5.45c, 5.45d)) and the modified version of Tolstoy's theory (Equations (5.45b, 5.47a–5.47d,5.47d, e and 5.49)), while leading to nearly identical effective impedance predictions for a hard rough surface formed by an almost-periodic array ($W = 0.95$) of 5 cm radius semi-cylinders on a hard surface at 20 cm mean centre-to-centre spacing, give rise to different predictions of effective impedance spectra for a nearly random array ($W = 0.1$) of identical semi-cylinders (see Figure 5.9).

5.4.2.2 *Rough Finite Impedance Surfaces*

Tolstoy considered sound propagation at the rough interface between two fluids [65] and his results have been used to predict the effective impedance of a rough finite impedance boundary [69,70]. Consider the general case involving a planar distribution of small fluid scatterers, N per unit area, embedded in a fluid half-space with density and sound speed ρ_3 and c_3 beneath a fluid half-space characterized by ρ_1 and c_1. The scatterers have density ρ_2 and c_2 , height h, centre-to-centre spacing l and occupy a total volume σ_V above the plane.

The effective admittance $\beta*$ of surfaces with 3-D scatterers is given by

$$\beta^* = -ik_0\varepsilon' + \beta_S \quad (5.50a)$$

where β_S represents the impedance of the imbedding plane, and

$$\varepsilon' = \varepsilon_{12} + \left(\frac{\rho_1 c_1^2}{\rho_3 c_3^2}\right)\varepsilon_{32} + \frac{\rho_1}{\rho_3}\left[1 - \frac{c_1^2}{c_3^2}\right]\varepsilon_{32}\delta_{32}$$

$$\varepsilon'\delta' = \varepsilon_{12}\delta_{12} + \left(\frac{\rho_1}{\rho_3}\right)\varepsilon_{32}\delta_{32}, \varepsilon_{ij} = a_{ij} - b_{ij}, \varepsilon_{ij}\delta_{ij} = a_{ij}, i = 1 \text{ or } 3, j = 2$$

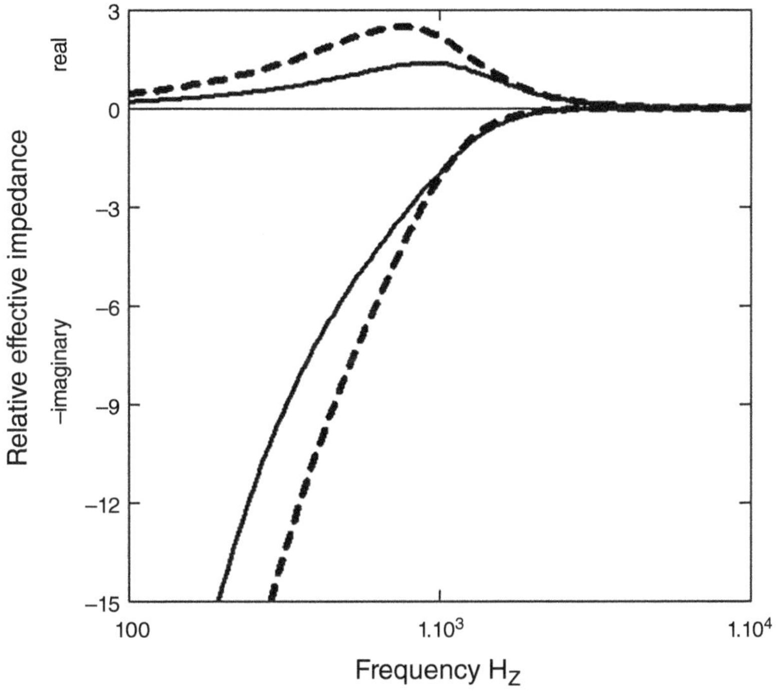

Figure 5.9 Effective impedance spectra predicted by equations (5.54) (continuous lines) and (5.45b', 5.47a, 5.47b, 5.47b, c, 5.47d, 5.47d, e and 5.49) (broken lines) of arrays of 5 cm radius semi-cylinders with mean centre to centre spacing of 0.2 m on a hard surface for incidence at 84.3° to the normal in a plane perpendicular to the (parallel) semi-cylinder axes and with **W** = 0.1.

$$a_{ij} = 3\sigma_V \left[\frac{\rho_j - \rho_i}{\rho_i + 2\rho_j}\right]\frac{s_{3D}}{v_{3D}}, s_{3D} = \left[\frac{\rho_i + 2\rho_j}{\rho_j + K\rho_i}\right]\frac{s_3}{2}, v_{3D} = 1 + \frac{3\pi\sigma_V}{8Nl^3}\left[\frac{\rho_j - \rho_i}{\rho_j + \rho_i / 2}\right]s_{3D}$$

$$b_{ij} = \sigma_V\left[1 - m_j / m_i\right], m_i = \frac{1}{\rho_i c_i^2}, m_j = \frac{1}{\rho_j c_j^2}. \tag{5.50b–j}$$

If $\rho_1 \ll \rho_3$ and $c_1 < c_3$, or $\rho_2 = \rho_3$ and $c_2 = c_3$, (5.50a–5.50b–j) reduces to

$$\beta_3^* = -ik_0\varepsilon_{12} + \beta_S. \tag{5.51}$$

For 2-D scatterers, the corresponding effective admittance is given by

$$\beta_2{}^* = -ik_0\cos^2(\theta)\varepsilon_{12}^* + \beta_S, \tag{5.52a}$$

where

$$\varepsilon_{12}{}^* = a_{12}^* - b_{12}^*, a_{12}^* = 2\sigma_V\left[\frac{\rho_2 - \rho_1}{\rho_2 + \rho_1}\right]\frac{s_{2D}}{v_{12}^*}, b_{12}^* = \sigma_V\left(1 - m_2 / m_1\right)$$

$$\upsilon_{12}^{*} = 1 + \frac{\pi^2}{3}\frac{h^2}{l^2}\frac{\rho_2 - \rho_1}{\rho_2 + \rho_1} s_2, s_{2D} = \frac{\rho_1 + \rho_2}{\rho_2 + K\rho_1} s_2, \nu_{2D} = 1 + \frac{2\pi\sigma_V}{3Nl^2}\left[\frac{\rho_2 - \rho_1}{\rho_2 + \rho_1}\right] s_{2D} \tag{5.52b–g}$$

and θ (= π/2–φ) is the angle between the source–receiver axis and the normal to the scatterer axis.

If the embedding material and scatterers are rigid and porous then they can be treated acoustically as if they are fluids but with complex densities and sound speeds. These complex quantities may be calculated from any of the models described in the previous section.

If $|\rho_2| >> \rho_1$, $s_2 = s_3 = \nu_2 = \nu_3 = 1$, Equations (5.51) and (5.52a–5.52b–g) are approximated by

$$\beta_3^* = -ik_0\frac{\sigma_V}{2} + \beta_S\left(1 - ik_S\sigma_V\right) \tag{5.53}$$

and by

$$\beta_2^* = -ik_0\cos^2(\theta)\sigma_V + \beta_S\left(1 - ik_S\sigma_V\right), \tag{5.54}$$

respectively, where k_S is the complex wave number within the lower half-space (i.e. the imbedding material).

As remarked earlier, Tolstoy's results for hard rough surfaces seem to ignore incoherent or non-specular scatter. A heuristic extension, using results obtained by Twersky, is to write the effective surface admittance of a porous surface containing 2-D roughnesses as

$$\beta_2^* = \eta - ik_0\cos^2(\theta)\varepsilon_{12}^* + \beta_S, \tag{5.55}$$

where η is given by Equation (5.49), s_2 and ν_2 are calculated from (5.53) and δ is calculated from $(2s_2/\nu_2)$.

For $\varphi = 0$, $\alpha \neq \pi/2$ (normal to scatterer axes, but non-grazing incidence),

$$\beta^* = \frac{k_0{}^3bV^2}{2}\left(1 - W^2\right)\left[1 + \left(\frac{\delta^2}{2}\right)\sin^2\alpha\right] - ikV\left[\delta\sin^2\alpha - 1 + \gamma\Omega\right] + \beta_s. \tag{5.56}$$

For $\alpha = \pi/2$, $\varphi \neq 0$ (grazing incidence but not normal to scatterer axes),

$$\beta^* = \frac{k_0{}^3bV^2}{2}\left(1 - W^2\right) - ikV\left[2(\delta - 1)\cos^2\varphi - 1 + \gamma\Omega\right] + \beta_s. \tag{5.57}$$

If the roughness is periodic, then $W = 1$ and only the imaginary part of admittance is changed. If the packing is random then $W = 0$ and the real part of the admittance changes by a factor that increases with f^3. For a randomly rough surface ($W = 0$), Equation (5.57), predicts the maximum roughness effect for a given value of Ω, V and β_s.

Roughness is predicted to be particularly significant for high flow resistivity or hard ground surfaces near grazing incidence. For 'hard' rough surfaces,

$$s_{2D} = s_2, v_{2D} = v_2 = 1 + \frac{2\pi V}{3Nb^2} s_2 \text{ and } \delta = 2s_2 / v_2. \tag{5.58}$$

The model for the effective admittance of the rough interface between two fluids was developed for semi-infinite media. However, for a high flow resistivity medium, the difference between the values of δ calculated from $2s_{2D}/\nu_{2D}$ and $2s_2/\nu_2$ is small.

This is illustrated in Figure 5.10.

Consequently, Equations (5.59) and (5.60) may be evaluated by (i) using any model for (smooth) surface admittance β_s (see Section 5.2) and (ii) adding the effective admittance for rough hard ground (see Section 5.4.2.1).

Figure 5.11 shows predictions of the impedance of various randomly rough finite impedance surfaces at a grazing angle of 5°. At low frequencies, and for the listed parameters, it is predicted that there is a reduction in the real part of impedance compared with the imaginary part. Such behaviour at low frequencies is also predicted in the impedance of smooth ground surfaces if there is layering [11]. For larger roughness and layered ground, more complicated relationships between real and imaginary parts of impedance may result. Note that the assumptions, $kh < kb \leq 1$, are violated above 1000 Hz for the largest roughness case considered in Figure 5.11.

The predicted high-frequency limit of the impedance of a smooth porous medium is determined by the ratio of tortuosity to porosity and, therefore,

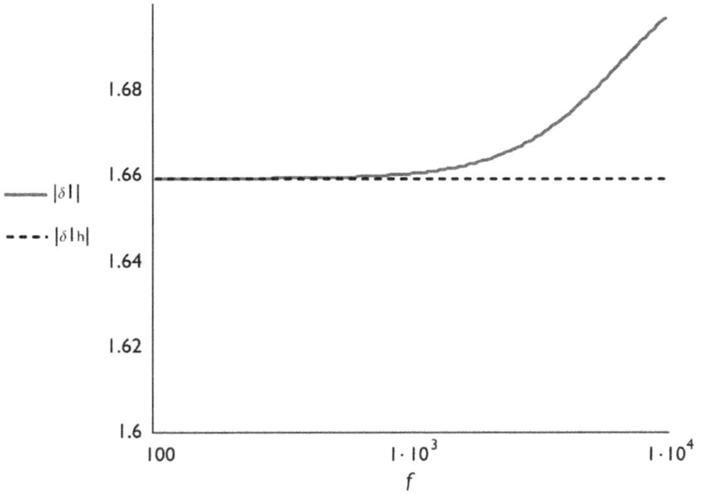

Figure 5.10 Comparison of $|\delta 1| = 2s_{2D}/\nu_{2D}$ and $\delta 1h = 2s_2/\nu_2$ for parameter values of flow resistivity 500 kPa s m^{-2}, porosity 0.4, tortuosity 2.5 and with semi-cylindrical roughness, radius 0.0075 m and mean spacing 0.03 m.

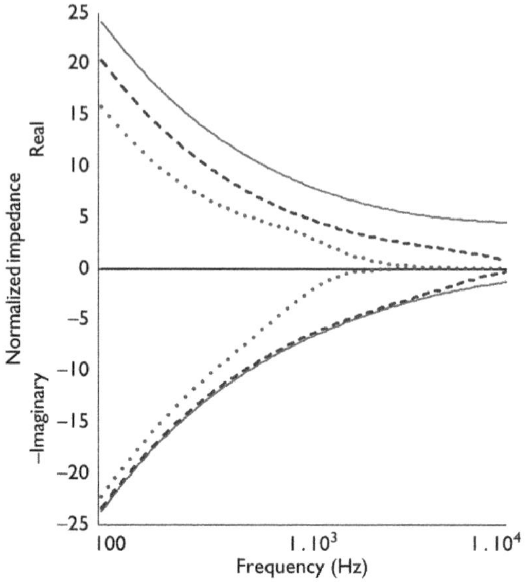

Figure 5.11 Predicted impedance of smooth and rough finite impedance surfaces. The porous medium is assumed semi-infinite with uniform triangular pores, flow resistivity 500 kPa s m^{-2}, porosity 0.4, and tortuosity = 2.5. The solid lines represent the predicted relative impedance of the smooth porous surface. The broken lines correspond to random semi-cylindrical roughness 0.0075 m high with mean spacing 0.03 m. The dotted lines correspond to random semi-elliptical roughness, height 0.025 m, eccentricity (*K*) 1.5 and mean spacing 0.06 m. The rough surface impedance is predicted for a grazing angle of 5°.

is always greater than 1. A common feature of predictions for the real part of impedance of rough finite impedance surfaces, is that, at high frequencies, they are less than would be expected for smooth surfaces with the same porosity and flow resistivity. The incoherent component of scatter ensures that the effective impedance of a randomly rough finite impedance surface tends to zero at high frequencies.

It is clear from Figure 5.11 that the presence of even relatively sparse small 2D roughness, less than 0.001 m in height and with mean spacing of 0.05 m, is predicted to cause a significant change in the smooth hard surface impedance.

Figure 5.12 shows that the predicted normalized surface impedance of a rough porous surface may be approximated by formula (5.2a and 5.2b) with an effective flow resistivity given by 0.8 × the actual flow resistivity. Clearly the prediction using the single-parameter impedance model is similar to that due to the more sophisticated rough finite impedance model after parameter adjustment. Predictions such as those shown in Figure 5.12 may be part of the explanation of why single-parameter models (Equations (5.2a) and (5.2b)) have been successful in modelling the impedance of outdoor ground surfaces.

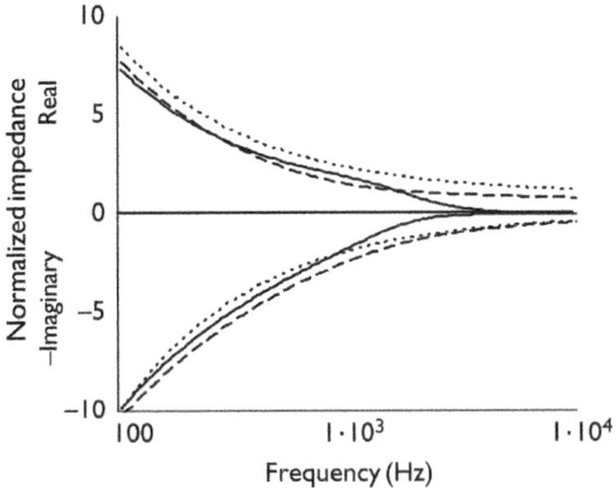

Figure 5.12 Predicted effect of increasing roughness parameter values on the impedance of a 'grass-like' 5 cm thick layer: flow resistivity 200 kPa s m^{-2}, porosity 0.4, tortuosity = 1/porosity, triangular pores. Continuous lines correspond to a smooth boundary. Dashed lines correspond to 2D roughness consisting of 0.01 m radius randomly spaced semi-cylinders, mean centre-to-centre spacing 0.04 m with axes normal to source-receiver axis. Dotted lines correspond to 2D roughness with 0.04 m radius, mean centre-to-centre spacing 0.16 m.

Also shown in Figure 5.12 are predictions of the formula

$$Z_r \approx Z_s - \left(\frac{HR_s}{\gamma\rho_0 c_0}\right)\left(\frac{2}{v}-1\right), \ \mathrm{Re}(Z_s) > \left(\frac{HR_s}{\gamma\rho_0 c_0}\right)\left(\frac{2}{v}-1\right), \tag{5.59}$$

where $\upsilon = 1+\frac{2}{3}\pi H$ and $\langle H\rangle$ is the rms roughness height. Equation (5.59) provides a low-frequency effective impedance model giving predictions in tolerable agreement with more exact calculations below 2 kHz. There are considerable differences between the predictions above 2 kHz, but such frequencies are seldom of importance in predicting outdoor sound.

Figure 5.12 shows that the predicted effect of roughness (small compared with wavelength) on an acoustically soft surface is to reduce real part of the effective surface impedance while leaving the imaginary part relatively unaffected and that a tolerably successful prediction of the effective normalized impedance of a rough surface is that of the smooth surface but with a reduced real part. The limitation on the domain of Equation (5.59) is important to avoid non-physical predictions, i.e. a negative real part of impedance, when the flow resistivity and roughness height are large.

Comparisons between data and impedance models are presented in Chapter 6.

Figure 5.13 shows the predicted effect of various roughness specifications on the impedance of a grass-like layer. Figure 5.14 shows the corresponding EA predictions. The predictions in Figure 5.13 show that the effective impedance

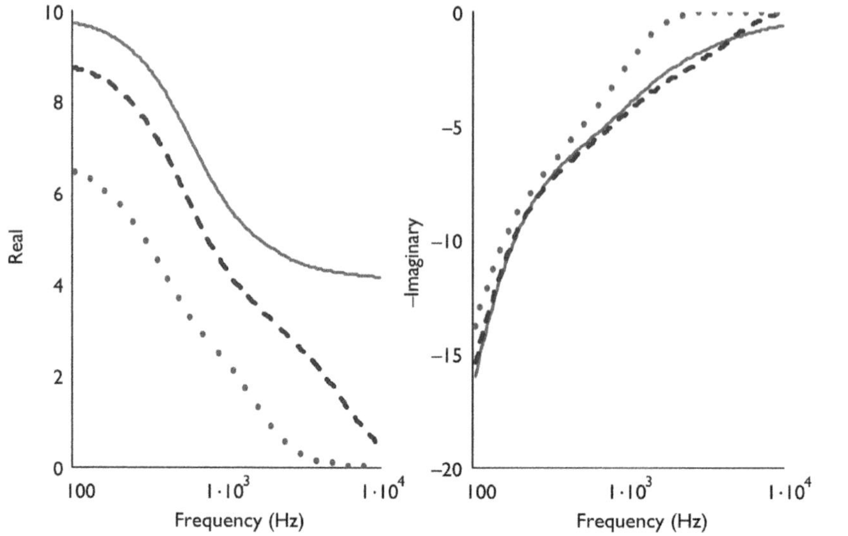

Figure 5.13 Predicted surface impedance of rough porous ground. Solid lines represent the predicted normalised surface impedance of a porous surface with 20/m semi-cylindrical roughness as specified in the caption for Figure 5.9. The dotted lines represent the result predicted by equation (3.2) with $\boldsymbol{R}_{se}$ = 0.8 × the actual flow resistivity. The dashed lines are predictions of equation (3.63).

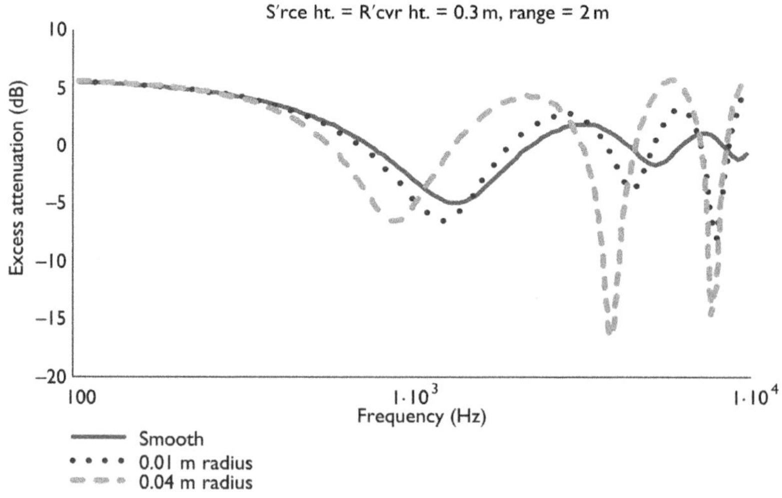

Figure 5.14 Predicted excess attenuation spectra corresponding to the impedance spectra shown in Figure 5.13.

changes corresponding to increase in mean roughness height from 1 cm to 4 cm cause the main ground effect to be shifted to lower frequencies. The predicted shifts in EA are substantial and may be important in short-range ground characterization.

5.4.2.3 *Modified 'Boss' and Empirical Models for Regularly Spaced Roughness Elements*

According to Brekhovskikh [71], the effective surface impedance of a 'comb-like' structure, a hard rough surface formed from periodically spaced thin rectangular strips, may be approximated by that of a thin fluid layer with the same thickness as the height of the strips, *h*,

$$Z_{eff} = \coth(-ikh). \tag{5.60}$$

Laboratory measurements of EA spectra measured over periodically rough hard surfaces shown in Chapter 6 and the effective impedance spectra deduced from these measurements, using a method described also in Chapter 6, depend on the roughness element spacing in a manner that is not predicted well by the 'boss' models (e.g. the modified Tolstoy Equations (5.47a–5.47d, e)). Based on results of measurements over various shapes of periodically spaced roughness elements, Bashir et al. [59] have proposed an empirical modification of Equations (5.47d, e),

$$v_2^{\text{mod}} = \frac{\pi V(1+K)(39b-0.07)}{3b}. \tag{5.61a}$$

Several of the maxima in EA spectra measured over periodically rough surfaces are associated with diffraction grating effects [59] and therefore not predicted when 'boss' models for effective impedance are used in classical theory for a point above a finite impedance boundary (see Chapter 2).

On the basis that the destructive interference observed in an EA spectrum at a higher frequency than predicted by 'boss' theory is similar to that associated with a thin layer type of surface impedance, Bashir et al. [59] have proposed an additional 'hard-backed layer' term, β_l for the effective rough surface impedance, i.e.

$\beta_{eff} = \beta_{RH}^{\text{mod}} + \beta_l$ where β_{RH}^{mod} is the result of using Equation (5.60a) in Equation (5.47a). Hence

$$\beta_l = 1/Z_l, Z_l = \coth(-ik_0 d_e(1+0.04i)), d_e = (0.5+4h)b \tag{5.61b, c, d}$$

where *h* is the roughness height.

Figure 5.15 compares predictions of Equations (5.45a, 5.45b, 5.45c, 5.45d, α = 84.3°, φ = 0), (5.60) and (5.61a, 5.61b, c, d) for the effective impedance of a surface formed by acoustically hard roughness elements (semi-cylinders or thin strips) of height 5 cm with almost regular spacing (W = 0.95, mean spacing of 0.2 m) on an acoustically hard surface. Example comparisons with data are in Chapter 6.

5.4.2.4 *Multiple Scattering Models*

If the 'bosses' have the form of semi-cylinders then a multiple scattering theory can be developed for calculating the field due to incident plane waves, or from a line source, above the resulting rough boundary [72]. A 2-D

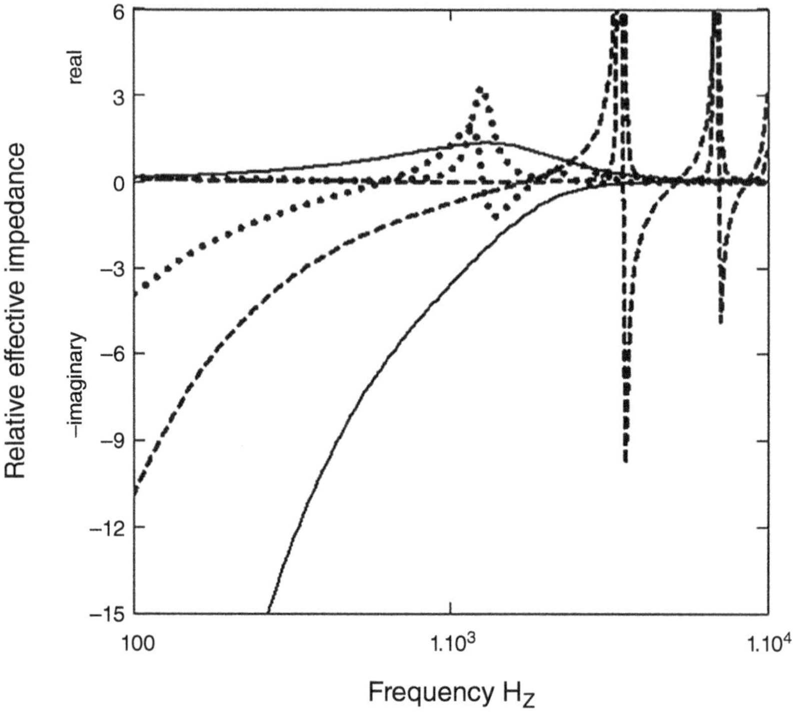

Figure 5.15 Effective impedance spectra predicted by equations (5.45a–5.45d), α = 84.3°, φ = 0, (solid lines), (5.60), (broken lines) and (5.61a and 5.61b, c, d), (dotted lines) for a surface formed by almost periodically-spaced (**W** = 0.95) acoustically hard roughness elements (semi-cylinders or thin strips) of height 5 cm and 20 cm mean spacing on an acoustically hard surface.

problem is considered in which each scatterer is represented by its circular cross section (see Figure 5.16(a)). The polar co-ordinates of the field point in the Cartesian reference frame (O_x, O_y) are represented by (r,θ), and the polar co-ordinates of the field point in the reference frame $(O_j x, O_j y)$ centred at the jth semi-cylinder centre $O_j(x_j, y_j)$ are represented by (r_j, θ_j).

Exterior to the semi-cylinders, the pressure field, P, satisfies

$$\nabla^2 P + k^2 P = 0. \tag{5.62a}$$

Interior to semi-cylinder j, the pressure field, P_j, satisfies

$$\nabla^2 P_j + k_j^2 P_j = 0. \tag{5.62b}$$

Consider an incident plane wave at angle β with respect to the $+x$-axis on an array of N infinitely long non-identical penetrable semi-cylinders embedded in a smooth hard surface (see Figure 5.16(a)).

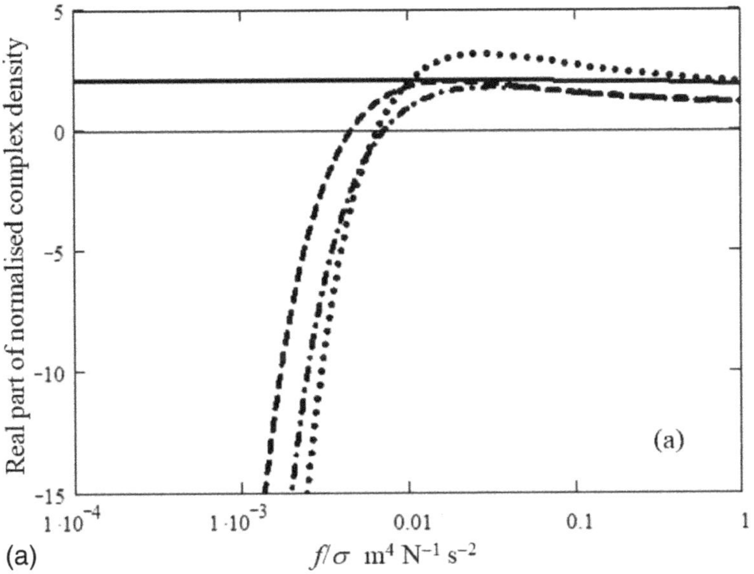

(a)

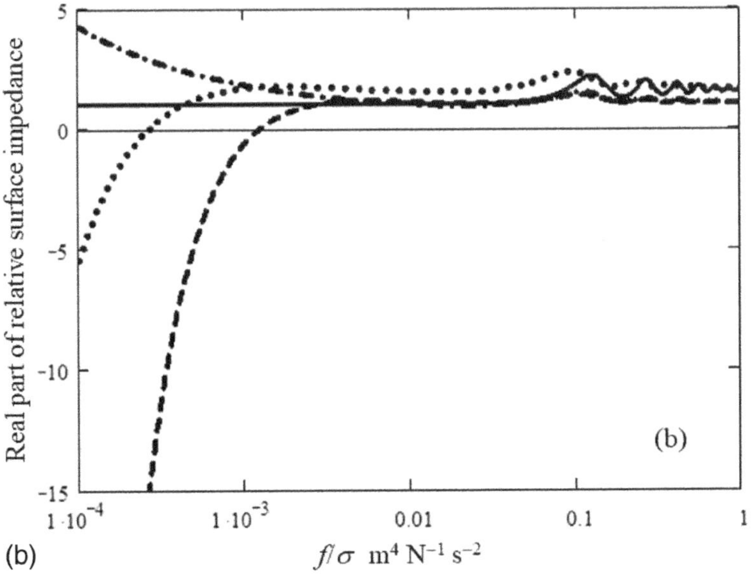

(b)

Figure 5.16 Cross-sections of two semi-cylinders and receiver locations considered for predicting multiple scattering by a rough boundary formed by finite impedance semi-cylinders on an acoustically hard plane for (a) plane wave incidence and (b) a line source.

$$P_{in} = e^{i\vec{k}\vec{r}} = e^{ikr\cos(\theta-\beta)} \tag{5.63}$$

The propagation vector $\vec{k}$ is considered to be perpendicular to the cylinder axes. When applying the boundary conditions, it is useful to express the incident wave in terms of the radial position, $r_{j,}$ of the jth semi-cylinder and the polar angle θ_j: $\vec{r} = \overrightarrow{OO_j} + \vec{r_j}$, where the dot product $\vec{k}.\vec{r}$ enables expression of the incident plane wave Equation (2.19) as

$$P_{in} = I_j e^{ikr_j\cos(\theta_j-\beta)}, \tag{5.64}$$

where I_j is a phase factor associated with semi-cylinder j defined as $I_j = e^{ikx_j \cos\beta}$.

The incident field is reflected and scattered by the array of semi-cylinders embedded on a smooth hard surface so the total field is the sum of incident, reflected and scattered waves

$$P = P_{in} + P_{refl} + P_{scat}. \tag{5.65}$$

As a result of the acoustically hard embedding plane, the reflected wave is the mirror reflection of the incident wave in the plane containing the semi-cylinder axes,

$$P_{refl} = e^{ikr\cos(\theta+\beta)}, \tag{5.66a}$$

or in terms of r_j and θ_j, the reflected wave is

$$P_{refl} = I_j e^{ikr_j\cos(\theta_j+\beta)}, \tag{5.66b}$$

where $I_j = e^{ik(x_j \cos\beta + y_j \sin\beta)}$ is a phase factor associated with the j_{th} cylinder.

The sum of the incident and reflected waves can be expanded as series of Bessel functions [73],

$$P_{in} + P_{refl} = 2I_j \sum_{n=-\infty}^{+\infty} J_n(kr_j) e^{in(\pi/2+\theta_j)} \cos(n\beta). \tag{5.67}$$

Around a given scatterer, the total scattered field P_{scat} is a sum of contributions from the N semi-cylinders. The scattering contribution from the jth semi-cylinder can be expresssed in the form of a cylindrical wave expansion using the basis function set, $e^{in\theta_j}$, for the polar angle contribution and Hankel functions of the first kind and order n, $H_n(kr_j)$, for the radial co-ordinates. Hence, the total scattered wave is

$$P_{scat} = \sum_{j=1}^{N} \sum_{n=-\infty}^{\infty} A_n^j Z_n^j H_n(kr_j) e^{in\theta_j} \tag{5.68}$$

The coefficients Z_n^j are chosen to be of a form that allows the cylinders to have different radii and finite surface impedance

$$Z_n^j = \frac{q_j J_n'(ka_j) J_n(k_j a_j) - J_n(ka_j) J_n'(k_j a_j)}{q_j H_n'(ka_j) J_n(k_j a_j) - H_n(ka_j) J_n'(k_j a_j)}. \tag{5.69}$$

The coefficients q_j are defined as $q_j = \frac{\rho_j c_j}{\rho_0 c_0}$ where ρ_j and c_j represent the density of fluid and the sound speed inside the jth cylinders, respectively. ρ_0 and c_0 are the density of air and the sound speed outside the cylinder. The limit $q_j \to \infty$ corresponds to acoustically hard semi-cylinders.

The solution to Equations (5.62a and 5.62b) have the form

$$P_j = \sum_{n=-\infty}^{\infty} B_n^j J_n(k_j r_j) e^{in\theta_j}, \tag{5.70}$$

where $J_n(k_j r_j)$ is a Bessel function of the first kind and order n.

The coefficients A_n^j and B_n^j are determined from the boundary conditions. Graf's addition theorem [68] for Bessel functions is used to express $H_n(kr_j)$ in terms of co-ordinates (r_s, θ_s) needed for the boundary conditions at the surface of cylinder s, and Equation (5.68) becomes

$$P_{scat} = \sum_{n=-\infty}^{\infty} A_n^s Z_n^s H_n(kr_s) e^{in\theta_s} + \sum_{j=1, j\neq s}^{N} \sum_{n=-\infty}^{\infty} A_n^j Z_n^j \sum_{m=-\infty}^{m=\infty} J_m(kr_s) H_{n-m}(kR_{js}) e^{im\theta_s} e^{i(n-m)\alpha_{js}}, \tag{5.71}$$

where α_{js} is 0 or π depending on the respective positions of the jth and s^{th} semi-cylinders. This equation is valid provided $r_s < R_{js}$,where R_{js} is the distance between the centres of cylinders j and s.

Equations (5.69)–(5.71) may be combined to give the total field at receiver by a finite array of acoustically soft semi-cylinders on a smooth acoustically hard ground:

$$P = 2I_j \sum_{n=-\infty}^{+\infty} J_n(kr_j) e^{in(\pi/2+\theta_j)} \cos(n\beta) + \sum_{n=-\infty}^{\infty} A_n^s Z_n^s H_n(kr_s) e^{in\theta_s} + \sum_{j=1, j\neq s}^{N} \sum_{n=-\infty}^{\infty} A_n^j Z_n^j \sum_{m=-\infty}^{m=\infty} J_m(kr_s) H_{n\ m}(kR_{js}) e^{im\theta_s} e^{i(n-m)\alpha_{js}}. \tag{5.72}$$

The boundary condition $P|_{rs=as} = P_s|_{rs=as}$on the sth semi-cylinder leads to expressions for the coefficients B_n^j in terms of A_n^j. After substituting this result in (5.72), the other boundary condition, $\left.\frac{1}{\rho_0}\frac{\partial P}{\partial r_s}\right|_{r_s=a_s} = \left.\frac{1}{\rho_s}\frac{\partial P_s}{\partial r_s}\right|_{r_s=a_s}$, gives the infinite system of equations

$$A_m^s + \sum_{\substack{j=1 \\ j \neq s}}^{N} \sum_{n=-\infty}^{\infty} A_n^j Z_n^j H_{n-m}(kR_{js}) e^{i(n-m)\alpha_{js}} = -2\cos(m\beta) I_s e^{im\pi/2},$$

$$m \in Z, s = 1, \ldots, N. \tag{5.73}$$

Note that the restriction $r_s < R_{js}$ associated with the use of Graf's addition theorem is met at the surface of the cylinders where the boundary conditions are applied.

To evaluate the unknowns A_n^j, the infinite system (5.73) is truncated to a system of $N(2M + 1)$ equations, where M is a truncation parameter, i.e.

$$A_m^s + \sum_{\substack{j=1 \\ j \neq s}}^{N} \sum_{n=-M}^{M} A_n^j Z_n^j H_{n-m}(kR_{js}) e^{i(n-m)\alpha_{js}} = -2\cos(m\beta) I_s e^{im\pi/2},$$

$$m = -M \ldots + M, s = 1..N. \tag{5.74}$$

This system is solved numerically. Setting the truncation parameter M equal to 5 has been found to give sufficient prediction accuracy between 100 Hz and 10 kHz [72].

A similar approach yields the pressure field above the same array of semi-cylindrical bosses due to cylindrical waves incident from a line source. The Helmholtz Equation (5.62a) is solved using the same co-ordinate system as was used for plane waves. The incident pressure amplitude is written as

$$P_{in} = H_0(k\rho_1), \tag{5.75a}$$

where ρ_1 is the source–receiver distance (see Figure 5.16(b)). It is useful for subsequent development to express $H_0(k\rho_1)$ in terms of the co-ordinates (r_s, θ_s) by again using Graf's addition theorem [68],

$$P = \sum_{n=-\infty}^{+\infty} J_n(kr_s) H_n(kS_{s1}) e^{-in\sigma_{s1}} e^{in\theta_s} \tag{5.75b}$$

with the restriction $r_s < S_{s1}$, where S_{s1} is the radial distance between the cylinder centre s and the source 1.

To develop an expression for the wave scattered by a finite array of non-identical finite impedance semi-cylinders embedded in a smooth hard surface, the effect of the hard surface embedding the semi-cylinders is taken into account by assuming an image source and hence a reflected wave,

$$P_{refl} = H_0(k\rho_2), \tag{5.76a}$$

where ρ_2 is the image source to receiver distance (see Figure 5.16(b)).

If the reflected wave (Equation (5.76a)) is expressed in terms of r_s and θ_s as in (5.75a) and (5.75b), the total field outside the semi-cylinders becomes

$$P = \sum_{n=-\infty}^{+\infty} J_n(kr_s)e^{in\theta_s}\left[H_n(kS_{s1})e^{-in\sigma_{s1}} + H_n(kS_{s2})e^{-in\sigma_{s2}}\right]$$
$$+\sum_{n=-\infty}^{\infty} A_n^s Z_n^s H_n(kr_s)e^{in\theta_s}$$
$$+\sum_{j=1,j\neq s}^{N}\sum_{n=-\infty}^{\infty} A_n^j Z_n^j \sum_{m=-\infty}^{m=\infty} J_m(kr_s)H_{n-m}(kR_{js})e^{im\theta_s}e^{i(n-m)\alpha_{js}}, \qquad (5.76b)$$

provided that $r_s < S_{s1}$ and $r_s < S_{s2}$.

The boundary conditions are as before and, after some algebra, lead to the infinite system of equations

$$A_m^s + \sum_{\substack{j=1\\ j\neq s}}^{N}\sum_{n=-\infty}^{\infty} A_n^j Z_n^j H_{n-m}(kR_{js})e^{i(n-m)\alpha_{js}}$$
$$= -H_m(kS_{p1})e^{-im\sigma_{p1}} - H_m(kS_{p2})e^{-im\sigma_{p2}} \qquad (5.77)$$

with unknowns A_n^j and $m \in Z$, $s = 1, \ldots, N$.

Note that the restrictions $r_s < S_{s1}$ and $r_s < S_{s2}$ associated with Graf's addition theorem applied to the source terms are met at the surfaces of the cylinders where the boundary conditions are applied. As for plane wave incidence, to evaluate the coefficients A_n^j, the infinite system of equations (5.77) is truncated to a system of $N(2M + 1)$ equations

$$A_m^s + \sum_{\substack{j=1\\ j\neq s}}^{N}\sum_{n=-M}^{M} A_n^j Z_n^j H_{n-m}(kR_{js})e^{i(n-m)\alpha_{js}}$$
$$= -H_m(kS_{p1})e^{-im\sigma_{p1}} H_m(kS_{p2})e^{-im\sigma_{p2}} \qquad (5.78)$$

with $m = -M \ldots +M$ and $s = 1 \ldots\ldots N$.

The system of Equations (5.78) is solved numerically. Note that only the right-hand side of Equation (5.78) differs from the result for plane waves (Equation (5.74)).

Although restricted to semi-cylindrical roughness, the multiple scattering theory described has been found to be more accurate than the corresponding boss models which are long-wavelength approximations. Moreover, the boss models are applicable only to non-deterministic random distributions, since the positions of the roughness elements are not included in these theories and the roughness distributions are characterized only by their size and number per unit area.

5.4.2.5 A Roughness Spectrum Model

By referring to an expression for the coefficient of reflection of the vertical component of electromagnetic waves from a perfectly conducting rough boundary derived using the small perturbation method (SPM) [74], Faure [75] has proposed an effective admittance for coherent sound reflection from a randomly rough acoustically hard surface

$$\beta_{eff} = \int_{-\infty}^{\infty} \frac{d\kappa'}{k_0 k_z(\kappa')} \left(k_0^2 - \kappa\kappa'\right)^2 \Re\left(\kappa - \kappa'\right), \tag{5.79}$$

where $k_0^2 = k_z^2 - \kappa^2$, k_z and $\kappa = k_0 \sin(\alpha)$ are the horizontal and vertical components, respectively, of the incident (at angle α) plane wave number and $\Re$ is the roughness spectrum.

The integral has to be evaluated numerically taking care of the singularity at $k_z(\kappa') = 0$.

The assumption of a Gaussian roughness spectrum

$$\Re(k) = \frac{\sigma_h l_c}{2\sqrt{\pi}} e^{\frac{-k^2 l_c^2}{4}} \tag{5.80}$$

introduces two roughness parameters; σ_h which is the mean roughness height and l_c is the correlation length.

Since the model is for random roughness, the mean of many realizations has to be used to obtain an effective admittance for calculating point-to-point propagation. As with the boss model, the effective impedance is predicted to be a function of the angle of incidence. Also, similar to the results in Section 5.4.2.2, Faure [75] has proposed a model for the effective admittance of a rough acoustically soft boundary which is the result of adding the roughness spectrum admittance (5.79) to the admittance of the corresponding smooth acoustically soft surface, as was the case with the boss model (see 5.4.2.2).

5.4.3 Propagation over Rough Seas

5.4.3.1 Effective Impedance of Rough Sea Surfaces

The ISO 9613-2 prediction scheme (see Chapter 12) assumes that any water surface is acoustically hard. While this may be true during exceptionally calm conditions when the sea surface is flat and smooth, more typically a sea surface is continuously in motion as a result of waves generated by winds and currents. In a similar manner to airborne propagation over rough ground, impulsive sounds will be modified during near-grazing propagation above a rough sea surface. When predicting impulsive sound propagation, such as from explosions, sonic booms or piling, the sea surface profile can be considered to be stationary in the same way that the atmospheric turbulence can be considered to be 'frozen' when predicting the influence of turbulence on outdoor sound (see Chapter 10).

The BEM can be used to predict propagation over arbitrarily rough boundaries. Short-range predictions of EA spectra over profiles formed by randomly spaced intersecting acoustically hard parabolas, intended to represent the various instantaneous sea states, have been made using BEM [76]. The wave crests are sharper and the wave troughs flatter in 'parabolic' waves than in sinusoidal ones [77]. The ratio of height H and wavelength λ of the parabolic roughness was fixed at $H/\lambda = 0.1$ which is below the value (0.14) beyond which water waves break [78]. Figure 5.17(a) shows an example profile and the source and receiver positions assumed for the BEM simulations. The length of rough sea surface in the BEM calculations was limited

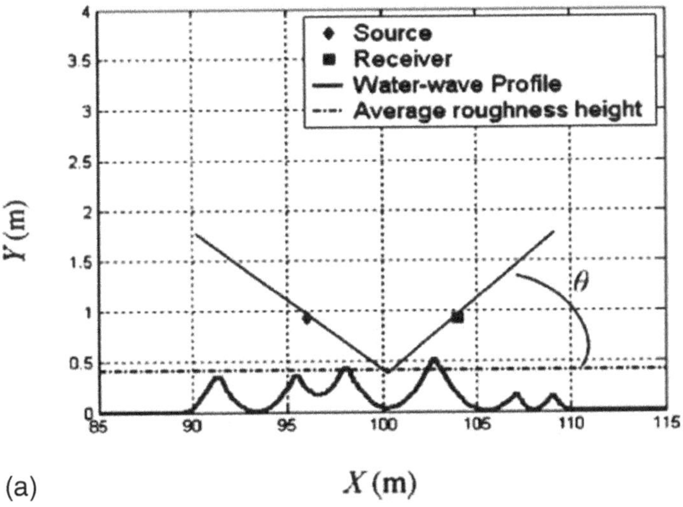

(a)

(b)

Figure 5.17 (a) Assumed source receiver locations over sea state 2 waves represented by randomly-spaced intersecting parabolas for BEM simulations of EA spectra [76]. Reproduced from P. M. Boulanger and K. Attenborough, Effective impedance spectra for rough sea effects on atmospheric impulsive sounds, *J. Acoust. Soc. Am.* *117*(2) 751–762 (2005) with the permission of the Acoustical Society of America. (b) Snapshot of an FDTD simulation of impulse propagation over a rough sea state 2 [80]. Reproduced from T. Van Renterghem, D. Botteldooren and L. Deckoninck, Airborne sound propagation over sea during offshore wind farm piling, *J. Acoust. Soc. Am.* *135* 599–609 (2014) with the permission of the Acoustical Society of America.

to 20 m but the rough surface was extended on each side of source and receiver by 90 m of flat hard surface to avoid edge effects which means that the total modelled surface length was 200 m.

The BEM EA predictions were fitted for effective plane surface admittance using a winding number integral method. BEM-deduced effective impedance spectra have been fitted in turn using simple polynomials,

$$\mathrm{Re}(Z) = \alpha f^{-m} + \delta \tag{5.81a}$$

$$Im(Z) = \alpha' f^{-m'} + \delta', \tag{5.81b}$$

in which it was assumed that $m = 1$ and $m' = 0.5$. Best-fit values of the other coefficients have been found for sea states from 2 to 7 (corresponding to rms wave heights from 0.1 to 9 m and wind speeds between 3 and 27.5 m/s) and five incidence angles [76]. For sea state 2 (mean wave height 0.25 m, crest heights between 0.1 m and 0.4 m), while values of α and α' were found to be more or less independent of incident angle (θ), the approximate Gaussian relationships

$$\delta = 3.75 - 7.5\exp\left(-\theta^2 / 0.008\right) \tag{5.82a}$$

$$\delta' = 3.5 - 4.5\exp\left(-\theta^2 / 0.002\right) \tag{5.82b}$$

were found to fit the angle dependence of δ and δ' (see Figure 5.18) indicating a stronger dependence on angle near grazing incidence than for smaller angles of incidence.

Predictions of long-range continuous tonal sound propagation over a rough sea allowing for complicated refraction conditions below a height of 1 km have been made using a Green's Function Parabolic Equation (GFPE) Model [72]. The GFPE allows for atmospheric turbulence as well as refraction but assumes flat terrain, so it was assumed that the rough sea surface can be represented by a locally reacting smooth finite impedance plane. To compare predictions with measurements of propagation over sea from 30 m high sound sources of 80 Hz, 200 Hz and 400 Hz signals on a lighthouse located 9 km from shore to a 1.7-m high receiver array on land approximately 10 km from the lighthouse, the effectively finite impedance of the rough sea was calculated by using the one-parameter Delany and Bazley semi-infinite model (see Equation (5.1a), (5.1b) and Table 5.1), but with effective flow resistivity values adjusted to fit the data at each frequency. Even though the effective impedance used in the calculations represents a relatively crude approximation, predictions were found to be in reasonable agreement with data [79].

To predict airborne sound propagation from wind farm piling operations, Equations (5.81a and 5.81b), with m and m' adjusted along with the other

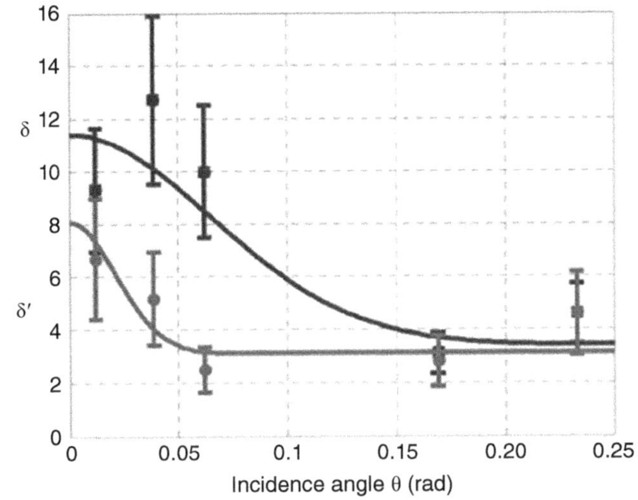

Figure 5.18 Angle dependence of coefficients in polynomial impedance models (equations (5.81a) and (5.81b)) based on fitting BEM EA simulations. The error bars indicate the uncertainty associated with the choice of surface profiles used to obtain the averages represented by the squares and circles. The continuous lines represent best fit Gaussian curves (equations (5.82a) and (5.82b)) [76]. Reproduced from P. M. Boulanger and K. Attenborough, Effective impedance spectra for rough sea effects on atmospheric impulsive sounds, *J. Acoust. Soc. Am.* 117(2) 751–762 (2005) with the permission of the Acoustical Society of America.

coefficients, have been used to represent the effective surface impedance [80]. The coefficient values required for this impedance model were obtained by matching predictions of EA spectra over short range with fixed source and receiver at 0.5 m above mean sea level and separated by 5 m, corresponding to a fixed grazing angle of 0.199 rad (11.42°), over a sea surface profile corresponding to sea states 2, 3 and 4 (see Figure 5.17(b) for sea state 2) using Equations (2.44), to predictions of a FDTD model (see Chapter 4). The sea surface wave profile was obtained by Fourier transforming a spectral density function which assumes that the waves are fully in equilibrium with the wind and that the trough to crest wave height of the highest third of the waves, H_S, is related to the wind speed at 10 m height by $H_S = 0.002466u_{10m}^2$. The maximum wave height was assumed to be $2H_S$. The assumed source and receiver locations are below the maximum wave height and a sufficiently large computational domain was included to avoid edge effects.

Figure 5.19 compares the predicted effective impedance spectra for sea state 2 according to three effective impedance models [76–80]. Solid and dash-dotted lines represent impedance spectra deduced from the methods based on fitting BEM and FDTD simulations of short-range EA spectra, respectively. Broken lines represent the uncertainty limits of the BEM-based method, taken to be ±25% for α and α' and ±50% for δ and δ' [76]. The

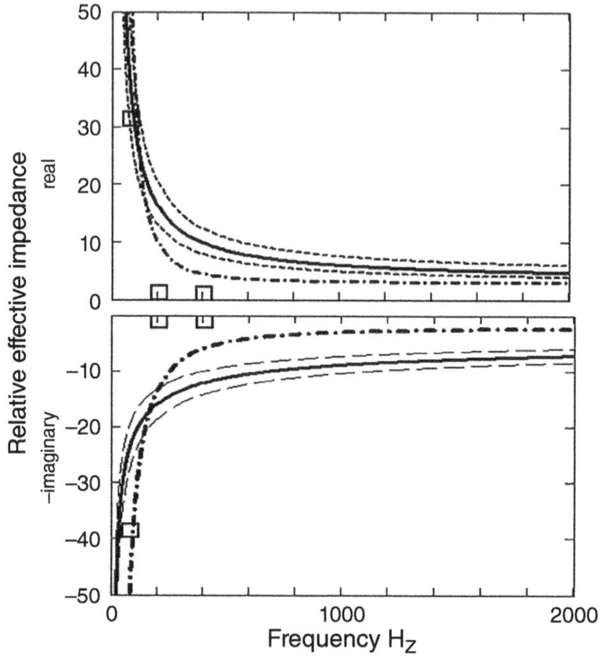

Figure 5.19 Effective impedance spectra for sea state 2 predicted according to three effective impedance models: solid and dash-dot lines represent impedance spectra deduced from the methods based on fitting EA spectra calculated by BEM [76] and FDTD [80] respectively. Broken lines represent the uncertainty limits of the BEM based method, taken to be ±25% for α and α' and ±50% for δ and δ' [76]. Open squares represent predictions of the Delany and Bazley model at 80, 200 and 400 Hz [79].

open squares represent predictions of the Delany and Bazley model at 80, 200 and 400 Hz with effective flow resistivity values of 2×10^5 kPa s m^{-2} at 80 Hz and 200 kPa s m^{-2} at 200 Hz and 400 Hz [79]. While the predicted values are in close agreement near 80 Hz, impedance values predicted by the Delany and Bazley model are significantly smaller at higher frequencies than those predicted by equations (5.81a, 5.81b) with parameter values deduced from fitting BEM and FDTD EA calculations. The BEM approach results in higher impedance values than the FDTD approach.

5.4.3.2 Predicted Propagation of Offshore Piling Noise

There is a trade-off between the increase in levels predicted during downwind sound propagation (relative to the levels predicted in a still atmosphere) and the extra attenuation due to sea surface roughness (relative to a flat surface). While downward refraction becomes stronger as the wind speed from source to receiver increases, at the same time the sea surface between source and receiver becomes rougher.

Impulsive sound propagation from wind turbine piling operations under downward-refraction conditions has been predicted [80]. A source power spectrum was measured, during pile driving (one of four steel pin piles) into the seabed to form a jacket foundation for supporting a 6.15 MW wind turbine. The corresponding equivalent source power level, obtained by integrating the acoustic energy, starting after a so-called 'soft start' procedure and ending when the top of the pile became submersed, is shown in Figure 5.20. An average source height (more precisely the hammer impact point) of 15 m was used in the simulations.

Numerical simulations have been made [80] with the GFPE method assuming that the rough sea can be represented by an equivalent impedance plane (see Section 5.4.3.1) but allowing also for atmospheric refraction.

In principle, a General Terrain Parabolic Equation (GTPE) model could be used for predicting long-range propagation in refraction conditions over rough seas (see e.g. [75]). But the GTPE method is restricted to local slopes of less than 30° [81] which makes it relatively unsuitable for detailed modelling of propagation over sea waves and justifies alternative effective impedance plane approaches. While it was found important to include turbulence effects in predictions from fixed single-frequency sources in predominantly upward-refraction conditions [79], it is less important to include turbulence when making predictions in downward-refraction conditions since,

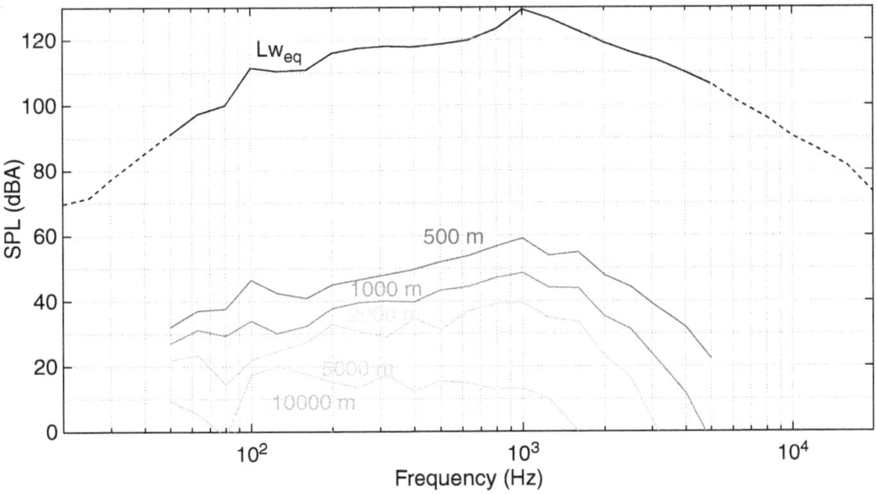

Figure 5.20 Predicted (equivalent) sound pressure level spectra (thinner lines) at different distances (receiver height 2 m) during piling as part of offshore wind farm construction, for sea state 3 (neutral marine boundary layer). The thick line represents the measured equivalent source power spectrum $L_{w,eq}$. Only the frequency range between 50 Hz and 5 kHz has been considered in the propagation simulations [80]. Reproduced from T. Van Renterghem, D. Botteldooren and L. Deckoninck, Airborne sound propagation over sea during offshore wind farm piling, *J. Acoust. Soc. Am. 135* 599–609 (2014) with the permission of the Acoustical Society of America.

typically, the refraction dominates over atmospheric scattering and, in any case, upward-refraction leads to much lower levels which are of less concern.

Predictions of the total A-weighted exposure levels at a receiver height of 2 m above sea level due to piling, up to 10 km distance, are shown in Figure 5.21. The sea surface undulations in combination with their corresponding wind profiles (downwind sound propagation) generally lead to lower sound pressure levels than in case of a still and thus perfectly flat sea surface.

Surface scattering is likely to be the dominant effect for long-distance sound propagation over sea, especially when the atmosphere is neutral or unstable. In a stable marine atmospheric boundary layer, assuming a Monin–Obukhov length (see Chapter 1) L_{MO} of 100 m, while the effects of sea surface scattering are still more important, they are somewhat less pronounced than in neutral or unstable conditions. Overall, sea states 2–4 give rather similar sound pressure level predictions.

Additional simulations [80], accounting only for wind refraction over a flat and rigid sea surface, thereby neglecting the rough sea surface scattering that would be typical for a given wind speed, have suggested an increase in level of between 8 and 10 dBA at 10 km relative to sea state 0 (no refraction and perfectly flat sea surface). If the sea surface scattering effect is considered separately, neglecting refraction of sound, although the differences between sea

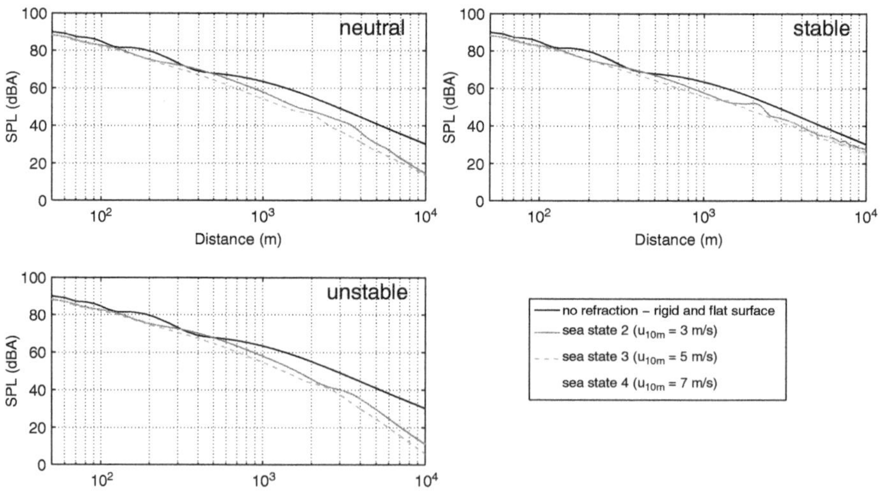

Figure 5.21 Predicted total (equivalent) airborne sound pressure level with distance from a piling location for wind farm construction during different sea states. The black line represents a reference prediction for no refraction and a perfectly flat, rigid sea surface (sea state 0) [80]. Reproduced from T. Van Renterghem, D. Botteldooren and L. Deckoninck, Airborne sound propagation over sea during offshore wind farm piling, *J. Acoust. Soc. Am. 135* 599–609 (2014) with the permission of the Acoustical Society of America.

states 2–4 are small within 2 km from the source, at larger distances, these differences increase rapidly. At 10 km, the sea surface representative for sea state 4 is predicted to lead to a decrease in level of 10 dBA relative to sea state 3, and to a decrease of more than 15 dBA relative to sea state 2.

A combined analysis of sea state and wind profile is needed to make realistic simulations of sound propagation over rough seas. The predicted sound pressure level spectra at selected distances in case of sea state 3, under neutral conditions, are depicted in Figure 5.20. These calculations show that the piling exposure levels are predominantly low frequency after long-distance propagation, as both atmospheric absorption and sea surface wave scattering attenuate the higher frequencies strongly.

Note that the simulations presented in this section all rely on the Monin–Obukhov similarity theory (see Chapter 1) to model the refractive state of the marine atmospheric boundary layer due to wind, neglect humidity profiles above the sea surface, neglect turbulent scattering and assume that the sound fields stay fully coherent during long-distance propagation. A constant standard environmental temperature lapse of –6.5°C/km was assumed in all simulations.

5.4.3.3 Predicted Rough Sea Effects on Sonic Booms

Increasing the mean sea-wave height is predicted to decrease the peak overpressure and increases the rounding of sonic boom wave profiles (see Figure 5.22), thereby indicating increased absorption. However, the most significant

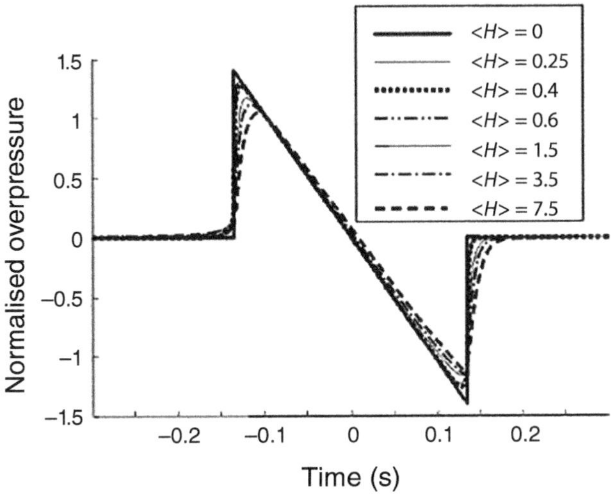

Figure 5.22 Sonic boom N-wave profiles predicted for various mean sea-wave heights up to 7.5 m [76]. Reproduced from P. M. Boulanger and K. Attenborough, Effective impedance spectra for rough sea effects on atmospheric impulsive sounds, *J. Acoust. Soc. Am.* 117(2) 751–762 (2005) with the permission of the Acoustical Society of America.

effect is on rise times which are predicted to be larger, during propagation over rough seas, than those induced by molecular relaxation or atmospheric turbulence. This suggests that sonic boom predictions in coastal areas should take account of sea surface roughness [76].

5.5 PREDICTING EFFECTS OF GROUND ELASTICITY

5.5.1 Coupling from Airborne Sound to Structures and Ground Vibration

Explosions and the shooting of heavy weapons create low-frequency impulsive sound waves which propagate over long distances and can create a major environmental problem for military training fields. Such impulsive sound sources tend to disturb neighbours more through the vibration and rattling induced in buildings, than by the audible airborne sound itself [82,83]. Human perception of whole body vibration includes frequencies down to 1 Hz [84] and the fundamental natural frequencies of buildings are in the range 1–10 Hz. Planning tools to predict and control such activities must therefore be able to handle sound propagation down to these low frequencies.

Despite their valuable role in many sound propagation predictions, impedance models that assume rigid-porous ground have the intrinsic shortcoming that they cannot account for air-to-ground coupling, i.e. the ability of atmospheric sound waves to cause seismic waves in the ground.

Air-ground coupling has been of considerable interest in geophysics, both because of the measured 'ground-roll' caused by intense acoustic sources and the possible use of air sources in ground layering studies. Theoretical analyses have been carried out for spherical wave incidence on a ground consisting either of a fluid or solid layer above a fluid or solid half-space [85]. However, to describe the phenomenon of acoustic-to-seismic coupling accurately it has been found that the ground must be modelled as an elastic *porous* material [86–87].

The resulting theory is relevant not only to predicting the ground vibration induced by low-frequency acoustic sources but also, as discussed later, when predicting the propagation of low-frequency sound above the ground.

The classical theory for a porous and elastic medium predicts the existence of three wave types in the porous medium: two dilatational waves and one rotational wave. In a material consisting of a dense solid frame with a low-density fluid saturating the pores, the first kind of dilatational wave, which often is called the 'fast' wave, has a velocity very similar to the dilatational wave (or geophysical 'P' wave) travelling in the drained frame. The attenuation of the first dilatational wave type is, however, higher than that of the P wave in the drained frame. The extra attenuation comes from the viscous forces in the pore fluid acting on the pore walls. This first dilatational wave has negligible dispersion and its attenuation is proportional to the square of the frequency, as is the case with the rotational wave. The viscous coupling leads to some of the energy in this propagating wave being carried in the pore fluid as the second type of dilatational wave.

In air-saturated soils, the second dilatational wave, which is often called the 'slow' wave, has a much lower velocity than the first. The attenuation of the 'slow' wave stems from viscous forces acting on the pore walls and from thermal exchange with the pore walls. The rigid-frame limit of dynamic poroelasticity theory predicts a single wave which is very similar to that predicted in the pores of a rigid-porous solid [88]. The 'slow' wave is the only wave type considered in the previous discussions of ground effect which have assumed that porous ground is rigid-framed. When the 'slow' wave is excited, most of the energy in this wave type is carried in the pore fluid. However, the viscous coupling at the pore walls leads to some propagation within the solid frame thereby creating the 'fast' wave. The attenuation for the 'slow' wave is higher than that of the 'fast' wave in most soil-like materials. At low frequencies, the real and imaginary parts of the 'slow' wave propagation constant are nearly equal, so the slow wave is more like a diffusion process than a wave and is analogous to heat conduction.

The rotational wave has a very similar velocity to the rotational wave carried in the drained frame (or the 'S' wave of geophysics). Again there is some extra attenuation due to the viscous forces associated with differential movement between solid and pore fluid. The fluid is unable to support rotational waves, but is driven by the solid.

5.5.2 Biot-Stoll Theory

A theory for dynamic poroelasticity was summarized in Chapter 4 in the context of time-domain calculations. The propagation of each wave type is determined by many parameters relating to the elastic properties of the solid and fluid components. Considerable efforts have been made to identify these parameters and determine appropriate forms for specific materials.

The most widely implemented frequency-domain formulation is described here. The coupled equations governing the propagation of dilatational waves can be written as [89–87]

$$\nabla^2 (He - C\xi) = \frac{\partial^2}{\partial t^2}(\rho e - \rho_f \xi) \tag{5.83a}$$

$$\nabla^2 (Ce - M\xi) = \frac{\partial^2}{\partial t^2}(\rho_f e - m\xi) - \frac{\eta}{\kappa}\frac{\partial \xi}{\partial t} F(\lambda), \tag{5.83b}$$

where e = ∇ .u is the dilatation or volumetric strain vector of the skeletal frame; $\xi = \Omega\, \nabla \cdot (u - U)$ is the relative dilatation vector between the frame and the fluid; u is the vector displacement of the frame, U is the vector displacement of the fluid, $F(\lambda)$ is the viscosity correction function, $\rho = (1 - \Omega)\rho_s + \Omega\rho_f$ is the bulk density of the medium, ρ_f is the fluid density, ρ_s is the solid density, η is dynamic fluid viscosity, Ω is the porosity and κ is the permeability.

The second term on the RHS of Equation (5.63b), $\frac{\eta}{\kappa}\frac{\partial \xi}{\partial t}F(\lambda)$, allows for the damping through viscous drag as the fluid and matrix move relative to one another. $m = \frac{T\rho_f}{\Omega}$, where T, the tortuosity, is a dimensionless parameter accounting for the fact that not all of the fluid flows along the axis of pores and in the direction of macroscopic pressure gradient, and Ω is the porosity.

H, C and M are elastic constants that can be expressed in terms of the bulk moduli K_s, K_f and K_b of the grains, fluid and frame, respectively, and the shear modulus, μ, of the frame.

Assuming that e and ξ vary as $e^{-i\omega t}$, $\partial/\partial t$ can be replaced by $-i\omega$ and Equation (5.63b) can be written as

$$\nabla^2 (Ce - M\xi) = -\omega^2 (\rho_f e - \rho(\omega)\xi), \tag{5.83c}$$

where $\rho(\omega) = m - \frac{i\eta}{\omega k}F(\lambda)$ is the dynamic fluid density.

The original formulation of $F(\lambda)$ (and hence of $\rho(\omega)$) was a generalization from the specific forms corresponding to slit-like and cylindrical forms but assumed identical pores [18]. As discussed in Section 5.2, expressions for $F(\lambda)$ are available also for triangular and rectangular pore shapes and for more arbitrary microstructures.

If plane waves of the form $e = Aexp(i(k_i x - \omega t)$ and $\xi = Bexp(i(k_i x - \omega t)$ are assumed, where k_i are wave numbers and i is an integer corresponding to each wave type, then the dispersion equations for the propagation constants of the dilatational waves may be derived. These are as follows:

$$A(k_i^2 H - \omega^2 \rho) + B(\omega^2 \rho_f - k_i^2 C) = 0 \tag{5.84a}$$

and

$$A(k_i^2 C - \omega^2 \rho_f) + B(\omega^2 \rho(\omega) - k_i^2 M) = 0. \tag{5.84b}$$

A non-trivial solution of Equations (5.82a, 5.82b) exists only if the determinant of the coefficient vanishes, i.e.

$$\begin{vmatrix} k_i^2 H - \omega^2 \rho & \omega^2 \rho_f - k_i^2 C \\ k_i^2 C - \omega^2 \rho_f & \omega^2 \rho(\omega) - k_i^2 M \end{vmatrix} = 0 \tag{5.85a}$$

or, introducing phase velocities $v_i = \omega/k_i$,

$$(\rho_f^2 - \rho\rho(\omega))v_i^4 + (H\rho(\omega) + \rho M - 2\rho_f C)v_i^2 + (C^2 - HM) = 0. \tag{5.85b}$$

This quartic dispersion equation has two complex roots from which both the attenuation and phase velocities of two corresponding dilatational waves are calculated.

The three (complex) moduli H, C and M are determined from the solid and fluid bulk moduli of the constituent parts [90].

$$H = \left[\frac{(K_s - K_b)^2}{D - K_b} + K_b + {}^4\!/_3\,\mu\right], C = k_s\frac{(k_s - k_b)}{D - k_b}, M = \frac{k_s^2}{D - k_b},$$

$$D = K_s\left[1 + \Omega\left(\frac{K_s}{K_f}\right) - 1\right], \tag{5.86a, b, c, d}$$

where K_s is the bulk modulus of the solid grains, K_b is the bulk modulus of the drained solid matrix and K_f is the bulk modulus of the pore fluid.

If either $K_b = 0$ (pure fluid) or $K_f = 0$ (elastic limit) then Equation (5.83b) has only one solution and the corresponding dispersion equation for a fluid or for a non-porous elastic medium is retrieved. In general, the fluid bulk modulus is complex and frequency dependent as a result of the thermal drag experienced by the fluid in the pores.

While the frequency-dependent thermal effect is not particularly important when pores are water-filled, it is significant in air-filled pores.

If the solid displacement vector (**u**) and relative fluid displacement vector (**w**) are expressed in terms of scalar potentials φ_s, φ_f and vector potentials χ_1 and χ_2 as

$$\boldsymbol{u} = \nabla\varphi_s + \nabla\times\chi_1, \tag{5.87a}$$

$$\boldsymbol{w} = \Omega(\boldsymbol{u} - U) = \nabla\varphi_f + m_3\nabla\times\chi_2, \tag{5.87b}$$

the equations of motion for the rotational motion are

$$\mu\nabla^2\chi_1 = -\omega^2(\rho\chi_1 - \rho_f\chi_2) \tag{5.88a}$$

$$0 = (\rho\chi_1 - \rho_f\chi_2) - \frac{i\eta}{\kappa\omega}F(\lambda)\chi_2, \tag{5.88b}$$

where μ is the rigidity modulus and m_3 is the ratio between fluid and solid rotational motion.

Equations (5.86a, b, c, d) lead to an expression for the velocity of shear waves in a poroelastic medium

$$v_S^2 = \frac{\mu}{\rho - \rho_f^2 / \rho(\omega)}. \tag{5.88c}$$

which will be close to the shear wave speed in the (non-porous) elastic solid if the second term in the denominator of the right-hand side of Equation (5.88c) is much smaller than the first.

At the interface between different porous elastic media, six boundary conditions may be applied. These are as follows:

1) Continuity of total normal stress
2) Continuity of normal displacement
3) Continuity of fluid pressure
4) Continuity of tangential stress
5) Continuity of normal fluid displacement
6) Continuity of tangential frame displacement

At an interface between a fluid and the poroelastic layer (such as the surface of the ground) the first four boundary conditions apply. The resulting equations and those for a greater number of porous and elastic layers are solved numerically [22,89,91].

5.5.3 Numerical Calculations of Acoustic–Seismic Coupling

5.5.3.1 *Fast Field Program for Layered Air-Ground Systems (FFLAGS)*

In the FFLAGS [87,91], each of two inhomogeneous media in contact (e.g. a fluid above a poroelastic ground) is considered to consist of vertically stratified homogeneous layers. The system is assumed to be bounded from above by homogeneous fluid half-space and from below by a homogeneous solid half-space. The wave equation in each layer, assuming a time dependence of exp(-$i\omega t$) and a cylindrical system of co-ordinates (r, z), is given by

$$\nabla^2\Psi_i(r,z)+k_i^2\Psi_i(r,z)=\delta_i(r,z), \tag{5.89}$$

where Ψ_i are the scalar displacement potentials for various wave types propagating in the medium, $k_i = \omega/c_i$ are the wave numbers in layer i and, since it is possible to consider multiple sources, δ_i represent source terms. One compressional wave propagates in the fluid. Two compressional waves and one shear wave propagate in each porous elastic ground layer. The wave numbers for the waves in the ground layers are determined from the dispersion equations (Equations (5.85a) and (5.85b)) for the dilatational waves). The subscripts i are used to identify the wave types: i = 0 denotes the fluid wave; in each poroelastic layer i =1, 2 denote the two compressional waves and i=3 the shear wave.

Noting that there is radial symmetry, to separate the radial and vertical variables in the equation and, thus, to reduce this partial differential equation to an ordinary one, a pair of Hankel transform integrals are used to represent the potentials:

$$\Psi_i(r,z) = \int_0^\infty \psi_i(z,k_r) J_0(k_r r) k_r dk_r \text{ and } \psi_i(z,k_r) = \int_0^\infty \Psi_i(r,z) J_0(k_r r) r dr, \tag{5.90a, b}$$

where $J_0(z)$ is the zero-order Bessel function and k_r, the variables of integration, can be thought of as the horizontal and radial components of the wave number, respectively. From (5.87b) we obtain the transformed Helmholtz Equation:

$$\frac{\partial^2}{\partial z^2}\psi_i + \beta_i^2 \psi_i = S_i \delta(z), \tag{5.91a}$$

where the right-hand side represents the source term and

$$\beta_i^2 = k_i^2 - k_r^2 \tag{5.91b}$$

In this way, the problem of determining the wave amplitudes is reduced to one of solving a set of ordinary differential equations.

The boundary condition equations (BCE) are put in the form of a Global Matrix equation

$$\mathrm{A} \cdot \mathrm{X} = \mathrm{B}, \tag{5.92}$$

where **X** is a vector containing the wave amplitudes (A_i in the porous medium and R_i in the fluid), **A** is an $N \times N$ matrix containing the coefficients from the BCE's and **B** is the source term vector. The order of the matrix, N, is related to the number of fluid layers, n_f, and the number of solid layers, n_s, both including the half-space, by

$$N = 6(n_s - 1) + 2n_f + 2. \tag{5.93}$$

The resulting matrix Equation (5.90a, b) can be solved by a variety of methods including Gaussian elimination with pivoting. Subsequently, the forward Hankel transform is applied to obtain the full wave solutions. Once the Green's functions (the range-independent ψ_i) are known as a function of k_i, the transform can be replaced by a Fast Fourier Transform. This may be calculated in the far field ($k_r r >> 1$ which restricts the model to ranges greater than a couple of wavelengths from the source) by substituting a large argument approximation for the Bessel function. The integral can be evaluated very quickly and efficiently using discrete Fourier transforms.

The wave potential in the first fluid layer may be expressed as

$$\psi_0 = R^{\uparrow} e^{j(z-h_1)\beta_0} + R^{\downarrow} e^{-j(z-h_2)\beta_0}, \tag{5.94a}$$

where, h_1 and h_2 denote the vertical co-ordinates of the lower and upper fluid layer boundaries ($h_2 > h_1$) and z is positive moving away from the fluid–solid interface.

Similarly for each poroelastic layer, there are three potentials (each potential consists of upgoing and downgoing terms) given by

$$\psi_1 = A_1^{\downarrow} e^{j(z-d_1)\beta_1} + A_1^{\uparrow} e^{-j(z-d_2)\beta_1} + A_2^{\downarrow} e^{j(z-d_1)\beta_2} + A_2^{\uparrow} e^{-j(z-d_2)\beta_2} \tag{5.94b}$$

$$\psi_2 = m_1\left[A_1^{\downarrow} e^{j(z-d_1)\beta_1} + A_1^{\uparrow} e^{-j(z-d_2)\beta_1}\right] + m_2\left[A_2^{\downarrow} e^{j(z-d_1)\beta_2} + A_2^{\uparrow} e^{-j(z-d_2)\beta_2}\right] \tag{5.94c}$$

$$\psi_3 = A_3^{\downarrow} e^{j(z-d_1)\beta_3} + A_3^{\uparrow} e^{-j(z-d_2)\beta_3}, \tag{5.94d}$$

where, d_1 and d_2 denote the upper and lower solid boundaries ($|d_2| > |d_1|$), z is positive downwards from the interface, $R^{\updownarrow}$ and $A_n^{\updownarrow}$ ($n = 1,2,3, \beta = \downarrow, \uparrow$) are the amplitudes to be determined from the BCE and m_n are the appropriate ratios of solid-borne wave to pore-borne wave since the two compressional wave types can exist simultaneously in solid and pore fluid phases and the potentials are a linear superposition of the two wave solutions.

The boundary conditions listed in Section 5.4.2 involve solid and fluid displacements and stresses. The fluid displacement and pressure are given, respectively, by

$$\nabla\Psi_0 = \left(\frac{\partial \Psi_0}{\partial r}, \frac{\partial \Psi_0}{\partial z}\right), p = \rho\omega^2\Psi_0. \tag{5.95a, b}$$

Using

$$u = \nabla\Psi_1 + \nabla \times \chi_1, \tag{5.95c}$$

the radial and vertical components of solid displacement are given by

$$u_r = \frac{\partial \Psi_1}{\partial r} + \frac{\partial^2 \Psi_3}{\partial r \partial z}, u_z = \frac{\partial \Psi_1}{\partial z} - \frac{1}{r}\frac{\partial}{\partial r}\left(r\frac{\partial \Psi_3}{\partial r}\right). \tag{5.95d}$$

However, it is more convenient to use the relative fluid motion and its components defined by

$$w = \nabla\Psi_2 + \nabla \times \tilde{\chi}_2, w_r = \frac{\partial \Psi_2}{\partial r} + m_3\frac{\partial^2 \Psi_3}{\partial r \partial z}, w_z = \frac{\partial \Psi_2}{\partial z} - m_3\frac{1}{r}\frac{\partial}{\partial r}\left(r\frac{\partial \Psi_3}{\partial r}\right), \tag{5.96a, b, c}$$

where $\chi_{1,2}$ are vector potentials representing the transverse motion (see Equations (5.78), (5.91a) and (5.91b)), and m_3 is the ratio between fluid and solid rotational motion.

The wave amplitude F at a point with cylindrical co-ordinates (r_m, z) due to a unit source at $(0, z_0)$ for a given angular frequency ω is derived from Hankel transform of the one-dimensional solution of the wave equation, Γ:

$$F(r_m, z) = \frac{\Delta k\sqrt{N}}{2\pi\sqrt{m}}\left(e^{-i\pi/4}\sum_{n=0}^{N-1}\frac{\Gamma(k_n, z)}{\sqrt{n}}e^{2i\pi mn/N} + e^{i\pi/4}\sum_{n=0}^{N-1}\frac{\Gamma(k_n, z)}{\sqrt{n}}e^{2i\pi mn/N}\right), \tag{5.97}$$

where N is the n index upper bound for the finite summation evaluating the Hankel's transform, m is the index for the range discretization, $r_m = m\Delta r = \dfrac{2\pi m}{N\Delta k}$ is the range, $\Delta k = \dfrac{k_{\max}}{N-1}$ is the horizontal wave number increment and $k_n = n\Delta k$. $\Gamma(k_n, z)$ is the product of k_n and the Green's function for the problem which is detailed elsewhere [87]. Two Fast Fourier Transforms and correction factors are used to evaluate the sums in Equation (5.80) which is based on a large argument approximation of Bessel function and the replacement of the integration by a finite sum over index n in the Hankel transform. The corrections are needed to allow for the truncation of the infinite integral associated with the Hankel transforms and for the presence of poles on the real axis. The integration contour is displaced by $\varepsilon\Delta k$ and a function $A(1 - exp(-\eta k_N))$ is subtracted from the integrand in the Hankel transform, where A can be expressed in terms of N, ε, η and the integrand. Typical values of the correction parameters ε and η used for the calculations are 1.1 and $3/k_{max}$ where k_{max} is the upper limit of the integration.

To allow a time-dependent acoustic source pulse, $f_S(t)$, as an input to the continuous wave model described in Section 5.5.3.1, the source pulse is Fourier transformed in the time domain to obtain a pulse spectrum

$$F_s(\omega) = \frac{1}{\sqrt{2\pi}} \int_{t=0}^{\infty} e^{i\omega t} f_S(t)\,dt. \tag{5.98a}$$

The predicted pulse spectrum is evaluated using equation (5.90a, b) and the predicted time-domain pulse is obtained from the inverse Fourier transform of the pulse spectrum as

$$\Psi_R(t) = \frac{1}{\sqrt{2\pi}} \int_{\omega=0}^{\infty} e^{-i\omega t} F_s(\omega) F(r_m, z, \omega)\,d\omega. \tag{5.98b}$$

5.5.3.2 Example Predictions of Low-Frequency Effects

The spectrum of the ratio between the normal component of the soil particle velocity at the surface of the ground and the incident sound pressure, the acoustic-to-seismic coupling ratio or transfer function, is influenced strongly by discontinuities in the elastic wave properties within the ground. At frequencies corresponding to peaks in the transfer function, there are local maxima in the transfer of sound energy into the soil [86]. These are associated with constructive interference between down- and up-going predominantly frame-borne waves within each soil layer. Consequently, there is a relationship between near-surface layering in soil and the peaks or 'layer resonances' that appear in the measured acoustic-to-seismic transfer function spectrum: the lower the frequency of the peak in the spectrum, the deeper the associated layer.

A numerical theory (FFLAGS) may be used to predict the ground impedance at low frequencies [92]. In Figure 5.23, the predictions for the surface

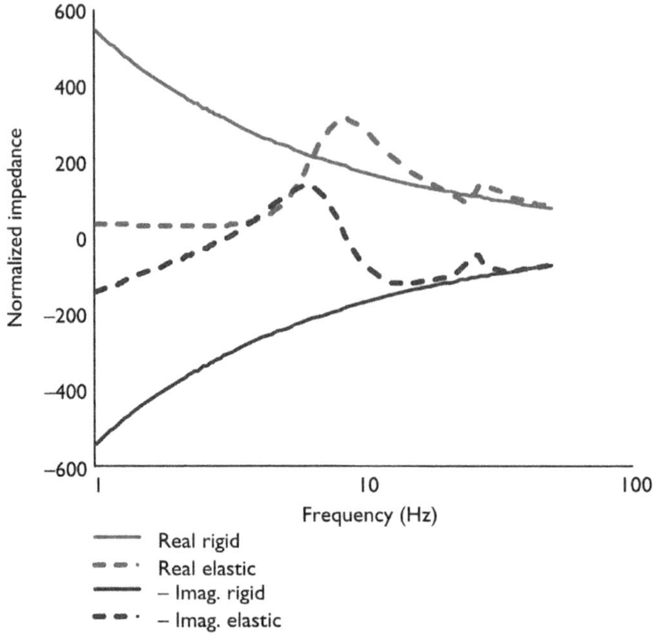

Figure 5.23 Predicted surface impedance at a grazing angle of 0.018° for poro-elastic and rigid porous ground (4-layer system, Table 5.8)

Table 5.9 Ground profile and parameters used in the calculations for Figure 5.23

Layer	*Flow resistivity (kPa s m⁻²)*	*Porosity*	*Thickness (m)*	*P-wave speed (m s⁻¹)*	*S-wave speed (m s⁻¹)*	*Damping*
1	1740	0.3	0.5	560	230	0.04
2	1740	0.3	1.0	220	98	0.02
3	1,740,000	0.01	150	1500	850	0.001
4	1,740,000	0.01	150	1500	354	0.001
5	1,740,000	0.01	Halfspace	1500	450	0.001

impedance at a grazing angle of 0.018° are shown as a function of frequency for the layered porous and elastic system described by Table 5.9 and compared with those for a rigid-porous ground with the same surface flow resistivity and porosity.

The influence of ground elasticity is to reduce the magnitude of the impedance considerably below 50 Hz. Potentially this is very significant for predictions of low-frequency airborne noise, for example, blast noise, at long range.

Figure 5.24 shows that the surface impedance of this four-layer poroelastic system varies between grazing angles of 5.7° and 0.57° but remains more or less constant for smaller grazing angles. The predictions show two resonances. The lowest frequency resonance is the most angle dependent. The

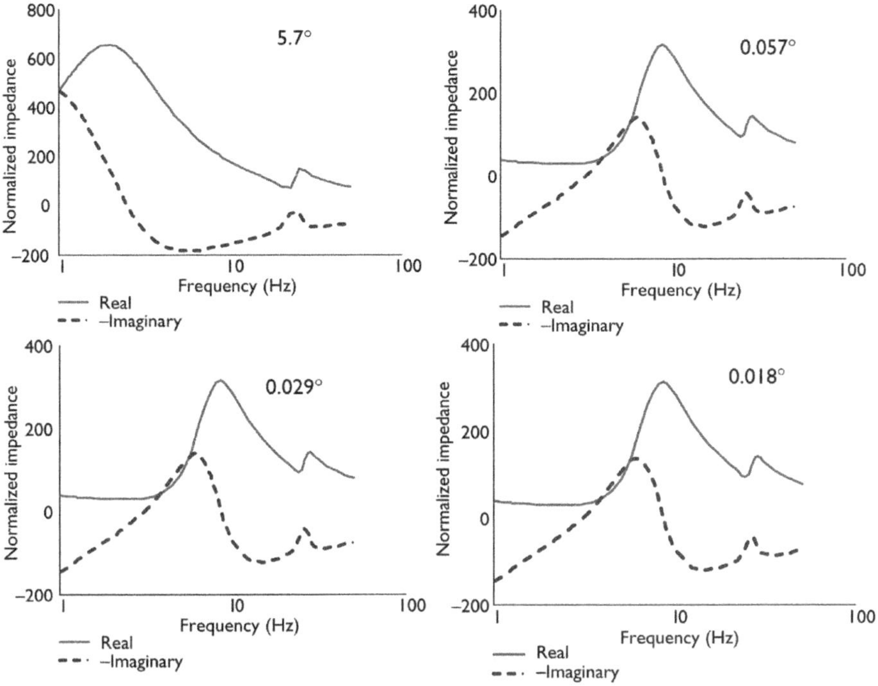

Figure 5.24 Normalised surface impedance predicted for the 4-layer structure, sound speed in air = 329 m/s (corresponding to an air temperature of -4°C) for grazing angles between 0.018 and 5.7 degrees

peak in the real part changes from 2 Hz at 5.7 degrees to 8 Hz at 0.057 deg. On the other hand, the higher frequency resonance peak near 25 Hz remains relatively unchanged with range.

The peak at the lower frequency may be associated with the predicted coincidence between the Rayleigh wave speed in the ground and the sound speed in air (Figure 5.25). Compared with the near-pressure doubling predicted by classical ground impedance models, the predicted reduction of ground impedance at low frequencies above layered elastic ground can be interpreted as the result of coupling of a significant fraction of the incident sound energy into ground-borne Rayleigh waves.

Numerical calculations have explored the consequences of this coupling for the EA of low-frequency sound above ground [92]. Figure 5.26 shows the EA predictions for 6.3, 7.2 and 12.6 km ranges. There are significant 'dips' in the predicted EA spectrum. According to the convention used here, these represent maxima in the EA. The depths of the dips increase with range. The depth of the dip predicted at 3.5 Hz is 9 dB and shows a small frequency-dependence with changing range. The enhanced (i.e. more than + 6dB) EA spectrum immediately before the 3.5 Hz dip in the prediction for 12.6 km range is symptomatic of a surface wave.

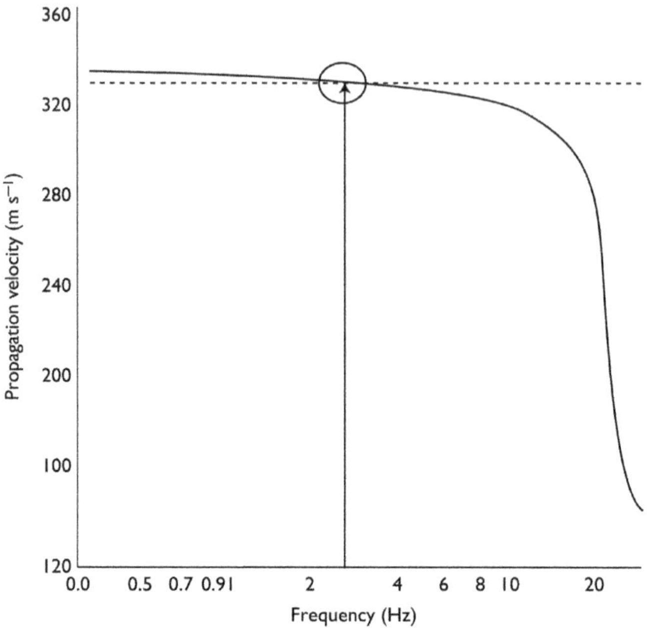

Figure 5.25 Rayleigh wave dispersion curve predicted for the system described by Table 5.8. Reprinted with permission from Elsevier.

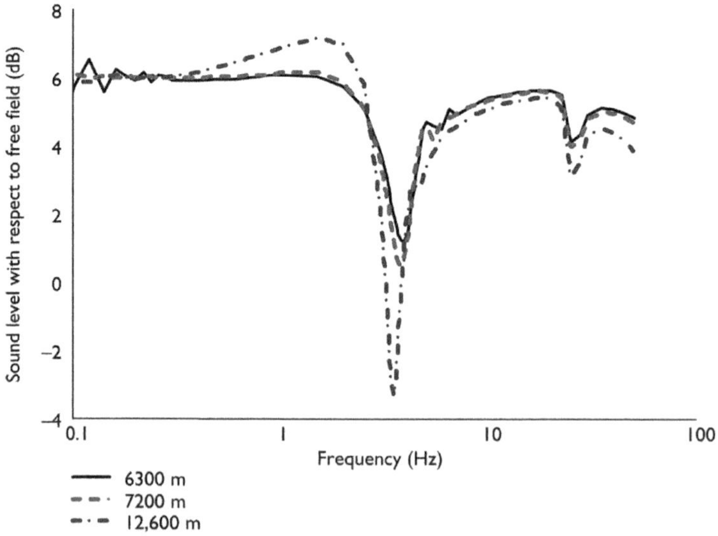

Figure 5.26 Excess attenuation spectra predicted for source and receiver heights of 2 m and 0.1 m and three ranges assuming a sound speed in air of 329 m/s. Reprinted with permission from Elsevier.

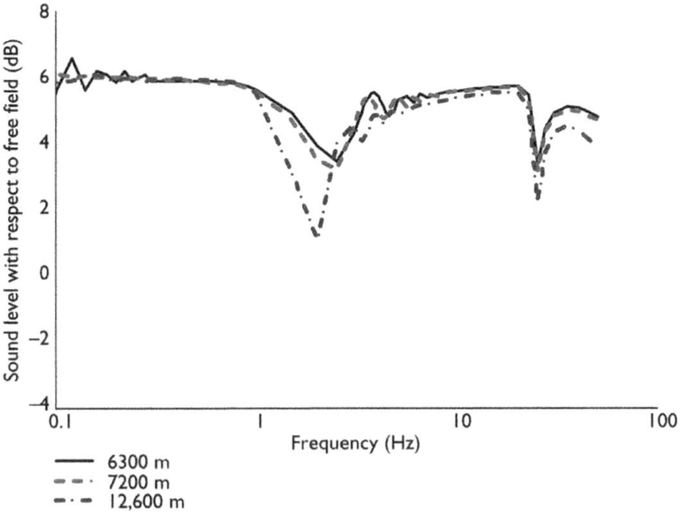

Figure 5.27 Excess attenuation spectra below 50 Hz predicted at ranges of 6300 m, 7200 m and 12600 m, source height 2.0 m, receiver height 1 m, with an assumed sound speed in air of 332 m/s.

However, as with the acoustic–seismic impedance predictions, the EA spectra predicted by FFLAGS for the four-layer poroelastic structure are rather dependent on the assumed sound speed in air. Figure 5.27 shows corresponding predictions with an assumed sound speed of 332 m/s. The lowest frequency dip is predicted to move to 2 Hz from 3.5 Hz and to become shallower. The dip between 20 and 30 Hz remains at the same frequency but becomes more pronounced.

It is important to consider whether it is possible to model the surface impedance of the ground to include elasticity effects in a way that is more accurate than classical rigid-porous impedance model predictions at low frequencies. Figure 5.28 shows the EA spectra predicted for source height 2 m, receiver height 0.1 m and horizontal range of 6.3 km over a layered ground profile corresponding to Table 5.8 (assumed sound speed in air of 329 m/s) and by classical theory for a point source above an impedance (locally reacting) plane (Equation (2.40)) using the impedance calculated for 0.018 degrees grazing angle (Zeff, broken lines in Figure 5.28).

The predictions show a significant extra attenuation between 2 and 10 Hz. Except for an enhancement near 2 Hz, the predictions also indicate that, for an assumed sound speed in air of 329 m/s, the EA spectrum might be predicted tolerably well by using modified classical theory instead of a full poroelastic layer calculation. On the other hand, Figure 5.29 shows the same information as Figure 5.28, but for an assumed sound speed of 332 m/s. In this case, the low-frequency dip in the EA spectrum is not predicted as well by assuming an effective impedance.

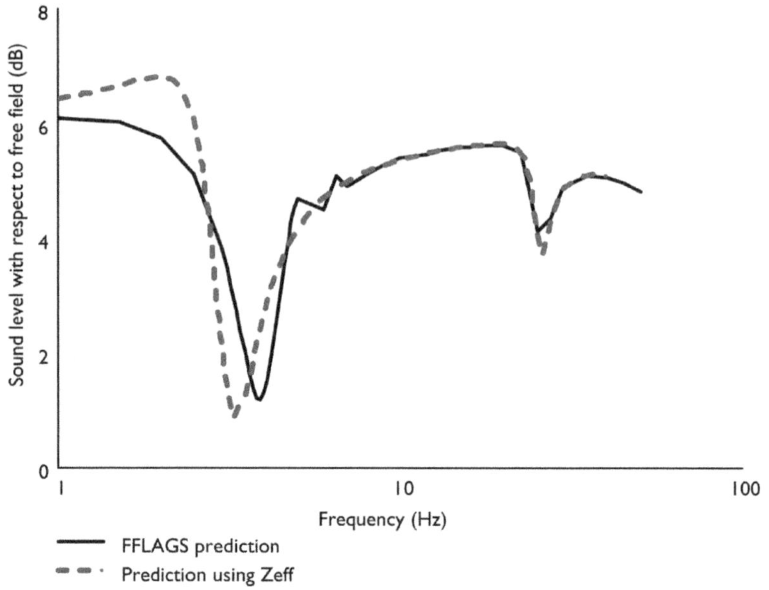

Figure 5.28 Excess attenuation spectra predicted for source height 2 m, receiver height 0.1 m and horizontal range of 6.3 km by FFLAGS (assumed sound speed in air of 329 m/s) and by classical theory using impedance calculated for 0.018 degrees grazing angle. Reprinted with permission from Elsevier.

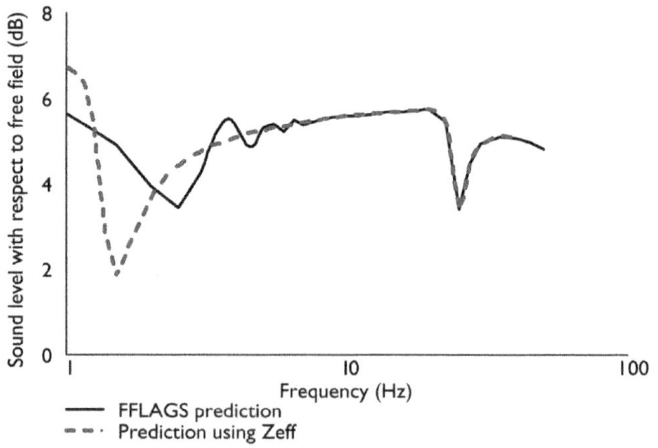

Figure 5.29 As for Figure 5.28 but with an assumed sound speed in air of 332 m/s.

Since it is difficult to measure the surface impedance of the ground at low frequencies, predictions of ground elasticity effects have been validated only by data for acoustic-to-seismic coupling, i.e. by measurements of the ratio of ground surface particle velocity relative to the incident sound pressure. Comparisons between predictions and data for acoustic-to-seismic coupling are discussed in Chapter 6.

REFERENCES

[1] ISO 9613-2:1996 Acoustics – Attenuation of sound propagation outdoors – Part 2: General method of Calculation, International Standards Organisation (1996).
[2] M. E. Delany and E. N. Bazley, Acoustical properties of fibrous absorbent materials, *Appl. Acoust.* **3**, 105–116 (1970).
[3] M. E. Delany and E. N. Bazley, Acoustical properties of fibrous absorbent materials, NPL Aero Report AC71 (1971).
[4] F. P. Mechel, Ausweitung der Absorberformel von Delany and Bazley zu teifel Frequenzen, *Acustica*, **35** 210–213 (1976).
[5] I. P. Dunn and W. A. Davern, Calculation of acoustic impedance of multilayer absorbers, *Appl. Acoust.*, **19** 321–334 (1986).
[6] Y. Miki, Acoustical properties of porous materials - Modifications of Delany and Bazley models, *J. Acoust. Soc. Japan* **11** 19–24 (1990).
[7] Y. Miki, Acoustical properties of porous materials generalizations of empirical models, *J. Acoust. Soc. Japan (E)* **11** 25–28 (1990).
[8] T. Komatsu, Improvement of the Delany–Bazley and Miki models for fibrous sound absorbing materials, *Acoust. Sci. Technol.* **29** 121–129 (2008).
[9] D. Dragna and P. Blanc-Benon, Physically Admissible Impedance Models for Time-Domain Computations of Outdoor Sound Propagation, *Acta Acust. united Ac.* **100** 401–410 (2014).
[10] K. B. Rasmussen, Sound propagation over grass covered ground, *J. Sound Vib.* **78**(2) 247–255 (1981).
[11] K. Attenborough, I. Bashir and S. Taherzadeh, Outdoor ground impedance models, *J. Acoust. Soc. Am.* **129**(5) 2806–2819 (2011).
[12] C. Zwikker and C. Kosten, *Sound absorbing materials*, Elsevier, New York (1949).
[13] S. I. Thomasson, Sound propagation above a layer with a large refractive index, *J. Acoust. Soc. Am.* **61** 659–674 (1977).
[14] J. F. Hamet and M. Bérengier, Acoustical characteristics of porous pavements: A new phenomenological model. Internoise'93, Leuven, Belgium, (1993). See also M. Bérengier, M. Stinson, G. Daigle and J.F. Hamet, Porous road pavements: acoustical characterization and propagation effects. *J. Acoust. Soc. Am.* **101** 155–162 (1997).
[15] M. R. Stinson, The propagation of plane sound waves in narrow and wide circular tubes and generalisation to uniform tubes of arbitrary cross-sectional shape, *J.Acoust. Soc. Am.* **89**(1) 550–558 (1991).
[16] K. Attenborough, On the acoustic slow wave in rigid framed porous media, *J. Acoust. Soc. Am.* **81** 93–102 (1987).
[17] M. A. Biot, Theory of elastic waves in a fluid-saturated porous solid. II High-frequency range, *J. Acoust. Soc. Am.* **28** 168–178 (1956).
[18] P. C. Carman, *Flow of gases through porous media*, Butterworths, London (1953).
[19] Y. Champoux and M. R. Stinson, On acoustical models for sound propagation in rigid frame porous materials and the influence of shape factors, *J. Acoust. Soc. Am.* **92**(2) 1120–1131 (1992).
[20] K. Attenborough, Models for the acoustical properties of air-saturated granular media, *Acta Acust. united Ac.* **1**, 213–226 (1993).
[21] D. L. Johnson, J. Koplik and R. Dashen, Theory of dynamic permeability and tortuosity in fluid-saturated porous materials, *J. Fluid Mech.* **176** 379–402 (1987).

[22] J. F. Allard, *Propagation of sound in porous media*, Elsevier Applied Science, London (1993).
[23] P. Allemon and R. Hazelbrouck, *Influence of pore size distribution on sound absorption of rubber granulates, Proc. InterNoise 96, publ.* IOA, Book **2** 927–930 (1996).
[24] O. Umnova, K. Attenborough and K. M. Li, Cell model calculations of dynamic drag parameters in packings of spheres. *J. Acoust. Soc. Am.* **107**(6) 3113–3119 (2000).
[25] P. Leclaire, M. J. Swift and K. V. Horoshenkov, Determining the specific area of porous acoustic materials from water extraction data, *J. Appl. Phys.* **84** 6886–6890 (1998).
[26] P. Leclaire, L. Kelders, W. Lauriks, C. Glorieux and J. Thoen, Determination of the viscous characteristic length in air-filled porous materials by ultrasonic attenuation measurements, *J. Acoust. Soc. Am.* **99**(4) 1944–1948 (1996).
[27] O. Umnova, K. Attenborough and K. M. Li, A cell model for the Acoustical properties of Packings of Spheres, *Acta Acust. united Ac.* **87** 226–235 (2001).
[28] Y. Champoux and J.-F. Allard, Dynamic Tortuosity and bulk modulus in air saturated porous media, *J. Appl. Phys.* **70** 1975–1979 (1991).
[29] D. Lafarge, P. Lemariner, J.-F. Allard and V. Tarnow, Dynamic compressibility of air in porous structures at audible frequencies, *J. Acoust. Soc. Am.* **102** 1995–2006 (1997).
[30] J. G. Berryman, Confirmation of Biot's theory, *Appl. Phys. Lett.* **37** 382–384 (1980).
[31] M. J. Swift, The physical properties of porous recycled materials, Ph.D. Thesis, University of Bradford, UK (2000).
[32] K. V. Horoshenkov and M. J. Swift, The acoustic properties of granular materials with pore size distribution close to log-normal, *J. Acoust. Soc. Am.* **110** 2371–2378 (2001).
[33] K. V. Horoshenkov, K. Attenborough and S. N. Chandler-Wilde, Pade approximants for the acoustical properties of rigid frame porous media with pore size distribution, *J. Acoust. Soc. Am.* **104** 1198–1209 (1998).
[34] T. Yamamoto and A. Turgut, Acoustic wave propagation through porous media with arbitrary pore size distributions, *J. Acoust. Soc. Am.* **83**(5) 1744–1751 (1988).
[35] K. Horoshenkov, J.-P. Groby and O. Dazel, Asymptotic limits of some models for sound propagation in porous media and the assignment of the pore characteristic lengths, *J. Acoust. Soc. Am.* **139**(5) 2463–2474 (2016).
[36] J.-F. Allard, B. Castagnede, M. Henry and W. Lauriks, Evaluation of tortuosity in acoustic porous materials saturated by air, *Rev. Sc. Instrum.* **65** 754–755 (1994).
[37] K. Attenborough, Ground parameter information for propagation modeling, *J. Acoust. Soc. Am.* **92** 418–427 (1992). see also R. Raspet and K. Attenborough, Erratum: 'Ground parameter information for propagation modeling', *J. Acoust. Soc. Am.* **92** 3007 (1992).
[38] K. Attenborough, Acoustical impedance models for outdoor ground surfaces, *J. Sound Vib.* **99** 521–544 (1985).
[39] R. Raspet and J. M. Sabatier, The surface impedance of grounds with exponential porosity profiles, *J. Acoust. Soc. Am.* **99**(1) 147–152 (1996).

[40] S. Taherzadeh, Private Communication (1997)
[41] D. K. Wilson, Relaxation-matched modeling of propagation through porous media including fractal pore structure, *J. Acoust. Soc. Am.* **94** 1136–1145 (1993).
[42] D. K. Wilson, Simple relaxation models for the acoustical properties of porous media, *Appl. Acoust.* **50** 171–188 (1997).
[43] D. K. Wilson, V. E. Ostashev, S. L. Collier, N. P. Symons, D. F. Aldridge and D. H. Marlin, Time-domain calculations of sound interactions with outdoor ground surfaces, *Appl. Acoust.* **68**(2) 173–200 (2007).
[44] S. R. Pride, F. D. Morgan and A. F. Gangi, Drag forces of porous medium acoustics, *Phys. Rev. B* **47** (9) 4964–4978 (1993).
[45] G. Taraldsen, The Delany–Bazley impedance model and Darcy's law, *Acta Acust. united Ac.* **91** 41–50 (2005).
[46] R. Kirby, On the modification of Delany and Bazley formulae, *Appl. Acoust.* **86** 47–49 (2014).
[47] R. Kirby and A. Cummings, Prediction of the bulk acoustic properties of fibrous material at low frequencies, *Appl. Acoust.* **56** 101–125 (1999).
[48] D. Dragna, K. Attenborough and P. Blanc-Benon, On the inadvisability of using single parameter models to represent the acoustical properties of ground surfaces, *J. Acoust. Soc. Am.* **138** (4) 2399–2413 (2015). https://doi.org/10.1121/1.4931447.
[49] V. E. Ostashev, S. L. Collier and D. K. Wilson, Padé approximation in time-domain boundary conditions of porous surfaces, *J. Acoust. Soc. Am.* **122**(1) 107–112 (2007).
[50] D. K. Wilson, V. E. Ostashev and S. L. Collier, Time-domain equations for sound propagation in rigid-frame porous media (L), *J. Acoust. Soc. Am.* **116**(4) 1889–1892 (2004).
[51] M. S. Howe, On the long-range propagation of sound over irregular terrain, *J. Sound Vib.* **98**(1) 83–94 (1985).
[52] I. Tolstoy, Smoothed boundary conditions, coherent low frequency scatter, and boundary modes, *J.Acoust. Soc. Am.* **75**(1) 1–22 (1984).
[53] T. Van Renterghem, S. Taherzadeh, M. Hornikx and K. Attenborough, Meteorological effects on the noise reducing performance of low parallel wall structure, *Appl. Acoust.*, **121** 74–81 (2017).
[54] M. A. Biot, Reflection on a rough surface from an acoustic point source, *J. Acoust. Soc. Am.* **29** 1193–1200 (1957).
[55] V. F. Twersky, On scattering and reflection of sound by rough surfaces, *J. Acoust. Soc. Am.* **29** 209–225 (1957).
[56] O Faure, *Analyse numerique de la propagation acoustique exterieure: effets du sol en presence d'irregularities de surface et methodes temporelles,* PhD Thesis, Ecole Polytechnique de Lyon (2014).
[57] K. Attenborough and S. Taherzadeh, Propagation from a point source over a rough finite impedance boundary, *J. Acoust. Soc. Am.* **98**(3) 1717–1722 (1995).
[58] P. M. Boulanger, K. Attenborough, T. Waters-Fuller and K. M. Li, Rough hard boundary ground effects, *J. Acoust. Soc. Am.* **104**(1) 1474–1482 (1998).
[59] I. Bashir, S. Taherzadeh and K. Attenborough, Diffraction assisted rough ground effect: models and data, *J. Acoust. Soc. Am.* **133** 1281–1292 (2013).
[60] S. Taherzadeh, I. Bashir, T. Hill, K. Attenborough and Maarten Honikx, Reduction of surface transport noise by ground roughness, *Appl. Acoust.* **83** 1–15 (2014). https://doi.org/10.1016/j.apacoust.2014.03.011.

[61] V. Twersky, Scattering and reflection by elliptically striated surfaces, *J. Acoust. Soc. Am.* **40** 883–895 (1966).
[62] V. Twersky, Multiple scattering of sound by correlated monolayers, *J. Acoust. Soc. Am.* **73** 68–84 (1983).
[63] V. Twersky, Reflection and scattering of sound by correlated rough surfaces, *J. Acoust. Soc. Am.* **73** 85–94 (1983).
[64] R. J. Lucas and V. Twersky, Coherent response to a point source irradiating a rough plane, *J. Acoust. Soc. Am.* **76**, 1847–1863 (1984).
[65] I. Tolstoy, Coherent sound scatter from a rough interface between arbitrary fluids with particular reference to roughness element shapes and corrugated surfaces, *J. Acoust. Soc. Am.* **72**(3) 960–972 (1982).
[66] H. Lamb, *Hydrodynamics*, Dover, New York, p. **85** (1945).
[67] H. Medwin, J. Bailie, J. Bremhorst, J. Savage and I. Tolstoy, The scattered acoustic boundary wave generated by grazing incidence at a slightly rough rigid surface, *J. Acoust. Soc. Am.* **66** 1131 (1979).
[68] H. Medwin, G. L. D'Spain, E. Childs and S. J. Hollis, Low-frequency grazing propagation over periodic steep-sloped rigid roughness elements, *J. Acoust. Soc. Am.* **76**(6) 1774 (1984).
[69] K. Attenborough and T. Waters-Fuller, Effective impedance of rough porous ground surfaces, *J. Acoust. Soc. Am.* **108**(3) 949–956 (2000).
[70] P. Boulanger, K. Attenborough and Q. Qin, Effective impedance of surfaces with porous roughness: models and data, *J. Acoust. Soc. Am.* **117**(3) 1146–1156 (2005).
[71] L. M. Brekhovskikh, *Waves in layered media* Translated by D. Liberman and R. T. Beyer. Academic press, London and New York, pp. 44–61 (1960).
[72] P. Boulanger, K. Attenborough, Q. Qin and C. M. Linton, Reflection of sound from random distributions of semi-cylinders on a hard plane: models and data, *J. Phys. D.* **38** 3480–3490 (2005).
[73] I. S. Gradshtein and I. M. Ryzhik *Tables of Integrals, Series, and Products*, Academic Press, New York (1980).
[74] Y. Brelet and C. Bourlier, SPM numerical results from an effective surface impedance for a one-dimensional perfectly conducting rough sea surface, *Prog. Electromagnet. Res.*, PIER **81** 413–436 (2008).
[75] O. Faure, B. Gauvreau, F. Junker, P. Lafon and C. Bourlier, Modelling of random ground roughness by an effective impedance and application to time-domain methods, *Appl. Acoust.* **119** 1–8 (2017).
[76] P. M. Boulanger and K. Attenborough, Effective impedance spectra for rough sea effects on atmospheric impulsive sounds, *J. Acoust. Soc. Am.* **117**(2) 751–762 (2005).
[77] E. Salomons, Computational study of sound propagation over undulating water, in Proceedings of 19th International Congress on Acoustics, Madrid, Spain (2007).
[78] M. Rousseau and F. Coulouvrat, Scattering of a high frequency acoustic wave by a sinusoidal swell: asymptotic formulation, numerical simulation and fluid motion influence, *Acustica* **86** 821–829 (2000).
[79] K. Bolin, M. Boué and I. Karasalo, Long range propagation over a sea surface, *J. Acoust. Soc. Am.* **126** 2191–2197 (2009).
[80] T. Van Renterghem, D. Botteldooren and L. Deckoninck, Airborne sound propagation over sea during offshore wind farm piling, *J. Acoust. Soc. Am.* **135** 599–609 (2014). https://doi.org/10.1121/1.4861244.

[81] T. Van Renterghem, D. Botteldooren and P. Lercher, Comparison of measurements and predictions of sound propagation in a valley-slope configuration in an inhomogeneous atmosphere, *J. Acoust. Soc. Am.* **121** 2522–2533 (2007).
[82] C. Madshus and N. I. Nilsen, *Low frequency vibration and noise from military blast activity -prediction and evaluation of annoyance, Proc. InterNoise 2000*, Nice, France (2000).
[83] E. Øhrstrøm, Community reaction to noise and vibrations from railway traffic. Proc. Internoise 1996, Liverpool UK (1996).
[84] ISO 2631-2, Mechanical vibration and shock – Evaluation of human exposure to whole body vibration – Part 2: Vibration in buildings International Standards Organization (1998).
[85] W. M. Ewing, W. S. Jardetzky and F. Press, *Elastic waves in layered media*, McGraw-Hill Book Company, New York (1957).
[86] J. M. Sabatier, H. E. Bass, L. M. Bolen and K. Attenborough, Acoustically induced seismic waves, *J. Acoust. Soc. Am.* **80** 646–649 (1986).
[87] S. Taherzadeh, *Sound propagation in inhomogeneous media*, PhD Thesis in Engineering Mechanics. The Open University, Milton Keynes, UK (1996).
[88] K. Attenborough, On the acoustic slow wave in air filled granular media, *J. Acoust. Soc. Am.* **81** 93–102 (1987).
[89] N. D. Harrop, *The exploitation of acoustic-to-seismic coupling for the determination of soil properties* Ph. D. Thesis, The Open University, Milton Keynes, UK (2000).
[90] R. D. Stoll, Theoretical aspects of sound transmission in sediments. *J. Acoust. Soc. Am.* **68** 1341–1350 (1980).
[91] S. Tooms, S. Taherzadeh and K. Attenborough, Sound propagation in a refracting fluid above a layered porous and elastic medium, *J. Acoust. Soc. Am.* **93**(1) 173–181 (1993).
[92] C. Madshus, F. Lovholt, A. Kaynia, L. R. Hole, K. Attenborough and S. Taherzadeh, Air–ground interaction in long range propagation of low frequency sound and vibration—field tests and model verification, *Appl. Acoust.* **66**(5) 553–578 (2005).
[93] Z. Fellah, C. Depollier and M. Fellah, Application of fractional calculus to the sound waves propagation in rigid porous materials: validation via ultrasonic measurements, *Acta Acust. united Ac.* **88** 34–39 (2002).
[94] M. Henry, P. Lemarinier, J. F. Allard, J. L. Bonardet and A. Gedeon, Evaluation of the characteristic dimensions for porous sound-absorbing materials, *J.Appl. Phys.* **77** 17 (1995).
[95] J-F. Allard, M. Henry and J. Tizianel, Pole contribution to the field reflected by sand layers, *J. Acoust. Soc. Am.* **111**(2) 685–689 (2002).
[96] M. Boeckx, G. Jansens, W. Lauriks and D. Albert, Modelling acoustic surface waves above a snow layer, *Acta Acust. united Ac.* **90**(2) 246–250 (2004).
[97] R. J. Donato, Impedance models for grass covered ground, *J. Acoust. Soc. Am.* **61** 1449–1452 (1977).

Chapter 6

Measurements of the Acoustical Properties of Ground Surfaces and Comparisons with Models

6.1 IMPEDANCE MEASUREMENT METHODS

6.1.1 Impedance Tube

Early measurements of reflection coefficients of outdoor ground surfaces were made at normal incidence and used adaptations of the standing wave or impedance tube technique which is a standard method to obtain the acoustical properties of building materials [1]. Initially, the probe microphone method was used which assumes plane wave incidence and requires the measurement of the ratio of the maximum sound pressure to the level of the first (or subsequent) pressure minimum (as counted from the surface of interest) as well as the distance from the surface to the first minimum at each frequency. The rather different acoustical, physical and biological properties of outdoor ground result in several problems not encountered often with artificial porous sound absorbents. These include the following: (i) the need for probe end and tube absorption corrections [2] when measuring the relatively high impedances associated with many ground surfaces, (ii) the difficulties of establishing the surface 'plane' location (particularly important in determining the phase (imaginary part) of the impedance), (iii) the changing micro-meteorological conditions and the formation of worm casts during lengthy measurements [3], (iv) the unrepresentative nature of the (necessarily) small tube sample in view of the spatial inhomogeneity typical of many grounds (discussed in Section 6.4) and (v) the destruction of the local ground structure while inserting the lower end of the tube into the ground to reduce flanking sound transmission. When using a probe microphone, there is the additional problem of devising a stable system for vertical probe traversal. Nevertheless, several useful data sets for impedance as a function of frequency have been obtained with a vertical probe microphone type of impedance [4,5]. Also, a probe microphone impedance tube method has proved particularly successful as a laboratory technique for measuring the acoustical properties of snow [6,7]. For this purpose, a horizontal tube was used with automatic tracking of the probe microphone and measurements both of hard-backed and quarter wavelength (air) backed finite length samples.

Both the normal surface impedance and the propagation constant are important when describing propagation over a low flow resistivity surface such as snow, which is potentially externally reacting.

A modification of the impedance tube method [8–11] measures the complex transfer function between two microphones positioned near to the sample surface. The method has the advantage that it is possible, by use of a broadband signal, to obtain impedance data over a wide frequency range simultaneously rather than at one frequency at a time as required by the probe microphone technique. The technique however is not as accurate as the probe method [9,10] and leads to spectral artefacts when the microphone spacing corresponds to integer numbers of quarter or half wavelengths. The microphone spacing (s) determines the frequency range of the measurement's validity according to

$$nf' < f < (n+1)f', \text{where} f' = c_0 / 2s$$

n is an integer and c_0 is the speed of sound in air. This technique has been used to determine the normal surface impedance of snow [12] and forest floors [13] (see also Section 6.4.2). Together with a system for sealing the contact between the tube and the surface of interest, it is used as a standard method for measuring the acoustical properties of road surfaces at normal incidence [14].

6.1.2 Impedance Meter

The normal surface impedance of the ground has been obtained from direct measurements of pressure and volume velocity at the ground surface by means of a mechanically driven cylindrical Helmholtz resonator chamber, which is pounded into the ground [15]. The sound pressure was measured by a microphone, mounted flush in the wall of the chamber and an optical monitor of the piston driver displacement permitted measurement of the phase angle between the volume velocity and the pressure. This enables deduction of the real and imaginary parts of impedance over the frequency range 300–1000 Hz. While this method overcomes several of the problems associated with outdoor use of the impedance tube, it shares the difficulties associated with small sample size and the inevitable destruction of soil structure caused by its insertion.

6.1.3 Non-Invasive Measurements

6.1.3.1 Direct Measurement of Reflection Coefficient

Dickinson and Doak [3] investigated a free field version of the impedance tube technique at normal incidence. Standing waves were set up along an axis between a loudspeaker source and the ground. The loudspeaker was mounted about 4 m high on an A-frame. The resulting interference pattern in the sound field was explored by a microphone traversing the axis and the normal surface impedance deduced in a similar manner to that used on impedance tube data.

As long as the measurements are made at normal incidence and the sum of the source and receiver heights is a sufficient number of wavelengths above the ground, then the total field is given with adequate accuracy by assuming plane rather than spherical reflection incidence in the analysis. In general, however, allowance should be made for the spherical spreading of the wave fronts.

Embleton et al. [4] adapted this free field technique to oblique incidence by measuring the interference field with a probe microphone along the track of the specularly reflected ray. Van der Heijden [16] avoided the need for using a probe microphone in the free field both at normal and oblique incidence by using up to 16 fixed microphone positions.

A free field adaptation of the two-microphone impedance tube technique developed initially for use in building acoustics [17] has been used successfully to measure the impedance of outdoor ground surfaces at normal and oblique incidence [18].

The typical geometry used a source height of 2.1 m, an upper microphone height of 135 mm, microphone spacing of 75 mm and an angle of incidence of 45°. With this geometry, data for the complex transfer function between the two microphones and hence for impedance were obtained over the frequency range 200–2000 Hz. The method was found to be very sensitive to meteorological conditions and to the assumed ground surface location. Background noise problems were alleviated by use of narrow band signals. At low frequencies it is difficult to achieve the source height required for the plane wave reflection coefficient assumption to be valid and to achieve the low background noise important to the free field standing wave technique.

Daigle and Stinson [19] and Sprague et al. [20] have measured the phase gradient between a pair of closely spaced, phase-matched microphones in the normal incidence interference pattern instead. This technique requires accurate measurements only in regions of the peaks and valleys in the phase gradient. However, it is extremely difficult to carry out, being extremely sensitive to unwanted reflections and to meteorological conditions. Using a loudspeaker source at a height of about 7 m and four different microphone spacings (between 10 and 80 cm) data have been obtained between 25 and 300 Hz.

Tone bursts or other impulse waveforms may be used as an alternative to continuous sound sources for impedance measurements. If the incident pulse is short enough then unwanted reflections may be isolated, and their effects may be eliminated. Tone bursts have been used to obtain the acoustical properties of ground surfaces brought into the laboratory [21]. In a series of outdoor measurements, Cramond and Don [22,23] used a blank-firing rifle as a source and deployed two microphones so that the ground-reflected signal at one arrived at the same time as the direct signal at the other. With appropriate choice of source and receiver heights (approximately 2 m) this enabled direct deduction of reflection coefficient as a function of frequency within the limitations of the source spectrum (400 Hz to 5 kHz). The method was found to be capable of distinguishing the impedance of a tyre track from that of the surrounding uncompacted soil. Also, it was used to show the existence of high frequency (>1 kHz) peaks in the surface impedance of a soil wetted so that a thin (1–2 cm thick) saturated layer was formed at the surface (see also Section 6.4.2). On the other hand, the frequency range of the method is limited by lack of source energy at low frequencies and by turbulent scattering at high frequencies.

6.1.3.2 Impedance Deduction from Short-Range Measurements

If a point source is located on a locally reacting surface and emits broadband sound, then for a reception point also on the surface $R_p = -1$ and Equation (2.40) is simplified considerably [4]. Specifically,

$$\text{total field} = 2F(w)e^{ik_0 r} / r \tag{6.1}$$

where, for $k_0 r \gg 1$,

$$F(w) \sim -1/2w^2 \tag{6.2}$$

and

$$w^2 \approx ik_0 r / Z^2. \tag{6.3}$$

This means that the excess attenuation (sound pressure level with respect to free field in dB) is given by

$$\text{E.A.} = 20\log\left|iZ^2/k_0 r\right| + 3. \tag{6.4}$$

Equation (6.4) enables deduction of the magnitude of the normal surface impedance as a function of frequency from measurements of the excess attenuation spectrum (or of excess attenuation as a function of range at each frequency of interest). This method has been used by Embleton et al. [4] on grassland. Results were found to be in reasonable agreement with both impedance tube and free field data obtained over the same surface. An elaboration of this technique has been used by Habault and Corsain [24]. The excess attenuation at grazing incidence was obtained as a function of range at several frequencies and the impedance was deduced from a minimization algorithm using a grazing-incidence approximation. Practically, the measurement technique was found to be restricted to frequencies above 500 Hz since the required range of measurement increases as the frequency of interest is decreased. Nevertheless, the resulting data shows remarkably little scatter.

More generally, least-squares or template fitting by means of theories described in Chapter 2 to measurements of the magnitude of the excess attenuation from a point source at non-grazing incidence may be used to yield impedance as a function of frequency [25]. Such a procedure may be regarded as a simple example of the matched field algorithms employed in underwater acoustics. It should be noted that the excess attenuation spectrum is independent of the source spectrum, consequently, in principle, any broadband source may be used as long as the measurements are taken far enough from the source that it may be regarded as a point source. On the other hand, the measurement range should be sufficiently small (<5 m) that meteorological influences are kept to a minimum [26]. Bolen and Bass [27] obtained impedances of grassland, soils and sands as a function of frequency from 40 Hz to 300 Hz using a large loudspeaker as a source. However, their deduced impedance spectra (even for a homogeneous sand surface) shows considerable scatter between successive frequencies. Probably, this is the result of the relatively large ranges (>15 m) used for their measurements and the fact that no constraint was imposed on the behaviour of the real and imaginary parts of impedance as separate functions of frequency. Such constraints may be imposed by the adoption of an appropriate model for the acoustical properties of the ground. This may involve fitting of either short-range measurements of excess attenuation, or the level difference between vertically separated microphones [28–32], using the classical theory for propagation from a point source over an impedance plane. As well as not requiring the assumption of plane waves, which is valid only near normal incidence if the source height is large and the frequency is high enough, the short-range propagation method near grazing incidence includes effects of small-scale surface roughness. Surface roughness alters the surface impedance, compared with a smooth surface with the same pore structure near grazing incidence but has less effect at normal incidence (see Section 5.3).

Wempen [33] measured and recorded a broadband signal in an anechoic chamber and then used it as a reference signal, transmitting it across several ground surfaces. In addition, the direct and ground reflected waves were synchronized by feeding them into an FFT analyser and adjusting the time delay between them. This enabled high signal-to-noise ratios to be obtained with relatively short average times.

An alternative to an anechoic chamber measurement of the free field spectrum level at the relevant range is to obtain a reference measurement after raising the source and receiver sufficiently far above the surface that ground effects may be ignored. This is more convenient in situ particularly when using a Maximum Length Sequence Signal Analysis (MLSSA) system. In this system, the source is controlled to give a pseudo-random sequence of impulses. The received signal is analysed for this emission sequence. The resulting analysis is robust to background noise and gives the time domain or impulse response, which may be used to eliminate unwanted portions of the time series. Examples of excess attenuation spectra obtained in this way

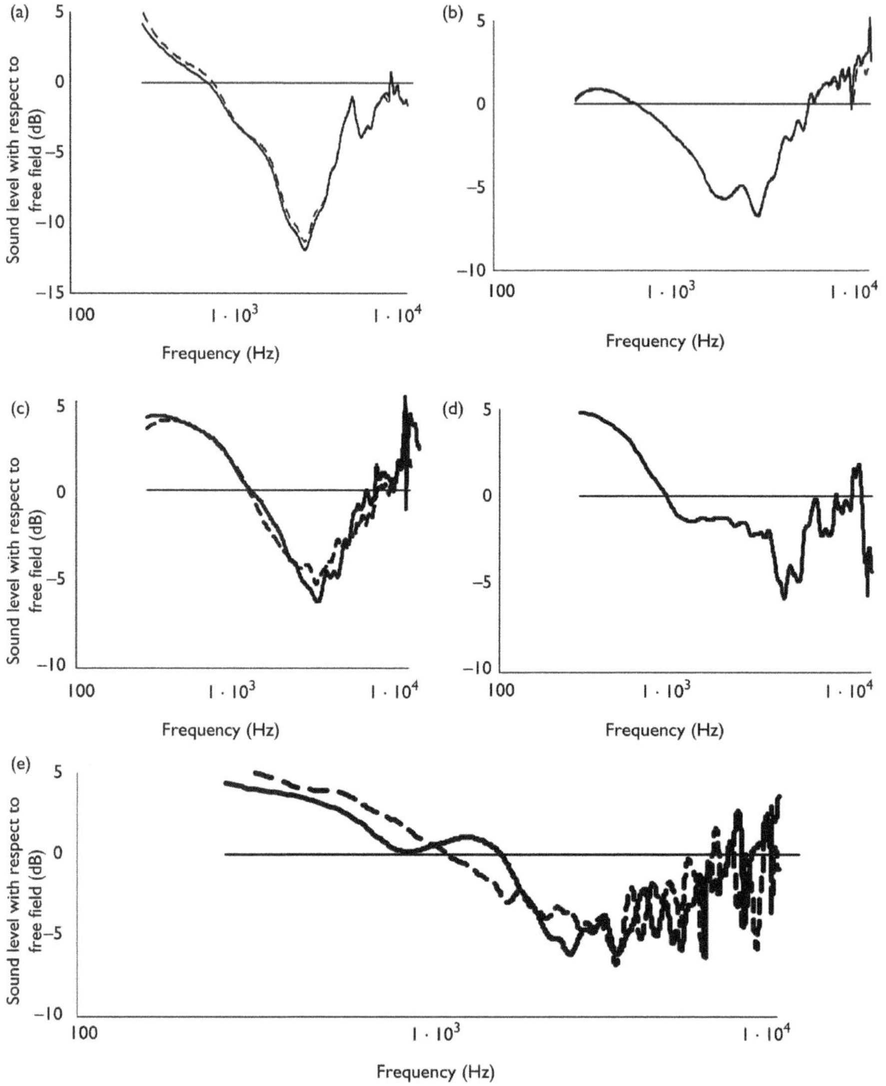

Figure 6.1 Excess attenuation spectra obtained with source and receiver at 0.1 m height and separated by 1 m over (a) grass 1, (b) soil, (c) grass 2, (d) bean plants and (e) wheat.The wheat was 0.55 m high. The different lines in each graph represent results of successive measurements.

at short-range outdoors are shown in Figure 6.1. It should be noted, however, that MLSSA is not reliable in a time-varying environment.

Rather than using the excess attenuation spectrum at a single point for the deduction of ground impedance which requires accurate knowledge of the free-field spectrum of the source, it is more convenient to use the difference in spectra between two separately located reception points, i.e. level difference spectra. The reception points may be separated either horizontally or vertically. If the source is omnidirectional in the plane containing source and receivers, then, like excess attenuation, level difference is independent of the source. The level difference between two vertically separated microphones at horizontal ranges of less than 3 m from a point source has been used extensively [34]. If the source–receiver distances are nearly the same for both receivers, the level difference spectrum may be calculated as the difference between the excess attenuation spectra at the two receivers. If the lower microphone is on the surface, the level difference spectrum corresponds closely in form to the excess attenuation spectrum at the upper microphone. However, locating the lower microphone at zero height may prove difficult because of the finite size of the microphone capsule and the roughness of the ground surface. Also, since there may be very large sound velocity gradients close to the ground, the lower microphone should not be on the ground surface but at a height small enough to ensure that the first interference minimum in the corresponding excess attenuation spectrum lies above the frequency range of interest. If source–receiver distances are less than 2 m, refraction and turbulence effects are likely to be small.

6.1.3.3 Model Parameter Deduction from Short-Range Propagation Data

Measurements of level difference spectra and excess attenuation spectra have been obtained at horizontal source–receiver ranges of 2 m or less, both in the laboratory and outdoors. These data have been fitted using some of the impedance models discussed in Chapter 5. Example laboratory measurements of excess attenuation spectrum over a sand surface, theoretical predictions and fits are shown in Figure 6.2.

Wempen [28] compared his measured excess attenuation spectra with predictions using several impedance models. A comprehensive set of outdoor excess attenuation measurements at short range (between 7 and 15 m), made by Embleton et al. [31], has been fitted by the single parameter Delany–Bazley model (Equations (5.1a and b), (5.2a and b) and Table 5.1) giving a wide range of effective flow resistivities for several surfaces (see Section 6.3). By making excess attenuation measurements at short range (4 m) and by making use of a phenomenological model (Equations (5.3) and (5.4a and b)) for the acoustical properties of a hard-backed porous layer, Thomasson [32] was able to deduce best-fit parameter values for grassland, hay and newly planted soil.

If the impedance model requires the fitting of more than two parameters, then problems of non-uniqueness will arise [34]. These can be overcome if one of the parameters can be fixed beforehand. For example, the porosity of the surface may be known to within ±20%. The deduction of parameters through the impedance-model-based approach offers a basis for remote

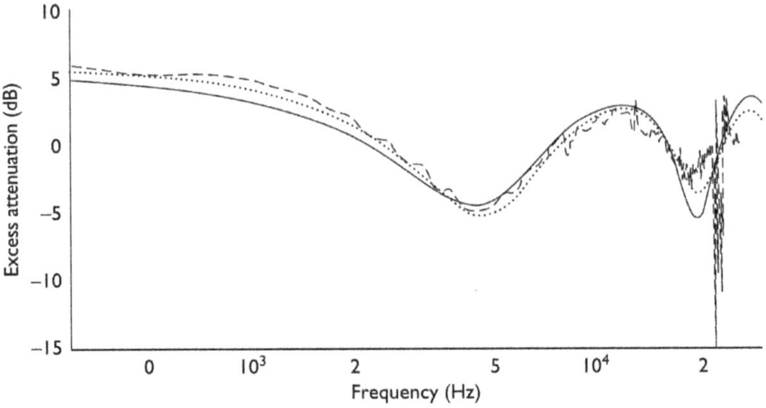

Figure 6.2 Measured excess attenuation spectrum with point source height = receiver height = 0.1 m, horizontal separation 1 m, above 0.3 m thick sand (broken line), compared with predictions of a two parameter model (Equation (5.38)) (solid line) and a prediction for a distribution of triangular pore sizes (dotted line) using measured flow resistivity (473 kPa s m^{-2}), assumed porosity (0.4), assumed tortuosity (2.5) and best-fit standard deviation of log-normal pore size distribution (1.3 ϕ units). Best-fit values for the two-parameter variable porosity model are σ_e = 499 kPa s m^{-2} and $\alpha_e = -185\ m^{-1}$.

acoustical monitoring of soils (see Section 6.9). Also, it is possible to deduce impedance model parameters for snow from pulse measurements [35].

If the ground surface is sloping, the source and receiver heights should be measured perpendicular to the surface. As discussed in Section 6.3, the presence of irregular or random roughness on porous surfaces changes the effective admittance. The effects may be predicted quantitatively if the irregularity or roughness is small compared with the wavelengths of interest and either the roughness elements have identifiable simple shapes or a known mean height and spacing. Also, this means that, if an independent measurement of flow resistivity is available, it may be possible to deduce ground roughness from acoustical measurements.

6.1.3.4 A Template Method for Impedance Deduction

An ANSI standard method includes the fitting of short-range measurements made with specified source–receiver geometries to *templates* of predicted excess attenuation or level difference spectra [36]. These templates are based on the likely range of ground surface parameters and the resulting variation in excess attenuation or level difference spectra predicted by classical theory for a point source above an impedance plane using two impedance models: the Delany - Bazley model and the variable porosity model (see Equation (5.38)). The best-fit parameters for the impedance model are deduced by eye from the closest 'template' excess attenuation prediction to the data. Templates are available for two different source–receiver geometries (Geometry A – Source height (h_s) = 0.325 m, Upper microphone height (h_t) = 0.46 m, Lower microphone height (h_b) = 0.23 m, Horizontal separation (d) = 1.75 m; Geometry B – Source height (h_s) = 0.20 m, Upper microphone height (h_t) = 0.20 m, Lower microphone height (h_b) = 0.05 m, Horizontal separation (d) = 1.0 m). Geometry A is intended to cover the broadest range of frequencies. Geometry B emphasizes ground effect at frequencies above 1000 Hz and is better suited for hard grounds. These geometries do not yield satisfactory results if the ground impedance is relatively high. But, in this case, accurate values for ground impedance should not be necessary. An example of the use of the ANSI template method based on predictions of the two-parameter variable porosity model (Equation (5.38)) is shown in Figure 6.3.

As long as the horizontal separation of the source and the receiver is not greater than 3.0 m or not less than 1.0 m and the angles of incidence are broadly the same as those for geometries A and B, the user of the standard has discretion in the choice of the geometry. The possibility of using a similar procedure with signals received by a 3-D microphone array such as that used for source location has been investigated [37].

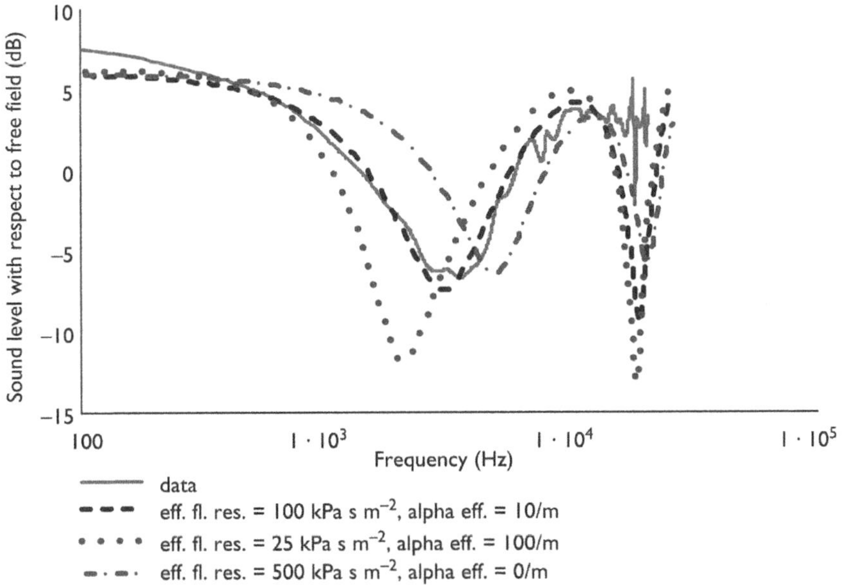

Figure 6.3 Excess attenuation spectrum measured (continuous line) at a microphone 0.1 m high and at a horizontal distance of 1 m from a point source loudspeaker 0.1 m above a sports field and predictions (dot-dash, broken, dotted lines) using various values of the parameters (see Legend) in a two-parameter impedance model. Clearly the prediction corresponding to an effective flow resistivity of 100 kPa s m^{-2} and effective alpha 10/m gives the best fit.

6.1.3.5 Effective Flow Resistivity Classification

A similar procedure for fitting level difference magnitude spectra is described in a NORDTEST ACOU 104 [38]. The method uses a single geometry (point source height 0.5 m, receiver heights at 0.5 and 0.2 m, separation 1.75 m). However, as mentioned in Section 6.3.1.4, the use of a single geometry restricts the range of ground impedance for which level difference magnitude spectra obtained using the chosen geometry are sufficiently sensitive.

The errors between predictions and data are calculated from

$$E = \sum_f \left| LD_M(f) - LD_C(f) \right| \tag{6.5a}$$

$$LD_C = EA(1) - EA(2) \tag{6.5b}$$

$$EA(1) = 20\log\left[\left|1 + f_b \frac{QR_2}{R_1} e^{k(R_2 - R_1)}\right| + \left(1 - f_b^{\,2}\right)\left|\frac{QR_2}{R_1}\right|^2\right] \tag{6.5c}$$

$$EA(2) = 20\log\left[\left|1 + f_b \frac{QR_4}{R_3} e^{k(R_4 - R_3)}\right| + \left(1 - f_b^{\,2}\right)\left|\frac{QR_4}{R_3}\right|^2\right]. \tag{6.5d}$$

In Equation (6.5a), LD_M are the measured level difference values and LD_C are the predicted level difference values between microphones at distances R_1 and R_3 from the source; R_2 and R_4 are the corresponding reflected ray path lengths. In Equations (6.5c) and (6.5d), Q is the spherical wave reflection coefficient (see Chapter 2, Equation (2.40)) and f_b is a factor related to third octave band averaging which is necessary if predictions at more closely spaced frequencies are compared with third octave band data. Adequate fits to a table of 'reference' error predictions based on the Delany - Bazley impedance model (or its hard-backed layer version, recommended particularly for snow) are used to place the site under investigation in a class of 12 effective flow resistivities between 10 and 20000 kPa s m^{-2} and one of four thickness classes (0.05–0.15 m). Several measurements are made at each site. A distinction is made between results at low (−20°C to + 5°C) and high temperatures (5–30°C). Level Difference data are not used for classification if the standard deviation of data in any frequency band exceeds 4 dB and/or the average error calculated from Equation (6.5a) is greater than 15 dB.

6.1.3.6 Direct Impedance Deduction

A method of deducing impedance directly from level difference or excess attenuation data has been proposed based on two-dimensional minimization of the difference between data and theory at each frequency [39]. This requires considerable computation and is relatively inefficient. An alternative numerical method is based on root-finding [40]. This technique takes advantage of the fact that the classical approximation to the spherical wave reflection coefficient is an analytic function of the impedance. The resulting saving in computation time can be up

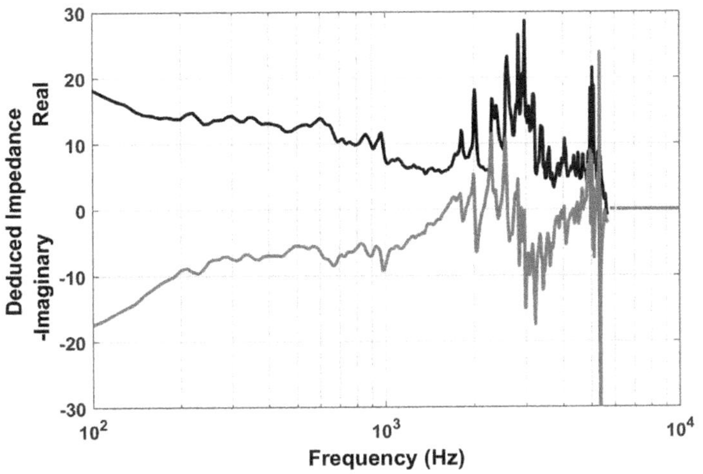

Figure 6.4 Complex impedance deduced from level difference amplitude and phase measured over bare-cultivated ground by placing source at height of 0.2 m, upper and lower microphone at heights of 0.2 m and 0.05 m respectively at a horizontal distance of 1 m from the source [41].

to a hundred-fold without any loss in accuracy. Examples of results obtained with this method are given in Sections 6.2 and 6.4. The ANSI standard [36] includes this method for deducing ground impedance directly and, therefore, independently of any impedance model. But it is suggested, nevertheless, that the outcomes of direct impedance deduction should be checked against the results of the template method (see Section 6.1.3.4).

Figure 6.4 shows the complex impedance deduced from **level difference** spectra measured over **bare-cultivated** ground after placing the point source at height of 0.2 m and upper and lower microphones at heights of 0.2 and 0.05 m, respectively, at a horizontal distance of 1 m from the source [41]. The deduced impedance spectra **tend to zero** above 5 kHz since it not possible to converge to a valid solution at higher frequencies. But this is unlikely to be of practical importance.

A two-microphone free-field method developed for characterizing sound absorbing materials [17] has been adapted for use on outdoor ground surfaces [42]. The method uses the definition of surface admittance in terms of the complex pressure gradient near the surface and approximates the pressure gradient by the complex level difference between vertically separated phase-matched microphones. Hence, if the vertical separation is denoted by Δz and the complex pressures at the microphones are p_1 and p_2, the surface admittance is given by

$$\beta = \left(\frac{i}{k_0}\frac{dp/dz}{z}\right)_{z=0} \approx \left(\frac{i}{k_0 \Delta z}\right)\left(\frac{p_2}{p_1} - 1\right) \tag{6.6}$$

An advantage of this method is that several pairs of vertically separated microphones can be deployed along a long propagation path to obtain ground impedance at several points simultaneously. The method has been validated through field trials at short and long range over different surfaces using a propane cannon as an impulsive source [42].

6.2 COMPARISONS OF IMPEDANCE SPECTRA WITH MODEL PREDICTIONS

Figures 6.5 and 6.6 compare data obtained by Cramond and Don using a pulse method (gun-shot source) [22,23] with predictions of the semi-empirical Delany - Bazley model (Equations (5.2)), and other models (Equations (5.11), (5.12) and (5.30)). The data are for compacted earth and for the same soil with the top 0.02 m loosened. There are significant differences between the measured impedance of compacted and loose soil. Although it is possible to obtain reasonable fits with the semi-empirical model in both cases, improved agreement is obtained by assuming a pore microstructure of identical tortuous slits (see Chapter 5).

Figure 6.7 Shows impedance spectra for a hard-backed sample of snow obtained by Buser [6] from measurements in an impedance tube. The measured flow resistivity and bulk density of this sample were 9.6 kPa s m^{-2} and 208 kg m^{-3}, respectively. Assuming a density of 913 kg m^{-3} for ice, the latter

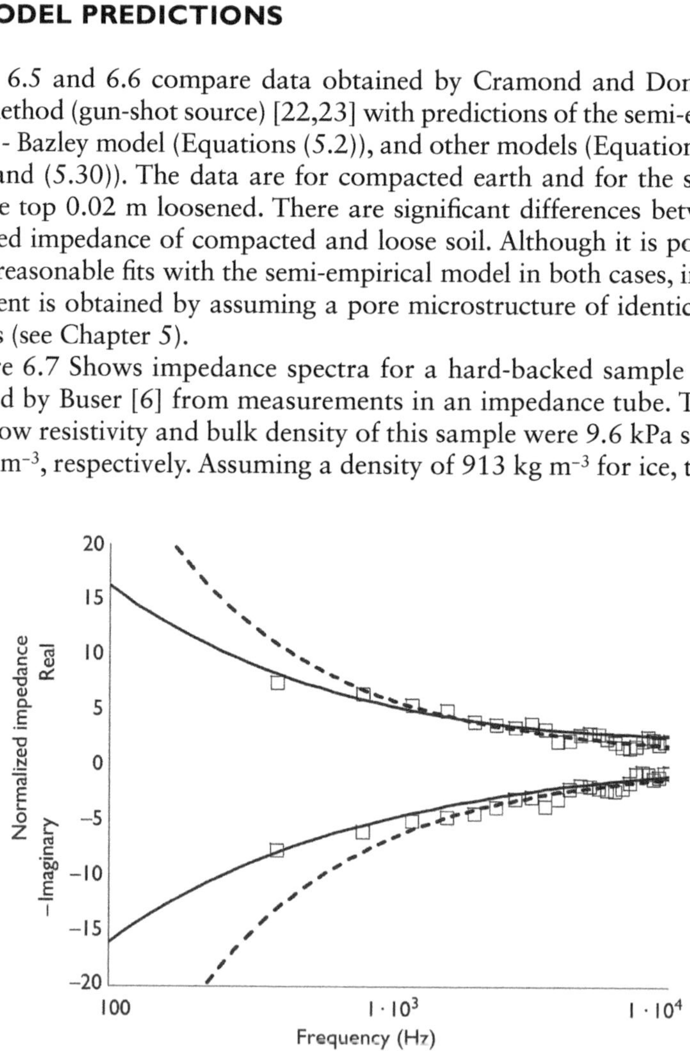

Figure 6.5 Impedance of compacted soil measured by Cramond and Don [20] and predictions using Equation (5.2) with effective flow resistivity 450 kPa s m^{-2} (dotted lines), Equations (5.11) and (5.12, and Equation (5.28)) (solid lines) with flow resistivity 275 kPa s m^{-2}, porosity 0.4 and tortuosity 2.

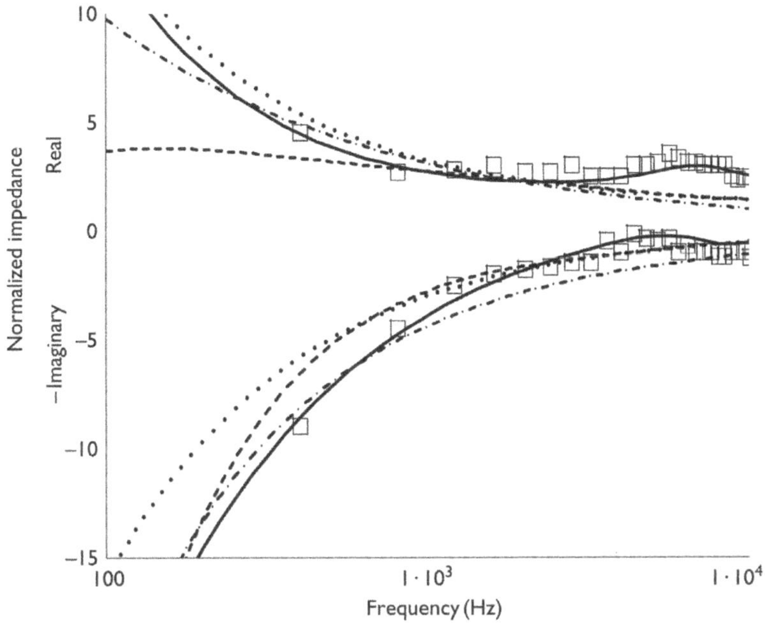

Figure 6.6 Impedance of loosened soil (0.02 m thick) above compacted soil measured by Cramond and Don [20] and predictions using Equation (5.2) with effective flow resistivity 450 kPa s m^{-2} (dotted lines); Equations (5.1), (5.2) and (5.3) (dash-dot lines) with layer thickness 0.02 m; Equations (5.11), (5.12) and (5.30) (solid lines), with upper layer flow resistivity 100 kPa s m^{-2}, porosity 0.4, tortuosity 1.5 and thickness 0.02 m; substrate flow resistivity 800 kPa s m^{-2}, porosity 0.2 and tortuosity 3 and equation (5.38) (broken lines) in which R_e = 50 kPa s m^{-2} and α_e = 133 m^{-1}.

value corresponds to a porosity of 0.772. The continuous lines in Figure 6.7 correspond to predictions of an identical slit-pore model assuming that tortuosity is given by the inverse of porosity. The dotted lines represent predictions of the Wilson relaxation model (Equations (5.42a, b)) with $s_B = 1$. The dashed lines represent predictions of the Hamet phenomenological model (Equation (5.5)) and the dash-dot lines represent predictions of the Delany–Bazley model (Equations (5.1), (5.2) and Table 5.1) with effective flow resistivity given by the product of measured flow resistivity and porosity.

The relaxation model predictions can be improved by adjusting the value of s_B (a value of 1.4 gives better agreement with data). However, this removes one of its advantages as a simple two-parameter model. The slit-pore predictions are obtained using the measured flow resistivity and porosity, without any adjustment, and assuming that tortuosity is given by the inverse of porosity. It is noticeable that use of an effective flow resistivity, given by the product of flow resistivity and porosity, in the Delany–Bazley model, gives relatively poor predictions for these data.

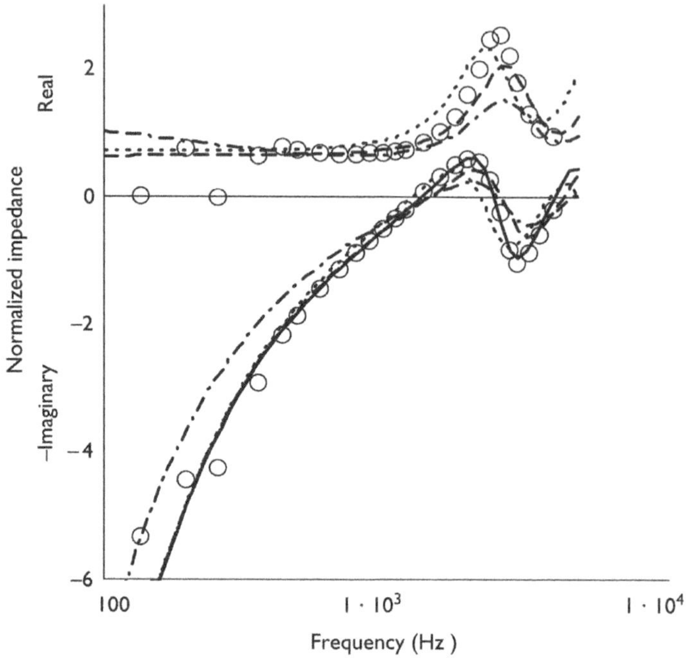

Figure 6.7 Measured and predicted impedance of a 0.05 m thick layer of snow with measured flow resistivity 9.6 kPa s m^{-2} and porosity 0.774. Predictions use identical slit pore model with measured flow resistivity and porosity and tortuosity = 1/porosity (solid lines), Hamet-Berengier (dashed), Wilson-relaxation (dotted) and Delany-Bazley (dash-dot) with $R_e = \Omega R_s$. For model details see Chapter 5.

Figure 6.8(a) shows example impedance values obtained by the direct deduction method (see Section 6.1.3.4) over a smoothed sand surface [43] together with fits based on Equations (5.10)–(5.12). Figure 6.8(b) shows impedance data obtained by direct deduction from complex excess attenuation measurements (see 6.1.3.6) over a rough sand surface [43] and predictions of Equation (5.59) using either the predicted or measured smooth surface impedance.

Figures 6.9 and 6.10 show level difference spectra obtained in the laboratory over layers of porous felt and polymer foam with 0.07 m high source and receiver separated by 0.7 m and the corresponding deduced impedance spectra [41].

6.3 FITS TO SHORT-RANGE PROPAGATION DATA USING IMPEDANCE MODELS

6.3.1 Short-Range Grassland Data and Fits

An extensive set of third-octave band short-range level difference data for several types of ground surface has been obtained [44–46] using the NORDTEST ACOU 104 standard procedure [38]. Also, narrow band data

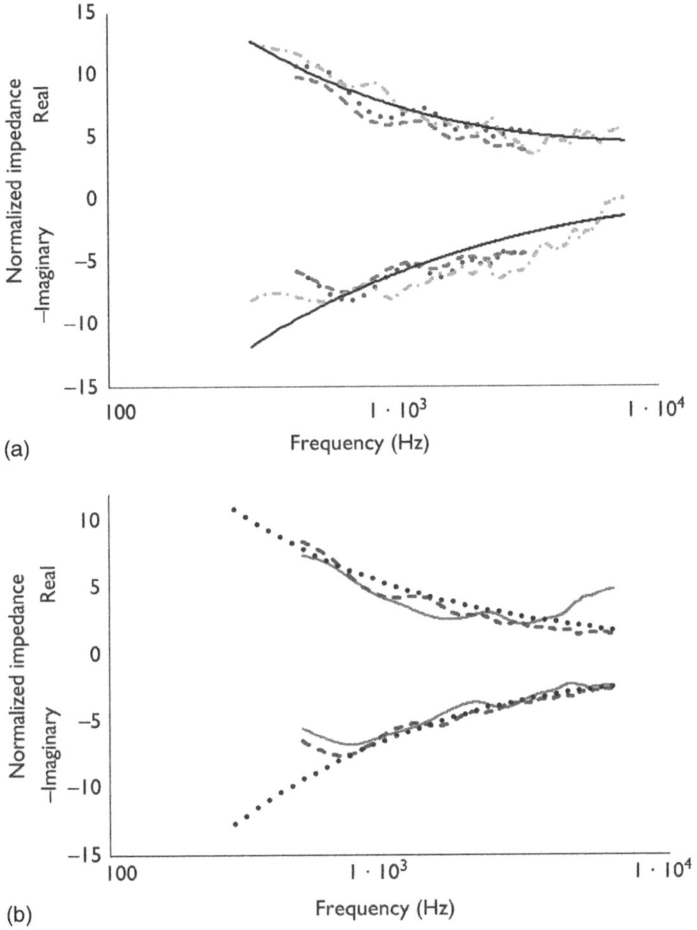

Figure 6.8 (a) Impedance spectra deduced from three complex excess attenuation measurements with a point source and receiver at 0.1 m height and separated by 1 m over smoothed sand (dotted broken and dot-dash lines). Also shown are predictions (solid lines) using a semi-infinite slit-pore model [7], measured flow resistivity 420 kPa s m^{-2}, assumed porosity = 0.4 and tortuosity = 1/porosity (b) impedance spectra deduced from a complex excess attenuation measurement (broken lines) with the same geometry as in (a) over a sand surface containing (approximately) semi-cylindrical roughness, height 0.008 m and centre-to-centre spacing 0.04 m. Also shown are predictions of equations (3.45) assuming semi-cylindrical roughness with given dimension and spacing and either the predicted (dotted line) or the measured (broken line) smooth surface impedance.

sets have been used [47] to validate the procedures in ANSI S1.18 [36]. Predictions using the Delany - Bazley single parameter model and the variable porosity, phenomenological, Taraldsen's models [44] and slit-pore two-parameter models have been fitted to these data using an automated numerical fitting procedure (Simplex Downhill algorithm).

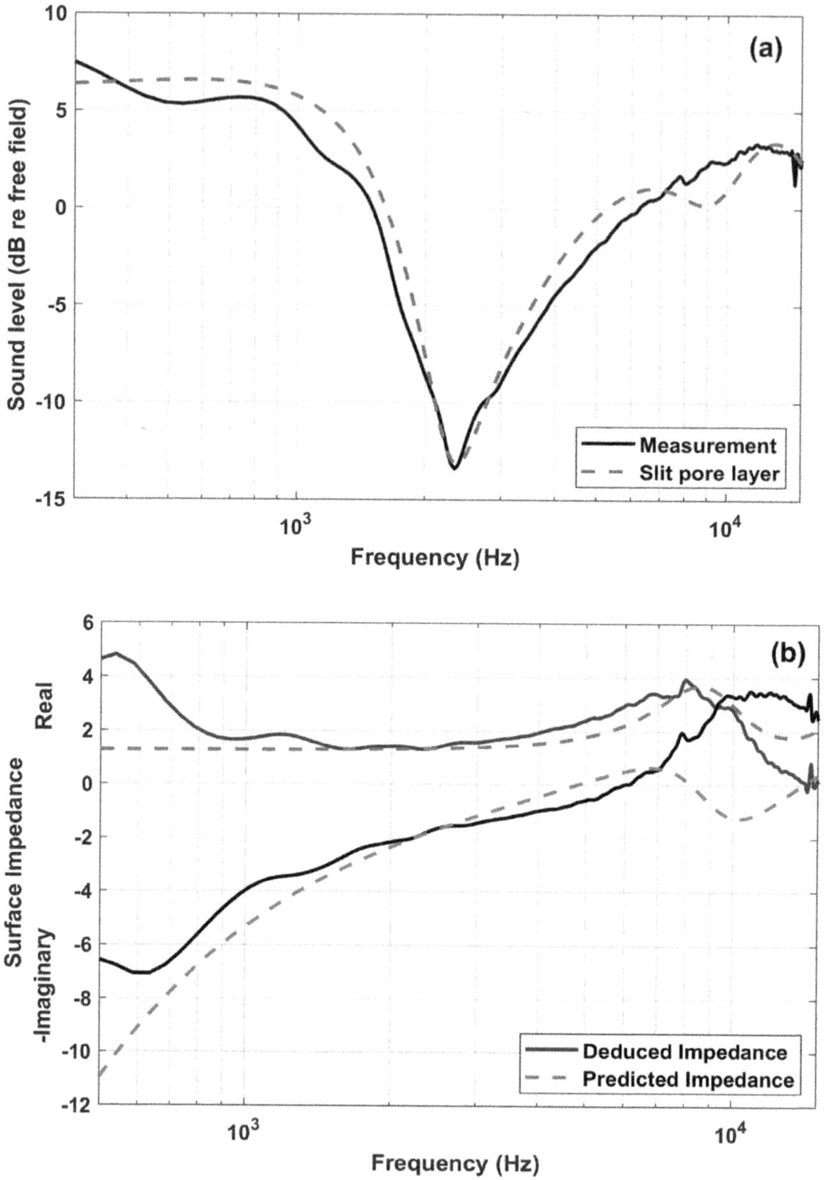

Figure 6.9 (a) Excess attenuation spectra measured with source and receiver at height of 0.07 m, distance between source and receiver 0.7 m (continuous line) and (b) deduced impedance spectra (continuous lines) over a 0.014 m thick layer of felt placed on MDF compared with predictions (broken lines) assuming a slit pore hard-backed layer impedance model with R_s = 85 kPa s m^{-2}, Ω = 0.5 and d = 0.014 m [41].

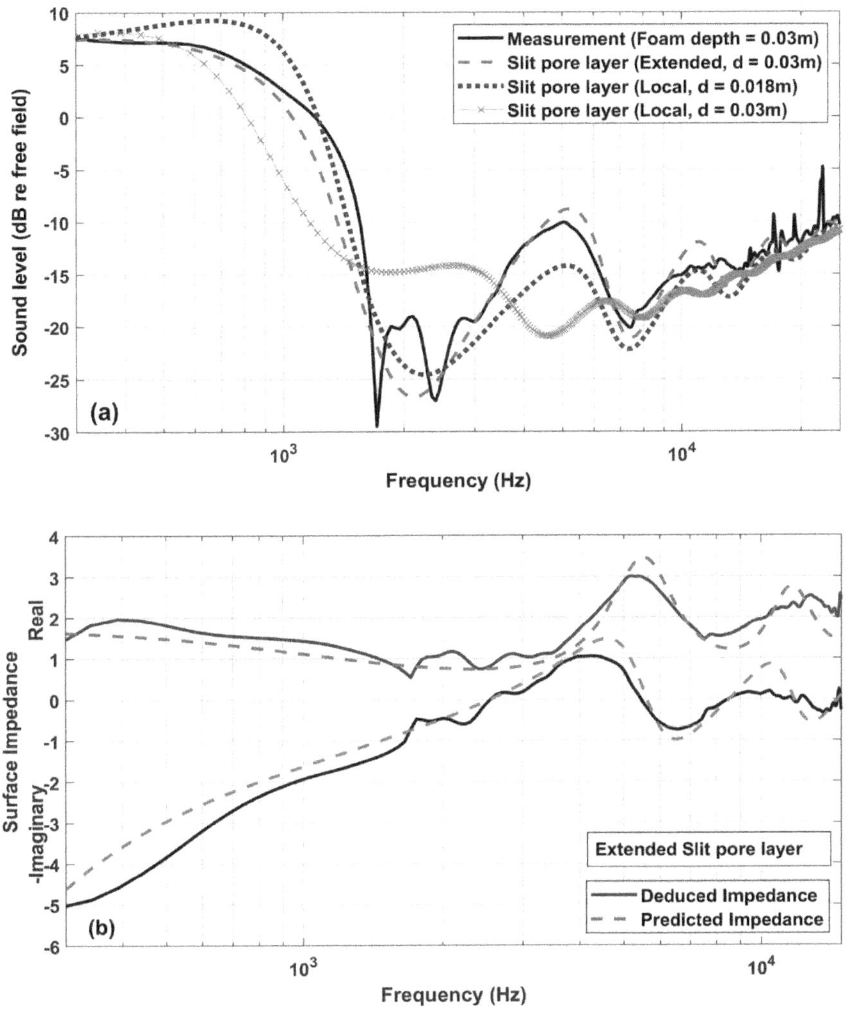

Figure 6.10 (a) Measured excess attenuation (continuous line) over a 0.03 m thick polymer foam layer placed on MDF with source and receiver at height of 0.015 m and separated by 0.7 m, and numerically obtained best fits assuming externally-reacting slit pore layer (broken line, Rs = 7.0 kPa s m^{-2}, Ω = 0.98, d = 0.03 m); locally-reacting slit pore layer (dotted line, Rs= 40.0 kPa s m^{-2}, Ω = 0.98, d = 0.018 m); and a locally-reacting slit pore layer (joined crosses, Rs = 40.0 kPa s m^{-2}, Ω = 0.98, d = 0.03 m); (b) Corresponding deduced (broken line) and predicted (solid line) impedance spectra assuming an externally reacting slit pore layer with the parameter values given in (a) [41].

The average error (Equation (6.5a)) from using the Delany - Bazley impedance model to fit level difference data from 26 grass-covered sites, all but one of which fits would qualify for Nordtest classification, is 9.3 dB. Since it was designed to give predictions similar to the Delany - Bazley model, Taraldsen's model [44] results in similar short-range propagation predictions and errors [47]. The corresponding average error using either the slit-pore or the phenomenological model is 8.7 dB and that from use of the variable porosity model is 6.7 dB. For eight grassland sites, the errors from using all four (semi-infinite) impedance models are comparable. Use of the variable porosity model provides better fits than other models for 24 out of 26 grassland sites. The variable porosity model also enables the best fits to narrow band data at three grassland sites used as examples in ANSI S1.18 [36].

Hard-backed-layer versions of the (semi-infinite) impedance models considered result in significantly smaller level difference fitting errors for 12 grassland sites. The mean fitting error (Equation 6.5a) using the hard-backed layer (i.e. two-parameter) version of Delany - Bazley model compared with data for nearly 50% of these sites is reduced from 11.8 to 7.5 dB. Similar reductions in mean fitting errors are achieved when using hard-backed-layer versions of the Taraldsen, slit-pore and phenomenological models. Nevertheless for three-quarters of these grassland sites, the two-parameter variable porosity model (Equation (5.38)), with the rate of change of porosity expressed as the thickness of an equivalent non-hard backed layer, i.e. $d_e = 4/\alpha_e$, gives fits that are better than those obtained with the two-parameter forms of Delany - Bazley or Taraldsen models (see Table 6.1).

Examples of NORDTEST data obtained at short range of two types of grassland and fits obtained using three impedance models (Delany - Bazley, Miki and modified Miki (see Section 5.2.1)) are shown in Figure 6.11. While the fits to the short-range level difference magnitudes over grassland obtained by using the Delany - Bazley model are adequate for NORDTEST classification, it should be noted that other models yielding at least as good fits to short-range level difference magnitude data (e.g. as shown in Figure 6.11) predict significantly different impedance spectra (see Figure 6.12).

Also, note that with the best-fit 'long grass' parameters, the Delany - Bazley model predicts that the real part of the surface impedance is negative below 150 Hz. Drawbacks of the Delany - Bazley and Miki models, including non-physical predictions at low frequencies, non-causality, less good fits to short-range data and the different impedance spectra obtained with best-fit parameters compared with physically admissible impedance models, are discussed in Chapter 5 and elsewhere [48a].

6.3.2 Fits to Data Obtained over Forest Floors, Gravel and Porous Asphalt

According to the NORDTEST criteria [36], the Delany - Bazley models (semi-infinite and hard-backed layer versions) fail to enable qualifying fits for the majority of short-range forest floor and gravel level difference data [44,46]. On the other hand, while giving best fits for grassland, the variable porosity model results in poorer fits to short-range data over relatively low

Table 6.1 Layer impedance model fits to third octave band data for twelve grassland sites [41,47]

	Delany - Bazley			Slit pore				Phenomenological				Taraldsen		
Site #	Effective flow resistivity kPa s m^{-2}	Effective layer depth m	Error (Eqn. 12a) dB	Effective flow resistivity kPa s m^{-2}	Effective porosity	Effective layer depth m	Error (Eqn. 12a) dB	Effective flow resistivity kPa s m^{-2}	Effective porosity	Effective layer depth m	Error (Eqn. 12a) dB	Effective flow resistivity kPa s m^{-2}	Effective layer depth m	Error (Eqn. 12a) dB
7	117.0	0.025	10.7	85.3	0.56	0.034	10.5	67.6	0.60	0.044	10.5	105.7	0.028	11.2
8	50.1	0.033	8.1	50.9	0.88	0.033	5.7	36.6	0.86	0.047	6.3	41.4	0.040	6.3
9	75.4	0.042	7.5	59.0	0.52	0.050	4.6	46.5	0.55	0.064	5.4	65.8	0.048	9.5
17	276.5	0.012	6.3	215.4	0.52	0.016	6.1	156.1	0.5	0.023	6.4	256.9	0.013	5.7
19	151.0	0.022	9.6	153.2	0.70	0.024	9.2	111.7	0.71	0.033	9.5	139.6	0.024	8.9
20	120.8	0.019	11.5	107.3	0.70	0.022	10.9	79.6	0.71	0.030	11.3	102.1	0.022	10.3
26	371.8	0.010	6.1	338.9	0.57	0.012	5.6	259.5	0.60	0.016	5.5	356.6	0.010	5.8
27	63.2	0.038	6.5	51.0	0.65	0.046	6.0	45.1	0.81	0.056	6.4	52.3	0.047	6.6
39	142.7	0.019	6.4	129.9	0.70	0.022	6.7	96.3	0.71	0.030	7.0	119.0	0.022	6.5
40	185.9	0.017	5.1	127.1	0.50	0.023	6.5	98.6	0.50	0.031	7.2	167.2	0.018	5.8
42	255.6	0.012	6.8	265.1	0.70	0.015	6.2	193.1	0.71	0.020	6.3	229.0	0.015	6.1
44	210.6	0.021	5.2	175.6	0.50	0.024	5.6	137.1	0.50	0.029	6.0	218.7	0.018	4.6

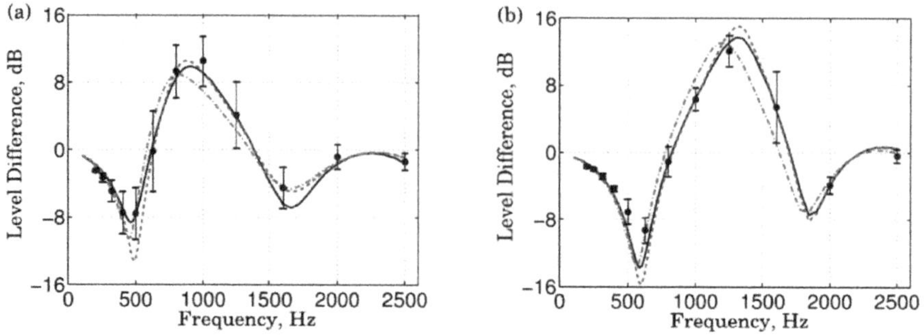

Figure 6.11 NORDTEST [38] third octave band level difference data (points) [error bars indicate 90% confidence limits (± 1.65 S.D.)] and best fit predictions [47] (a) for long grass using the variable porosity impedance model (solid line, effective flow resistivity 20 kPa s m^{-2} and porosity rate 50/m); the Delany - Bazley layer impedance model (broken line, effective flow resistivity 110 kPa s m^{-2} and effective layer depth 0.019 m); and the modified Miki layer impedance model (dash-dot line, effective flow resistivity 100 kPa s m^{-2} and effective layer depth 0.025 m) and (b) for lawn using the variable porosity model (solid line, effective flow resistivity 366.5 kPa s m^{-2} and porosity rate −79.5/m); the Delany - Bazley model (broken line, effective flow resistivity 746 kPa s m^{-2}) and the modified Miki model (dash-dot line, effective flow resistivity 565 kPa s m^{-2}) [48]. For details of the impedance models see Chapter 5.

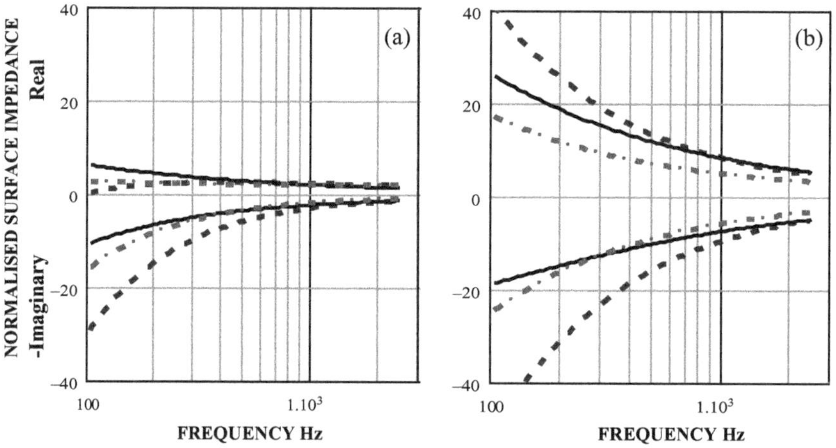

Figure 6.12 Impedance spectra corresponding to the best fits to short range level difference data over long grass and lawn shown in Figure 6.11. The line types correspond to the impedance models used for Figure 6.11.

flow resistivity forest floors and gravel than obtained with the two-parameter versions of the slit-pore and phenomenological models (see Table 6.2).

A three-parameter version of the Miki model also provides better fits to short-range data over low flow resistivity surfaces but, as discussed in Chapter 5 and elsewhere [48], this model is not physically admissible. The

Table 6.2 Comparison of best fit impedance model parameters and (Eqn. 12a) fitting errors based on third octave band data for five sites in a pine forest [41, 47].

	Delany - Bazley		*Slit pore*			*Zwikker and Kosten*			*Variable porosity*		
Site #	*Effective flow resistivity (kPa s m⁻2)*	*Error (Equation 6.5a) dB*	*Effective flow resistivity (kPa s m⁻2)*	*Effective porosity*	*Error (Equation 6.5a) dB*	*Effective flow resistivity (kPa s m⁻2)*	*Effective porosity*	*Error (Equation 6.5a) dB*	*Effective flow resistivity (kPa s m⁻2)*	*Effective rate of change of porosity (/m)*	*Error (Equation 6.5a) dB*
2	43	15.8	26.7	0.44	4.9	19.5	0.46	4.9	44.1	−80.9	13.5
3	38	11.3	33.03	0.56	7.1	23.9	0.57	6.9	23.9	−27.3	12.5
4	133	14.6	62.38	0.38	3.4	46.0	0.40	3.4	66.7	−75.1	6.9
5	51	18.7	22.75	0.43	3.8	16.4	0.44	3.7	47.9	−88.1	9.5
6	74	8.8	52.65	0.63	5.9	38.2	0.56	6.0	28.5	−18.3	10.3

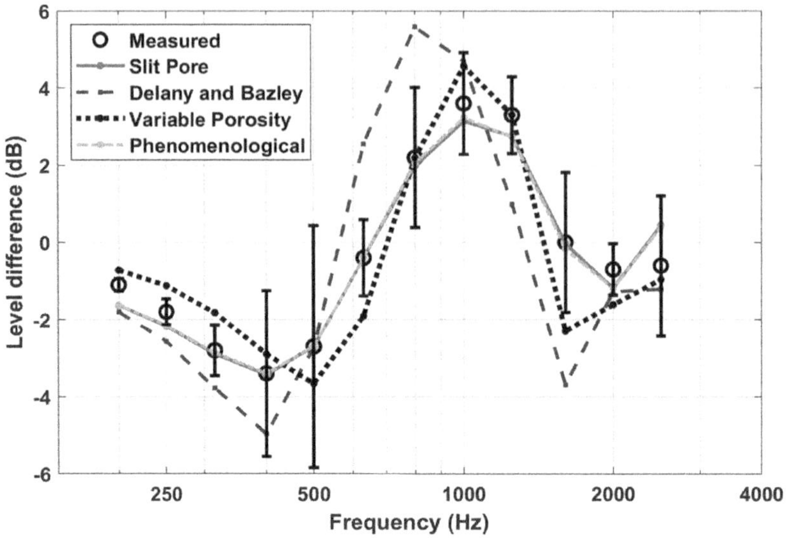

Figure 6.13 Short range level difference spectra (open circles; error bars indicate 90% confidence limits (±1.65 S.D.)) measured over a pine forest floor [38] and best fit predictions [47] using the slit-pore, variable porosity, phenomenological and Delany - Bazley impedance models [41].

example shown in Figure 6.13 demonstrates the better agreement with short-range propagation data over a forest floor that can be achieved with impedance models other than those of Delany - Bazley and Miki [41]. The corresponding parameter values and those resulting from fitting short-range level difference data over gravel are listed in Table 6.3.

Porous asphalt represents an artificial low flow resistivity outdoor ground surface. Independently (i.e. non-acoustically) measured flow resistivity values lie between 2 and 15 kPa s m^{-2}; independently measured porosity values are between 0.15 and 0.3, tortuosity values (deduced from absorption coefficient spectra) are between 2.5 and 3.3 and a typical layer thickness is 0.04 m [49]. The HARMONOISE prediction method [50] recommends the three-parameter Hamet model (Equations (5.5a, b)) for representing the acoustical properties of such a surface. Predictions of the three-parameter Hamet model and those of a five-parameter microstructural model, similar to the Johnson-Allard model (Equations (5.13) and (5.14)) but using cylindrical pore functions, have been found to give good agreement with narrow-band short-range propagation data over porous asphalt [49].

As part of a study related to potential revision of ACOU 104 [38], measurements were made above newly laid single-layer drainage asphalt using a source height of 0.42 m, distance between source and microphones of 4 m and microphone heights of 0.28 and 0.075 m [46]. The numerical best fit using the Delany - Bazley layer model results in a large fitting error (34.0 dB)

Table 6.3 Impedance model parameters giving best fits to short range level difference spectra measured over a pine forest floor and gravel in a pit [41,47]

Model	*Effective flow resistivity kPa s m⁻2*	*Porosity*	*Layer thickness m*	*Error dB*
	Pine forest floor NORDTEST site #5			
Delany - Bazley	51.0	-	-	18.7
Zwikker and Kosten	16.4	0.44	-	3.7
Slit pore	22.8	0.43	-	3.8
	Gravel NORDTEST site #38			
Slit pore layer	33.6	0.33	0.068	6.8
Delany - Bazley layer	55.0	-	0.04	26.6
Modified Miki layer	50.0	-	0.04	26.6
3-parameter Miki layer	13.8	0.462	0.057	14.4

and improbable best-fit values of effective flow resistivity (2687 kPa s m^{-2}) and layer thickness (0.097 m). Significantly better fits to these data (fitting errors of 3.9 and 7.4 dB, respectively) are obtained by using the slit-pore layer model with layer thickness 0.036 m, porosity 0.18 and effective flow resistivity 67.5 kPa s m^{-2} and the phenomenological layer model with layer thickness 0.04 m, porosity 0.18 and effective flow resistivity 86.9 kPa s m^{-2}. Figure 6.14 compares predictions using the slit-pore layer, Delany - Bazley layer and Taraldsen layer models with the short-range level difference spectra measured over porous asphalt [47]. The slit-pore layer predictions shown use the two-parameter form of the slit-pore model which assumes $T = 1/\Omega^{0.5}$. The latter relationship assumes spherical particles but, not surprisingly, since gravel particles are unlikely to be spherical, as shown also in Figure 6.14, $T = 1/\Omega^{0.64}$ yields improved agreement at higher frequencies [41].

6.3.3 Railway Ballast

Railway ballast consists of much larger stones and has a much lower flow resistivity than typical of gravel [51]. Heutschi [52] has made measurements (source height 0.5 m, receiver height 1.2 m and separation 7.5 m) and predictions of excess attenuation spectra above (Swiss) railway ballast surfaces. The Delany - Bazley model, Attenborough (4 parameter) model [47] and a model based on electrical network theory for the acoustical properties of railway

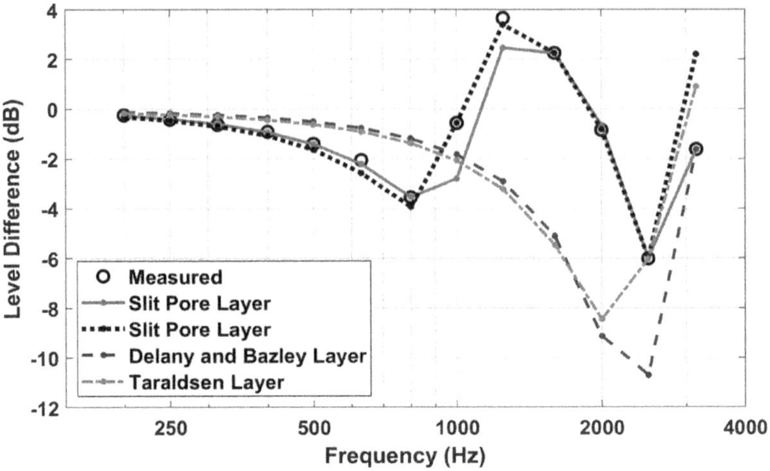

Figure 6.14 Numerically-obtained best fits to level difference data (open circles) over newly laid porous asphalt source height 0.42 m, distance between source and microphones 4 m, and microphone heights 0.28 m and 0.075 m (continuous line – slit pore layer R_s = 67.5 kPa s m^{-2}, h = 0.18, n' = 0.5, d = 0.036 m; dotted line - R_s = 61.2 kPa s m^{-2}, h = 0.22, n' = 0.64, d = 0.036 m); (a) broken line – Delany - Bazley layer, effective flow resistivity 2687 kPa s m^{-2}, d = 0.097 m; dash-dot line – Taraldsen layer, flow resistivity 1948 kPasm^{-2}, d = 0.14 m [41].

ballast were used. The electrical network model is based on an assumed microstructure of a rectangular lattice of spherical cavities (diameter d_p) connected by tubes (diameter d_k, vertical length l and horizontal length s). Because of the complicated track profiles, calculations of propagation loss were made using the Boundary Element Method. The best-fit parameter values were found to be: d_p = 0.0056 m, d_k = 0.00127 m, l = 0.00132 m, s = 0.006 m and a layer depth of 0.33 m, the latter value being consistent with the typical depth of the railway ballast which is expected to be between 30 and 60 cm. It was shown that excess attenuation predictions over ballast should allow for extended reaction. While it leads to predictions that compare well with excess attenuation data, the electrical network impedance model has the disadvantage that it is difficult to relate its parameters to measurable physical properties of the ballast.

Figure 6.15(a) shows that, use of the local reaction assumption and hard-backed-layer versions of the Delany - Bazley, slit-pore layer and phenomenological layer models result in reasonable fits to measured narrow band excess attenuation spectra [no railway tracks, see Figure 6 in Ref. 52]. However, the first ground effect dip magnitude and width are significantly over-predicted when using the Delany - Bazley layer model. Better fits to the first ground effect dip (see Figure 6.15(b)), similar to that obtained with the electrical network model, result from allowing for extended reaction [41,47].

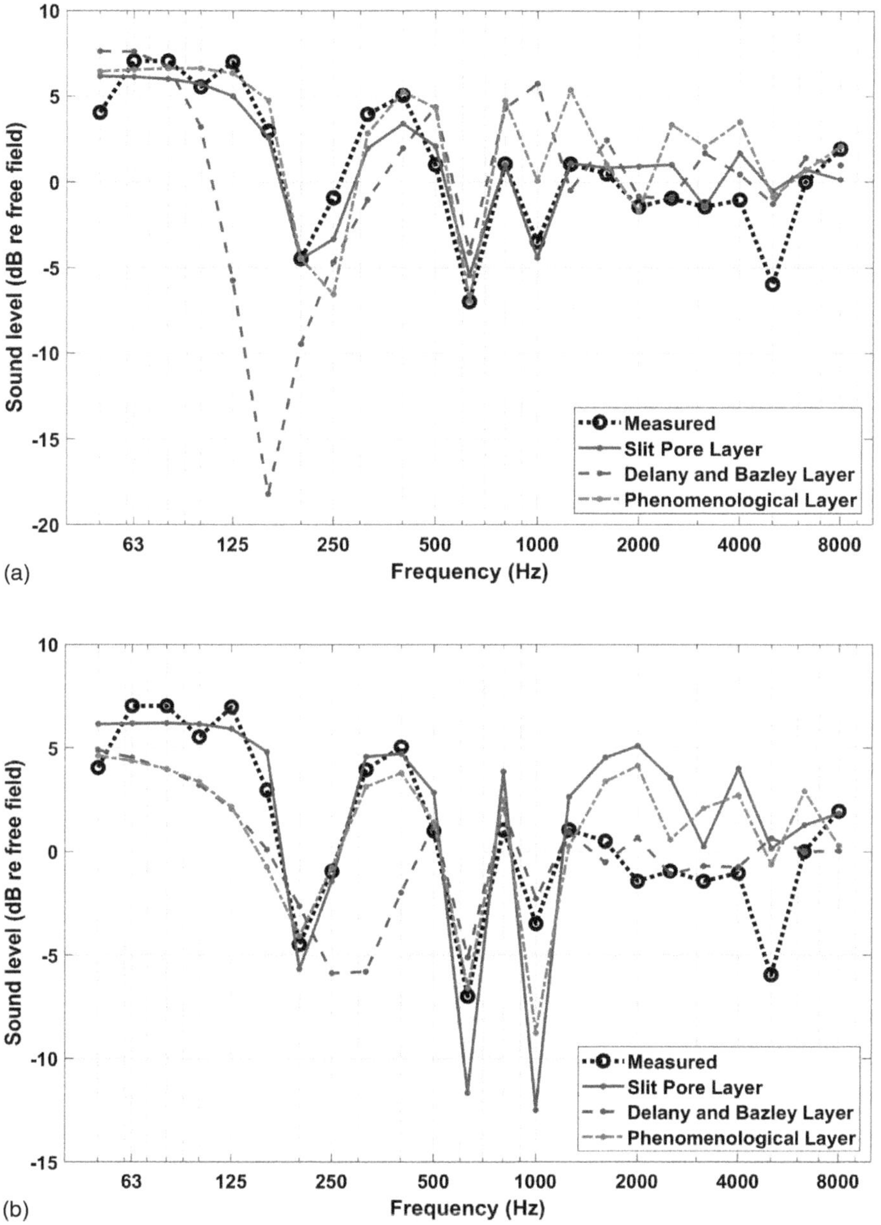

(a)

(b)

Figure 6.15 Comparison of data (black dotted line and circles) for excess attenuation spectra over railway ballast [ref. [53], Figure 6] and predictions (a) assuming local reaction and (b) assuming extended reaction using hard-backed-layer versions of the Delany - Bazley model (broken lines), the slit pore model (continuous lines) and the phenomenological model (dot-dash lines) with the parameter values given in Table 6.4 [41].

The corresponding fitted parameter values are listed in Table 6.4. However, the only realistic value of depth (0.3 m) is that resulting from fitting with the slit-pore layer model.

6.3.4 Measured Flow Resistivities and Porosities

Tables 6.5–6.7 show the range of flow resistivities and porosities that have been measured for outdoor ground surfaces [16,53–55]. The range of flow resistivities is particularly wide. According to Table 6.5, the flow resistivity of a ground described simply as 'grassland' could vary by a factor of five. According to Tables 6.6 and 6.7, the flow resistivity of other surfaces including soils, sands, snow and gravel could vary by a factor of more than 200. This is consistent with the variation in effective flow resistivity deduced from fitting-level difference spectra discussed earlier.

Table 6.4 Fitted parameter values for railway ballast corresponding to Figure 6.15 [41]

	Local Reaction			Extended Reaction		
Impedance Model	*Delany - Bazley Layer*	*Slit pore layer*	*Phenomenological Layer*	*Delany - Bazley Layer*	*Slit pore layer*	*Phenomenological Layer*
Effective flow resistivity ($Pa\ s\ m^{-2}$)	15	5409	161	10	100	66
Effective porosity	-	0.2	0.28	-	0.2	0.67
Effective layer depth (m)	0.274	0.2	0.23	0.7	0.3	1.445

Table 6.5 Measured flow resistivities and porosities of grassland

Ground type	*Flow resistivity ($kPa\ s\ m^{-2}$)*	*Porosity – (air-filled/ water-filled) or total*
Loamy sand beneath lawn (no roots)	677 ± 93	0.288/0.137
Grass covered compact sandy soil	463 ± 122	0.417/0.052
Grass-covered field	300	0.345/0.160
Loamy sand beneath lawn (0.06 m thick with roots)	237 ± 77	0.505
Grass	220	—
Grass root-filled layer	189 ± 91	—
Loamy sand with roots (mixed grass overgrowth)	114 ± 52	0.211/0.271

Table 6.6 Measured flow resistivities and porosities of other surfaces

Ground Type	*Flow resistivity (kPa s m⁻2)*	*Porosity total or (air-filled / water-filled)*
Wet sandy loam	1501	0.11
Compacted silt	1477	0.12
Mineral layer beneath mixed deciduous forest	540 ± 92	0.365/0.15
Sand (moistened)	479	0.37
Loamy sand on plain	422 ± 165	0.375/0.112
Hard clay field	400	-----
Sand (dry)	376	0.35
Bare sandy plain	366 ± 108	0.269/0.093
Dry sandy loam	259	0.5
Humus on pine forest floor	233 ± 223	0.581/0.161

Table 6.7 Measured values for low flow resistivity outdoor surfaces

Ground type	*Flow resistivity (kPa s m^{-2})*	*Porosity – (air-filled/ water-filled) or total*
Litter layer on mixed deciduous forest floor (0.02–0.05 m thick)	30 ± 31	—
Wet peat mul	24 ± 5	0.55/0.29
Beech forest litter layer (0.04–0.08 m thick)	22 ± 13	0.825
Snow (old)	16.4	0.574
Pine forest litter (0.06–0.07 m thick)	9 ± 5	0.389/0.286
Snow (new)	4.73	0.86
Gravel (mean max. grain dimension 9.02 mm)	1.648	0.38
Railway ballast [53]	0.2	0.491

Table 6.8 indicates that the presence of an acoustically soft organic *root layer* (the root zone) above otherwise rootless mineral soil reduces the flow resistivity near the surface by a factor of between 2 and 5 compared with that in the substrate [53]. The measured effects of roots on porosity are less dramatic.

6.3.5 Comparison of Template and Direct Deduction Methods over Grassland

In association with the development of ANSI S1.18 [36], measurements of complex level difference spectra have been made at three locations on a cricket field at the Open University (OU), Milton Keynes, UK and over institutional grass at the National Research Council (NRC) of Canada, Ottawa.

Table 6.8 Influence of root zones on flow resistivity and porosity

Ground type	*Flow resistivity (kPa s m^{-2})*	*Porosity (volume %)*
Bare loamy sand	422 ± 165	48.3 ± 1.7
Grass root layer in loamy sand	153 ± 91	47.9 ± 4.4
Grass root layer in loamy sand	237 ± 77	50.5 ± 9.3
Loamy sand beneath root-zone	677 ± 93	42.5 ± 1.7
Loamy sand with roots	1 14 ± 52	55.2 ± 4.5

Model parameter values have been deduced by using both the template method and by fitting impedance spectra resulting from the direct impedance deduction method. Figures 6.16 (a) and (b) show 'model-independent' impedance spectra deduced from the OU measurements (three sites × two geometries) and corresponding 'template' impedance model fits, respectively. Table 6.9 lists the parameters for the two-parameter variable porosity model which provides the best fits in good agreement with deduced impedance spectra.

6.4 SPATIAL AND SEASONAL VARIATIONS IN GRASSLAND IMPEDANCE

6.4.1 Predicted Effects of Spatial Variation

Ostashev et al. [56] have considered sound propagation above a homogeneous impedance plane with random properties as a function of range. They use the Wilson relaxation model (Equation (5.42)) to represent the surface impedance of the ground and assume log-normal distributions of flow resistivity, tortuosity and porosity. Statistics of the sound pressure level (SPL) at long range were computed using the analytical solution for a homogeneous plane. The main results of their study are (i) that the mean intensity, which corresponds to the mean SPL, is almost identical to the intensity obtained for the mean value of the ground characteristics and (ii) that the standard deviation of the intensity fluctuations increases with the distance, meaning that for a particular realization of the ground parameters, larger deviations of the SPL from the mean value can be obtained as the range increases. Another conclusion of their work was that, since it depends on the angle of incidence, propagation over randomly varying surfaces cannot be described through an effective impedance.

On the other hand, Dragna et al. [57] offer angle-dependent expressions for effective admittance in terms of the mean admittance with a correction depending on the correlation length of the spatial variation in surface

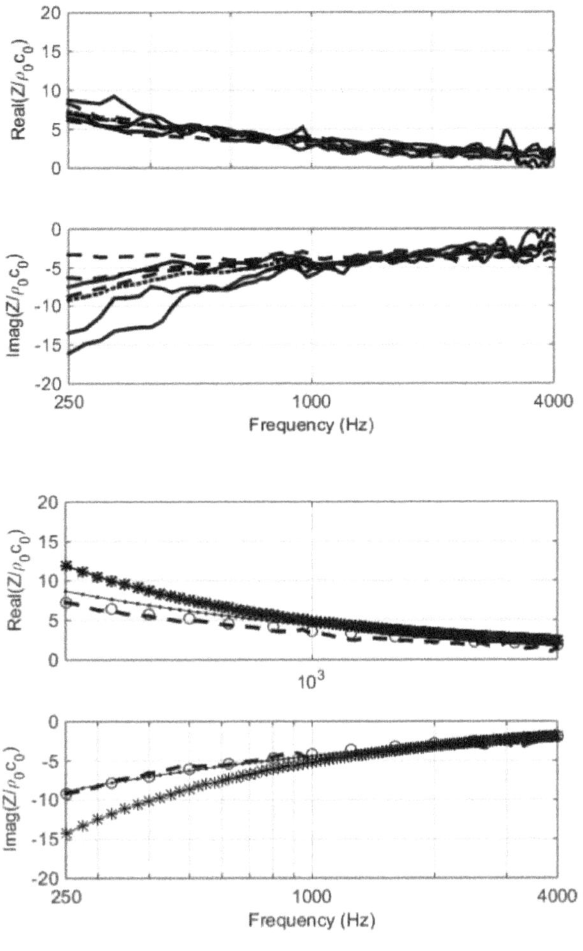

Figure 6.16 (a) Normalized specific acoustic impedance ratios deduced from six sets of averaged complex sound pressure ratios over a cricket field using the ANSI S1.18 procedure [36 and see section 6.1.3.4], solid lines represent deductions using geometry A and broken lines are from using geometry B: the dotted lines indicate the arithmetic means (b) Corresponding impedance model predictions; the broken line is the average deduced spectrum from Figure 6.16(a), asterisks and joined dots represent predictions from 'template' fitting of the one and two-parameter models respectively, the open circles represent best fit predictions of the two-parameter model to the average deduced impedance spectrum with values of σ_e = 70 kPa sm^{-2} and α_e = 25 m^{-1} (see also Table 6.9).

characteristics. Their numerical investigations indicate that (i) near grazing, the real part of the effective admittance decreases as the correlation length decreases, (ii) the coherence between direct- and ground-reflected components is reduced as the correlation length is decreased, (iii) the increased standard deviation of intensity fluctuations predicted by Ostashev et al. [56]

Table 6.9 Best fit parameters to short range level difference spectra over grassland using the variable porosity model

		Geometry A		Geometry B	
Location		*σeff* *kPa s m*$^{-2}$	*αe /m*	*σeff* *kPa s m*$^{-2}$	*αe /m*
		Best fit parameters using the template method			
OU	Area 1	100	50	100	3
	Area 2	100	3	100	100
	Area 3	100	3	100	50
NRC		100	50	100	50
		Best fit to directly deduced impedance spectra			
OU	Averaged over all	70	25	70	25
NRC		100	50	100	3

is particularly strong if the mean impedance corresponds to that of a thin hard-backed layer and (iv) that the amplitude of the surface wave over a spatially varying impedance is less than that predicted over a homogeneous impedance. Effects (i) and (ii) are consistent with those of surface roughness discussed elsewhere (Section 6.7 and Chapter 5) but effect (iii) is contrary to the predicted and measured effects of surface roughness.

6.4.2 Measured Effects of Varying Moisture Content

Several studies of the influence of moisture content on the acoustical properties of sands and soils [58–61] show that the higher the water content, the higher the measured acoustic surface impedance. Since it influences water retention in soils, organic matter plays a role also [61]. An interesting high-frequency resonance (~5 kHz) in the surface impedance spectrum has been found when applying water to a soil surface to form a thin surface film [58]. When impedance or admittance spectra deduced from impedance tube measurements have been fitted with single- or two-parameter models, increasing water content leads to an increase in the best-fit effective flow resistivity [59], at least until the saturation reaches 50%. However, when using the single parameter model, different values of effective flow resistivity were needed to fit the real and imaginary parts of surface admittance [59]. In contrast, a two-parameter model was found to enable reasonable fits to both real and imaginary parts. This supports the suggestion made elsewhere ([48], Chapter 5 and Section 6.6) that the single parameter empirical model is a poor choice for modelling the surface impedance of soils.

Independent acoustical influences of air and water contents have been investigated through impedance tube measurements on a sandy soil [60].

A larger air content than in the 'dry' state is found when there is 20% by weight of water. For a given water content, increase in air content results in higher absorption coefficients. On the other hand, for a fixed air content, increasing water content results in higher absorption also. This is attributed to increased particle cohesion associated with water content resulting in larger pores and a greater fraction of interconnected pores. Nevertheless, typically, an increase in water content is associated with a decrease in air content and hence with higher surface impedance and lower sound absorption [58,59,61].

The decrease in the noise shielding by a green roof due to increasing soil moisture in a vegetation (approximately 3-cm thick sedum plants)-covered substrate (7-cm thick, mineral fraction between 70% and 90%; organic content between 3% and 8%) has been investigated by measurements of the influence of natural rainfall and artificial wetting experiments [62]. This decrease in diffraction-induced noise shielding is related to the measured decrease in ground effect attenuation associated with the increase in moisture content, which is discussed in Section 6.4.4.

6.4.3 Influence of Water Content on 'Fast' and Shear Wave Speeds

Impedance tube and measurements of sound reflection from the ground involve mainly the 'slow' wave (see Sections 5.5.2 and 6.8) and, therefore, can be used to indicate the influence of water content on pore structure related parameters. However, moisture affects the other wave types in soil associated with frame elasticity. In soil physics, not only is moisture content considered in terms of the proportion of water by weight but also in terms of matric potential energy, often abbreviated to *matric potential or water potential*. This is a quantitative indication of the tendency for water to move from one area to another. A portion of the matric or water potential can be attributed to the attraction of the soil matrix for water. Over a large part of its range, the matric potential is due to capillary action, i.e. surface tension. However, as the water content decreases, there is a reduction in the proportion of water held in pores due to capillarity, compared to the proportion of water held directly on particle surfaces. A continuous set of measurements made in an instrumented trench on a university campus over a period of a little less than two years has shown that the 'fast' wave speed is related directly to matric potential [63]. An approximately exponential relationship (coefficient of determination 0.92) was found between 'fast' wave speed (V_p in m/s) and matric potential (ψ in kPa) given by

$$V_p = 161.6 + 4.72\psi^{0.49}. \tag{6.7}$$

The dependence during drying phases differed from that during wetting phases indicating hysteretic effects. When the moisture content was between

17% and 40%, the sound ('fast' wave) speed varied with the soil moisture content in an approximately linear fashion. When soil was relatively dry, i.e. moisture content <17%, the sound speed decreased non-linearly with the moisture content. Near the minimum soil moisture content, sound speed changed rapidly, i.e. independently of the soil moisture content indicating that the water potential rather than moisture content is the governing factor. The instrumented pit data showed not only how the 'fast' wave speed increased with depth as a result of overburden pressure but also that the depth dependence of the 'fast' wave speed is seasonal [63].

Measurements of shear wave speed have been made on samples of unconsolidated unsaturated sandy loam and clay soils in a triaxial cell and used to derive an empirical relationship between shear wave speed (V_s m/s), porosity (Ω), matric potential (ψ kPa) and overburden pressure (σ_s kPa) given by [64]:

$$V_s = 2.368\frac{(2.97-\Omega)^2}{1+\Omega}\left[\sigma_s^{\,0.746} + \psi\left(\frac{\psi}{\psi_{ae}}\right)^{-0.55}\right]^{7.085}, \tag{6.8}$$

where ψ_{ae} is the air entry potential, i.e. the maximum pressure potential at which soil starts to desaturate (starting at saturation), which is governed by the largest pores. This relationship was found to account of 95.6% of the observed variation independently of soil type.

Soil water status has been deduced empirically from measurements of seismic wave speeds using embedded sensors [64] but, as has been pointed out [65], several attempts to use acoustical measurements for delivering in situ information about water content are based on an uncertain theoretical foundation.

A more rigorous three-phase model of dynamic poroelasticity and consequent fitting of an extended set of parameters will be necessary to deduce moisture content in addition to pore structure parameters and soil-strength profiles by non-invasive acoustical methods (see also Section 6.8).

6.4.4 Measured Spatial and Seasonal Variations

A study of the spatial and seasonal variation of short-range propagation over three types of ground: an artificial grass football (soccer) pitch, a grass lawn and a 'natural' grass-covered ground has been reported by Guillaume et al. [66]. Over each ground measurements were made of spectra of the level difference between vertically separated microphones at heights of 0.6 and 0.0 m positioned 4 m from an omnidirectional source at height of 0.6 m and these spectra were measured at a series of locations. Measurement campaigns were carried out in summer and winter when the grounds were dry and moist, respectively. The resulting level difference data were fitted using the hard-backed layer versions of the Delany - Bazley and Miki models (Equations (5.1), (5.2) and Table 5.1), the fitted parameters being effective flow resistivity and effective layer thickness. The Delany - Bazley model fits for both the artificial grass and the grass lawn were found to lead to negative real parts of the fitted impedance at frequencies below 100 Hz. The Miki model was found to predict positive values for the real part of fitted impedance; however, the original Miki model coefficients and exponents were used (Table 5.1). As discussed in Chapter 5 and elsewhere [48,67], the original Miki model can lead to physically inadmissible predictions at low frequencies also. Nevertheless, the parameter values resulting from the Miki model fits are used in the subsequent analysis.

Table 6.10 lists mean, maximum and minimum Miki hard-backed layer model parameter values corresponding to (thirty) measured level difference spectra [66]. The variable porosity model has been shown to be physically admissible [67] and to give good fits to many short-range measurements over grassland [47]. Level difference spectra predicted using the Miki model for the measurement geometry used have been fitted also using the variable porosity model and the resulting parameter values are listed in Table 6.10 also. Note that, in this table, the porosity rate parameter (α_e) is replaced by effective layer thickness, d_e, using $\alpha_e = 4/d_e$.

The overall change in the short-range level difference spectra between Summer and Winter conditions is greater than the point-to-point variability at both sites. Furthermore, the parameters corresponding to the grass lawn show more seasonal variation than those for the 'natural' ground. The ranges of mean effective flow resistivity for the grass lawn and grass-covered natural ground, according to the variable porosity model (60–200 kPa s m^{-2} and 19.3–110 kPa s m^{-2}, respectively), are at the lower end of the range found elsewhere [47] for 26 grass-covered grounds (21.7–1296 kPa s m^{-2}).

Figure 6.17(a) shows an example of the short-range level difference spectra predicted by the Miki model with the mean winter grass lawn parameters listed in Table 6.10 and the best-fit predictions obtained using the variable porosity model using the parameter values in Table 6.10. Figure 6.17(b) shows the predictions of the corresponding impedance spectra. While the predicted level difference spectra are nearly identical, the predicted impedance spectra are significantly different.

6.5 GROUND EFFECT PREDICTIONS BASED ON FITS TO SHORT-RANGE LEVEL DIFFERENCE SPECTRA

Figures 6.18 and 6.19 show that a result of the different best-fit impedance spectra is that, at a long range, the variable porosity model results in predictions of larger ground effect maxima at lower frequencies than predicted when using the Miki model. The differences between the excess attenuation predictions based on the different impedance models are greater than those due to spatial variation and comparable with those due to seasonal

Table 6.10 Mean, maximum and minimum parameter values from fits to level difference spectra measured over grassland in summer and winter [66]

	Miki model parameter values				*Variable porosity model parameters*			
	Summer		*Winter*		*Summer*		*Winter*	
Season parameter	*Effective flow resistivity kPa s* m^{-2}	*Effective thickness m*	*Effective flow resistivity kPa s* m^{-2}	*Effective thickness m*	*Effective flow resistivity kPa s* m^{-2}	*Effective thickness m*	*Effective flow resistivity kPa s* m^{-2}	*Effective thickness m*
Grass lawn								
mean	354	0.0157	732	0.0058	80	0.035	200	0.011
maximum	510	0.0116	990[a]	0.0042	105	0.023	285	0.008
minimum	246	0.0138	409	0.0082	60	0.035	115	0.018
Natural ground								
mean	173	0.0183	243	0.0158	40	0.042	55	0.037
maximum	355	0.0148	460	0.0127	60	0.035	110	0.028
minimum	73	0.025	75	0.026	19.3	0.055	20	0.062

[a] this is the maximum effective flow resistivity permitted in the fitting process (990 kPa m^{-2}). It is stated as the 'best fit value on six occasions …. over the grass lawn but might indicate sub-optimal fitting'

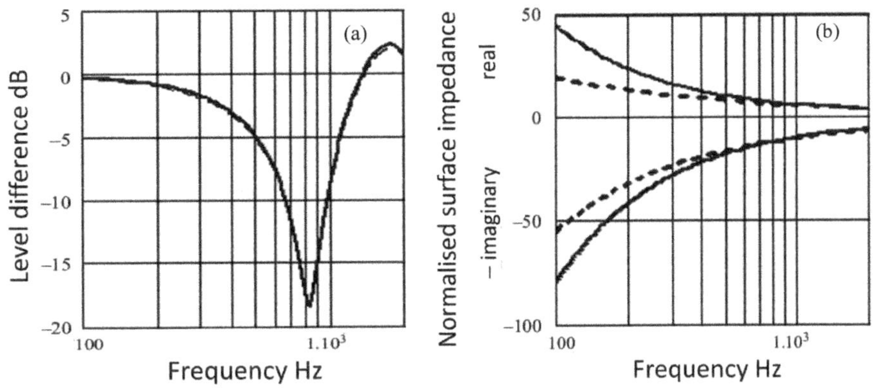

Figure 6.17 (a) Level difference spectra predicted for the short-range geometry used by Guillaume et al [57] parameter values for the mean winter grass lawn listed in Table 6.10 (b) the corresponding predicted impedance spectra (broken lines for Miki model and solid lines for variable porosity model).

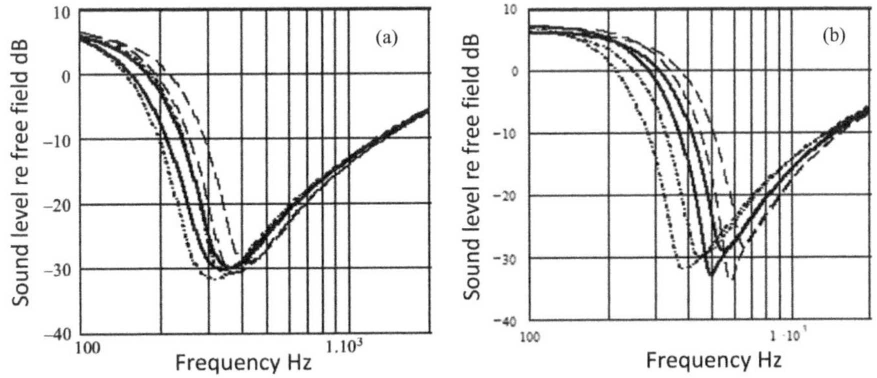

Figure 6.18 Predictions of the excess attenuation spectra between 2 m high source and receiver separated by 500 m above a grass lawn in (a) Summer and (b) Winter, using the Miki impedance model (black lines) and the variable porosity impedance model (grey lines), with the respective mean (solid lines), maximum (broken lines) and minimum (dash dot lines) parameter values listed in Table 6.10.

differences. The predictions of excess attenuation spectra in Figures 6.18 and 6.19 include very slight turbulence (Gaussian: mean variance 10^{-10}, outer scale length 1 m) (see Chapter 10).

However, even allowing for moderate turbulence, the fact that comparably good fits to short-range level difference magnitude data over grassland result from significantly different impedance spectra means that predictions of excess attenuation due to ground effect at longer range are significantly different also. Examples predictions are shown in Figure 6.20.

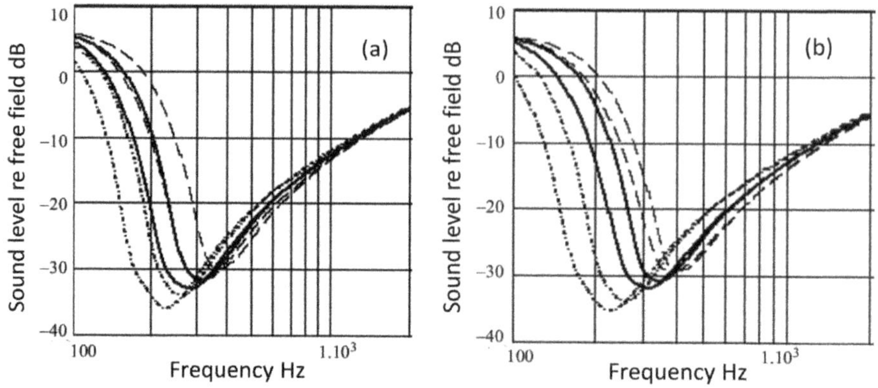

Figure 6.19 Predictions of the excess attenuation spectra between 2 m high source and receiver separated by 500 m above a natural ground in (a) Summer and (b) Winter, using the Miki impedance model (black lines) and the variable porosity impedance model (red lines), with the respective mean (solid lines), maximum (broken lines) and minimum (dash dot lines) parameter values listed in Table 6.10.

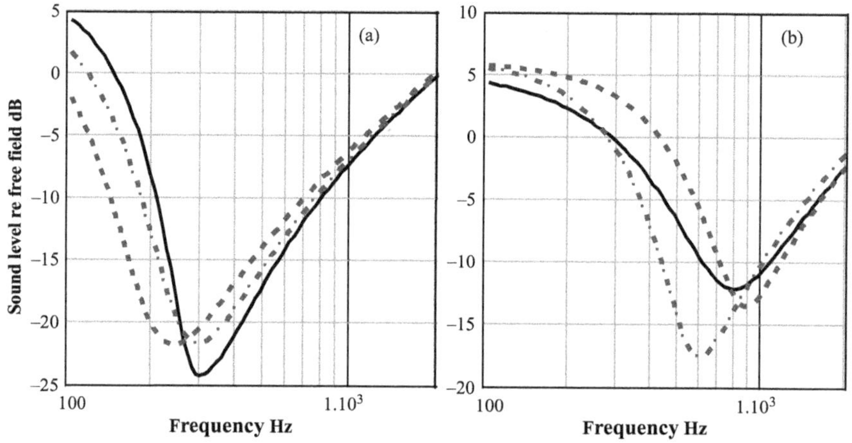

Figure 6.20 Comparisons of excess attenuation spectra in a moderately turbulent atmosphere ($\langle \mu^2 \rangle = 10^{-6}$, L_0 = 1 m) for (point) source height 1 m, receiver height 1.5 m and range 100 m predicted using impedance models with parameter values giving best fits to short range data obtained over (a) long grass, NORDTEST site #20 and (b) lawn, NORDTEST site #30 (see Figure 6.11). The line types correspond to those used in Figures 6.11 and 6.12.

According to Figure 6.20a, at 200 Hz for 100 m source–receiver separation, the predictions obtained using the physically admissible variable porosity model differ by about 10 dB from those obtained using the Delany - Bazley (layer) model, despite using parameter values that enable predictions, using both models, that are in close agreement to short-range level difference magnitudes over long grass. According to Figure 6.20b at 500 Hz, there would be at least 5 dB difference in the predictions over 'lawn'. The latter difference would be significant for A-weighted road traffic noise predictions. While the difference of 10 dB at 200 Hz shown in Figure 6.20a might not be particularly significant for predictions of A-weighted road traffic noise spectra, it would be important for predictions of levels from sources with substantial low-frequency content. Moreover, the difference is predicted to become greater as the range increases.

There can be significant differences between predictions for impulse propagation (in the time domain) in which allowance is made for atmospheric refraction [48b]. Figures 6.21(a) and (b) show example time domain calculations over a gravel surface (see Table 6.3) using three hard-backed layer impedance models (Delany - Bazley, 3-parameter Miki and slit pore) in downward and upward refraction conditions, respectively. A logarithmic sound-speed profile (see Chapter 10) is assumed with source and receiver heights of 1 and 2 m, respectively, separated by 100 m. Although the

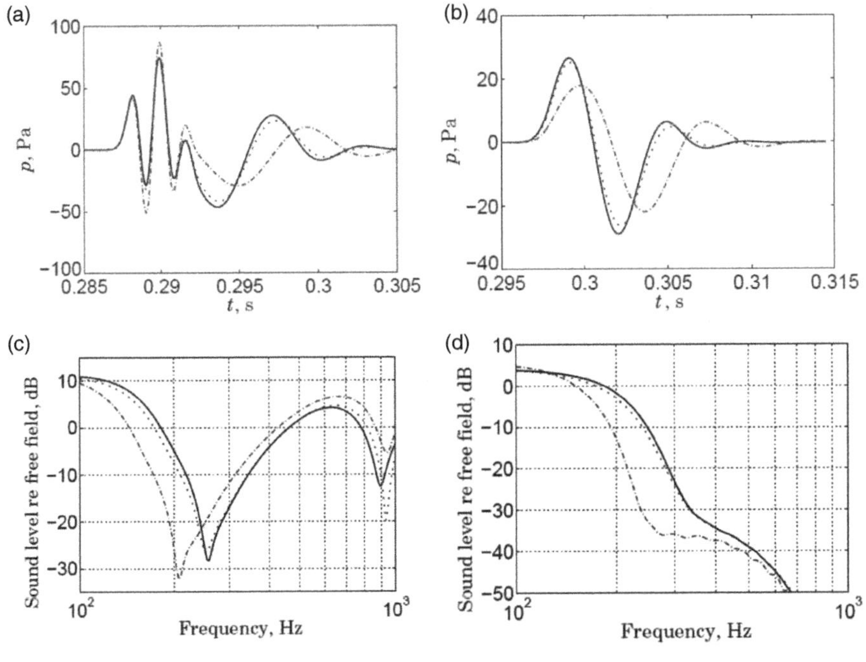

Figure 6.21 Predicted time series of acoustic pressure obtained at a 2 m high receiver at 100 m (a) for a downward-refracting and (b) an upward-refracting atmosphere using impedance models and parameters values (see Table 1) giving best fits to short range data obtained over gravel, NORDTEST #38 ((solid) slit pore, (dash-dotted) modified Miki and (dotted) three-parameter Miki impedance models. (c) and (d) show the corresponding predictions of excess attenuation spectra.

slit-pore and three-parameter Miki models lead to similar predictions, there are significant differences between these predictions and those using the Delany - Bazley model particularly in upward refraction conditions, i.e. 5 Pa differences in predicted peak pressures and at least 10 dB differences in the predicted EA spectra between 200 and 300 Hz.

6.6 ON THE CHOICE OF GROUND IMPEDANCE MODELS FOR OUTDOOR SOUND PREDICTION

The Delany - Bazley one-parameter semi-empirical model, developed for predicting the acoustical properties of highly porous fibrous materials, has been used widely as the ground impedance model for outdoor sound predictions since the 1970s. It is recommended as the default model for predicting outdoor ground impedance (except for porous asphalt) in the HARMONOISE prediction scheme and in NORDTEST ACOU 104. Moreover, effective flow resistivity classes based on the Delany - Bazley model feature in the European prediction scheme for noise mapping [68].

Despite the reported success of the Delany - Bazley model in fitting data obtained over many grass-covered surfaces, extensive fitting of short-range propagation data shows that other impedance models result in better fits than the Delany - Bazley model, albeit requiring at least one additional parameter. Furthermore, use of the Delany - Bazley model (either in its semi-infinite or hard backed layer versions) fails consistently to give good fits to short-range data for acoustically 'softer' surfaces such as those of forest floors, porous asphalt and gravel. Other models for the acoustical properties of rigid-porous media reviewed in Chapter 5 including the Hamet model, Wilson's relaxation model, the variable porosity model, the phenomenological (Zwikker and Kosten) model and identical tortuous pore models give better fits. By using a fixed relationship between tortuosity and porosity, two-parameter versions of the slit-pore, Wilson relaxation and Hamet impedance models can be obtained which give similar predictions and could be used as alternatives to single parameter models for ground characterization. In the frequency domain, the option of assuming slit-like pores is computationally convenient since the required hyperbolic tangent function is needed anyway when a hard-backed layer model gives better fits than a semi-infinite one. Also, if allowance is made for extended reaction [41,47], the slit-pore layer model has been found to enable a reasonable fit to excess attenuation data obtained over railway ballast with realistic parameter values.

On the basis of the mean fitting errors summarized in Table 6.11, the fits for porous asphalt and railway ballast that have been discussed, the non-physical predictions that can result from using the Delany - Bazley model and the differences in longer range predictions that can result (discussed in Section 6.5), that it should be used as the default impedance model [69] is

Table 6.11 Mean errors (eqn. (12a)) dB in fitting to third-octave band data using eight impedance models [48,49]

Model/ ground type	*Delany - Bazley*	*Delany - Bazley layer*	*Slit pore*	*Slit pore layer*	*Phenomenological*	*Phenomenological layer*	*Variable porosity*	*Taraldsen layer*
Grassland (26 sites)	9.3	-	8.7	-	8.7	-	6.7	-
Grassland (12 sites)	11.8	7.5	11.4	7.0	11.4	7.3	7.6	7.3
Pine forest floor (single forest, 5 sites)	13.8	-	5.0	-	5.0	-	10.5	13.8
Pine forest floors (2 sites) forests, 5 sites)	17.4	15.4	16.7	14.3	17.0	15.0	15.8	-
Beech forest floor (3 sites)	26.3	22.6	18.8	11.6	19.4	13.8	24.8	24.4
Gravel and sand (4 sites)	15.2	15.2	12.5	9.4	12.5	9.9	9.8	15.2

not a good recommendation and it should be replaced by two- or three- (if layer depth is included) parameter impedance models which are physically admissible. Despite the fact that original and modified three-parameter versions of the Miki model enable better fits to short-range data than the single parameter version of the Miki model or the Delany - Bazley model, for the same reasons, use of the Miki model in any of its forms for representing outdoor ground impedance is not to be recommended.

To encourage the use of alternative physically admissible impedance models when predicting outdoor sound, Tables 6.12–6.14 summarize ranges of best-fit parameter values for different ground surfaces. Table 6.12 lists parameters for grassland sites where the variable porosity model gave better fits to short-range data than the single-parameter models and at least as good fits as the hard-backed layer (two-parameter) version of the Delany - Bazley model. Table 6.13 lists parameters for sites where the two-parameter slit-pore model gave better fits to short-range data than the single parameter

Table 6.12 Parameter values giving best-fits to short range level difference spectra over 'grassland' using the variable porosity impedance model [48,49].

Grassland description	*Effective flow resistivity kPa s m^{-2}*	*Porosity rate m^{-1}*
Lawn	39	17
Natural ground (summer)	40	95
Heath	52	33
Natural ground (winter)	55	108
Sports field	66	58
Lawn	75	47
Sports field	78	158
Lawn (summer)	80	114
Long grass	167	48
Lawn (winter)	200	364
Lawn	367	−76
Pasture	383	43

Table 6.13 Parameter values giving best-fits to short range level difference spectra over various types of ground using the two-parameter slit pore model

Ground description	*Effective flow resistivity kPa s m^{-2}*	*Effective porosity*
Beech wood floor	14	0.51
Pine forest floor	27	0.44
gravel	34	0.33
Pine forest floor	62	0.38
Institutional grass	159	0.45

Table 6.14 Parameter values giving best-fits to short range level difference spectra measured over various types of ground using the hard-backed layer version of the slit pore model [48].

Ground description	*Effective flow resistivity kPa s m^{-2}*	*Effective porosity*	*Effective depth m*
Beech wood floor	32	0.35	0.07
gravel	34	0.33	0.07
Urban grassland	59	0.52	0.050
heath	175.6	0.50	0.024
pasture	338.9	0.57	0.012

Delany - Bazley model and the variable porosity model. Table 6.14 lists parameters for sites where the hard-backed layer version of the slit-pore model (i.e. a three-parameter model) gave the best fits. In contexts where further simplification might be preferred, such as for noise mapping, the three-parameter form of the slit-pore layer model could be reduced to two parameters by fixing the porosity at a value of 0.4.

6.7 MEASURED AND PREDICTED SURFACE ROUGHNESS EFFECTS

6.7.1 Roughness-Induced Ground Effect

Near grazing incidence, the presence of roughness even at small scales compared with the incident wavelengths reduces the effective acoustical impedance of a surface. The effects are most noticeable if the surface material is acoustically hard and increases with the mean roughness height. Roughness effects have been observed in laboratory measurements of excess attenuation from an elevated continuous point source of white noise in the frequency range 500 Hz–5 kHz over various rough hard and rough finite-impedance boundaries [43,69–74]. In comparing predictions based on Equation (5.54) with data, the assumed location of the effective admittance plane had to be adjusted to improve agreement with data at higher frequencies [43].

Laboratory measurements have been made of sound propagation over several small-scale rough surfaces formed by placing varnished wooden strips with various cross-sectional shapes including (approximately) semi-cylindrical (0.008 m high with 0.02 m long base), triangular (0.015 m high and 0.03 m long base), rectangular (sides 0.012 and 0.0285 m), (approximately) square (sides 0.018 and 0.02 m), with random or periodic spacing, on a glass sheet [41,74]. The (point) source and receiver were located at equal heights and the strip locations were centred on the point of specular reflection halfway between them. Figure 6.22 illustrates the different excess attenuation spectra that resulted from each random arrangement.

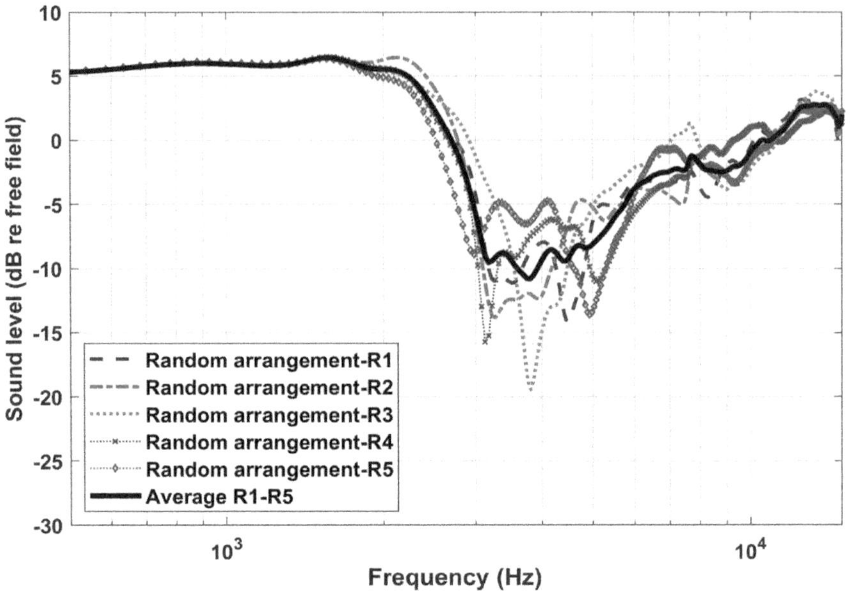

Figure 6.22 EA spectra measured over 15 randomly spaced parallel triangular strips (height 0.015 m, base 0.030 m) with mean centre-to-centre spacing of 0.05 m with source and receiver at a height of 0.07 m and separated by 0.7 m. The data correspond to five different random distributions (R1- dashed line, R2- dash-dot line, R3- dotted line, R4- joined crosses, R5- joined diamonds) and the average spectrum is shown also (thick solid line) [41].

Figures 6.23(a) and (b) compare the mean EA spectra measured over several types of 2-D roughness with that measured over the smooth hard glass sheet. For source and receiver heights of 0.07 m and a separation of 0.7 m, the first destructive interference above a smooth hard ground should occur at a frequency of 12.3 kHz. This is confirmed by the EA spectrum measured over the glass plate alone (i.e. a smooth surface). The measured EA maxima in the presence of identical randomly spaced strips with various cross-sectional shapes (semi-cylindrical, triangular, square, short rectangular and tall rectangular) are at lower frequencies. Moreover, for a given average spacing, these maxima increase in magnitude and become sharper as the roughness height increases. The first EA maximum occurs at about 2 kHz for the tallest (0.0285-m high) roughness elements. There is little difference between the mean EA spectra measured for average random spacings of 0.05 and 0.08 m.

6.7.2 Excess Attenuation Spectra for Random and Periodic Roughness

Measurements have shown that there are considerable differences between the ground effects caused by periodically and randomly spaced roughness with the same packing density and same height. Excess attenuation spectra

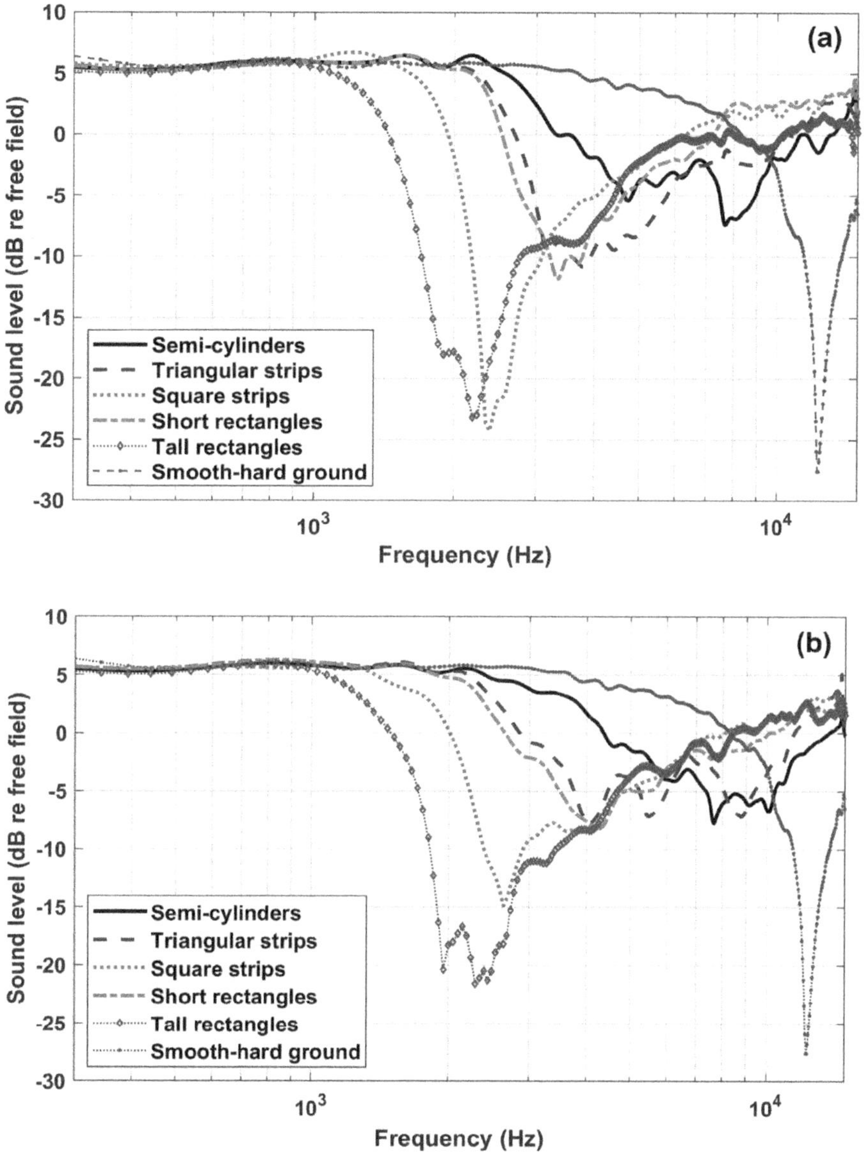

Figure 6.23 Averages of EA spectra measured over five random strip distributions with source and receiver height at 0.07 m separated by 0.7 m over surfaces composed of semi-cylinders (0.008 m high, 0.020 m base, solid line), triangular strips (dimensions as for Figure 6.22, dashed line), approximately square strips (0.018 m high, 0.020 m base, dotted line), short rectangular strips (0.012 m high, 0.0285 m base, dash-dot line) or tall rectangular strips (0.025 m high, 0.013 m base, joined diamonds) randomly spaced on a glass sheet with mean centre-to-centre spacing of (a) 0.05 m and (b) 0.08 m. In both cases EA spectra measured over the smooth hard glass sheet are shown by the joined asterisks [41].

measured over periodically spaced roughness elements with different cross-sectional shapes show multiple maxima and greater relative sound level minima [69,72]. Since a modified Tolstoy/Twersky theory (see Chapter 5) predicts a single excess attenuation maximum, it has been suggested that the additional maxima are caused by diffraction grating effects. An additional diffraction grating term has been incorporated by a heuristic modification of the classical analytical approximation for the propagation of sound from a point source over an impedance plane at near-grazing incidence [71]. However, an efficient alternative approach has been to incorporate the multiple edge diffraction effects through a heuristic modification of the effective admittance of a rough hard boundary [40,73]. Predictions of ground effect due to regularly spaced roughness elements based on an extended effective admittance model are sensitive to small deviations from exactly periodic spacing. Incoherent scattering has been shown to play an important role for the source–receiver geometries and roughness sizes studied. The effective admittance model has been generalized for arbitrary scatterer shape. While predictions of the resulting approximation agree with data obtained over wooden slats and triangular wooden rods on a flat hard surface to some extent, for the larger scatterers considered, the boundary element method gives better predictions.

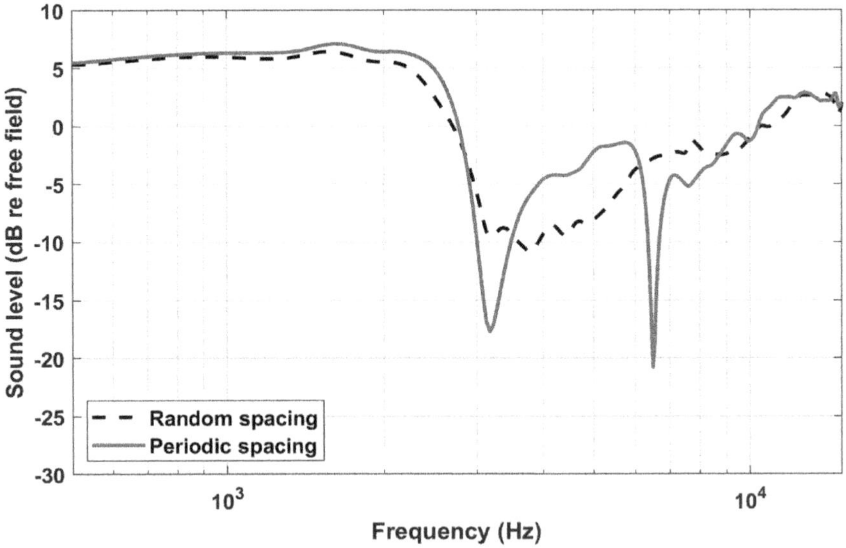

Figure 6.24 Comparison between measured EA spectra for propagation with source and receiver at a height of 0.07 m separated by 0.7 m over surfaces composed of 15 randomly- (broken line) and periodically- (continuous line) spaced triangular strips (dimensions as for Figure 6.22) mean centre-to-centre spacing of 0.05 m) on a glass sheet.The spectrum for random spacing is the result of averaging measurements over five random distributions [41].

Laboratory measurements of sound propagation over several small-scale rough surfaces composed from identical 2-D roughness elements, of various cross-sectional shapes, randomly and regularly spaced with a (mean) centre-to-centre spacing of between 0.03 and 0.08 m on a glass sheet [41,74] provide more evidence of the contrast between the excess attenuation spectra resulting from regular- and random arrangements of identical roughness elements with the same mean spacing. The lowest frequency of destructive interference observed over a regularly spaced roughness arrangement is close to that observed in the mean of spectra obtained over equivalent random arrangements. However periodic spacing results in deeper and additional excess attenuation maxima. Example excess attenuation spectra obtained over 15 triangular cross section strips with mean spacing of 0.05 m are shown in Figure 6.24.

As illustrated by Figure 6.25, the frequency of the first destructive interference observed due to a periodically rough hard surface depends on the spacing of the elements which is not predicted by Twersky/Tolstoy effective admittance models (see Chapter 5).

The frequency of the second excess attenuation magnitude maximum observed for periodic structures depends linearly on the spacing in a similar manner to the variation expected over a diffraction grating and corresponds to the first zero-crossing of the imaginary part of the deduced effective impedance (see Figure 6.26). In this respect, the effective impedance spectra

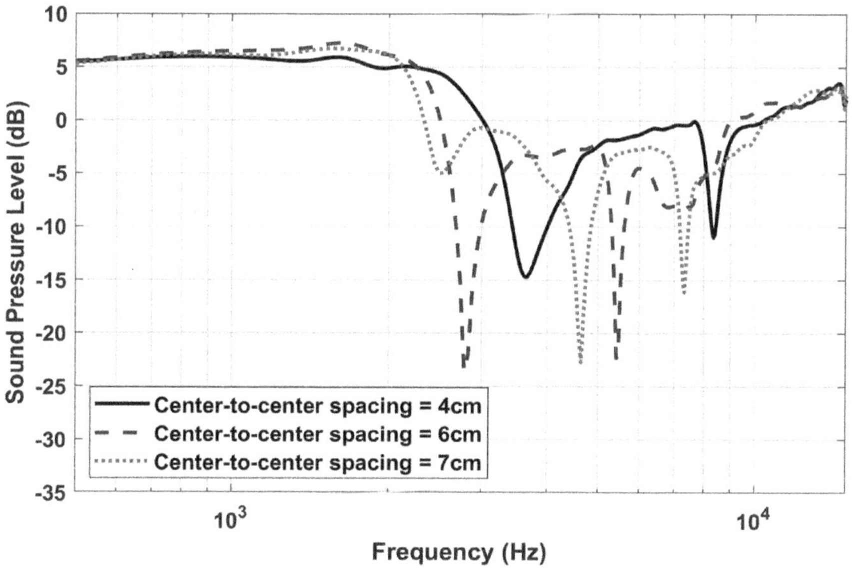

Figure 6.25 Measured EA spectra for source and receiver heights of 0.07 m separated by 0.7 m over surfaces consisting of triangular strips (dimensions as for Figure 6.22) regularly spaced on a glass sheet with mean centre-to-centre spacings of 0.04 m, 0.06 m and 0.07 m [41].

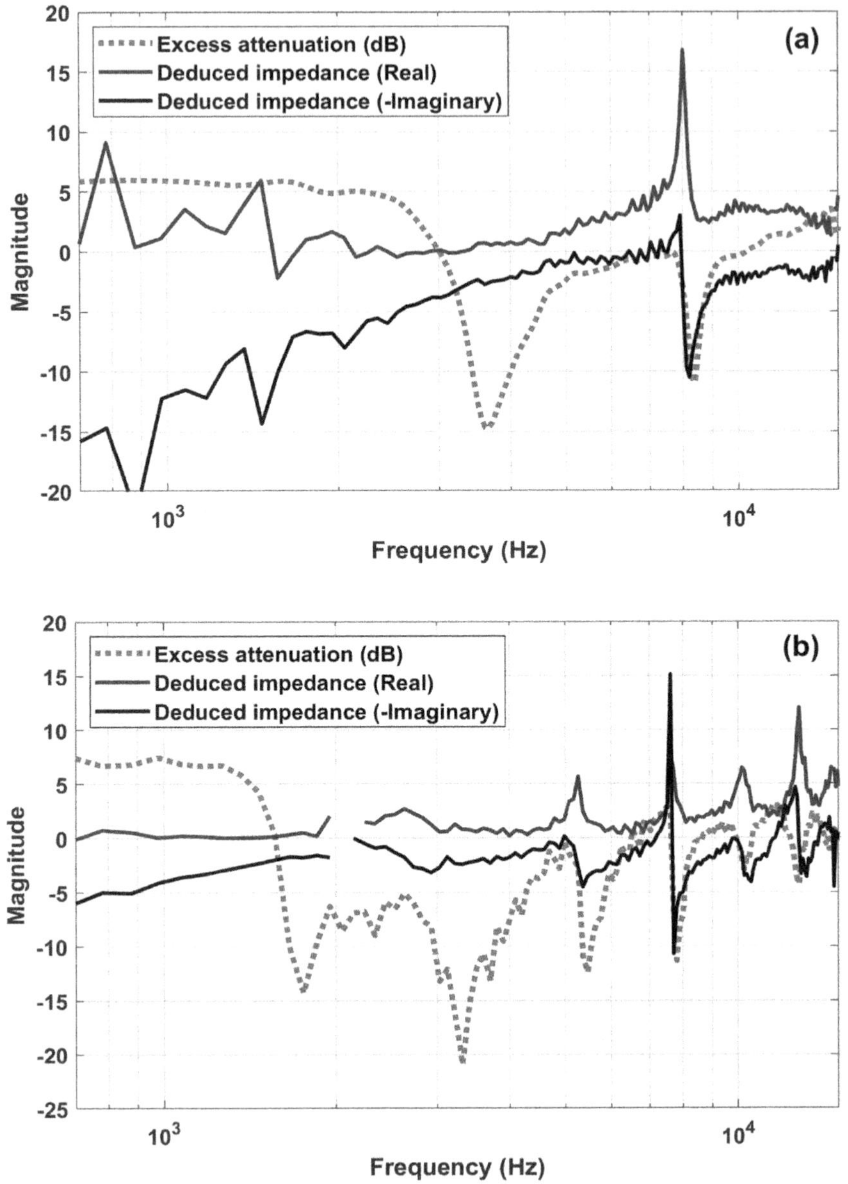

Figure 6.26 Measured EA spectra (dotted lines) due to strips on a glass sheet with source and receiver heights of 0.07 m separated by 0.7 m and spectra for real and imaginary parts of impedance (solid lines) deduced from the complex EA data for (a) 15 Triangular Strips (dimensions as for Figure 6.22) with centre-to-centre spacing of 0.04 m and (b) 9 tall rectangular strips (dimensions as for Figure 6.23) with centre-to-centre spacing of 0.08 m [41].

resemble those expected for a hard-backed layer of porous material. The second EA maximum frequency corresponds to a half-wavelength resonance in the 'effective layer'.

Data such as those shown in Figures 6.24–6.26 support the empirical modifications to Tolstoy effective admittance theory shown in Equations (5.61). Further support for these modifications is supplied by the comparisons between laboratory data and predictions shown in Figure 6.27 [41].

Higher frequency excess attenuation maxima (other than the second order maxima), such as the maximum observed near 9 kHz in Figure 6.27(a), depend on the proportion of the surface not occupied by the roughness elements.

6.7.3 Roughness-Induced Surface Waves

Thin porous polymer foam layers with very high porosity support two types of surface waves. One is similar to a dispersive Rayleigh wave and is associated mainly with elliptical *solid* particle motion in the frame [75]. In polymer foam layers it is the counterpart of the air-coupled surface wave in the ground, known as ground roll, observed following volcanic eruptions [76]. The other has a phase speed slightly lower than the speed of sound in air [77–79]. It is associated mainly with elliptical motion of *air* particles and results when the imaginary part (reactance) of the surface impedance is greater than the real part (resistance) (see Chapter 2). It spreads cylindrically along the surface but decays exponentially with height above the surface. It is an evanescent wave. However, of interest in this section is the fact that the 'thin layer' character of the effective surface impedance of a hard boundary with small-scale roughness results in the formation of an air-coupled surface wave also.

Brekhovskikh [80] studied surface wave propagation over a comb-like structure and suggested that the structure has an effective surface impedance equivalent to that of a layer of air with the same thickness as the height of the comb. Medwin et al. [81–83] carried out laboratory studies of the 'boundary' wave formed over several different types of roughness and, as a result of comparisons of these data with Tolstoy's theory (see Chapter 5), suggested empirical modifications to the theory.

Donato [84] used continuous sound from a point source to study the propagation of surface waves over a rectangular lattice (lighting diffuser) placed on a wooden board and showed that the measured vertical and horizontal attenuation rates of the acoustically induced surface waves were consistent with theoretical predictions. Others [85,86] have studied propagation of tone bursts and pulses over lattice surfaces. Separation of the surface wave from the main arrival in the time domain has confirmed that the surface wave travels slightly slower than the speed of sound in air. The measured properties of surface waves have been found to be in good agreement with theoretical predictions. As is the case with the surface impedance of a

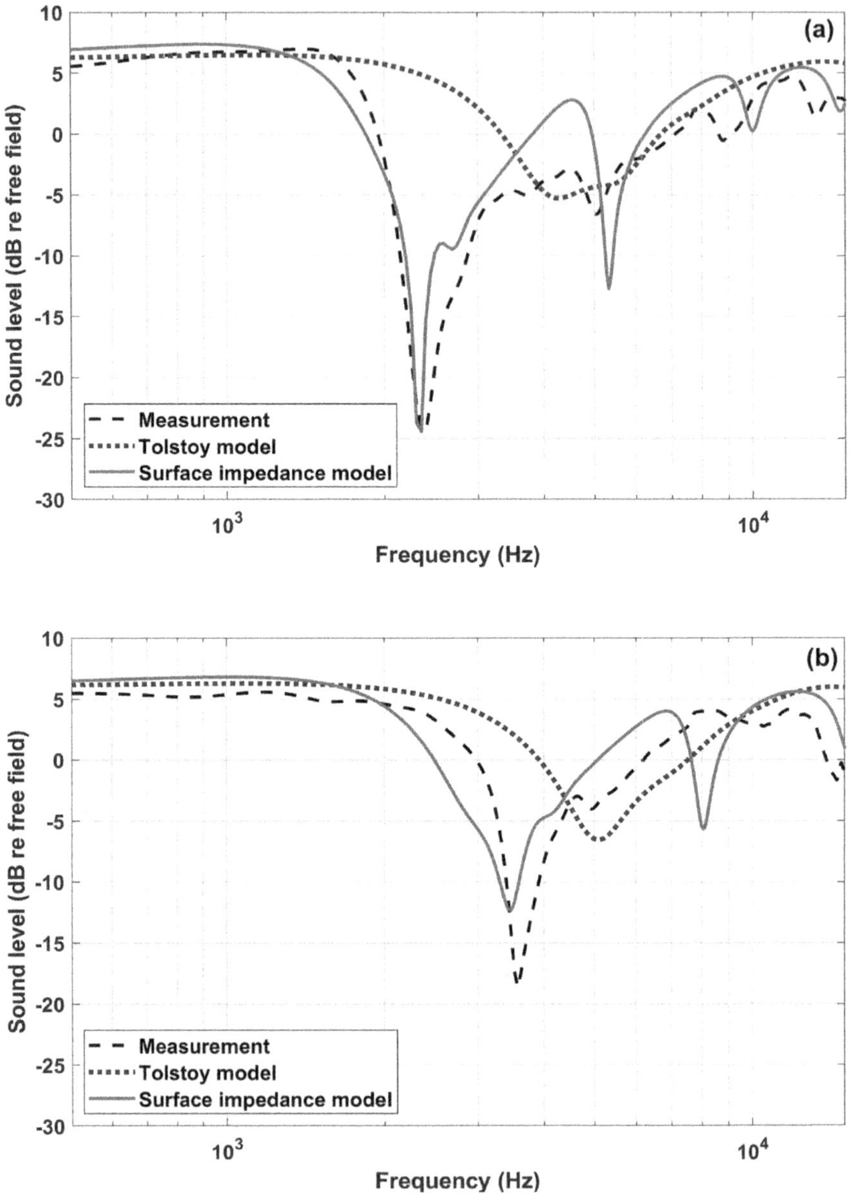

Figure 6.27 EA spectra measured with source and receiver at a height of 0.07 m and separated by 0.7 m over a glass sheet on which were placed (a) 13 'square' strips with regular centre-to-centre spacing of 0.06 m and (b) 15 rectangular (0.0285 m wide × 0.012 m high) strips with regular centre-to-centre spacing of 0.04 m, compared with predictions using the Tolstoy effective admittance model (Eqs. (5.47)) and the heuristic surface impedance model (Eqs. (5.60) and (5.61)) [41].

thin hard-backed porous layer, the measured impedance of the lattice surface shows a reactance that is much larger than the resistance. Measured excess attenuation spectra magnitudes that exceed more than +6 dB at certain frequencies indicate the existence and propagation of surface waves [41]. The 'greater than +6 dB' phenomenon in the excess attenuation spectra can be observed in Figures 6.29(b), 6.32 and 6.33 which are discussed later.

Zhu et al. [87] have studied reflection and diffraction and the dispersion and formation of surface waves over a comb-like surface and presented an analytical solution for the plane-wave scattering by a comb-like grating. Subsequently, Zhu et al. [88] have carried out laboratory measurements on the air coupled surface wave due a point source above lattice and mixed impedance surfaces and investigated surface wave generation and amplification by lattice strips of different widths.

The formation and propagation of ultrasonic surface waves have been studied over triangular grooves [89], rectangular grooves [90], a rectangular lattice [91] and honeycombs [92]. A modal theory for sound propagation over periodically rough surfaces, adapted from electromagnetic wave propagation theory, was found to yield better agreement with data than Tolstoy's model (see Chapter 5).

Systematic measurements have been made in an anechoic chamber of excess attenuation spectra due to a point source over surfaces composed of single, double and triple layers of a plastic light diffuser lattice placed on medium density fibreboard (MDF) board [41,93]. The measurement arrangement with a single lattice layer is shown in Figure 6.28.

The lattice layer has square pores or cells with depth of 12.63 mm, centre-to-centre spacing of 14.04 mm, wall thickness of 1.85 mm and side of 12.19 mm. To form double and triple lattice layers two or three lattice sheets were

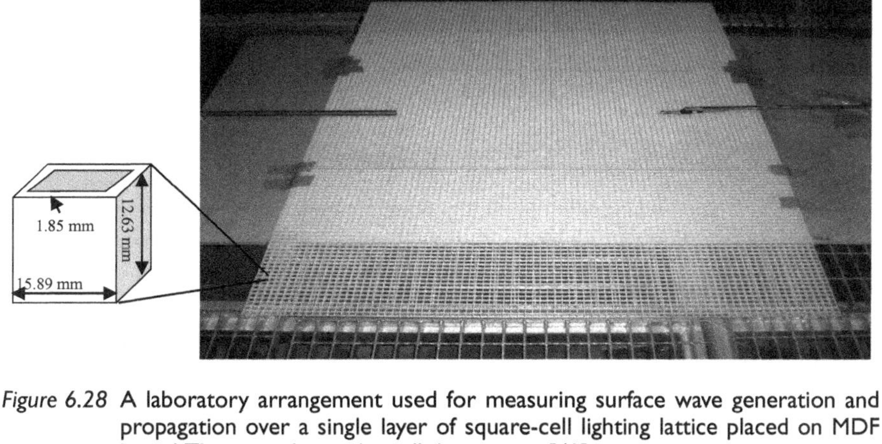

Figure 6.28 A laboratory arrangement used for measuring surface wave generation and propagation over a single layer of square-cell lighting lattice placed on MDF board. The inset shows the cell dimensions [41].

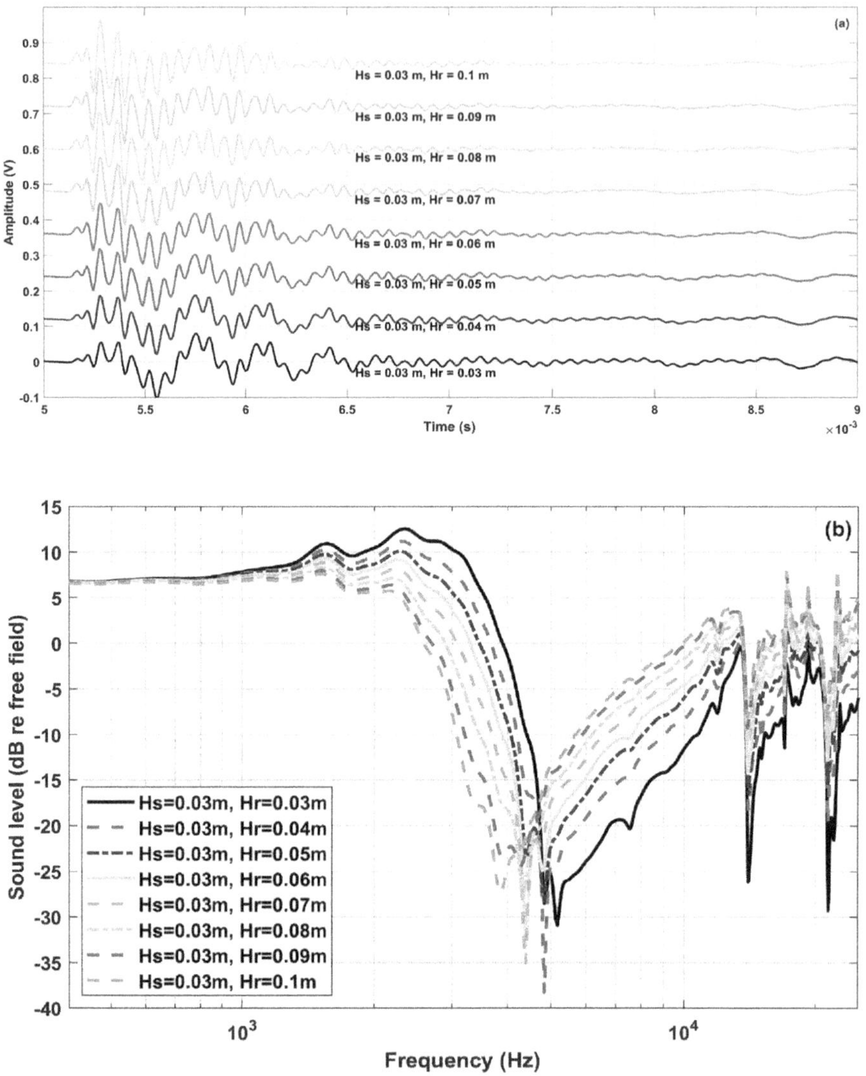

Figure 6.29 (a) Measured time domain signals and (b) corresponding excess attenuation spectra over single lattice layer placed on a MDF board with source at height of 0.03 m and receiver is placed at different heights of 0.03 m, 0.04 m, 0.05 m, 0.06 m, 0.07 m, 0.08 m, 0.09 m and 0.1 m. The horizontal separation of source and receiver was 0.7 m [41].

stacked carefully so as to align the cell walls directly above each other. The measured layer depths of double and triple layer lattices are 25.26 and 37.89 mm, respectively.

Figure 6.29 (a) compares time domain signals obtained with receiver heights of between 0.03 and 0.10 m over a surface composed of a single lattice layer placed on an MDF board. The source and receiver were separated by 0.7 m. For these and subsequently reported data the source and receiver heights were measured from the MDF board surface. The time domain plots corresponding to the lower receiver heights show a strong surface wave, whereas this feature does not appear when the receiver is in the higher locations. Figure 6.29 (b) shows measured excess attenuation spectra corresponding to the time domain data in Figure 6.28(a).

Over a smooth hard boundary below the first destructive interference frequency, the direct and reflected waves reinforce each other and the excess attenuation (EA) is +6 dB. Over the lattice the EA is greater than +6 dB between 1.5 kHz and 3 kHz; indeed, for source and receiver at a height of 0.03 m, the peak EA spectra magnitude is 14 dB. This indicates the spectral redistribution of reflected sound energy into a surface wave. Figure 6.29(b) indicates that the magnitude of the extent to which the EA exceeds +6 dB, i.e. the strength of the surface wave, depends on the receiver height. There is a strong surface wave when the receiver is at height of 0.03 m, but there is little or no surface wave when the receiver is at a height of 0.10 m. The data in Figures 6.29(a) and (b) are consistent with the expected exponential decrease with height required for a surface wave. Figure 6.30(a) shows an example time domain signal of the surface wave measured over single lattice layer placed on a MDF board with source and receiver at height of 0.03 m and separated by 0.7 m. The two instants on the waveform labelled in Figure 6.30(a) were used in corresponding waveforms obtained at a series of receiver heights for determining the rates of decay shown in Figures 6.30(b) and (c).

The exponential fits shown also were obtained using the curve fitting available in the Matlab® toolbox and based, respectively, on the time instants shown in Figure 6.30(a).

Lattice surfaces can be considered either as rigid porous layers with square pores or as cellular roughness. Similarly, arrangements of closely and regularly spaced parallel low walls or strips may be considered either as rigid porous layers with slit-like pores or as periodically rough or grooved surfaces. A series of laboratory experiments on propagation over parallel aluminium strips (width 0.0126 m, height 0.0253 m) spaced regularly at edge-to-edge intervals of between 0.003 and 0.0674 m on a varnished MDF have shown that the resulting surfaces may be modelled as a locally reacting slit-pore layer until their spacing approaches about 50% of the layer depth, i.e. the strip height. At larger separations they behave as periodically rough surfaces and propagation over such surfaces is better modelled using BEM [41,93]. Figure 6.31 shows two example laboratory arrangements with source and receiver at heights of 0.045 m and separated by 0.7 m over 0.0253 m high and 0.013 m wide aluminium strips with edge-to-edge spacing of 0.0124 and 0.0674 m, respectively. Figure 6.32 compares the resulting data with predictions in which the surfaces are modelled as slit-pore layers with the parameters listed in Table 6.13 (layer thickness 0.0253 m)

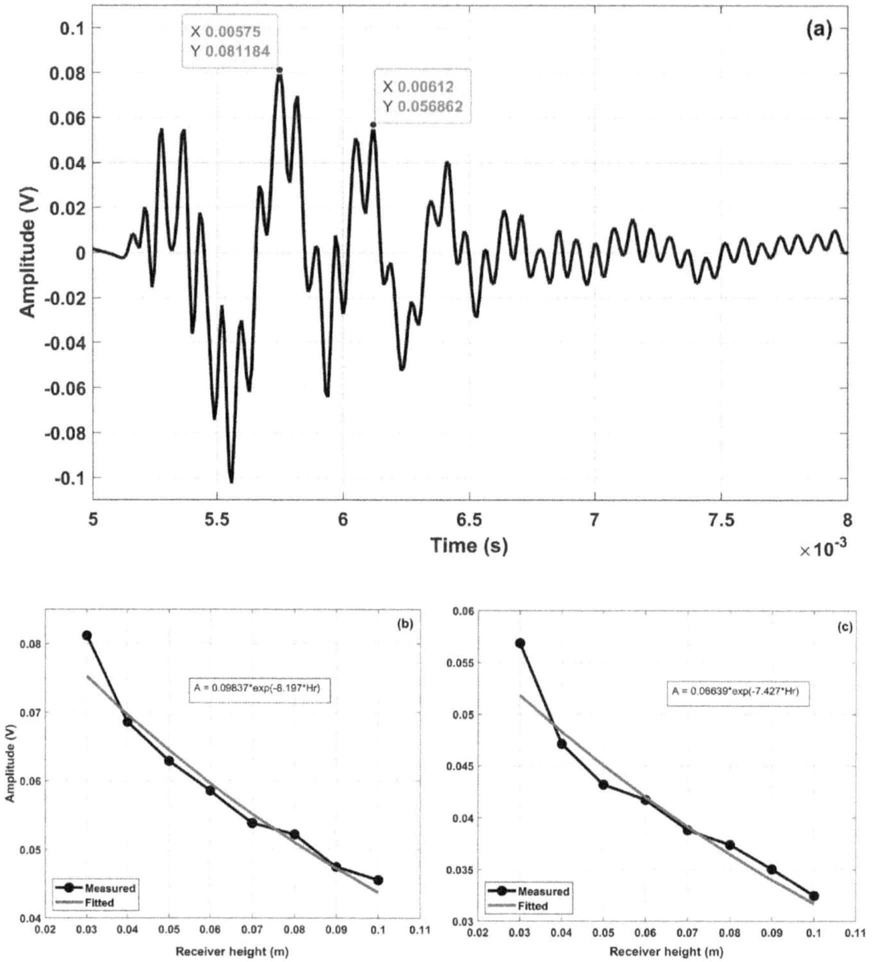

Figure 6.30 (a) Example time domain signal of the surface wave measured over single lattice layer placed on MDF with source and receiver at 0.03 m high and separated by 0.7 m. The amplitudes at the two instants labelled on the wave trace were used in determining the rate of decay: (b) and (c) show measured surface wave decay rates and exponential fits based on the times instants shown in (a) [41].

and with BEM predictions. Better predictions, particularly for the wider spacing, are obtained using BEM.

Other laboratory measurements [41] have shown (a) that a periodic spacing produces slightly higher amplitudes of surface waves than do random spacings of identical roughness elements with the same mean spacing and (b) that the frequency content of the surface wave above periodically rough surfaces depends on the spacing of the elements.

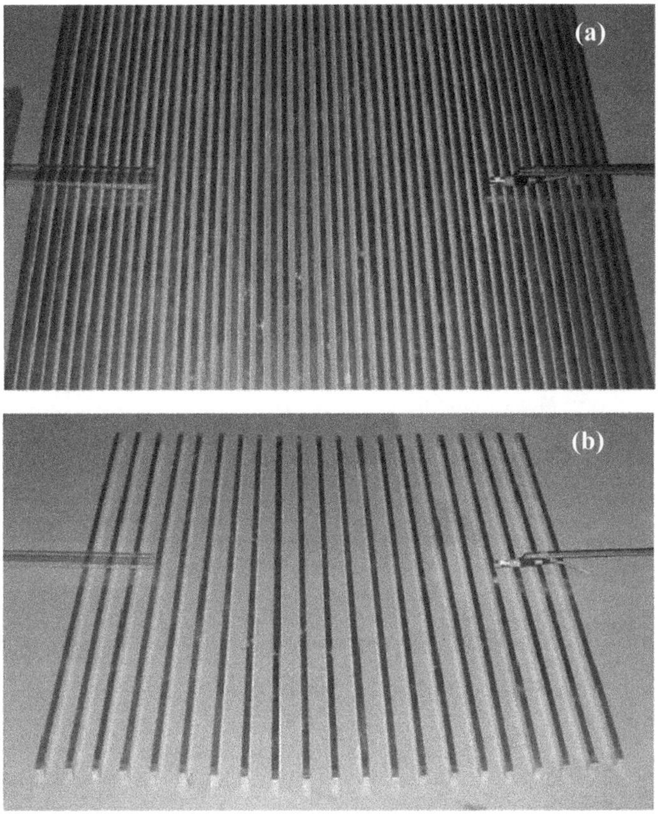

Figure 6.31 Laboratory arrangements for measuring excess attenuation from point source to a receiver at a height of 0.045 m separated by 0.7 m over 0.0253 m high aluminium strips placed on MDF with edge-to-edge spacing of (a) 0.0124 m and (b) 0.0674 m [41].

Figures 6.33(a) and (b) show excess attenuation spectra measured over periodically and randomly spaced triangular strips and Figure 6.33(c) shows two corresponding time domain plots.

Figure 6.34 compares the EA spectra measured with source and receiver heights of 0.045 m, separated by 0.7 m over aluminium strips placed on MDF board with edge-to-edge spacings of 0.003, 0.0124 and 0.0674 m, respectively. Although surface waves with similar amplitudes are evident for each of these configurations, the frequency content of the surface waves moves to lower frequencies as the edge-to-edge spacing is increased. The change in surface wave characteristics is proportionately larger if the spacing is small.

Multiple scattering simulations and laboratory measurements of propagation over regularly spaced cylinders on an acoustically hard plane have

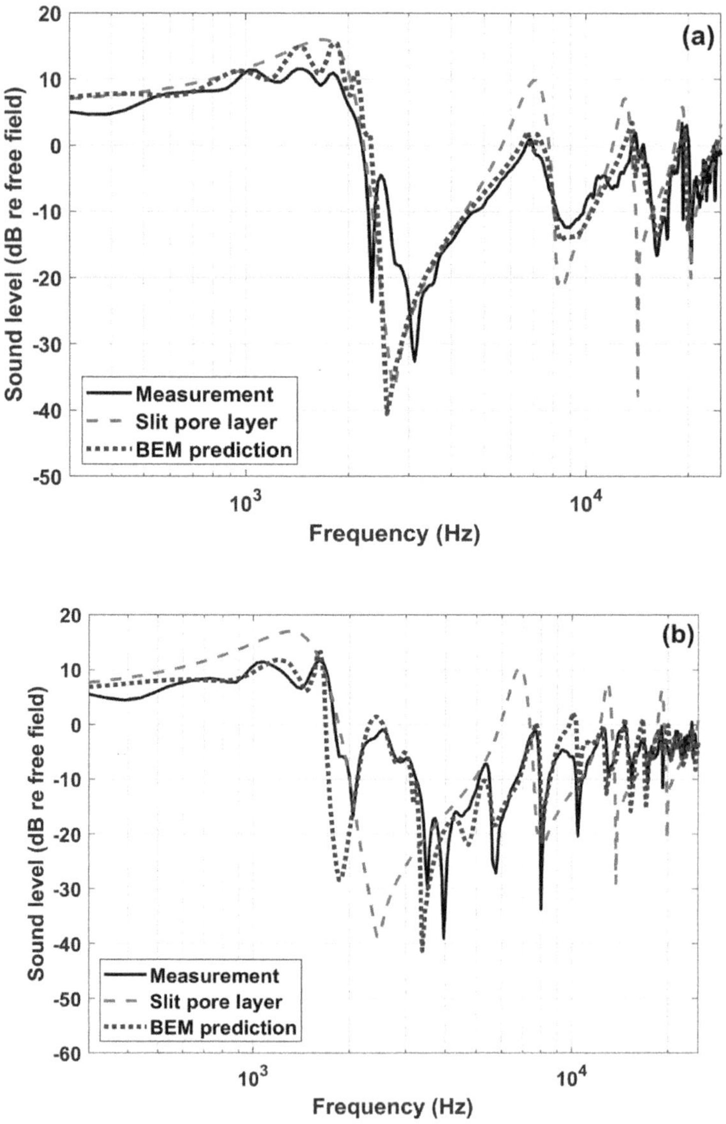

Figure 6.32 Comparisons between excess attenuation spectra measured (black continuous line) with source and receiver at 0.045 m height and 0.7 m separation over aluminium strips placed on MDF board with edge-to-edge spacing of (a) 0.0124 m and (b) 0.0674 m with BEM calculations (dotted lines) and predictions assuming a slit-pore layer impedance (dashed lines). The parameter values used for the slit-pore predictions (broken line) are listed in Table 6.13. The BEM predictions assume an MDF board impedance given by the 2-parameter variable porosity model (see Chapter 5) with effective flow resistivity 10 MPa s m^{-2} and effective porosity rate 1.0 m^{-1}. The source and receiver heights are measured with respect to the MDF base [41].

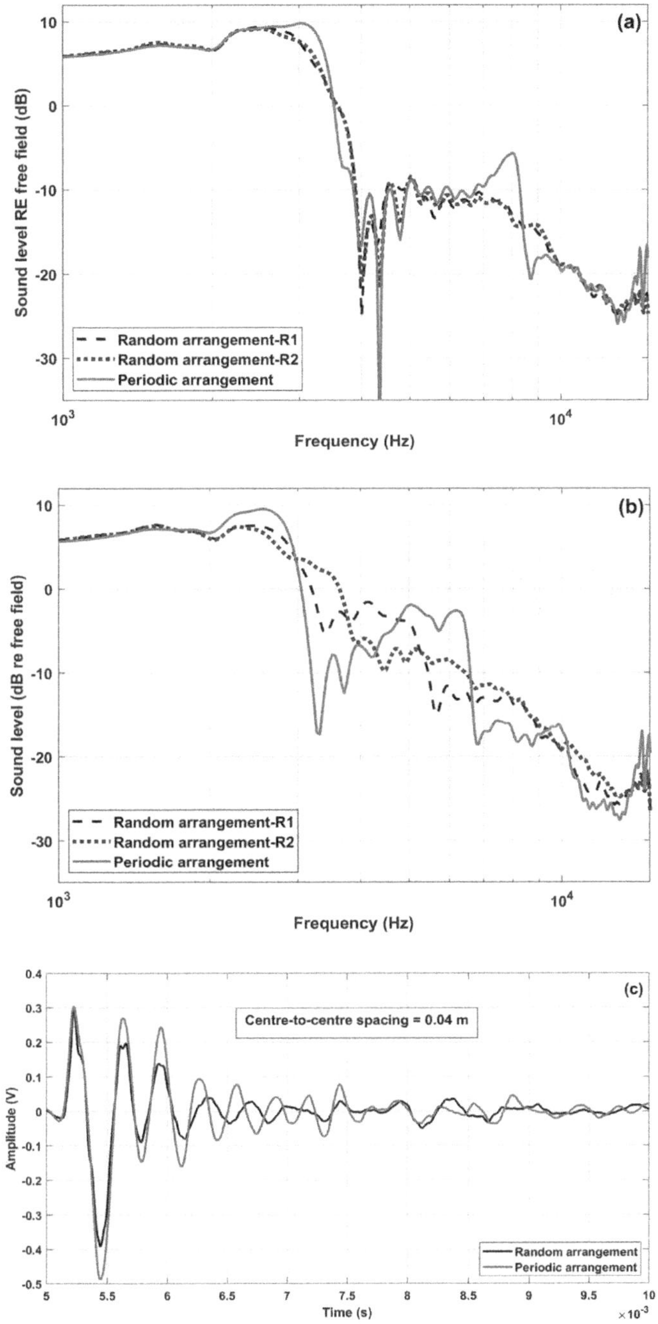

Figure 6.33 Measured EA spectra for source and receiver heights of 0.02 m separated by 0.7 m over 15 triangular strips with either random or periodic distributions (a) with (mean) centre-to-centre spacing of 0.04 m (b) with (mean) centre-to-centre spacing of 0.06 m (c) Time domain plots for the periodic arrangement with centre-to-centre spacing of 0.04 m and random arrangement R1 with mean centre-to-centre spacing of 0.04 m [41].

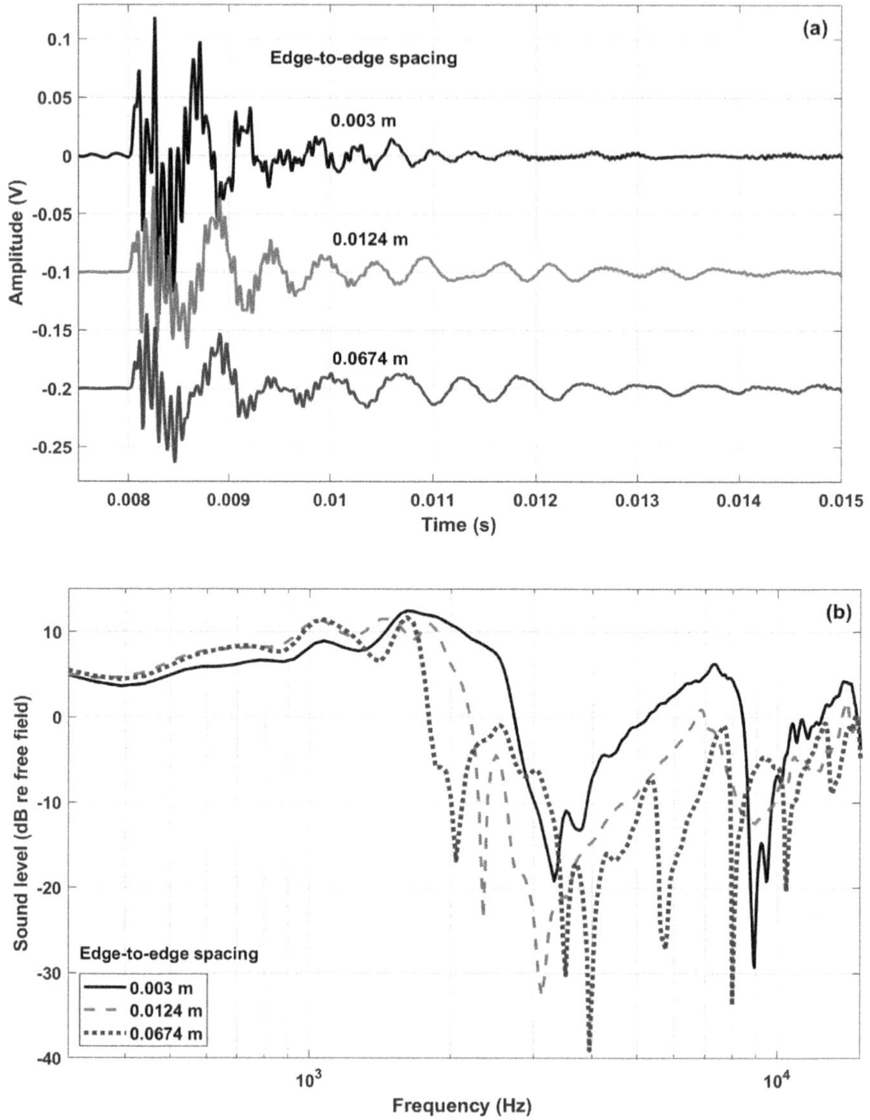

Figure 6.34 Time domain data and EA spectra with 0.045 m high source separated horizontally by 0.7 m from a receiver at heights of 0.003 m, 0.0124 m and 0.0674 m respectively showing surface waves over arrays of parallel aluminium strips placed on MDF board with edge-to-edge spacing of 0.0124 m. The source and receiver heights are measured with respect to the MDF base [41].

explored the relative importance of contributions to excess attenuation spectra from Bragg scattering, 'organ pipe' resonances in the gaps and roughness-induced finite effective surface impedance as the size of the cylinders and their spacing are varied [94]. For closely spaced cylinders, surface wave creation seems to involve the merging of 'organ pipe' resonances between the cylinders.

According to laboratory measurements, roughness-induced surface waves penetrate the shadow region beyond the apex of a convex surface [95, 96]. This suggests that roughness-induced surface waves may increase the usefulness of airborne sound in detecting low flying aircraft using 'nap-of-the-earth' tactics to avoid radar.

Most of the published studies of air-coupled surface wave propagation have been in the laboratory. However, using an impulsive source (gun shots), Albert [97] has observed audio-frequency surface waves outdoors over snow covered ground, successfully verifying the surface wave properties such as exponential decay and a phase speed slower than the speed of sound. Subsequently, these outdoor data have been fitted using a rigid-porous model for the acoustical properties of snow [98]. Also, as mentioned in Chapter 2, Section 2.5, tails in the received waveforms associated with air-coupled surface waves have been observed in long-range measurements of propagation from impulsive sources.

6.7.4 Outdoor Measurements of the Influence of Roughness on Ground Effect

While many laboratory measurements have shown the effects of roughness over rough hard surfaces, much of the published in situ data obtained outdoors over porous ground include the combined effects of roughness and flow resistivity. Aylor [99] measured a considerable change in excess attenuation over approximately 50 m after disking a soil but without any significant change in meteorological conditions. Other measurements before and after ploughing have demonstrated the additional effects of roughness [100]. Figures 6.35 and 6.36 show spectra of the difference in sound pressure levels received by vertically separated microphones over ploughed and sub-soiled heavy (boulder) clay, respectively. The microphones were at 1 and 0.1 m heights and located 30 m from the Electro-Voice loudspeaker source (centre at 1.65 m height) on a sunny day with light wind. As discussed earlier, vertical-level difference spectra are surrogates for excess attenuation spectra since they are independent of the source spectrum. There are clear differences in the ground effects before and after cultivation.

Also, Figures 6.35 and 6.36 show theoretical predictions for the level-difference spectra at 30 m. Predictions are shown both for rough (dotted lines) and smooth (broken lines) surfaces in each case. Although some of the change in ground effect between smooth and ploughed or sub-soiled conditions will result from changes in flow resistivity, improved agreement with data is obtained by including roughness effects and by attributing different mean roughness heights to each ground state. For the conditions and source–receiver geometry obtaining during these measurements, turbulence affects only the higher frequencies (see Chapter 10) and is not

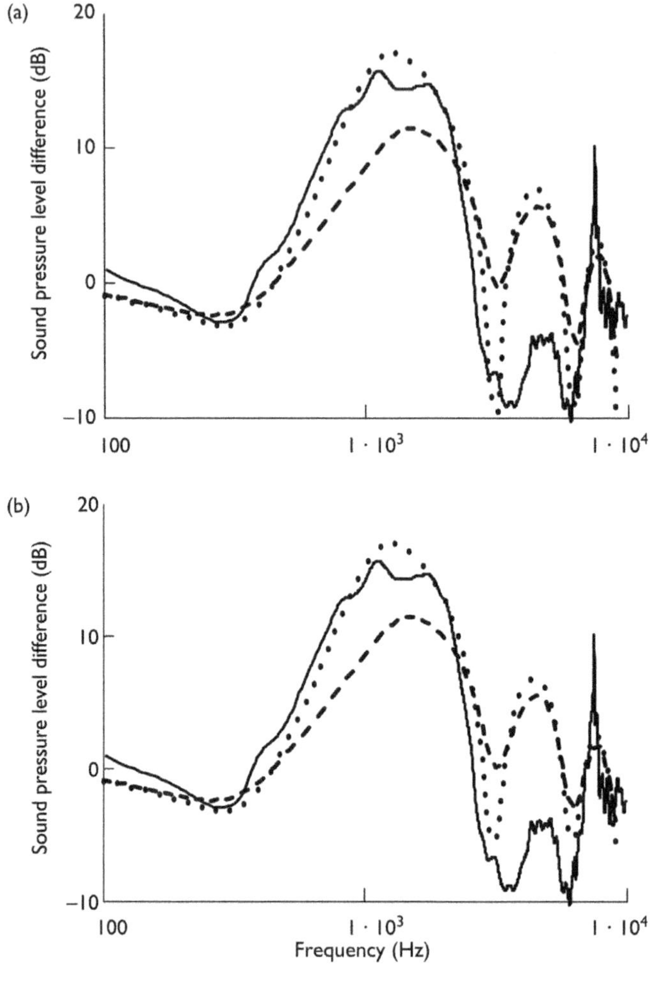

Figure 6.35 Average level difference spectrum obtained with a loudspeaker source at 1.65 m height and vertically separated microphones at 1 and 0.1 m height at a range of 30 m over sub-soiled ground (solid line). Predictions assume either a hard-backed slit pore medium with smooth surface (flow resistivity 80 kPa s m^{-2}, porosity 0.4, tortuosity 2.5, thickness 0.08 m, broken line) or a slit pore medium with these parameters plus 0.05 m high close-packed but partially-random (randomness factor 0.75) semi-cylindrical roughness (dotted line) on its surface. Predictions are (a) without and (b) with turbulence characterized by $\langle \mu^2 \rangle = 10^{-5.75}$, $L_0 = 1$ m (see Chapter 10).

responsible for the behaviour of the data around 1 kHz which seem consistent with the predicted effects of surface roughness. Also, in keeping with the fact that sub-soiling produces less surface roughness than ploughing, the roughness size assumed to fit the data over ploughed ground is greater than that assumed for the sub-soiled ground. The fitted thickness (0.175 m) for the surface layer of the ploughed ground is consistent with

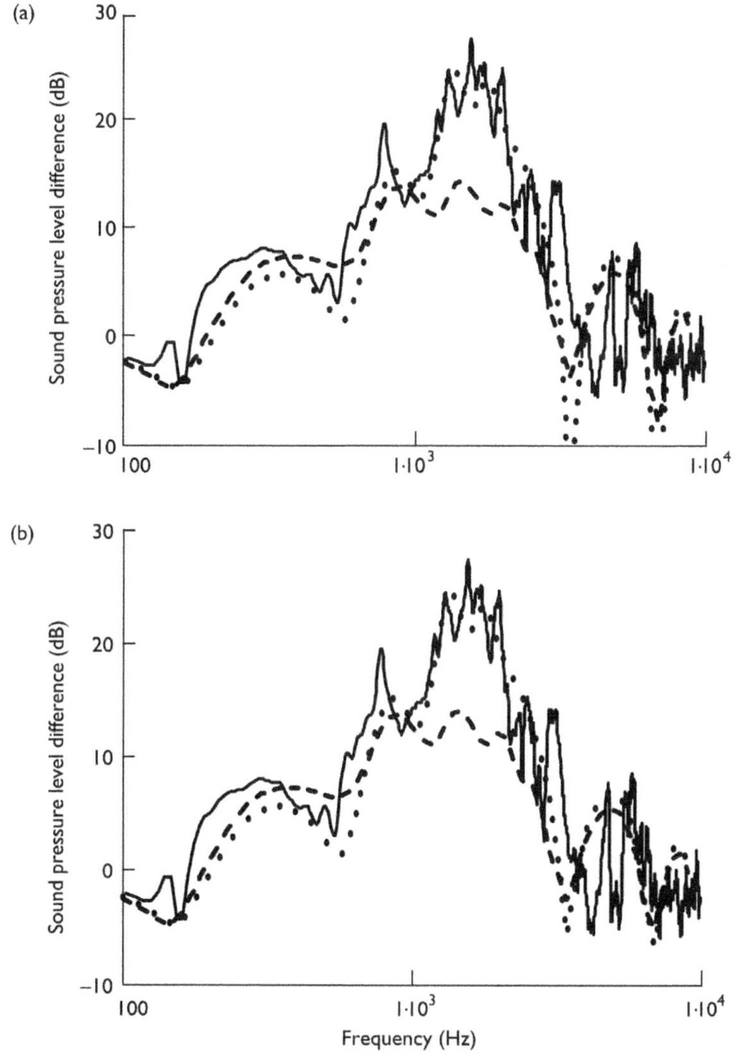

Figure 6.36 Average level difference spectrum obtained at a range of 30 m over ploughed ground (solid line). Predictions assume either the smooth surface of a hard-backed slit pore medium with flow resistivity 10 kPa s m^{-2}, porosity 0.4, tortuosity 2.5 and thickness 0.175 m (broken line) or a rough surface with these parameters plus 0.08 m high, close-packed, semi-elliptical cylinder (eccentricity 1.6) roughness, (dotted line). The turbulence parameters assumed are the same as for Figure 6.35.

the expected depth of the plough pan. Figure 6.37 shows impedance spectra predicted using the parameters that enable fitting of the level difference data in Figures 6.35 and 6.36. The impedance spectrum predicted for the ploughed field shows layer resonance behaviour (see Figure 5.15), whereas that predicted for the sub-soiled ground has the ‘rough finite impedance’

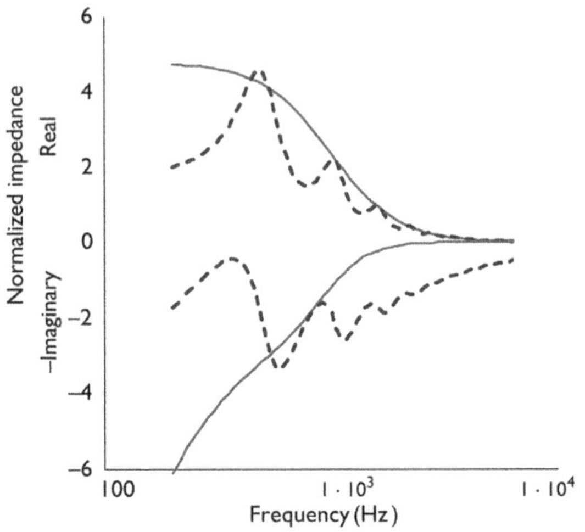

Figure 6.37 Surface impedance of sub-soiled (continuous lines) and ploughed clay soil (broken lines) predicted using the parameters obtained by fitting the data shown in Figures 6.33 and 6.34.

form in which both real and imaginary parts of the impedance tend to zero (see Figures 5.11 and 5.12).

6.8 MEASURED AND PREDICTED EFFECTS OF GROUND ELASTICITY

6.8.1 Elasticity Effects on Surface Impedance

As discussed in Chapter 5, Section 5.4, ground may be described as a layered poroelastic solid. Typically, the stiffness and rigidity of soils are sufficiently low that their seismic wave speeds may be less than the speed of sound in air. Ground elasticity is predicted to have its greatest influence on the surface impedance at low frequencies [20]. Moreover 'frame' resonances have been observed below 500 Hz in admittance spectra deduced from impedance tube measurements on moist sands and soils [59,60]. However, as remarked already, the measurement of surface impedance in situ at low frequencies is challenging. It is difficult to achieve much sound wave penetration using acoustic (loudspeaker) sources above ground. Moreover, the influence of ground elasticity depends strongly on the attenuation of the elastic waves in the surface layer for which there is a paucity of data. Consequently, the available measurements of impedance are rather inconclusive in respect of the effects of elasticity.

Figure 6.38 shows predictions of rigid- and elastic-framed models for the surface impedance of a soil at Bondville, Il, USA characterized as a porous layer over a non-porous elastic intermediate layer and a non-porous elastic substrate and by the parameters listed in Table 6.14 [20]. The impedance is predicted to be slightly less at low frequencies if the soil elasticity is taken into account.

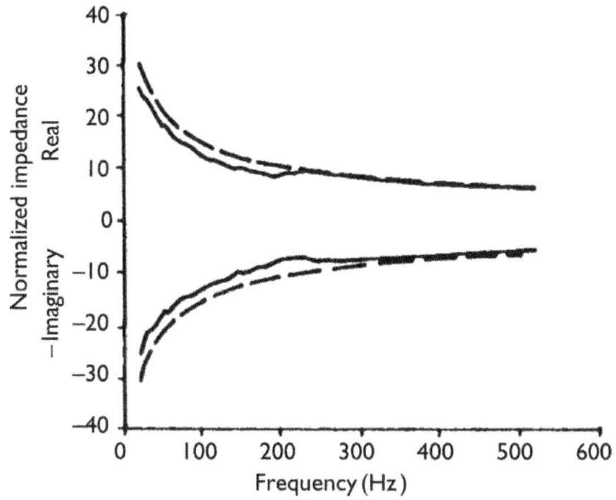

Figure 6.38 Predicted surface impedance at low frequencies for a soil at Bondville, Il., USA [3.68]. The continuous lines represent predictions including effects of elasticity whereas the broken lines represent predictions that ignore these effects. A double layer structure was found by a shallow seismic survey and the parameters (listed in Table 6.15) were found by direct measurement and by fitting data for the acoustic-to-seismic coupled spectrum.

6.8.2 Ground Vibrations Due to Airborne Explosions

Data for the ratio of air pressure 0.1 m above the ground to the vertical soil particle velocity measured at a collocated geophone buried just beneath the ground surface obtained as the result of C4 explosions have been reported [101]. These data were obtained at ranges between 2 and 17 km at Finnskogen in Norway. One objective was to derive an improved ground impedance model for low frequencies that takes ground elasticity into account. Calculations of the impedance of a multi-layered poroelastic system at an angle of incidence of 90° (grazing incidence) have been obtained by means of a plane wave model (MULTIPOR) [102]. This model was used also to compute the acoustic-seismic coupling ratio (A/S), i.e. the ratio of incident pressure to the component of solid particle velocity normal to the surface. This ratio has the same units as impedance (Pa s m^{-}) and may be termed an 'acousto-seismic impedance'. It was shown that the computed quantities are sensitive to ground elasticity. Moreover, it was found possible, using a plane wave model, to match the computed acoustic-seismic coupling ratio to the data measured at ranges of 2 km and above during the blast propagation trials in Norway. The seismic profile deduced on site (see Table 6.16) using the Spectral Analysis of Surface Waves (SASW) method is simplified to enable such data fitting (see Table 6.17).

The plane wave model predicts results that are independent of range. On the other hand, it is noticeable that the A/S impedance data have a significant spread below 7 Hz at horizontal ranges between 2 and 17 km. Plane wave predictions show a significant minimum in the acoustic-to-seismic coupling ratio at 2.5 Hz which is the frequency at which the (non-porous

Table 6.15 Layer parameters measured and deduced for Bondville site

Layer	*Thickness m*	*P-wave speed m/s*	*S- wave speed m/s*	*Density kg/m³*	*porosity*
1	0.15	114	30	900	0.6
2	1.98	260	120	2650	0
3	∞	1800	340	2650	0

Table 6.16 Layering deduced by Spectral Analysis of Surface Waves (SASW) at Finnskogen in Norway [100].

Layer	*Thickness (m)*	*P-wave speed m/s*	*S-wave speed m/s*	*Wave attenuation constant*
1	0.5	560	230	0.04
2	1.0	220	90	0.02
3	3.0	415	170	0.002
4	3.0	1500	170	0.001
5	4.0	1500	200	0.001
6	5.0	1500	400	0.001
7	halfspace	1500	450	0.001

Table 6.17 Parameters used for fitting Acoustic-to-Seismic data at Finnskogen, Norway [100].

Layer	*Thickness m*	*Flow resistivity kPa s m⁻²*	*Porosity*	*P-wave speed m/s*	*S-wave speed m/s*	*Wave attenuation constant*
1	0.8	1740	0.3	560	230	0.04
2	1.2	1740	0.3	220	98	0.02
3	150	17400	0.1	150	850	0.001
4	halfspace	174000	0.01	1500	354	0.001

elastic) Rayleigh wave speed predicted from the dispersion curve for the profile given in Table 6.16, coincides with the sound speed in air. Minima in the A/S impedance data appear between 1.5 and 3 Hz but it is noticeable, that with one possible exception, the data do not show minima as sharp as predicted by a plane wave model [102].

Calculations for propagation from a point source that take ground elasticity, porosity and layering into account may be made using FFLAGS (Fast Field program for Layered Air-Ground Systems). Details of FFLAGS are given elsewhere [103]. FFLAGS enables atmospheric refraction to be included also but for the calculations presented here the atmosphere is assumed to be homogeneous and stationary. The values of flow resistivity and porosity that have been used in the calculations reported here are shown in Table 6.18. Although the assumed porosities are not consistent with the assumed densities, the inconsistency has little effect on the predictions.

Table 6.18 Ground parameters used for predictions (Figures 6.43–6.45). Adapted from [103].

Parameter		*Hard soil*	*Soft soil*
Speed of sound in air deduced from time of flight (m s^{-1})		358.5	358.5
No. of porous elastic layers (excluding substrate)		1	1
Layer	flow resistivity (σ Pa s m^{-2})	927,000	127,000
	Porosity (Ω)	0.17	0.37
	Pore shape factor (s_p)	0.3	0.3
	Grain shape factor (n')	0.5	0.5
	P-wave speed (ms^{-1})	600	490
	S-wave speed (ms^{-1})	400	290
	Soil density (ρ kg m^{-3})	1700	1900
	Layer thickness (m)	1.5	2.7
	Wave attenuation constant (α)	0.02	0.02
Substrate	flow resistivity (Pa s m^{-2})	1,600,000	1,600,000
	Porosity	0.07	0.07
	Pore shape factor (s_p)	0.3	0.3
	Grain shape factor ()	0.5	0.5
	P-wave speed (m s^{-1})	2040	2040
	S-wave speed (m s^{-1})	1020	1020
	Soil density (kg m^{-3})	2600	2600
	Wave attenuation constant (α)	0.05	0.05

Example comparisons between predicted and measured acoustic-seismic impedance (i.e. the ratio of sound pressures at microphones divided by vertical soil particle velocities measured at collocated geophones) are shown in Figure 6.39(a). Points (circles, boxes and crosses) in Figure 6.39(a) represent data. Predictions by FFLAGS at ranges of 6.3, 7.2 and 12.6 km for source height 2 m, receiver heights 0.1 m (microphone) and –0.05 m (geophone) and the four-layer ground profile (Table 6.19) are shown in Figures 6.39(b) and 6.39(c). The FFLAGS predictions are in tolerable agreement with data, similar to the plane wave predictions [102] but, unlike these, show a range dependence that is consistent with the differences in data at different ranges. Compared with the corresponding plane wave predictions, the main minimum in the acoustic-seismic coupling spectrum below 5 Hz predicted by FFLAGS is relatively shallow and depends on range.

Comparison of predictions in Figures 6.39(b) and 6.39(c) indicates that predictions have a significant dependence on the assumed sound speed in air. Figure 6.40 shows predictions obtained at 6300 m range with four different air sound speeds. Sound speeds in air of 329, 332, 335 and 343 m/s correspond to temperatures of – 4°C, 1°C, 6°C and 20°C, respectively, at sea level.

In particular, the predicted A/S impedance below 5 Hz is much altered by different air sound speed values. This is consistent with the interpretation

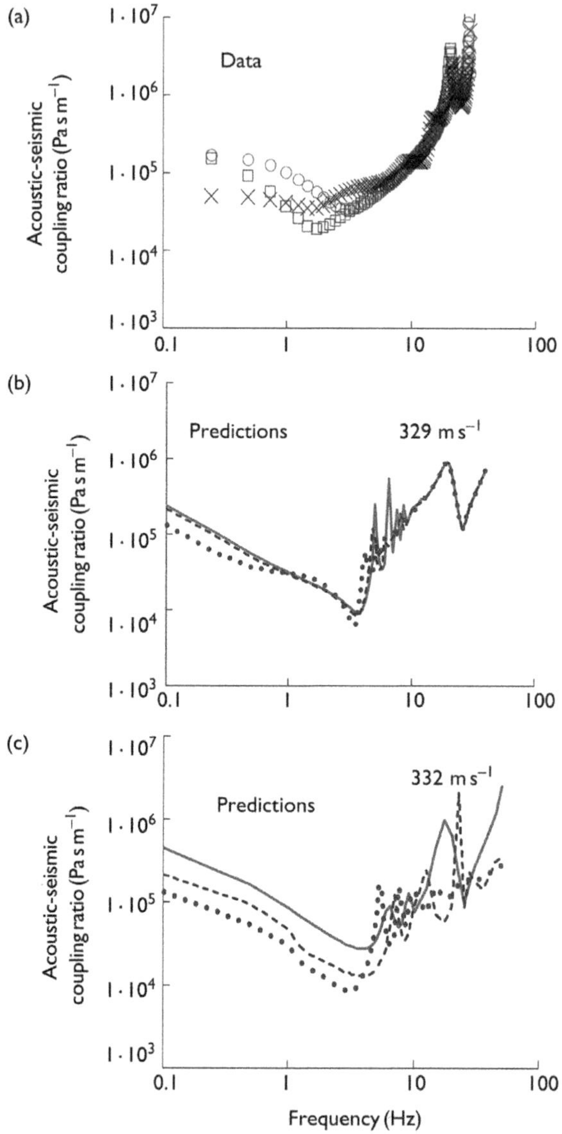

Figure 6.39 (a) Acoustic-seismic coupling ratios deduced from measurements [100] at ranges of 6.3 km (circles), 7.2 km (boxes) and 12.6 km (crosses) (b) Predictions of FFLAGS for sound speed in air of 329 m/s and (c) Predictions of FFLAGS for sound speed in air of 332 m/s. The predictions for 6.3 km are represented by solid lines, for 7.2 km by broken lines and for 12.6 km by dotted lines.

that an air-coupled Rayleigh wave is propagating below 5 Hz since the slope of the predicted Rayleigh wave dispersion curve is small near 329 m/s.

The data at a given range display some variability [101]. However, there will have been some sources of variability, e.g. meteorology, topography and changing conditions with range, which are not taken into account by FFLAGS.

Table 6.19 The flow resistivity and porosity corresponding to an array of strips (0.0253 m high, 0.013 m wide) with edge-to-edge spacings of either 0.0124 m or 0.0674 m.

Edge-to-edge spacing 'a' (m)	*Flow resistivity (Pa s m−2)* $R_s = \dfrac{2\mu T s_o}{\Omega r_h^2}$ $\mu = 1.811\times 10^{-5}, T = 1.0, s_o = 1.5, r_h = a/2$	*Porosity 'Ω'*
0.0124	2.85	0.496
0.0674	0.06	0.8425

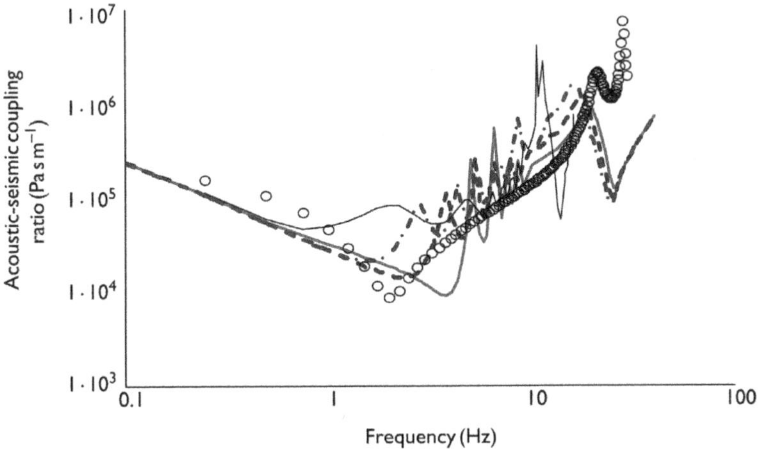

Figure 6.40 Predicted Acoustic-seismic coupling ratio at 6300 m assuming the layered ground system given by Table 6.17 and sound speeds in air of 329 m/s (thick solid line), 332 m/s (broken line), 335 m/s (dash-dot line) and 343 m/s (thin solid line) respectively.

Predictions of the surface impedance of layered poroelastic ground were shown and discussed in Chapter 5 (Figure 5.18). As well as A/S coupling and surface impedance, FFLAGS may be used to predict excess attenuation spectra (i.e. the sound pressure spectra relative to free field) including the effects of elasticity. Such predictions were discussed in Chapter 5.

Seismic signatures produced by above-ground near-surface explosions measured in a variety of ground and vegetative conditions for propagation distances from 8 to 565 m are consistent with the two propagation paths illustrated in Figure 6.39 where C2 and C1 represent the amplitude coupling coefficients between the blast wave arrivals close to and distant from the source, respectively [104].

At a vibration sensor distant from the explosion buried in a ground possessing seismic wave speeds greater than the speed of sound in air, 'precursor' waves induced by the blast overpressure near the explosion arrive first. They are followed by seismic waves that result from acoustic-to-seismic coupling close to the receiver from the arrival of the airborne sound wave from the blast. Ground vibration amplitudes caused by the atmospheric wave arrivals at a distant seismic sensor are significantly greater than those associated with the precursor seismic waves. Typically, the acoustic-to-seismic coupling ratio C1 is constant with respect to distance and maximum pressure at a given location but varies from site to site over the range from 1 to 13 $\mu m\ s^{-1}\ Pa^{-1}$. The waveforms shown in Figure 6.40 were recorded by a PCB pressure sensor and by the vertical and radial components of a triaxial geophone located on a concrete surface at 90 m from a 1.5 m high explosion of 0.57 kg C4. The largest peak amplitudes of 1.2 kPa, 5.5 mm/s and 1.5 mm/s, respectively, are registered at about 0.25 seconds after the blast and correspond to the acoustic arrival (Path A in Figure 6.41). A smaller precursor seismic arrival at 0.05 seconds is visible on the radial sensor. Later arrivals at about 0.35 seconds are associated with acoustic reflections from a wall located near the test site.

Figure 6.43 shows pressure (recorded by a microphone near the surface) and seismic waveforms recorded by a geophone buried at 1 cm depth in a 'hard' sandy soil with some vegetation (grass, weeds, scattered small trees) at 100 m from a C4 explosion.

The air-coupled seismic signals induced by the air wave arrival are about 30 ms in duration, about as long as the waveform duration recorded by the pressure sensor. The maximum amplitudes of both seismic components are nearly equal, and the maximum seismic frequency is about 175 Hz.

It is possible to distinguish geophone signals associated with soils in which the P- and S-wave speeds are greater than the speed of sound in air, which

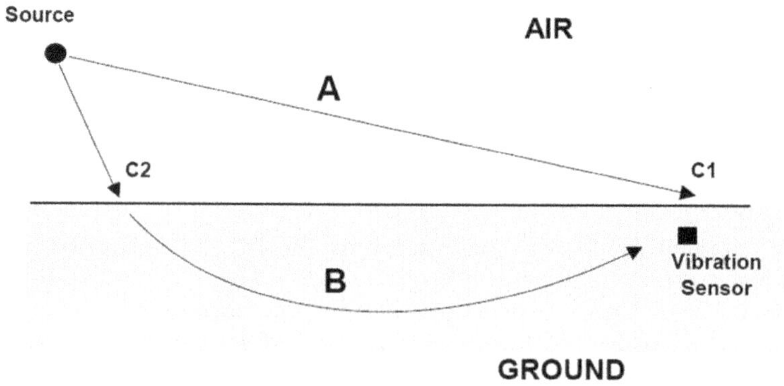

Figure 6.41 Hypothetical propagation paths to a buried vibration sensor produced by an explosive source in the air or on the ground: an airborne path (A) and a predominantly subsurface (seismic) path (B). C2 and C1 represent the air-to-ground coupling strengths for each path [104]. Reprinted from *Appl.Acoust.*, D. G. Albert, S. Taherzadeh, K. Attenborough, P. Boulanger and S. Decato, Ground vibrations produced by surface and near-surface explosions, *74*: 1279 – 1296 (2013) with permission from Elsevier.

are termed 'hard' soils, from 'soft' soils in which, while the P-wave speed is greater than the speed of sound in air, the shear wave speed is less than that in air. 'Soft' soils result in air-coupled Rayleigh waves. Figure 6.44 shows data recorded using the same geometry at a site with sparsely vegetated (grass and weeds) 'soft' soil. While the 30-ms-long acoustic waveform in Figure 6.44 is similar to that in Figure 6.43 (but more rounded, indicating higher ground attenuation for this soil), the seismic responses include long, low-frequency mono-chromatic wave trains over 80 ms long generated after the air wave arrival durations and phase shifted with respect to each other. These are consistent with air-coupled Rayleigh waves.

All sensors recorded maximum energy at a frequency of about 50 Hz and the coupling occurs at this frequency when the acoustic velocity and Rayleigh wave velocity (controlled by the shallow soil stratigraphy and seismic velocity structure) are identical.

Predictions shown also in Figures 6.42–6.44 have been made using PFFLAGS (a time domain version of FFLAGS; see Chapter 5, section 5.5.3). The source acoustic pulse waveform was deduced from other C4 measurements [104] at a distance assumed to be beyond that involving non-linear

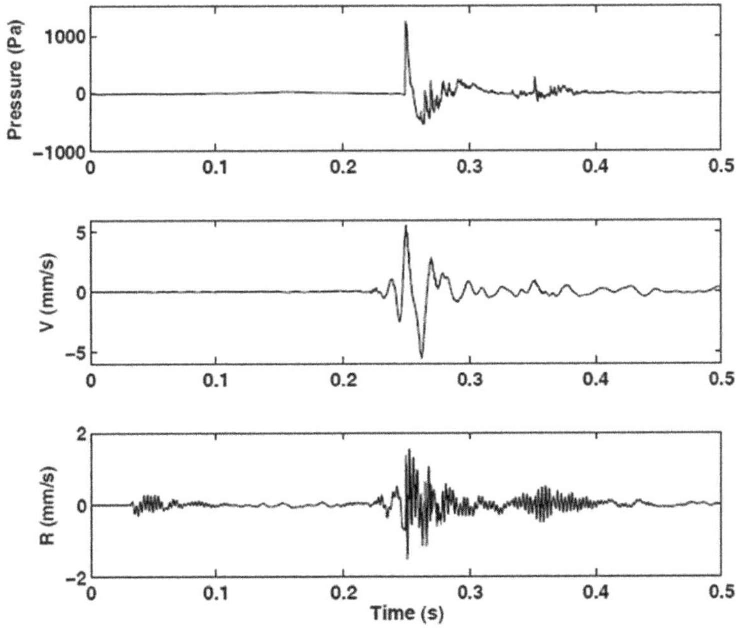

Figure 6.42 Waveforms at a pressure sensor at the surface (top), vertical component geophone (centre), and radial component geophone (bottom), located 90 m from the detonation of 0.57 kg of C4 at a height of 1.5 m above a concrete pad [104]. Reprinted from *Appl. Acoust.*, D. G. Albert, S. Taherzadeh, K. Attenborough, P. Boulanger and S. Decato, Ground vibrations produced by surface and near-surface explosions, *74*: 1279 – 1296 (2013) with permission from Elsevier.

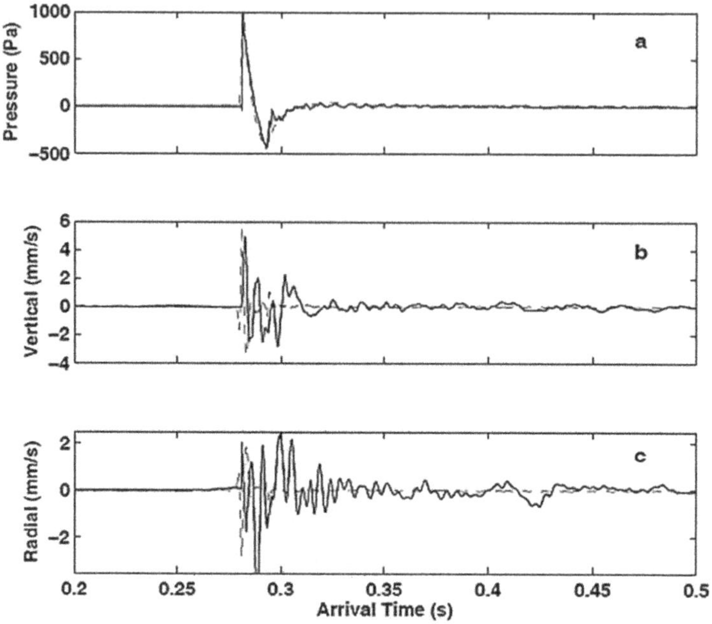

Figure 6.43 Measured (continuous lines) and predicted (broken lines) waveforms of (a) acoustic pressure, (b) soil vertical particle velocity and (c) soil radial particle velocity at 100 m from a C4 explosion. The microphone recording signal (a) is close to the surface and the geophone recording signals (b) and (c) is buried at a depth of 1 cm in 'hard' soil [104]. The parameters used for the predictions are listed in Table 6.18. Reprinted from *Appl. Acoust.*, D. G. Albert, S. Taherzadeh, K. Attenborough, P. Boulanger and S. Decato, Ground vibrations produced by surface and near-surface explosions, *74*: 1279 – 1296 (2013) with permission from Elsevier.

interaction and the amplitude and pulse length have been adjusted to fit the acoustic data obtained over 'hard' soil at 100 m. In the absence of seismic refraction information for the sites, a simple ground structure consisting of a single porous and elastic layer over a semi-infinite porous and elastic substrate has been assumed.

The P- and S-wave speeds and thickness of the upper porous and elastic layers, determined by trial and error fitting of the seismic data starting with typical values and fixing the pore-related parameters at values that give good fits to the acoustic pulse data, are listed in Table 6.18. The differences in spectra are consistent with lower flow resistivity and elastic wave speeds deduced for the 'soft' soil compared with the 'hard' soil.

Probably the late arrivals observed at about 0.52 s in the data for both vertical and radial seismic components but not in the predictions are the result of additional layers below the topsoil not included in the model. Other potential causes of discrepancies between predictions and data include (i)

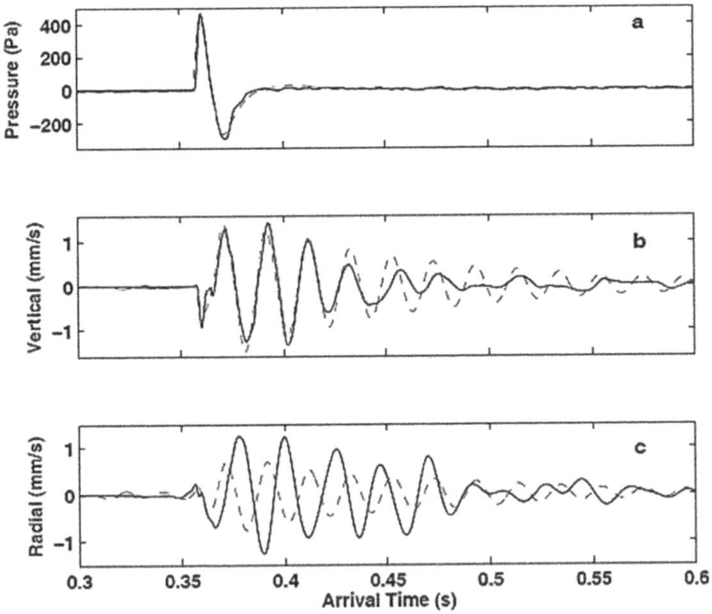

Figure 6.44 Measured (continuous line) and predicted (dashed line) waveforms of (a) acoustic pressure close to the surface, (b) soil vertical particle velocity and (c) soil radial particle velocity waveforms at a depth of 1 cm in 'soft' soil located 120 m from the source [104]. The parameters used for the predictions are listed in Table 6.18. Reprinted from *Appl. Acoust.*, D. G. Albert, S. Taherzadeh, K. Attenborough, P. Boulanger and S. Decato, Ground vibrations produced by surface and near-surface explosions, 74: 1279 – 1296 (2013) with permission from Elsevier.

non-linear effects near the source (ii) range-dependent topography and lithography and (iii) atmospheric turbulence.

Charge sizes of between 0.28 and 4.5 kg of C4 were detonated either on the surface or at a height of 1.5 m above the (frozen) ground level and the resulting signatures were measured using a digital seismograph at propagation distances of 30–100 m over a very low density and low strength thin snow layer. Spectra measured 100 m from the C4 explosions and predictions of PFFLAGS with the parameters listed in Table 6.20 are shown in Figure 6.45. The measurements were conducted in an approximately 200 × 200 m demolition area, free of vegetation but with surface height variations between 30 and 50 cm. Because the snow cover was very shallow, the geophones were installed by drilling mounting holes into the frozen soil beneath the snow and freezing them in place with the consequence that the seismic signals are likely to be dominated by propagation in the frozen soil rather than through the snow. Before testing, the snow cover thickness was between 10 and 15 cm and had an extremely low density of 130–330 kg m^{-3}. Snow characterization was carried out after the testing by which time a

Table 6.20 Parameters used for the predictions shown in Figure 6.49 of plane reflection coefficient as a function of sound pressure on a 15 cm thick hard-backed layer of lead shot

Parameter	*value*	*How obtained*
flow resistivity ($R_s(0)$, Pa s m^{-2})	1373	measured
Porosity (Ω)	0.385	measured
Viscous characteristic dimension (Λ m)	5×10^{-4}	Cell model [105]
Tortuosity (m)	1.6	$1/\sqrt{(\text{porosity})}$
Forchheimer's parameter (ξ)	3.7	measured

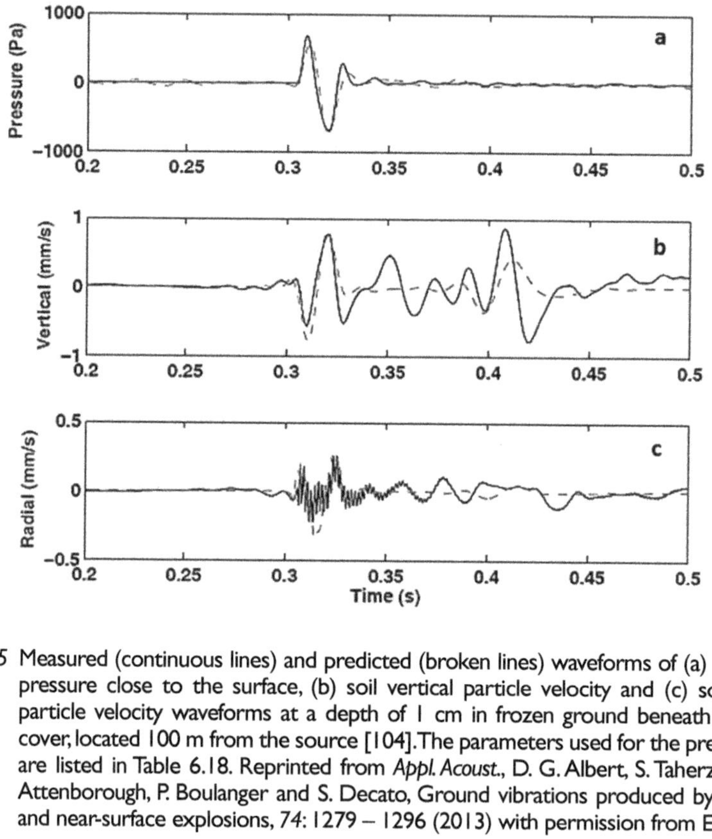

Figure 6.45 Measured (continuous lines) and predicted (broken lines) waveforms of (a) acoustic pressure close to the surface, (b) soil vertical particle velocity and (c) soil radial particle velocity waveforms at a depth of 1 cm in frozen ground beneath a snow cover, located 100 m from the source [104]. The parameters used for the predictions are listed in Table 6.18. Reprinted from *Appl. Acoust.*, D. G. Albert, S. Taherzadeh, K. Attenborough, P. Boulanger and S. Decato, Ground vibrations produced by surface and near-surface explosions, *74*: 1279 – 1296 (2013) with permission from Elsevier.

considerable area near the charge location was clear of snow. The calculated peak amplitudes were, respectively, 0.9, 2.1 and 1.1 times as large as the measured peak amplitudes. The (scaled) source function used for the theoretical predictions is shown near 0.21 seconds in Figure 6.45(a). There are significant secondary arrivals and a minor precursor in the vertical seismic component. There is a clear secondary arrival and a significant precursor in the radial seismic component. The best fit of the acoustic pulse data (Figure 6.45a) is obtained by assuming that the snow has a surface crust consistent with the measured density profile. The total assumed thickness of the snow layer (15 cm) corresponds to the deepest section measured but the predictions are not very sensitive to the assumed thickness of the lower snow layer. The assumed flow resistivity of the snow layer beneath the crust is based on the snow core data. The measured permeability of cylindrical cores extracted near the 60 m sensor location was between 35 and 40×10^{-0} m^{-} corresponding to a flow resistivity of between 4.47 and 5.1 kPa s m^{-}. The geophone depth is assumed to be 16 cm, i.e. just within the frozen ground layer.

The values listed in Table 6.20 have been used to obtain the predictions of the main and secondary arrivals in the radial seismic component shown in Figure 6.45c. However, to fit the precursor data shown in Figure 6.46b required a P-wave speed of 490 m/s (rather than 690 m/s as in Table 6.20). Moreover, the measured radial seismic component precursors are predicted only if the geophone is assumed to be at 14 cm depth or less, i.e. within the

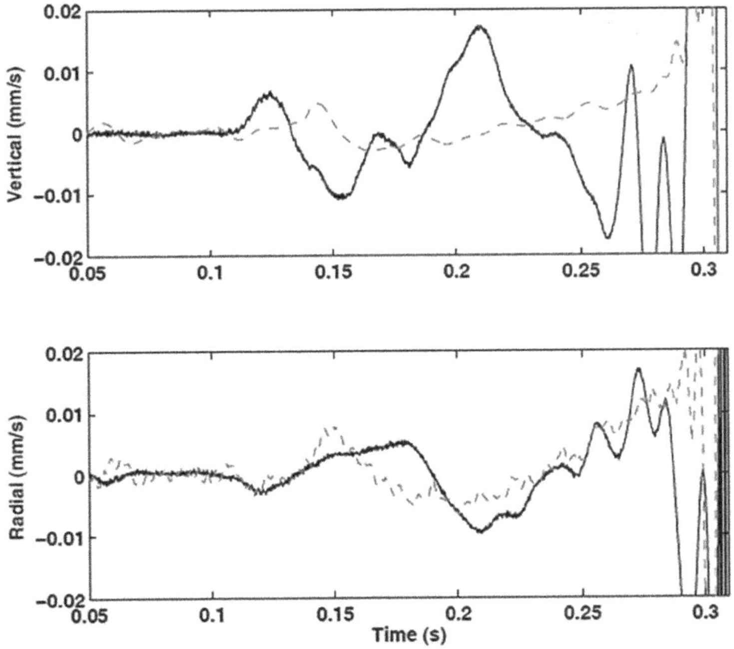

Figure 6.46 Measured (continuous lines) and predicted (broken lines) (a) soil vertical particle velocity and (b) soil radial particle velocity waveforms (precursors) arriving before the acoustically-coupled wave at a geophone buried at a depth of 1 cm in frozen ground beneath a snow cover and 100 m from the source [104]. The parameters used for the predictions are listed in Table 6.18. Reprinted from *Appl. Acoust.*, D. G. Albert, S. Taherzadeh, K. Attenborough, P. Boulanger and S. Decato, Ground vibrations produced by surface and near-surface explosions, *74*: 1279 – 1296 (2013) with permission from Elsevier.

snow layer. The high-frequency 'jitter' in the main and secondary arrivals in the radial component signal (Figure 6.45c) may be due to sensor resonance. Nevertheless, it is predicted to some extent if the geophone is assumed to be inside the snow layer.

6.9 NON-LINEAR INTERACTION WITH POROUS GROUND

Near an airborne explosion close to the ground, the sound level may be sufficiently high to involve non-linear interaction with a porous ground surface. For above ground propagation and assuming that the porous ground has a rigid frame, then the non-linearity results from the increase of flow resistivity with increasing air particle velocity associated with the increased acoustic pressure. Figure 6.47 shows an example of laboratory measurements of the increase of the flow resistivity with flow velocity for 1.89 mm radius lead shot [105]. The increase is approximately linear, i.e.

$$R_s(v) \approx R_s(0)[1 + \xi\Omega v] \tag{6.9}$$

where ξ, is Forchheimer's non-linearity parameter. The straight line fit to the data shown in Figure 6.47 corresponds to ξ = 3.7 s/m.

A model which combines Forchheimer's non-linearity with an equivalent fluid model has been validated against measured impedance and reflection coefficient spectra for three types of hard-backed rigid-porous layers (lead shot, porous aluminium and porous concrete) at amplitudes up to 500 Pa over a wide range of frequencies [105].

At frequencies near the quarter wavelength hard-backed layer resonance, both growth and decrease of the plane wave reflection coefficient with

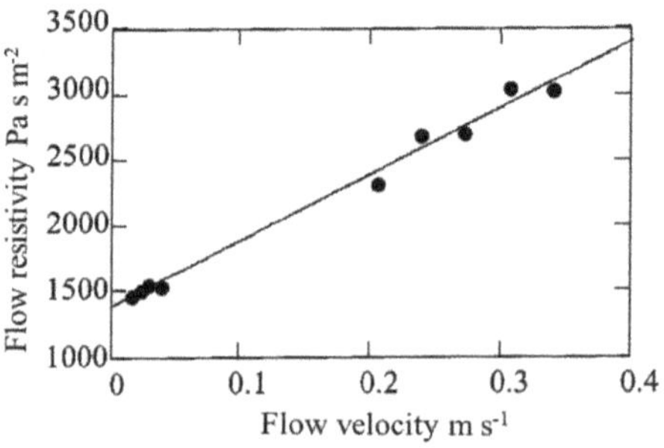

Figure 6.47 The measured dependence of flow resistivity on flow velocity for 1.89 mm radius lead shot and a straight-line fit [105]. Reproduced from O. Umnova, K. Attenborough, E. Standley and A. Cummings, Behavior of Rigid-Porous Layers at High Levels of Continuous Acoustic Excitation: Theory and Experiment, *J. Acoust. Soc. Am.* *114*(3): 1346–1356 (2003) with the permission of the Acoustical Society of America.

pressure amplitude are predicted and measured [105]. Whether the reflection coefficient increases or decreases, depends on the frequency, the material parameters and the thickness of the layer. An approximate condition for a decrease may be expressed in terms of the parameters introduced by the Johnson-Champoux-Allard-Umnova model (Equations (5.13), (5.14), (5.17)–(5.19)) for a porous layer of thickness d [106], i.e.

$$\frac{R_s^2\Omega^2\Lambda^2}{4T\rho_0\eta} < \frac{\Lambda c_0}{\sqrt{T}\delta_{res}d}\ln\left[\frac{1+\dfrac{T}{\Omega}}{\sqrt{\dfrac{\sqrt{T}}{\Omega}-1}}\right] - \omega_{res}\frac{\Lambda}{L_1}, \tag{6.10}$$

where $\delta_{res} = \sqrt{2\omega_{res}/\eta\rho_0}$ is the boundary layer thickness at the porous layer resonant frequency, ω_{res}, $1/L_1 = 1/\Lambda + (\gamma-1)/\sqrt{N_{PR}\Lambda'}$, $\Lambda' = 3\Lambda/\left[2\sqrt{T}(1-\Theta)\right]$ and $\Theta \approx 0.675(1 - \Omega)$.

A 15-cm thick hard-backed layer of lead shot (mean radius 1.89 mm) satisfies this condition. Figures 6.48(a) shows the measured and predicted linear quarter wavelength layer resonance exhibited by the normal incidence pressure reflection coefficient. Figures 6.48(b) and (c) show this resonance in the real and imaginary parts of admittance. The resonance is at just above 415 Hz. Figure 6.49 shows the behaviour of the reflection coefficient as a function of incident pressure at frequencies between 385 Hz and 500 Hz revealed by measurements in a shock tube [106]. Also shown are predictions using the parameters listed in Table 6.20. Both data and predictions show that, in this frequency range, the plane wave reflection coefficient of the 15-cm thick hard-backed lead shot layer passes through a minimum as the incident sound pressure is increased.

Typically, soils have flow resistivities that are too high for any significant change in reflection coefficient with increasing acoustic pressure. On the other hand, the non-linear acoustic response of 'pea' gravel has been predicted to be instrumental and exploitable in reducing the sound levels due to airborne explosions [106–108].

6.10 DEDUCTION OF SOIL PROPERTIES FROM MEASUREMENTS OF A/S COUPLING

Pore structure and soil strength are important in determining root growth and thereby crop yields. Figure 6.50 shows a measurement arrangement for using acoustic-to-seismic coupling non-invasively to determine parameters related to pore structure and elasticity in soils [109]. In this arrangement, signals at two vertically separated microphones and a Laser Doppler Vibrometer (LDV) provide three transfer functions which are used in an optimized fitting procedure to find a parameter set minimizing a cost

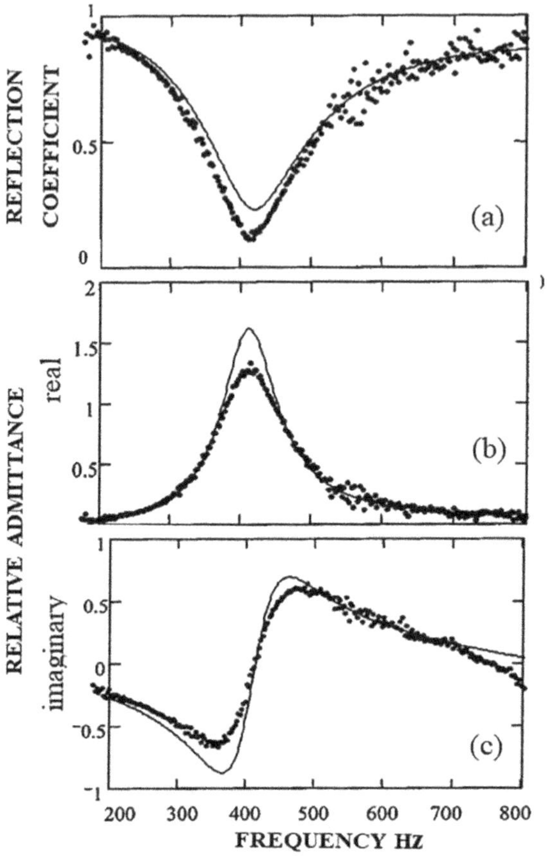

Figure 6.48 The frequency dependence of reflection coefficient and surface admittance for lead shot subject to low amplitude sound. Points - white noise data, Line - Johnson/Champoux /Allard/Lafarge/Umnova linear model [105]. Reproduced from O. Umnova, K. Attenborough, E. Standley and A. Cummings, Behavior of Rigid-Porous Layers at High Levels of Continuous Acoustic Excitation: Theory and Experiment, *J. Acoust. Soc. Am.* *114*(3): 1346–1356 (2003) with the permission of the Acoustical Society of America.

function (CF) based on measured and simulated transfer functions and Equation (6.11):

$$CF = \sum_{p=1}^{3}\left(\sum_{f}\left(\left|TF_{meas,p}(f)\right| - \left|TF_{sim,p}(f)\right|\right)^2 / \sum_{f}\left|TF_{meas,p}(f)\right|^2\right). \tag{6.11}$$

The subscripts *meas* and *sim* denote measurement data and the results of simulations using a wave propagation model, respectively. f is the frequency, and the type of transfer function is indicated by the subscript p. The wave propagation model is based on Biot-Stoll theory and the FFLAGS numerical

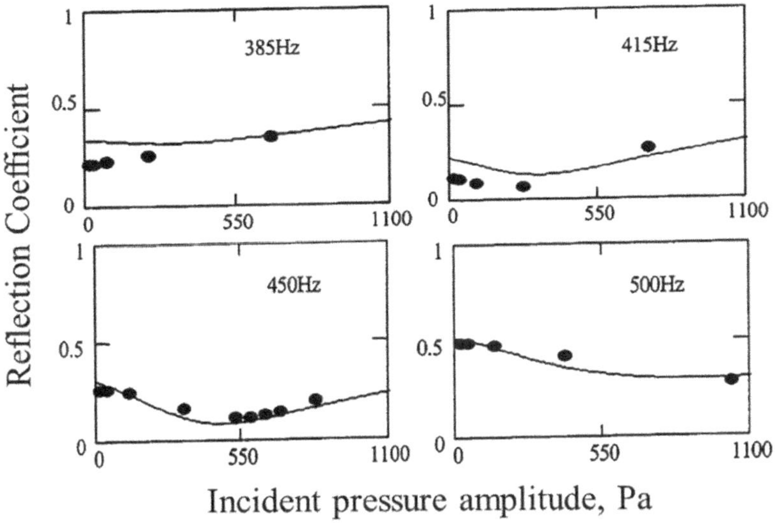

Figure 6.49 Plane wave pressure reflection coefficient of 15 cm thick hard-backed layer of 1.89 mm radius lead shot as a function of the pressure amplitude of the incident wave at four different frequencies. Points – data from shock tube measurements, lines – predictions at (a) 385 Hz, (b) 415 Hz, (c) 450 Hz, (d) 500 Hz using parameter values listed in Table 6.20 [105]. Reproduced from O. Umnova, K. Attenborough, E. Standley and A. Cummings, Behavior of Rigid-Porous Layers at High Levels of Continuous Acoustic Excitation: Theory and Experiment, *J. Acoust. Soc. Am.* 114(3): 1346–1356 (2003) with the permission of the Acoustical Society of America.

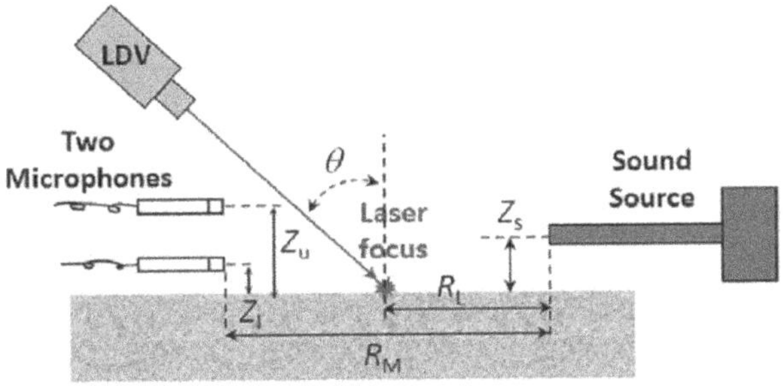

Figure 6.50 Schematic diagram of a configuration for non-contact measurement of acoustic-to-seismic coupling showing an acoustic point source, two microphones and an LDV [109, 110]. Z_s, Z_l and Z_u represent the heights of the source and the lower and upper microphones. The horizontal distances from the source to the laser beam focus and to the microphones are denoted by R_L and R_M respectively. The incidence angle of the laser beam is denoted by ϑ.

code (described in Chapter 5) and parameters for up to four layers were considered. The cost function uses the magnitudes of the complex transfer functions. Using the difference of the magnitudes improves the convergence of the optimization compared with using the complex values. The transfer functions include the acoustic level difference (p_u/p_l) and two acoustic-to-seismic (AS) coupling ratios (either v/p_u or v/p_l), where p_u and p_l are the acoustic pressures measured by the upper and lower microphones, respectively, and v is the surface particle velocity measured by the LDV.

It is not essential that the microphones are vertically separated as long as their positions are known. Furthermore, the Laser Doppler Vibrometer beam can be at a known azimuthal angle to the vertical plane through source and receivers. Figure 6.51 shows an example measured transfer function magnitude spectrum (LDV and upper microphone signals) at a sandy loam site growing winter wheat [110].

Figures 6.52(a) and (b) compare soil strength profiles obtained from penetrometer readings with those deduced from AS coupling measurements at two experimental sandy loam sites growing winter wheat [111]. The measurement geometries differed slightly but the sound source was at a height of about 10 cm and the microphone were at heights of about 5 cm and 15 cm and about 45 cm from the source. The laser beam elevation angle was about 35°. In these plots, the means of five penetrometer readings are represented by the broken black lines, the grey region boundaries represent the minimum and maximum readings, the continuous black lines represent the mean of soil strength profiles deduced from AS data, the continuous grey

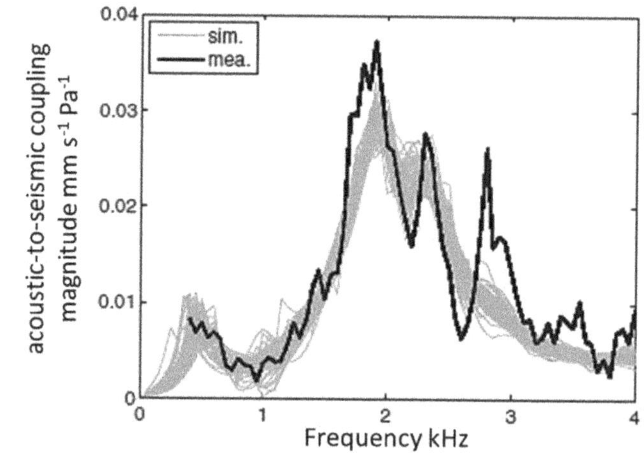

Figure 6.51 Measured (black) and simulated (grey) acoustic-to-seismic (AS) coupling transfer function magnitude spectra (soil particle velocity divided by incident acoustic pressure) obtained at a sandy loam site growing winter wheat using the signals from the laser Doppler vibrometer and the upper of a pair of vertically-separated microphones. The range of simulated results correspond to 49 numerical optimisations of fits assuming four layers with a hardpan at a variable depth (approximately 0.4 m) [110]. Reproduced from H.-C. Shin, R. Whalley, C. Watts, K. Attenborough and S. Taherzadeh, Non-invasive estimation of the depth profile of soil strength using acoustic-to-seismic coupling measurement in the presence of crops, *Eur. J. Soil. Sci.*, *68*: 758–768 (2017) with permission from Wiley.

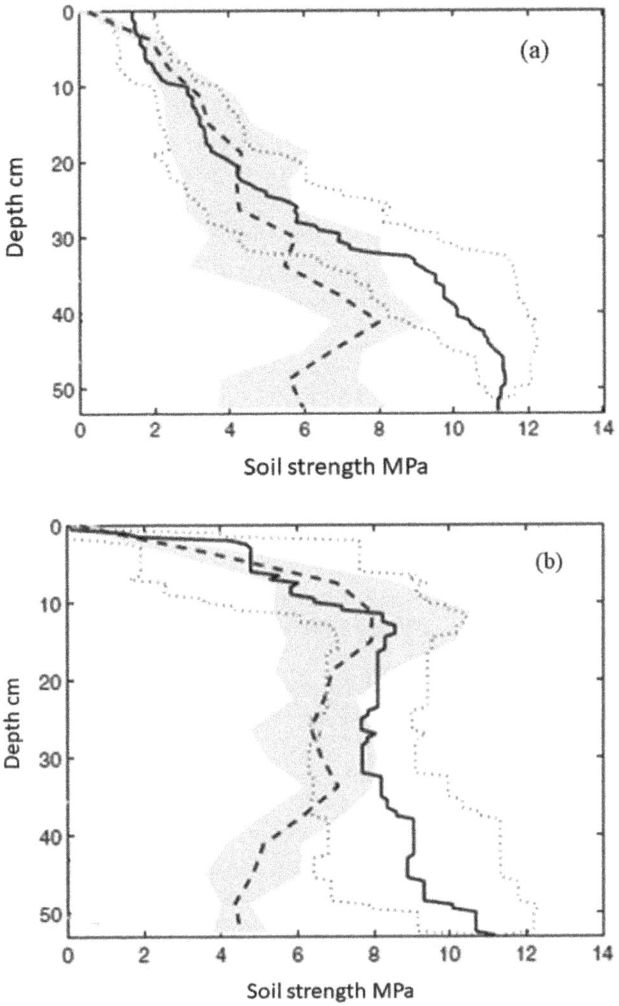

Figure 6.52 Soil strength profiles measured by a penetrometer and deduced through numerically optimized fitting of the acoustic-to-seismic (AS) coupling measurements at a two sandy loam experimental plots growing a variety of winter wheat [110]. Reproduced from H.-C. Shin, R. Whalley, C. Watts, K. Attenborough and S. Taherzadeh, Non-invasive estimation of the depth profile of soil strength using acoustic-to-seismic coupling measurement in the presence of crops, *Eur. J. Soil. Sci.*, *68*: 758–768 (2017) with permission from Wiley.

lines are the medians below 30 cm depth and the dotted lines indicate the minimum and maximum of the estimates from AS data. Although each profile deduced from AS data allowed for only four layers, the plots are based on several profiles giving similar goodness of fits, but different deduced layer thicknesses, as well as the best-fit profile. At the site associated with Figure 6.49(a), two of the penetrometer readings were not possible beyond depths of 41.3 and 52.5, respectively, and 40 strength profiles deduced from AS data were used. At the site associated with Figure 6.49(b), five penetrometer readings were available, but nine strength profiles only were deduced from AS data. More AS profiles were used when the goodness of fit was relatively poor.

The agreement between the penetrometer readings and AS-deduced soil strength is tolerable down to 30 cm depth. Beyond this depth the agreement is less encouraging partly because penetrometer measurements were not uniformly successful at those depths and partly due to lack of a sufficient low-frequency output from the sound source used in these AS measurements.

REFERENCES

[1] American Society for the testing of materials, ASTM Standard test method for impedance and absorption of acoustical materials by the tube method, 348–377 (1977).

[2] Y. Ando. The directivity and acoustic centre of a probe microphone, *J. Acoust. Soc. Japan* **24** 334–342 (1968).

[3] P. J. Dickinson and P. E. Doak, Measurement of the acoustical impedance of ground surfaces, *J. Sound Vib.* **13** 309–322 (1970).

[4] T. F. W. Embleton, J. E. Piercy and N. Olson, Outdoor sound propagation over a ground of finite impedance, *J. Acoust. Soc. Am.* **59** 267–277 (1976).

[5] R. Talaske, *The acoustic impedance of a layered forest floor*, M.S. Thesis Pennsylvania State University (1978).

[6] O. Buser, A rigid frame model of porous media for the acoustic impedance of snow. *J. Sound Vib.* **111** 71–92 (1986).

[7] American Society for the Testing of Materials ASTM E1050. Standard testing method for impedance and absorption of acoustical materials using a tube, two microphones and a digital frequency analyzer (1986).

[8] J. Y. Chung and D. A. Blaser, Transfer function method of measuring a duct acoustic properties: experiment, *J. Acoust. Soc. Am.* **68** 907–921 (1980).

[9] F. J. Fahy, Rapid method for the measurement of sample acoustic impedance in a standing wave tube, *J. Sound Vib.* **97**: 168–170 (1984).

[10] W. T. Chu, Further experimental studies on the transfer technique for impedance tube measurements. *J. Acoust. Soc. Am.* **83** 2255–2260 (1988).

[11] ISO 10534-2:1998 *Acoustics - Determination of sound absorption coefficient and impedance in impedance tubes – Part 2: Transfer-function method.*

[12] P. Datt, J. C. Kapil, A. Kumar and P. R. Srivastava, Experimental measurements of acoustical properties of snow and inverse characterization of its geometrical parameters, *Appl. Acoust.* **101** 15–23 (2016).

[13] G. M. Heissler, O. McDaniel and M. Dahl, Measurements of normal impedance of six forest floors by the tube method, Proc. 2nd symposium on long range sound propagation and acoustic/seismic coupling vol. 2, 408–427, University of Mississippi PARGUM (1985).

[14] ISO 13472-2: 2010 Acoustics – Procedure for measuring sound absorption properties of road surfaces in situ – Part 2: Spot method for reflective surfaces.

[15] A. J. Zuckerwar, Acoustic ground impedance meter, *J. Acoust. Soc. Am.* 73 2180–2186 (1983).
[16] L. A. M. van der Heijden, *The influence of vegetation on acoustic properties of soils*. Ph.D. Thesis, University of Nijmegen, The Netherlands (1984).
[17] J. F. Allard and B. Sieben, Measurements of acoustic impedance in a free field with two microphones and a spectrum analyzer, *J. Acoust. Soc. Am.* 77 1617–1618 (1985).
[18] D. Waddington, *Acoustical impedance measurement using a two-microphone transfer function technique*, Ph.D. Thesis. University of Salford (1990).
[19] G. A. Daigle and M. R. Stinson, Impedance of grass covered ground at low frequencies using a phase difference technique, *J. Acoust. Soc. Am.* **81**: 62–68 (1987).
[20] M. W. Sprague, R. Raspet, H. E. Bass and J. M. Sabatier, Low frequency acoustic ground impedance measurement techniques, *Appl. Acoust.* **39** 307–325 (1993).
[21] L. A. M. van der Heijden, J. G. E. M. de Bie and J. Groenewoud, A pulse method to measure the impedance of semi-natural soils, *Acustica* **51** 193–197 (1982).
[22] A. J. Cramond and C.G. Don, Reflection of impulses as a method of determining acoustic impedance. *J. Acoust. Soc. Am.* 77 382–389 (1984).
[23] C. G. Don and A. J. Cramond, Soil impedance measurements by an acoustic pulse technique, *J. Acoust. Soc. Am.* 77 1601–1609 (1985).
[24] D. Habault and G. Corsain, Identification of the acoustical properties of a ground surface. *J. Sound Vib.* **100** 169–180 (1985).
[25] C. Hutchinson-Howorth, K. Attenborough and N.W. Heap, Indirect in-situ and free-field measurement of impedance model parameters or surface impedance of porous layers, *Appl. Acoust.* **39** 77–117 (1993).
[26] K. Attenborough, A note on short-range ground characterization. *J. Acoust. Soc. Am.* **95** 3103–3108 (1994).
[27] L. N. Bolen and H. E. Bass, Effects of ground cover on the propagation of sound through the atmosphere, *J. Acoust. Soc. Am.* **69** 950–954 (1981).
[28] J. M. Sabatier, H. M. Hess, P. A. Arnott, K. Attenborough, E. Grissinger and M. Romkens, In-situ measurements of soil physical properties by acoustic techniques, *Am. J. Soil Sci.* **54** 658–672 (1990).
[29] H. M. Hess, K. Attenborough and N. W. Heap, Ground characterization by short-range measurements of propagation, *J. Acoust. Soc. Am.* **87** 1975–1985 (1990).
[30] H. M. Hess, *Acoustical determination of physical properties of porous grounds*, Ph.D. Thesis, The Open University, Milton Keynes, UK (1988).
[31] T. F. W. Embleton, J. E. Piercy and G. A. Daigle, Effective flow resistivity of ground surfaces determined by acoustical measurements, *J. Acoust. Soc. Am.* **74** 1239–1244 (1983).
[32] S.-I. Thomasson, Sound propagation above a layer with a large refractive index. *J. Acoust. Soc. Am.* **61** 659–674 (1977).
[33] J. Wempen, Ground effect on long range propagation. Proc. I.O.A. (September 1987).
[34] J. M. Sabatier, R. Raspet and C.K. Frederickson, An improved procedure for the determination of ground parameters using level difference measurements, *J. Acoust. Soc. Am.* **94** 396–399 (1993).

[35] D. G. Albert, Acoustic waveform inversion with application to seasonal snow covers, *J. Acoust. Soc. Am.* **109**, 91–101 (2001).

[36] ANSI/ASA S1.18-2018 *American National Standard Method for Determining the Acoustic Impedance of Ground Surfaces*

[37] W. C. K Alberts and K. J. Sanchez. Deduction of the acoustic impedance of the ground via a simulated three-dimensional microphone array, *J. Acoust. Soc. Am.* **134** EL471–EL476 (2013).

[38] NORDTEST ACOU 104. 1998. Ground Surfaces: determination of acoustic impedance. http://doutoramento.schiu.com/referencias/outras/NT%20ACOU%2014%20-%20Ground%20Surfaces%20Determination%20of%20acoustic%20impedance.%201999.pdf (last viewed 08/07/20).

[39] C. Nocke, V. Mellert, T. Waters-Fuller, K. Attenborough and K. M. Li, Impedance deduction from broad-band, point-source measurements at grazing angles, *Acta Acust. united Ac.* **83** 1085–1090 (1997).

[40] S. Taherzadeh and K. Attenborough, Deduction of ground impedance from measurements of excess attenuation spectra. *J. Acoust. Soc. Am.* **105** 2039–2042 (1999).

[41] I. Bashir, *Acoustical exploitation of rough, mixed impedance and porous surfaces outdoors*, Ph.D. dissertation, The Open University, Milton Keynes, UK (2013).

[42] J. H. Soh, K. E. Gilbert, W. G. Frazier, C. L. Talmadge and R. Waxler, A direct method for measuring acoustic ground impedance in long-range propagation experiments. *J. Acoust. Soc. Am.* **128** EL286–EL293 (2010).

[43] K. Attenborough and T. F. Waters-Fuller, Effective impedance of rough porous ground surfaces. *J. Acoust. Soc. Am.* **108** 949–956 (2000).

[44] G. Taraldsen and H. Jonasson. Aspects of ground effect modelling, *J. Acoust. Soc. Am.* **129** 47–53 (2011).

[45] M. Sohlman, H. Jonasson and A Gustafsson, Using satellite data for the determination of the acoustic impedance of ground. SP Swedish National Testing and Research Institute Report 2004 - 12 ISSN 0284-5172 (2004).

[46] H. Jonasson and S. Storeheier, Revision of NT ACOU 104 for the Measurement of the Acoustic Impedance of Ground, Nordic Innovation Centre project number 04145 (2006). http://ftp.aip.org/epaps/j_acoust_soc_am/E-JASMAN-129-023105/JonassonStoreheier06NTImpedance06.pdf (last viewed 08/07/20).

[47] K. Attenborough, I. Bashir and S Taherzadeh, Outdoor ground impedance models, *J. Acoust. Soc. Am.* **129** 2806–2819 (2011).

[48] A. D. Dragna, K. Attenborough and P. Blanc-Benon, On the inadvisability of using single parameter models for representing outdoor ground impedance, *J. Acoust. Soc. Am.* **138** 2399–2413 (2015); b. D. Dragna, private communication.

[49] M. C. Berengier, M. Stinson, G. A. Daigle and J.-F. Hamet, Porous road pavements: Acoustical characterization and propagation effects, *J. Acoust. Soc. Am.* **101** 155–162 (1997).

[50] Harmonoise WP 3 Engineering method for road traffic and railway noise after validation and fine-tuning, Technical Report HAR32TR-040922-DGMR20, January 2005.

[51] K. Attenborough, P. Boulanger, Q. Qin and R. Jones, Predicted influence of ballast and porous concrete on rail noise. Proc. InterNoise 2005, Rio de Janeiro, Paper 1583 (2005).

[52] K. Heutschi, Sound propagation over ballast surfaces, *Acta Acust. united Ac.* **95** 2006–2012 (2009).
[53] M. J. M. Martens, L. A. M. van der Heijden, H. H. J. Walthaus and W. J. J. M. van Rens, Classification of soils based on acoustic impedance, air flow resistivity and other physical soil parameters*J.Acoust. Soc. Am.* **78** 970–980 (1985).
[54] T. Ishida, Acoustic properties of snow, *Contrib. Intensity Low Temp. Sci. Series A* **20** 23–63 (1965).
[55] W. H. T. Huisman, *Sound propagation over vegetation-covered ground*, Ph.D. Thesis, University of Nijmegen, The Netherlands (1990).
[56] V. E. Ostashev, D. K. Wilson and N. Vecherin, Effect of randomly varying impedance on the interference of the direct and ground-reflected waves, *J. Acoust. Soc. Am.* **130** 1844–1850 (2011).
[57] D. Dragna and P. Blanc-Benon, Sound propagation over the ground with a random spatially- varying surface admittance, *J. Acoust. Soc. Am.* **142** 2058–2072 (2017).
[58] A. J. Cramond and C. G. Don, Effects of moisture content on soil impedance, *J. Acoust. Soc. Am.* **82** 293–301 (1987).
[59] K. V. Horoshenkov and M. H. Mohamed, Experimental investigation of the effects of water saturation on the acoustic admittance of sandy soils, *J. Acoust. Soc. Am.* **120**(4) 1910–1921 (2006).
[60] T. Oshima, Y. Hiraguri and Y. Okuzono, Distinct effects of moisture and air contents on acoustic properties of sandy soil, *J. Acoust. Soc. Am. Express Lett.* **138**(3) EL258–EL263 (2015).
[61] K. Suravi, H-C. Shin, K. Attenborough, S. Taherzadeh and R. Whalley, The influence of organic matter on acoustical properties of soil, Proc. ICA Aachen, Germany (2019).
[62] T. Van Renterghem and D. Botteldooren, Influence of rainfall on the noise shielding by a green roof, *Build. Environ.* **82** 1–8 (2014).
[63] Z. Lu and J. M. Sabatier, Effects of soil water potential and moisture content on sound speed, *Soil Sci Soc Am J.* **73**(5) 1614–1625 (2009).
[64] R. Whalley, M. Jenkins and K. Attenborough, The velocity of shear waves in unconsolidated soils, *Soil Till. Res.* **125** 30–37 (2012).
[65] H.-C. Shin, W. R. Whalley, K. Attenborough and S. Taherzadeh, On the theory of Brutsaert about elastic wave speeds in unsaturated soils, *Soil Till. Res.* **156** 155–165 (2016).
[66] G. Guillaume, O. Faure, B. Gavreau, F. Junker and M. Berengier, Estimation of impedance model input parameter from in situ measurements: principles and applications, *Appl. Acoust.* **95** 27–36 (2015).
[67] D. Dragna and P. Blanc-Benon, Physically admissible impedance models for time-domain computations of outdoor sound propagation, *Acta Acust. united Ac.* **100** 401–410 (2014).
[68] Annex to Commission Directive 2015/996 in Official Journal of the European Union L168 (2015).
[69] J. P. Chambers, J. M. Sabatier and R. Raspet, Grazing incidence propagation over a soft rough surface, *J. Acoust. Soc. Am.* **102** 55–59 (1997).
[70] K. Attenborough and S. Taherzadeh, Propagation of sound from a point source over a rough finite impedance surface, *J. Acoust. Soc. Am.* **98** 1717–1722 (1995).
[71] P. M. Boulanger, K. Attenborough, S. Taherzadeh, T. Waters-Fuller and K. M. Li, Ground effect over hard rough surfaces, *J. Acoust. Soc. Am.* **104** 1474–1482 (1998).

[72] P. Boulanger, K. Attenborough and Q. Qin, Effective impedance of surfaces with porous roughness: models and data, *J. Acoust. Soc. Am.* **117** 1146–1156 (2005).
[73] P. Boulanger, K. Attenborough, Q. Qin and C. M. Linton. Reflection of sound from random distributions of semi-cylinders on a hard plane: models and data, *J. Phys. D.* **38** 3480–3490 (2005).
[74] I. Bashir, S. Taherzadeh and K. Attenborough, Diffraction assisted rough ground effect: models and data, *J. Acoust. Soc. Am.* **133** 1281–1292 (2013).
[75] J. F. Allard, G. Jansens, G. Vermeir and W. Lauriks, Frame-borne surface waves in air-saturated porous media, *J. Acoust. Soc. Am.* **111** 690–696 (2002).
[76] F. Press and M. Ewing, Ground roll coupling to atmospheric compressional waves, *Geophysics* **16** 416–430 (1951).
[77] W. Lauriks, L. Kelders and J. F. Allard, Poles and zeros of the reflection coefficient of a porous layer having a motionless frame in contact with air, *Wave Motion* **28** 59–67 (1998).
[78] L. Kelders, W. Lauriks and J. F. Allard, Surface waves above thin porous layers saturated by air at ultrasonic frequencies, *J. Acoust. Soc. Am.* **104** 882–889 (1998).
[79] M. R. Stinson, Surface wave formation at an impedance discontinuity, *J. Acoust. Soc. Am.* **102** 3269–3275 (1997).
[80] L. M. Brekhovskikh, *Waves in layered media* Translated by D. Liberman and R. T. Beyer. Academic press, London and New York, pp. 44–61 (1960).
[81] H. Medwin, J. Bailie, J. Bremhorst, J. Savage and I. Tolstoy, The scattered acoustic boundary wave generated by grazing incidence at a slightly rough rigid surface, *J. Acoust. Soc. Am.* **66** 1131–1134 (1979).
[82] H. Medwin, G. L. D'Spain, E. Childs and S. J. Hollis, Low-frequency grazing propagation over periodic steep-sloped rigid roughness elements, *J. Acoust. Soc. Am.* **76** 1774–1790 (1984).
[83] H. Medwin and G. L. D'Spain, Near-grazing, low-frequency propagation over randomly rough, rigid surfaces, *J. Acoust. Soc. Am.* **79** 657–665 (1986).
[84] R. J. Donato, Model experiments on surface waves, *J. Acoust. Soc. Am.* **63** 700–703 (1978).
[85] C. Hutchinson-Howorth and K. Attenborough, Model experiments on air-coupled surface waves, *J. Acoust. Soc. Am.* **92**(A) 2431 (1992).
[86] G. A. Daigle, M. R. Stinson and D. I. Havelock, Experiments on surface waves over a model impedance plane using acoustical pulses, *J. Acoust. Soc. Am.* **99**(4) 1993–2005 (1996).
[87] W. Zhu, M. R. Stinson and G. A. Daigle, Scattering from impedance gratings and surface wave formation, *J. Acoust. Soc. Am.* **111**(5) (A) 1996 (2002).
[88] W. Zhu, G. A. Daigle and M. R. Stinson, Experimental and numerical study of air-coupled surface waves generated above strips of finite impedance, *J. Acoust. Soc. Am.* **114**(3) 1243–1253 (2003).
[89] W. Lauriks, L. Kelders and J. F. Allard, Surface waves above gratings having a triangular profile, *Ultrasonics* **36**(8) 865–871 (1998).
[90] L. Kelders, J. F. Allard and W. Lauriks, Ultrasonic surface waves above rectangular-groove gratings, *J. Acoust. Soc. Am.* **103**(5) 2730–2733 (1998).
[91] J. F. Allard, L. Kelders and W. Lauriks, Ultrasonic surface waves above a doubly periodic grating, *J. Acoust. Soc. Am.* **105**(4) 2528–2531 (1999).

[92] J. Tizianel, J. F. Allard and B. BROUARD, Surface waves above honeycombs, *J. Acoust. Soc. Am.* **104**(4) 2525–2528 (1998).
[93] I. Bashir, S. Taherzadeh and K. Attenborough, Surface waves over periodically-spaced strips, *J. Acoust. Soc. Am.* **134**(6) 4691–4697 (2013).
[94] D. Berry, S. Taherzadeh and K. Attenborough, Acoustic surface wave generation over rigid cylinder arrays on a rigid plane, *J. Acoust. Soc. Am.* **146**(4) 2137–2144 (2019).
[95] A. Whelan and J. P. Chambers, A note on the effects of roughness on acoustic propagation past curved rough surfaces, *J. Acoust. Soc. Am.* **125**(6) EL231–EL235 (2009).
[96] Q. Qin, K. Attenborough, S. Ollivier and P. Blanc-Benon, Effects of surface roughness and turbulence on propagation of shock waves above a curved surface, Proc. CFA-DAGA Strasberg (2009).
[97] D. G. Albert, Observations of acoustic surface waves in outdoor sound propagation, *J. Acoust. Soc. Am.* **113**(5) 2495–2500 (2003).
[98] M. Boeckx, G. Jansens, W. Lauriks and D. Albert, Modelling acoustic surface waves above a snow layer, *Acta Acust. united Ac.* **90**(2) 246–250 (2004).
[99] D. Aylor, Noise reduction by vegetation and ground. *J. Acoust. Soc. Am.* **51**(1) 197–205 (1972).
[100] J. P. Chambers and J. M. Sabatier, Recent advances in utilizing acoustics to study roughness of agricultural surfaces, *Appl. Acoust.* **63** 795–812 (2002).
[101] C. Madshus, F. Løvholt, A. Kaynia, L. R. Hole, K. Attenborough and S. Taherzadeh, Air–ground interaction in long range propagation of low frequency sound and vibration. Field tests and model verification, *Appl. Acoust.* **66** 553–578 (2005).
[102] A. M. Kaynia, F. Lovholt and C. Madshus, Effects of a multi-layered poroelastic ground on attenuation of acoustic waves and ground vibration, *J. Sound Vib.* **330** 1403–1418 (2011). https://doi.org/10.1016/j.jsv.2010.10.004.
[103] S. Taherzadeh, *Sound propagation in inhomogeneous media*, PhD Thesis in Engineering Mechanics. The Open University: Milton Keynes, UK (1996).
[104] D. G. Albert, S. Taherzadeh, K. Attenborough, P. Boulanger and S. Decato, Ground vibrations produced by surface and near-surface explosions, *Appl. Acoust.* **74** 1279–1296 (2013). https://doi.org/10.1016/j.apacoust.2013.03.006.
[105] O. Umnova, K. Attenborough, E. Standley and A. Cummings, Behavior of Rigid-Porous Layers at high levels of continuous acoustic excitation: theory and experiment, *J. Acoust. Soc. Am.* **114**(3) 1346–1356 (2003). https://doi.org/10.1121/1.1603236.
[106] E. Vedy and F. V. D. Eerden, Propagation of shock waves from source to receiver, *Noise Cont. Eng. J.* **53** 87–93 (2005).
[107] K. Attenborough, P. Schomer, F. van der Eerden and E. Védy, Overview of the theoretical development and experimental validation of blast sound absorbers, *Noise Cont. Eng. J.* 53 70–80 (2005).
[108] P. Schomer and K. Attenborough, Basic results of tests at Fort Drum, *Noise Cont. Eng. J.* **53** 94–109 (2005).
[109] F. V. D. Eerden and E. Carton, *Mitigation of open-air explosions using blast absorbing barriers and foam, Proc.* InterNoise 2012, *New York* (2012).
[110] H.-C. Shin, S. Taherzadeh, K. Attenborough, W. R. Whalley and C. W. Watts, Non-invasive soil parameter deduction using acoustic-seismic coupling and

linear Biot-Stoll theory, *Eur. J. Soil. Sci.* **64** 308–323 (2013). https://doi.org/10.1111/ejss.12000.

[111] H.-C. Shin, R. Whalley, C. Watts, K. Attenborough and S. Taherzadeh, Non-invasive estimation of the depth profile of soil strength using acoustic-to-seismic coupling measurement in the presence of crops, *Eur. J. Soil. Sci.* **68** 758–768 (2017). https://doi.org/10.1111/ejss.12462.

Chapter 7

Influence of Source Motion on Ground Effect and Diffraction

7.1 INTRODUCTION

Chapters 2, 5 and 6 have involved considerations of the sound field due to a stationary monopole source above an absorbing ground. The asymptotic solution for a stationary monopole was derived in Chapter 2. However, powered vehicles are noise sources that are neither stationary nor monopole in nature. This chapter investigates analytical ways of predicting the effect of source motion on sound propagation outdoors. It makes use of the results for the sound field due to directional sources in Chapter 3.

Previous analyses have simplified the problem of modelling a moving source either by treating it as a quasi-stationary point source or, in the case of road traffic noise, as a continuous line source. While such simplifications are valid for sources travelling slowly, they are unlikely to be adequate for sources travelling at a non-negligible Mach number, such as a high-speed trains, helicopters or aircraft during overflights, landing or taking-off. A better physical understanding of the effects of source motion should enable better designs of suitable noise control devices.

The many theoretical and experimental efforts to study the effects of motion on aerodynamic noise have been restricted to either an unbounded medium or propagation above a rigid half-plane [1, 2]. This chapter investigates the effect of motion of the source on propagation of sound above an absorbing ground by means of analytical approximations. The task can be considered as an extension to the problem of predicting the sound pressure due to a source moving in a free field which has a long history and has been addressed by many authors. In his seminal paper, Lowson [3] pointed out the importance of stating whether the theoretical derivation is based on the wave equation for the velocity potential or on the wave equation for the perturbation pressure. Similarly, Graham and Graham [4] have distinguished two different types of models for a moving source: a conventional moving source that injects fluid as it moves and a modified source that can be depicted as an expanding and contracting balloon in uniform motion. Despite the considerable previous theoretical work, there are relatively few in-depth studies of outdoor sound propagation from sources moving at high speeds.

This chapter provides a theoretical basis for such developments based on the wave equation for the velocity potential of the sound field due to a moving monopole source [3]. The use of velocity potential is consistent with the fundamental theory of fluid mechanics, whereas an approach based on the wave equation for the perturbation pressure has little physical significance.

7.2 A MONOPOLE SOURCE MOVING AT CONSTANT SPEED AND HEIGHT ABOVE A GROUND SURFACE

Consider a point monopole source of unit strength moving at a constant speed u (=cM) along the x-axis where c is the speed of sound and M is the Mach number. The source is moving at a constant height, z_s above an impedance ground with a specific admittance β. The intention is to determine an analytical expression for the sound field due to a moving source above an impedance ground in the homogeneous atmosphere.

The governing equation is the space-time wave equation given by

$$\nabla^2\varphi - \frac{1}{c^2}\frac{\partial^2\varphi}{\partial t^2} = -e^{-i\omega_0 t}\delta(x - cMt)\delta(y)\delta(z - z_s), \tag{7.1a}$$

where ω_0 is the angular frequency of the source, (x, y, z) are rectangular coordinates, t is time and φ is the velocity potential relating to the acoustic pressure p by

$$p = \partial\varphi / \partial t \tag{7.1b}$$

The boundary condition, at $z = 0$, is determined by

$$\frac{1}{c}\frac{\partial\varphi}{\partial t} - \frac{1}{\beta}\frac{\partial\varphi}{\partial z} = 0 \tag{7.2}$$

The analysis can be simplified considerably by using a Lorentz transformation [5]:

$$\left.\begin{aligned} x_L &= \gamma^2(x - cMt) \\ y_L &= \gamma y \\ z_L &= \gamma z \\ t_L &= \gamma^2(t - Mx/c) \end{aligned}\right\} \tag{7.3}$$

$$\gamma = 1/\sqrt{1 - M^2} \tag{7.4}$$

This transformation reduces the governing wave equation and the boundary condition (at $z_L = 0$) to

$$\nabla_L^2\phi - \frac{1}{c^2}\frac{\partial^2\phi}{\partial t_L^2} = -\gamma^2 e^{-i\omega_0(t_L + Mx_L/c)}\delta(x_L)\delta(y_L)\delta(z_L - h_s) \tag{7.5a}$$

$$\frac{1}{c}\frac{\partial \varphi}{\partial t_L} - M\frac{\partial \varphi}{\partial x_L} - \frac{1}{\gamma\beta}\frac{\partial \varphi}{\partial z_L} = 0, \tag{7.5b}$$

where $h_s = \gamma z_s$.

The resulting Equation (7.5a) is analogous to that of a stationary source except that the source strength and the time-dependent factor are modified. In addition, the boundary condition (7.5b) is somewhat different from that of a stationary source above an impedance plane. Use of a Lorentz transformation leads to an equation which is relatively simple compared with that resulting from use of a Galilean transformation [6, 7].

The interpretation of different terms in the analytic expression will be left until Section 7.5 after deriving the corresponding expression for a source moving perpendicularly to the ground surface in Section 7.3.

The time-dependent factor, $e^{-i\omega_0 t_L}$, can be factored out by separating the variables in the solution of (7.6) into spatial and temporal domains as

$$\phi = e^{-i\omega_0 t_L} G_L(x_L), \tag{7.6}$$

where $G_L(x_L)$ is the required Green's function for the Helmholtz equation with the source located at $(0,0,h_s)$. Then equations (7.5a and 7.5b) become

$$\nabla_L^2 G_L + k_0^2 G_L = -\gamma^2 e^{-ik_0 M x_L}\delta(x_L)\delta(y_L)\delta(z_L - h_s) \tag{7.7a}$$

$$k_0 G_L + M\frac{\partial G_L}{\partial x_L} + \frac{1}{\gamma\beta}\frac{\partial G_L}{\partial z_L} = 0, \tag{7.7b}$$

where $k_0 = \omega_0/c$ is the wavenumber. A Fourier transformed pair for the Green's function in the Lorentz space [*cf* Equations (2.5a) and (2.5b) of Chapter 2] is

$$\hat{G}_L(k_x,k_y,z_L) = \int_{-\infty}^{\infty}\int_{-\infty}^{\infty} G_L(x_L)e^{-ik_x x_L - ik_y y_L}\,dx_L dy_L \tag{7.8a}$$

and

$$G_L(x_L) = \frac{1}{4\pi^2}\int_{-\infty}^{\infty}\int_{-\infty}^{\infty} \hat{G}_L(k_x,k_y,z_L)e^{ik_x x_L + ik_y y_L}\,dk_x dk_y \tag{7.8b}$$

Using the same analysis as described in Chapter 2,

$$\hat{G}_L(k_x,k_y,z_L) = \frac{i}{2k_z}\left\{e^{ik_z|z_L - h_s|} + \frac{[k_z - \gamma\beta(k_0 + Mk_x)]}{[k_z + \gamma\beta(k_0 + Mk_x)]}e^{ik_z(z_L + h_s)}\right\}. \tag{7.9}$$

Substitution of (7.9) into (7.8b) gives an integral representation of the sound field which can be estimated by evaluating the integrals asymptotically. The solution for (7.9) is well known for the case of a stationary source and it is sometimes referred to as the Weyl-Van der Pol formula. The method of steepest descent is used to derive the asymptotic solution. The total field is given by a sum of three components: a direct wave term G_1, a contribution due to an image source G_2 and a boundary wave term G_b. The boundary wave term is particularly important for the near-grazing propagation above a ground of finite impedance. The same analytical approach can be applied to the corresponding result for a moving source and leads to an accurate asymptotic solution. It follows that

$$G_1 = \frac{\gamma^2 \exp[ik_0 d_1]}{4\pi d_1} \tag{7.10}$$

$$G_2 = \frac{\gamma^2 \exp[ik_0 d_2]}{4\pi d_2}, \tag{7.11}$$

where d_1 and d_2 are the direct and reflect path lengths in Lorentz space:

$$d_1 = \sqrt{x_L^2 + y_L^2 + (z_L + h_s)^2}, \quad d_2 = \sqrt{x_L^2 + y_L^2 + (z_L + h_s)^2} \tag{7.12}$$

The boundary wave term, G_b, can be expressed as

$$G_b = -\frac{\gamma^2}{4\pi^2}\int_{-\infty}^{\infty}\int_{-\infty}^{\infty} \frac{\gamma\beta(k_0 + Mk_x)\exp\{i[k_x x_L + k_y y_L + k_z(z_L + h_s)]\}}{ik_z[k_z + \gamma\beta(k_0 + Mk_x)]} dk_x dk_y \tag{7.13}$$

The evaluation of the above integral can be simplified considerably by using spherical polar co-ordinates centred on the source and the concept of effective admittance such that

$$\beta_L = \gamma\beta(1 + M sin\theta_L \cos\psi_L), \tag{7.14}$$

where θ_L and ψ_L are, respectively, the polar and the azimuthal angles of the reflected wave in the Lorentz space. The polar angle θ_L may be interpreted as the angle of incidence of the reflected wave also. Using the procedure detailed in Chapter 2, the boundary wave term can be evaluated to give

$$G_b = 2i\gamma^2\sqrt{\pi}\frac{\beta_L}{\beta_L + \cos\theta_L} w_L e^{-w_L^2} erfc(-iw_L)\frac{e^{ik_0 d_2}}{4\pi d_2}, \tag{7.15a}$$

where

$$w_L = \frac{1}{2}(1+i)\sqrt{k_0 d_2}(\cos\theta_L + \beta_L) \tag{7.15b}$$

Noting the relationship (7.6) and summing the contributions due to the direct wave term, image source term and the boundary wave term,

$$\phi = \frac{\gamma^2 e^{-i\omega_0 t_L}}{4\pi}\left\{\frac{e^{ik_0 d_1}}{d_1} + [R_L + (1 - R_L)F(w_L)]\frac{e^{ik_0 d_2}}{d_2}\right\}, \tag{7.16a}$$

where

$$R_L = \frac{\cos\theta_L - \beta_L}{\cos\theta_L + \beta_L} \tag{7.16b}$$

$$F(w_L) = 1 + i\sqrt{\pi} w_L e^{-w_L^2} erfc(-iw_L) \tag{7.16c}$$

For a stationary source ($M = 0$), Equations (7.16a, 7.16b and 7.16c) reduce to the classical Weyl-Van der Pol formula (2.40). The ground wave term in (7.16a) was ignored in earlier work [5–7]. However, since the ground wave term is particularly important for a stationary source at low frequencies near the ground effect dip, it can be expected that the same will apply to a moving source above an impedance plane. Omitting the ground wave term will lead to erroneous predictions especially at frequencies near the ground effect dip.

Equations (7.16a–7.16c) provide an asymptotic expression for the sound field due to a source moving at constant speed parallel to the ground. To arrive at a physical interpretation of each term in (7.16a), first consider the amplitudes of the direct and reflected waves. Since they have similar characteristics, it is enough to study the reflected wave term. In general, sounds heard at different positions are heard at different times. It is convenient to use the 'emission time geometry', which was used already to describe the sound field due to a stationary source in the homogeneous atmosphere. Ignoring the constant factor, $1/4\pi$, the virtual distance travelled is $d_2/\gamma^2, = r_2(t)$. Replacing the Lorentz co-ordinates with the emission time co-ordinates, the virtual distance is

$$r_2(t) = \sqrt{(x - cMt)^2 + (1 - M^2)\left[y^2 + (z + z_s)^2\right]} \tag{7.17a}$$

The instantaneous distance, $R_2(t)$, between the source image and receiver is given by

$$R_2(t) = \sqrt{(x - cMt)^2 + y^2 + (z + z_s)^2} \tag{7.17b}$$

Suppose τ_2 is the emission time of sound heard at x at time t. An implicit relationship between t and τ_2 can be expressed as

$$c(t - \tau_2) = R_2(\tau_2) = \sqrt{(x - cM\tau_2)^2 + y^2 + (z + z_s)^2}, \tag{7.18a}$$

which can be solved to obtain τ_2 explicitly for any observer locations,

$$\tau_2 = \frac{ct - Mx \pm \sqrt{(x - cMt)^2 + (1 - M^2)\left[y^2 + (z + z_s)^2\right]}}{c(1 - M^2)} \tag{7.18b}$$

The spurious roots have been introduced as a result of squaring both sides of (7.18a) to obtain a solution for τ_2^*. Since τ_2^* must be real and less than t, only the root (in the numerator) with the negative sign satisfies these conditions. To write the virtual distance r_2 in terms of R_2, it is useful to combine (7.18a) and (7.18b). With some simple algebraic manipulations, we obtain

$$r_2(t) = c(t-\tau_2) - M(x - cM\tau_2) \tag{7.19}$$

The above equation can be simplified with the introduction of Doppler factor, $1-M_{r2}$, such that

$$M_{r2} = M(x - cM\tau_2)/R_2(\tau_2) = M sin\theta cos\psi, \tag{7.20}$$

where θ and ψ are the polar angle and azimuthal angle between the source and observer's position at emission time. Hence, it follows from (7.19) and (7.20) that the virtual distance is given by

$$r_2(t) = c(t-\tau_2)\left[1 - \frac{M(cM\tau_2 - x)}{c(t-\tau_2)}\right] = R_2(\tau_2)(1-M_{r2}) \tag{7.21}$$

The virtual distance, r_2, may be interpreted as the distance between the image source and receiver's position at the emission time τ_2. This distance is further modified by the well-known Doppler factor $(1 - M_{r2})$. The emission time τ_2 may be regarded as the retarded time as a result of the time delay for the sound propagation from the image source to the receiver's position. To represent the phase of the reflected wave in terms of the retarded time, the phase function of the reflected wave, $\bar{\omega}_2$ can be transformed back from the Lorentz space to the emission time co-ordinates as follows:

$$\begin{aligned}\bar{\omega}_2 &= -i(\omega_0 t_L - k_0 d_2) = \\ &-i\omega_0\gamma^2\left[(t - Mx/c) - \frac{1}{c}\sqrt{(x-cMt)^2 + (1-M^2)\left[y^2 + (z+z_s)^2\right]}\right] = -i\omega_0\tau_2.\end{aligned} \tag{7.22}$$

Similarly, the plane wave reflection coefficient R_p and the numerical distance w_L can be written in the emission time co-ordinates as

$$R_L = R_p = \frac{\cos\theta - \beta}{\cos\theta + \beta} \tag{7.23a}$$

$$w_L = \frac{1}{2}(1+i)\sqrt{k_0 d_2}\left(\cos\theta_L + \beta_L\right)$$

or

$$w_L = \frac{1}{2}(1+i)\sqrt{k_0 R_2(\tau_2)/(1-M_{r2})}\left(\cos\theta + \beta\right) = w/\sqrt{1-M_{r2}}, \tag{7.23b}$$

where w is the numerical distance and R_p is the plane wave reflection coefficient in the emission time geometry.

Identities (7.24a and 7.24b), adapted from Morse and Ingard [3], viz.

$$1 - M_{r2} = 1 - M sin\theta cos\psi = \frac{1}{\gamma^2\left(1 + M sin\theta_L \cos\psi_L\right)} \tag{7.24a}$$

and

$$\cos\theta_L = \frac{\cos\theta}{\gamma\left(1 - M_{r2}\right)} \tag{7.24b}$$

have been used in deriving (7.23a and 7.23b).

The source motion has no effect on the plane wave reflection coefficient, but the numerical distance has been modified by a factor $(1 - M_{r2})^{-1/2}$. The same transformation can be applied to the direct wave term, although R_p and w_L are not required in this situation. Denoting the Doppler factor for the direct wave by $1 - M_{r1}$, (7.16a–7.16c) can be expressed in the emission time co-ordinates:

$$\phi = \frac{e^{-i\omega_0 t}}{4\pi}\left\{\frac{e^{ik_0 R_1}}{R_1\left(1 - M_{r1}\right)} + \left[R_p + \left(1 - R_p\right)F\left(w/\sqrt{1 - M_{r2}}\right)\right]\frac{e^{ik_0 R_2}}{R_2\left(1 - M_{r2}\right)}\right\} \tag{7.25}$$

In principle, the sound field can be obtained by differentiating (7.25) because it is related to the velocity potential, φ through the use of (7.2). However, this leads to a somewhat lengthier evaluation [8] than an alternative method for obtaining an asymptotic form for p, which is to differentiate both sides of the space-time equation.

Noting (7.2) and expanding the right-hand side (7.1a), we obtain

$$\begin{aligned}\nabla^2 p - \frac{1}{c^2}\frac{\partial^2 p}{\partial t^2} &= i\omega_0 e^{-i\omega_0 t}\delta\left(x - cMt\right)\delta\left(y\right)\delta\left(z - z_s\right)\\ &+ cMe^{-i\omega_0 t}\delta'\left(x - cMt\right)\delta\left(y\right)\delta\left(z - z_s\right),\end{aligned} \tag{7.26a}$$

where the prime denotes the derivative respect to its argument. Similarly, the boundary condition [*c.f.* (7.2)] can be re-stated in terms of the sound pressure as

$$\frac{1}{c}\frac{\partial p}{\partial t} - \frac{1}{\beta}\frac{\partial p}{\partial z} = 0 \tag{7.26b}$$

A close examination of (7.26a) reveals that there are two source terms consisting of a monopole and horizontal dipole in the space time equation. The sound field corresponding to each individual component source is found and then, by applying the principle of superposition, the total sound field is the sum of these two contributions. From the analysis earlier in this section, the monopole sound field, p_m can be written as

$$p_m = -\frac{i\omega_0 e^{-i\omega_0 t}}{4\pi}\left\{\begin{array}{c}\dfrac{e^{ik_0 R_1}}{R_1\left(1-M_{r1}\right)}+ \\ \left[R_p+\left(1-R_p\right)F\left(w/\sqrt{1-M_{r2}}\right)\right]\dfrac{e^{ik_0 R_2}}{R_2\left(1-M_{r2}\right)}\end{array}\right\}. \quad (7.27)$$

The asymptotic solution for the sound field due to the horizontal dipole is available from Chapter 3. The sound field due to a moving horizontal dipole, p_d can be found by starting from the solution for a moving monopole. The asymptotic solution is obtained by differentiating the appropriate solution for the moving monopole to give

$$p_d = \frac{cMe^{-i\omega_0 t_L}}{4\pi}\left\{\begin{array}{c}\left(1-ik_0 d_1\right)x_L\dfrac{e^{ik_0 d_1}}{d_1^3} \\ +\left[R_L+\left(1-R_L\right)F\left(w_L\right)\right]\left(1-ikd_2\right)x_L\dfrac{e^{ik_0 d_2}}{d_2^3}\end{array}\right\}. \quad (7.28)$$

This can be transformed back to the emission time co-ordinates by using the same approach as was used for the sound field due to a monopole source.

After some tedious algebraic manipulation, the total sound field from combining (7.27) and (7.28) results in

$$p = -\frac{i\omega_0 e^{-i\omega_0 t}}{4\pi}\left\{\begin{array}{c}\left[1+\dfrac{1}{ik_0 R_1}\left(\dfrac{M^2-M_{r1}}{1-M_{r1}}\right)\right]\dfrac{e^{ik_0 R_1}}{R_1\left(1-M_{r1}\right)^2} \\ +\left[R_p+\left(1-R_p\right)F\left(w/\sqrt{1-M_{r2}}\right)\right]\left[1+\dfrac{1}{ik_0 R_1}\left(\dfrac{M^2-M_{r2}}{1-M_{r2}}\right)\right]\dfrac{e^{ik_0 R_2}}{R_2\left(1-M_{r2}\right)}\end{array}\right\}. \quad (7.29)$$

This is the asymptotic solution for the sound field due to a monopole source moving at constant speed parallel to a ground surface. It may be called the 'Dopplerized Weyl–Van der Pol formula'. As expected, the solution can be reduced to the classical form (2.40) when the speed of the source vanishes.

7.3 THE SOUND FIELD OF A SOURCE MOVING WITH ARBITRARY VELOCITY

In the last section, the use of a simple transformation served to bring the source at rest relative to the ground surface in the frame of Lorentz space. But this is possible only when the source is moving at a constant height above the ground surface. Generalization to other situations faces considerable geometrical complexities, even for the relatively simple case when the source moves vertically upwards at constant speed. This is because the ground surface would be moving away from the source after the transformation, thereby rendering subsequent intractable analyses. Given that it is rather difficult to extend the method of Lorentz transformation to more general situations, a new approach is required to enable the derivation of the required asymptotic formulae.

First, it is instructive to consider the classical prediction of the free field due to a moving source. The space-time wave equation is

$$\nabla^2\phi - \frac{1}{c^2}\frac{\partial^2\phi}{\partial t^2} = -e^{-i\omega_0 t}\delta\left(r - r_s(t)\right), \tag{7.30}$$

where $r \equiv (x, y, z)$ is the field point and $r_s \equiv (x_s, y_s, z_s)$ is the instantaneous position of the source. Differentiation of r_s with respect to t gives the instantaneous velocity, $u = cM \equiv (cM_x, cM_y, cM_z)$ of the source. There is a subtle way to derive an exact solution for the sound field due to a moving source without resorting to the use of method of Fourier transformation. The field generated by a source (or even an extended source) of strength $q(r, t)$ is given by the well-known integral representation [9] of

$$\phi(r,t) = -\int \frac{q(y,\tau)}{4\pi|r - y|} d^3y, \tag{7.31}$$

where y is the source's location and τ is the retarded time which varies over the source. It is given by $\tau = t - |r - y|/c$.By the use of generalized function, $\delta(t - \tau - |r - y|/c)$, (7.31) is identical to

$$\phi(r,t) = -\int \frac{q(y,\tau)}{4\pi|r - y|}\delta\left(t - \tau - |r - y|/c\right)d^3y d\tau \tag{7.32}$$

Now, for a point moving harmonic source, the source strength is simply $q(r,t) = -\exp(-i\omega_0 t)\delta(r - r_s(t))$. Hence, the sound field given in (7.32) may be rewritten as

$$\phi(r,t) = -\int \frac{e^{-i\omega_0 t}\delta\left(y - r_s(\tau)\right)}{4\pi|r - y|}\delta\left(t - \tau - |r - y|/c\right)d^3y d\tau \tag{7.33}$$

Changing the order of integration, the triple integral with respect to y can be evaluated straightforwardly and exactly because of the δ-function and its solution is

$$\phi(r,t) = -\int \frac{e^{-i\omega_0 t}}{4\pi|r - r_s(\tau)|}\delta\left(t - \tau - |r - r_s(\tau)|/c\right)d\tau \tag{7.34}$$

The integral with respect to τ can be evaluated also. Using a property of δ-functions

$$\int_{-\infty}^{\infty} f(\tau)\delta[g(\tau)]d\tau = \left[\frac{f(\tau)}{|g'(\tau)|}\right]_{\tau=\tau*}, \tag{7.35}$$

where f and g are any arbitrary functions and τ^* is a zero of g, that is any solutions for $g(\tau) = 0$, and the prime denotes the derivative with respect to τ. It is possible to identify that $f = e^{-i\omega_0 \tau}/4\pi|\mathrm{r} - \mathrm{r_s}(\tau)|$ and $g = t - \tau - |\mathrm{r} - \mathrm{r_s}(\tau)|/c$. Differentiation of g with respect to τ gives

$$\begin{aligned} g' &= -1 + \frac{(x-x_s)\dfrac{dx_s}{d\tau} + (y-y_s)\dfrac{dy_s}{d\tau} + (z-z_s)\dfrac{dz_s}{d\tau}}{c|r - r_s(\tau)|} \\ &= -1 + \frac{(x-x_s)M_x + (y-y_s)M_y + (z-z_s)M_z}{c|r - r_s(\tau)|} \end{aligned} \tag{7.36a}$$

For a subsonic source speed,

$$|g'| = 1 - M_r, \tag{7.36b}$$

where

$$M_r = \frac{(x-x_s)M_x + (y-y_s)M_y + (z-z_s)M_z}{|r - r_s(\tau)|}$$

By comparing with (7.21), it is possible to see that M_r is the component of the source Mach number in the direction of the observer. With the aid of the above analysis, the integral with respect to τ in (7.34) can be evaluated exactly to yield

$$\phi(r,t) = \frac{\exp(-i\omega_0 \tau *)}{4\pi R(\tau *)(1 - M_r)}, \tag{7.37a}$$

where $R(\tau*) = |r - r_s(\tau*)|$ and $\tau*$ satisfies

$$c(t - \tau^*) = |r - r_s(\tau^*)| \tag{7.37b}$$

The sound field due to a source moving above a ground surface with an arbitrary velocity is governed by the space-time wave equation given in (7.30) subject to the impedance boundary condition (7.2). As expected, the sound field is composed of two terms: a direct wave φ_1 and a reflected wave φ_2. With the above analysis, the solution for the direct wave can be written as

$$\varphi_1(r,t) = e^{-i\omega_0 t} \frac{e^{ik_0 R_1(\tau_1)}}{4\pi R_1(\tau_1)(1 - M_{r1})}, \tag{7.38a}$$

where

$$R_1(\tau_1)=\sqrt{(x-x_s(\tau_1))^2+(y-y_s(\tau_1))^2+(z-z_s(\tau_1))^2}, \tag{7.38b}$$

τ_1 is a zero of

$$c(t-\tau_1)=R_1(\tau_1) \tag{7.38c}$$

and M_{r1} is the component of the source Mach number in the direction of the observer.

The same method is readily adaptable to the reflected wave by replacing the source strength $q(r,t)$ in (7.32) in a heuristic manner, i.e. by

$$q(r,t)=-e^{-i\omega_0 t}Q\delta(r-r_s(t)), \tag{7.39a}$$

where Q is the spherical wave reflection coefficient given by

$$Q=R_p+(1-R_p)F(w) \tag{7.39b}$$

Using the same procedure, an asymptotic formula for the reflected wave is

$$\varphi_2(r,t)=e^{-i\omega_0 t}Q\frac{e^{ik_0R_2(\tau_1)}}{4\pi R_2(\tau_2)(1-M_{r2})}, \tag{7.40a}$$

where

$$R_2(\tau_2)=\sqrt{(x-x_s(\tau_2))^2+(y-y_s(\tau_2))^2+(z+z_s(\tau_2))^2}, \tag{7.40b}$$

τ_2 is a zero of

$$c(t-\tau_2)=R_2(\tau_2) \tag{7.40c}$$

and M_{r2} is the component of the image source Mach number in the direction of the observer. The spherical wave reflection coefficient Q in (7.40a) is calculated at the emission time τ_2. For the special case of a source moving at a constant speed in the x-direction, the reflection wave term derived in this section is same as that derived in Section 7.2, except that the numerical distance w differs by a correction term $1/\sqrt{(1-M_r)}$. The correction term, introduced by approximating the source strength in (7.39a), is small if the source traverses at low subsonic speed. To allow a uniform asymptotic solution for two derivations, it is assumed that the numerical distance should be modified by the same correction term as derived in Section 7.2. Hence, summing the contribution of the direct and reflected wave terms, the velocity potential for a source moving at arbitrary velocity is

$$\phi=\frac{e^{-i\omega_0 t}}{4\pi}\left\{\frac{e^{ik_0R_1}}{R_1(1-M_{r1})}+\left[R_p+(1-R_p)F\left(w/\sqrt{1-M_{r2}}\right)\right]\frac{e^{ik_0R_2}}{R_2(1-M_{r2})}\right\}. \tag{7.41}$$

To obtain an asymptotic expression for the acoustic pressure, the velocity potential should be differentiated with respect to t. It is useful just to consider the direct wave, p_1

$$p_1 = \frac{\partial \varphi_1}{\partial t} = \frac{1}{4\pi}\frac{\partial}{\partial t}\left\{\frac{e^{-\omega_0(t-R_1/c)}}{R_1\left(1-M_{r1}\right)}\right\} \tag{7.42}$$

Making use of the following identities:

$$\begin{cases} \dfrac{\partial R_1}{\partial t} = -\dfrac{cM_{r1}}{1-M_{r1}} \\ \dfrac{\partial \tau_1}{\partial t} = \dfrac{1}{1-M_{r1}} \\ \dfrac{\partial M_{r1}}{\partial t} = -\dfrac{M_r}{R_1}\dfrac{\partial R_1}{\partial t} - \dfrac{cM^2}{R_1}\dfrac{\partial \tau_1}{\partial t} + \dot{M}_r\dfrac{\partial \tau_1}{\partial t} \end{cases},$$

where $c\dot{M}_{r1}$ is the component of the source acceleration in the direction of the observer, the direct wave is given by

$$p_1 = -i\omega_0\left[\left(1-M_{r1}+i\dot{M}_{r1}/\omega_0\right)+\frac{M^2-M_{r1}}{ik_0R_1}\right]\frac{e^{-i\omega_0(t-R_1/c)}}{4\pi R_1\left(1-M_{r1}\right)^3} \tag{7.43}$$

Equation (7.43) is the free field sound pressure due to a source moving at arbitrary velocity. An extra term is introduced as a result of the acceleration of the source, which is significant when $\dot{M}_{r1} \gg \omega_0\left(1-M_{r1}\right)$. Morse and Ingard [3] considered the sound field due to a source moving at a constant speed. As well as correcting a minor typographical error Equation (11.2.15) of [3], this furnishes a generalization of their result.

For a distant slow source, variations of the spherical coefficient Q with emission time t are generally small. Hence, the reflected wave term can be derived in an analogous manner as that for the direct wave. The solution is

$$\begin{aligned} p_2 &= \frac{\partial \varphi_2}{\partial t} = \frac{1}{4\pi}\frac{\partial}{\partial t}\left\{\frac{e^{-\omega_0(t-R_2/c)}}{R_2\left(1-M_{r2}\right)}\right\} \\ &= -i\omega_0 Q\left[\left(1-M_{r2}+i\dot{M}_{r2}/\omega_0\right)+\frac{M^2-M_{r2}}{ik_0R_2}\right]\frac{e^{-i\omega_0(t-R_2/c)}}{4\pi R_1\left(1-M_{r2}\right)^3}. \end{aligned} \tag{7.44}$$

The total sound field is obtained by summing the contributions of the direct and reflected waves to give

$$p_2 = -i\omega_0 \left\{ \begin{array}{c} \left[\left(1 - M_{r1} + i\dot{M}_{r1} / \omega_0\right) + \dfrac{M^2 - M_{r1}}{ik_0 R_1}\right] \dfrac{e^{-i\omega_0 (t - R_1/c)}}{4\pi R_1 \left(1 - M_{r1}\right)^3} \\ + \\ Q\left[\left(1 - M_{r2} + i\dot{M}_{r2} / \omega_0\right) + \dfrac{M^2 - M_{r2}}{ik_0 R_2}\right] \dfrac{e^{-i\omega_0 (t - R_2/c)}}{4\pi R_1 \left(1 - M_{r2}\right)^3} \end{array} \right\} \tag{7.45}$$

The attenuation associated with ground effect for stationary sources is usually expressed using Excess Attenuation (EA), i.e., the ratio of the total pressure field to the direct field in dB (see Chapter 2). For sources in motion, it is usual to refer to the *instantaneous* EA which is that observed at the emission time t_e for which the interfering direct and reflected fields are calculated by means of the retarded times t_L and t'_L, respectively. An example numerical prediction of the instantaneous EA for a source moving above a porous ground at a height of 3 m at Mach number 0.3 in uniform circular motion with radius 2 m about a vertical axis is shown in Figure 7.1 [10]. The impedance of the porous ground is modelled by using (5.38) with

140 kPa s m⁻⁻ and $^{-1}\alpha_e = 35$ m

Equation (7.45) can be reduced to (7.29) for the special case of a monopole source moving parallel to the ground surface at constant speed. The solutions for a moving source ((7.45) and (7.29)) have a similar form to that for a

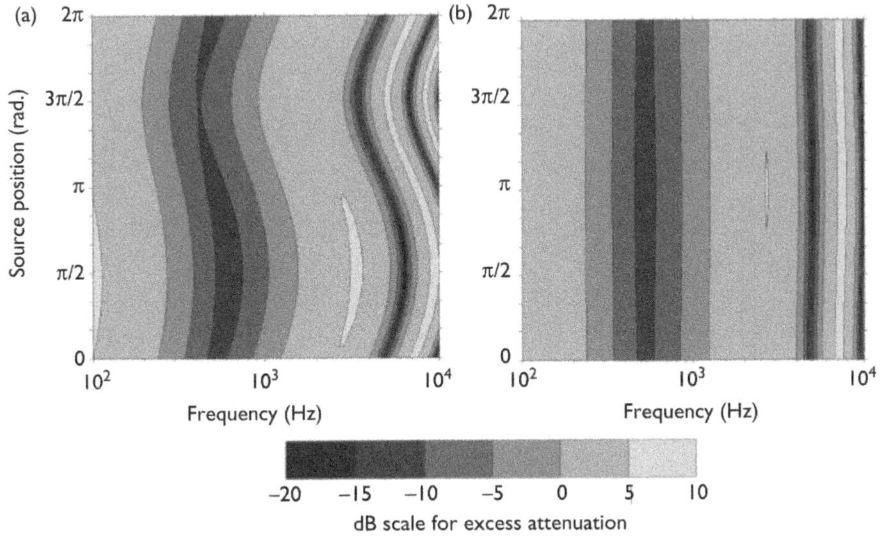

Figure 7.1 (a) Predicted Instantaneous Excess Attenuation for a source in circular motion about the vertical axis. (b) Excess Attenuation for the corresponding stationary source.

stationary source. The only difference is that the magnitudes of the direct and reflected waves are modified due to the effect of source motion. The apparent impedance of the ground surface is unaffected by the convection of the source but there is a small correction to the numerical distance. Since this alters the boundary loss factor only slightly, it has a relatively small effect on the overall sound field especially for a relatively slow-moving source.

In the next section, the 'exact' formulation derived in this section is compared with another scheme based on heuristic approximations.

7.4 COMPARISON WITH HEURISTIC CALCULATIONS

As a first approximation for low source speed, the effects of motion are sometimes ignored. For example, a road or railway track may be modelled as a coherent line source [11]. This assumption produces a simple solution since the asymptotic solution for a line source (see Section 2.7) can be used. An alternative heuristic approach is to model the problem as the sound field due to a quasi-stationary point source with the source frequency adjusted by the Doppler factor, $(1 - M_r)^{-1}$. The sound field can be computed by using the Weyl-Van der Pol formula (2.40) with source frequency $f_1 \equiv f_0(1 - M_r)^{-1}$. Moreover, the ground admittance can be calculated using the Dopplerized frequency for the interfering reflection. Numerical calculations of the pressure field due to a source in uniform motion above an impedance plane have been carried out [10] for Mach numbers up to 0.3 (370 km/h). The modified Weyl-Van der Pol calculation (7.16a–7.16c) is compared with the heuristic approach for the calculation of the sound field in Figure 7.2.

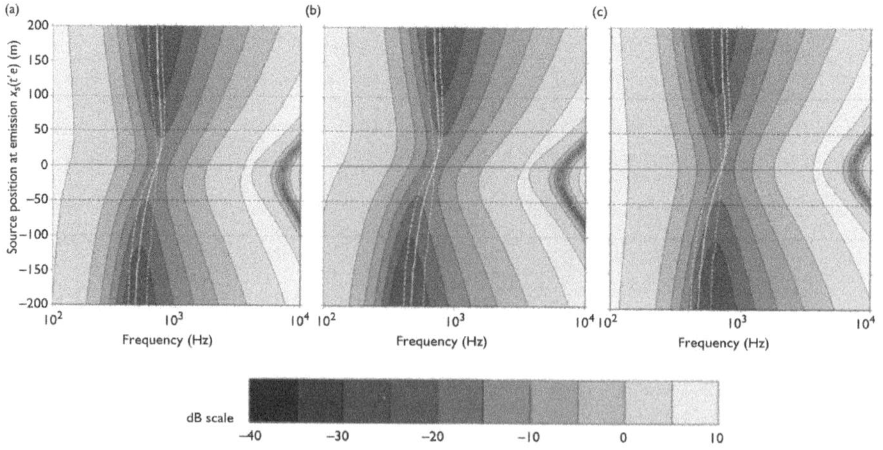

Figure 7.2 Predicted instantaneous excess attenuation at a receiver 1.2 m high and 50 m from an approaching source moving parallel to a horizontal flat ground at a height of 1 m (a) using (7.16a–7.16c) (b) using the heuristic approach with Dopplerised frequency and (c) assuming a stationary source.

The source is 1 m above the ground and is travelling along the x-axis. The receiver height is 1.2 m. The ground admittance is calculated by means of the variable porosity model (see (5.48)) with parameters σ_e= 140 kPa s m⁻⁻ and α_e= 35 m^{-1}. Predictions of the two models for a moving point source are compared with the calculation for a stationary source in Figures 7.2 and 7.3. The varying time parameter used here and subsequently relates to the position of the source corresponding to the emission of the interfering reflection. For Figure 7.2, the separation distance between the source path and the receiver is 50 m. With the chosen geometry, there is near-grazing incidence for all source positions, so the values taken by the emission times for interfering direct and reflected waves are indistinguishable. The observation time t varies non-linearly with the emission position.

Figure 7.3(a) and (b) show that the ground effect dip is shifted towards the low frequencies when the source approaches the receiver (the ground is seen as softer) and towards higher frequencies when the source is receding (the ground is seen as harder). However, the heuristic, i.e. Dopplerized frequency, calculation shows greater sensitivity to the motion than the modified Weyl-Van der Pol calculation (7.16a–7.16c). Although effects due to source motion are observable in the frequency domain, the influence on the magnitude of the ground effect dip is only a few dB. For the chosen geometrical and ground parameters, at Mach number 0.3, there is a shift of about 100 Hz according to the modified Weyl-Van der Pol formula (solid lines) and 170 Hz with the heuristic calculation (dashed lines).

Figure 7.4 shows the frequency shift of the ground effect dip due to source motion at Mach numbers 0.1 (white), 0.2 (grey) and 0.3 (black). The calculation of the frequency of the ground effect dip is carried out by locating the first minimum in the instantaneous EA. Because this calculation has been carried out numerically, the resulting curves are not very smooth. Nevertheless, this confirms the difference between predictions of the modified Weyl-Van der Pol formula (solid lines) and those of the heuristic calculation (dashed lines).

For higher source speeds, the shift of the ground effect dip is predicted to have a significant influence on the resulting sound level, particularly for a source whose spectrum peaks in the frequency range where the maximum attenuation is observed.

7.5 POINT SOURCE MOVING AT CONSTANT SPEED AND HEIGHT PARALLEL TO A RIGID WEDGE

7.5.1 Kinematics

Consider a semi-infinite rigid wedge with arbitrary top angle $T = 2\pi - T_1 - T_2$ and constant cross section in the y-z plane and a sound source S is moving along the x-axis, parallel to the edge of the wedge. As shown in Figure 7.5, the cylindrical coordinate system $[\xi, r, x]$ is introduced and $\mathrm{r_S} \equiv [\xi_S, r_S, x_S]$ represents the source coordinates.

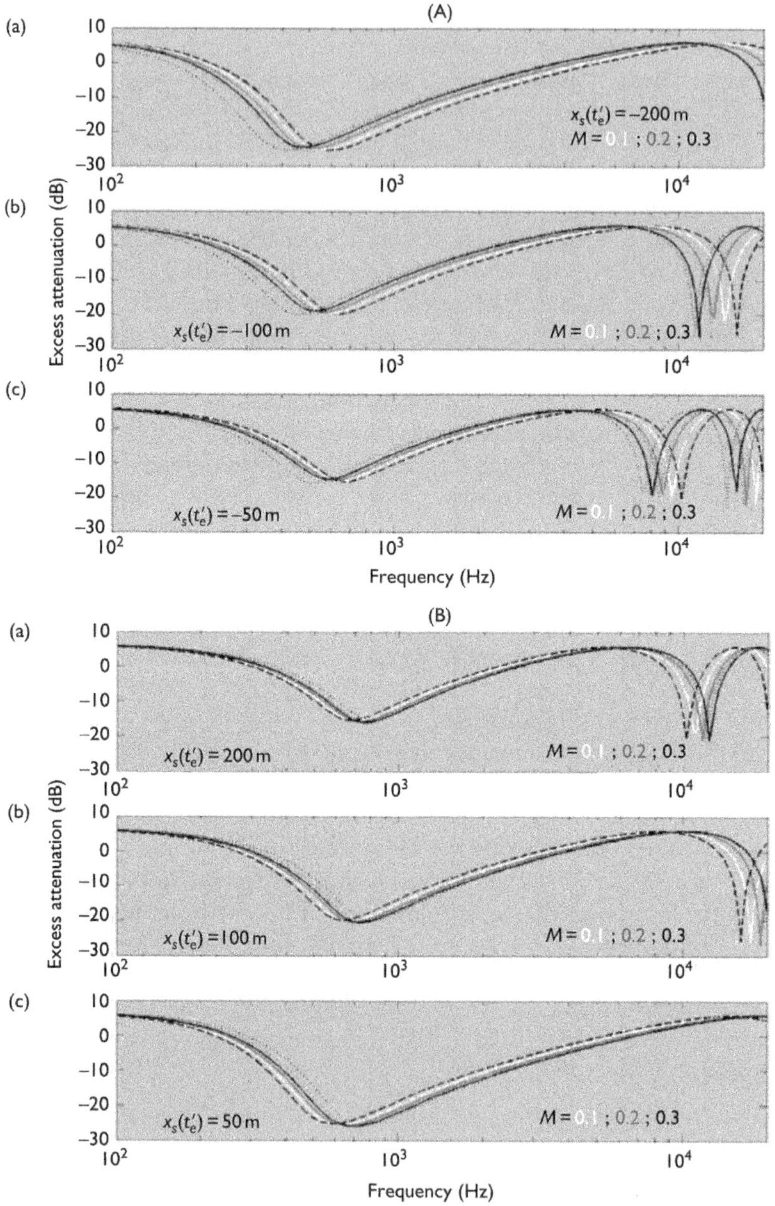

Figure 7.3 (a) Predicted instantaneous excess attenuation spectra predicted at 1 m high receiver when a 1 m high source is moving at three different speeds $M = 0.1$, $M = 0.2$, $M = 0.3$, (a) approaching the receiver at three instants (i) $x_s\left(t_e'\right) = -200\,\text{m}$; (ii) $x_s\left(t_e'\right) = -100\,\text{m}$; (iii) $x_s\left(t_e'\right) = -50\,\text{m}$,; (b) receding from the receiver at three equivalent instants (i) $x_s\left(t_e'\right) = 200\,\text{m}$, (ii) $x_s\left(t_e'\right) = 100\,\text{m}$, (iii) $x_s\left(t_e'\right) = 50\,\text{m}$. Solid lines represent predictions of the modified Weyl-Van der Pol formula (7.16a–7.16c); dashed lines represent predictions using the Heuristic Approach (i.e. Dopplerised frequency). The dotted lines represent predictions for a stationary source.

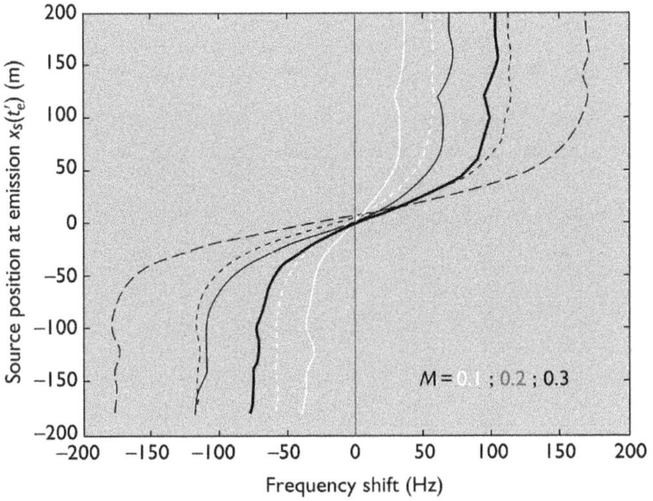

Figure 7.4 Frequency shift of the ground effect dip due to source motion. The configuration is the same as that for Fig.7.3. Solid lines represent predictions using the modified Weyl-Van der Pol formula. Broken lines represent predictions using the heuristic approach.

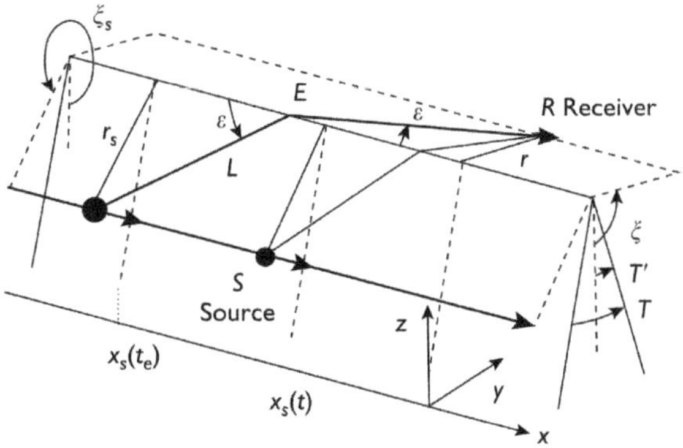

Figure 7.5 Geometry describing a source in uniform motion along a semi-infinite wedge.

For uniform source motion at Mach number M, the only varying parameter is the source–receiver offset $x - x_s$. If the initial position of the source is at $x = 0$ and c_0 denotes the speed of sound,

$$x_S(t) = c_0 M t \tag{7.46}$$

The projection of the geometry in the cross-sectional plane remains unchanged with source motion so there is an analogy with a stationary source in a 2-D situation. According to the geometrical theory of diffraction, the diffracted sound ray is a broken segment linking the source S to the edge of the wedge E to the receiver R. The diffraction angle ε is the same on both source and receiver side [12]. If the receiver is at $x = 0$,

$$x_E = \frac{r}{\tan \varepsilon} \tag{7.47}$$

where t_e is the time of emission of the diffracted wave reaching the receiver at time t and L the total length of the diffracted ray. Also, from the geometry,

$$c_0(t - t_e) = L = \sqrt{(r + r_S)^2 + (c_0 M t_e)^2} \tag{7.48}$$

and

$$L = \frac{\Lambda - M(c_0 M t)}{1 - M^2} \tag{7.49a}$$

where

$$\Lambda = \sqrt{(c_0 M t)^2 + (1 - M^2)(r + r_S)^2} = L(1 - M_E), \tag{7.49b}$$

$M_E = M \cos \varepsilon$ being the component of the Mach number along the diffracted ray.

7.5.2 Diffracted Pressure for a Source in Uniform Motion

The solution for wedge-diffracted wave from a stationary point source is well known. The result given by Pierce [12], based on the electromagnetic theory, is used here. Morse and Ingard [3] have derived the sound field due to a monopole in uniform motion in the free field by means of a Lorentz transformation. Buret et al. [13] have used this transformation to formulate the Doppler-Weyl-Van der Pol formula (7.30).

The coordinates and time (x_L, y_L, z_L, t_L) in the Lorentz space are given by (7.3).

In the Lorentz space, it can be shown that the acoustic pressure field can be expressed as the sum of contributions from two stationary sources: a monopole and a dipole oriented in the direction of source motion [14]. Denoting these relative contributions by $p^{(0)}$ and $p^{(1)}$, respectively, the total pressure p due to a source of strength P_0 can be expressed as [13]:

$$p = P_0\left(\gamma^4 p^{(0)} - \gamma^4 \frac{M}{ik_0} M p^{(1)}\right) \tag{7.50}$$

With constant Mach number, the Lorentz transformation involves uniform expansion along the three dimensions, as well as a time-dependent translation along the direction x of source motion. As a result, the transformed geometry is analogous to that for a stationary source in the physical space. A cylindrical coordinate system $[\xi_L, r_L, x_L]$ is introduced and is related to that in the physical plane by

$$x_L = \gamma^2 (x - c_0 M t) \tag{7.51}$$

The polar angle is unaffected by the transformation. As a result, the top angle of the wedge is the same in both the transformed and the physical space.

The expression for the monopole component of the diffracted wave is deduced straightforwardly from Pierce's formulation [3] as

$$p^{(0)} = \frac{e^{-i\omega t_L}}{4\pi}\frac{1+i}{2}\frac{e^{ik_0 L_L}}{L_L}\left[A_D(X_{L+}) + A_D(X_{L-})\right], \tag{7.52a}$$

where L_L is the diffracted path in the transformed coordinates and A_D is the diffraction integral given by

$$A_D(X_{L\pm}) = \frac{1}{\pi\sqrt{2}}\int_{-\infty}^{\infty}\frac{e^{-u^2}}{(\pi/2)^{1/2}X_{L\pm} - e^{-i\pi/4}u}du \tag{7.52b}$$

$$X_{L\pm} = \Gamma_L M_V(\alpha_\pm); \quad \alpha_- = \xi + \xi_S - 2T_1; \quad \alpha_+ = \xi - \xi_S \tag{7.52c}$$

$$\Gamma_L = \sqrt{K_0 r_L \rho_S / \pi L_L}; \quad M_\nu(\alpha_\pm) = \frac{\cos\nu\pi - \cos\nu\alpha_\pm}{\nu\sin\nu\pi} \tag{7.52d}$$

where $\nu = \pi/(2\pi - T)$ is the wedge index and $\rho_S = \gamma r_s$.

Buret et al. [15] have derived the expression for the 3-D dipole field diffracted by a wedge and they have validated this expression through laboratory measurements. For a dipole lying along the x_L-axis,

$$\begin{aligned} p^{(1)} &= \frac{e^{-i\omega t_L}}{4\pi}\frac{1+i}{2}\cos\varepsilon_L\frac{1-ik_0L_L}{L_L}\frac{e^{ik_0L_L}}{L_L}\left[A_D(X_{L+}) + A_D(X_{L-})\right] \\ &- \frac{e^{-i\omega t_L}}{4\pi}\frac{1-i}{4}\frac{\cos\varepsilon_L}{L_L}\frac{e^{ik_0L_L}}{L_L}\left\{(X_{L+} + X_{L-}) - \pi\left[X_{L+}^2 A_D(X_{L-}) + X_{L-}^2 A_D(X_{L+})\right]\right\} \end{aligned} \tag{7.53}$$

In (7.53), dipole orientation cosines missing from Equation (14a) of Ref. [15] have been re-introduced, i.e. $(\ell_x, \ell_y, \ell_z) = (\cos\varepsilon_L, 0, 0)$. For long diffraction paths and high frequencies, X_{L+} and X_{L-} take small values and the second term in (7.53) can be neglected [15]. This approximation is particularly sensible for source motion along the wedge, as the diffracted path length increases with the time-dependent source–receiver offset x–x_S. The dipole component of the pressure then takes a form similar to that for the monopole pressure given in Equations (7.52a–7.52d), except for a directivity factor and a strength factor:

$$p^{(1)} = \frac{e^{-i\omega t_L}}{4\pi}\frac{1+i}{2}\cos\varepsilon_L\frac{1-ik_0L_L}{L_L}\frac{e^{ik_0L_L}}{L_L}\left[A_D(X_{L+}) + A_D(X_{L-})\right] \tag{7.54}$$

The total diffracteccg Equations (7.52a–7.52d) and (7.54) into Equation (7.50). Transformation back into the physical space requires use of

$$L_L = \sqrt{(\gamma r + \gamma r_S)^2 + x_L^2} = \gamma^2 L(1 - M\cos\varepsilon) \tag{7.55a}$$

$$t_L - \frac{L_L}{c_0} = t_e = t - \frac{L}{c_0} \tag{7.55b}$$

and

$$\cos\varepsilon_L = \frac{x_L}{L_L} = \frac{\cos\varepsilon - M}{1 - M\cos\varepsilon} \tag{7.55c}$$

Hence, the diffracted acoustic pressure wave due to a monopole in uniform motion in parallel to a rigid wedge can be formulated as

$$p = P_0 \frac{e^{-i\omega t}}{4\pi} \frac{e^{ik_0 L_L}}{L(1 - M\cos\varepsilon)^2}$$

$$\left\{\frac{1+i}{2}\left[1 - \frac{1}{ik_0 L}\frac{M\cos\varepsilon - M^2}{1 - M\cos\varepsilon}\right]\left[A_D\left(\frac{X_+}{\sqrt{1 - M\cos\varepsilon}}\right) + A_D\left(\frac{X_-}{\sqrt{1 - M\cos\varepsilon}}\right)\right]\right\}, \tag{7.56}$$

where A_D is defined as in (7.52b) with

$$\Gamma = \sqrt{k_0 r r_S / \pi L} \tag{7.57}$$

For a receiver in the shadow zone of the obstacle, the total pressure field is given by (7.56). However, for arbitrary source or receiver position, it involves potential contributions from a direct pressure wave, as well as reflections on the sides of the wedge. Although the formulation of the total pressure in the transformed space using Equation (7.45) together with the expression for diffraction of sound from arbitrarily positioned monopoles has been derived [12] and has been extended to dipoles by Buret et al. [15], it is not reproduced here.

In Figure 7.6, predictions of the diffracted wave are plotted for discrete frequencies 100 Hz, 1 kHz and 5 kHz, for a point source moving at Mach number 0.3 in the vicinity of a semi-infinite rigid half-plane ($T = 0$; $\nu = ½$) with source, receiver and obstacle edge at heights 1, 1.2 and 2 m, respectively. The shortest horizontal separation from the obstacle to the source is 10 m and that to the receiver on the other side is 20 m. Solid lines show the calculation by means of (7.56). Dotted lines show the calculation by means of the complete dipole component as in (7.53) and are indistinguishable from the latter. The approximation used to derive (7.56) is valid along the whole source path. Broken lines show the calculation for the corresponding stationary omnidirectional point source located at the position $x_S(t_e)$ of emission.

Comparison of the predictions for a stationary source with those for a moving source enables assessment of the effects of source motion. As the moving source approaches the receiver ($x_S\left(t_e'\right) \leq 0$), the Sound Pressure Levels (SPLs) are predicted to be higher than for a stationary source. When the source is receding ($x_S\left(t_e'\right) \geq 0$), the SPLs are predicted to be lower than for a stationary source. In particular, the maximum in SPL is predicted to occur when the

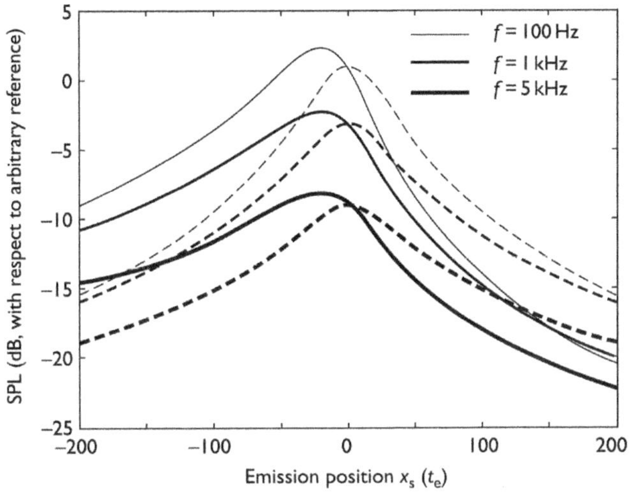

Figure 7.6 Diffracted pressure predicted at discrete frequencies 100 Hz (——), 1 kHz () and 5 kHz () for a source in uniform motion at Mach 0.3 along a half-plane. The source path is 1 m below the edge, at a horizontal separation of 10 m. The receiver is 0.8 m below the edge and at a horizontal separation of 20 m. Solid lines correspond to the approximate solution, dotted lines (undistinguishable) correspond to predictions without approximation (i.e. using the full dipole component). Dashed lines correspond to predictions for the corresponding stationary source at the point of emission.

source is approaching the receiver, i.e. before the position of closest approach. These effects are related to the Doppler factor $(1 - M \cos \varepsilon)^{-1}$ and the convection coefficient $(1 - M \cos \varepsilon)^{-2}$ in (7.56). Both are larger than unity on approach and smaller than unity on recession. These observations are consistent with previous results for a source in the free field [3].

7.6 SOURCE MOVING PARALLEL TO A GROUND DISCONTINUITY

7.6.1 Introduction

A semi-empirical formula for the sound field due to a stationary source in the presence of a ground discontinuity has been derived by De Jong by considering the superposition of two half-planes with different admittance [16]. This model was shown to be valid in 3-D situations for which the propagation path is not necessarily perpendicular to the discontinuity [16] and was later extended to dipole sources [15]. Derivation of the acoustic pressure field for source motion in parallel to the discontinuity is then straightforward (see Equation (7.45)). Consider a point source in uniform motion at height z_s parallel to an impedance jump which coincides with the x-axis (see Figure 7.7).

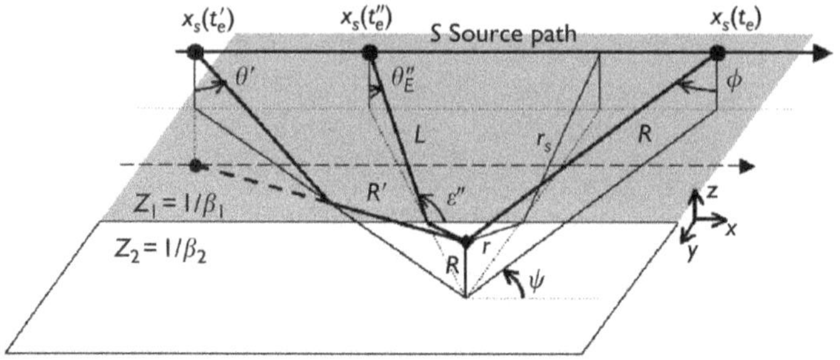

Figure 7.7 Geometry describing a source in uniform motion parallel to a discontinuity in ground impedance.

The ground on the source side is characterized by its specific admittance β_1 and that on the receiver side by β_2. The direct wave reaching the receiver R at time t has been emitted at instant t_e, following path $R = c_0(t - t_e)$ with azimuthal angle ψ and elevation angle ϕ. The reflected wave reaching R, at instant t, and denoted subsequently by dashed parameters in the physical space, has been emitted at retarded time $t_e^{'}$ and has followed path $R' = c_0\left(t - t_e^{'}\right)$ with azimuthal and elevation angles $\psi' = \psi\left(t_e^{'}\right)$ and $\theta' = \theta\left(t_e^{'}\right)$, respectively. For brevity, the expressions for R, ψ, ϕ and R', ψ', θ are not detailed here but their derivation is straightforward [14, 15].

The diffracted wave follows path L defined by (7.48) and its time of emission $t_e^{''}$ is calculated from (7.48) and (7.49a and 7.49b). During the source motion, the point of specular reflection remains on a line parallel to the admittance discontinuity and the source path. Consequently, the proportion of each type of ground covering on the source–receiver path also remains constant during the motion of the source. This is of interest if there are strips of different grounds, parallel to the source path. Some models for such a situation involve tedious calculations whereby the diffracted waves at each discontinuity are summed [17–19]. However, Nyberg [20] has shown that for an infinite alternation of strips with respective admittance β_1 and β_2, the ground effect could be modelled using the area-averaged admittance

$$\bar{\beta} = \left(a\beta_1 + b\beta_2\right) / \left(a + b\right) \tag{7.58a}$$

with a and b the spatial periods for each strip. This theory is valid for short periods compared with the incident wavelength λ [20], i.e. for

$$a + b << \lambda. \tag{7.58b}$$

This condition might not be fulfilled for source motion parallel to the strips, since the effective width of the strips along the sound propagation path is increased by a factor $1/\cos\psi$ due to the source–receiver offset. However, some tolerance in the accuracy of the model is acceptable for source positions at large offsets for which the distance attenuation is large.

Moreover, Nyberg's theory has shown to give good agreement with measurements beyond its limit of approximation [18]. The acoustic pressure field for a source moving along periodic strips of ground is calculated by substituting the area-averaged impedance $\bar{\beta}$ into the Doppler-Weyl-Van der Pol formula (7.16a–7.16c).

7.6.2 Uniform Motion Parallel to a Single Discontinuity

After Lorentz-transforming De Jong's solution for a stationary source [16], the monopole component of the acoustic pressure near a single impedance discontinuity is

$$p^{(0)} = \frac{e^{-i\omega t_L}}{4\pi}\left[\frac{e^{ik_0R_L}}{R_L} + Q_G'\frac{e^{ik_0R_L'}}{R_L'} + (Q_1 - Q_2)\frac{1+i}{2}\frac{e^{ik_0L_L}}{L_L}D_L\right], \tag{7.59}$$

where L_L is the diffraction path in the Lorentz space for a horizontal half-plane whose edge coincides with the ground discontinuity and $D_L = A_D(X_{L+}) + A_D(X_{L-})$, i.e. the sum of the two diffraction integrals given in (7.52a). Q_G is the spherical wave reflection coefficient at the point of specular reflection, with G ≡ 1 or 2 depending on the geometry, Q_1 and Q_2 are the spherical wave reflection coefficients for each type of ground. They are calculated by using the Doppler-Weyl-Van der Pol formula (7.16a–7.16c), i.e.

$$Q_i = R_{Pi} + (1 - R_{Pi})F\left(\frac{w_i}{\sqrt{1 - M_R'}}\right), \tag{7.60a}$$

where M_R', the component of the Mach number in the direction of propagation of the reflected wave, is

$$M_R' = M\cos\psi'\sin\theta' \tag{7.60b}$$

and $R_{P,i}$, the plane wave reflection coefficient, is given by

$$R_{P,i} = \frac{\cos\theta' - \beta_i}{\cos\theta' - \beta_i} \tag{7.60c}$$

The expression for the boundary loss factor F is

$$F\left(\frac{w_i}{\sqrt{1 - M_R'}}\right) = 1 + i\sqrt{\pi}\frac{w_i}{\sqrt{1 - M_R'}}\exp\left(-\frac{w_i^2}{1 - M_R'}\right)erfc\left(-i\frac{w_i}{\sqrt{1 - M_R'}}\right), \tag{7.60d}$$

where erfc denotes the complementary error function and w_i the numerical distance is given by

$$w_i = \left(-\frac{1}{2}ik_0R'\right)^{1/2}(\cos\theta' + \beta_i) \tag{7.60e}$$

When the point of specular reflection coincides with the admittance step, $R_L' = L_L$ and $X_{L+} = 0$. Since $A_D(0) = -(1-i)/2$, the continuity of the monopole component of the acoustic pressure across the jump is ensured.

The dipole component is also expressed as the sum of a direct, a reflected and a diffracted pressure wave to which are applied the appropriate reflection coefficients. The formulations for the direct and the reflected fields are given in Li et al. [21] and the form of the diffracted wave is given in Buret et al. [15]. Hence, the total dipole contribution may be expressed as

$$p^{(1)} = \frac{e^{-i\omega t_L}}{4\pi}\left\{\begin{aligned} &\Theta_R \frac{1-ik_0R_L}{R_L}\frac{e^{ik_0R_L}}{R_L} + \Theta_R' \frac{1-ik_0R_L'}{R_L'}\frac{e^{ik_0R_L'}}{R_L'}Q_G' \\ &+\frac{1+i}{2}\Theta_L \frac{1-ik_0L_L}{L_L}\frac{e^{ik_0L_L}}{L_L}\left[\left(Q_1-Q_2\right)+\frac{\partial\left(Q_1-Q_2\right)}{\partial x_L}\right]D_L \end{aligned}\right\} \tag{7.61a}$$

In (7.61a),

$$\begin{cases} \Theta_R = \cos\psi_L \sin\varphi_L \\ \Theta_R' = \cos\psi_L \sin\theta_L' \\ \Theta_E = \cos\psi_L \sin\theta_{E,L} \end{cases} \tag{7.61b}$$

represent the directivity factors where ψ_L, ϕ_L and θ_L are the azimuthal angle and the elevation angles for the direct and reflected ray in the Lorentz space and $\theta_{E,L}$ is the elevation angle of the ray linking the source to the admittance step. Corresponding angles in the physical space are shown in Figure 7.7. It should be noted that, when the point of specular reflection lies on the impedance jump, $\Theta_R' = \Theta_E = \cos\varepsilon_L$ and continuity of the acoustic pressure across the jump is ensured.

Since variations of the spherical wave coefficient in the horizontal directions are known to be small [15], the last term in Equation (7.61a) can be neglected and (7.61a) can be simplified to an expression analogous to (7.59) for the monopole component. The total pressure field is derived by substituting (7.59) and (7.60a–7.60d) into (7.45). Transformation back into the physical space is straightforward by using (7.55a–7.55c) and noting that [13]:

$$R_L = \gamma^2 R(1 - M\cos\psi\sin\varphi);\ t_L - \frac{R_L}{C_0} = t_L\frac{R}{C_0};\ \Theta_R = \frac{\cos\psi\sin\varphi - M}{1 - M\cos\psi\sin\varphi} \tag{7.62a}$$

$$R_L' = \gamma^2 R'(1 - M\cos\psi'\sin\varphi');\ t_L - \frac{R_L'}{C_0} = t_L\frac{R'}{C_0};\ \Theta_R' = \frac{\cos\psi'\sin\theta' - M}{1 - M\cos\psi'\sin\theta'} \tag{7.62b}$$

and

$$\Theta_E = \frac{\cos\varepsilon'' - M}{1 - M\cos\varepsilon''} = \frac{\cos\psi''\sin\theta_E'' - M}{1 - M\cos\psi''\sin\theta''} \tag{7.62c}$$

A double dash denotes the parameter in the physical space (see Figure 7.7).

The pressure field for a source in uniform motion along an admittance discontinuity is then expressed as:

$$p = P_0 \frac{e^{-i\omega t}}{4\pi}\left[p_{dir} + Q_G p_{refl} + \left(Q_1 - Q_2\right) p_{diffr}\right], \tag{7.63a}$$

where

$$p_{dir} = \frac{e^{ik_0 R}}{R\left(1 - M\cos\psi\sin\varphi\right)^2}\left[1 - \frac{M}{ik_0 R}\left(\frac{\cos\psi\sin\varphi - M}{1 - M\cos\psi\sin\varphi}\right)\right] \tag{7.63b}$$

$$p_{refl} = \frac{e^{ik_0 R'}}{R'\left(1 - M\cos\psi'\sin\varphi'\right)^2}\left[1 - \frac{M}{ik_0 R'}\left(\frac{\cos\psi'\sin\varphi' - M}{1 - M\cos\psi'\sin\varphi'}\right)\right] \tag{7.63c}$$

$$p_{diffr} = \frac{1+i}{2}\frac{e^{ik_0 L}}{L\left(1 - M\cos\psi''\sin\theta_E''\right)^2}\left[1 - \frac{M}{ik_0 L}\left(\frac{\cos\psi''\sin\theta_E'' - M}{1 - M\cos\psi''\sin\theta_E''}\right)\right]$$
$$\left[A_D\left(\frac{X_+}{\sqrt{1 - M\cos\psi''\sin\theta_E''}}\right) + A_D\left(\frac{X_-}{\sqrt{1 - M\cos\psi''\sin\theta_E''}}\right)\right]. \tag{7.63d}$$

Figure 7.8(a) shows the predictions of the SPL according to (7.63a–7.63d) as a function of the source position at emission of the reflected wave $x_S\left(t_e'\right)$ with a 33%-hard ground mix. The source is supposed to be moving at Mach number 0.3 with its track located 10 m away from the impedance jump at height 1 m above hard ground. The receiver is located 20 m away from the

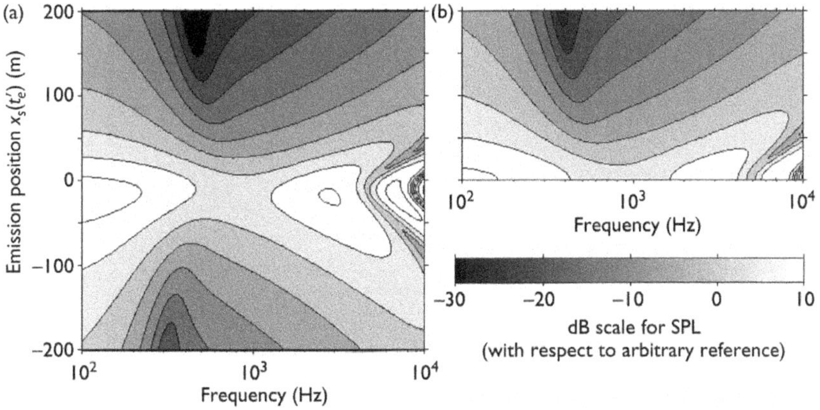

Figure 7.8 (a) Sound Pressure Level (SPL, re. Arbitrary reference) predicted for a source moving at Mach number 0.3 parallel to an admittance discontinuity, 10 m away from it, at height 1 m. The receiver is 20 m away from the step-line at height 1.2 m. The ground on the source side (33%) is hard, the ground on the receiver side is characterised by σ_e= 140 kPa s m^{-2} and α_e= 35 m^{-1}; (b) SPL for a stationary source in the same geometry as in a), but with offset from 0 to 200 m.

discontinuity, 1.2 m above soft ground characterized by the two-parameter impedance model (5.38) with flow resistivity σ_e = 140 kPa s m^{-2} and porosity rate α_e= 35 m^{-1}. Figure 7.8(b) shows the corresponding SPL predictions for a stationary source at offset locations between 0 m and 200 m. Comparison of these Figures enables the effects of source motion to be assessed.

Figure 7.9 shows the same calculations but for the source track and receiver 20 m and 10 m away from the impedance jump, respectively (66% of hard ground).

Corresponding predictions of instantaneous EA, 10 *log* $(|p(t)|/|p_{dir}(t)|)$, are shown in Figure 7.10, for the source at an offset of 100 m on approach,

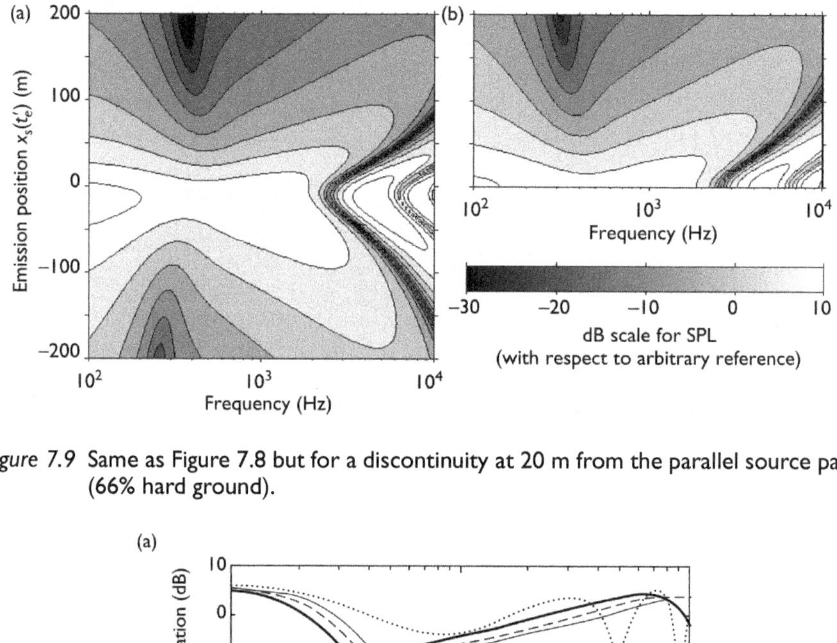

Figure 7.9 Same as Figure 7.8 but for a discontinuity at 20 m from the parallel source path (66% hard ground).

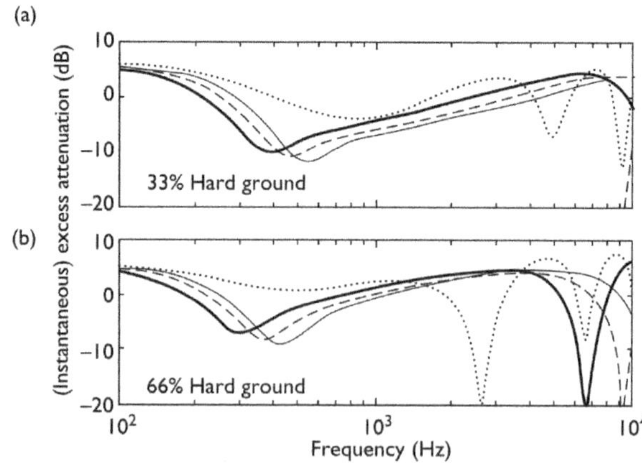

Figure 7.10 Instantaneous Excess Attenuation for the same configurations as in (a) Figure 7.8, (b) Figure 7.9. Approach $x_S(t_e') = -100$ m, —— Recession $x_S(t_e') = +100$ m, ---- stationary source at offset x_S = 100 m, moving source at closest separation ($x_S(t_e')$ = 0), stationary source at closest separation (undistinguishable from the latter).

$(x_S(t_e') = -100\,\text{m}$, thick solid line) and recession $(x_S(t_e') = +100\,\text{m}$; thin solid line) and for the corresponding stationary source (broken line). Thick and thin dotted lines represent the predictions at the closest position $(x_S(t_e') = 0)$ for moving and stationary source, respectively, and are undistinguishable from one another.

As observed for homogenous ground in Section 7.4 (see also [13]), the predicted effects of motion are small. The predicted SPL contours are asymmetric along the source motion direction due to the Doppler shift and the convection factor that affect the location and magnitude of the ground effect dip. This results in lower SPLs being predicted than for a stationary source when the source approaches the receiver $(x_S(t_e') \le 0)$ up to the frequency of the ground effect dip and higher levels being predicted from the dip up to the higher frequency destructive interference. This trend in the predictions is reversed when the source recedes from the receiver. As the source approaches, the ground effect dip is predicted to be slightly sharper and deeper than that for a stationary source, whereas, as the source recedes it is predicted to be slightly broader and shallower. Deformation of the SPL pattern and hence sensitivity to motion is predicted to be stronger on approach than on recession. On the other hand, change of the proportion in the ground mix covering the source–receiver path is predicted to have little effect.

7.7 SOURCE MOVING AT CONSTANT HEIGHT PARALLEL TO A RIGID BARRIER ABOVE THE GROUND

7.7.1 Barrier over Hard Ground

Sound propagation in the presence of a rigid barrier above the ground is a classical problem. In the case of hard ground on both sides of the barrier, the solution is straightforward, by means of mirror images. When there is no direct sound and no reflection on the barrier, the total sound field is the sum of four diffracted waves, corresponding to the paths linking the source S and its image with respect to the ground S' to the receiver R and its image R' via the edge of the barrier (Figure 7.11).

The paths including a ground reflection on the source side are denoted with subscript 1 and subscript 2 is used to indicate the paths for which there is a ground reflection on the receiver side. The total field is given by

$$p = P_0\left(p_0 + p_1 + p_2 + p_{12}\right), \tag{7.64}$$

where p_0 stands for the diffracted wave obtained for path 0, linking source to edge to receiver with no ground reflection.

Using the same reference systems as in Section 7.5, the coordinate vectors of the Image Source and the Image Receiver are $\mathbf{r}_S'$ and $\mathbf{r}'$, respectively. In the Cartesian and the cylindrical coordinates (x,y,z) and $[r,\xi,x]$ they are expressed as

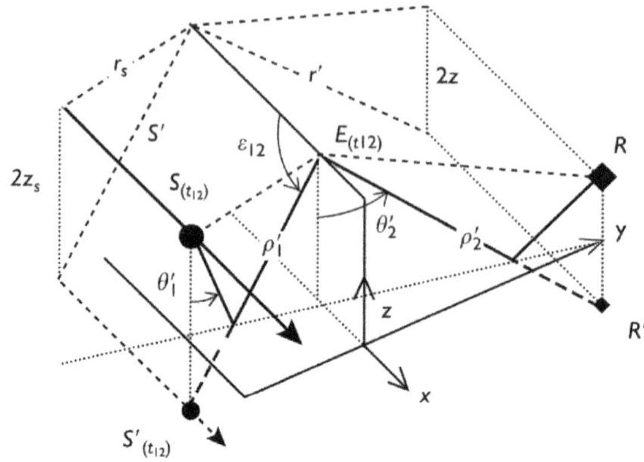

Figure 7.11 Geometry describing a source in uniform motion parallel to a thin barrier above the ground. The diffraction path "12" with ground reflections on both sides of the barrier is represented by thick lines.

$$\mathbf{r}_S'(t) \equiv (-c_0Mt, y_S, -z_S) \equiv \left[r_S', 2\pi - \varphi_S, -c_0Mt\right] \tag{7.65a}$$

$$\mathbf{r}'(t) \equiv (0, y, -z) \equiv \left[r', \varphi, x\right] \tag{7.65b}$$

where φ_S and φ are the elevation angles of the paths linking the projection of the Image Source and Image Receiver to the top of the barrier (see Figure 7.11). If z_E denotes the barrier height,

$$r_s' = r_s\sqrt{1 - 4\frac{z_E}{r}\left(\cos\xi_s - \frac{z_E}{r}\right)}; \quad r' = r\sqrt{1 - 4\frac{z_E}{r}\left(\cos\xi - \frac{z_E}{r}\right)} \tag{7.66}$$

The four diffracted paths have different lengths and as a result, each contribution in (7.64) has been emitted at a different time t_K , which is a solution of

$$L_K = c_0(t - t_K), \tag{7.67a}$$

where the L_K's are the diffracted path lengths:

$$\begin{cases} L_0 = \sqrt{(r_S + r)^2 + (c_0Mt_0)^2} \\ L_1 = \sqrt{(r_S' + r)^2 + (c_0Mt_1)^2} \\ L_2 = \sqrt{(r_S + r')^2 + (c_0Mt_2)^2} \\ L_{12} = \sqrt{(r_S' + r')^2 + (c_0Mt_{12})^2} \end{cases} \tag{7.67b}$$

The subscripts 0, 1, 2 and 12 have the same meanings as in (7.64). The components of the Mach number in the direction of the diffracted rays M_K are determined from the diffraction angles ε_K by:

$$M_K = M\cos\varepsilon_K = -M\frac{c_0 M t_K}{L_K} \tag{7.68}$$

After solving the system of Equations (7.67b), each pressure wave component is calculated from (7.56) by substituting the appropriate (image) source and (image) receiver coordinates together with the corresponding diffracted path and angle, L_K and ε_K.

The total pressure is then calculated by substituting the component pressures into (7.64).

Figure 7.12(a) shows predictions of the sound field for the same configuration as used for Figure 7.6, but above hard ground (at zero height). The SPL predicted for the corresponding stationary source locations is given for comparison in Figure 7.12(b).

Again, the pressure field is predicted to be asymmetric along the source travel direction. Lower levels and hence stronger barrier attenuation are predicted as the source recedes from the receiver ($x_S(t_0) \geq 0$). On the other hand, sensitivity of the sound field to source motion is predicted to be more important on source approach ($x_S(t_0) \leq 0$), for which the deformation of the SPL pattern is stronger. Strong interference between the four components of the acoustic pressure is predicted at high frequencies.

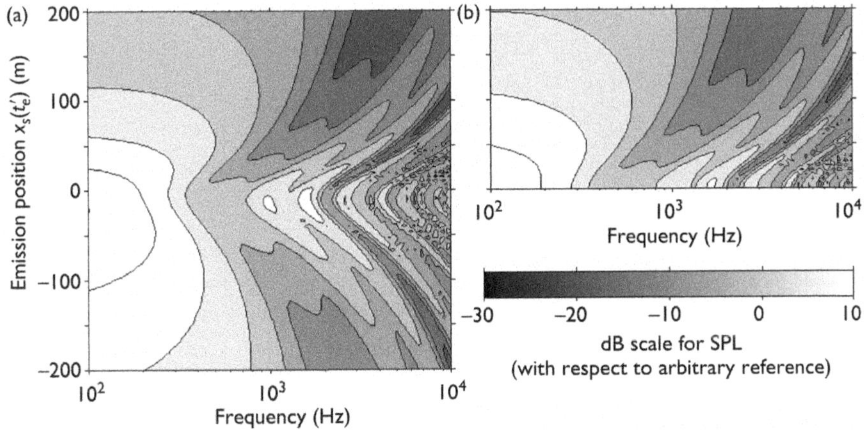

Figure 7.12 Predicted sound field due to a source moving at Mach 0.3 along a thin barrier of height 2 m above hard ground. The source path is 10 m from the barrier, height 1 m. The receiver is in the shadow zone 20 m from the barrier at height 1.2 m.

7.7.2 Barrier over Impedance Ground

If the ground on either or both source and receiver side of the barrier has finite impedance, the corresponding reflection coefficients differ from 1 (the value for acoustically hard ground). Equation (7.64) must be amended to account for the ground reflections [22]. It is then necessary to consider separately the monopole and the dipole component of the pressure field given in (7.45). β_1 and β_2 denote the admittance of the ground on the source and receiver sides, respectively. As in the previous section, subscript 1 denotes a ground reflection on the source side, whereas subscript 2 corresponds to a ground reflection on the receiver side.

The monopole component is obtained by transforming to the Lorentz space the sound field due to a stationary monopole in the presence of a barrier over the ground [22]:

$$p^{(0)} = p_0^{(0)} + Q_1 p_1^{(0)} + Q_2 p_2^{(0)} + Q_1' Q_2' p_{12}^{(0)} \tag{7.69}$$

The primes on the reflection coefficients in the last term denote the fact that the emission time for p_{12} is different from those for p_1 and p_2. The diffracted monopole pressures $p_K^{(0)}$ are calculated from (7.52a–7.52d). Using Huygens's principle and considering that the diffracted wave is emitted by a secondary source located at the edge of the obstacle, the spherical wave reflection coefficients are calculated for the paths joining the image source and the image receiver to the edge of the barrier [23]. Hence,

$$\rho_2' = \frac{r'}{r' + r_S'} L_{12} \tag{7.70}$$

Using the Doppler-Weyl-Van der Pol formula (7.16a–7.16c), with $i \equiv 1, 2$ and $M_i = M\cos\varepsilon_i$,

$$Q_i = R_{Pi} + (1 - R_{Pi}) F\left(\frac{w_i}{\sqrt{1 - M_i}}\right) \tag{7.71a}$$

$$R_{P,i} = \frac{\cos\theta_i - \beta_i}{\cos\theta_i - \beta_i} \tag{7.71b}$$

$$F\left(\frac{w_i}{\sqrt{1-M_i}}\right) = 1 + i\sqrt{\pi}\frac{w_i}{\sqrt{1-M_i}} \exp\left(-\frac{w_i^2}{1-M_i}\right) erfc\left(-i\frac{w_i}{\sqrt{1-M_i}}\right) \tag{7.71c}$$

$$w_i = \left(-\frac{1}{2} i k_0 \rho_i\right)^{1/2} (\cos\theta_i + \beta_i) \tag{7.71d}$$

$$\cos\theta_1 = (z_S + z_E)/\rho_1; \; \cos\theta_2 = (z + z_E)/\rho_2 \, ; \tag{7.71e}$$

(7.71a–7.71e) can be used as appropriate also to calculate Q_1' and Q_2'.

To obtain the dipole component of the pressure field expressed in the Lorentz space, use is made of the solution for the stationary dipole [15].

For a dipole orientated along the x_L-axis,

$$p^{(1)} = p^{(1)}_{S-R} + Q_1 p^{(1)}_{S'R} + Q_2 p^{(1)}_{SR'} + Q_1 Q_2 p^{(1)}_{S'R'} + \frac{\partial Q_1}{\partial x_S} p^{(1)}_{S'R}, \qquad (7.72)$$

where the diffracted dipole pressures $p_K^{(1)}$ are calculated using (7.54).

The last term in (7.72) can be neglected due to the small variations of the spherical wave reflection coefficient in the horizontal plane. Hence, after summing the monopole and the dipole contributions and transforming back into the physical space:

$$p = p_0 + Q_1 p_1 + Q_2 p_2 + Q_1' Q_2' p_{12} \qquad (7.73a)$$

$$p_K = P_0 \frac{e^{-i\omega t}}{4\pi} \frac{e^{ik_0 L_K}}{L_K (1 - M\cos\varepsilon_K)^2} \times \left\{ \frac{1+i}{2} \left[1 - \left(\frac{1}{ik_0 L_K} \right) \left(\frac{M\cos\varepsilon_K - M^2}{1 - M\cos\varepsilon_K} \right) \right] \left[\begin{array}{l} A_D\left(\dfrac{X_{K+}}{\sqrt{1 - M\cos\varepsilon_K}} \right) \\ +A_D\left(\dfrac{X_{K-}}{\sqrt{1 - M\cos\varepsilon_K}} \right) \end{array} \right] \right\}$$

$$K \equiv 0, 1, 2, 12 \qquad (7.73b)$$

Figure 7.13 shows predictions of the pressure field for the same source and barrier geometries as were used for Figure 7.11 but with hard ground on the source side and soft ground on the receiver side. Figure 7.14 shows the corresponding predictions for soft ground on both sides.

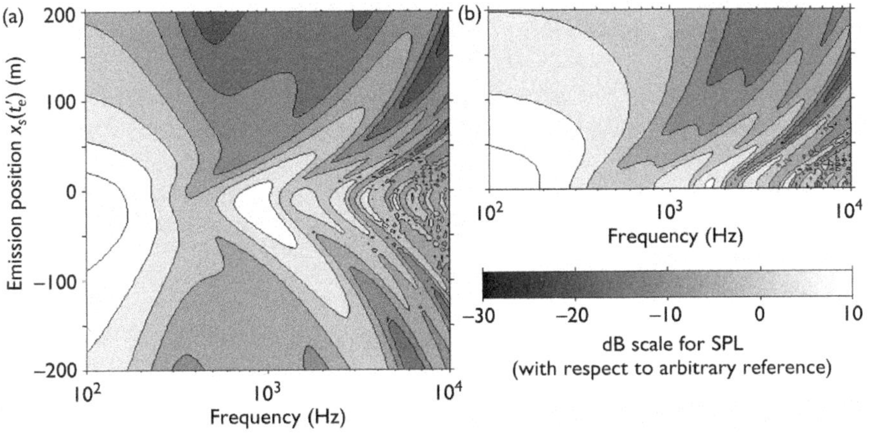

Figure 7.13 Predictions for the same situation as in Figure 7.12 but with hard ground on the source side and soft ground σ_e= 140 kPa s m^{-2} and α_e= 35 m^{-1}) on the receiver side.

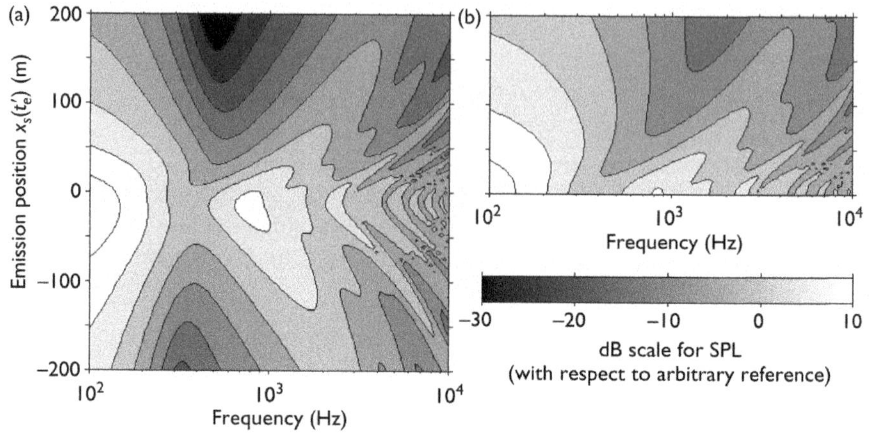

Figure 7.14 As for Figures 7.12 and 7.13 but with soft ground on both sides.

As observed in Figure 7.11, lower levels are predicted when the source is receding ($x_S(t_0) \geq 0$). Two dips related to the reflections on each side of the barrier are identifiable, particularly at large offsets and for source recession. As before, sensitivity to motion is predicted to be stronger on source approach. Also, source motion is predicted to be more important for soft ground. This was not so obvious in the predictions involving a ground discontinuity as the proportion of hard ground was varied and suggests that the source motion effect is related to the fact that, in the presence of a barrier, the angles of incidence of the reflected waves take higher values.

7.8 SOURCE MOVING OVER EXTERNALLY REACTING GROUND

The assumption that the ground is locally reacting, and, therefore, represented as an impedance plane, may not be valid for many acoustically 'soft' grounds such as snow- or gravel-covered ground or forest floors. Li and Tao [24] have investigated sound propagation from a stationary source above an externally reacting ground and have derived a generalization of (2.40) and (7.29) in the form,

$$p(x,y,z) \approx \frac{e^{ikR_1}}{4\pi R_1} + \left[V_p + (1 - V_p) F(w_a)\right] \frac{e^{ikR_2}}{4\pi R_2}, \tag{7.74a}$$

in which the plane wave reflection coefficient, R_p, is replaced by

$$V_p = \frac{\cos\theta - \chi_p}{\cos\theta + \chi_p}, \tag{7.74b}$$

where V_p has been called the 'poles-of-reflection-coefficient' [25], the apparent admittance, χ_p, is related to the (surface wave) pole location, μ_p, by

$$\cos(\mu_p) = -\chi_p, \tag{7.74c}$$

$$\sin\mu_p = +\sqrt{1-\chi_p^2} \tag{7.74d}$$

Also

$$F(w_a) = 1 + i\sqrt{\pi}w_a \exp(-w_a^2)\,erfc(-iw_a) \tag{7.74e}$$

$$w_a^2 = ikR_2\left(1 + \chi_p\cos\theta - \sqrt{1-\chi_p^2}\sin\theta\right), \tag{7.74f}$$

θ being the angle of incidence and erfc() the complementary error function (see Chapter 2). Expressions for χ_p corresponding to different types of externally reacting ground surface are listed in Table 7.1.

Wang et al. [26] have used these results when calculating the sound field due to a line source moving horizontally at constant speed, c_0M, where M is the Mach number, over externally reacting ground. Figure 7.15 shows the assumed geometry.

Their expression for the total field has the form of a Dopplerized Weyl Van der Pol formula (see (7.29)), i.e.

$$p(x_L, z_L, t) \approx p_-(R_-, \theta_-, t) + \left[V_\theta + A(1-V_\theta)F(w_p)\right]p_+(R_+, \theta_+, t) \tag{7.75a}$$

$$V_\theta = \frac{\cos\theta - \beta_e}{\cos\theta + \beta_e},\; p_\mp(R_\mp, \theta_\mp, t) = \frac{\rho_0\omega_0}{4}D_\mp^{\frac{3}{2}}\sqrt{2/i\pi k_s R_\mp}\,e^{-i\omega_0(t - R_\mp/c_0)} \tag{7.75b}$$

$$A = \frac{r_\beta/r_w}{\Delta},\; r_\beta = \frac{D_p\varsigma_p\sqrt{n_p^2 - \sin^2\mu_p}}{D + \varsigma + \sqrt{n_+^2 - \sin^2\theta_+}},$$

$$r_w = 2\sqrt{D_p/D_+\left\{\sin\frac{1}{2}(\mu_p - \theta_+)/(\cos\theta_+ - \beta_e)\right\}}, \tag{7.75c}$$

Table 7.1 Apparent surface admittance of different porous structures in which the upper layer has characteristic impedance ρ_1c_1 and where $\zeta = \rho_0/\rho_1$, $n = k_1/k_0$, $k_1 = \omega/c_1$, $N = \frac{n^2-1}{1-\zeta^2}$, and the thickness of the upper layer is d.

Type of surface	*Apparent admittance* (β_e)
Locally reacting semi-infinite	ρ_1c_1
Externally reacting semi-infinite	$(n^2 - 1)/(1 - \zeta^2)$
Externally reacting hard backed	$-i\zeta N\tan[k_0dN]$
Externally reacting non-hard-backed layer, substrate admittance $\beta_2 = \rho_2c_2$	$\frac{\beta_2 - i\zeta N\tan[k_0dN]}{1 - i\beta_2\zeta N\tan[k_0dN]/(\zeta N)}$

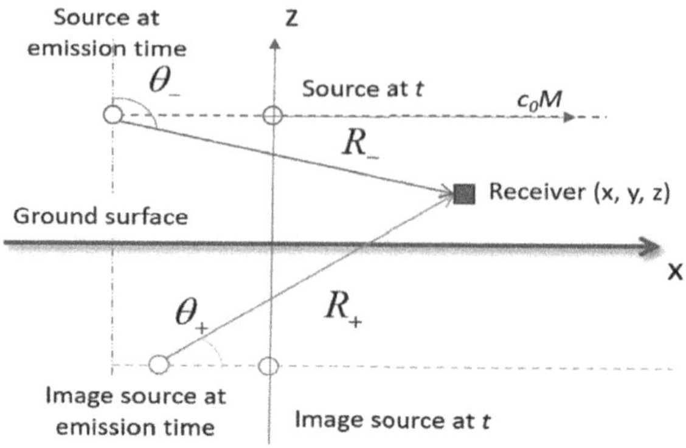

Figure 7.15 Assumed source-receiver geometry with a point source moving parallel to an externally reacting plane surface [26].

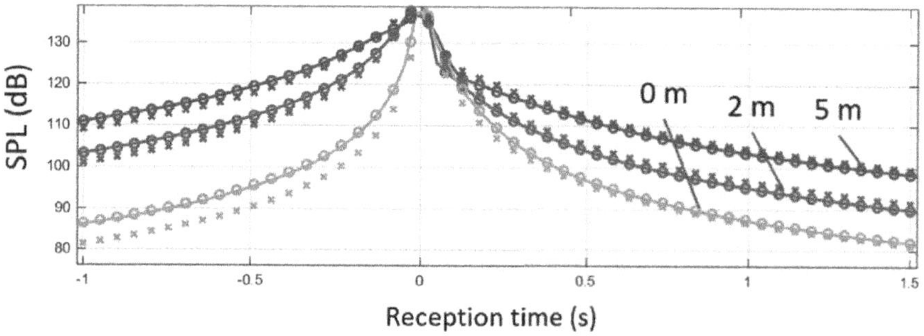

Figure 7.16 Comparisons between asymptotic solutions assuming locally reacting (α) and non-locally reacting ground (open circles) with time-domain finite difference solutions (solid lines). The assumed harmonic source has a frequency 300 Hz, sound pressure of 20 mPa (120 dB), moves at Mach number 0.3 at a height of 0.5 m above an externally reacting hard-backed layer of thickness 0.15 m. The receiver height is 0, 2, or 5 m above the ground. ***x*** coordinates of source and receiver are both 0 when reception time is 0 [26].

where $\gamma = \sqrt{1-M^2}$, $(R_{\mp},\theta_{\mp})$ are the polar coordinates in the emission time geometry and $D_{\mp} \equiv D(\theta_{\mp})$ are the corresponding Doppler factors.

Figure 7.16 shows examples of predictions that assume a 0.5 m high 300 Hz single-frequency source moving at Mach 0.3 parallel to a porous hard backed layer with acoustical properties given by the Hamet model (see Chapter 5) with parameters of flow resistivity 10 kPa s m^{-2}, porosity 0.5 and tortuosity 1.6. Predictions of analytical approximations for either local- or extended-reaction are compared with numerical results for extended reaction obtained using an FDTD method [27 and Chapter 4]. The predictions of the analytical approximation and numerical method allowing for extended reaction agree well. Both indicate that the largest difference (as much as 3 dB) between local- and extended-reaction predictions occur when the source is approaching a receiver on the surface.

REFERENCES

[1] D. G. Crighton, Scattering and diffraction of sound by moving bodies, *J. Fluid. Mech.* **72** 209–227 (1975).
[2] A. P. Dowling, Convective amplification of real simple sources, *J. Fluid Mech.* **74** 529–546 (1976).
[3] P. M. Morse and K. U. Ingard, *Theoretical Acoustics*, Princeton University Press, Princeton, New Jersey, Chapter 11 (1986).
[4] M. V. Lowson, Sound field for singularities in motion, *Proc. Royal Soc. Series A* **286** 559 (1965).
[5] E. W. Graham and B. B. Graham, Theoretical acoustical sources in motion, *J. Fluid Mech.* **49** 481 (1971).
[6] T. D. Norum and C. H. Liu, Point source moving above a finite impedance reflecting plane - experiment and theory, *J. Acoust. Soc. Am.* **63** 1069–1073 (1978).
[7] S. Oie and R. Takeuchi, Sound radiation from a point source moving in parallel to a plane surface of porous material, *Acustica* **48** 123–129 (1981).
[8] S. Oie and R. Takeuchi, Sound radiation from a point source moving vertical to a plane surface of porous material, *Acustica* **48** 137–142 (1981).
[9] G. Rosenhouse and N. Peled, Sound field of moving sources near impedance surfaces, *International Conference on Theoretical and Computational Acoustics*, Mystic, Connecticut, USA, 5–9 July 1993, Vol. I, 377–388 (1993).
[10] A. P. Dowling and J. E. Ffowcs Williams, *Sound and Sources of Sound*, Ellis Horwood Ltd., Chichester, *Chapter* 7 (1983).
[11] M. Buret, *New Analytical Models For Outdoor Moving Sources of Sound*, Ph.D. Thesis, The Open University, UK (2002).
[12] A. D. Pierce, *Acoustics, An Introduction to its Physical Principles and Applications*, Acoustical Society of America, New York (1989).
[13] M. Buret, K. M. Li and K. Attenborough, Optimisation of ground attenuation for moving sound sources, *Appl. Acoust.* **67** 135–156 (2006).
[14] K. M. Li, M. Buret and K. Attenborough, The propagation of sound due to a source moving at high speed in a refracting medium, *Proc. Euro-Noise* **98**, 2, 955–960 (1998).
[15] M. Buret, K. M. Li and K. Attenborough, Diffraction of sound from a dipole source near to a barrier or an impedance discontinuity, *J. Acoust. Soc. Am.* **113**(5) 2480–2494 (2003).
[16] B. A. De Jong, A. Moerkerken and J. D. Van der Toorn, Propagation of sound over grassland and over an earth barrier, *J. Sound Vib.* **86**(1) 23–46 (1983).
[17] D. C. Hothershall and J. N. B. Harriot, Approximate models for sound propagation above multi-impedance plane boundaries, *J. Acoust. Soc. Am.* **97**(2) 918–926 (1995).
[18] P. Boulanger, T. Waters-Fuller, K. Attenborough and K. M. Li, Models and measurements of sound propagation from a point source over mixed impedance ground, *J. Acoust. Soc. Am.* **102**(3) 1432–1442 (1997).

[19] M. R. Bassiouni, C. R. Minassian and B. Chang, Prediction and experimental verification of far field sound propagation over varying ground surface, *Proc. Internoise* **83**(1 287–290 (1983).

[20] C. Nyberg, The sound field from a point source above a striped impedance boundary, *Acta Acust. united Ac.* **3** 315–322 (1995).

[21] K. M. Li, S. Taherzadeh and K. Attenborough, Sound propagation from a dipole source near an impedance plane, *J. Acoust. Soc. Am.* **101**(6) 3343–3352 (1997).

[22] H. G. Jonasson, Sound reduction by barriers on the ground, *J. Sound Vib.* **22**(1) 113–126 (1972).

[23] P. Koers, Diffraction by an absorbing barrier or by an impedance transition, *Proc. Inter-Noise* **83**(1) 311–314 (1983).

[24] K. M. Li and H. Tao, Heuristic approximations for sound fields produced by spherical waves incident on locally and non-locally reacting planar surfaces, *J. Acoust. Soc. Am.* **135**(1) 58–66 (2014).

[25] J. F. Allard, M. Henry, J. Tizianel, J. Nicolas and M. Yasushi, Pole contribution to the field reflected by sand layers, *J. Acoust. Soc. Am.* **111** 685–689 (2002).

[26] Y. Wang, K. M. Li, D. Dragna and P. Blanc-Benon, On the sound field from a source moving above non-locally reacting grounds, *J. Sound Vib.* **464** 114975 (2020).

[27] D. Dragna, P. Pineau and P. Blanc-Benon, A generalized recursive convolution method for time-domain propagation in porous media, *J. Acoust. Soc. Am.* **138** 1030–1042 (2015).

Chapter 8

Predicting Effects of Mixed Impedance Ground

8.1 INTRODUCTION

In many situations where outdoor sound propagation is to be predicted, the nature of the ground is likely to change from acoustically hard to acoustically soft. For example, when considering sound propagation from vehicles on a road, the sources move over an acoustically hard surface such as hot-rolled asphalt, but the receivers might be above acoustically soft grassland adjacent to the road. Other situations might involve propagation over a strip of acoustically hard ground such as a service road or over a strip of acoustically soft ground supporting arable crops or a belt of trees. In urban locations there might be several areas of widely differing impedance. Even if the propagation is entirely over naturally occurring acoustically soft outdoor ground surfaces, then, as discussed in Chapter 6, the ground impedance varies from place to place along the propagation path. How do these variations influence propagation? Can the influence be predicted using an average or effective impedance and the models described in Chapters 2, 3 and 5? Current engineering prediction schemes for outdoor sound allow for mixtures of hard and soft ground either by reducing the attenuation due to ground effect in accordance with the proportion of soft ground between source and receiver [1–3] or by computing only the soft ground contribution [4].

This chapter concerns models for predicting the effects of impedance changes along propagation paths and the resulting understanding of these effects. First, models for propagation over a single discontinuity are outlined. Subsequently, models for propagation over multiple discontinuities and ways of including atmospheric refraction are discussed. Laboratory measurements over different impedance distributions are described and these laboratory data are compared with predictions. Then predictions of the broadband attenuation that could result from various ground treatments including low parallel walls and lattices near surface transport corridors are presented and discussed. Next, the influences of wind-induced refraction and turbulence on the attenuation due to arrays of low parallel walls between the source and receiver are investigated. The final section discusses how propagation over a concrete-to-grassland impedance transition is affected by a downward refracting atmosphere.

8.2 SINGLE IMPEDANCE DISCONTINUITY

8.2.1 De Jong's Semi-Empirical Method

Various analytical methods of calculating sound propagation over a two-impedance boundary have been proposed. These range from an exact solution in the form of a triple integral [5] to an approximate solution of the boundary integral [6]. Semi-empirical formulae have been proposed by Koers [7] and by De Jong et al. [8]. The former presented a model based on Kirchoff diffraction theory in which a plane impedance transition is represented by a wedge with a top angle of 180° having a different impedance on each side. De Jong et al. [8] also considered the limiting case of sound diffraction from a wedge with different surface acoustic impedances on either side [9] and derived expressions for the propagation of sound over an impedance discontinuity as an empirical combination of the formulae for the diffraction of spherical waves at a rigid half plane and the field due to a point source over an impedance plane. This uses the spherical wave reflection coefficient calculation described in Chapter 2 and, thereby, offers a particularly useful and relatively straightforward computation.

Consider the situation shown in Figure 8.1.

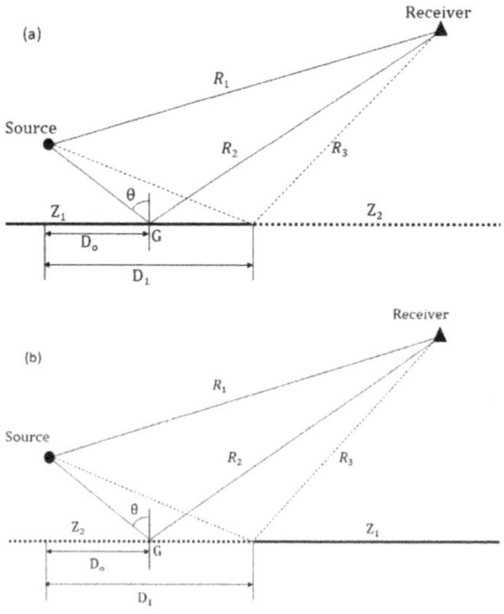

Figure 8.1 Propagation from a point source to a receiver showing specular reflection and reflection from the point of impedance discontinuity (a) from hard to soft (b) from soft to hard.

According to the De Jong model, the excess attenuation over the single discontinuity between portions of ground with impedance Z_1 and Z_2, the source being over Z_1, is given by

$$EA = 20\log\left|\frac{P}{P_1}\right|, \tag{8.1}$$

where

$$\frac{P}{P_1} = 1 + \frac{R_1}{R_2} Q_{1,2}\exp\{ik(R_2 - R_1)\} + (Q_2 - Q_1)\frac{\exp\left(\frac{-i\pi}{4}\right)}{\sqrt{\pi}}\frac{R_1}{R_3}\{F_{d1} \pm F_{d2}\exp\{ik(R_2 - R_1)\}\} \tag{8.2}$$

$Q_{1,2}$ is replaced by Q_1 the spherical wave reflection coefficient for the portion of the ground with impedance Z_1 and the positive sign in the curly brackets is used when the point of specular reflection falls on that portion of ground. Conversely, $Q_{1,2}$ is replaced by Q_2, the spherical wave reflection coefficient for the portion of the ground with impedance Z_2 and the negative sign in the curly brackets is used when the point of specular reflection falls on that portion of ground.

$F_{d1} \equiv F\left[\sqrt{k(R_3 - R_1)}\right]$ and $F_{d2} \equiv F\left[\sqrt{k(R_3 - R_2)}\right]$ are Fresnel integrals of the form $F(x) = \int_x^\infty \exp(iw^2)dw$ and the path lengths are defined in Figure 8.1. R_3 is the path from source to discontinuity to receiver.

According to Daigle et al. [10] and Hothersall and Harriott [11], the De Jong model fails at grazing incidence. On the other hand, it agrees remarkably well with data obtained when source and receiver are elevated (see Section 8.4).

8.2.2 Modified De Jong Method

Lam and Monazzam [12] have pointed out that, although the De Jong formulation is correct if the source is over hard ground and the receiver is over soft ground, it is incorrect if the source is above soft ground and the receiver is above hard ground, i.e. it violates the reciprocity condition as a result of an inconsistency in the use of the ± sign inside the square brackets in Equation (8.2). They have proposed a modified De Jong model, using additional parameters μ and γ, given by

$$\frac{P}{P_1} = 1 + \frac{R_1}{R_2} Q_G e^{ik(R_2 - R_1)} + (Q_2 - Q_1)e^{-i\pi/4}\frac{1}{\sqrt{\pi}}\frac{R_1}{S_1}X \times\left[\mu F_2\left(\sqrt{k(S_1 - R_1)}\right) + \gamma F_2\left(\sqrt{k(S_1 - R_2)}\right)e^{ik(R_2 - R_1)}\right] \tag{8.3}$$

The value of γ depends on the location of the impedance discontinuity relative to the point of specular reflection, i.e. the distances D_0 and D_1 shown in Figure 8.1(a). If $D_0 < D_1$, then the value of γ is +1 and, if $D_0 > D_1$, then the value of γ is −1. The value of μ depends on the relative magnitudes of impedance across the discontinuity. If the impedance on the source side is greater than that on the receiver side, i.e. propagation from 'hard' to 'soft' ground, then $\mu = -1$. Conversely for propagation from 'soft' to 'hard' (see Figure 8.1(b)), $\mu = +1$.

8.2.3 Rasmussen's Method

Rasmussen [13] has suggested a numerical method for determining the sound field over a plane containing an impedance discontinuity. A hypothetical planar source 40 wavelengths wide and 20 wavelengths tall is placed above the discontinuity. The planar source is discretized into an array of point sources a fifth of a wavelength apart. The relative strength of each source is calculated using the usual point source theory for propagation over infinite impedance Z_1. The received field is calculated as the sum of the contributions from each of the constituent planar sources over an infinite Z_2. However, comparisons between predictions of this relatively numerically intensive method, the De Jong semi-empirical formula, and data indicate that the De Jong formula is adequate for engineering purposes [13]. The method developed by Rasmussen is also known as the *substitute sources method* and has been developed further [14].

Based on this approach, Rasmussen [15] has suggested a numerical integration (see Equation (8.4)) for calculating the velocity potential, ψ, over an impedance discontinuity (see Figure 8.2).

$$\psi = x\sqrt{(8\pi k)}\frac{e^{-i\pi/4}}{16\pi^2}\int_0^\infty \left[\begin{array}{l} \dfrac{e^{ik(R_1+R_2)}}{\sqrt{R_3^3R_1(R_1+R_3)}} + Q_2\dfrac{e^{ik(R_1+R_4)}}{\sqrt{R_3^3R_1(R_1+R_3)}} \\ + Q_2\dfrac{e^{ik(R_1+R_3)}}{\sqrt{R_3^3R_2(R_2+R_3)}} + Q_1Q_2\dfrac{e^{ik(R_2+R_4)}}{\sqrt{R_4^3R_2(R_2+R_4)}} \end{array}\right] dh \tag{8.4}$$

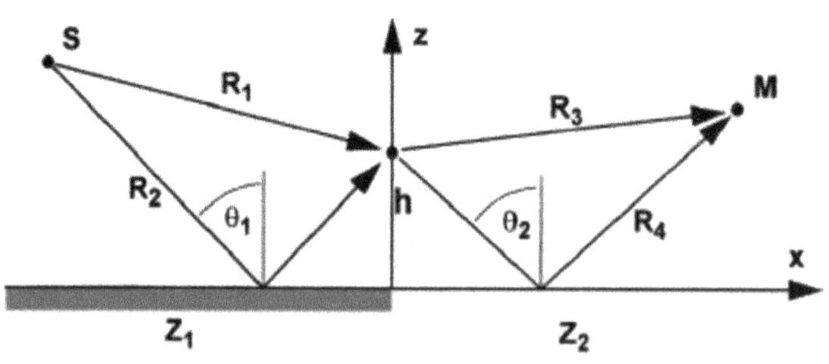

Figure 8.2 Source-receiver-discontinuity geometry used by the Rasmussen calculation.

8.3 MULTIPLE IMPEDANCE DISCONTINUITIES

8.3.1 An Extended De Jong Method

Hothersall and Harriott [11] have extended the De Jong model to predict propagation over two impedance discontinuities, for example, over a 'soft' strip in a 'hard' ground surface. However, Boulanger et al. [16] have pointed out an error in sign convention for the multi-impedance De Jong model proposed by Hothersall and Harriott [11] and have extended and corrected De Jong multi-impedance model for a ground surface composed of different impedance surfaces (see Figure 8.3). The pressure at the receiver relative to the free field in the case of multiple strips of impedance Z_1 imbedded in a plane of impedance Z_2 is given by

$$\frac{P}{P_1} = 1 + \frac{R_1}{R_2} Q_{1,2} e^{ik(R_2 - R_1)}$$
$$+(Q_2 - Q_1) e^{-i\pi/4} \frac{1}{\sqrt{\pi}} R_1 \sum_{j=1}^{n} \left\{ \begin{array}{l} \dfrac{1}{R_{b_j}} \left[F_{b,1} + \gamma_j F_{b,2} e^{ik(R_2 - R_1)} \right] + \\ \dfrac{1}{R_{a_j}} \left[-F_{a,1} + \delta_j F_{a,2} e^{ik(R_2 - R_1)} \right] \end{array} \right\}, \qquad (8.5)$$

where j is the index characterizing the various strips and n is the total number of strips between source and receiver.

In accordance with previous definitions, $F_{a,1} \equiv F(R_{aj} - R_1)$. The spherical wave reflection coefficients Q_1 or Q_2 are used when the specular reflection point falls on the ground of impedance Z_1 or Z_2, respectively. The sequences of values $(Q_{1,2}, \gamma_j, \delta_j)$ appear in Equation (8.5) according to the following combinations:

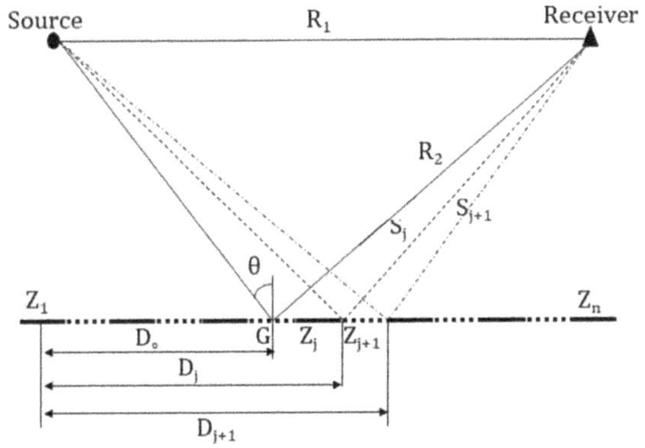

Figure 8.3 Point source and receiver over a plane containing multiple impedances showing diffraction at and distances to each discontinuity.

$(Q_2, 1, -1)$ if the specular reflection point is on the source side of the jth strip,
$(Q_2, -1, 1)$ if the specular reflection point is on the receiver side of the jth strip,
$(Q_2, -1, -1)$ if the specular reflection point is inside the jth strip.

However, this form of extended multi-impedance De Jong model is incorrect for the 'soft' to 'hard' ground transitions along a propagation path.

8.3.2 The nMID (Multiple Impedance Discontinuities) Method

As well as correcting the De Jong model for a single 'soft' to 'hard' ground transition, Lam and Monazzam [12] have proposed a model for propagation from a point source over a ground with n sections of differing impedance and hence $(n - 1)$ impedance discontinuities.

Their Multi-Impedance Discontinuities Model (nMID) is given by

$$\frac{P}{P_1} = 1 + \frac{R_1}{R_2} Q_G e^{ik(R_2 - R_1)} + \sum_{j=1}^{n-1} (Q_{j+1} - Q_j) e^{-i\pi/4} \frac{1}{\sqrt{\pi}} \frac{R_1}{S_j} \left[\begin{array}{c} \mu_j F_2\left(\sqrt{k(S_j - R_1)}\right) \\ +\gamma_j F_2\left(\sqrt{k(S_j - R_2)}\right) e^{ik(R_2 - R_1)} \end{array} \right] \tag{8.6a}$$

If $D_0 < D_j$, then the value of γ_j is equal to +1 and if $D_0 > D_j$, then the value of γ_j is equal to –1. When the admittance $\beta_{j+1} > \beta_j$ then $\mu_j = 1$ and when $\beta_{j+1} < \beta_j$ then $\mu_j = -1$. The different path lengths used in Equation (8.5) are shown in Figure 8.3.

Similarly, for a mixed impedance ground surface having periodic impedance discontinuities created with strips of impedance Z_1 and Z_2, the excess attenuation term can be written as

$$\frac{P}{P_1} = 1 + \frac{R_1}{R_2} Q_G e^{ik(R_2 - R_1)} + (Q_2 - Q_1) \frac{e^{-i\pi/4}}{\sqrt{\pi}} \sum_{k=1}^{m} \left\{ \begin{array}{c} \frac{R_1}{S_{2k-1}} \left[\begin{array}{c} F_2\left(\sqrt{k(S_{2k-1} - R_1)}\right) + \\ \gamma_{2k-1} F_2\left(\sqrt{k(S_{2k-1} - R_2)}\right) e^{ik(R_2 - R_1)} \end{array} \right] \\ + \frac{R_1}{S_{2k}} \left[\begin{array}{c} F_2\left(\sqrt{k(S_{2k} - R_1)}\right) + \\ \gamma_{2k} F_2\left(\sqrt{k(S_{2k} - R_2)}\right) e^{ik(R_2 - R_1)} \end{array} \right] \end{array} \right\} \tag{8.6b}$$

Although predictions of the nMID model agree with boundary element method predictions for many situations involving multiple impedance strips, it is not as accurate when source and receiver are close to the ground surface. Moreover, it gives poor agreement with BEM results when the widths of the strips are small. Predictions of nMID are compared with those of a more accurate Boundary Element method in Section 8.3.5. Comparisons with data are discussed further in Section 8.4.

8.3.3 Nyberg's Method

Nyberg [17] has solved the Helmholtz equation with the boundary condition for a point source above an infinite plane surface consisting of strips with alternating impedance by using Cartesian coordinates and a Fourier transform technique. The ground effect due to a two-valued, infinitely periodic (i.e. striped) impedance (see Figure 8.4) may be determined, approximately, from that predicted for a point source (see Chapter 2). Hence, the expression for the field at height z, due to a source at height z_0, is

$$\Psi\left(x,y,z\right) = -\frac{e^{ikR_1}}{4\pi R_1} - \frac{e^{ikR_2}}{4\pi R_2} + \frac{1}{2\pi}\int_{-\infty}^{+\infty} L_0'\left(\alpha,y,h\right)e^{i\alpha x}d\alpha, \tag{8.7}$$

where a, b are the periods of the two types of strips, $L_0' = \frac{\beta_0 b + \beta_1 a}{a+b}L_0$ and L_0 is the usual integral for a point source (see Chapter 2) using the area-averaged admittance $\frac{\beta_0 b + \beta_1 a}{a+b}$. Also α is the horizontal wavenumber and $R_{1,2} = \sqrt{x^2 + y^2 + \left(z \pm z_0\right)^2}$.

This approximation requires $a + b \ll \lambda$.

Use of an area-averaged impedance does not lead to accurate predictions for sound propagation over a single discontinuity. Figure 8.5 compares predictions of excess attenuation for two source-receiver-discontinuity geometries using the De Jong formula, each of the corresponding continuous

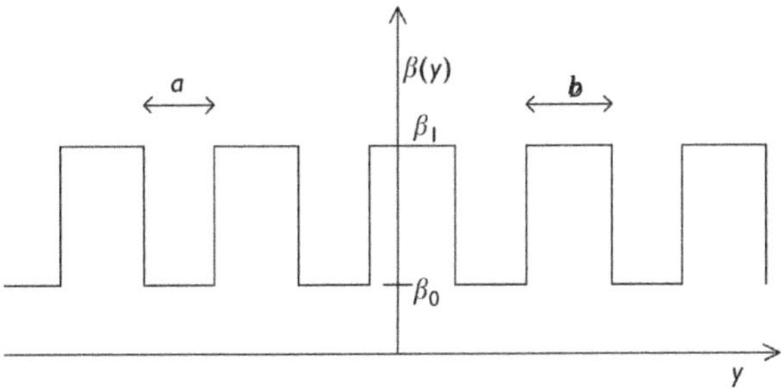

Figure 8.4 An admittance which varies regularly between values of β_0 and β_1 with periods **a** and **b** respectively, as considered by Nyberg [16].

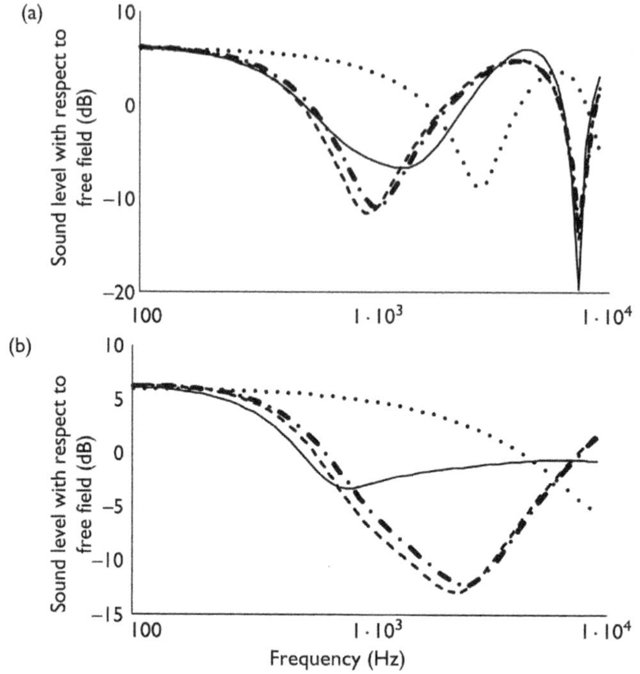

Figure 8.5 Predictions of excess attenuation due to the ground from a point source over ground containing a single discontinuity (solid lines), continuous impedance 1 (dotted lines), continuous impedance 2 (broken lines) and the area-averaged impedance (dot-dash lines) for receiver height 1.2 m, separation 25 m, distance to discontinuity 3.5 m and two source heights above impedance 1 (a) 0.5 m, (b) 0.1 m. The impedances are predicted by the two-parameter model (3.13) with σ_e= 10000 kPa s m^{-2} and 100 kPa s m^{-2} respectively (α_e = 100 m^{-1}).

impedances and with the continuous averaged impedance representing Nyberg' approximate result.

On the other hand, predictions of Nyberg's approximate theory have been found yield reasonable agreement with laboratory data for propagation over a spatially periodic impedance variation formed from alternative strips of hardwood and sand [16].

Nevertheless, Nyberg theory does not distinguish between the excess attenuation due to periodic impedance ground for a 'hard' or 'soft' surface at the point of specular reflection whereas measured EA spectra show a difference between these conditions (e.g. see Figure 8.14(c)). Also, Boundary Element calculations show such difference (see Section 8.3.5).

8.3.4 *Fresnel-Zone Methods*

Following a suggestion by Slutsky and Bertoni [18], Hothersall and Harriott [11] have developed an approach based on Fresnel diffraction theory. If the oscillations due to diffraction are ignored, the excess attenuation with range from the source varies between the values appropriate to point-to-point propagation over each impedance alone. The transition is within a well-defined region around the point of specular reflection between source and receiver. This may be defined as a Fresnel zone (see Figures 8.5 and 8.6) at the boundary of which the path length is some fraction (F) of a wavelength (e.g. $\lambda/3$) greater than the specularly reflected path. Hothersall and Harriott [10] have shown that path length differences between $\lambda/3$ and $\lambda/4$ are acceptable. The locations P' of points on the surface of the Fresnel volume define the boundary of an ellipsoid with foci S' and R and its major axis along S'R with semi-major axis $a = \dfrac{R_2 + F\lambda}{2}$ and semi-minor axis $b = \sqrt{\dfrac{R_2 F\lambda}{2} + \left(\dfrac{F\lambda}{2}\right)^2}$.

The Fresnel ellipsoid cuts the ground plane at an ellipse (see Figure 8.6) with the equation

$$\frac{x^2}{b^2} + \frac{(y\cos\theta - c)^2}{a^2} + \frac{y^2\sin^2\theta}{b^2} = 1, \tag{8.8a}$$

where $c = \dfrac{R_2}{2} - SP$, SP is the path length between source S and specular reflection point P and the origin of the coordinate system is the point of specular reflection.

The area of this ellipse may be computed from the locations of the points $x_{1,2}$ and $y_{1,2}$ shown in Figure 8.7 given by:

$$x_{1,2} = \pm b\sqrt{1 - \frac{(y_m\cos\theta - c)^2}{a^2} + \frac{{y_m}^2\sin^2\theta}{b^2}} \tag{8.8b}$$

$$y_{1,2} = -\frac{B}{A} \pm \sqrt{\frac{1}{A} - \left(\frac{c\sin\theta}{Aab}\right)^2}, \tag{8.8c}$$

where $A = \left(\dfrac{\cos\theta}{a}\right)^2 + \left(\dfrac{\sin\theta}{b}\right)^2$, $B = \dfrac{c\cos\theta}{a^2}$ and $y_m = \dfrac{cb^2\cos\theta}{a^2\sin^2\theta + b^2\cos^2\theta}$.

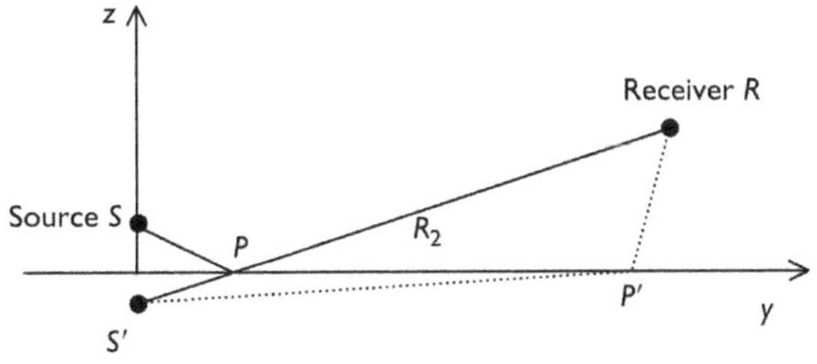

Figure 8.6 Source-receiver geometry, specular reflection point P, equivalent specularly-reflected path (S′R) and path (S′P′R) via point P′ on the Fresnel zone boundary so that **S′P′** + **RP′** − **R**$_2$ = **F**λ.

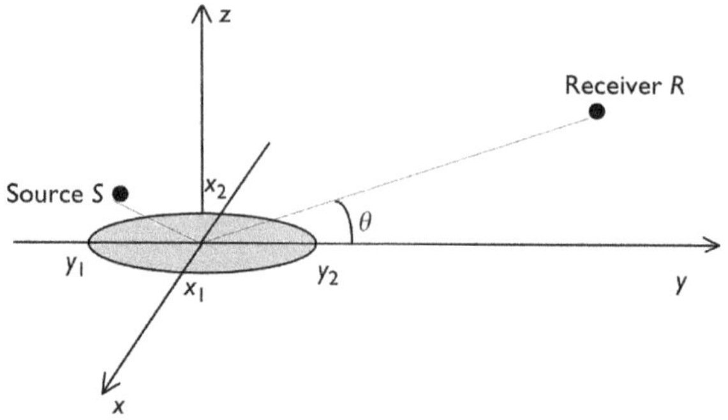

Figure 8.7 The elliptical boundary of the Fresnel zone in the ground plane.

The expression for $y_{1,2}$ differs from that given by Hothersall and Harriott [11], since they interchange variables x and y without maintaining a consistent notation. Moreover, their corresponding expression is not dimensionally correct.

The area of the ellipse is $\pi x_1 y_1$. The frequency-dependent and geometry-dependent elliptical area defined in this way may be interpreted as the area of ground between point source and receiver which is important in determining the ground effect.

The reduction in Fresnel-zone size with increasing frequency is demonstrated in Figures 8.7–8.9 for source and receiver at 1 m height above the ground, separated horizontally by 10 m and $F = 1/3$. Note that the major axis of the ellipse exceeds the source–receiver separation (10 m) at 100 Hz (see also the 63 Hz ellipse in Figure 8.10). This implies that the area of

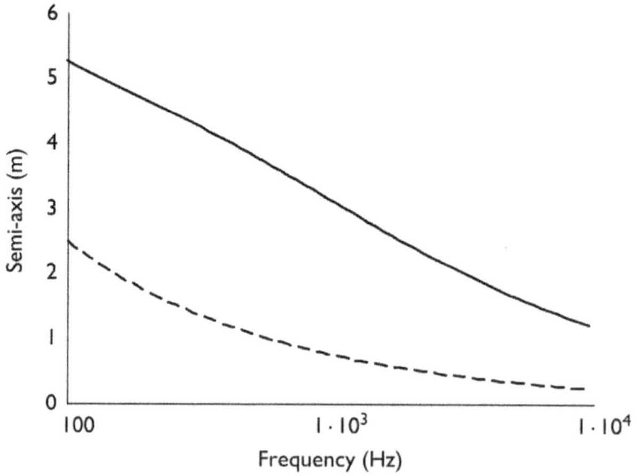

Figure 8.8 Semi-axes of elliptical Fresnel zone (8.3) as a function of frequency calculated for $h_s = h_r = 1$ m, $r = 10$ m and $F = 1/3$. The solid line represents the semi-major axis and the broken line represents the semi-minor axis.

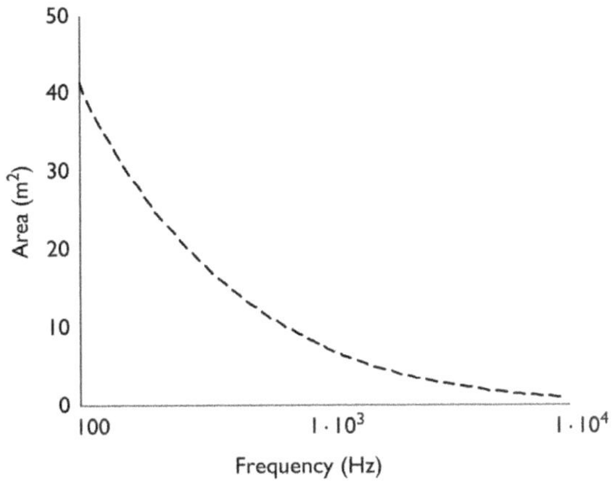

Figure 8.9 Area of elliptical Fresnel zone (8.3) calculated for $h_s = h_r = 1$ m, $r = 10$ m and $F = 1/3$.

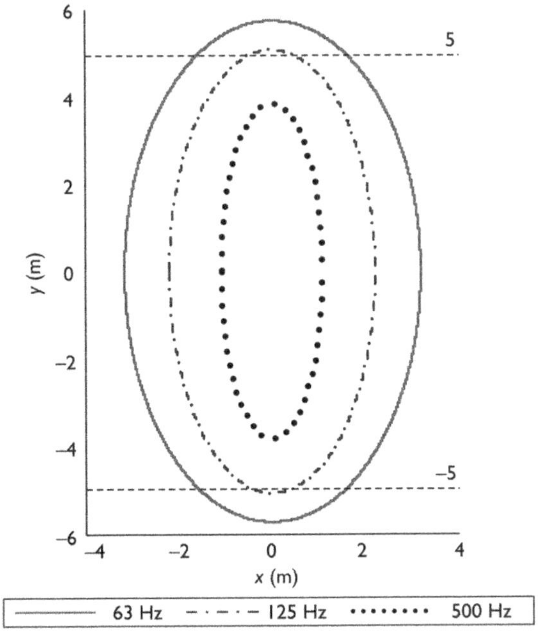

Figure 8.10 Fresnel ($F = 1/3$) zones centred at the specular reflection point (5 m from source or receiver along source-receiver axis) at three frequencies for source and receiver at 1 m height and separated by 10 m. The projections of the source and receiver positions on the (x, y) plane are at (0,–5) and (0,5) respectively.

ground of interest at this frequency extends beyond that between source and receiver. The elliptical area becomes circular at normal incidence.

If it is assumed that the excess attenuation is linearly dependent on the proportion of the different surfaces (μ) along the line between source and receiver representing the Fresnel zone, then

$$\text{EA} = \mu 20\log\left|1+\frac{R_1}{R_2}Q_1e^{ik(R_2-R_1)}\right| + (1-\mu)20\log\left|1+\frac{R_1}{R_2}Q_2e^{ik(R_2-R_1)}\right| \quad (8.9a)$$

The line representing the Fresnel zone is defined by the intersection between the Fresnel-zone ellipse (Equation (8.8a) and Figure 8.7) and the vertical plane through source, receiver and specular reflection point.

Apart from the factors μ and $(1 - \mu)$, the terms on the right-hand side of Equation (8.9a) represent the excess attenuation over uniform boundaries with surface impedance Z_1 or Z_2, respectively. A linear combination of excess attenuation magnitudes, as in Equation (8.7), is implicit in standard prediction schemes [1,2].

An alternative approach is to assume that the excess attenuation components are proportional to the areas of the Fresnel zone occupied by each impedance. Further possibilities are that either the pressure or the intensity is proportional to μ. For example, if a linear interpolation of pressure is used [16],

$$\text{EA} = 20\log\left|\mu\left|1+\frac{R_1}{R_2}Q_1e^{ik(R_2-R_1)}\right| + (1-\mu)\left[1+\frac{R_1}{R_2}Q_2e^{ik(R_2-R_1)}\right]\right|, \quad (8.9b)$$

where μ is calculated from the areas of each impedance inside the Fresnel zone.

Predictions of this modification of the Fresnel-zone method (Equation (8.9b)) have been found to be in better agreement with data than the method suggested by Hothersall and Harriott (Equation (8.9a)) [16]. Nevertheless, the Fresnel-zone method cannot differentiate between single and multiple impedance discontinuities. Consider a situation for a ground surface with single impedance discontinuity of equal proportion of both ground type inside the Fresnel zone. Similarly, consider another situation with multiple impedance discontinuities of equal proportion of both ground type inside the Fresnel zone. For both cases, the value of μ is equal to 0.5. The Fresnel-zone model does not predict the spectral oscillations associated with diffraction effects but gives a reasonable estimate of the excess attenuation spectra.

8.3.5 The Boundary Element Method

Increasingly powerful computational resources make the use of numerical schemes such as those based on the Boundary Element Method (BEM) more attractive. The computational complexity of the BEM approach is reduced considerably if it is used only for two-dimensional problems. Let a line source produce a time-harmonic sound field, $\phi(r,z)$, in a medium, **D**, bounded by a locally reacting impedance surface, **S**. This impedance plane can have features such as impedance discontinuities, etc.

The Boundary Integral form of the wave equation (BIE) can be written as

$$\epsilon\phi(r,z) = G(\mathbf{r},\mathbf{r}_0) - \int_S \left[G(\mathbf{r},\mathbf{r}_s)\frac{\partial\phi(r_s,z_s)}{\partial \mathbf{n}(\mathbf{r}_s)} - \phi(r_s,z_s)\frac{\partial G(\mathbf{r},\mathbf{r}_s)}{\partial \mathbf{n}(\mathbf{r}_s)} \right] ds, \tag{8.10}$$

where $\mathbf{r}_s$ is the position vector of the boundary element ds, and $\mathbf{n}$ is the unit normal vector out of ds. The parameter, ε, depends on the position of the receiver [19] being equal to 1 for **r** in the medium, ½ for **r** on the flat boundary and equal to the Ω/2π at edges where Ω is the solid angle. The Green's function, $G(\mathbf{r},\mathbf{r}_0)$, is the solution of the wave equation in the domain without the effect of scatterers.

The integral is then the contribution of the scatterer elements to the total sound field at a receiver position. This integral formulation (first derived by Kirchhoff in 1882) is called the Helmholtz–Kirchhoff wave equation. It is the mathematical formulation of Huygen's principle. For receiver points on the boundary, one obtains an integral equation for the field potential at the boundary. This BIE is a Fredholm integral equation of the second kind. Once solved, the contribution of the scatterers can be determined by evaluating the integral in equation (8.10) and calculating the total field for any point in the entire domain, **D**. This is the main BIE for the acoustic field potential in the presence of non-uniform boundary.

The BEM represents the acoustic propagation in a medium by the boundary integral equation and solves this integral equation numerically.

The locally reacting impedance boundary condition is

$$\frac{d\varphi}{dn} - ik_0\beta\varphi = 0, \tag{8.11}$$

where β, the admittance, can be a function of the position on the boundary. This applies to the surface of the plane boundary and to the surfaces of the scatterers.

So, the derivative term of the unknown potential can be written in terms of the potential itself. Hence,

$$\epsilon\phi(r,z) = G(\mathbf{r},\mathbf{r}_0) - \int_S \phi(r_s,z_s)\left[ik_0\beta G(\mathbf{r},\mathbf{r}_s) - \frac{\partial G(\mathbf{r},\mathbf{r}_s)}{\partial \mathbf{n}(\mathbf{r}_s)} \right] ds, \mathbf{r}, \mathbf{r}_s \in \mathrm{S} \tag{8.12}$$

The integral Equation (8.12) is solved numerically after discretizing the boundary and assuming that the unknown potential is constant in each element, thereby reducing the integral equation to a set of linear equations [19–22]. One numerical integration method involves a quadrature technique (Simpson's rule or Gauss) which is similar to a *finite difference* formulation.

To minimize the number of elements, the Green's function includes reflection from the flat impedance surface. Consider a two-dimensional problem with an infinitely long line source radiating cylindrical waves in the medium.

In this case, the boundary integral is a line integral and its evaluation by numerical means is a simple matter. In this case, the Green's function takes the form [19]

$$G_2(\mathbf{r},\mathbf{r}_0) = -\frac{i}{4}\left\{H_0^{(1)}\left(k\left|r_0 - r\right|\right) + H_0^{(1)}\left(k\left|r_0' - r\right|\right)\right\} + P_\beta(r,r_0) \tag{8.13a}$$

With

$$P_\beta(\mathbf{r},\mathbf{r}_0) = \frac{i\beta}{2\pi}\int_{-\infty}^{+\infty}\left\{\frac{e^{ik\left[(z+z_0)\sqrt{(1-s^2)}-(x-x_0)s\right]}}{\sqrt{(1-s)}\left(\sqrt{(1-s^2)}-\beta\right)}\right\}, \ \mathrm{Re}(\beta)>0 \tag{8.13b}$$

In the above expressions, $H^{(1)}(\ )$ is the Hankel function of the first kind, $\mathbf{r}$, $\mathbf{r}_0$ and r_0' are the receiver, source and image source position vector, respectively.

The wavenumber k and the complex admittance β are dependent on the frequency. The function P_β represents the ground wave term. The derivatives of the Green's function in the x and z directions are also available [19].

Although more computationally intensive, BEM calculations are more accurate than predictions using the De Jong semi-empirical formula (8.2) and its modification (8.3) for propagation over a single impedance discontinuity. Figure 8.11(a)–(c) compare predictions of BEM, the original De Jong model (8.2) and the modified De Jong model (8.3) for propagation over a hard to soft impedance discontinuity located at distances of 0.2, 0.4 and 0.6 m, respectively, from the source [23]. As expected, the original and modified De Jong models predict identical EA spectra for hard to soft impedance discontinuity. The De Jong model gives good agreement with the numerically predicted EA spectra for an impedance discontinuity at distances of 0.2 and 0.6 m (see Figure 8.11(a) and (c), respectively). However, the agreement between De Jong model and BEM is less good if the impedance discontinuity is near the specular reflection point (see Figure 7.4(b)). Figure 8.11(d)–(f) compare predictions of BEM, the original De Jong model and the modified De Jong model for a soft to hard impedance discontinuity located at distances of 0.2, 0.4 and 0.6 m, respectively. As expected, the original and modified De Jong models predict different EA spectra for a soft to hard impedance discontinuity. The modified De Jong model gives better agreement with the BEM predictions than the original De Jong model for a soft to hard impedance discontinuity.

Consider a single strip of absorbing material, impedance Z_2 lying perpendicular to the line between source and receiver in a plane of less absorbing material with impedance Z_1 (see Figure 8.12).

The appropriate Boundary Integral Equation formulation developed by Chandler-Wilde and Hothersall [19] may be written as

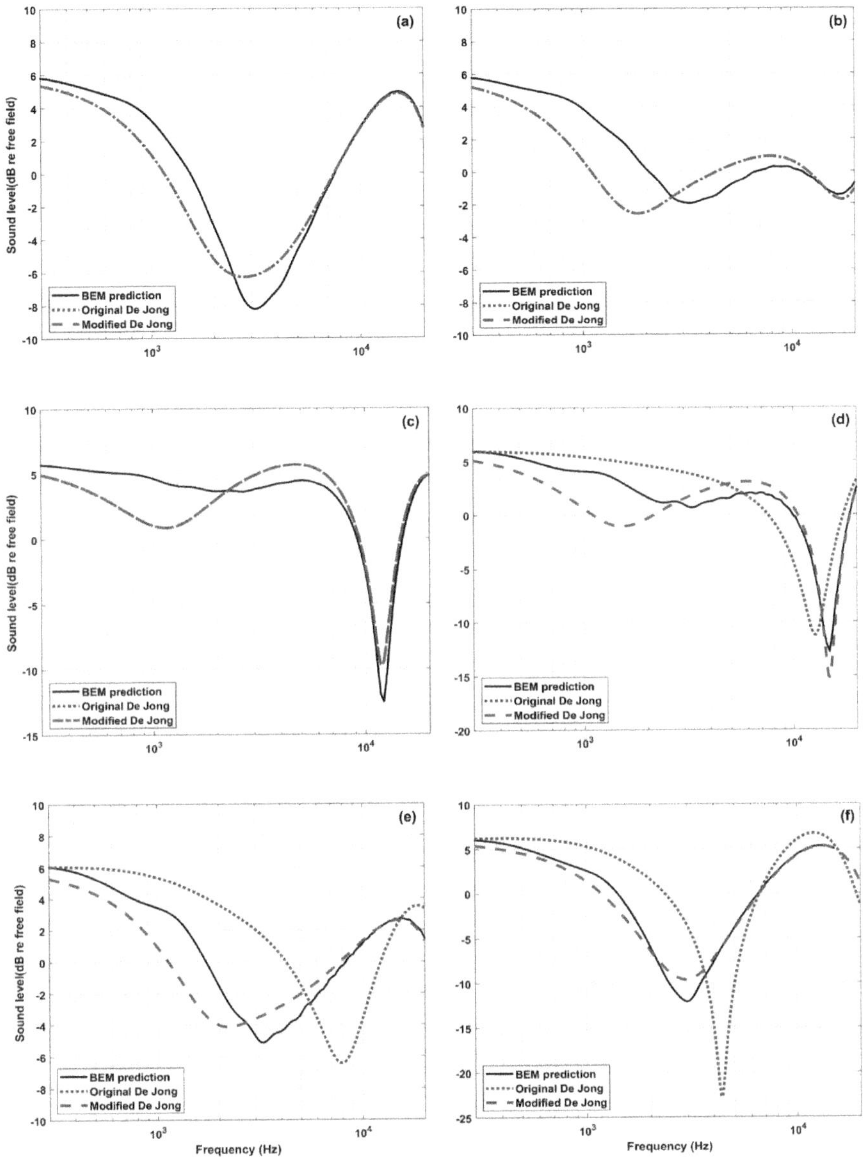

Figure 8.11 Excess attenuation spectra predicted by BEM, the original De Jong model and the modified De Jong model for EA over a ground surface containing a single impedance discontinuity assuming equal source and receiver height of 0.07 m and separation 0.7 m. The curves in Figure 8.11(a)–(c) are predicted for source over 'hard' and receiver over 'soft'. The curves in Figure 8.11 (d)–(f) are for source over 'soft' and receiver over 'hard'. The discontinuity is located at distances of ((a) and (d)) 0.2 m, ((b) and (e)) 0.4 m and ((c) and (f)) 0.6 m from the source. The impedance spectra are modelled using the variable porosity model (see Chapter 5) with σ_e=30 kPa s m^{-2} and α_e= 15 m^{-1} (soft) and σ_e=100 MPa s m^{-2} and α_e=15 m^{-1} (hard) respectively [23].

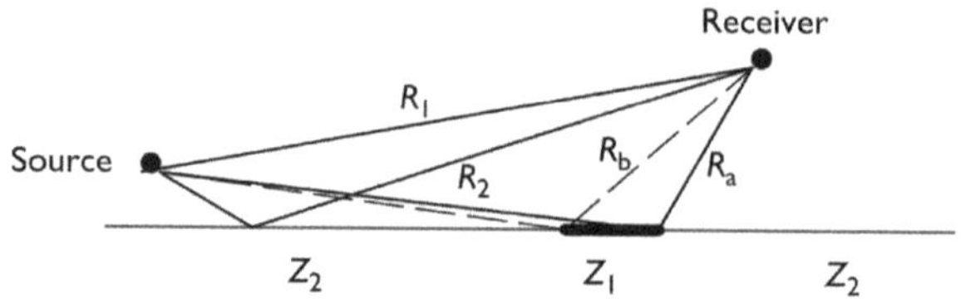

Figure 8.12 Path lengths for propagation over an impedance strip.

$$P(\bar{r},\bar{r}_0) = P_1(\bar{r},\bar{r}_0) - ik\left(\frac{1}{Z_2} - \frac{1}{Z_1}\right)\int_s G_{Z_2}(\bar{r}_S,\bar{r})P(\bar{r}_S,\bar{r}_0)ds(\bar{r}_S), \qquad (8.14)$$

where $P_1(\bar{r},\bar{r}_0)$ is the pressure at the receiver for a surface with homogeneous impedance Z_1, $\bar{r}_S, \bar{r}, \bar{r}_0$ are the position vectors of the source, the receiver and a point in the boundary, respectively; $G_{Z_2}(\bar{r}_S,\bar{r})$ is the Green's function associated with propagation over a boundary of impedance Z_2(see, for example, Equation (8.13a)); and S is the surface of the strip.

The integral can be calculated numerically by a standard boundary element technique once the pressure $P_1(\bar{r},\bar{r}_S)$ is known at point $\bar{r}_S$ in the strip. If the source and receiver are restricted to a vertical plane perpendicular to the impedance discontinuity, then the surface integral is replaced by a line integral which saves computation time.

8.4 COMPARISONS OF PREDICTIONS WITH DATA

8.4.1 Single Impedance Discontinuity

Rasmussen [13] has compared predictions of the De Jong model and the numerical model described in Section 8.2.3 with short-range outdoor data taken under light wind conditions (<2 m/s). The source height was 1 m, the receiver height was 0.1 m and the horizontal separation was 10 m (source over 'hard' and receiver over 'soft'). This corresponds to a grazing angle of about 5.7°. The agreement between predictions and data was good.

While there are few *outdoor* studies of propagation over impedance discontinuities, there have been many comparisons of predictions with laboratory data. In one laboratory study, 'model' discontinuous surfaces have been constructed using varnished hardwood board and 0.3 m deep sand contained within a box measuring 1.1 × 2 × 0.3 m placed on the grid floor of an anechoic chamber [16]. Excess attenuation spectra were measured between a point source at 0.1 m height and a receiver also at 0.1 m height and separated by 1 m for different distances from the source to the discontinuity and with the source over the 'hard' and 'soft' surfaces, respectively. The measurement geometry corresponds to grazing angles of about 11°. Predictions have been made with the De Jong model (Equation (8.2)) using the average of impedance spectra deduced from complex excess attenuation spectra obtained over the continuous sand surface [24] (see Figure 8.16(a)). The hardwood board was assumed to be acoustically hard. Figures 8.13(b) and (c) illustrate the good agreement between predictions and data obtained for the laboratory-scale geometry with hardwood and sand as the two impedances [16].

8.4.2 Impedance Strips

Laboratory measurements have been made also of sound propagation over surfaces formed from alternating rectangular strips of MDF and felt with

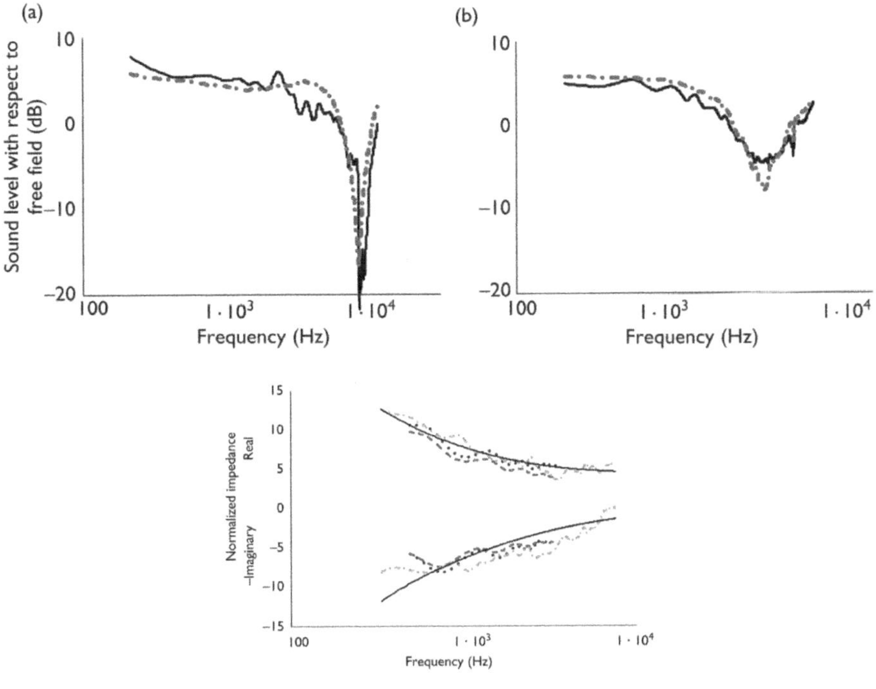

Figure 8.13 Comparison of De Jong *et al* [7] model predictions with laboratory measurements over a single surface discontinuity between hardwood and sand [16]: source and receiver at 0.1 m height separated by 1 m. (a) source over wood, discontinuity at 0.8 m from source (b) source over sand, discontinuity at 0.6 m from source. An average sand impedance spectrum was used in the calculations deduced from excess attenuation data (range 1 m, source height = receiver height = 0.1 m (dotted and dot-dash lines); source height = receiver height = 0.05 m, (broken lines)) [24]. The deduced impedance spectra are shown in Figure 8.13(c) together with best-fit predictions using a semi-infinite slit pore model (see Chapter 5) with (measured) flow resistivity of 420 kPa s m^{-2}, assumed porosity of 0.4 and tortuosity = 1/porosity [23].

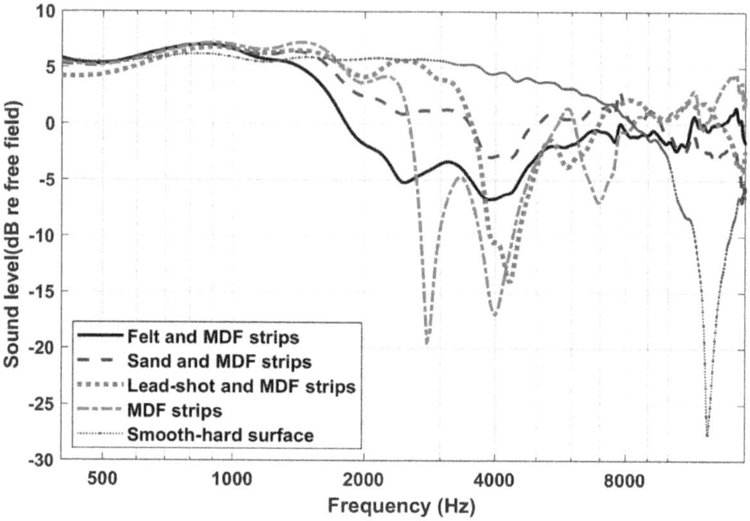

Figure 8.14 Excess attenuation spectra measured with source and receiver height 0.07 m and separation 0.7 m over alternate 'hard' and 'soft' strips (continuous black line felt, broken line sand, dotted line lead shot), smooth hard surface (continuous brown line MDF) and a hard rough surface formed by regularly-spaced MDF strips on an MDF board [23].

equal widths (2.85 cm) and heights (1.2 cm) [23]. The strips were tightly packed to avoid gaps at the impedance discontinuities and the resulting mixed impedance surfaces were plane. Five measurements were carried out over each surface at different source and receiver heights. These measurements were repeated after removing the felt strips and replacing them with other 'soft' surfaces composed either of sand or lead shot. Figure 8.14 compares excess attenuation spectra obtained with source and receiver at a height 0.07 m and separated by 0.7 m over three 'striped' impedance surfaces (MDF strips alternating with felt, sand and lead shot, respectively), a periodic rough surface formed by regularly spaced MDF strips on an MDF board and the MDF board alone. Compared to the maximum EA for the smooth hard surface, the maxima are at lower frequencies in the EA spectra for both mixed impedance and rough surfaces. In the mixed impedance spectra, the EA maxima are broader, and their magnitudes are less than those observed in the EA spectrum obtained over the rough surface. The frequencies and breadth of the EA maxima depend on the impedance of the 'soft' material. The EA maxima obtained over the surface composed of felt and MDF strips are at lower frequency and broader than observed over alternating strips of sand or lead shot and MDF.

Figure 8.15(a) and (b) compare EA spectra measured over a surface composed of felt and MDF strips with Fresnel-zone and BEM predictions, respectively [23]. To make the predictions the surface impedances have been calculated using the variable porosity model (see Chapter 5) with values of

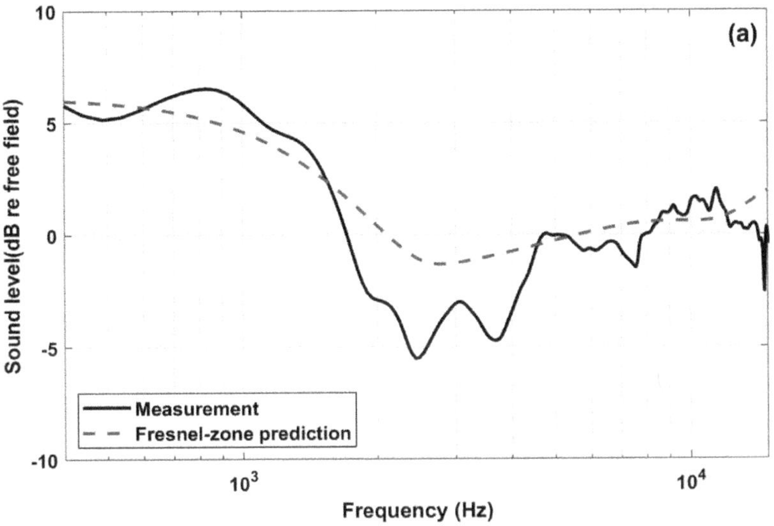

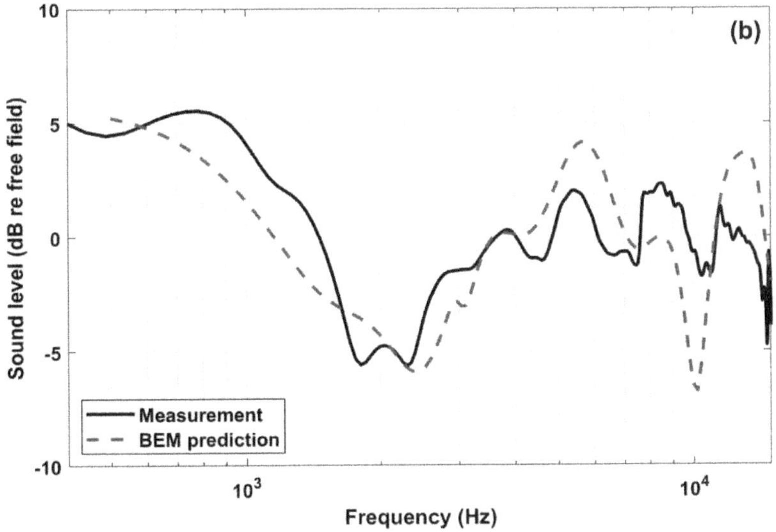

Figure 8.15 Excess attenuation spectra measured with source and receiver separated by 0.7 m above alternate MDF and felt strips at an equal height of (a) 0.07 m and (b) 0.12 m compared respectively with Fresnel zone and BEM predictions.

(Continued)

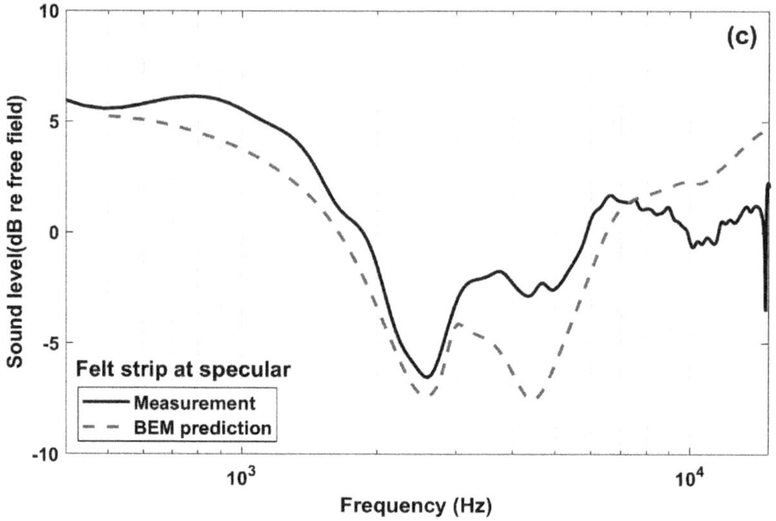

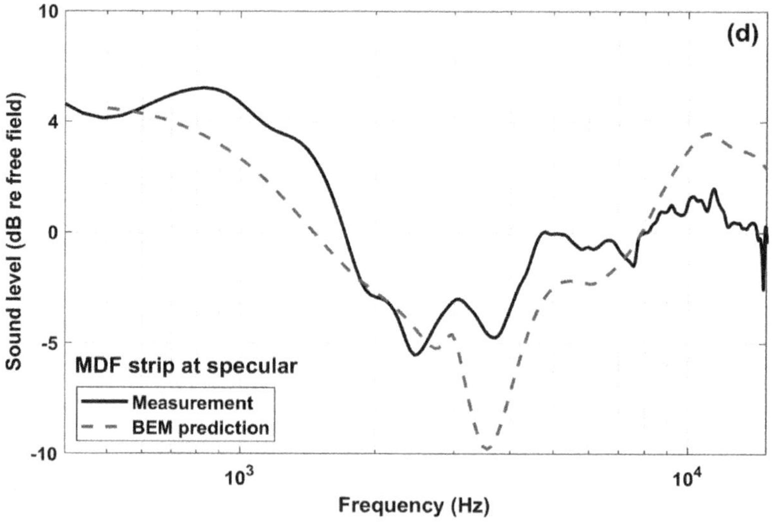

Figure 8.15 (Continued) (c) and (d) show measured spectra with source and receiver at 0.07 m and the specular reflection point on an MDF strip and felt strip respectively compared with BEM predictions [23].

flow resistivity and porosity change rate for felt-on-MDF of 30 kPa s m^{-2} and 15 m^{-1} and, for MDF alone, 100 MPa s m^{-2} and 15 m^{-1}, respectively. Since it does not account for diffraction at the impedance discontinuities, the Fresnel-zone predictions compare less well than BEM with the data. Moreover, the Fresnel-zone method does not distinguish between the presence of different impedance (either an acoustically hard (MDF) strip or a soft (felt) strip) whereas BEM predictions do (see Figure 8.15(c) and (d)).

8.5 REFRACTION ABOVE MIXED IMPEDANCE GROUND

The Parabolic Equation method is a useful technique for incorporating both a refracting medium and range-dependent impedance. For example, Robertson et al. [24] have used a PE routine to investigate low-frequency propagation over an impedance mismatch, and You and Yoon [25] have used a polar PE formulation to investigate the effect of road-grass verge discontinuity on sound field in a refracting medium. Berengier et al. [26] have used the PE method successfully to model propagation involving impedance variation, topography and wind speed gradients. Here, we describe heuristic extensions of De Jong's formulation to allow simultaneously for refraction and range-dependent impedance and a numerical test of their validity by comparing predictions of the modified De Jong formulation with those of an alternative hybrid numerical procedure based on the Boundary Integral Equation and Fast Field Program [27].

Equation (8.2) can be extended to allow for atmospheric refraction using ideas of geometrical optics. In downwind or temperature inversion conditions there may be multiple ray arrivals and rays may have multiple reflections from the ground.

The second term in Equation (8.2) becomes a sum over all rays with $Q_{a,b}$ replaced by

$$Q_{l,m} = Q_a^l \times Q_b^m, \tag{8.15}$$

where l is the number of the bounces the ray makes at the region a and m is the number of bounces in region b. There will be more than one diffracted ray path from the source to the point of discontinuity and then to the receiver. In general, if the numbers of ray paths from source to the discontinuity and from the discontinuity to the receiver are m and l, respectively, then the total number of diffracted ray paths is $m \times l$. Since there is more than one R_3, the term inside the brackets that includes the Fresnel function should also be replaced by a sum over all possible R_3.

Furthermore, the term under the square root in the argument of the Fresnel function can now have negative value, but this is unrealistic. To avoid this, we revert to the original expression by Pierce [9] and introduce a modulus function. The expression for the total field becomes

$$P_t = \frac{e^{ikR_1}}{R_1}\left\{1 + (Q_b - Q_a)\frac{e^{-i\pi/4}}{\sqrt{\pi}} \cdot \sum_{j=1}^{M} \frac{R_1}{R_{d,j}} F\left(\sqrt{k\left|R_{d,j} - R_1\right|}\right)\right\}$$
$$+ \sum_{n=2}^{N} \frac{e^{ikR_n}}{R_n} \left\{ \begin{aligned} & Q_{l,m} + (Q_b - Q_a)\sum_{j=1}^{M} sgn\left[\cos\left(\frac{\mu_{1,j} + \mu_{2,j}}{2}\right)\right] \\ & \frac{e^{-i\pi/4}}{\sqrt{\pi}} \cdot \frac{R_n}{R_{d,j}} F\left(\sqrt{k\left|R_{d,j} - R_n\right|}\right) \end{aligned} \right\} \tag{8.16}$$

In this expression, R_n are sound ray trajectories from source to the receiver with $n = 1$ taken as the direct ray path and M, the total number of possible ray paths, is equal to $m \times l$. The parameters, $\mu_{1,j}$ and $\mu_{2,j}$ are the polar angles of the incident and diffracted wave at the point of discontinuity for any R_d, respectively, and sgn(x) is the sign function. Its value is −1 for a negative x and +1 otherwise. The evaluation of $R_{d,j}$ requires computation of all possible combinations of ray trajectories from the source to the point of discontinuity at the ground plus trajectories from the point of discontinuity to the receiver. This will involve multiple values of R_d in the downwind case. In the presence of a positive gradient, caused by a temperature inversion or downwind, the rays that go through n reflections on the ground can be determined from the following *quartic* equation:

$$n(n+1)x^4 - (2n+1)Rx^3 + \left[b_r^2 + (2n^2-1)b_s^2 + R^2\right]x^2 - (2n-1)b_s^2 Rx + n(n-1)b_s^4 = 0 \quad (8.17)$$

where

$$b_i^2 = \frac{z_i}{a}(2 + az_i) \text{ for } i = r \text{ or } s \quad (8.18)$$

and x is the distance from the source to the first reflection from the ground. Only real roots of the equation are to be considered. This equation is to be solved for all n until no real solutions exist.

A more useful parameter is the angle of incidence of any particular ray. This is given in terms of x by

$$\tan\mu_0 = \frac{2x}{a\left[x^2 + b_s^2\right]}, \quad (8.19)$$

where x is determined from Equation (8.18).

All other ray parameters can easily be determined from this angle.

The horizontal distances from the source, receiver or the point of reflection at the ground to the turning point of the ray path are

$$T_1 = \frac{2n}{a\tan\mu_0} \quad (8.20)$$

$$T_2 = \frac{\cos\mu_s}{a\sin\mu_0} \quad (8.21)$$

$$T_3 = \frac{\cos\mu_r}{a\sin\mu_0}, \quad (8.22)$$

where

$$\cos\mu_i = \sqrt{1-(1+a^2b_i^2)\sin^2\mu_0}\,.\ \mathrm{i = s\ or\ r} \quad (8.23)$$

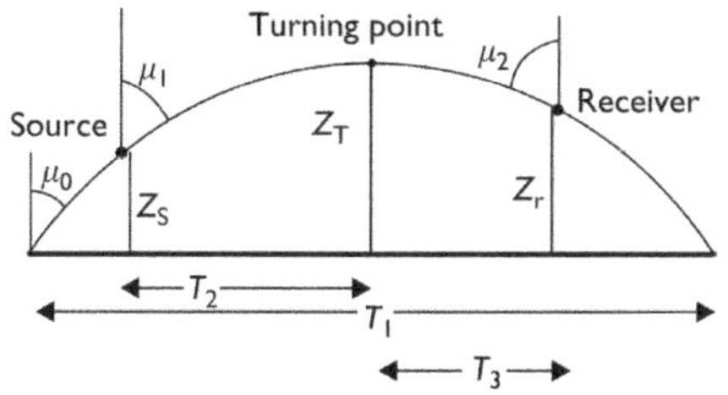

Figure 8.16 Definitions of the terms used to describe a ray path in the De Jong model extended to account for refraction.

The angle μ_i is the polar angle (the angle that the tangent to the ray makes with the positive z-axis) of the ray at the source or receiver position. These distances and angles are shown in Figure 8.16. Once the angle of incidence for a particular ray is known the phase and amplitude of each ray can be evaluated.

There are four cases that can be distinguished.

1. *Range* = $T1 + T2 + T3$

$$R_a = \frac{2n}{a}\ln\left(\frac{1+\cos\mu_0}{\sin\mu_0}\right) + \frac{1}{a}\ln\left(\frac{1+\cos\mu_s}{\sin\mu_s}\right) + \frac{1}{a}\ln\left(\frac{1+\cos\mu_r}{\sin\mu_r}\right) \quad (8.24)$$

$$R_g = \frac{1}{a\sin\mu_0}\left[(n+1)(\pi - 2\mu_0) - (\mu_r + \mu_s - 2\mu_0)\right] \quad (8.25)$$

2. *Range* = $T1 + T2 - T3$

$$R_a = \frac{2n}{a}\ln\left(\frac{1+\cos\mu_0}{\sin\mu_0}\right) + \frac{1}{a}\ln\left(\frac{1+\cos\mu_s}{\sin\mu_s}\right) - \frac{1}{a}\ln\left(\frac{1+\cos\mu_r}{\sin\mu_r}\right) \quad (8.26)$$

$$R_g = \frac{1}{a\sin\mu_0}\left[(n)(\pi - 2\mu_0) + (\mu_r - \mu_s)\right] \quad (8.27)$$

3. *Range* = $T1 - T2 + T3$

$$R_a = \frac{2n}{a}\ln\left(\frac{1+\cos\mu_0}{\sin\mu_0}\right) - \frac{1}{a}\ln\left(\frac{1+\cos\mu_s}{\sin\mu_s}\right) + \frac{1}{a}\ln\left(\frac{1+\cos\mu_r}{\sin\mu_r}\right) \quad (8.28)$$

$$R_g = \frac{1}{a\sin\mu_0}\left[(n)(\pi - 2\mu_0) - (\mu_r - \mu_s)\right] \quad (8.29)$$

4. *Range* = $T1 - T2 - T3$

$$R_a = \frac{2n}{a}\ln\left(\frac{1+\cos\mu_0}{\sin\mu_0}\right) - \frac{1}{a}\ln\left(\frac{1+\cos\mu_s}{\sin\mu_s}\right) - \frac{1}{a}\ln\left(\frac{1+\cos\mu_r}{\sin\mu_r}\right) \quad (8.30)$$

$$R_g = \frac{1}{a\sin\mu_0}\left[(n-1)(\pi - 2\mu_0) + (\mu_r + \mu_s - 2\mu_0)\right] \quad (8.31)$$

The geometrical path length, R_g, of the ray is simply the arc length of the ray path and can be determined by the integral

$$R_g = \int_{\mu_1}^{\mu_2} R_c d\mu = \frac{1}{a \sin \mu_0}(\mu_2 - \mu_1), \tag{8.32}$$

where μ_1 and μ_2 are the polar angles of the beginning and the end points, respectively and R_c is the radius of curvature of the ray. R_g determines the amplitude along the ray. The phase of the ray is determined by its acoustic path length, R_a. This takes into account the change of the speed of sound along the ray path and is given by

$$R_a = c_0 \int_{\mu_1}^{\mu_2} \frac{R_c}{c(z)} d\mu = \int_{\mu_1}^{\mu_2} \frac{1}{a sin \mu_0 \sin \mu} d\mu \tag{8.33}$$

It is important to note that for $n > 0$ more than one option may be valid. In fact, for $n = 1$ there may be one or three rays present, and for each $n > 1$ there may be two or four rays up to a maximum value of n depending on the gradient and the range.

Figure 8.17(a) and (b) compare the predictions of the extended De Jong model (8.16) and a numerical hybrid BIE/FFP code [27] in a downward-refraction condition corresponding to a positive wind speed gradient of 0.25 m/s/m over an impedance discontinuity at frequencies of 200 Hz and 1 kHz. Source and receiver are assumed at 5 m height.

For the predictions at 200 Hz shown in Figure 8.17(a), the discontinuity is assumed to be 200 m from the source, the ground is assumed acoustically hard from the source to the discontinuity and to be absorbing thereafter. In Figure 8.17(b), for a higher frequency (1 kHz), the hard section is assumed

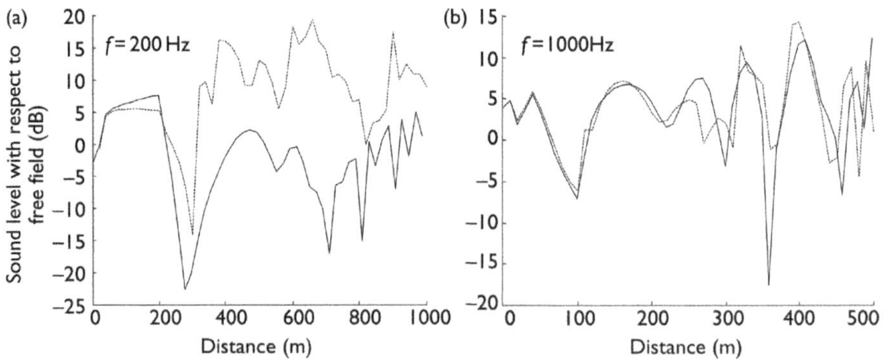

Figure 8.17 (a) Predictions, obtained by using the extended De Jong model (broken line) and the BIE/FFP (solid line) at a frequency of 200 Hz, of the sound field in a downwind condition over an impedance discontinuity at 200 m. The source and receiver heights are 5 m. (b) As for (a) but at a frequency of 1000 Hz and with the impedance jump at only 100 m from the source.

to extend only up to 100 m from the source. While the two models agree out to 500 m range at 1 kHz, the predictions of the extended De Jong model are not as good at 200 Hz.

8.6 PREDICTING EFFECTS OF GROUND TREATMENTS NEAR SURFACE TRANSPORT

8.6.1 Roads

8.6.1.1 Sound Propagation from a Road over Discontinuous Impedance

Consider the situation illustrated in Figure 8.18(a) in which sound propagates from a two-lane urban road to receivers located over soft ground. Vehicles are represented by three source heights of 0.01, 0.3 and 0.75 m on each lane. The lowest source height represents the tyre/road contact source. The 0.3 m high source corresponds to a light vehicle engine and the 0.75 m high source represents a heavy vehicle engine. The centres of the two lanes are 3.5 m apart. The traffic on the road is assumed to consist of 95% light and 5% heavy vehicles moving at an average speed of 50 km/h. The power spectrum at each source height is calculated and then these spectra are added incoherently [23]. The traffic flow is assumed to be 833 vehicles per hour, i.e. 20,000 vehicles per 24-hour day. The receiver is assumed to be at 50 m from

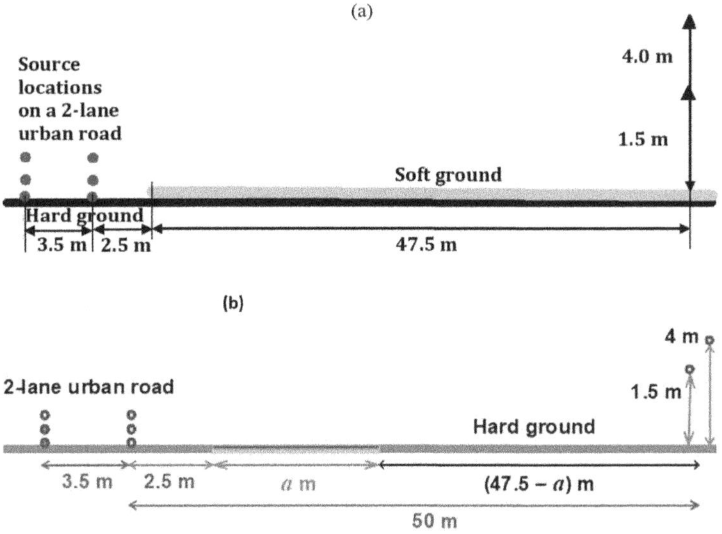

Figure 8.18 Schematics of assumed vehicle source positions on a two-lane urban road and two adjacent ground conditions (a) continuous acoustically soft ground and (b) a single wide acoustically soft strip in otherwise acoustically hard ground [23].

the centre of the nearest lane and at a height of either 1.5 or 4.0 m. There is a single hard/soft impedance discontinuity between the source and the receiver. The soft ground starts at 5.0 m from the nearest source and extends up to the receiver.

Figure 8.19 compares the EA spectra predicted using the De Jong method and BEM for the single impedance discontinuity case (see Figure 8.18(a)). The soft ground impedance is assumed to be described by a two-parameter model (see Chapter 5) with values of flow resistivity and porosity of 104 kPa s m^{-2} and 0.36, respectively. The De Jong and BEM predictions are in reasonable agreement.

The cases shown in Figure 8.18(b) and (c) introduce two or more impedance discontinuities. Figure 8.20 compares BEM and modified De Jong model (nMID) predictions of EA spectra at 50 m from the source in the presence of a single 10 m wide strip starting 2.5 m from the nearest source. The strip impedance is assumed to be that given by a semi-infinite two-parameter

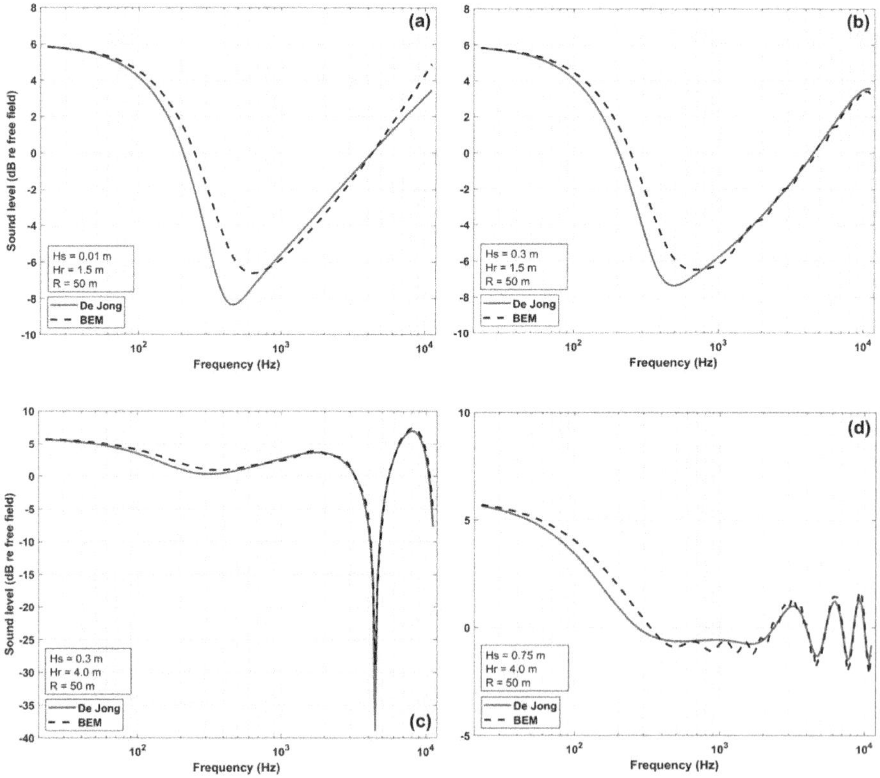

Figure 8.19 Predictions of the de Jong model and BEM for excess attenuation spectra in the presence of a single hard-soft impedance discontinuity (Figure 8.18(a)) for various source and receiver heights and separations [23].

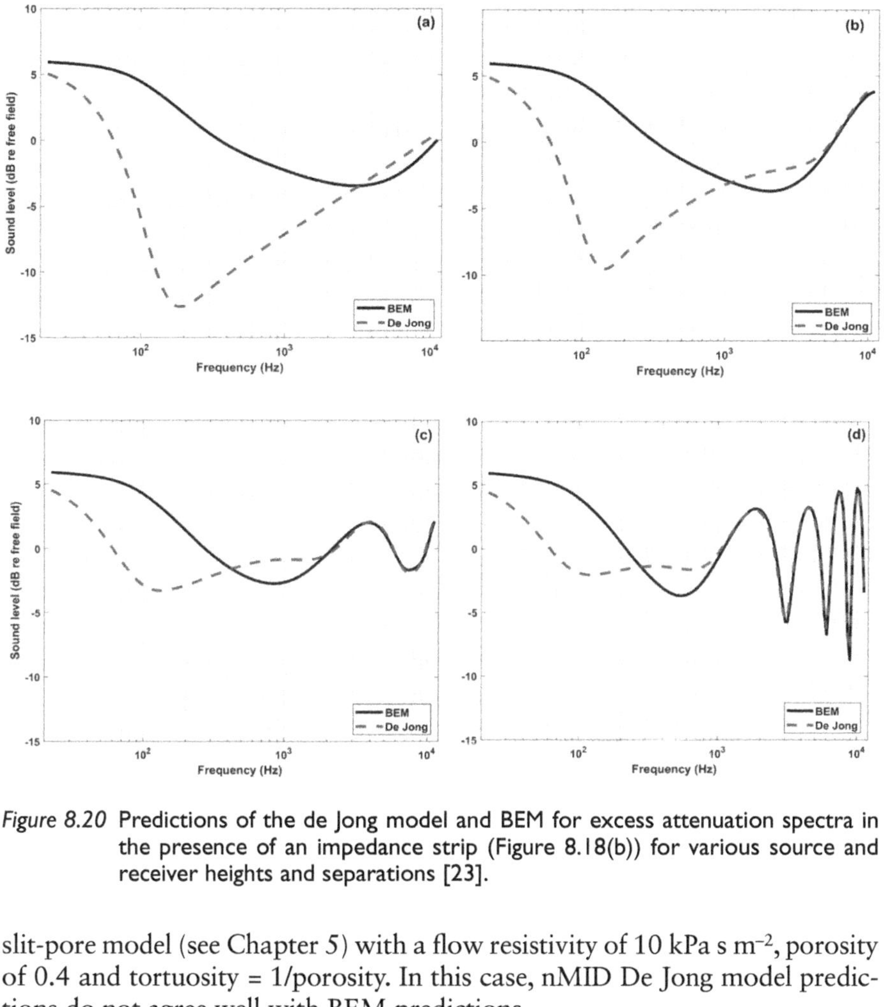

Figure 8.20 Predictions of the de Jong model and BEM for excess attenuation spectra in the presence of an impedance strip (Figure 8.18(b)) for various source and receiver heights and separations [23].

slit-pore model (see Chapter 5) with a flow resistivity of 10 kPa s m^{-2}, porosity of 0.4 and tortuosity = 1/porosity. In this case, nMID De Jong model predictions do not agree well with BEM predictions.

Consequently, while the De Jong model gives sufficiently accurate predictions for situations involving a single hard/soft impedance discontinuity and is used in these types of calculations reported in the next section, the BEM should be used in situations involving more than one discontinuity.

8.6.1.2 Predicted Effects of Replacing 'Hard' by 'Soft' Ground Near a Road

Table 8.1 shows predictions of the insertion loss resulting from replacing acoustically hard ground by acoustically soft ground, represented by the two-parameter slit-pore impedance model (either 'grass' with flow resistivity 150 kPa s m^{-2} and porosity 0.4 or 'gravel' with flow resistivity 10 kPa s m^{-2} and porosity 0.5), near a two-lane urban road obtained by using the De

Table 8.1 Insertion losses associated with replacing hard ground by soft ground near a two-lane urban road predicted for the configurations shown in Figure 8.18 and others.

	Reductions (dB) compared with smooth hard ground					
	Hr = 1.5 m			*Hr = 4 m*		
Surface description	*Lane 1*	*Lane 2*	*Combined lanes*	*Lane 1*	*Lane 2*	*Combined lanes*
Continuous 'grass'	8.9	8.3	8.6	4.0	2.5	3.2
25 m wide strip of 'grass'	8.1	7.5	7.8	4.0	2.6	3.3
25 m wide strip of 'gravel'	9.5	8.6	9.1	3.8	2.3	3.0
9 m wide strip of 'grass'	5.1	4.6	4.9	3.9	2.7	3.3
15 m wide strip of gravel	8.2	7.3	7.7	4.0	2.5	3.2
25 × 1 m wide alternating strips of 'grass' and 'hard'	6.1	5.6	5.8	3.5	2.3	2.9
25 × 1 m wide alternating strips of 'gravel' and 'hard'	8.0	7.1	7.5	3.6	2.2	2.8

Jong model and assuming the configurations shown in Figure 8.18 and another involving multiple impedance strips. According to these results, replacing a 45 m wide strip of acoustically hard ground by any acoustically soft ground will decrease levels by at least 5 dB at a 1.5 m high receiver and by between 1 and 3.5 dB at a 4 m high receiver. These calculations have assumed that the receiver is located over soft ground, but additional calculations show that if the receiver is over hard ground and within 5 m of the nearest soft ground, then there should be similar noise reductions. Note that the predicted insertion losses from replacing only 25 m of hard ground next to a road with low flow resistivity 'grass' are comparable to those that result from replacing a greater width (47.5 m) by 'grass'.

The flow resistivity chosen for 'grass' in these calculations is relatively low. Figure 8.21 shows the significant differences in the predicted A-weighted sound level spectra at a 1.5 m high receiver 50 m from a two-lane urban road, implying 2.6 dB difference in predicted insertion loss that result from using the potentially wide range of grassland flow resistivity (see Chapter 6) [23]. The flow resistivity of grassland will be relatively low if it is not regularly maintained by mowing and rolling. Since 'gravel' it has an even lower flow resistivity than the lowest 'grass' flow resistivity, strip configurations including 'gravel' are predicted to result in larger insertion losses than the same configurations with 'grass'.

The results of calculations for a wider set of configurations including the additional insertion loss due to crops are given elsewhere [23,28–30].

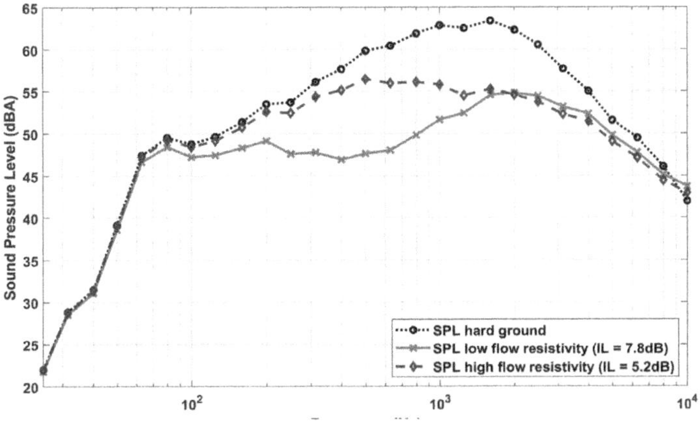

Figure 8.21 Predictions of A-weighted sound pressure level spectra at a 1.5 m high receiver 50 m from the nearest lane of a two-lane urban road carrying 95% cars and 5% HGV at a mean speed of 50 km/h (see Figure 8.18(a)). The dotted line represents the prediction over hard ground. The continuous line is the prediction for a low flow resistivity 'grass' (R_s= 104.0 kPa s m^{-2}, Ω = 0.36) and the dashed line is for a high flow resistivity 'grass' (R_s= 664.0 kPa s m^{-2}, Ω = 0.22) [23].

8.6.1.3 Predicting Effects of Low Parallel Walls and Lattices

As discussed in Chapter 5, Section 5.4 and Chapter 6, Section 6.7, if a surface is artificially or naturally rough, it does not reflect sound perfectly since the roughness causes scattering. The distribution of the scattered sound depends on the roughness topology, the ratio of the roughness dimensions to the incident wavelength and the relative locations of source and receiver. If enough of the reflected sound retains a phase relationship with the incident sound, i.e. there is significant coherent scattering and specular reflection, there can be a significant change in ground effect.

One method of deliberately introducing roughness outdoors is to construct an array of low parallel walls. Low walls of heights 0.3 m (approximately the wavelength of sound in air at 1 kHz) or less can be considered as a form of artificial ground roughness. The potential usefulness of such an arrangement for road traffic noise reduction was demonstrated by outdoor measurements over sixteen 0.21-m high parallel walls with 0.2 m edge-to-edge spacing in 1982 [31] and it was suggested that the creation and subsequent attenuation of surface waves was the main mechanism of noise reduction. Although surface wave creation is one of the consequences of placing a low parallel wall array on an acoustically hard ground, the array has a significant influence on ground effect over a wider range of frequencies than those affected directly by the surface wave. While, quarter wave resonance in a 0.3 m high array, at about 280 Hz, is unlikely to be important in A-weighted traffic noise reduction since it occurs at relatively low frequency, diffraction-grating effects and the interference between direct and multiply reflected paths between adjacent walls contribute to diffraction-assisted ground effect [32].

In a series of measurements deploying low parallel brick walls and brick lattices on car parks [33], a two-brick high rectangular lattice was found to offer a similar insertion loss to regularly spaced parallel wall arrays of the same height but twice the total width. Part of the insertion loss due to such configurations is the result of transfer of incident sound energy to surface waves which without treatment would make undesirable contributions above pressure doubling at low frequency. But they can be reduced by introducing wall absorption or material absorption in the form, for example, of gravel between the walls or in the lattice cells [23,28].

Finite length effects have been predicted using a PseudoSpectral Time-Domain Method (PSTD), which models the complete 3D roughness profile. It is concluded from measurements and predictions that since a lattice design has less dependence on azimuthal source–receiver angle than parallel wall configurations, it is more suitable for reducing noise from moving vehicles [33].

The acoustical effects of infinitely long parallel wall arrays can be predicted using a 2D BEM after discretizing the complete wall array profile [23,33]. However, rather than having to carry out a full discretization, it is possible to predict the acoustical performance of low parallel wall arrays and lattices using a 2D BEM with the walls or lattices represented by raised surfaces having impedance spectra predicted by a two-parameter slit-pore layer model [23,28].

Table 8.2 lists the predicted reduction in noise from a two-lane urban road due to low walls and lattices for the type of configuration shown Figure 8.22. A 12 m wide 0.3 m high lattice is predicted to give up to 10 dB reduction at a 1.5 m high receiver.

Table 8.2 Insertion losses compared with smooth hard ground predicted for various parallel wall and lattice constructions (see Figure 8.22) for a receiver either 50 m or 100 m from the nearside road edge and at a height of either 1.5 m or 4 m. The constructions are assumed to start 2.5 m from the nearest traffic lane on a two-lane urban road (95% cars, 5% lorries travelling at 50 km/h) [33]

Receiver distance from road edge m	*50*		*100*	
height m	*1.5*	*4*	*1.5*	*4*
Ground treatment	*Insertion loss dB*			
1.65 m wide array of 9 parallel walls 0.3 m high, 0.05 m thick, 0.2 m centre-to-centre spacing	5.8	5.4	5.2	5.7
3.05 m wide array of 16 parallel walls, 0.05 m thick, 0.2 m centre-to-centre spacing	6.6	5.6	6.1	6.5
12.05 m wide array of 61 Parallel walls 0.3 m high, 0.05 m thick, 0.2 m centre-to-centre spacing	8.6	5.1	8.5	8.2
1.53 m wide 0.3 m high 0.2 m square cell lattice	5.9	5.6	5.3	5.8
3.05 m wide 0.3 m high 0.2 m square cell lattice	7.2	6.1	6.5	7.0
12.05 m wide 0.3 m high 0.2 m square cell lattice	10.5	6.1	10.0	9.6

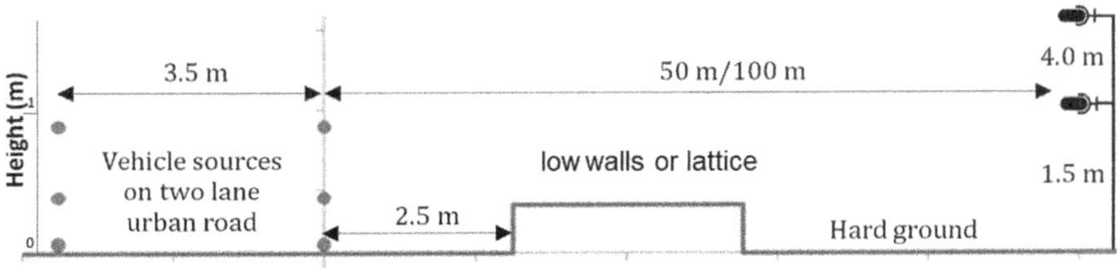

Figure 8.22 Two-lane urban road sources and receiver locations assumed for the predictions in Table 8.2 [23].

8.6.2 Tramways

Measurement made near tram tracks in Grenoble [29] have shown that replacing acoustically hard surfaces alongside the tracks by grass reduces tram noise levels at a 1.5 m high receiver located about 4 m from the nearest track by between 1 and 10 dBA with an average of about 3 dBA. BEM calculations made by assuming that vertical arrays of point sources at heights of 0.05, 0.3 and 0.5 m represent noise sources on each wheel, a standard tram noise spectrum and 'grass' type impedance spectra suggest reductions of between 1 and 6 dBA at a 1.5 m high receiver, 4 m from the nearest track when trams are on the further track and of between 0.5 and 4.5 dBA when trams are on the nearer track [23].

Predictions can be made of the acoustical effects of other types of 'soft' ground around tram tracks [23]. Figure 8.23(a) shows the assumed A-weighted source spectrum. Consider the situation shown schematically in Figure 8.23(b) in which each tram is represented by two 'wheel' sources and 'hard' ground within and between tram tracks is replaced by a gravel. The separation between two tram-wheel sources is 1.45 m and that between the tramways is 1.6 m. The receivers are assumed to be at heights of 1.5 and 4.0 m and distances of 52.5 and 102.5 m from the nearest track. The gravel impedance is calculated using the two-parameter slit-pore model (see Chapter 5) with a flow resistivity of 10 kPa s m^{-2} and porosity of 0.4. Predicted tramway noise reductions are listed in Table 8.3. As the noise emission is very close to the tram track, i.e. at a height of 0.05 m, replacing hard ground by gravel is predicted to be very effective for reducing the noise levels and insertion losses of 5 dB and 10 dB are predicted for tracks 1 and 2, respectively, with different receiver positions and heights.

BEM predictions have been made for the effects of introducing various other ground treatments near tram tracks, including parallel wall and lattice arrays (see Figure 8.23(b)) and example results are listed in Table 8.3 also. A 0.3 m high, 5.85 m wide lattice is predicted to give 10 dB insertion loss at 1.5 or 4 m high receivers located 52.5 m from the nearest tram track.

8.6.3 Railways

8.6.3.1 Porous Sleepers and Porous Slab Track

A likely consequence of noise mapping near major railways, in accordance with the European Directive on Environmental Noise (Commission of the European Communities, 2002), is that additional noise mitigation will be required. Methods for controlling railway noise that have been explored include barriers, cuttings and tunnels. Less attention has been paid to options relating to track design. These options include optimizing the characteristics and deployment of railway ballast and the use of porous materials in sleepers and slab track. Such options will be particularly important in respect of wheel/rail noise.

A boundary element code (see Section 8.3.5) has been used to predict the likely effects on wheel/rail noise of introducing porous sleepers and porous

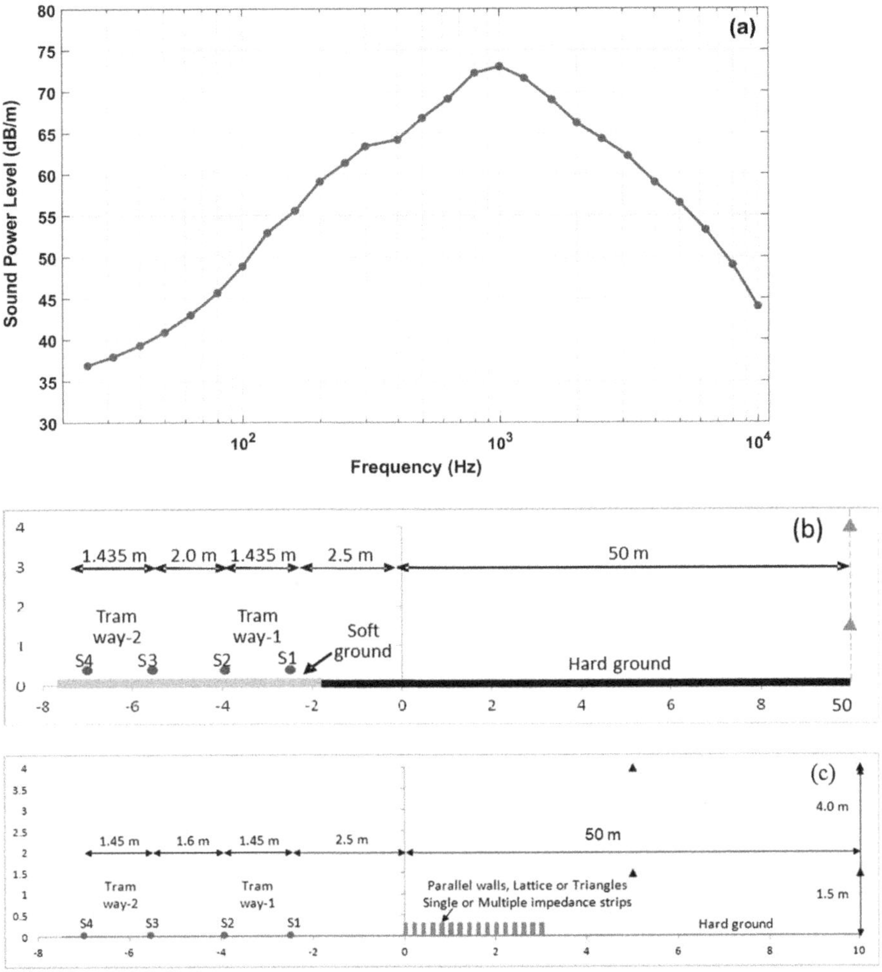

Figure 8.23 (a) Assumed A-weighted tramway source power spectrum, (b) and (c) ground treatment configurations for predictions in Table 8.3 [23].

Table 8.3 Predicted insertion losses at a receiver 52.5 m from the nearest tram track due to various treatments between and alongside tram tracks.

receiver height m	*1.5*	*4*
Ground treatment	*Insertion loss dB*	
replacing 'hard' ground by gravel	5.0	5.1
5.85 m wide and 0.3 m wide array of 30×0.05 m thick parallel walls centre-to-centre spacing 0.2 m	9.0	8.9
5.85 m wide and 0.3 m high square lattice	10.2	10.2
25.0 m wide array of 1.0 m wide alternating hard and gravel strips	8.9	3.9

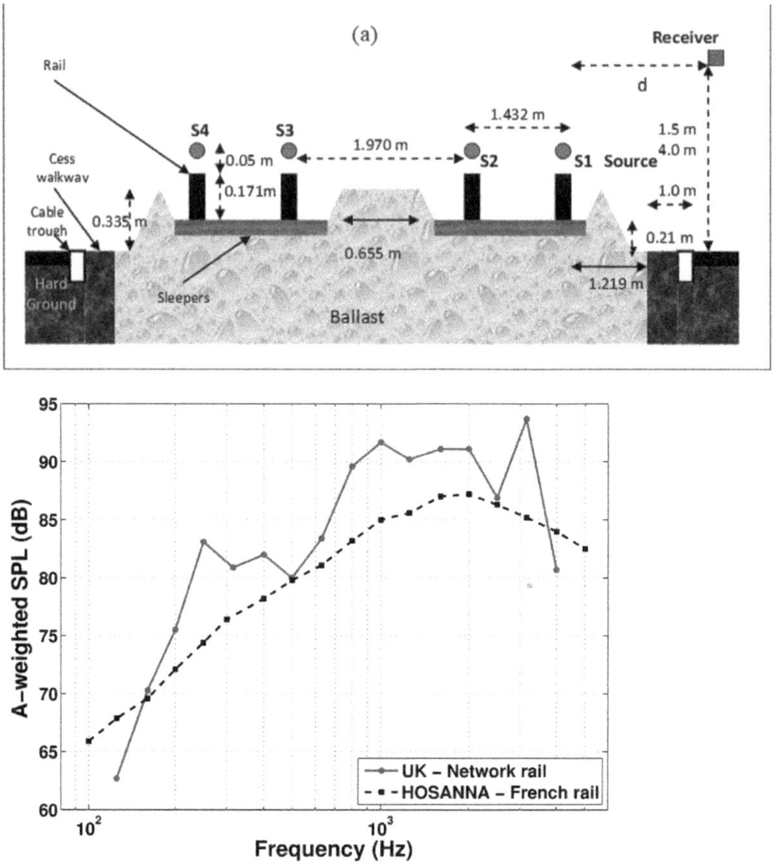

Figure 8.24 Assumed railway track profile and source spectra [23].

slab track [23,34]. Figure 8.24(a) details the assumed UK track profile. As well as two tracks, there is a Cess walkway (an area either side of the railway immediately off the ballast shoulder on which workers stand when trains approach) and a cable trough. The walkway and cable trough are assumed to be acoustically hard. The edges on either side of the railway track are assumed to be flat and grass covered. In this UK track profile, ballast is heaped to a height of 0.335 m above the surrounding ground plane on both sides of the track and in the centre. Since the track consists of areas of ballast and sleepers it offers a mixed impedance situation in respect of ground effect.

The rail/wheel excitation is modelled by a coherent line of monopole sources in the vertical plane perpendicular to the track. Two monopole sources are assumed on each track, i.e. four sources altogether, located 0.05 m above the top of the 0.171 m high rails. A separate simulation is made for each

Table 8.4 Parameter values used for calculating the acoustical properties of railway ballast, porous concrete and grass [23].

Material	*Flow resistivity (kPa s m⁻²)*	*Porosity*	*Tortuosity*	*Viscous Characteristic length (m)*
Railway Ballast	0.2	0.491	1.3	0.01
Porous concrete	3.619	0.3	1.8	2.2×10^{-4}
Grass	125	0.5	1.85	0.001

source position and the resulting predicted noise levels are added logarithmically in pairs assuming that the sources are incoherent. The two railway tracks are treated independently to calculate the insertion loss because at any given time only one train might be passing by. Two trains are only present together only for a short period of time when they are crossing over.

The acoustical properties of the ballast (assumed 30 cm deep), porous concrete and grass have been calculated using the Johnson-Allard-Umnova model [35] (also see Chapter 5) and the parameter values listed in Table 8.4. The parameter values for grass are consistent with those used by Morgan et al. [36].

The receiver height is assumed to be 1.5 m above a grass surface and the horizontal source/receiver ranges are denoted by d_i, where $i = 1$ to 4 corresponding to each of the four rails. The assumed values are 28.3 m, 26.9 m, 24.9 m and 23.5 m, respectively.

A 2D BEM has been used to calculate the excess attenuation at a 1.5 or 4.0 m high receivers at 50 m from the nearest track in the reference railway track profile (see Figure 8.24) assuming only an acoustically hard surface at the side of the track. Another excess attenuation calculation has been made assuming the various abatement configurations shown in Figure 8.25. The predicted excess attenuation spectra have been combined with the source spectra (Figure 8.24(b)) to predict sound pressure levels (SPL) and corresponding insertion losses have been calculated by subtracting the predicted SPL with these abatement configurations from that predicted with an acoustically hard surface only. The predicted insertion losses are listed in Table 8.5. Although reference source spectra are shown for UK and French trains in Figure 8.24(b), since both are found to lead to similar predictions only results assuming the UK train spectrum are shown. The predictions assume the ballast to be locally reacting despite its very low flow resistivity. Comparisons with predictions that allow for extended reaction [37,38] show differences only below 400 Hz where the assumed A-weighted source spectrum has little energy [23]. For convenience, therefore, only predictions assuming local reaction are listed.

Figure 8.26 shows the track profile assumed for predicting the effect of using porous concrete instead of sealed concrete for slab track.

BEM-predicted A-weighted SPL spectra in vertical planes through porous concrete and hard slab are shown in Figure 8.27 for both railway

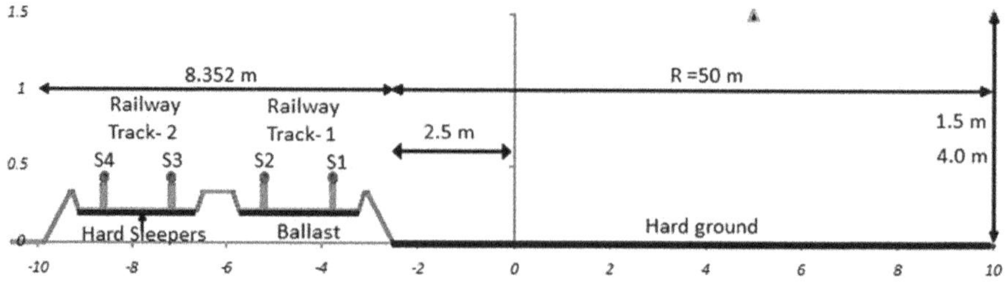

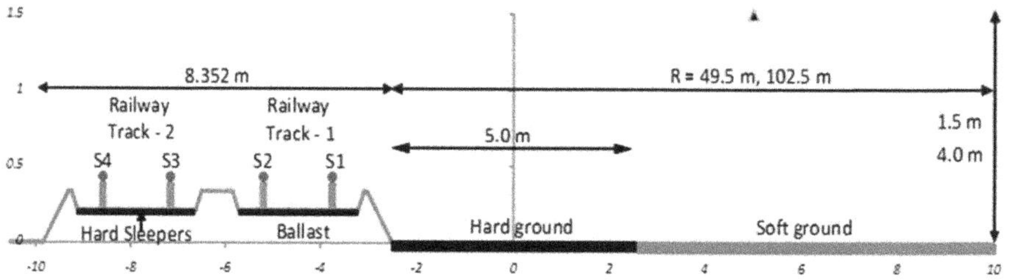

(c)

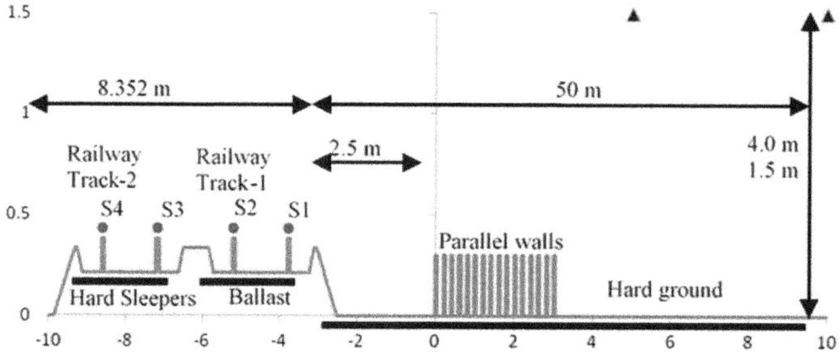

(d)

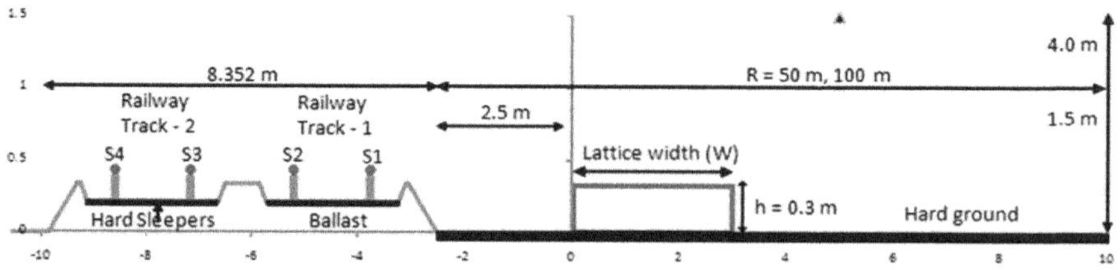

Figure 8.25 Geometries of sources, receivers and intervening ground assumed for BEM predictions of insertion losses from the railway track sources (Figure 8.24) (a) reference hard ground (b) soft ground (c) low parallel walls and (d) lattice [23].

Table 8.5 Predicted insertion loss compared with hard ground (Figure 8.25(a)) at 1.5 m or 4.0 m high receivers 50 m from nearest railway track due to continuous grass (Figure 8.25(b)), low parallel walls (Figure 8.25(c)) and a lattice (Figure 8.25(d)) [23].

			Insertion Loss (dB) (Frequency range: 25–10 kHz)					
Configuration	*Range (m)*	*Receiver height (m)*	*Source-1*	*Source-2*	*Combined 1&2*	*Source-3*	*Source-4*	*Combined 3&4*
Grass (flow resistivity 176 kPa s m^{-2}, porosity 0.5) (Figure 8.25(b))	49.5	1.5	7.2	7.2	7.2	5.0	5.4	5.2
		4.0	1.1	1.1	1.1	0.4	0.3	0.4
5.85 m wide array of 30 × 0.05 m thick 0.3 m high walls (Figure 8.25(c))	50	1.5	8.9	7.2	8.2	5.4	5.0	5.2
		4.0	1.8	2.1	1.9	0.4	0.6	0.5
5.85 m wide, 0.3 m high lattice (Figure 8.25(d))	50	1.5	8.5	7.0	7.9	5.4	5.1	5.3
		4.0	2.1	2.4	2.2	0.5	0.7	0.6

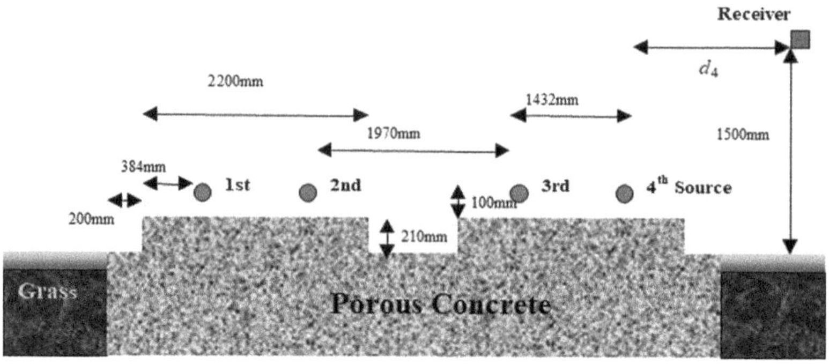

Figure 8.26 Idealised 'slab track' railway profile and source/receiver geometry used for BEM calculations of the effect of porous slab track [23].

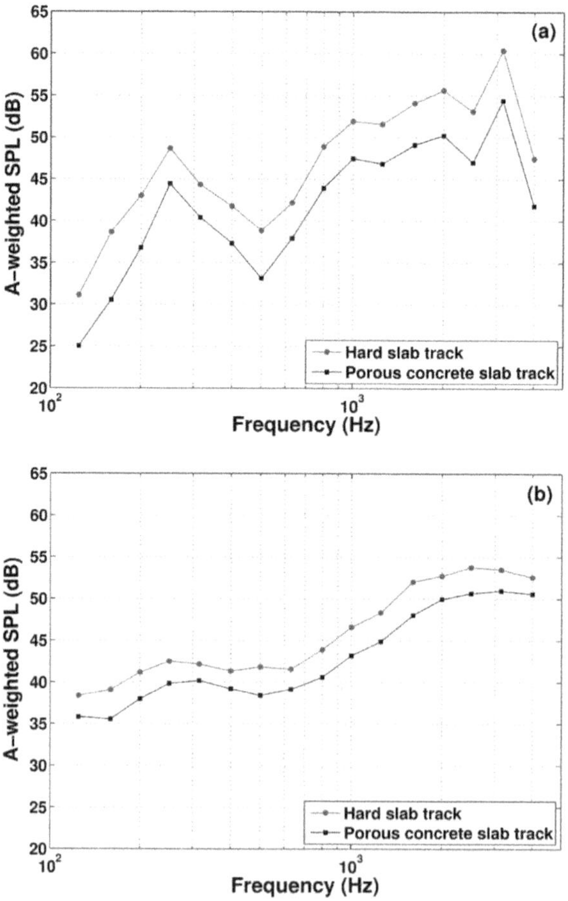

Figure 8.27 BEM-predicted A-weighted SPL spectra over continuous slab tracks with over porous and non-porous slab tracks (a) for the UK railway noise spectrum and (b) for the French railway noise spectrum (see Figure 8.24) [23].

source spectra (Figure 8.24(a) and (b)). The porous concrete properties assumed are listed in Table 8.4. The predicted effects of replacing acoustically hard slab track by porous concrete slab are to reduce the A-weighted sound level above 200 Hz, and by nearly 8 dB at 2 kHz. Predicted effects of other ground and track treatments, including the replacing of acoustically hard sleepers by porous concrete sleepers, are to be found elsewhere [23,34].

8.7 PREDICTING METEOROLOGICAL EFFECTS ON THE INSERTION LOSS OF LOW PARALLEL WALLS

8.7.1 Configuration and Geometry

The sound pressure level reductions that can be achieved with an array of low parallel walls (LPW) between a sound source and a receiver (see Section 8.6.1.3) will be influenced by atmospheric effects, particularly at higher frequencies. For example, turbulence will limit the depth of destructive interferences. Downward refraction of sound will lead to multiple sound paths interacting at a single receiver.

This section investigates the effects of meteorology on the acoustical performance of the raised parallel wall structure and the source–receiver geometry depicted in Figure 8.28(a). The source located at (0, 0.01) m (in a two-dimensional coordinate system (x, z)) is representative of the rolling tyre/road noise source. A receiver is located at 50 m from the source at 4 m height. All surfaces, including those of the LPW, are assumed to be acoustically rigid. The LPW (Figure 8.28(b)) consists of 24 × 0.2 m high, 0.065 m thick walls, with a regular centre-to-centre spacing of 0.26 m. The edge of the nearest wall in the array facing the source is 2.5 m from the source. The furthest face of the last wall is 8.545 m from the source.

8.7.2 Numerical Methods

Calculations have been performed with the Boundary Element method (BEM), the finite-difference time-domain (FDTD) method and the PSTD method. FDTD and PSTD enable investigation of the effect of wind flow on sound propagation. Furthermore, the FDTD model has been used to investigate the influence of a turbulent atmosphere by applying the turbule approach (see Section 10.7.12). If the wind direction is assumed to be downwind along the x-axis, i.e. orthogonal to the low parallel walls, full three-dimensional predictions of the insertion loss of the LPW structure agree very well with those resulting from two-dimensional simulations [39]. Similarly, results of using the effective sound speed approach agree well with those resulting from explicitly solving those terms in the linearized Eulerian equations where the wind velocity appears [39].

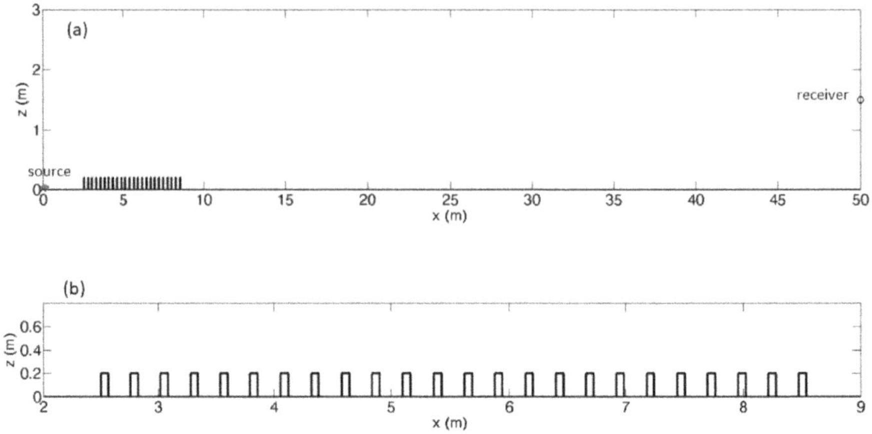

Figure 8.28 (a) A schematic of a low parallel wall structure (LPW) with assumed source height of 0.01 m at the origin and receiver height of 1.5 m 50 m from the source; (b) details of the LPW [39].

8.7.3 Meteorological Effects

A logarithmic wind speed profile, $u_z = \frac{u_*}{\kappa}\ln\left(\frac{z}{z_0}\right)$, representative of a neutral atmosphere is assumed, where u_z is the wind speed at height z, u^* is the friction velocity, κ is the von Kármán constant (= 0.4) and z_0 is the roughness length (assumed to be 0.01 m). The wind speed is directed parallel to the surface and orthogonal to the LPW. Friction velocities of u^* = 0.4 m/s (moderate wind) and 0.8 m/s (strong wind) are assumed. To a much less extent than a conventional noise barrier (see Chapter 9), LPW increase the ground roughness compared with grassland but slight increases in z_0 to account for this effect do not change the sound propagation results [39].

FDTD simulations, including turbulence but in a non-moving atmosphere, have been performed by assuming weak (represented by a temperature structure factor C_T^2 = 0.05 K^2/m$^{2/3}$) or strong (C_T^2= 2 K^2/m$^{2/3}$) temperature-related atmospheric turbulence, following a Kolmogorov spectral density function (see Chapter 10). New realizations were added until the energetically averaged insertion losses converged. The results of no-turbulence predictions without wind and with two wind conditions are shown in Figure 8.29.

The predicted spectra of insertion loss due to LPW have a pronounced frequency-dependent behaviour. At frequencies, sufficiently low that their wavelengths are much larger that the wall dimensions, the predicted sound propagation is similar to that over smooth flat rigid ground. Between 100 Hz and 200 Hz, the LPW excites a surface wave (see Chapter 6, Section 6.7.3) that results in negative insertion loss, i.e. an increase in sound pressure level over and above that due to the constructive interference (> +6 dB relative to

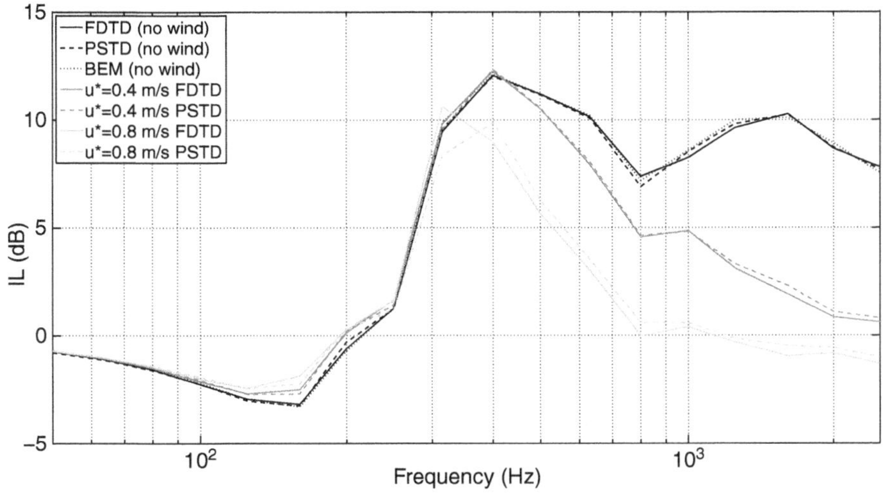

Figure 8.29 2D numerical predictions of insertion loss spectra due to an LPW (see Figure 8.28) without wind and with moderate and strong winds from source to receiver but in the absence of turbulence [39].

free field sound propagation) above the reference situation of a flat acoustically hard surface. Note, however, that it is straightforward to mitigate such surface waves by adding absorption to walls or between them, or by small variations in their height [23,33]. At higher sound frequencies, and in the absence of wind, the LPW is predicted to result in significant insertion losses, exceeding 10 dB in the 315 Hz octave band.

While the predicted effect of wind refraction is negligible at low frequencies, including the surface wave zone, with increasing wind speed, compared with those predicted in a still atmosphere, the insertion losses above 300–400 Hz are reduced. In strong wind, no insertion loss is predicted above about 1 kHz, although the steep increase in IL that is predicted between 200 and 300 Hz is not affected.

In the absence of wind, the agreement between predictions of the full-wave techniques BEM, FDTD and PSTD is excellent. Also, although small differences are inevitable since frequency domain techniques are opposed to time-domain approaches and very different spatial and temporal resolutions have been used, when including wind-induced refraction, predictions between FDTD and PSTD are very much alike.

The predicted effects of temperature-related turbulence on the LPW insertion loss are shown in Figure 8.30. The effects of turbulent scattering are very small at low sound frequencies but become more pronounced at higher frequencies. In a weakly turbulent atmosphere, the average insertion loss is predicted to be very similar to that predicted for no turbulence. In strong turbulence, on average, a loss in insertion loss above about 400 Hz is predicted. On the other hand, as illustrated by the extent of the error bars on

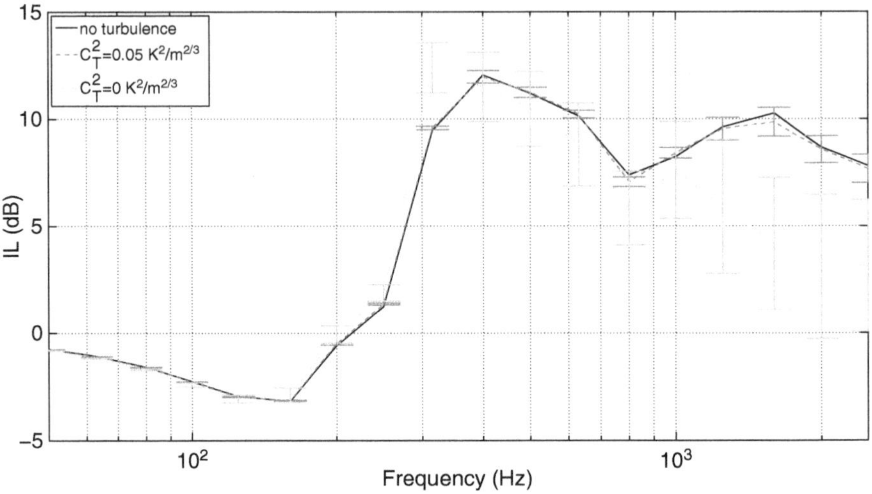

Figure 8.30 2D FDTD (no wind) predictions of insertion loss spectra above an LPW (see Figure 8.28) without turbulence, and with weak (15 realizations) and strong (40 realizations) temperature related turbulence. The error bars have a total length of two times the standard deviations based on the individual realizations [39].

the predictions, there may be sufficiently large fluctuations in insertion loss, that, at some frequencies in strong turbulence (e.g. in the 315 Hz 1/3 octave band), the insertion loss might be higher than predicted without turbulence, even after the time-averaging.

In conclusion, the noise reduction due to LPW will be sensitive to atmospheric effects, especially at higher frequencies. For a typical road traffic noise spectrum (vehicle speed of 70 km/h, assuming 15% heavy vehicles) and a receiver at 4 m height, the predicted insertion loss for the LPW in Figure 8.28 is 5.7 dBA without wind, 2.6 dBA in moderate wind and less than 0.5 dBA in strong wind [39]. As with conventional noise barriers, assessments of the noise reducing potential of LPW need to account for meteorological effects.

8.8 PREDICTING EFFECTS OF VARIABILITY IN DOWNWARD-REFRACTION AND GROUND IMPEDANCE

8.8.1 Introduction

Outdoor sound levels can be influenced significantly by changes in meteorological and ground conditions. During upwind sound propagation, an acoustical shadow zone is formed and, thereby, contributions from the source are limited. When the propagation is downwind, sound levels are increased since multiple sound paths arrive at the receiver. Also, in downwind conditions, the influence of the ground might be stronger due to multiple interactions with it. Given this potential volatility, the choice of the duration and location for immission measurements at moderate distances from a source should not be random.

Although measuring in absence of wind might be attractive, this would not lead to an adequate exposure assessment near a continuous noise source, since downwind episodes might dominate long-term exposure metrics like the equivalent sound pressure level. Good practices, like the ones described in ISO 1996-2 [40], prescribe that sound pressure level measurements should be made during (moderately) downwind episodes. This choice is logical, since, during upwind sound propagation, the start of the shadow zone is rather sensitive to the source and receiver height, the magnitude of the gradients in the sound speed profile and the turbulent state of the atmospheric boundary layer. During downwind sound propagation, levels are expected to be less volatile and the influence of turbulent scattering less strong [41]. Nevertheless, even when under moderately downwind sound propagation conditions at limited distance (<250 m), there will still be substantial variation. Quantifying this variability is the goal of this section.

Simulations have been performed in keeping with the restricted data available for typical industrial noise monitoring or assessment [42]. It is unlikely that detailed wind speed profiles and ground impedance measurements at the site under study will be available. Even when ground appears to be flat uniform grassland, there could be a wide range of surface impedance [43]. The combination of variations in sound speed gradients and in ground impedance will lead to variations in exposure level that are not captured by typical monitoring campaigns.

8.8.2 Meteorological Data and Processing

To explore the variability of sound levels at downwind receivers, sound propagation simulations have been carried out [42], making use of detailed meteorological data collected from a tower over a full year, consisting of 10-minute averages of wind speed at 5 heights (24, 48, 69, 78 and 114 m above the ground) and of air temperature at 5 heights (8, 24, 48, 78 and 114 m). Wind direction was monitored at 3 heights (24, 69 and 114 m). Hourly relative humidity data from a nearby observation point 1.5 m above the ground were also used.

Sound propagation simulations were carried out for 'favourable' periods within which the wind direction was between 135 and 315 degrees including the dominant south-west wind direction of 225 degrees, measured at 24 m height. Wind alone was used for defining favourable sound propagation conditions since wind speed and direction are relatively easy to measure and estimate. However, since temperature profiles are accounted for in the simulations, this means that some sound speed profiles in the periods, selected as representing favourable propagation, might have contained temperature-gradient-related upward refracting zones as well. Linear-logarithmic curves with coefficients a_{log} and a_{lin} (see Chapter 1) were used to give effective sound speed profiles.

To limit the number of calculations, downwind conditions were divided into 146 classes (with class widths of 0.25 m/s and 0.025 1/s for a_{log} and a_{lin}, respectively). Note that refraction strength is governed by the combination (a_{log}, a_{lin}) and not simply by a_0 (which is fixed at 340 m/s). As an additional constraint, periods with wind speeds above 5 m/s at a (virtual) receiver height of 4 m were removed from the database, thereby restricting to 'moderately' downwind sound propagation. This prevents excessive wind-induced microphone noise in outdoor monitoring campaigns. Relative humidity from the nearby monitoring point and the air temperature from the sensor at 8 m height were used to calculate the frequency-dependent atmospheric absorption following ISO 9613-1 [44].

The distribution of best-fit (a_{log}, a_{lin}) pairs for the selected favourable propagation periods is shown in Figure 8.31.

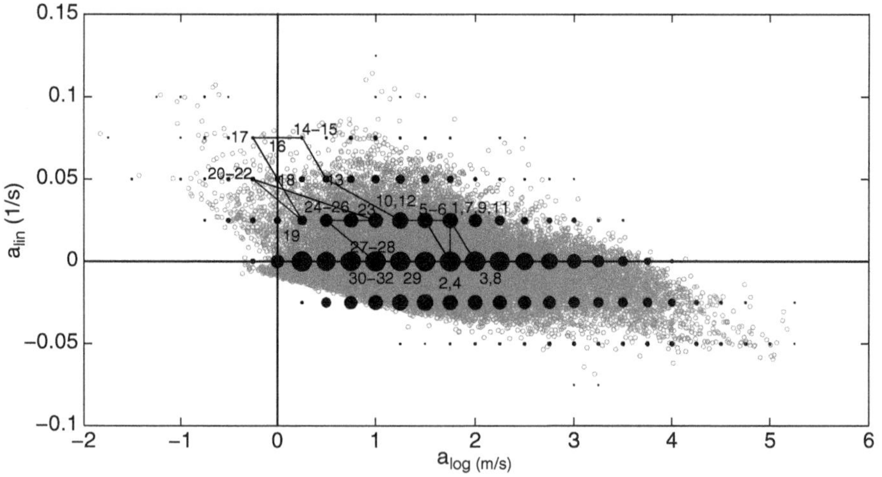

Figure 8.31 Scatter plot of (a_{log},a_{lin}) pairs providing best-fit linear-logarithmic sound speed profiles (see Chapter 1) for selected 10-minute periods of moderate downwind sound propagation based on meteorological tower data obtained during a particular year. The grey circles indicate all fits. The filled black circles represent 146 classes of (a_{log},a_{lin}) combinations (with radii proportional to the number of occurrences in each class). The numbers (1 to 32) and lines track the changes in the refractive state of the atmospheric boundary layer during the selected 320-minute period.

8.8.3 Grassland Impedance Data

Short-range measurements [43] show a large variety in the surface impedance of grounds that could be categorized as 'grassland'. Seven types of grasslands for which multiple measurements are available (classified as 'arable', 'heath', 'lawn', 'long grass', 'pasture', 'sports field' and 'urban') have been considered in the current calculations. For each of these types, average values of the effective flow resistivity and effective porosity have been used (see Table 8.6).

Although other model choices could have been made (see Chapter 5), the phenomenological impedance model [45] is used, since it enables reasonably good fits to measured short-range level difference spectra over these types of ground surface [43]. Each grassland-type is combined with all valid meteorological conditions. For a single source height, this leads to 146 times 7 (= 1022) sound propagation calculations. The effect of soil moisture is not considered explicitly, although it is likely that variation due to water content is already present in the reported measurements (and their parameter fits) [43].

To represent a situation that might be typical of noise from industry, the first 20 m between source and receiver is modelled as rigid ground (e.g. concrete) beyond which the ground is assumed to be grass-covered. The transition from the rigid source zone to grass-covered land involves a significant impedance discontinuity.

8.8.4 Sound Propagation Modelling and Numerical Parameters

The axi-symmetric Green's function Parabolic Equation (GFPE) method [46,47] has been used for the sound propagation calculations, assuming that the ground between source and receiver is flat. Atmospheric refraction has been simulated using the effective sound speed approach. Especially for

Table 8.6 Average best fit effective flow resistivities and effective porosities used in the phenomenological model to simulate the impedance of various types of grassland [42].

Type of grassland	*Effective flow resistivity (kPa s/m²)*	*Effective porosity*
Arable	742	0.453
Heath	226	0.856
Lawn	216	0.763
Long grass	57	0.650
Pasture	418	0.833
Sports field	514	0.240
Urban grass	35	0.610

sound propagation over flat ground, vertical gradients in the horizontal component of the wind speed are the main drivers for refraction, and these are represented well by effective sound speed profiles. Even in more complex cases, the effective sound speed approach proves to be quite accurate [48,49]. The refractive state of the atmosphere is assumed to be range-independent here and turbulent scattering is not considered.

GFPE needs a rather fine discretization in the vertical direction (1/10th of the wavelength), while, in the horizontal direction, spatial discretization constraints are much more relaxed. Forward stepping is performed at 5 times the wavelength, however, with a maximum of 5 m to allow sufficient spatial resolution when presenting results. An 8th order starter function [47] has been used to initiate the calculations from the assumed point source. An absorbing layer consisting of 500 wavelengths, exceeding the minimum advised thickness of 200 wavelengths [50], was used. The 1/3 octave bands between 50 Hz and 4 kHz were considered, and 10 sound frequencies were explicitly calculated to constitute each band.

8.8.5 Detailed Analysis of a Temporal Sequence

A sequence of 32 successive 10-minute averages of meteorological conditions is represented by (a_{log}, a_{lin}) values plotted in Figure 8.31 to illustrate the temporal changes in sound speed profiles one might observe. The lines track trajectories in (a_{log}, a_{lin}) space during the selected 320 minutes.

Figures 8.32–8.35 show examples of the impact of the varying sound speed profiles on the predicted spatial behaviour of excess attenuation in the 1/3 octave bands with centre frequencies 250 Hz and 1000 Hz, for sound propagation from a 2 m high point source over 'long grass' (see Table 8.6). In the 250 Hz band, there is a distinct zone of destructive interference associated with ground effect (see Figure 8.32) which is predicted to be very sensitive to the changing meteorological conditions.

Figure 8.34 indicates that the minimum in this zone is predicted to move by roughly 100 m during the selected 320 minutes. When the effective sound speed profiles become more linear (i.e. the a_{log} value approaching zero while a_{lin} increases, corresponding to periods from 13 to 18 in Figure 8.31), the destructive interference zone is predicted to become less pronounced since there are less strong sound speed gradients near the ground. Excess attenuation in the 1000 Hz 1/3 octave band is predicted to have a more complex spatial behaviour (see Figures 8.33 and 8.35), with clearly identifiable zones of constructive interference. When the sound speed profile becomes more linear or slightly upward refracting, a small shadow zone is predicted very close to the ground (e.g. during periods 20–22). Although all of these conditions could be categorized as downwind, some remarkable features are predicted. At receiver heights on the order of 2 m or less, going further away from the source will not necessarily lead to a decrease in sound pressure level.

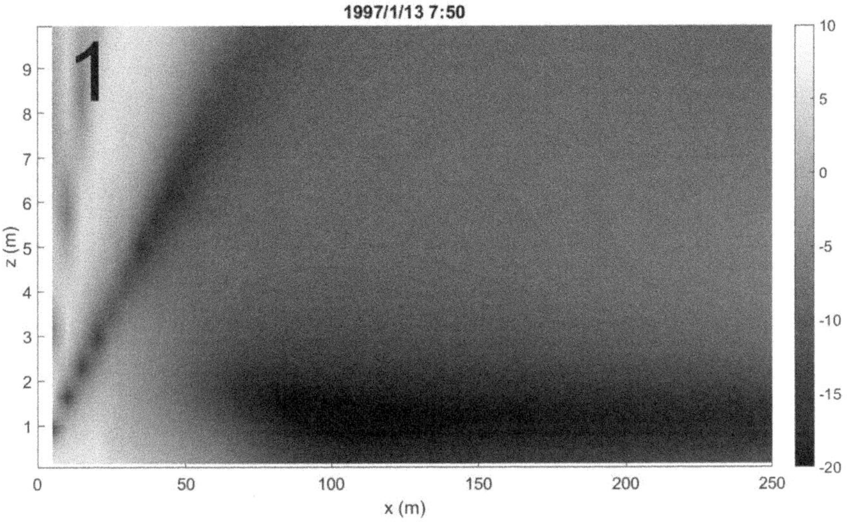

Figure 8.32 Predicted distribution of sound levels relative to free field in the 250 Hz 1/3 octave band, for a source height of 2 m and propagation over "long grass" (see Table 8.6). The axes are not to scale. For the significance of the '1' in the top left hand of this and Figure 8.33, see Figure 8.31.

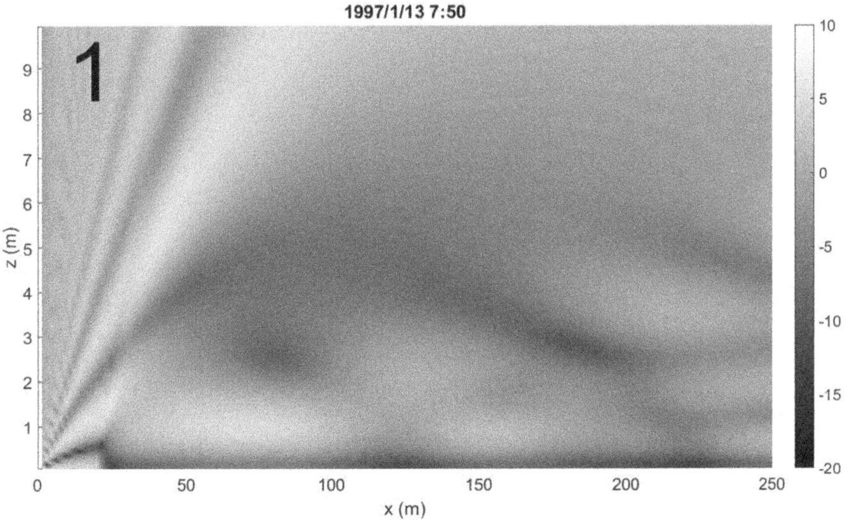

Figure 8.33 Predicted distribution of sound levels relative to free field in the 1000 Hz 1/3 octave band, for a source height of 2 m and propagation over "long grass" (see Table 8.6). The axes are not to scale.

Figure 8.34 A sequence of plots of the spatial distribution of sound pressure level relative to free field propagation in the 250 Hz 1/3-octave band, in 32 successive 10-minute intervals of moderate downwind conditions. The upper left figure is identical to Figure 8.32, and similar axes were used. The numbers in each subplot correspond to those in Figure 8.31.

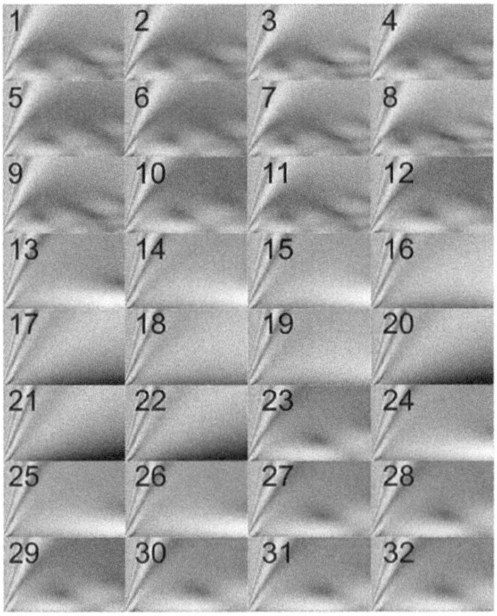

Figure 8.35 A sequence of plots of the spatial distribution of sound pressure level relative to free field propagation in the 1000 Hz 1/3-octave band, in 32 successive 10-minute intervals of moderate downwind conditions. The upper left figure is identical to Figure 8.33, and similar axes were used. The numbers in each subplot correspond to the ones in Figure 8.31.

8.8.6 Statistical Analysis of Temporal Variation over a Full Year

8.8.6.1 Spectral Variation

This section looks at the predicted variation in sound propagation resulting from all possible combinations of ground parameters (Table 8.6) and moderate downward refraction conditions obtained from the meteorological tower data over a full year.

Given that the (temporal) distribution of sound pressure levels at a single downwind location is predicted to be far from normally distributed, a distribution-independent variation descriptor is used, namely the 97.5 percentile minus the 2.5 percentile value. This range contains 95% of the values and would be equivalent to 4 times the standard deviation if there were to be a normal distribution. Additional graphs are shown later for 68% of the distribution values, equivalent to 2 times the standard deviation in a normal distribution.

Figure 8.36 shows typical patterns of the predicted 95% and 68% variations in twenty 1/3 octave bands between 50 Hz and 4 kHz for a source height of 2 m, receiver heights of 1.5 m and 4 m and ranges to 250 m. Below 100 Hz an approximately linear increase in the variation with distance is predicted.

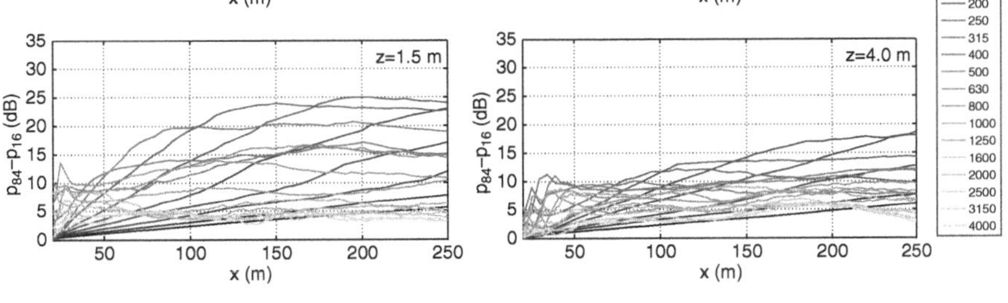

Figure 8.36 95% (upper two plots) and 68% (lower two plots) variation margins in sound pressure level in twenty 1/3 octave bands at receiver heights of 1.5 m and 4 m from a 2 m high source, predicted due to changes in the (moderately) downwind refractive state of the atmospheric boundary layer throughout the year and impedance changes corresponding to 7 types of "grassland" [42]. Reproduced from T. Van Renterghem, D. Botteldooren, Variability due to short distance favorable sound propagation and its consequences for immission assessment, *J. Acoust. Soc. Am.*, **143**: 3406–3417 (2018) with the permission of the Acoustical Society of America.

Above 100 Hz, the predicted variation increases rapidly until it plateaus at distances that are closer to the source as frequency increases. Near 315 Hz, the magnitude of this plateau starts to decrease, while above 1000 Hz the predicted variation is maximum within 50 m of the source. When the source height is 2 m and the receiver height is 1.5 m, the predicted 95% variation in 1/3 octave bands from 100 Hz to 200 Hz is about 30 dB. With increase in receiver height to 4 m, this variation reduces to about 25 dB. For an elevated source height of 20 m and a receiver height of 4 m the predicted 95% variation reduces to 17 dB [51].

The strong frequency dependence of the variation between 100 and 1000 Hz is consistent with the influence of refraction on the destructive interference zone associated with ground effect as discussed in Section 8.8.5 (see also Chapters 2, 5 and 6). Downward refraction of sound causes significant changes in path length and leads to multiple paths contributing to the level at a receiver. The combination of different types of grassland impedance and the wide range of refraction conditions is predicted to result in particularly strong variations between about 100 Hz and 1000 Hz which can be pronounced over several tens of metres.

8.8.6.2 Variation in A-Weighted Pink Noise

Although variation in individual bands might be important when considering tonal noise components, they will be partly levelled out when considering broadband noise typical of industrial sources, so the variation in A-weighted pink noise is analysed separately. Figure 8.37 shows the 95%

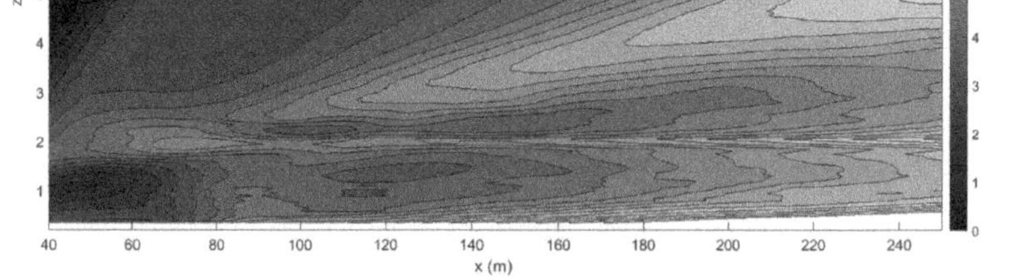

Figure 8.37 Spatial distribution of P97.5 - P2.5 predicted due to changes in moderately downwind refractive state of the atmosphere throughout the year for A-weighted pink noise propagating over "grassland" when the source height is 2 m [42]. Reproduced from T. Van Renterghem, D. Botteldooren, Variability due to short-distance favorable sound propagation and its consequences for immission assessment, *J. Acoust. Soc. Am.*, **143**: 3406–3417 (2018) with the permission of the Acoustical Society of America.

variation in A-weighted pink noise predicted from the combination of the changing moderately downwind conditions over a full year, and the influence of seven types of 'grassland', for a source height of 2 m. The predicted overall variation for A-weighted pink noise is much less than predicted for individual 1/3 octave bands. Moreover, in general, the variation is predicted to increase with increasing distance. However, distinct zones with higher variability are predicted for specific height ranges. For a source height of 2 m (see Figure 8.37), such a zone is predicted very close to the ground. Two other zones of high variability are predicted near the source height and at heights of between 4 and 5 m. This has the important practical consequence for performing short-term (snapshot) measurements, that it might be possible to minimize the variability and hence uncertainty by careful choice of microphone height. Specifically, measuring at a height of 2 or 4 m would be less likely to produce a representative level in the short term, or would need a longer period to reduce measurement uncertainty. The predicted variation at these receiver heights could be roughly halved by different choice of microphone location.

Localized steep gradients in the predictions are not encountered if the source is significantly more elevated (see Figure 8.38). The variation increases relatively smoothly when moving closer to the ground and when going further away from the source. Overall, the predicted variation is much less than that predicted for the low source height.

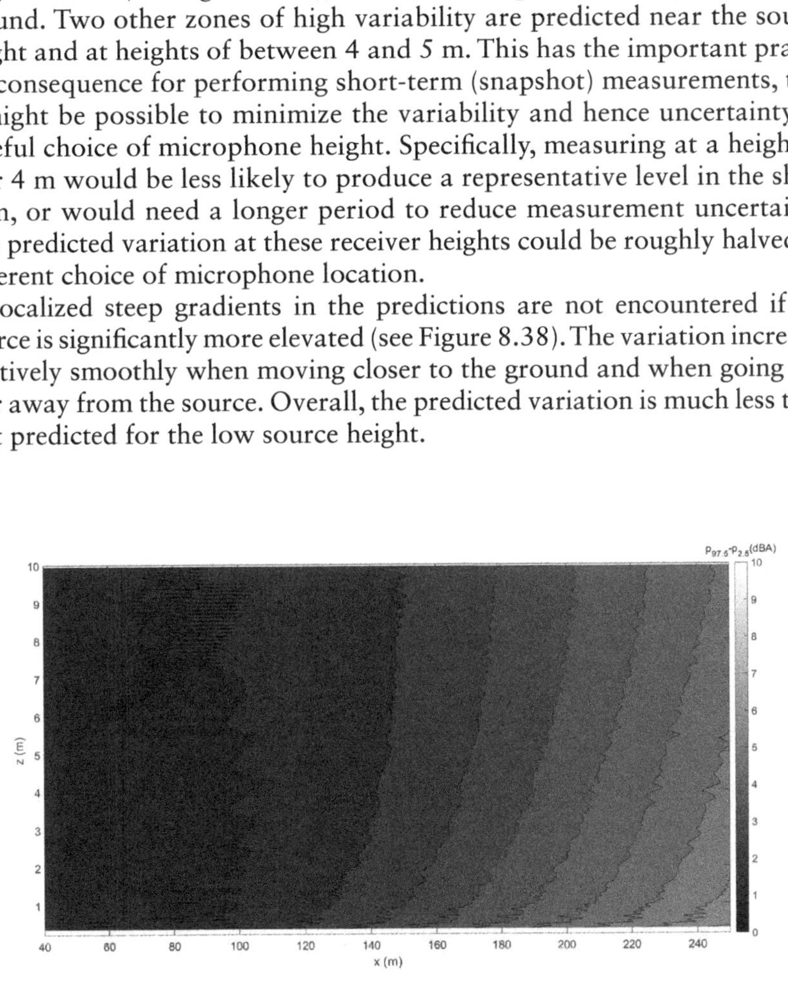

Figure 8.38 Spatial distribution of P97.5 - P2.5 due to changes in moderately downwind refractive state of the atmosphere throughout the year for A-weighted pink noise propagating over "grassland" when the source height is 20 m [42]. Reproduced from T. Van Renterghem, D. Botteldooren, Variability due to short-distance favorable sound propagation and its consequences for immission assessment, *J. Acoust. Soc. Am.*, **143**: 3406–3417 (2018) with the permission of the Acoustical Society of America.

8.8.6.3 Convergence to Yearly L_{Aeq}

Given the predicted variations (for pink noise), what measurement duration is needed to establish long-term equivalent sound pressure levels? To investigate this, the yearly L_{Aeq} for (continuous) pink noise is calculated by including an increasing number of 10-minute periods with moderately downwind sound propagation conditions (over 'long grass'), in a chronological way. This means that the nth datapoint considers all moments up to n to calculate the equivalent sound pressure level. Note that more advanced sampling approaches could be used (like Monte Carlo, Latin hypercube or importance sampling [51,52]) to account for variability.

Figure 8.39 shows the predicted convergence at propagation distances of 50, 100 and 200 m, for receiver heights of 1.5 and 4 m. Since the starting period of the integration will influence the convergence towards the yearly L_{Aeq}, 50 starting moments were defined, uniformly spread over the valid downwind episodes throughout the year under study. The meteorological data set was treated cyclically to allowing capturing a full year of data for each starting moment. The convergence curves for the additional starting

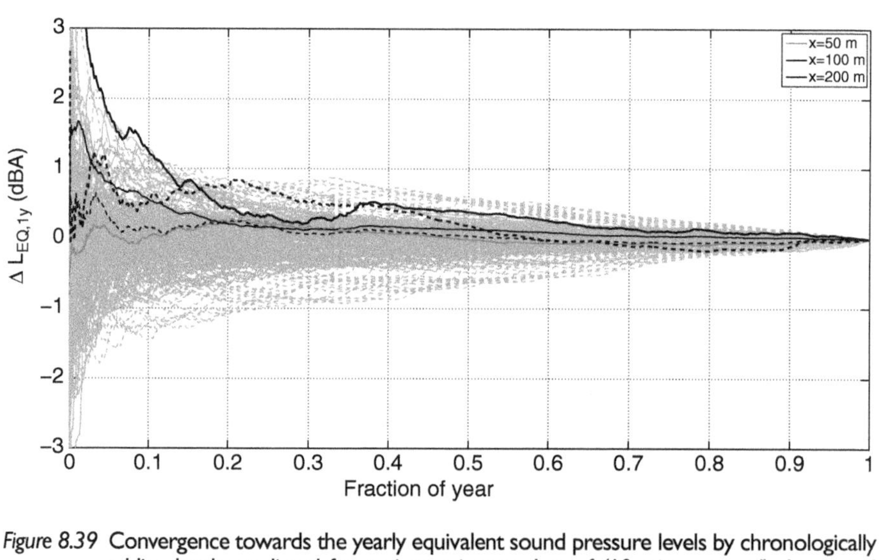

Figure 8.39 Convergence towards the yearly equivalent sound pressure levels by chronologically adding levels predicted for an increasing number of (10-min averaged) downwind episodes (starting from the beginning of the year). $\Delta L_{EQ,1y}$ is the difference between the short-term (cumulatively averaged) equivalent sound pressure level and its yearly value. Three propagation distances (50 m, 100 m, and 200 m) are considered at two receiver heights (full lines 1.5 m, dashed lines 4 m) for a specific type of grassland ("long grass"). The additional set of (thin grey) background lines are for shifted starting points uniformly distributed over the year. The source is positioned at a height of 2 m and emits pink noise [42]. Reproduced from T. Van Renterghem, D. Botteldooren, Variability due to short-distance favorable sound propagation and its consequences for immission assessment, *J. Acoust. Soc. Am.*, **143**: 3406–3417 (2018) with the permission of the Acoustical Society of America.

moments are indicated by the thin grey curves in Figure 8.39. While not intended to present a detailed analysis, nevertheless they show the full range of deviations one could obtain.

When the receiver is located further away from the source, a longer measurement period is needed to stay within a preset maximum allowable deviation from the long-term averaged level. A non-smooth transition towards zero deviation from the long-term equivalent level is observed, especially for the receiver at 200 m from the source (see Figure 8.39). However, the initial strong deviation at the low receiver height of 1.5 m (exceeding 3 dBA) rapidly decreases when including more periods. At 50 and 100 m propagation distance, a receiver height of 4 m seems beneficial when only a short measurement period is planned. However, this advantage relative to a receiver height of 1.5 m is less obvious when aiming at a more stringent convergence criterion of, e.g. 0.5 dBA, due to the undulating nature of the convergence curves as shown in Figure 8.39. Close to the source (e.g. 50 m), the influence of the receiver height is very limited and a convergence criterion of 0.5 dBA is rapidly obtained. For a higher source position (20 m), the convergence curves (not shown) become smoother and convergence towards the yearly averaged equivalent sound pressure level is obtained with less samples taken [42]. The choice of receiver height then has a negligible effect on the simulated convergence to the long-term level. Note that in Figure 8.39, setting the starting point at the first relevant downwind episode leads to an initial over estimation of the yearly level since this particular year started with rather strong downward refracting atmospheres. Considering the convergence curves for all starting moments, an overall symmetry is found. If only a few episodes are considered, then the probabilities of overestimation or underestimation are similar. At all receiver heights and distances considered here and for a low source position, about 10% of the year must be covered to stay within ±1 dBA relative to the converged yearly L_{Aeq}.

8.8.6.4 Conclusions

The variation between short-term measurements of noise from broadband sources during moderately downwind conditions over grassland, even when adhering to measurement standards, is predicted to depend strongly on the acoustic frequency, propagation distance, source height and receiver height. Even for relatively short propagation distances (less than 250 m), this variation might be large; the 95% margins on the variation might reach 30 dB in individual 1/3 octave bands, and roughly 10 dBA in the overall level for pink noise. Generally, the predicted variation increases with increasing propagation distance but is less for more elevated sources.

A good choice of receiver height might lead to a decrease in the variation that would result from short-term measurements and enable a more rapid convergence to the long-term equivalent sound pressure level when assessing continuous noise sources. In general, the predictions show that long-term monitoring is necessary and that 'snapshot' measurements do not result in realistic exposure assessments.

The current analysis did not consider influences from inexact knowledge of source and receiver positions, turbulent scattering and non-flat grounds. These are relevant sources of uncertainty, in addition to incomplete knowledge of meteorological conditions and ground impedance. Ground undulations could have a strong effect on sound propagation close to the ground. Turbulent scattering is expected to be mainly relevant in individual 1/3 octave bands, although less important under the downwind 'favourable' conditions considered than in upwind conditions. However, more research is needed to identify the dominant causes of uncertainty in outdoor sound propagation.

REFERENCES

[1] ISO9613-2 Acoustics - Attenuation of sound during propagation outdoors - Part 2: A general method of calculation (1996).

[2] The Calculation of Road Traffic Noise, Department of Environment 1985 (HMSO).

[3] Annex to Commission Directive 2015/996 in Official Journal of the European Union L168 (2015).

[4] K. J. Marsh, The CONCAWE model for calculating the propagation of noise from open-air industrial plants, *Appl. Acoust.* **15** 411–428 (1982).

[5] B. O. Enflo and P. H. Enflo, Sound wave propagation from a point source over a homogeneous surface and over a surface with an impedance discontinuity, *J. Acoust. Soc. Am.* **82** 2123–2134 (1987).

[6] J. Durnin and H. Bertoni, Acoustic propagation over ground having inhomogeneous surface impedance, *J. Acoust. Soc. Am.* **70** 852–859 (1981).

[7] P. Koers, Diffraction by an absorbing barrier or by an impedance transition, *Proc. Internoise*, 311–314 (1983).

[8] B. A. De Jong, A. Moerkerken and J. D. Van Der Toorn, Propagation of sound over grassland and over an earth barrier, *J. Sound Vib.* **86** 23–46 (1983).

[9] A. D. Pierce, Diffraction of sound around corners and over wide barriers, *J. Acoust. Soc. Am.* **55** 941–955 (1974).

[10] G. A. Daigle, J. Nicolas and J.-L. Berry, Propagation of noise above ground having an impedance discontinuity, *J. Acoust. Soc. Am.* **77** 127–138 (1985).

[11] D. C. Hothersall and J. N. B. Harriott, Approximate models for sound propagation above multi-impedance plane boundaries, *J. Acoust. Soc. Am.* **97** 918–926 (1995).

[12] Y. W. Lam and M. R. Monazzam, On the modeling of sound propagation over multi-impedance discontinuities using a semiempirical diffraction formulation, *J. Acoust. Soc. Am.* **120** 686–698 (2006).

[13] K. B. Rasmussen, Propagation of road traffic noise over level terrain, *J. Sound Vib.* **82** 51–61 (1982).

[14] J. Forssen, Calculation of noise barrier performance in a turbulent atmosphere by using substitute-sources above the barrier, *Acta Acust. united Ac.* **86** 269–275 (2000).

[15] K. B. Rasmussen, A note on the calculation of sound propagation over impedance jumps and screens, *J. Sound Vib.* **84** 598–602 (1982).
[16] P. Boulanger, T. Waters-Fuller, K. Attenborough and K. M. Li, Models and Measurements of sound propagation from a point source over mixed impedance ground, *J. Acoust. Soc. Am.* **102** 1432–1442 (1997).
[17] C. Nyberg, The sound field from a point source above a striped impedance boundary, *Acta Acust. united Ac.* **3** 315–322 (1995).
[18] S. Slutsky and H. L. Bertoni, Analysis and programs for assessment of absorptive and tilted parallel barriers, Transportation Research Record 1176, Transportation Research Record, National Research Council, Washington, DC, 13–22 (1987).
[19] S. N. Chandler-Wilde and D. C. Hothersall, Sound propagation above an inhomogeneous impedance plane, *J. Sound Vib.* **98** 475491 (1985).
[20] S.N. Chandler-Wilde, *Ground effects in environmental sound propagation*, Ph.D. Thesis, University of Bradford (1988).
[21] R. P. Shaw, Boundary Integral Equation Methods Applied to Wave Problems, in *Developments in Boundary Element Methods-1*, Eds. P.K. Banerjee and R. Butterfield, Applied Science Publishers, London, 1980.
[22] D. F. Mayers, Quadrature Methods for Fredholm Equations of The Second Kind, in *Numerical Solution of Integral Equations*, Eds. L. M. Delves and J. Walsh, Clarendon Press, Oxford (1974).
[23] I. Bashir, *Acoustical exploitation of rough, mixed impedance and porous surfaces outdoors*, PhD Thesis, Engineering and Innovation, The Open University (2014).
[24] J. S. Robertson, P. J. Schlatter and W. L. Siegmann, Sound propagation over impedance discontinuities with the parabolic approximation, *J. Acoust. Soc. Am.* **99** 761–767 (1996).
[25] C. S. You and S. W. Yoon, Predictions of traffic noise propagation over a curved road having an impedance discontinuity with refracting atmospheres, *J. Korean Phy. Soc.* **29** 176–181 (1996).
[26] M. C. Bérengier, B. Gauvreau, P. Blanc-Benon and D. Juvé, Outdoor sound propagation: a short review on analytical and numerical approaches, *Acta Acust. united Ac.* **89** 980–991 (2003).
[27] S. Taherzadeh, K. M. Li and K. Attenborough, A hybrid BIE/FFP scheme for predicting barrier efficiency outdoors, *J. Acoust. Soc. Am.* **110** (2) 918–924 (2001).
[28] K. Attenborough, I. Bashir and S. Taherzadeh, Exploiting ground effects for surface transport noise abatement, *Noise Mapping J.* **3** 1–25 (2016).
[29] K. Attenborough, S. Taherzadeh, I. Bashir, J. Forssen, B. V. D. Aa and M. Männel, Porous ground, crops and buried resonators, chapter 7 in *Environmental methods for transport noise reduction*, ed. M.E. Nillson, J. Bengtsson and R. Klaeboe, CRC Press, an imprint of Taylor and Francis, London (2015).
[30] I. Bashir, S. Taherzadeh, H.-C. Shin and K. Attenborough, Sound propagation over soft ground with and without crops and potential for surface transport noise reduction, *J. Acoust. Soc. Am.* **137** 154–164 (2015).
[31] L. A. M. van der Heijden and M. J. M Martens, Traffic noise reduction by means of surface wave exclusion above parallel grooves in the roadside, *Appl. Acoust.* **15** 329–339 (1982).
[32] I. Bashir, S. Taherzadeh and K. Attenborough, Diffraction-assisted rough ground effect: models and data, *J. Acoust. Soc. Am.* **133** 1281–1292 (2013).

[33] S. Taherzadeh, I. Bashir, T. Hill, K. Attenborough and M. Hornikx, Reduction of surface transport noise by ground roughness, *Appl. Acoust.* **83** 1–15 (2014).
[34] K. Attenborough, P. Boulanger, Q. Qin and R. Jones, Predicted influence of ballast and porous concrete on railway noise, Proc. InterNoise 2005, Rio de Janeiro, Paper IN05_1583 (2005).
[35] O. Umnova, K. Attenborough, E. Standley and A. Cummings, Behavior of rigid-porous layers at high levels of continuous acoustic excitation: theory and experiment, *J. Acoust. Soc. Am.* **114** (3) 1346–1356 (2003).
[36] P. A. Morgan and G. R. Watts, Investigation of the screening performance of low novel railway noise barriers. Unpublished project report, TRL, *PR SE/705/03* (2003).
[37] K. M. Li, T. Waters-Fuller and K. Attenborough, Sound propagation from a point source over extended-reaction ground, *J. Acoust. Soc. Am.* **104** 679–685 (1998).
[38] K. M. Li and H. Tao, Heuristic approximations for sound fields produced by spherical waves incident on locally and non-locally reacting planar surfaces, *J. Acoust. Soc. Am.* **135** 58–66 (2014).
[39] T. Van Renterghem, S. Taherzadeh, M. Hornikx and K. Attenborough, Meteorological effects on the noise reducing performance of a low parallel wall structure, *Appl. Acoust.* **121** 74–81 (2017).
[40] ISO 1996-2 (2017) Acoustics – description, measurement and assessment of environmental noise. Part 2: Determination of sound pressure levels. International Organisation for Standardisation, Geneva, Switzerland (1996).
[41] E. Salomons, *Computational atmospheric acoustics*, Kluwer, Dordrecht, The Netherlands, 335 pages (2001).
[42] T. Van Renterghem and D. Botteldooren, Variability due to short-distance favorable sound propagation and its consequences for immission assessment, *J. Acoust. Soc. Am.* **143** 3406–3417 (2018).
[43] K. Attenborough, I. Bashir and S. Taherzadeh, Outdoor ground impedance models, *J. Acoust. Soc. Am.* **129** 2806–2819 (2011).
[44] ISO 9613-1 Acoustics – attenuation of sound during propagation outdoors – Part 1. International Organisation for Standardisation, Geneva, Switzerland (1996).
[45] C. Zwikker and C.W. Kosten, *Sound Absorbing Materials*, Elsevier, New York (1949).
[46] K. Gilbert and X. Di, A fast Green's function method for one-way sound propagation in the atmosphere, *J. Acoust. Soc. Am.* **94** 2343–2352 (1993).
[47] E. Salomons, Improved Green's function parabolic equation method for atmospheric sound propagation, *J. Acoust. Soc. Am.* **104** 100–111 (1998).
[48] R. Blumrich and D. Heimann, Numerical estimation of atmospheric approximation effects in outdoor sound propagation modelling, *Acta Acust. united Ac.* **90** 24–37 (2004).
[49] T. Van Renterghem, D. Botteldooren and P. Lercher, Comparison of measurements and predictions of sound propagation in a valley-slope configuration in an inhomogeneous atmosphere, *J. Acoust. Soc. Am.* **121** 2522–2533 (2007).
[50] J. Cooper and D. Swanson, Parameter selection in the Green's function parabolic equation, *Appl. Acoust.* **68** 390–402 (2007).
[51] C. Pettit and D. K. Wilson, Proper orthogonal decomposition and cluster weighted modeling for sensitivity analysis of sound propagation in the atmospheric surface layer, *J. Acoust. Soc. Am.* **122** 1374–1390 (2007).
[52] D. K. Wilson, C. Pettit, V. Ostashev, S. Vecherin, Description and quantification of uncertainty in outdoor sound propagation calculations, *J. Acoust. Soc. Am.* **136** 1013–1028 (2014).

Chapter 9

Predicting the Performance of Outdoor Noise Barriers

9.1 INTRODUCTION

The erection of a noise barrier is popular as a means of reducing noise from surface transport. As long as the direct transmission of sound through it is negligible and it intercepts the line-of-sight from the noise source, a barrier reduces the noise from the source arriving at a receiver located on the other side of the barrier. The acoustic field in the shadow region created by the barrier is dominated by the sound diffracted over and around it.

As well as conventional barriers with straight top edges, many different types of noise barriers have been developed. These include a top-bended or 'cranked' barriers, inclined barriers, louvred barriers and barriers with multiple edges. Some noise barriers are in the form of earth berms. There is increasing interest also in barriers made from caged stones (gabions) and in barriers consisting of periodically arranged scattering elements, i.e. so-called sonic crystal noise barriers.

To design noise barriers with optimum acoustic performance, it is important to have accurate methods for predicting their sound reduction.

The aims of this chapter are

(i) to introduce the basic principles of the diffraction of sound by noise barriers and identify the important parameters,
(ii) to discuss analytical and empirical models for studying the acoustic performance of noise barriers,
(iii) to determine the degradation of performance due to presence of gaps in barriers,
(iv) to explore the effectiveness of a barrier in screening a directional source,
(v) to investigate predictions and measurements on gabion barriers,
(vi) to describe the influence of meteorological conditions (particularly for downwind receivers) on barrier performance,
(vii) to compare the acoustical performances of berms and conventional fence type barriers,

(viii) to describe aspects of the design of sonic crystal noise barriers and
(ix) to investigate the consequences of assuming a fixed source spectrum when calculating the performance of a noise barrier.

9.2 ANALYTICAL SOLUTIONS FOR THE DIFFRACTION OF SOUND BY A BARRIER

Diffraction of spherical, cylindrical or plane waves by a thin half plane is of interest in optics as well as in acoustics. It has been studied, experimentally and theoretically, since the 18th century. In the optical context, it was suggested that the diffraction pattern behind a half plane is the result of the superposition of waves scattered by the edge and the unobstructed portion of the incident waves. The first mathematically rigorous solution of a half-plane diffraction problem was formulated in 1896 by Sommerfeld who considered the 2-D problem of a plane wave incident on a thin, perfectly reflecting, half plane and solved the corresponding partial differential equations [1,2]. The resulting solutions have two main components. The first component is the direct wave, expressed exactly according to the principles of geometrical acoustics and the second component is the contribution of the diffracted wave, expressed in terms of Fresnel integrals. Subsequently, Carslaw [3,4] and MacDonald [5] extended Sommerfeld's approach to the generalized problems of the diffraction of cylindrical and spherical incident waves by wedges. In a similar manner to Sommerfeld's solution, their solutions are expressed as the sum of two integrals, associated with the incident and scattered or diffracted fields, respectively.

An alternative to seeking a formal solution for the governing partial differential equations that arise when solving diffraction problems directly is an integral formulation. Copson [6] and Levine and Schwinger [7,8] have used this approach to diffraction problems and obtained exact solutions by applying the Wiener–Hopf method, which is a standard technique for solving certain types of linear partial differential equations subject to mixed boundary conditions and involving semi-infinite geometries [9,10].

Tolstoy [11,12] has obtained an alternative exact and explicit solution for sound waves diffracted by wedges. In his solution, the diffracted sound field can be represented by a sum of an infinite series. The coefficient of each term of the series is given by a set of linear equations which can be solved by a simple recursion scheme. Edge diffractions are included exactly without the need for an asymptotic approximation of integrals. However, application of this formulation is limited by the slow convergence of the series particularly at high frequencies.

9.2.1 Formulation of the Problem

Consider first a geometrical acoustics interpretation of the problem of sound propagating past a thin screen. The edge of the screen may be considered to act as a 'secondary' line source, so it is convenient to use a cylindrical polar coordinate system to describe the relative position of source and receiver from this edge. Let the origin be located at the edge and (r,θ,y) be the cylindrical polar coordinates. The y-axis coincides with the edge of the semi-infinite half plane, and the initial line of the polar coordinates lies on the right-hand surface of the half plane. Hence, the thin plane is made of two surfaces at $\theta = 0$ (right surface) and $\theta = 2\pi$ (left surface) with zero thickness and it has an exterior angle of 2π. With this coordinate system, all radial distances are measured from the edge and all angular positions are measured in the counterclockwise direction as shown in Figure 9.1.

Without loss of generality, consider a point source placed in front of the left surface of the thin plane at (r_o,θ_o,y_0), i.e. $2\pi \geq \theta_0 \geq \pi$ and a receiver located at (r_r,θ_r,y_r) where $2\pi \geq \theta_r \geq 0$.

According to geometrical acoustics, the sound field consists of a diffracted field, p_d and a geometrical solution that combines the incident and reflected waves, p_i and p_r. The dashed lines at $\theta = 3\pi - \theta_0$ and $\theta = \theta_0 - \pi$ in Figure 9.1, subdivide the field into three separate regions, I, II and III. The ray that strikes the top edge of barrier is reflected backwards along the path represented by a broken line, $\theta = 3\pi - \theta_0$.

Again, according to geometrical acoustics, none of the reflected rays can penetrate region II and III and they are confined to region I. The dividing line at $\theta = 3\pi - \theta_0$ separates region I from region II and creates a shadow boundary B_r for the reflected wave. Similarly, the direct wave cannot penetrate region III because of the presence of the thin plane preventing direct

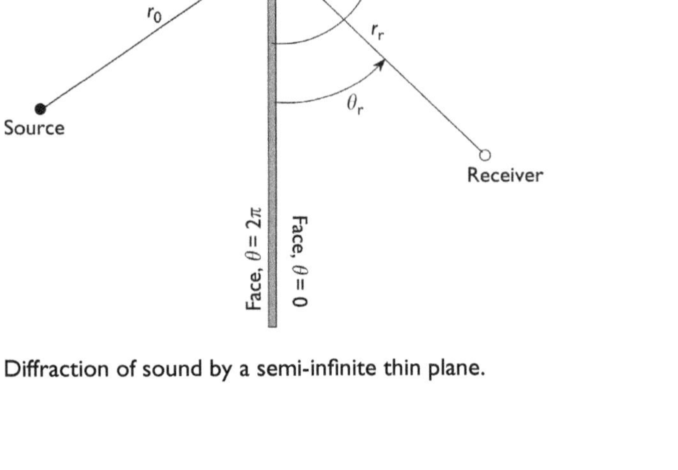

Figure 9.1 Diffraction of sound by a semi-infinite thin plane.

line-of-sight contact between source and receiver. Region III is called the shadow zone. The dividing line at $\theta = \theta_0 - \pi$ separates region II from region III and it is called the shadow boundary, B_i, for the direct wave. If the source faces the right-hand side of the thin plane, then there is a wave reflected on the $\theta = 0$ face. The occurrence of each wave type can be summarized as follows:

1) a direct wave where $\pi - |\theta_r - \theta_0| \geq 0$;
2) a wave reflected from the $\theta = 0$ face when $\pi - (\theta_r + \theta_0) \geq 0$;
3) a wave reflected from the $\theta = 2\pi$ face when $(\theta_r + \theta_0) - 3\pi \geq 0$; and
4) a diffracted wave from edge of plane when $0 \leq \theta_0, \theta_r \leq 2\pi$.

Items (2) and (3) above are mutually exclusive as the source cannot 'see' both surfaces of the half plane simultaneously.

The analytical solution for the total sound field in the vicinity of a semi-infinite half plane can be decomposed into three terms. The first term takes account of the contribution of a direct wave, p_i; the second term accounts for the contribution of a reflected wave from the surface of the semi-infinite half-plane, p_r; and the last term deals with the contribution of the diffracted wave at the edge of the half plane, p_d. Except the diffracted wave term, the occurrence of the direct and reflected wave terms depends very much on the relative position of source, receiver and the thin plane. In each region, the total sound field, p_T, is as given below:

1. Region I

$$p_T = p_i + p_r + p_d; \tag{9.1a}$$

2. Region II

$$p_T = p_i + p_d; \tag{9.1b}$$

and

3. Region III

$$p_T = p_d \tag{9.1c}$$

The diffracted sound field may be treated as having two components which are associated with the direct and reflected fields, p_{di} and p_{dr}, respectively. The geometrical solution becomes discontinuous at the shadow boundary of the reflected wave because the reflected wave term vanishes in region II. However, the associated component of the diffracted wave for the reflected fields p_{dr} also becomes discontinuous on crossing the shadow boundary of the reflected wave. As a result, the total sound field remains continuous. Similarly, the geometrical solution at the shadow boundary of the direct wave becomes discontinuous due to the absence of the direct wave term, but the diffraction solution associated with the direct field p_{di} becomes discontinuous also and that leads to a continuous solution throughout.

9.2.2 The MacDonald Solution

Consider the sound field due to a point monopole source which is placed near to a rigid half plane. The source has unit strength and the time-dependent factor, $e^{-i\omega t}$, is omitted for simplicity. Using the cylindrical polar coordinate system described in the last section, the distance from the source and its image (see Figure 9.2) to the receiver, R_1 and R_2, can be determined according to the following equations:

$$R_1 = \sqrt{r_0^2 + r_r^2 - 2r_0 r_r \cos(\theta_0 - \theta_r) + (y_0 - y_r)^2} \tag{9.2a}$$

and

$$R_2 = \sqrt{r_0^2 + r_r^2 - 2r_0 r_r \cos(\theta_0 + \theta) + (y_0 - y_r)^2} \tag{9.2b}$$

The shortest distance from the source (or image source) to receiver after diffraction from the half plane, i.e. the shortest source-edge-receiver path is

$$R' = \sqrt{(r_0 + r)^2 + (y_0 - y)^2} \tag{9.3}$$

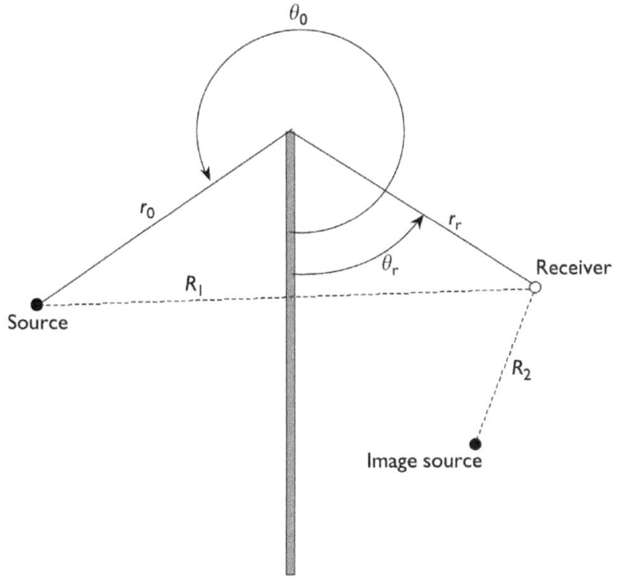

Figure 9.2 The geometry of source, image source, and receiver in the proximity of a semi-infinite plane.

Following Sommerfeld [1], MacDonald [5] developed a solution in a spherical polar coordinate system. This was subsequently recast in the cylindrical polar system by Bowman and Senior [13] to give the total field as a sum of two contour integrals as follows:

$$p_T = \frac{ik}{4\pi}\int_{\varsigma_1}^{\infty}\frac{H_1^{(1)}\left(kR_1+s^2\right)}{\sqrt{s^2+2kR_1}}ds+\frac{ik}{4\pi}\int_{\varsigma_2}^{\infty}\frac{H_1^{(1)}\left(kR_2+s^2\right)}{\sqrt{s^2+2kR_2}}ds, \tag{9.4}$$

where k is the wave number of the incident wave and $H_1^{(1)}(u)$ is the Hankel function of the first kind. The limits of the contour integrals are determined according to

$$\varsigma_1 = \mathrm{sgn}\left(\left|\theta_0-\theta_r\right|-\pi\right)\sqrt{k\left(R'-R_1\right)}, \tag{9.5a}$$

and

$$\varsigma_2 = \mathrm{sgn}\left(\theta_0+\theta_r-\pi\right)\sqrt{k\left(R'-R_2\right)}, \tag{9.5b}$$

where sgn(u), the sign function, takes the value of +1 or −1 depending on the sign of the argument. Note that both terms in Equation (9.4) contain integrals that are related to the integral representation of spherical wave where

$$\frac{e^{ikR}}{4\pi R}=\frac{ik}{4\pi}\int_{\varsigma_1}^{\infty}\frac{H_1^{(1)}\left(kR+s^2\right)}{\sqrt{s^2+2kR}}ds \tag{9.6}$$

The integral solution of Equation (9.4) can be spilt into two parts: a geometrical and diffraction contribution. The total sound field in regions I, II and III can be expressed in the form given in Equations (9.1a–9.1c) with the direct wave, reflected wave and diffracted wave determined, respectively, by

$$p_i = \frac{e^{ikR_1}}{4\pi R_1}, \tag{9.7a}$$

$$p_r = \frac{e^{ikR_2}}{4\pi R_2}, \tag{9.7b}$$

and

$$p_d = \frac{ik\,sgn(\varsigma_1)}{4\pi}\int_{|\varsigma_1|}^{\infty}\frac{H_1^{(1)}\left(kR_1+s^2\right)}{\sqrt{s^2+2kR_1}}ds+\frac{ik\,sgn(\varsigma_2)}{4\pi}\int_{|\varsigma_2|}^{\infty}\frac{H_1^{(1)}\left(kR_2+s^2\right)}{\sqrt{s^2+2kR_2}}ds \tag{9.7c}$$

A limiting case that should be checked is the high-frequency limit, for which $k \to \infty$. Both $|\varsigma_1|$ and $|\varsigma_2|$ tend to infinity in this limit. Hence, the diffraction integrals in Equation (9.7c) become negligibly small and the geometrical solution is recovered as it should. Another limiting case of interest is the situation where the source or the receiver (or both) are close to the edge of the half plane such that $k(R' - R_1) \ll 1$ and $k(R' - R_2) \ll 1$ where R' is defined by (9.3c). The integrands can be expanded by means of Taylor series and the resulting expression integrated to yield an approximate solution for the total pressure as

$$p_T = \frac{e^{ikR'}}{4\pi R'} + \frac{ik}{2\pi} H_1^{(1)}(kR')\sqrt{\frac{r_0 r_r}{R'}}\cos(\theta_0/2)\cos(\theta_r/2) \tag{9.8}$$

The last case is when both the source and receiver are far from the edge and the shadow boundaries, i.e. $kR' \gg 1$. MacDonald [5] derived an asymptotic solution for this case in which the contribution from the diffracted wave can be written in terms of Fresnel integrals as follows:

$$p_d = \frac{ke^{-i\pi/4}}{4\pi\sqrt{kR'}}\left\{\frac{\text{sgn}(\varsigma_1)e^{ikR_1}}{\sqrt{k(R'+R_1)}}G\left(\sqrt{2N_1}\right) + \frac{\text{sgn}(\varsigma_2)e^{ikR_2}}{\sqrt{k(R'+R_2)}}G\left(\sqrt{2N_2}\right)\right\} \tag{9.9}$$

The function $G(u)$ is defined as

$$G(u) = \int_u^{\infty} e^{iu^2}d\mu = F_r(\infty) - F_r(u), \tag{9.10a}$$

where

$$F_r(u) = C(u) + iS(u), \tag{9.10b}$$

and the Fresnel integrals $C(u)$ and $S(u)$ are defined [14] by

$$C(u) = \int_0^u \cos\left(\frac{\pi t^2}{2}\right)dt \tag{9.11a}$$

and

$$S(u) = \int_0^u \sin\left(\frac{\pi t^2}{2}\right)dt \tag{9.11b}$$

The corresponding Fresnel numbers of the source and image source are denoted, respectively, by N_1 and N_2. They are defined as follows:

$$N_1 = \frac{R' - R_1}{\lambda/2} = \frac{k}{\pi}(R' - R_1) \tag{9.12a}$$

and

$$N_2 = \frac{R' - R_2}{\lambda/2} = \frac{k}{\pi}(R' - R_2) \tag{9.12b}$$

If the Fresnel numbers are sufficiently large ($N_1 \gg 0.125$ and $N_2 \gg 0.125$), the asymptotic expansions for the Fresnel integrals can be used in Equation (9.9) to yield an approximate solution for the diffracted sound:

$$p_d = i\frac{e^{-i\pi/4}}{8\pi\sqrt{2\pi kR'}}\frac{e^{ikR'}}{\sqrt{r_0 r_r}}\left\{\sec\left(\frac{\theta_0 - \theta_r}{2}\right) + \sec\left(\frac{\theta_0 + \theta_r}{2}\right)\right\} \tag{9.13}$$

The approximation in Equation (9.13) is invalid at the shadow boundaries. In this case, the exact integral representation of the total field [c.f. Equation (9.4)] should be used instead. Thus, the total sound field at the shadow boundary of the reflected wave is

$$p_T = \frac{e^{ikR_1}}{4\pi R_1} + \frac{e^{ikR_2}}{8\pi R_2} - \frac{ik}{4\pi}\int_{\sqrt{k(R'-R_1)}}^{\infty} \frac{H_1^{(1)}\left(kR_1 + s^2\right)}{\sqrt{s^2 + 2kR_1}}ds \tag{9.14}$$

When the receiver is located at the shadow boundary of the direct wave, the solution becomes

$$p_T = \frac{e^{ikR_1}}{8\pi R_1} + \frac{ik}{4\pi}\int_{\sqrt{k(R'-R_2)}}^{\infty} \frac{H_1^{(1)}\left(kR_2 + s^2\right)}{\sqrt{s^2 + 2kR_2}}ds \tag{9.15}$$

9.2.3 The Hadden and Pierce Solution for a Wedge

The last section considered the diffraction of sound by a thin half plane. Next to be investigated is sound diffraction around wedges for arbitrary source and receiver locations. An accurate integral representation of the diffracted wave from a semi-infinite wedge has been given by Hadden and Pierce [14]. Consider the geometrical configuration in Figure 9.3 of diffraction by a wedge with an exterior angle of ϕ. In a similar fashion to diffraction by a thin screen, the solution is composed of a geometrical solution of a direct wave term, a wave reflected from the right face of the wedge and a wave reflected from the left face. The direct wave is zero unless the receiver can 'see' the source.

There are no contributions from the reflected waves unless one can construct specularly reflected waves from either faces of the wedge. In addition to the geometrical solutions, the total field consists of a diffracted wave which can be decomposed into four terms:

$$p_d = \sum_{i=1}^{4} V(\varsigma_i), \tag{9.16}$$

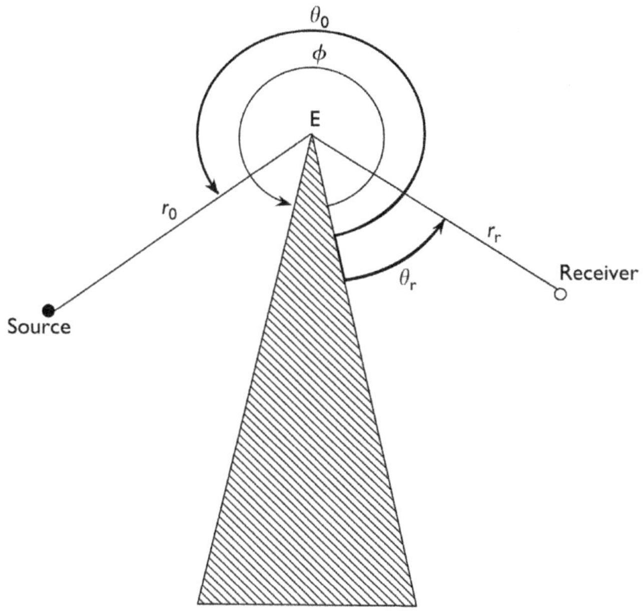

Figure 9.3 Diffraction of a spherical sound wave by a wedge.

where

$$\varsigma_1 = |\theta_r - \theta_0| \tag{9.17a}$$

$$\varsigma_2 = 2\varphi - |\theta_r - \theta_0| \tag{9.17b}$$

$$\varsigma_3 = \theta_r + \theta_0 \tag{9.17c}$$

$$\varsigma_4 = 2\varphi - (\theta_r + \theta_0) \tag{9.17d}$$

Each of the diffracted wave terms correspond to the sound paths between the source S_0, its image in the barrier S_0', the receiver R_0 and its image in the barrier R_0', i.e. paths $S_0ER_0(i=1)$, $S_0'ER_0\,(i=2)$, $S_0ER_0'\,(i=3)$ and $S_0'ER_0'\,(i=4)$ for the wedge shown in Figure 9.4. For each path, the diffracted field $V(\varsigma_i)$ can be calculated from

$$V(\varsigma_n) = -\left(1/4\pi^2\right)A_n\left(e^{ikL}/R'\right)F(\varsigma_n), \tag{9.18a}$$

where

$$A_n \equiv A(\varsigma_n) = (\upsilon/2)(-\beta - \pi + \varsigma_n) + \pi H(\pi - \varsigma_n), \tag{9.18b}$$

$$F(\varsigma_n) = \int_0^\infty \frac{kR'}{kR' + iy}\left(1 + \frac{i}{kR' + iy}\right)q_n e^{-y}dy \tag{9.18c}$$

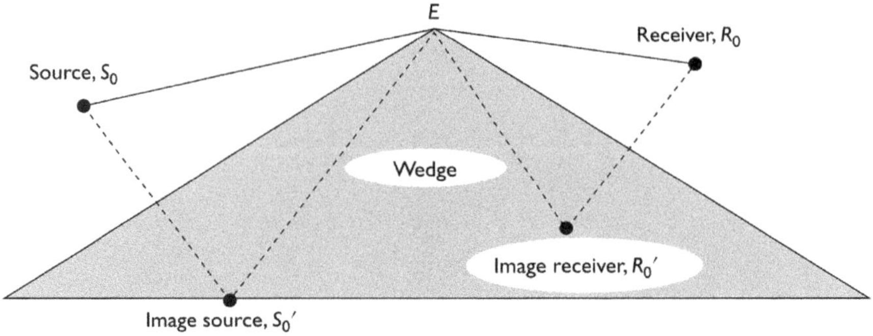

Figure 9.4 Source, image source, receiver, and image receiver for a wedge.

$$q_n = \frac{1}{|A_n|}\tan^{-1}\left(\tan|A_n|\tanh(\upsilon X_n)\right) \tag{9.18d}$$

$$\sinh X_n = \sqrt{\frac{y}{\alpha}\frac{i}{2} - \frac{y}{4kR'}} \tag{9.18e}$$

$$\alpha = kr_0r_r/R' \tag{9.18f}$$

and R', the shortest diffraction path from source to receiver, is given in (9.3c). In Equation (9.18b), the function $H(\ )$ is the Heaviside step function which is 1 for positive argument and zero otherwise. The parameter υ is the wedge index with $\upsilon = \frac{1}{2}$ for a thin screen and $\upsilon = \frac{2}{3}$ for a right-angle wedge.

The diffracted field can be computed by using Equations (9.18a–9.18f). As the integral is dominated by the e^{-y} factor, the rest of the integrand is non-oscillatory, and its magnitude is bounded by unity. The formulation allows evaluation of the sound fields for arbitrary point source locations in the vicinity of a rigid wedge. This representation of the sound field has been shown to agree well with experiment. Unlike the MacDonald solution, the formulation allows for reflections of the source and receiver on the faces of the barrier (wedge). As a result, it is possible to consider the effect of barrier surfaces with finite impedance by incorporating appropriate spherical wave reflection coefficient(s) for the paths from the image sources. However, instead of pursuing this idea here, a special case is explored of a thin screen with either the source or receiver located at a distance a few wavelengths from the edge of the screen such that the wedge index, $\upsilon = \frac{1}{2}$ and the length ratio $r_0r_r/R'^2 \to 0$ but $\alpha(\equiv kr_0r_r/R')$ remains finite. In such a case, the integral given in Equation (9.18c) can be simplified and expressed in terms of Fresnel integrals. Furthermore, the four diffraction terms can be grouped leading to the compact formula,

$$p_d = \left[(1+i)/2\right]\left[e^{ikR'}/4\pi R'\right]\left[A_D(X_+) + A_D(X_-)\right], \tag{9.19a}$$

where

$$X_+ = X(\theta_o + \theta_r), \tag{9.19b}$$

$$X_- = X(\theta - \theta_o), \tag{9.19c}$$

$$X(\Theta) = -2\sqrt{2r_o r_r / \lambda R'} \cos(\Theta / 2), \tag{9.19d}$$

$$A_D(X) = \mathrm{sgn}(X)\left[f(|X|) - ig(|X|)\right] \tag{9.19e}$$

The function $A_D(X)$ is known as the diffraction integral and can be expressed in terms of the auxiliary Fresnel integrals $f(x)$ and $g(x)$ given by

$$f(x) = \left[\frac{1}{2} - S(X)\right]\cos\left(\frac{\pi X^2}{2}\right) - \left[\frac{1}{2} - C(X)\right]\sin\left(\frac{\pi X^2}{2}\right) \tag{9.20a}$$

and

$$g(x) = \left[\frac{1}{2} - C(X)\right]\cos\left(\frac{\pi X^2}{2}\right) + \left[\frac{1}{2} - S(X)\right]\sin\left(\frac{\pi X^2}{2}\right) \tag{9.20b}$$

The Fresnel integrals $C(X)$ and $S(X)$ are defined in (9.11a) and (9.11b), respectively. Note that the first and fourth terms (with $i = 1$ and 4) in (9.16) are combined to give the term $A_D(X_-)$ in (9.19a). On the other hand, the second and the third terms in (9.16) are grouped to yield the term $A_D(X_+)$.

Figure 9.5 shows a comparison of the two exact solutions with experimental data.

The insertion loss (IL) of the screen (sometimes known as the attenuation, Att) defined by

$$IL = Att = 20\log_{10}\left(\left|\frac{p_{w/o}}{p_w}\right|\right), \tag{9.21}$$

where p_w and $p_{w/o}$ are the total sound fields with or without the barrier, respectively, is a useful method of assessing the barrier's acoustic performance since it accounts for attenuation due to other causes, e.g. soft ground effect, before the construction of the barrier which may be affected by the barrier's construction.

9.2.4 Approximate Analytical Formulation

Many approximate analytical solutions for diffraction by a half plane include the physical interpretation of diffraction suggested by Young and Fresnel. The most common one, the Fresnel-Kirchhoff approximation, is a

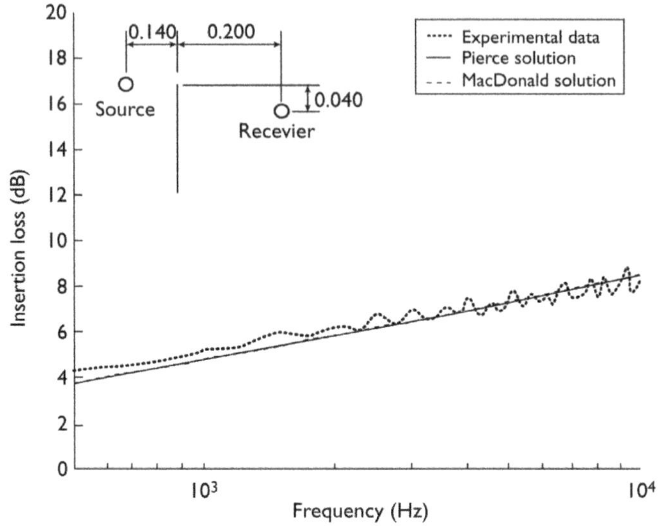

Figure 9.5 Comparison of predictions of the MacDonald solution and Hadden and Pierce formulation with laboratory data. The insertion loss of the thin half plane is plotted against the Fresnel Number, *N*1.

mathematical representation of the Huygens-Fresnel principle and is used widely in optics. While, typically, the wavelengths of optical light are smaller than obstacles such as screens, the Fresnel-Kirchhoff approximation represents a high-frequency approximation in acoustics. It results from solving the Helmholtz equation with the aid of Green's theorem. The sound field behind a screen is expressed as the surface integral over the open aperture above the screen. Alternatively, by applying Babinet's principle, the same result can be obtained if the surface integral is performed on the barrier surface. The details of Young's and Fresnel's studies and the Fresnel-Kirchhoff approximation can be found in various textbooks by Hecht [15] and Born and Wolf [2].

Rubinowics [16] transformed Kirchhoff's solution to a new diffraction formula, which is now known as Rubinowics-Young formula. He intended to investigate waves diffracted by a screen containing an aperture. The diffracted field was expressed as a line integral along the aperture edge instead of a surface integral as stated in the Kirchhoff's solution. In addition, Rubinowics was able to show that the Kirchhoff solution for plane or spherical incident waves can be decomposed into two portions: the unobstructed portion of incident wave and the edge-scattered wave. Embleton [17] derived a formula for sound diffracted by a 2-D barrier based on the Rubinowics-Young formula. In his studies, Embleton assumed that the line integral was along the edge of the barrier and a semicircular arc at infinity that connects the two ends of the barrier edge. Embleton's formula is convenient for numerical implementation because the integration variable is reduced to one

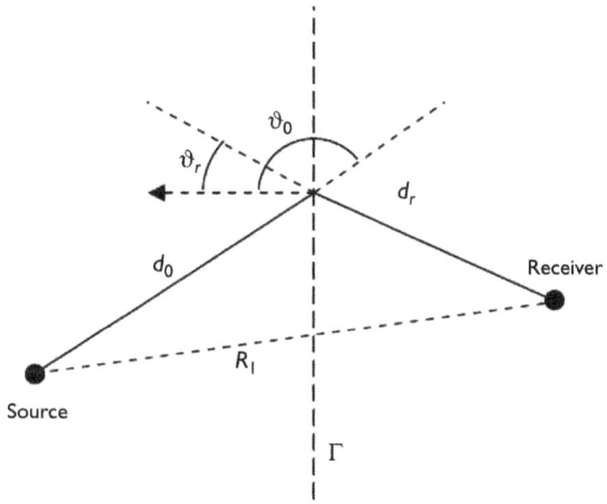

Figure 9.6 Geometry and notation used for the Fresnel-Kirchhoff approximation for the sound field with a transparent screen in front of a source.

dimension only. In this section, we present the Fresnel-Kirchoff approximation scheme only.

Consider the configuration shown in Figure 9.6.

The dashed line represents an infinite plane surface between the source and receiver. Again, the time-dependent factor $e^{-i\omega t}$ is understood and suppressed throughout. According to the Fresnel-Kirchoff diffraction formula [2], the sound pressure at the receiver can be written as

$$p_T = \frac{ik}{16\pi^2} \iint_\Gamma \frac{(\cos\vartheta - \cos\vartheta_0)}{d_0 d_r} \exp\left[ik(d_0 + d_r)\right] dA \tag{9.22}$$

where the integral is to be evaluated over a surface Γ of infinite extent.

The variables d_0 and d_r are the distances from a point on the surface Γ to the source and to the receiver, respectively. The angles ϑ and ϑ_0 are defined in Figure 9.6. The integral of Equation (9.22) can be evaluated exactly to yield the free field sound pressure as

$$p_T = \frac{e^{ikR_1}}{4\pi R_1} \tag{9.23}$$

Suppose that a semi-infinite thin screen is interposed between the source and receiver as shown in Figure 9.7. The screen is represented by the surface Γ_2 and the 'aperture' is represented by the surface Γ_1.

The advantage of adopting Equation (9.22) to calculate the sound field is that the effect of a screen can be modelled by assuming that there is no contribution to the pressure at the receiver point from areas of the surface

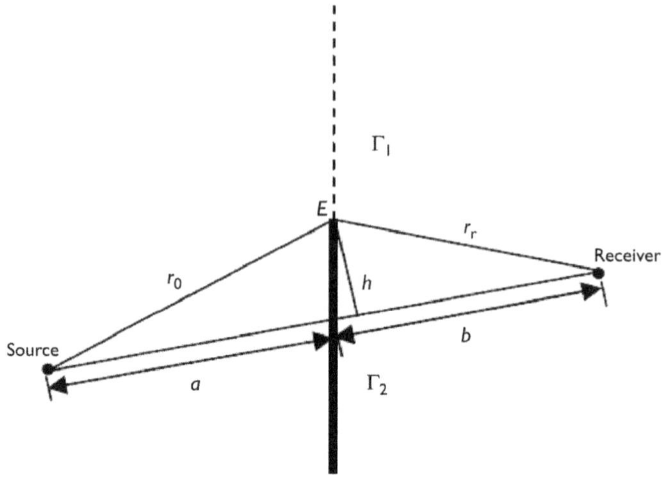

Figure 9.7 Geometry and notation used for the Fresnel-Kirchhoff approximation for the sound field with an aperture above a rigid screen.

corresponding to the screen. For the situation in Figure 9.7 the pressure at the receiver can be calculated approximately from

$$p_T = \frac{ik}{16\pi^2} \iint_{\Gamma_1} \frac{(\cos\vartheta - \cos\vartheta_0)}{d_0 d_r} \exp\left[ik(d_0 + d_r)\right] dA \tag{9.24}$$

Invoking the Fresnel-Kirchoff diffraction integral for an approximate solution, we implicitly assume that the thin screen transmits and reflects no sound. The expression (9.23) can be evaluated readily by standard numerical quadrature techniques that provide a far field approximation of the total field. If the opening is small compared to the distances of the surface Γ from the source and the receiver, the factor $(\cos\vartheta - \cos\vartheta_0)/d_0 d_r$ does not change appreciably over the surface and can be moved out of the integral. Under this condition, Equation (9.24) can be simplified and expressed in terms of the Fresnel integrals as

$$p_T = \frac{e^{ik(R_1+R_2)}}{8\pi(R_1+R_2)}(1+i)\left[F_r(\infty) + F_r\left(\sqrt{2/\lambda\left(a^{-1}+b^{-1}\right)}h\right)\right], \tag{9.25}$$

where the distances a, b and h are defined in Figure 9.7.

When h in (9.25) is positive, the receiver is in the illuminated zone.

An interesting case is when $h \to -\infty$, *i.e.* the sound field in the absence of the screen. Then, according to Fresnel-Kirchhoff model [cf (9.25)], the total (free) sound pressure is $ie^{ik(a+b)}/4\pi(a+b)$, which differs from the exact solution by a phase factor of $\pi/2$. Although the condition for a small aperture in the study of Fresnel diffraction is violated in the problem of the half plane, the approximation given in (9.24) appears to give reasonable predictions of barrier attenuation. However, there are many reported discrepancies between predictions using the Fresnel-Kirchhoff model and both those using other more accurate theoretical models and data [18].

Using (9.22) and (9.24), the insertion loss of the screen, defined by (9.21) can be calculated from

$$IL = 10\log 2 - \log\left\{F_r(\infty) - F_r\left(\sqrt{2\left(a^{-1}+b^{-1}\right)/\lambda h}\right)\right\}. \tag{9.26}$$

To give a better perspective of different approximate solutions, the comparison of the Fresnel-Kirchhoff formula with data will be deferred until the next section because it is more desirable to discuss a well-known empirical scheme first.

9.3 EMPIRICAL FORMULATIONS FOR STUDYING THE SHIELDING EFFECT OF BARRIERS

The most direct way to investigate the acoustical performance of a noise barrier is to conduct full-scale field measurements. Alternatively, it can be determined through indoor scale model experiments. The first known graph for sound attenuation in the shadow zone of a rigid barrier due to a point source was developed by Redfearn in the 1940s [19]. The noise attenuation was given as a function of two parameters in his graphs, namely, (i) the effective height of barriers normalized by the wavelength and (ii) the angle of diffraction. Nearly 30 years later, Maekawa [20] measured the attenuation of a thin rigid barrier, with different source and receiver locations. In his experimental measurements, a pulsed tone of short duration was used as the noise source. He presented his experimental data in a chart where the attenuation was plotted against a single parameter known as the Fresnel number. The Fresnel number is the numerical ratio of path difference (the difference between the lengths of the diffracted path and the direct path of sound) to the half of a sound wavelength. In the same period, Rathe presented an empirical table for the sound attenuation by a thin rigid barrier due to a point source [21]. His empirical table was based on a large number of data sets. The attenuation was given in octave frequency bands normalized by the reference frequency with the Fresnel number of 0.5. Kurze and Anderson [22,23] reviewed diffraction theory from Keller [24] and used Maekawa's and Redfearn's experimental data to derive empirical formulae for the barrier attenuation. The attenuation is expressed as the function of relative locations of source and receiver, including the diffracted angles at the source and receiver side. There are some common features of the experimental investigations of these studies. First, a point source was used to generate incident waves in these early experimental studies. Second, the Fresnel number is an elementary parameter for determining the barrier attenuation. As a result of their comparative simplicity, Kurze–Anderson's formulae and Maekawa's chart are used extensively for engineering calculations.

Although there are several other empirical models and charts to predict the acoustic performance of a thin barrier, only the Maekawa chart and other related developments will be discussed in the following sections. Maekawa described the attenuation of a screen using an empirical approach based on the important parameters affecting the screening. An important parameter which appears in the treatments described above is the difference in path lengths from the source to the receiver via the top of the barrier ($r_0 + r_r$ in Figure 9.7) and the direct path length from source to receiver, $R_1 = a + b$. This is a measure of the depth of shadow produced by the screen for the given source and receiver positions. The path difference, δ_1, is given by

$$\delta_1 = (r_0 + r_r) - R_1 \tag{9.27}$$

The other important parameter is the wavelength of the sound, λ. The longer the wavelength the greater is the diffracted wave amplitude.

These parameters may be combined into the Fresnel number N_1 associated with the source, i.e.

$$N_1 = \frac{2\delta_1}{\lambda} \tag{9.28}$$

In the 1960s, Maekawa obtained a comprehensive set of data for the insertion loss, which corresponds to the attenuation in the absence of ground. The measurements were made in a test room using a spherically spreading pulsed tone of short duration. The solid line in Figure 9.8 shows predictions according to Maekawa's chart for the attenuation of sound by a semi-infinite plane screen as a function of positive N_1 compared with data from measurements made in an anechoic chamber. Various geometrical configurations were used, and the frequency range varied from 500 Hz to 10 kHz. Despite the wide range of Fresnel numbers associated with these data, they are consistent with the measurements shown in Maekawa's chart [20] and hence with the design curve. Also shown are predictions of the MacDonald solution (9.9) and the Fresnel-Kirchoff approximation (9.26).

Maekawa obtained data not only in the shadow zone but also in the illuminated zone where the source can see the receiver. In his notation, a negative value of N_1 signifies that the receiver is located in the illuminated zone although the same definition (9.28) is used. For $N_1 > 1$ the abscissa is logarithmic. Below this value, it has been adjusted to make the design curve linear. For $N_1 = 0$ the attenuation is 5 dB. At this point the source is just visible over the top of the barrier. For $N_1 < 0$, the receiver is in the illuminated zone and the attenuation quickly drops to zero. In Figure 9.8, we only show N_1 ranging from 0.01 to 9. A function which fits this curve quite well is

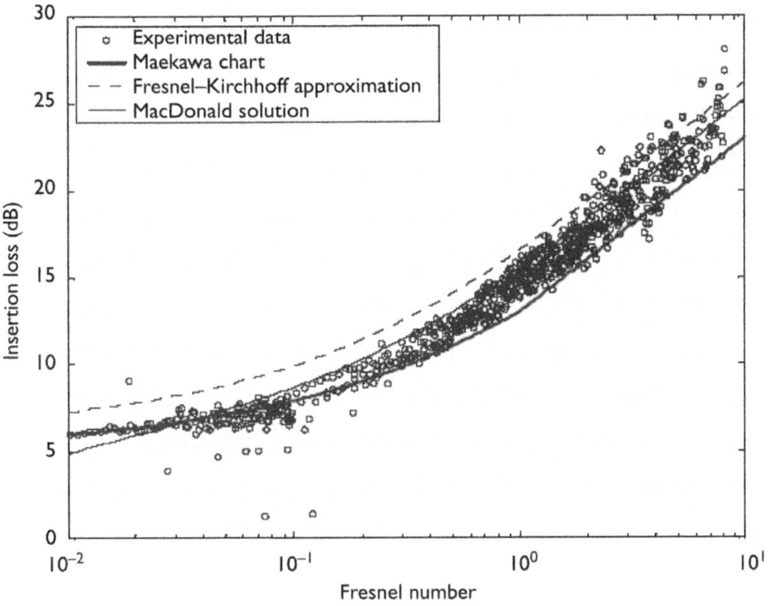

Figure 9.8 Comparison between laboratory data for the sound attenuation by a semi-infinite hard screen in free space and predictions of the Maekawa Chart (thick solid line), the Fresnel Kirchoff approximation (broken line) and the MacDonald solution (thin solid line).

$$Att = 10\log(3 + 20N)\ \text{dB} \tag{9.29}$$

This formula was originally defined only for $N > 0$ but is often used for $N_1 > -0.05$ [25]. More accurate formulae have been proposed by Yamamoto and Takagi [26]. Based on Maekawa's original chart, they developed four different approximations.

$$Att = \begin{cases} 10\log\left[1 + G(N_1)(N_1 + 0.3)\right] & \text{for } N_1 > -0.3 \\ 0 \text{ for } N_1 < -0.3 \end{cases} \tag{9.30a}$$

$$Att = \begin{cases} 10\log N_1 + 13 \text{ for } N_1 > 1 \\ 5 + \dfrac{8N_1}{|N_1|^{0.55+0.143|N_1|}} \text{ for } -0.3 < N_1 < 1 \\ 0 \text{ for } N_1 < -0.3 \end{cases} \tag{9.30b}$$

$$Att = \begin{cases} 10\log N_1 + 13 \text{ for } N_1 > 1 \\ 5 \pm 8|N_1|^{0.438} \text{ for } -0.3 < N_1 < 1 \\ 0 \text{ for } N_1 < -0.3 \end{cases} \tag{9.30c}$$

$$Att = \begin{cases} 10\log N_1 + 13 \text{ for } N_1 > 1 \\ 5 \pm 9.07674 \times \sinh^{-1}|N_1|^{0.485} \text{ for } -0.3 < N_1 < 1. \\ 0 \text{ for } N_1 < -0.3 \end{cases} \tag{9.30d}$$

where the function $G(N_1)$ in (9.30a) is given by

$$G(N_1) = 3.621\left\{\tan\left(\frac{N_1 - 5\times10^{-3}}{1.45\times10^{-2}}\right) + \frac{\pi}{2}\right\} + 6.165\left\{1 - e^{-0.205(N_1+0.3)}\right\} + 2.354. \tag{9.30e}$$

Any one of above expressions gives good agreement with the data obtained from Maekawa's chart. The maximum difference is less than 0.5 dB in (i) and is no more than 0.3 dB in cases (ii), (iii) and (iv). Although formula (i) leads to a marginally greater 'error', it has the advantage that only one formula is required to describe the whole chart. Using the data shown in Figure 9.8 again, predictions of the Yamamoto and Takagi formulae and the MacDonald solution are compared in Figure 9.9.

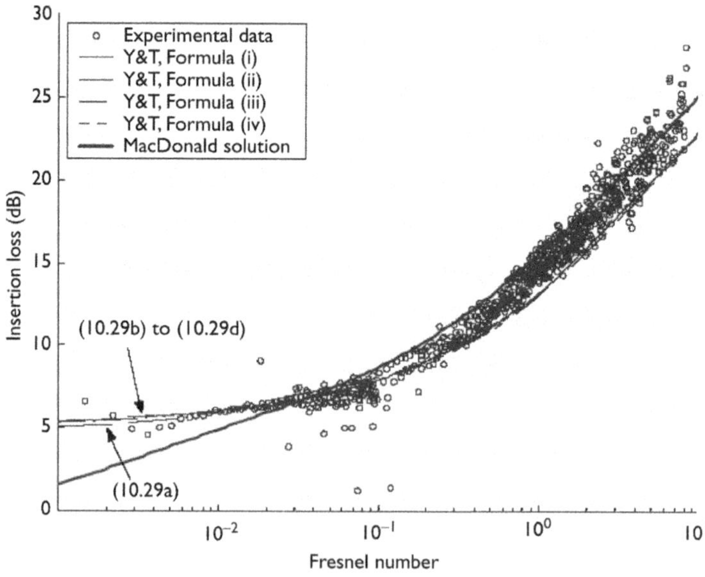

Figure 9.9 Comparison between laboratory data and predictions of the different empirical formulae of Yamamoto and Takagi (Y&T) and of the MacDonald solution (9.29). The insertion loss of the thin screen is plotted against the Fresnel Number, $\mathbf{N}_1$.

Kurze and Anderson [22] have derived another simple formula that has been used widely. Maekawa's curve can be represented mathematically by

$$Att = 5 + 20\log\frac{\sqrt{2\pi N_1}}{\tanh\sqrt{2\pi N_1}} \tag{9.31}$$

The empirical formulae associated with Maekawa's chart only predict the amplitude of the attenuation and no wave interference effects are predicted [20]. The next section explains how they can be extended to model the interference effects of contributions from different diffracted wave paths.

Menounou [27] has modified Maekawa's chart from a single curve with one parameter to a family of curves with two Fresnel numbers. The first Fresnel number is the conventional Fresnel number associated with the relative position of the source to the barrier and the receiver. The second Fresnel number represents the relative position of the image source and the receiver. Menounou also modified the Kurze–Anderson empirical formula and the Kirchhoff solution. Unlike earlier studies, Menounou considered plane, cylindrical and spherical incident waves. Her study combines simplicity of use with the accuracy of sophisticated diffraction theories. An improved Kurze–Anderson formula that allows a better estimation of the barrier attenuation by including the effect of image source on the total field is

$$Att = Att_s + Att_b + Att_{sb} + Att_{sp} \tag{9.32a}$$

where

$$Att_s = 20\log\frac{\sqrt{2\pi N_1}}{\tanh\sqrt{2\pi N_1}} - 1, \tag{9.32b}$$

$$Att_b = 20\log\left[1 + \tanh\left(0.6\log\frac{N_2}{N_1}\right)\right], \tag{9.32c}$$

$$Att_{sb} = \left(6\tanh\sqrt{N_2} - 2 - Att_b\right)\left(1 - \tanh\sqrt{10N_1}\right), \tag{9.32d}$$

$$Att_{sp} = -10\log\left[\frac{1}{\left(\frac{R'}{R_1}\right)^2 + \left(\frac{R'}{R_1}\right)}\right] \tag{9.32e}$$

The term Att_s is a function of N_1 which is a measure of the relative position of the receiver from the source. The second term depends on the ratio of N_2/N_1 and is a measure of the proximity of either the source or the receiver to the half plane.

The third term is only significant when N_1 is small and is a measure of the proximity of the receiver to the shadow boundary. The last term is a

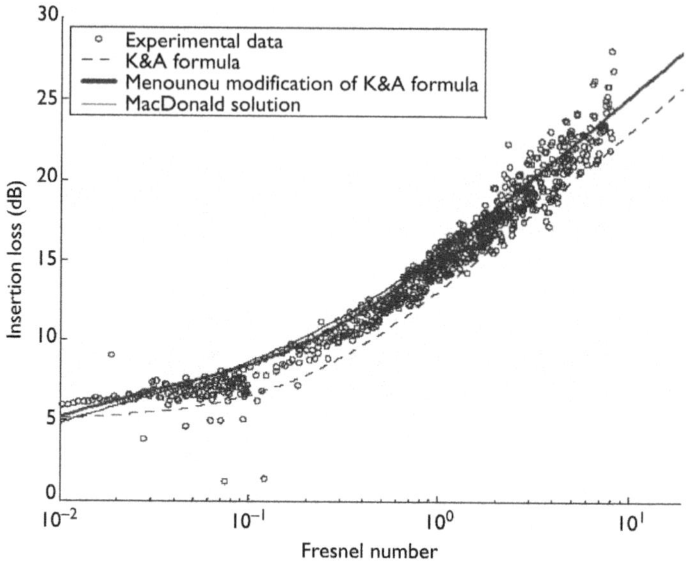

Figure 9.10 Comparison between laboratory data and predictions of the Kurze and Anderson (K&A) empirical formula. Predictions of Menounou's modification on the Kurze and Anderson formula is shown also. The insertion loss of the thin half plane is plotted against the Fresnel Number, $\boldsymbol{N}_1$.

function of the ratio R'/R_1 which is used to account for the diffraction effect due to spherical incident waves.

Figure 9.10 shows the predicted attenuation according to the Kurze and Anderson formula and the Menounou modification. Again, both formulae appear to give predictions in agreement with the data for a thin screen.

9.4 THE SOUND ATTENUATION BY A THIN PLANE ON THE GROUND

The diffraction of sound by a semi-infinite rigid plane is a classical problem in wave theory, dating back to the 19th century. There was renewed interest in the problem between the 1970s and early 1980s in connection with the acoustic effectiveness of outdoor noise barriers. Since outdoor noise barriers are always on the ground, the attenuation is different from that discussed in Sections 9.2 and 9.3 for a semi-infinite plane screen. Extensive literature reviews on the pertinent theory and details on precise experimental studies have been carried out by Embleton and his co-workers [28,29].

Figure 9.11 shows the situation of interest. A source S_g is located at the left side of the barrier, a receiver R_g at the right side of the barrier and O is

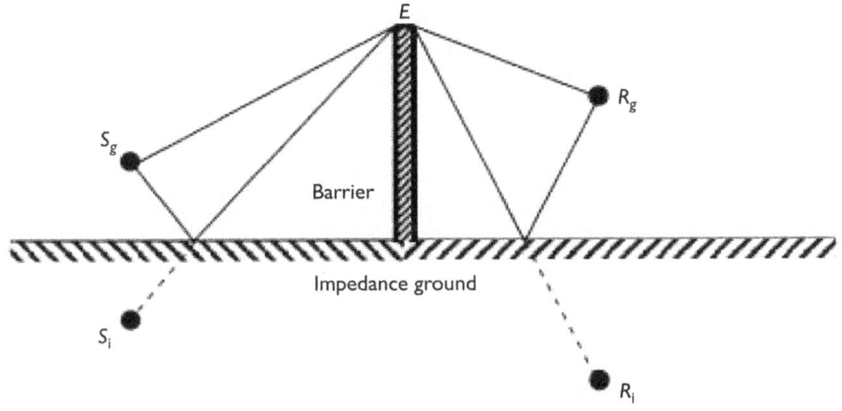

Figure 9.11 Ray paths for a thin barrier on the ground between the source and receiver.

the diffraction point on the barrier edge. The sound reflected from the ground surface can be described by the source's 'mirror' image, S_i. On the receiver side, sound waves will also be reflected from the ground and this effect can be considered in terms of the image of the receiver, R_i. The pressure at the receiver is the sum of four terms which correspond to the sound paths S_gOR_g, S_iOR_g, S_gOR_i and S_iOR_i.

If the surface is a perfectly reflecting ground, the total sound field is the sum of the diffracted fields of these four paths,

$$P_T = P_1 + P_2 + P_3 + P_4 \tag{9.33a}$$

where

$$P_1 = P(S_g, R_g, O), \tag{9.33b}$$

$$P_2 = P(S_i, R_g, O), \tag{9.33c}$$

$$P_3 = P(S_g, R_i, O), \tag{9.33d}$$

$$P_4 = P(S_g, R_i, O), \tag{9.33e}$$

and $P(S, R, O)$ is the diffracted sound field due to a thin barrier for given positions of source S, receiver R and the point of diffraction at the barrier edge O. If the ground has a finite impedance (such as grass or a porous road surface) then the pressure corresponding to rays reflected from these surfaces will need to be multiplied by the appropriate spherical wave reflection coefficient(s) to allow for the change in phase and amplitude of the wave on reflection as follows:

$$P_T = P_1 + Q_s P_2 + Q_R P_3 + Q_s Q_R P_4, \tag{9.34}$$

where Q_s and Q_R are the spherical wave reflection coefficient at the source and receiver side, respectively. The spherical wave reflection coefficients can be calculated according to Equation (2.40c) for different types of ground surfaces and source/receiver geometries.

For a given source and receiver position, the acoustic performance of the barrier on the ground is normally assessed by use of either the excess attenuation (EA) or the insertion loss (IL). They are defined as follows:

$$EA = SPL_f - SPL_b \tag{9.35a}$$

$$IL = SPL_g - SPL_b, \tag{9.35b}$$

where SPL_f is the free field noise level, SPL_g is the noise level with the ground present and SPL_b is the noise level with the barrier and ground present. Note that, in the absence of a reflecting ground, the numerical value of EA (see also Att defined by (9.21) in Section 9.3) is the same as IL. If the calculation is carried out in terms of amplitude only, as described in [20], then the attenuation Att_n for each ray path can be directly determined from the appropriate Fresnel number F_n for that path. The excess attenuation of the barrier on a rigid ground is then given by

$$EA = -10\log\left(10^{-\left|\frac{Att_1}{10}\right|} + 10^{-\left|\frac{Att_2}{10}\right|} + 10^{-\left|\frac{Att_3}{10}\right|} + 10^{-\left|\frac{Att_4}{10}\right|}\right) \tag{9.36}$$

The attenuation for each path can either be calculated by the empirical or analytical formulae listed in Sections 9.2 and 9.3 depending on the complexity of the model and the required accuracy. If the calculation demands an accurate estimation on both the amplitude and phase of the diffracted waves, then the MacDonald solution or the Hadden and Pierce solution should be used although the latter solution will give a more accurate solution for both source and receiver locating very close to the barrier [29]. Nevertheless, the former solution is equally accurate for most practical applications where both the source and receiver are far from the barrier edge.

Figure 9.12(a) displays predicted insertion loss spectra for propagation from a point source for the source–receiver-barrier geometry indicated. The sound pressure was calculated using the four sound paths described above and the MacDonald expression (9.9). Curves plotted with a thin line is for a rigid ground surface and a dashed line is for an absorbing surface such as grassland. For the grassland impedance, the Delany and Bazley one-parameter model with an effective flow resistivity of 300 kPa s m^{-2} has been used. The oscillations are due to interference between the sound waves along the four ray paths considered. The curve plotted as a thick line is the result using Maekawa's method to determine the attenuation of each ray path. The total insertion loss is obtained by summing all four paths incoherently by

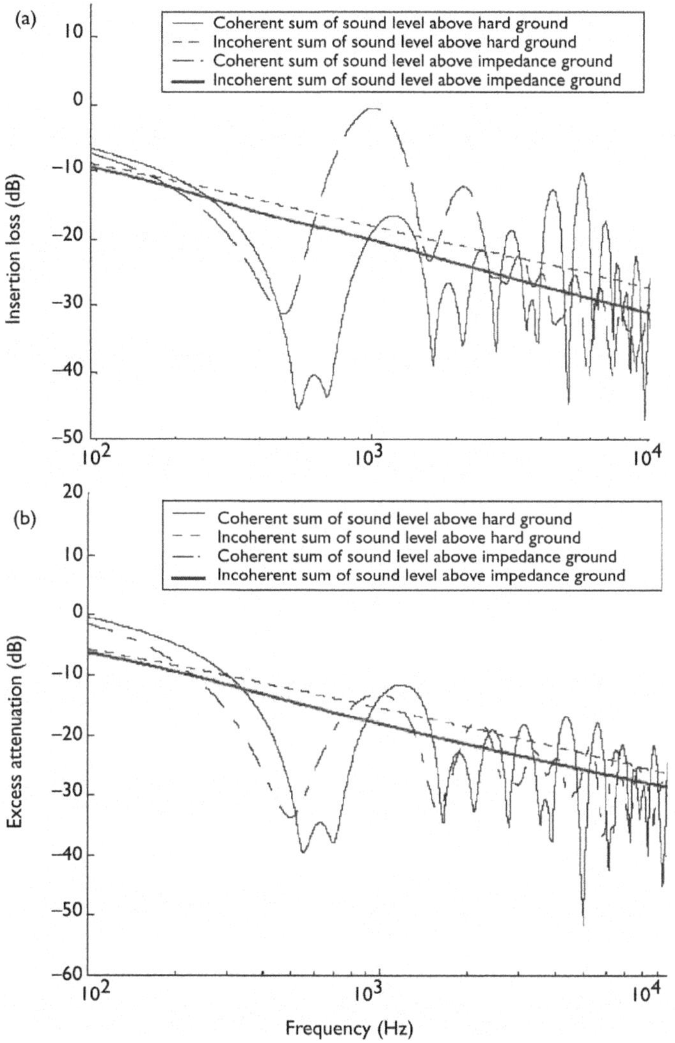

Figure 9.12 Comparison of predictions resulting from summing the contributions of different ray paths incoherently and coherently for the sound field (a) insertion loss due to a point source and (b) excess attenuation from a coherent line source behind an outdoor noise barrier placed on a ground surface.

assuming the point source is located above a rigid ground. The same method is used to predict the insertion loss above the absorbing surface. The predicted results (dotted line) are presented in Figure 9.12(a). Neither of the incoherent predictions oscillates since the phases of the waves are not considered. Nevertheless, these lines give a reasonable approximation to the results of more complex calculations [28].

Figure 9.12(b) shows results for a similar geometry except that in this case a line of *incoherent* point sources, separated by approximately 1.5 m along the nominal road line, has been assumed and the calculation has been carried out for each source. The averaging over all the predictions for the different sources has smoothed the curves. Note that the *IL* for the grassland case becomes negative around 500 Hz indicating that the 'soft' ground attenuation is greater than the screening effect of the barrier for these conditions.

In both Figures 9.12(a) and (b), the height of the barrier is taken as 3 m. The source is assumed to be 5 m from the barrier and 0.3 m above the ground. The receiver is assumed to be 30 m from the barrier and 1.2 m above the ground.

Lam and Roberts [30] have suggested a relatively simple approach for modelling full wave effects. In their model the amplitude of the diffracted wave, Att_1, may be calculated using a method such as that described in Section 9.3 [34]. However, the phase of the wave at the receiver is calculated from the path length via the top of the screen assuming a phase change in the diffracted wave of $\pi/4$. It is assumed that the phase change is constant for all source, barrier and receiver heights and locations. For example, the diffracted wave for the path S_gOR_g would be given by

$$P_1 = Att_1 e^{i\left[k(r_0 + r_r) + \pi/4\right]} \tag{9.37}$$

It is straightforward to compute the contributions for other diffracted paths and hence the total sound field can be calculated using Equation (9.34) for impedance ground and Equation (9.33a) for a hard ground. This approach provides a reasonable approximation for many conditions normally encountered in practical barrier situations where source and receiver are many wavelengths from the barrier and the receiver is in the shadow zone. The scheme works well for a barrier located on a hard ground but less so if the ground has finite impedance. To illustrate this point, Figure 9.13 compares data obtained in an anechoic chamber near a thin barrier placed on an absorbing surface of thickness 0.015 m with predictions of the Hadden and Pierce formula and Lam and Roberts approximate scheme.

Predictions obtained by using the Hadden and Pierce formula agree well with the experimental observations. On the other hand, while the Lam and Roberts model predicts the general trend of the insertion loss spectrum it does not predict its magnitude as well, particularly in frequency intervals where the ground effects are strong.

9.5 NOISE REDUCTION BY A FINITE-LENGTH BARRIER

All outdoor barriers have finite length and, under certain conditions, sound diffracting around the vertical ends of the barrier may be significant. The MacDonald solution, (see Section 9.2.2) has been adapted to include the

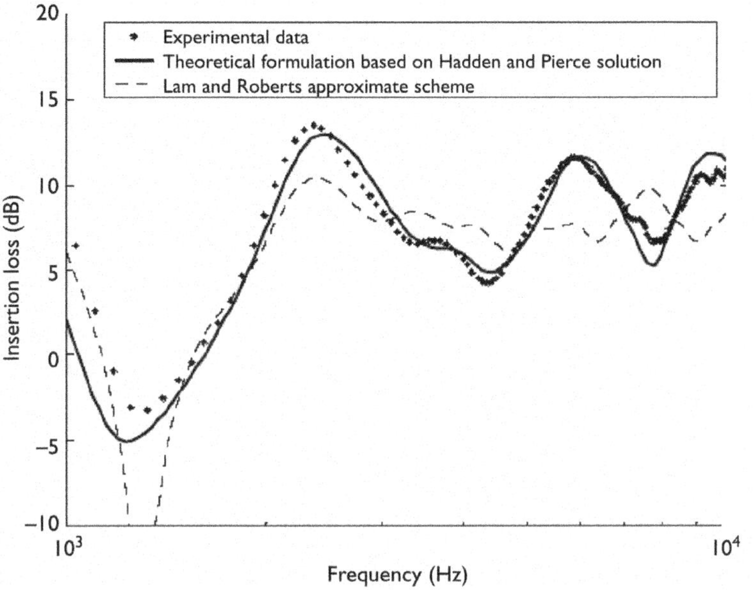

Figure 9.13 Comparison of laboratory data (points) with predictions obtained from the Hadden and Pierce solution (solid line) and the Lam and Roberts approximate scheme (broken line) for a barrier place on an impedance ground (Delany and Bazley hard backed layer (Eqns. 5.1 and 5.2) with effective flow resistivity 9000 Pa s m^{-2} and (known) layer thickness of 0.015 m). The source is 0.355 m from the 0.3 m high barrier and 0.163 m above the ground. The receiver is 0.342 m from the barrier and 0.198 m high.

sound fields due to rays diffracted around the two vertical edges of a finite-length barrier [31]. There are other diffraction theories for the vertical edge effects which have met with varying degrees of success when compared with data [32,33]. In this section, a more practical method is described [30]. The accuracy of this approach has been verified [34].

Figure 9.14 shows eight diffracted ray paths contributing to the total field behind a finite-length barrier. In addition to the four 'normal' ray paths associated with diffraction at the top edge of the barrier (see Figure 9.7), four more diffracted ray paths from the vertical edges, including two ray paths each from either side, have been identified. In principle, there are reflected–diffracted–reflected rays also which reflect from the ground twice, but these are ignored. This is a reasonable assumption if the ground is acoustically soft. The two rays at either side are, respectively, the direct diffracted ray and the diffracted–reflected ray from the image source. The reflection angles of the two diffracted–reflected rays are independent of the barrier position. These rays will either reflect at the source side or on the receiver side of the barrier depending on the relative positions of the source, receiver and barrier. Both possibilities are illustrated in Figure 9.14. The total field is given by

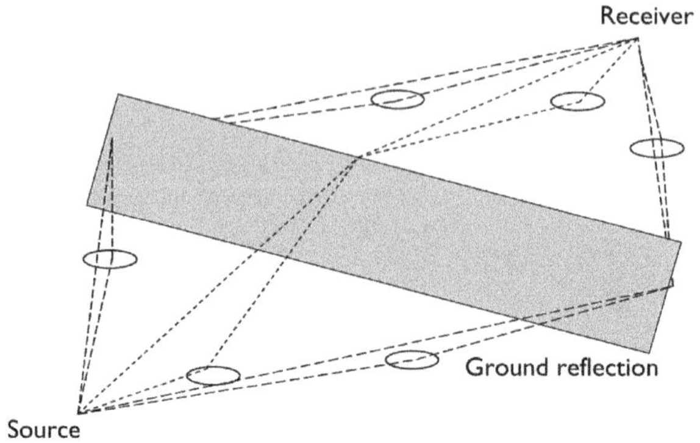

Figure 9.14 Eight ray paths associated with sound diffraction by a finite length barrier.

$$P_T = P_1 + Q_s P_2 + Q_R P_3 + Q_s Q_R P_4 + P_5 + Q_R P_6 + P_7 + Q_R P_8, \tag{9.38}$$

where $P_1 - P_4$ are those given in Equations (9.33a–9.33d) for diffraction at the top edge of the barrier. The contributions from the two vertical edges E_1 and E_2 are

$$P_5 \equiv P(S_g, R_g, E_1),\ P_6 \equiv P(S_i, R_g, V_1),\ P_7 \equiv P(S_g, R_g, V_2)$$

and

$$P_8 \equiv P(S_i, R_g, E_2),$$

respectively. The spherical wave reflection coefficient Q_R is used for P_6 and P_8 as the reflection is assumed to take place at the receiver side but Q_s should be used instead if the reflection happens at the source side.

Although we can use any of the more accurate diffraction formulae described in Section 9.2 to compute P_1–P_8, a simpler approach, following Lam and Roberts [30], is to assume that each diffracted ray has a constant phase shift of $\pi/4$ regardless the position of source, receiver and diffraction point. Consequently, the empirical formulations described in Section 9.3 can also be used to compute the amplitudes of the diffracted rays. Indeed, Muradali and Fyfe [34] compared the use of the Maekawa chart, the Kurze–Anderson empirical equation and the Hadden–Pierce analytical formulation with a wave-based boundary element method (BEM). Predictions of the ray-based approaches show excellent agreement with those of the relatively computationally intensive BEM formulation.

In Figure 9.15, data obtained by Lam and Roberts [Figure 3 of Ref. 29] are compared with predictions using their proposed approximation scheme. In their studies, the measurements were conducted over a hard ground with the barrier height of 0.3 m and length of 1.22 m. The source was located at

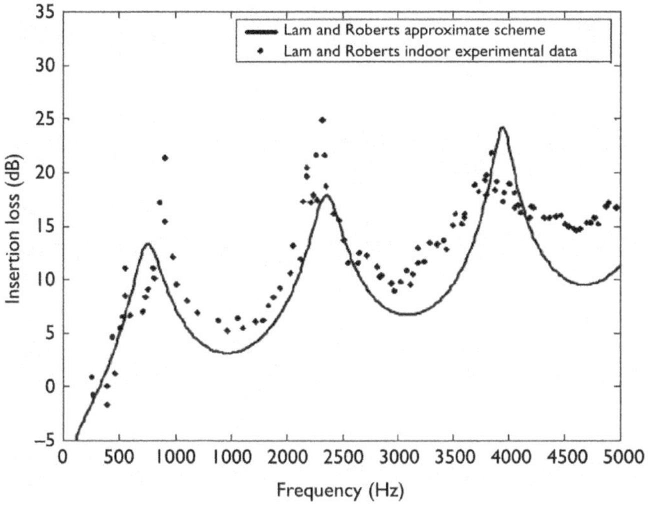

Figure 9.15 Comparison between laboratory data for the insertion loss of a barrier (0.3 m high and 1.22 m long) on hard ground and predictions using the Lam and Roberts approximate scheme [30]. The source was 0.033 m high and 1.008m from the barrier. The receiver was on the ground and 1.491 m from the barrier.

1.009 m from the barrier at the source side and 0.033 m above the ground. The receiver was situated at 1.491 m from the barrier at the receiver side with the microphone placed on the ground. Predictions based on the other numerical schemes detailed in Sections 9.2 and 9.3 are not shown because they tend to give similar results to the Lam and Roberts predictions.

9.6 ADVERSE EFFECT OF GAPS IN BARRIERS

When the transmission of sound through the barrier is negligible, the acoustic field in the shadow region is mainly dominated by the sound diffracted around the barrier. However, in some cases, leakage will occur due to shrinkage, splitting and warping of the panels, and weathering of the acoustic seals. The problems of shrinkage and splitting are particularly acute for noise barriers made of timber. Sometimes gaps are unavoidable since spaces are required, for example, for the installation of lamp posts in urban districts. Watts [35] has investigated the resulting sound degradation of screening performance due to such leakage. He used a 2-D numerical model based on the BEM to predict the sound fields behind barriers of various heights with different gap widths and distributions. The A-weighted sound leakage increases as the gap size increases. Also, he reported (a) that sound leakage is more significant in the region close to the barrier and (b) that there are no significant leakage effects of noise from heavy vehicles. It should be noted that only horizontal gaps can be included in his 2-D boundary element model. The theory developed by Wong and Li [36] is outlined here and it is shown that the resulting predictions of the effects of barrier leakages agree reasonably well with data from laboratory and outdoor experiments.

Consider the barrier diffraction problem shown in Figure 9.16.

Based on the Helmholtz integral formulation, Thomasson [37] derived an approximate scheme for the prediction of the sound field behind an infinitely long barrier. The solution for a thin rigid screen is given by

$$\varphi(\mathbf{r}_R,\mathbf{r}_S) = \varphi_L(\mathbf{r}_R|\mathbf{r}_S) - 2\iint_{S_B} \varphi_L(\mathbf{r}_0|\mathbf{r}_S)\frac{\partial \varphi_L(\mathbf{r}_R|\mathbf{r}_0)}{\partial x_R}\, dS, \tag{9.39}$$

where dS is the differential element of the barrier surface; $\mathbf{r}_S$, $\mathbf{r}_R$ and $\mathbf{r}_0$ are, respectively, the position of source, receiver and barrier surface. The symbol $\phi_L(\mathfrak{R}/\mathfrak{I})$ signifies the sound field at receiver points $\mathfrak{R}$ (either $\vec{r}_R$ or $\vec{r}_0$) due to the source located at $\mathfrak{I}$ (either $\vec{r}_s$ or $\vec{r}_0$). The computation of $\phi_L(\mathfrak{R}/\mathfrak{I})$ is straightforward if the Weyl-Van der Pol formula is used (see Equation (2.40a)). The sound field from source to receiver in the absence of a barrier can be determined by noting that the sound field vanishes when the rigid screen, Γ_∞ say, occupies the infinite plane of $y = 0$.

Then

$$\varphi_L(\mathbf{r}_R|\mathbf{r}_S) = 2\iint_{\Gamma_\infty} \varphi_L(\mathbf{r}_0|\mathbf{r}_S)\frac{\partial \varphi_L(\mathbf{r}_R|\mathbf{r}_0)}{\partial x_R}\, dS \tag{9.40a}$$

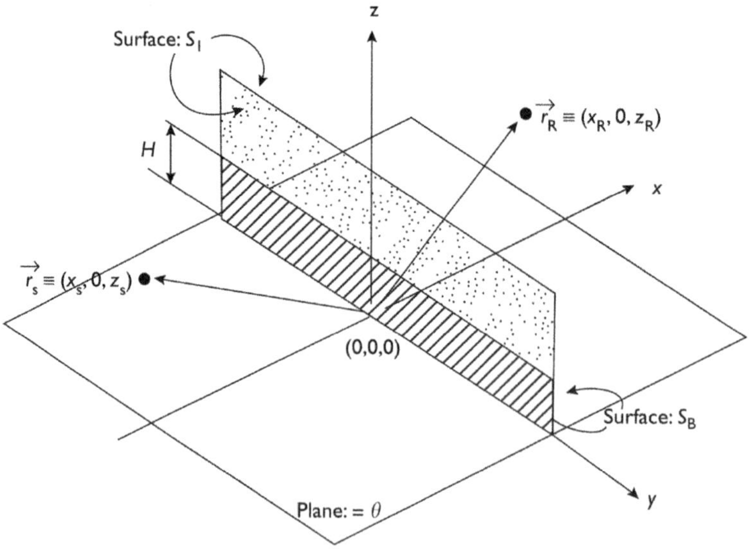

Figure 9.16 The co-ordinate system used in the Helmholtz integral formulation for a sound diffracted by a rigid infinitely long barrier of height H.

or

$$\varphi_L(\mathbf{r_R}|\mathbf{r_S}) = 2\int_0^{\infty}\int_{-\infty}^{\infty}\varphi_L(\mathbf{r_0}|\mathbf{r_S})\frac{\partial\varphi_L(\mathbf{r_R}|\mathbf{r_0})}{\partial x_R}dydz \tag{9.40b}$$

The sound field behind a rigid screen can be computed by substituting Equation (9.40a) into Equation (9.39) to yield

$$\varphi(\mathbf{r_R},\mathbf{r_S}) = 2\iint_{\Gamma_\infty-\Gamma_B}\varphi_L(\mathbf{r_0}|\mathbf{r_S})\frac{\partial\varphi_L(\mathbf{r_R}|\mathbf{r_0})}{\partial x_R}dS \tag{9.41}$$

Although the Thomasson approach (9.41) provides an accurate method for predicting the sound field behind a rigid screen, the use of Equation (9.39) in conjunction with the Weyl-Van der Pol formula is preferable because the area required for integration is generally smaller than that required in Equation (9.41). Hence, less computation is required for the evaluation of the sound fields behind a thin barrier. Since (9.39) is the most general formula for the computation of a sound field behind a thin barrier, it can be generalized to allow any arbitrary gaps in the barrier. Specifically, if there are gaps in the barrier, then their areas should not be included in the computation of the integral. For instance, suppose there is a vertical gap of width $2d$ in an otherwise infinitely long barrier of height H. Without loss of generality, the gap is assumed to extend from $y = -d$ to $y = d$. The sound field behind such a barrier can be computed by

$$\begin{aligned}\varphi(\mathbf{r_R},\mathbf{r_S}) = \varphi_L(\mathbf{r_R}|\mathbf{r_S}) - 2\int_0^{H}\int_{-\infty}^{-d}\varphi_L(\mathbf{r_0}|\mathbf{r_S})\frac{\partial\varphi_L(\mathbf{r_R}|\mathbf{r_0})}{\partial x_R}dydz \\ -2\int_0^{H}\int_{d}^{\infty}\varphi_L(\mathbf{r_0}|\mathbf{r_S})\frac{\partial\varphi_L(\mathbf{r_R}|\mathbf{r_0})}{\partial x_R}dydz\end{aligned} \tag{9.42}$$

Note that the term $\frac{\partial\varphi_L(\mathbf{r_R}|\mathbf{r_0})}{\partial x_R}$ can be regarded as the contribution due to horizontal dipoles on the surface of the barrier. The asymptotic solution for a horizontal dipole above an impedance plane is detailed in Chapter 3.

In many practical situations, the thin barrier may be assumed to be infinitely long and to have a constant cross section along the y-direction. In view of Equation (9.39), the sound field can therefore be expressed as

$$\varphi(\mathbf{r_R},\mathbf{r_S}) = \varphi_L(\mathbf{r_R}|\mathbf{r_S}) - 2\int_{l_B}\int_{-\infty}^{\infty}\varphi_L(\mathbf{r_0}|\mathbf{r_S})\frac{\partial\varphi_L(\mathbf{r_R}|\mathbf{r_0})}{\partial x_R}dydz, \tag{9.43}$$

where l_B is the constant barrier 'height'. The y-integral of the second term can be evaluated asymptotically to yield a closed form analytical expression.

The evaluation of the integral involves tedious algebraic manipulations but, eventually, the sound field can be simplified to

$$\varphi\left(\mathbf{r}_R,\mathbf{r}_S\right) = \varphi_L\left(\mathbf{r}_R|\mathbf{r}_S\right) - X_R\sqrt{(8\pi k)}\frac{e^{-i\pi/4}}{16\pi^2}\int_{l_B}\left[\begin{array}{l}\dfrac{e^{ik(d_1+d_3)}}{\sqrt{d_3^3 d_1\left(d_1+d_3\right)}}+\dfrac{Q_R e^{ik(d_1+d_4)}}{\sqrt{d_4^3 d_1\left(d_1+d_4\right)}}\\ +\dfrac{Q_s e^{ik(d_2+d_3)}}{\sqrt{d_3^3 d_2\left(d_2+d_3\right)}}+\dfrac{Q_s Q_R e^{ik(d_2+d_4)}}{\sqrt{d_4^3 d_2\left(d_2+d_4\right)}}\end{array}\right] dz, \tag{9.44}$$

where d_1,d_2,d_3 and d_4 are defined in Figure 9.17 and X_R is the shortest distance measured from the receiver to the barrier plane. The spherical wave reflection coefficients Q_s and Q_R are calculated on the basis of ground impedance on the source side and receiver side, respectively. In our case, the ground impedance is the same for both sides but, as demonstrated by Rasmussen [38], they can be different.

The integral limits in (9.44) specify the 'height' of the thin barrier. If there are horizontal gaps in the barrier, the height of the barrier can be adjusted accordingly. Consider the thin barrier of height H with a horizontal gap of size $(h_2 - h_1)$ shown in Figure 9.18.

The integral term in (9.44) is broken down into two parts where the integral limits for l_B range from $z = 0$ to $z = h_1$ and from $z = h_2$ to $z = H$ as follows:

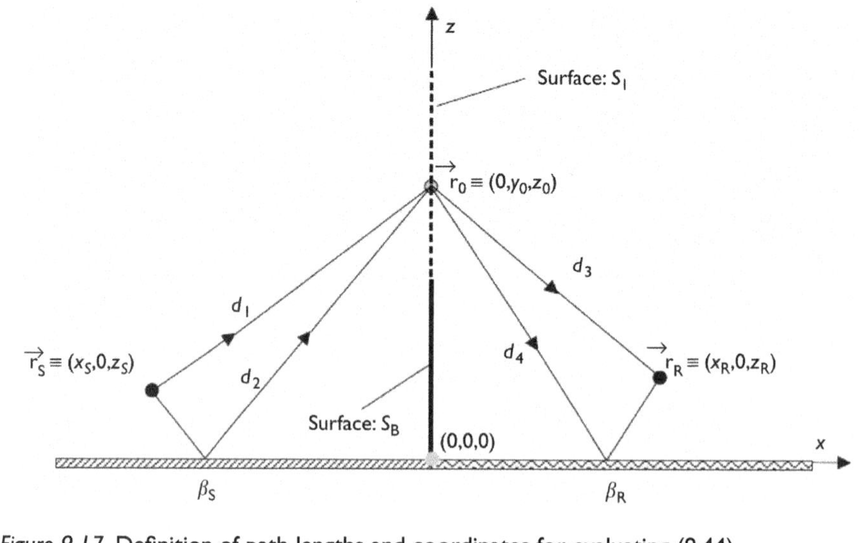

Figure 9.17 Definition of path lengths and coordinates for evaluating (9.44).

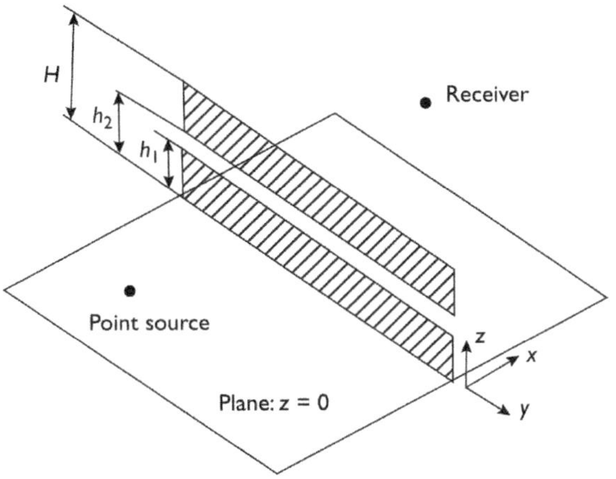

Figure 9.18 An infinitely long thin barrier with a horizontal gap.

$$\varphi(\mathbf{r}_R,\mathbf{r}_S) = \varphi_L(\mathbf{r}_R|\mathbf{r}_S) - X_R\sqrt{(8\pi k)}\frac{e^{-\frac{i\pi}{4}}}{16\pi^2}\left\{\int_0^{h_1} + \int_{h_2}^{H}\right\} \left[\begin{array}{l} \dfrac{e^{ik(d_1+d_3)}}{\sqrt{d_3^3 d_1(d_1+d_3)}} + \dfrac{Q_2 e^{ik(d_1+d_4)}}{\sqrt{d_4^3 d_1(d_1+d_4)}} \\ + \dfrac{Q_1 e^{ik(d_2+d_3)}}{\sqrt{d_3^3 d_2(d_2+d_3)}} + \dfrac{Q_1 Q_2 e^{ik(d_2+d_4)}}{\sqrt{d_4^3 d_2(d_2+d_4)}} \end{array}\right] d \tag{9.45}$$

It is straightforward to generalize ((9.45) to allow for multiple horizontal gaps in the barrier by splitting the integral into appropriate smaller intervals representing the barrier surface.

It is important to note that the sound field behind a barrier with horizontal gaps can also be calculated by means of a simple ray method, thus eliminating the need for the time-consuming numerical integration procedure. Either the analytical or the empirical formulation of the diffraction theory outlined earlier can be used to evaluate the required sound field in the presence of horizontal gaps. $P_T(H)$ may be regarded as the asymptotic solution for the right-hand side of Equation (9.44) and, combining the two terms in the equation, the integral is

$$\int_H^{\infty} \left[\begin{array}{l} \dfrac{e^{ik(d_1+d_3)}}{\sqrt{d_3^3 d_1(d_1+d_3)}} + \dfrac{Q_R e^{ik(d_1+d_4)}}{\sqrt{d_4^3 d_1(d_1+d_4)}} \\ + \dfrac{Q_s e^{ik(d_2+d_3)}}{\sqrt{d_3^3 d_2(d_2+d_3)}} + \dfrac{Q_s Q_R e^{ik(d_2+d_4)}}{\sqrt{d_4^3 d_2(d_2+d_4)}} \end{array}\right] dz = \frac{16\pi^2 e^{i\pi/4}}{X_R\sqrt{(8\pi k)}} P_T(H) \tag{9.46}$$

For a barrier of height H with a horizontal gap starting from $z = h_1$ and extending to $z = h_2$ (see Figure 9.18), the sound field can be calculated by using Equation ((9.45) to yield

$$\varphi(r_R,r_S) = X_R\sqrt{(8\pi k)}\frac{e^{-\frac{i\pi}{4}}}{16\pi^2}\left\{\int_H^{\infty}+\int_{h_1}^{\infty}-\int_{h_2}^{\infty}\right\}\left[\begin{array}{l}\dfrac{e^{ik(d_1+d_3)}}{\sqrt{d_3^3d_1(d_1+d_3)}}+\dfrac{Q_2e^{ik(d_1+d_4)}}{\sqrt{d_4^3d_1(d_1+d_4)}}\\+\dfrac{Q_1e^{ik(d_2+d_3)}}{\sqrt{d_3^3d_2(d_2+d_3)}}+\dfrac{Q_1Q_2e^{ik(d_2+d_4)}}{\sqrt{d_4^3d_2(d_2+d_4)}}\end{array}\right]dz \tag{9.47}$$

Using this equation, the sound field behind a barrier that has a horizontal gap is computed from

$$\varphi_T(\mathbf{r}_R,\mathbf{r}_S) = P_T(H)+P_T(h_1)-P_T(h_2) \tag{9.48}$$

The first term of Equation (9.48) corresponds to the sound field diffracted by a thin rigid barrier without any gaps. Grouping the second and third terms, we can interpret (9.48) as the sound field due to the leakage through the barrier gap. That is to say, the 'leakage' of sound through a single gap can be represented by the difference of two barriers with different heights, h_1 and h_2

$$\varphi_{gap}(\mathbf{r}_R,\mathbf{r}_S) = P_T(h_1)-P_T(h_2) \tag{9.49}$$

The total sound field behind a barrier with n horizontal gaps can be generalized:

$$\varphi_T(\mathbf{r}_R,\mathbf{r}_S) = \varphi(\mathbf{r}_R,\mathbf{r}_S)+\sum_{i=1}^{n}\varphi_{gap}(\mathbf{r}_R,\mathbf{r}_S), \tag{9.50}$$

where $\varphi(\mathbf{r}_R,\mathbf{r}_S)$ is the sound field behind the thin barrier (with no gaps) and $\sum_{i=1}^{n}\varphi_{gap}(\mathbf{r}_R,\mathbf{r}_S)$ is the leakage of sound through gaps. The leakage of sound at each gap can be calculated by using Equation (9.50) with the appropriate barrier 'heights'.

Figure 9.19 shows data from a laboratory experiment in which the barrier top edge was 0.275 m from the ground with a horizontal gap of 0.017 m starting from the ground level [36]. The total area of the gap was approximately 6% of the total area of the barrier surface. The source height was located at 0.07 m above the ground and at 0.49 m from the barrier. The

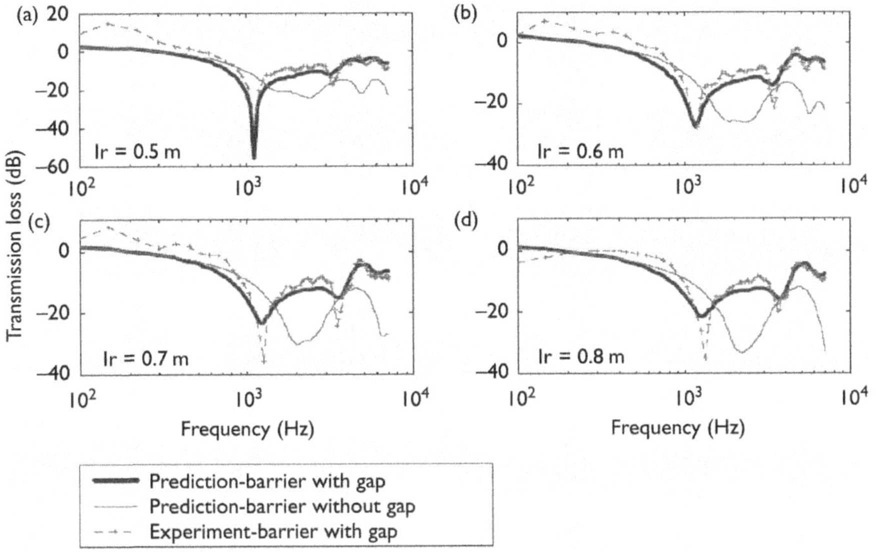

Figure 9.19 Comparison of measured (broken line) and predicted transmission loss for a 0.275 m high barrier with a 0.017 wide gap at ground level (thick solid line). Also shown is the predicted transmission loss for a barrier of the same dimensions but with no gap (thin solid line). The receiver is located at horizontal distances of (a) 0.5 m, (b) 0.6 m, (c) 0.7 m, and (d) 0.8 from the barrier.

receiver was located at 0.11 m above the ground, and at horizontal distances of (a) 0.5 m, (b) 0.6 m, (c) 0.7 m and (d) 0.8 from the barrier. Measurements were conducted on a hard ground. The resulting data, denoted as Transmission Loss (TL) in Figure 9.19, are presented as the relative sound pressure levels with respect to the reference sound pressure measured at 1 m in free field.

Numerical predictions are in general agreement with these experimental results. Hence, it is possible to use the proposed model to assess the degradation of acoustic performance of the barrier due to presence of a horizontal gap.

9.7 THE ACOUSTIC PERFORMANCE OF AN ABSORPTIVE SCREEN

The application of sound absorbent materials on barrier surfaces for increasing the insertion loss of a barrier has been a subject of interest in the past few decades. Although the theory of the sound diffracted by a rigid half plane has been studied with considerable success, there are still uncertainties about the usefulness of covering barriers with sound absorption materials. An early theoretical study [39] has suggested that an absorbent strip of one wavelength wide should be adequate to provide a similar insertion loss as the corresponding barrier totally covered by the same sound absorption material. However, this study does not provide the required information about the effect of placing this absorbent strip on either the source side or the receiver side of the barrier. In addition, the experimental and theoretical studies of Isei [40] indicated that the absorptive properties of the barrier surface did not significantly increase the acoustic performance of barriers. However, his conclusion is in contrast with previous experimental and theoretical studies by Fujiwara [41].

In this section, an analytical solution which is based on a heuristic extension of MacDonald's solution is outlined along with a heuristic solution based on Hadden and Pierce [42]. A corresponding heuristic extension of the line integral approach of Embleton [40] is not pursued here.

Fujiwara [41] suggested the use of a complex reflection coefficient, R_p into the second term of the MacDonald solution, see Equation (9.9) to obtain an approximate solution for the sound field behind an absorbent thin screen. Clearly, the diffracted field could be modified to allow for reflection of spherical wavefronts. Hence,

$$p_d = \frac{ke^{-i\pi/4}}{4\pi\sqrt{kR'}}\left\{\frac{\operatorname{sgn}(\varsigma_1)e^{ikR_1}}{\sqrt{k(R'+R_1)}}G\left(\sqrt{2N_1}\right)+Q_b\frac{\operatorname{sgn}(\varsigma_2)e^{ikR_2}}{\sqrt{k(R'+R_2)}}G\left(\sqrt{2N_2}\right)\right\}, \tag{9.51}$$

where Q_b is the spherical wave reflection coefficient for the façade of the barrier facing the source. No account is taken of the impedance on the side of the barrier facing the receiver. Note that neither of the diffraction models developed by Isei [40] and Fujiwara [41] satisfy the reciprocity condition: i.e. the predicted sound fields are different if one exchanges the position of source and receiver behind an absorbent barrier with different impedance on the source and receiver sides.

To satisfy the reciprocity condition, L'Espérance et al. [42] extended Hadden and Pierce diffraction model to include consideration of the impedance on both sides of the barrier in the model. In the presence of a rigid wedge, the diffracted field is given by Equation (9.16), restated here as

$$p_d = V_1 + V_2 + V_3 + V_4, \tag{9.52}$$

where $V_i \equiv V(\varsigma_i)$ for $I = 1, 2, 3$ and 4. The values of V_1, V_2, V_3 and V_4 can be obtained by evaluating the integrals of Equation (9.18c) numerically using the standard Laguerre technique. These four terms can be interpreted physically as the contributions from the diffracted rays travelling from the source to the edge to the receiver, from the image source to the edge to the receiver, from the source to the edge to the image receiver and from the image source to the edge to the image receiver, respectively. As a result, the effect of the impedance boundary conditions on both sides of the wedge can be incorporated into the Hadden and Pierce formulation as

$$p_d = V_1 + Q_b V_2 + Q_f V_3 + Q_b Q_f V_4, \tag{9.53}$$

where $Q_b \equiv Q\left(r_0 + r_r, \beta_b, \frac{\pi}{2} - \theta_0\right)$ and $Q_f \equiv Q\left(r_0 + r_r, \beta_f, \frac{\pi}{2} - \theta_r\right)$ are, respectively, the spherical wave reflection coefficient of the barrier façades facing the source and receiver. The corresponding parameters β_b and β_f are the effective admittances of the source and receiver sides of the wedge. The formula specified in Section 9.4 can be used to determine the insertion loss of an absorbent barrier resting on an impedance ground. The implementation is straightforward, but the details will not be given here.

Figure 9.20 shows the comparison between laboratory data obtained by L'Espérance et al. [scaled from Figure 4(b) of Ref. 10.41] with predictions of the heuristic modification of the Hadden and Pierce model. In the experiments, the height of barrier was 0.577 m, the source was placed at 1.2 m from the barrier and at 0.011 m above the ground. The receiver was located at 0.6 m from the barrier and 0.006 m above the ground. A 0.04 m thick fibreglass layer was used to cover either the source side or both sides of the barrier surface. In Figure 9.20, the data corresponding to these two cases are

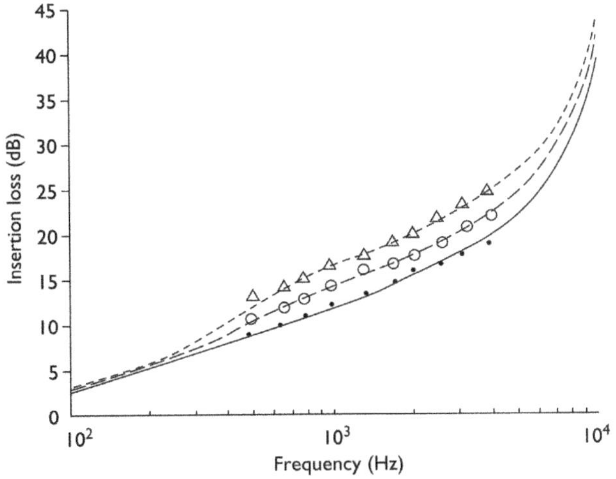

Figure 9.20 Comparison of laboratory data [42] with the predictions of a heuristic modification of the Hadden and Pierce model [14] for an absorptive screen. The solid circles and solid line represent theoretical and experimental results, respectively, for the hard barrier. The open circles and short-dashed line represent theoretical and experimental results, respectively, when the surface of the barrier facing the source is covered with fiberglass. The open triangles and long-dashed line represent theoretical and experimental results, respectively, when both surfaces of the barrier are covered with fiberglass.

shown as the open circles and open triangles, respectively. Also shown are the measured results for a rigid barrier (small solid circles). The impedance of the fibreglass layer was predicted by using the Delaney and Bazley one-parameter model (3.2) and assuming an effective flow resistivity of 40 kPa s m^{-2}. Good agreement between the predicted and measured insertion losses is evident.

In this way, L'Espérance et al. have shown that the application of sound absorbing materials on the barrier façade can lead to a significant improvement in the barrier's performance. Although Hayek [43] has suggested that barrier absorption is less effective in reducing the diffracted noise in the shadow zone where absorbing ground might be a more dominant factor in determining the acoustic performance of the barrier, it remains important to investigate the effectiveness of using absorptive screens in protecting receivers from excess noise particularly in urban environments where multiple reflections and scattering between surfaces are prevalent.

9.8 GABION BARRIERS

9.8.1 Numerical Predictions of Comparative Acoustical Performance

Gabions consist of cages made of steel wire filled with stones. Traditionally, they have been used for foundations, for slope stabilization or to prevent erosion (e.g. of riverbanks). Increasingly, they replace vegetative hedges in gardens to define land property borders. But this section looks at their increasing use as noise barriers along roads because of their natural appearance and their low maintenance requirements.

Compared with traditional fence like noise barriers, gabions barriers are relatively 'leaky' at low frequencies since sound waves travel through the channels formed between the stones. On the other hand, sound penetrates the outer surface of the gabion, so the pressure doubling that would be associated with reflection at a rigid, thick, non-porous noise barrier, does not occur. Although reflections from traditional environmental noise barriers can be reduced, to some extent, by adding sound absorbing materials to their surfaces, barrier reflections remain an issue at low frequencies. In contrast, gabion barriers do not reflect much sound towards the source.

Two full wave methods (namely the BEM and the finite-difference time-domain (FDTD) method (see Chapter 4)) may be used to compare predictions of insertion loss, defined as the sound pressure level in absence of any barrier (i.e. for sound propagation from a line source over rigid flat ground) minus the sound pressure level in presence of gabion barriers or a rigid thick barrier of the same outer dimensions.

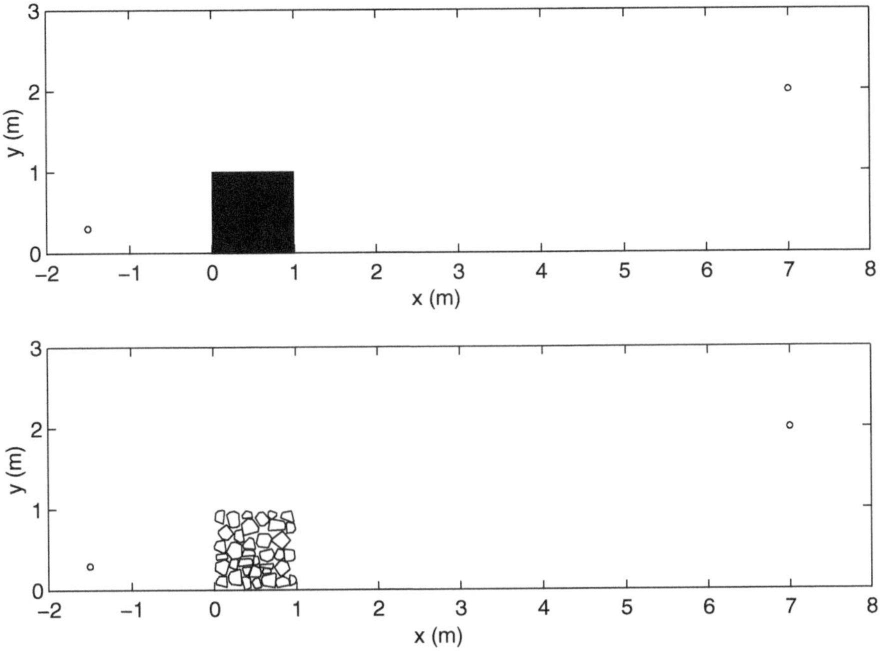

Figure 9.21 Geometry assumed for numerical predictions of acoustical insertion loss due to a rigid full thick barrier and a gabion barrier of similar outer dimensions (from [117]).

In the example presented here, the assumed source height of 0.3 m corresponds to the location of a light vehicle engine and the receiver is assumed to be 2 m high and 8.5 m from the source (see Figure 9.21).

The example gabion (see Figure 9.22) consists of relatively large stones the shapes of which have been modelled in detail. Neighbouring stones are positioned in such a way that none of the voids between them are blocked and, thereby, prevent sound waves to pass. A distinction is made between gabions made with fully rigid non-porous stones and those made with porous stones of the same shape and position as the rigid stones.

Corresponding predictions using BEM are shown in Figure 9.23 and those using FDTD are shown in Figure 9.24. In making the FDTD predictions, the acoustical properties of porous clay stones are characterized using the Zwikker and Kosten phenomenological model [44], whereas in making the BEM calculations the Hamet model [45 and Chapter 5] has been used. The assumed parameter values are flow resistivity = 82.2 kPa s m^{-2}, porosity = 0.758 and tortuosity or structure factor = 1.37.

Below about 700 Hz, both numerical methods predict that the rigid barrier provides a significantly higher acoustic insertion loss than the rigid-stone gabion. However, at higher frequencies, the predicted insertion losses are similar. Note that, for this specific source–receiver configuration, the dip

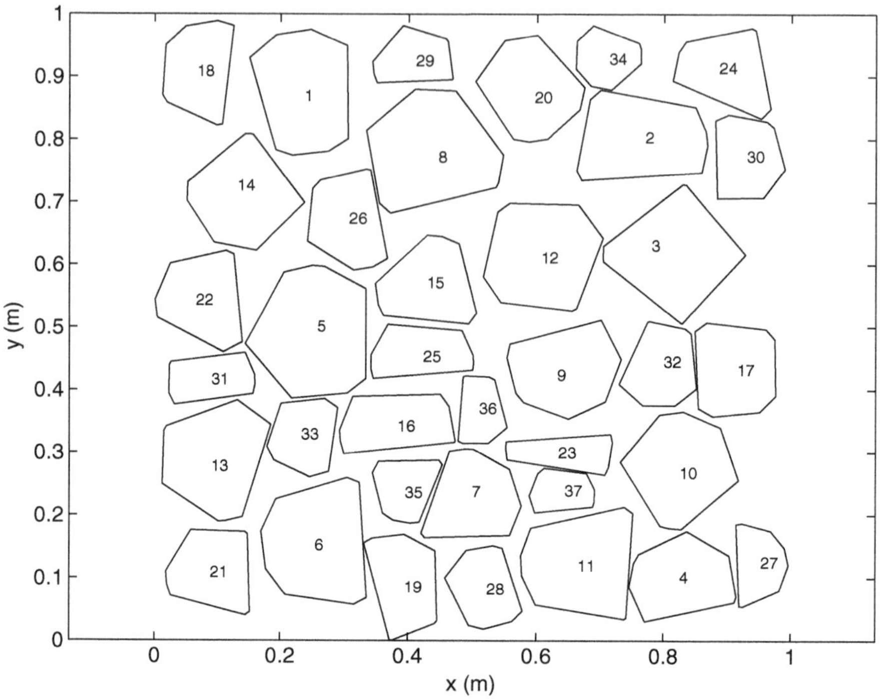

Figure 9.22 Locations and shapes of the 37 stones in the gabion barrier (from [117]).

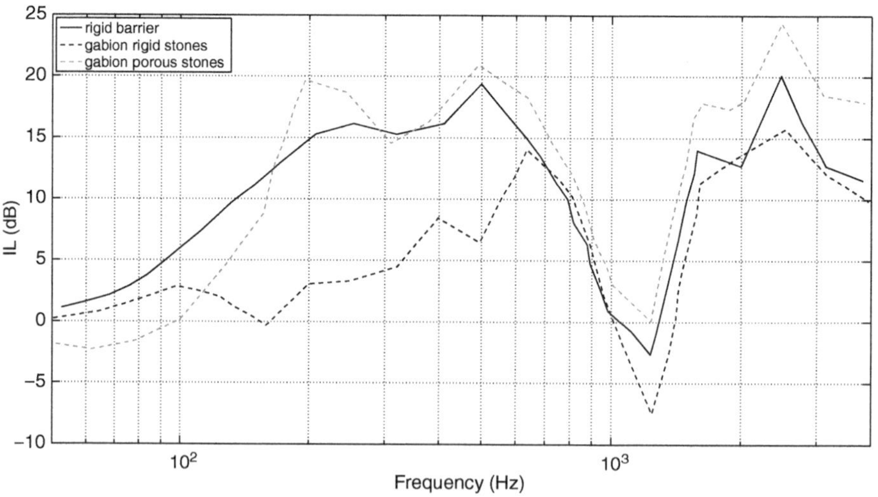

Figure 9.23 BEM predictions of the insertion loss of a 1 m high, 1 m wide rigid barrier and of gabion barriers of the same dimensions filled with either rigid stones or porous stones. All barriers are assumed to be above rigid ground (from [117]).

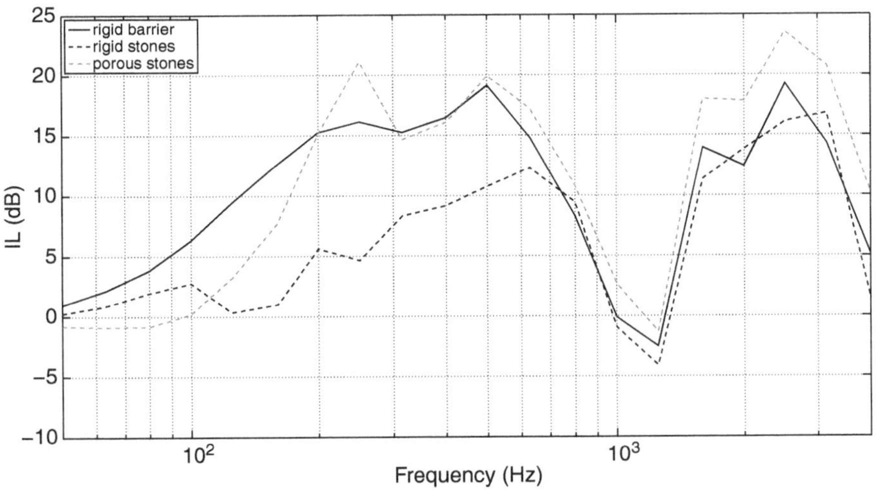

Figure 9.24 FDTD predictions of the insertion loss of a 1 m high, 1 m wide rigid barrier and of gabion barriers of the same dimensions filled either with rigid stones or porous stones. The barriers are assumed to be above rigid ground (from [117]).

in insertion loss near 1 kHz originates from the pronounced destructive interference between the direct and ground-reflected sound path predicted for the reference rigid flat ground case.

With porous stones, a very low insertion loss is predicted below 150 Hz. Indeed, both numerical methods predict the insertion loss due to porous stone gabions to be slightly negative below 100 Hz. But, as the frequency increases, the predicted insertion loss of porous stone gabions grows rapidly, exceeding that for corresponding rigid-stone gabions and becoming at least comparable with that predicted for the full rigid barrier. Above 1.5 kHz, the high insertion loss predicted for the porous stone gabion is due to the absorption provided by the stones. It should be noted that the validity of these numerical simulations is supported by the fact that the predicted insertion loss spectra in Figures 9.23 and 9.24 are very similar, despite resulting from different numerical techniques.

9.8.2 Laboratory Measurements on Porous-Stone Gabions

According to the predictions discussed in 9.8.1, the use of porous stones rather than non-porous ones increases the shielding efficiency of gabion barriers over a significant frequency range. Although sound propagates between the stones, part of the sound energy enters the stones and, effectively, is absorbed there, leading to less transmission of sound through the gabion. The diffracted and reflected sound components are reduced as well.

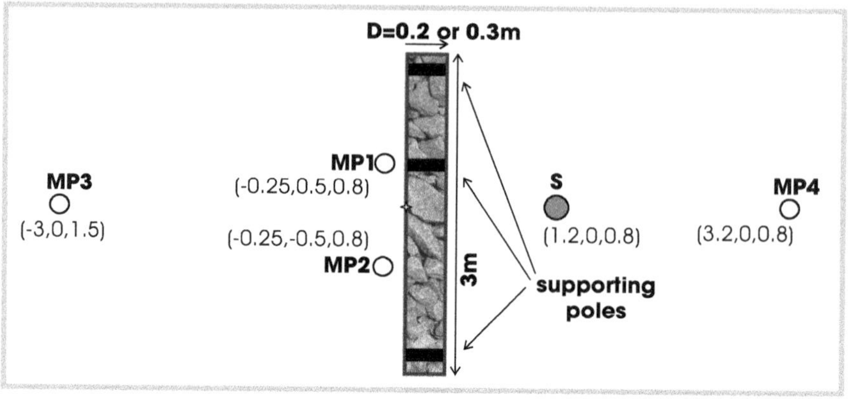

Figure 9.25 A plan view of source (S), microphone (MP) and gabion barrier positions in the semi-anechoic room during the full-scale experiment. The thick lines inside the rectangle representing the stone-filled gabion barrier indicate the positions of three supporting poles [46].

A full-scale experiment has been carried out in a semi-anechoic room (with a rigid floor) to study the effect of stone type on the acoustic performance of a gabion [46]. The gabion barriers were 1.4 m or 1.6 m high, 20 cm or 30 cm thick and were composed of either fully rigid stones or porous lava stones. The stones were contained in a 2 m high metal cage, consisting of metal wires of diameter between 3 mm and 4 mm in the form of a square lattice with sides of 5 cm. There were three supporting poles. Near the bottom was a 10 cm high metal U-profile. Sound pressure level measurements were made in response to test signals at fixed positions, both on the source side and on the opposite side of the barrier (see Figure 9.25). The largest rigid-stone dimensions were between 4 and 6 cm, while those of the porous stones were between 4 and 8 cm (see Figure 9.26).

Measured insertion loss spectra with the source close to the gabion are shown in Figure 9.27. For comparison, the measured acoustic effect of the empty cage (including the three supporting poles) is shown as well.

At microphone positions MP1, MP2 and MP3 and at frequencies below 1 kHz, the gabions provide less than 5 dB insertion loss. Above about 1 kHz, the shielding strongly increases up to 25 dB in the considered frequency range. Although MP1 and MP2 were positioned symmetrically relative to the source and barrier, they show rather different insertion loss spectra which may be the result of non-uniform filling of stones in the cages, non-symmetric positioning of the poles and potentially small microphone positioning errors. Note that the empty container has a non-negligible influence on the insertion loss since the poles and the U-profile partly shield and scatter sound. When the gabion barriers were formed from porous lava stones rather than rigid non-porous stones, the shielding is enhanced particularly above 1 kHz. Also, when the gabion stones were porous, increasing the gabion thickness from

Figure 9.26 Porous Lava stones used in full scale gabion barrier experiment in an anechoic chamber [46].

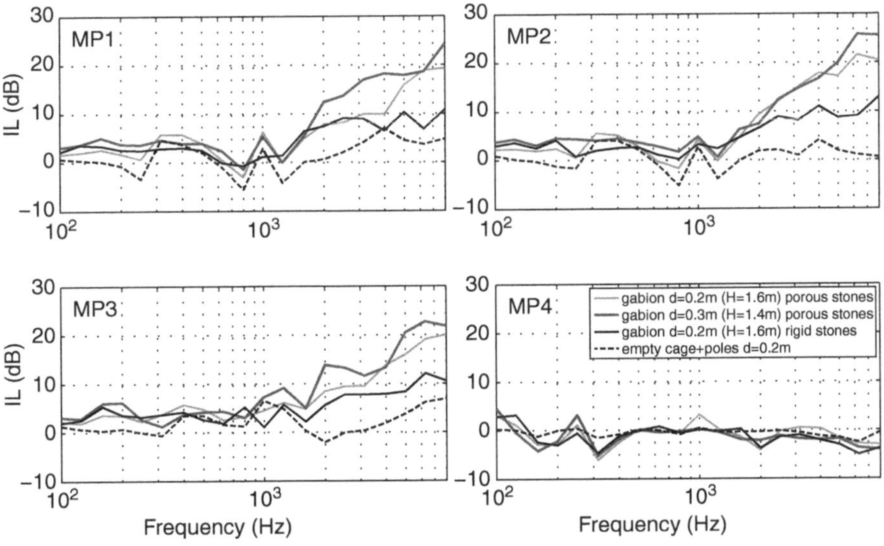

Figure 9.27 Measured insertion loss spectra at the different microphone positions (see Figure 9.25) for different gabion barrier formations [46].

0.2 to 0.3 m leads to higher sound shielding over a wide frequency range, even though stones within the 30-cm thick barrier were piled to a lower height. This means that larger thickness seems to be more important for the shielding than small differences in the height of the stone filling. The insertion losses at microphone position MP4 on the same side as the source due to the

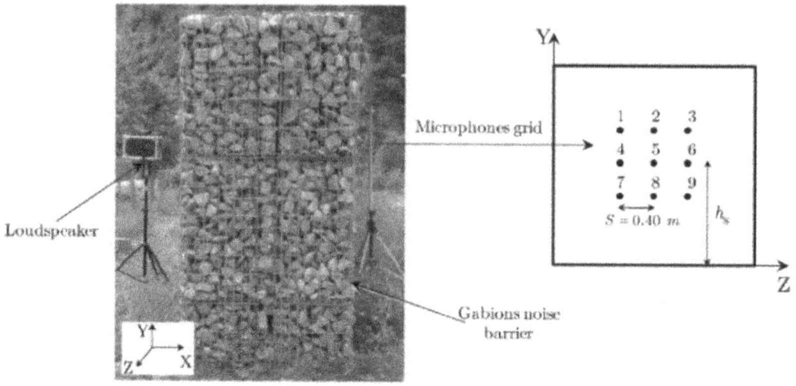

Figure 9.28 An outdoor gabion noise barrier and source and microphone arrangement for measuring transmission loss [47].

empty cage and poles are close to zero but slightly negative when the gabion cage is filled with stones, from which one can conclude that the stones are mainly responsible for reflections coming from the gabion. Differences in the gabion height and width do not influence these reflections significantly.

9.8.3 Outdoor Measurements on a Gabion Barrier

Outdoor measurements of sound reflection and sound transmission have been made at a gabion barrier 3 m high and 1.1 m thick, with rigid stones of sizes between 7 and 15 cm and a porosity near 40% [47]. The barrier and the arrangement for measuring sound transmission are shown in Figure 9.28. The octave band sound reflection and insulation spectra for this specific gabion are shown in Figure 9.29. The corresponding single-number ratings for in situ sound reflection (DL_{RI}) and sound insulation (DL_{SI}) [48], including weighting by a standard road traffic noise spectrum [49], are 5 dB and 20 dB, respectively. Such a barrier meets the requirements of class A2 'moderate' [50] for its absorbing properties, and B2 'moderate' [51] for its acoustic insulation properties.

9.8.4 Optimizing Gabion Barriers for Noise Reduction

Based on the measurements and simulations reported in this section, a gabion barrier must be quite wide to achieve a useful sound insulation. A barrier which is only 20–30 cm wide might be too 'leaky', especially at low frequencies. Height seems less important, so 1 m high gabion barriers might be useful in some situations. On the other hand, the type and size of stones are important. Porous stones result in improved insertion loss compared

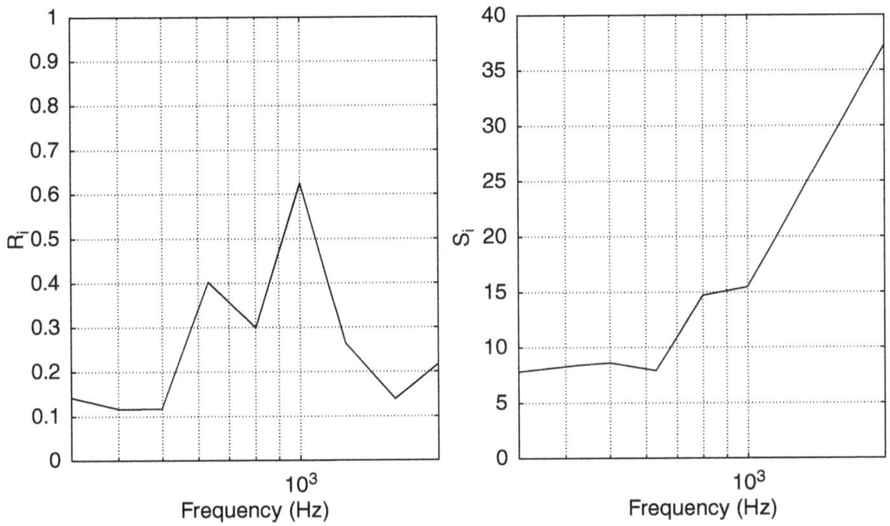

Figure 9.29 Measured octave band sound reflection R_i and sound insulation S_i of the outdoor gabion noise barrier shown in Figure 9.28 [47].

with non-porous rigid stones. Ensuring that the stones have dimensions close to the dominant wavelengths in the source spectra is another way to improve the acoustic efficiency [46]. Scale-model measurements and numerical calculations have shown that the acoustic performance of gabions could be further improved by filling the gabion in a non-uniform manner, thereby creating a layered system in which waves travelling through the gabion are partially reflected at each change in the effective medium properties [46].

9.9 OTHER FACTORS IN BARRIER PERFORMANCE

9.9.1 Barrier Shape

Most published theoretical and experimental studies are focused on a straight edge barrier or a wedge. In fact, the shapes of barrier edges can affect their shielding efficiencies significantly. Variations in the barrier shapes may be used to improve the shielding performance without increasing the height of barriers.

(a) Ragged-edge barriers

Wirts [52] introduced barriers with non-straight top edges called 'Thnadners'. Saw-tooth Thnadners barrier consists of large triangular panels placed side by side. Flat top Thnadners are similar to saw-tooth Thnadners but each of the sharp corners at the saw-tooth top is flattened. Wirts suggested using the simplified Fresnel theory to calculate the sound fields behind Thnadners barriers. Results of his scale model experiments agreed with the proposed theory. An interesting outcome is the finding of additional improvements in shielding performance when some parts of the barrier are removed. Improvements are obtained at some receiver locations even if over 50% of the barrier surface areas are removed.

Ho et al. [53] conducted many scale model experiments and concluded that the acoustic performance of a barrier can be improved at high frequencies by introducing a random fluctuation to the height of the barrier, producing the so-called 'ragged-edge' barrier. However, the poorer performance of the ragged-edge barrier at low frequencies was not explained. An empirical formula for the insertion loss of the ragged-edge barrier was obtained from the experimental results. The formula involves four parameters which are the frequency of the incident sound, the Fresnel number, the maximum height fluctuation from the average barrier height and the horizontal spacing between the height fluctuations.

Shao et al. [54] have proposed a 2-D model for a random edge barrier. Their models were based on the Rubinowics-Young formula which is a line integral method. The integration is performed along the random edge. They also conducted scale model experiments to validate their theoretical models. They concluded that the improvement could be obtained by introducing a random edge profile, especially at high frequencies. The insertion loss is controlled by the average barrier height and the variations of the randomness. In some cases, an improvement in the insertion loss can be obtained if the average height of a random edge barrier is smaller than the original straight edge barrier.

(b) Multiple-edge barriers

After reviewing relevant previous experimental studies by May and Osman [55,56] and Hutchins et al. [57,58], Hothersall et al. [59] have used the BEM approach to study T-profile, Y-profile and arrow-profile barriers theoretically (see Figure 9.30). They concluded from their numerical simulation results and previous work that a multiple-edge profile barrier has better acoustical performance than a normal thin plane barrier of the same height. However, in most conditions, the Y-profile and arrow-profile barriers

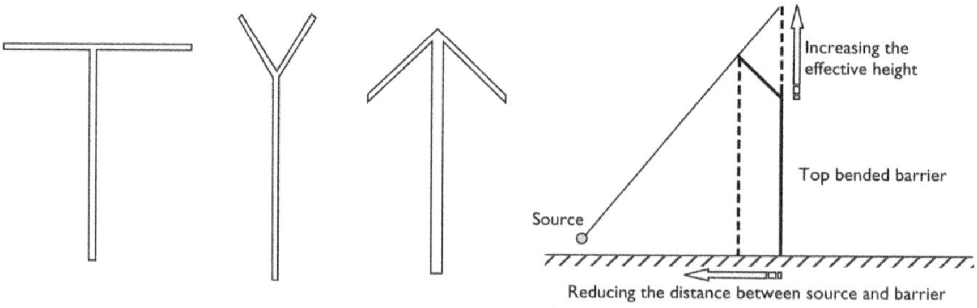

Figure 9.30 T-profile, Y-profile, arrow-profile and cranked (top-bended) barriers.

perform less efficiently than the T-profile barrier. The use of T-profile barrier can benefit regions close to the barrier and close to ground. On the other hand, the height of barrier is a dominant factor if the receiver located far away from the barrier and, in this case, the top profile plays a less important role. The application of absorbent materials on the cap of T-profile has also been studied.

A T-profile barrier with an absorptive cap can increase the insertion loss if the absorbent layers are thin. An excessively thick layer of sound absorbing materials can lead to a less efficient shielding. At present, only numerical tools, such as BEM, are used to study the acoustic performance of the T-profile and multiple edge barriers. Consequently, most of the theoretical studies are restricted to two-dimensions.

Another type of multiple edge barrier is the 'cranked' or top-bended barrier (see Figure 9.30) used increasingly in 'high rise' cities. Jin et al. [60] have proposed an analytical method for the prediction of sound behind a top-bended barrier. They used a theoretical model based on the uniform geometrical theory of diffraction [61]. The total diffracted sound is calculated by summing multiple diffractions occurring at the top and corner points, at which convex and concave edges are formed. They concluded that the third and higher orders of multiple diffraction did not contribute significantly to the total diffracted sound and so these higher-order diffracted terms were ignored in their model. They compared the predicted insertion loss spectra between finite- and infinite-length barriers, but some ripples caused by edge effects are evident in their predictions for the finite-length barrier. Jin et al. used a commercially available BEM programme to compare with their analytical solutions for the 2-D case and obtained good agreement.

Okubo and Fujiwara [62] have suggested another design in which an acoustically soft waterwheel-shaped cylinder is placed on the top of a barrier. They studied its effects by using a 2-D BEM approach and conducted 3-D scale model experiments. They concluded that the insertion loss of the barrier is improved by the cylinder, but the effect is strongly dependent on the source frequency. Since the depth and opening angle of each channel influence the centre frequency, a suitable variation of the channel depth can flatten the frequency dependency and improve the overall barrier performance for A-weighted noise.

(c) Other barrier types

Wassilieff [63] introduced the concept of picket barrier suggesting that traditional Fresnel-Kirchhoff diffraction is not applicable to a picket barrier at low frequencies because the wavelengths of the low frequency sound are comparatively larger than the widths of the vertical gaps in the picket barrier. He added terms for a 'mass-layer' effect to the Fresnel-Kirchhoff diffraction theory to increase the accuracy of the model at low frequencies. Nevertheless, the original Fresnel-Kirchhoff diffraction theory gives better accuracy at the high-frequency region. He concluded that a careful adjustment of the picket gap can result in an improvement at certain low frequencies compared with a solid barrier. However, the overall acoustic performance of barriers is reduced at other frequencies.

Lyons and Gibbs [64] have studied the acoustic performance of open screen barriers, i.e. the performance of a louver type barrier, at low frequency. Their model barrier consisted of vertical pickets, having a sound absorbing surface on one side, which were in two offset rows. The cavity left between two offset pickets is designed for better ventilation. They concluded that the side with the open area has less influence on transmission loss at low-frequency sound. They also derived an empirical formula for the calculation of transmission loss of this louver-type barrier with the dependent parameters being the overlapped area and the cavity volume between two offset rows of pickets. They found that the leakage loss through two offset rows of pickets is relatively small at low frequency. As a refinement of this study, Viverios et al. [65] have developed a numerical model that was based on the Fresnel-Kirchoff approximate scheme to estimate the transmission loss of a louver barrier. After incorporating the amplitude and phase changes due to the absorbing materials within the louvre bladders, their model agreed reasonably well with measurements. Watts et al. [66] have used the BEM formulation to study the acoustic performance of louvred noise barriers. Nevertheless, it seems that more theoretical and experimental studies are required to develop a simple scheme for predicting the sound field behind open screen barriers.

A comprehensive summary of the efficiency of noise barriers of different shapes can be found in the paper by Ishizuka and Fujiwara [67]. The authors used the BEM and a standard traffic noise spectrum to determine the broadband efficiency of several popular noise barrier shapes with respect to that predicted for an equivalent 3 m high plain noise screen. Their work shows that a complex noise barrier edge, e.g. a Y-shape barrier or a barrier with a cylindrical edge, can increase in the insertion loss by between 4 and 5 dB. This is important when increasing the barrier height is neither practical nor impossible. Figure 9.31 illustrates the change in the mean insertion loss relative to a standard 3 m high, plane screen noise barrier.

A laboratory study using a spark source has compared the efficiency of several types of barrier tops and indicated that T- shape barriers are superior to jagged or random edge barriers [68].

If a source of noise is close to the barrier, then the positive effect of the complex barrier shape can become small in comparison with the effect of treating a barrier surface with absorbing materials. This phenomenon has been investigated, experimentally and numerically, by Hothersall et al. [69] who studied the performance of railway noise barriers. Reflecting and absorbing railway noise barriers of different shapes were positioned as close as possible to the train structure, within the limitations of the structure gauge (see Figure 9.32). The barrier shapes included a plain screen (a), a cranked screen (b), a parabolic screen (c), a multiple edge noise screen (d) and a corrugated barrier (e).

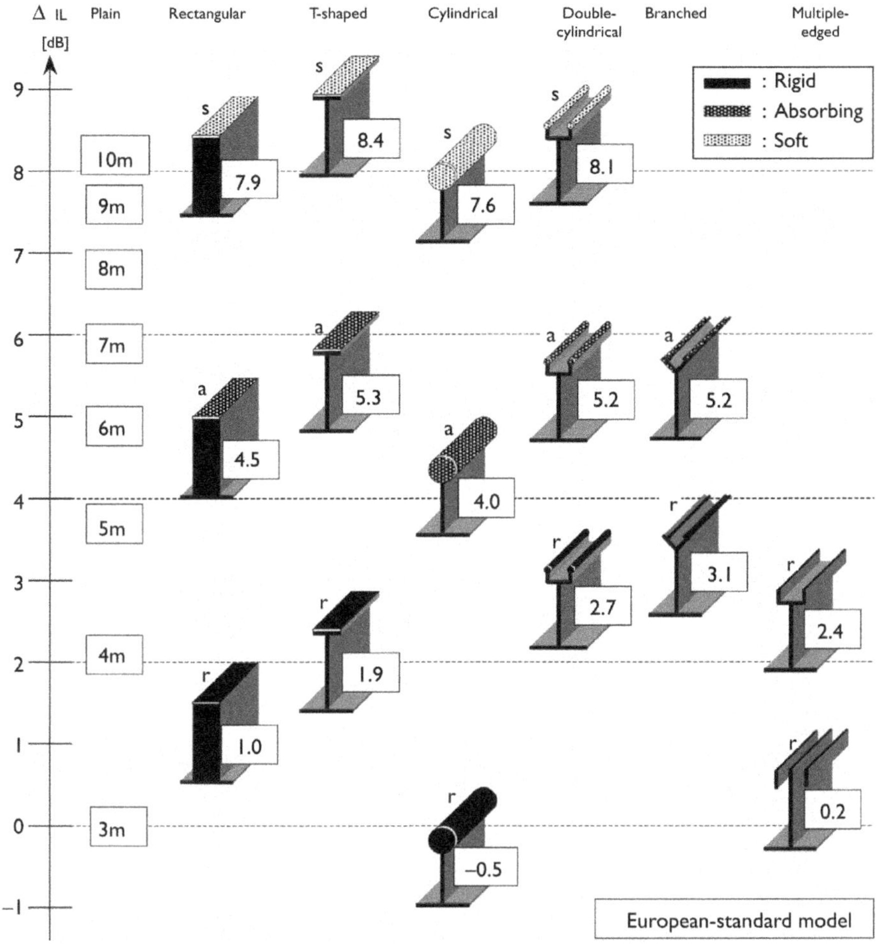

Figure 9.31 Predicted acoustic performance of traffic noise barriers of various shapes assuming a standard traffic noise spectrum (reproduced from reference [67]). The numerical labels indicate the predicted improvement in the mean insertion loss relative to that for a 3 m high plane screen.

It was found that the insertion loss values for a rigid screen were between 6 dB and 10 dB lower than those for a similar screen with absorbing treatment. In a particular case, where a 1.5 m high plain screen was installed 1.0 m away from a passing railway carriage, the maximum predicted improvement in the average insertion loss due to the absorbing barrier treatment was 14 dB. On the other hand, none of the modifications in the barrier shape, resulted in more than 4 dB gain in the insertion loss compared with that of the basic plain screen of equivalent height [69].

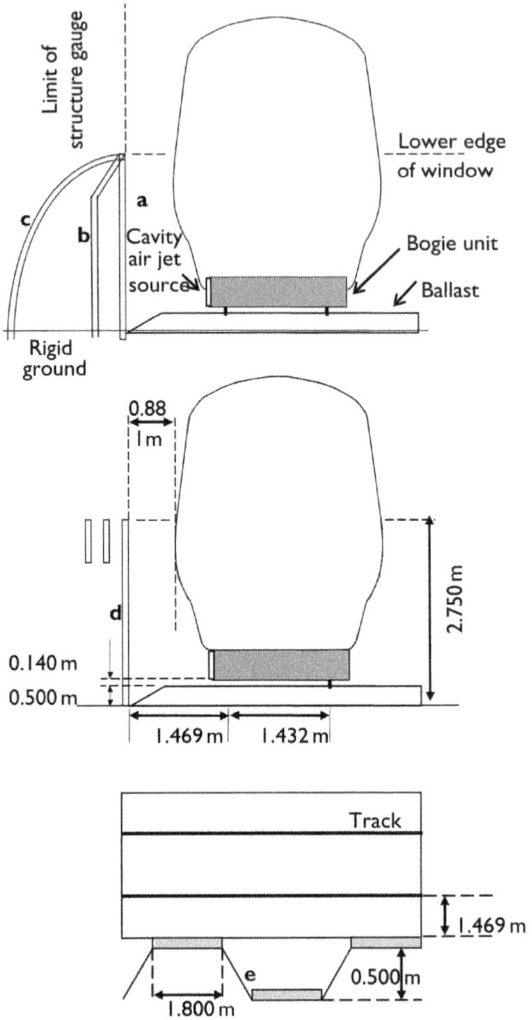

Figure 9.32 Configuration of the model and barrier cross-sections in [68].

9.9.2 Meteorological Effects on Barrier Performance

In the barrier diffraction models described earlier, atmospheric conditions have been assumed to be steady and uniform. But wind velocity gradients play a very important role in determining the acoustic performance of barriers. Not only the refraction that would occur above unscreened ground but also the specific flow fields near a noise barrier, leading to additional wind speed gradients, are important. Barrier effectiveness is much reduced for receivers downwind. By means of full-scale measurements conducted in the 1970s, Scholes et al. [70] demonstrated such wind effects on the acoustic performance of a noise barrier. Salomons [71] was among the first to study the acoustical performance of a barrier in a refracting medium by using a Parabolic Equation (PE) formulation.

Rasmussen [72] studied the diffraction of barrier over an absorbing ground and in an upwind condition both theoretically and experimentally. He used an approach, rather similar to the Fast Field formulation, to estimate effective distances from source to the barrier edge and from the barrier edge to receiver in an upwind condition. Then the uniform theory of diffraction [61] was used to compute the diffracted sound field in the shadow zone. It is questionable whether the barrier diffraction can be 'de-coupled' from the wind gradient effects in this way. Nevertheless, Rasmussen's numerical results agreed reasonably well with his 1:25 scale model experiments. Muradali and Fyfe [73] have proposed a heuristic method for modelling barrier performance in the presence of a linear sound speed profile. Their model combines the effects of the sound refracted over a barrier and the sound diffracted by the barrier. Geometrical ray theory is used to determine the ray paths and the diffracted wave effect is incorporated by using the modified Hadden–Pierce formula. Li and Wang [74] have proposed a method to simulate the effects of a noise barrier in a refracting atmosphere. They mapped the wave equation in a medium where the speed of sound varies exponentially with height above a flat boundary to that for propagation in an iso-velocity field above a curved surface. Propagation over a barrier in an upward refracting medium is then solved by applying the standard BEM over a curved surface in a homogeneous medium. Taherzadeh et al. [75] developed a hybrid scheme for predicting barrier performance in downwind conditions. Their numerical scheme is based on a combination of the boundary integral and FFP techniques in which the Green's function required in the boundary integral equations is evaluated by using the FFP formulation.

Atmospheric turbulence is an important factor in barrier performance also since it causes scattering of sound into the shadow zone. Although experimental studies of the effects of atmospheric turbulence on the acoustic performance of the barrier are rare, theoretical studies have been conducted by Daigle [76], Forssen and Ogren [77] and others quoted in these references. Some progress in modelling the acoustic performance of barriers in a turbulent atmosphere using a substitute sources method has been reported by Forssen [78,79]. Hybrid models have been proposed, involving combinations of the BEM with the PE, which enable effects of barriers, range-dependent impedance and turbulence to be taken into account [80,81]. Rasmussen and Arranz [82] have applied the Gilbert et al. [83] approach, assuming weak atmospheric turbulence with a simple Gaussian distribution in the fluctuating velocity field, in a PE formulation to predict the sound field behind a thin screen in a scale model experiment. The indoor experiments based upon a 1:25 scaling ratio were conducted in a wind tunnel in which the agreement of numerical results and experimental measurements was very good. They concluded that the flow pattern associated with a specific noise barrier could be an important design parameter for improving the acoustic performance of barriers. Improvements in computational speeds and resources have encouraged wider use of time-dependent computations including computational fluid dynamics [84,85]. To date, except as discussed in the next section, there are few measurements or accurate predictions of the velocity profiles and associated turbulence in the region close to a barrier on the ground and it remains a challenge to combine fluid dynamics and diffraction models to establish an accurate scheme for predicting the acoustic effectiveness of screens under the influence of wind.

9.9.3 Rough and Soft Berms

The diffracting edge on a berm with a triangular cross section is further away from the source than that of a noise barrier of the same height positioned at the foot of the berm. While this puts such a berm at a disadvantage in terms of shielding efficiency, to some extent it can be compensated by making the surfaces of the berm acoustically soft or rough [86–89]. A scale model experiment mimicking a typical highway setup [86] showed that the optimal choice between a noise barrier and a berm strongly depends on the material with which the berm is made. A 4-m high noise barrier reduced total A-weighted road traffic noise levels better than a triangular berm consisting of packed, i.e. relatively acoustically hard, earth with similar height and diffracting edge position. However, when the berm was acoustically soft, the noise shielding by the berm was slightly greater than that of the noise barrier. Decreasing the slope of an acoustically soft berm while retaining its height was shown to increase shielding. In contrast, for the packed earth berm, angle of slope was not an important parameter. Also, the general shape of a berm can be optimized; for example, flat-topped (trapezoidal shaped) berms perform better than triangular ones [88].

As discussed in Chapters 5, 6 and 8, deliberate roughening of hard ground between source and receiver can lead to useful reductions in surface transport noise. There is some evidence also that rough tops of noise barriers can enhance their insertion loss [89,90]. Boundary Element Methods have been used to give numerical predictions of the extent to which roughening the sides of a trapezoidal section berm increases its insertion loss compared with that for a smooth-sided berm of the same (outer) dimensions with a finite impedance surface represented by the Delany and Bazley model (see Chapter 5) with effective flow resistivity 300 kPa s m^{-2} [89,91]. Example numerical results for a trapezoidal section berm (4 m high, 10 m wide base, 3 m wide top) alongside a 4-lane motorway (Figure 9.33) with various roughness profiles (Figure 9.34) are shown in Figure 9.35. Additional insertion losses of between 4.1 dBA and 7.2 dBA are predicted; the smallest gain in insertion loss is for the irregular roughness cross section shown in the second from

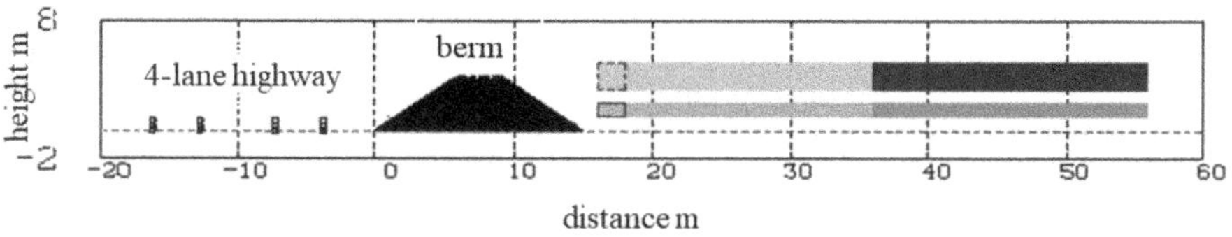

Figure 9.33 The geometry assumed for BEM calculations of berm insertion loss with 12 source locations (4 lanes with 3 sources per lane), trapezoidal section berm and two heights (1.5 m and 4 m) of receiver locations with three zones at each height.

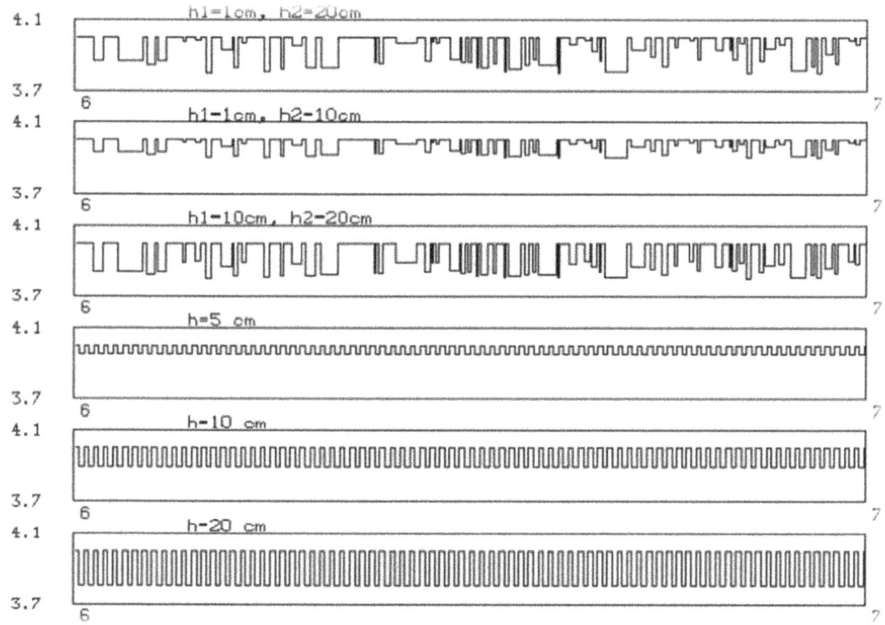

Figure 9.34 1 m long sections of six hypothetical roughness profiles on the top of a berm [91].

top panel of Figure 9.34 and the maximum is predicted for the regular roughness cross section shown in the lowest panel of Figure 9.34.

As a result of the predictions [89,91], several conclusions can be reached:

(i) For the geometry in Figure 9.33, adding roughness is predicted to increase insertion loss by up to 10 dBA compared with a smooth berm of the same dimensions.

(ii) The greatest increase in insertion loss is predicted for the highest edge density and deepest irregular or regular profiles (such as the profile in the bottom panel of Figure 9.34) but there is not much difference.

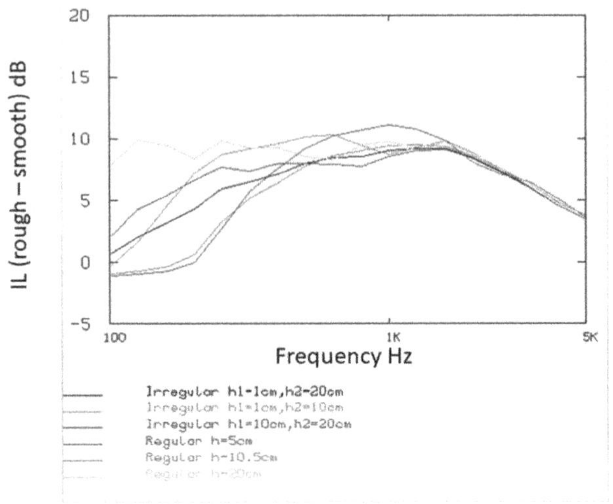

Figure 9.35 Spectra of the extra insertion loss relative to that for a smooth berm predicted for the geometry shown in Figure 9.33 with various roughness profiles (Figure 9.34) on the top of the berm [91].

(iii) The predicted low insertion loss at low frequencies is evidence of surface wave generation, but this is avoided if only the side furthest from the source is roughened.

(iv) Roughening an acoustically soft surface on a berm is predicted to generate less surface wave. This is consistent with predictions (see Chapter 5) that roughness increases the real part rather than imaginary part of effective impedance.

(v) Rough hard berms or smooth berms with acoustically soft surfaces perform better than a thin noise barrier of the same height placed where berm starts or at nearest highest point on berm.

9.9.4 Berms vs Barriers in Wind

Typically, highway noise barriers are positioned where little or no shelter from wind is provided by surrounding objects. If enough land is available for a berm, then the choice between a noise barrier and an earth berm as a noise abatement solution near surface transport infrastructure should be considered carefully. The different ways in which berms and barriers behave in downwind conditions is important, since downwind episodes result in the most important contributions to long-term equivalent sound levels.

As was discussed in detail in Section 9.9.2, the acoustical efficiency of a traditional noise barrier is largely reduced if receivers are downwind. In contrast, because berms are aerodynamically smoother, the screen-induced refraction of sound by wind is weaker. A detailed analysis of the spatial distribution of the wind speed gradients predicted near barriers and berms of different shapes (see Figure 9.36) is presented in Figure 9.37.

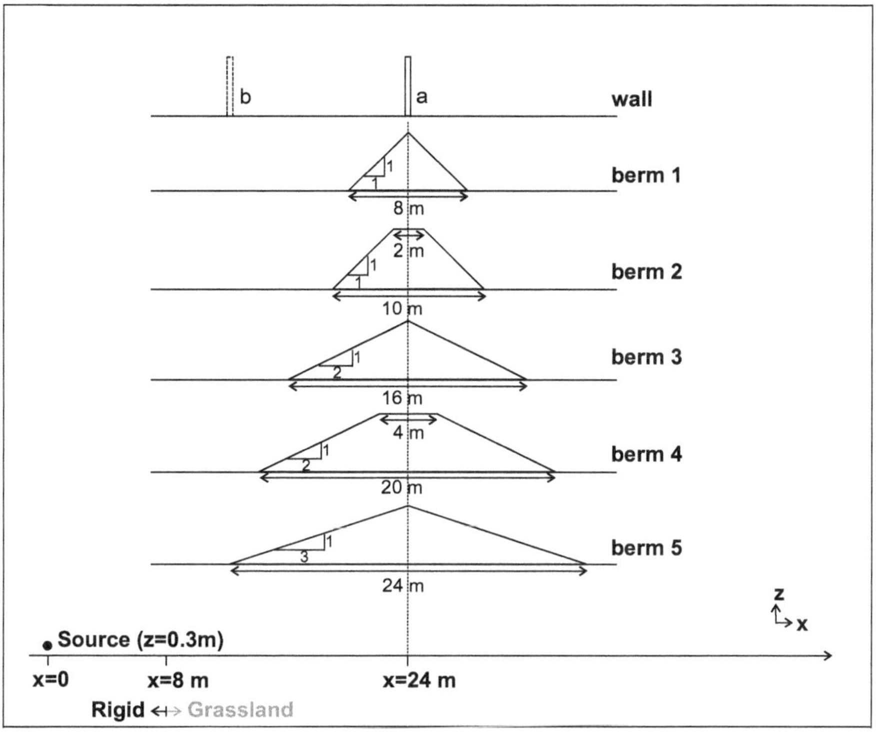

Figure 9.36 Position(s) of the noise wall(s) and variously shaped berms relative to the source position assumed for Figure 9.37 [88].

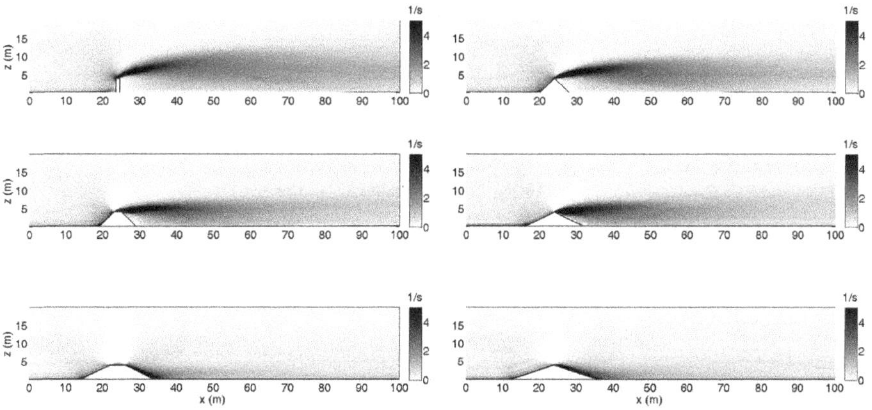

Figure 9.37 Plots of the vertical gradients in the horizontal component of the wind velocity near wall a and berms 1–5 shown in Figure 9.36. The incident friction velocity equals u* = 0.8 m/s (strong wind), the aerodynamic roughness length z_0 equals 0.01 m. A neutral atmosphere is assumed [88].

Strong positive gradients in the wind speed are predicted near a noise barrier in position a and near berms with steeper sides (i.e. berms 1–3). Typically, the zones in which there are strong gradients, thereby causing strong downward refraction, start at the diffracting edge and occupy an area just above the zone where the acoustic shadow is expected in absence of wind. Wind speed gradients are predicted to be less pronounced near berms with smaller slopes. Berms 4 and 5 are predicted to cause a zone with pronounced downward refraction only at their lee sides and very close to the surface.

The predicted impacts of wind-induced flow fields on noise shielding by a barrier and berm 5 (with a 'forest floor' type of impedance cover – see Chapter 5) are shown by the contours of insertion loss of total A-weighted road traffic noise, assuming light vehicles driving at 100 km/h (single traffic lane), for both a moderate and strong wind in Figure 9.38. For comparison, the predicted spatial distribution of the insertion loss in absence of wind is depicted as well. For these plots, the insertion loss is defined as the sound pressure level in absence of a noise barrier, in a still atmosphere, minus the sound pressure level due to the barrier configuration of interest in wind.

While, in absence of wind, a similar spatially distributed insertion loss is obtained for the barrier and the shallow berm, the shielding performance of the barrier is predicted to decrease rapidly in wind, even with moderate wind speed (u^*=0.4 m/s), creating zones with negative insertion loss, starting at 150 m from the source. Such zones are predicted to develop even closer to the noise barrier in strong wind. In contrast, the insertion loss of a small-slope berm is

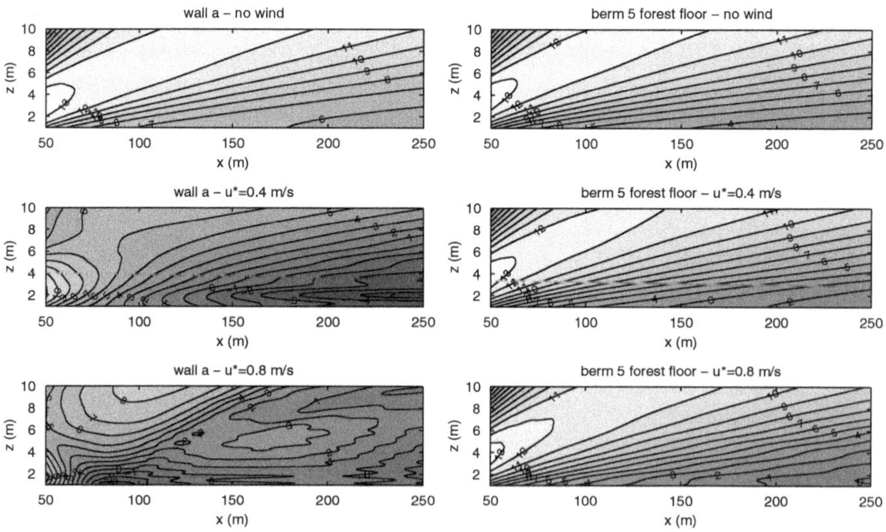

Figure 9.38 Contour plots of the total A-weighted road traffic noise insertion loss predicted from a road carrying light vehicles at 100 km/h, for wall a and berm 5 (with 'forest floor' cover), in absence of wind (first row), in case of moderate wind (second row) and in case of strong wind (third row) [88].

much more resilient to this negative action by wind. In case of moderate wind, the insertion loss is almost fully preserved, while the degradation in efficiency in case of strong wind is only a few decibels. Note that in the current analysis, only refraction is considered and not turbulent scattering.

To illustrate the importance of accounting for wind effects, a long-term prediction of the overall shielding efficiency of barrier in position b and berm 4 (with a 'forest floor' type of ground cover) has been made using numerical simulations. Note that position b for the barrier is at the foot of berm 5 (see Figure 9.36) and 2 m closer to the source than the foot of berm 4. Periods of downwind sound propagation often dominate long-term equivalent noise levels. Periods with upwind sound propagation can be assumed not to contribute significantly to these long-term equivalent levels, especially for sources and receivers close to the ground surface. Moreover, a measurement campaign has shown that the downwind effect behind a noise barrier is not strictly limited to the periods where the wind is blowing exactly normal to the barrier [92]. Similar effects are observed for deviations of up to ±45 degrees relative to the normal to the barrier. So, if it is assumed that the wind direction is uniformly distributed, downwind conditions would occur 25% of the time. Similarly, upwind sound propagation would occur 25% of the time as well. During the remaining 50%, there is no clear downwind or upwind sound propagation, and the 'no wind' insertion loss should be a good estimate [93]. This leads to a predicted long-term berm insertion loss of 11.1 dBA (with $u^* = 0.4$ m/s in case of downwind sound propagation and assuming a light vehicle source moving at 100 km/h) averaged over a zone between $z = 1$ m and $z = 10$ m, and distances between $x = 50$ m and $x = 250$ m.

For a noise barrier at position b, the corresponding predicted value is 8.3 dBA. For this downwind receiver case, the acoustically soft berm with a small slope is predicted to outperform the noise barrier. On the other hand, if wind effects are ignored, the predicted insertion losses are 12.8 dBA in case of barrier b and 10.2 dBA in case of berm 4. On this basis, neglect of wind effects would give rise to a non-optimal choice.

An additional acoustical benefit is that a small slope berm will limit sound reflection towards receivers on the source side of the barrier. Although reflections from conventional noise barriers can be reduced by making their source side surfaces absorbing, this increases their cost significantly, and their acoustical efficiency remains limited at low frequencies.

The additional influence of vegetation either on the berm or near the barrier has not been considered in this analysis. Note that in Section 10.6, it is explained how the downwind efficiency loss of a noise barrier can be mitigated by the presence of a row of trees behind it.

An earth berm has various non-acoustical advantages. Often, traditional noise barriers are perceived as visually intrusive, except in highly urbanized regions. An earth berm can be integrated into the environment more easily than a noise barrier since it can preserve the feeling of 'openness' that might be important in the rural or suburban landscape. Furthermore, berms provide spaces where vegetation can develop and thereby may be considered as a green noise reducing measure. An additional advantage of vegetated berms is the development of an acoustically highly porous soil (forest floor). Ecological impacts of noise barriers like wildlife habitat fragmentation and bird strikes are expected to be much less pronounced near gently sloped earth berms [94]. Traditional noise barriers might last only 20–30 years whereas berms have a relatively unlimited life span. Moreover, a berm is less likely to encourage graffiti than a barrier. A berm can be constructed using soil removed during nearby construction. Typically, the cost of a berm is much lower than that of a noise barrier of comparable height. The difference in cost can be even more pronounced when excess material from other locations or constructional works can be used.

A major drawback of a berm, especially if it is wide, is that it requires more land than a noise barrier of comparable height. If erecting a berm requires purchase of extra land along road infrastructure, then any cost benefits over a traditional barrier could be lost.

9.10 SONIC CRYSTAL NOISE BARRIERS

There has been a lot of interest in the potential of sonic crystal noise barriers (SCNB) as alternatives to conventional noise barriers. A sonic crystal is a periodic arrangement of solid elements, which, frequently, are vertical cylinders. If the cylinders are closely and regularly spaced, then coherent multiple scattering (sometimes called Bragg scattering) causes constructive and destructive interference effects. The results of destructive interference in the frequency domain are called band gaps and represent frequency ranges with strongly reduced transmission. If a sound wave is normally incident on a regular scatterer array with a scatterer-to-scatterer separation (lattice constant) of d, the frequencies at which the band gaps occur are given by

$$f = \frac{nc}{2d}, \tag{9.54}$$

where c is the speed of sound in the host medium, e.g. 344 m/s if the host medium is air, and n is 1,2 3...etc.. The first band gap corresponds to $n = 1$; the frequencies of subsequent higher order band gaps are obtained by substituting higher values of n. Constructive interferences or pass bands occur between the band gaps. The magnitude of the transmission loss peaks associated with the band gaps depends on the filling fraction. In a square array of cylinders of radius R with lattice constant d the filling fraction is given by

$$\frac{\pi R^2}{d^2} \tag{9.55}$$

Laboratory investigations of the acoustical performance of sonic crystals [95], have shown that triangular (i.e. part of hexagonal) lattices produce wider attenuation bands than square ones for similar filling fractions. It has been suggested that the maximum number of rows needed for a SCNB is four [96]. However, for the larger filling fractions, diffraction effects due to the finite array height, could limit the potential attenuation [97].

The mixture of stop and pass bands in transmission through an array of rigid cylinders means that the acoustical performance of such an SCNB is poor compared with that of a conventional noise barrier of the same height and thickness. If the cylinders are made absorbing [98–100] or local resonances are introduced, for example, by split hollow cylinders [100–103] or elastic shells [104,105], or by combining different types of resonator elements or absorption and resonance [106,107], then the transmission loss spectra of SCNB are increased and fluctuate less with frequency.

A further way of improving SCNB performance is to optimize the spatial distribution of scatterers. A design involving a range of scatterer sizes, scatterer forms and a fractal spatial distribution has been developed [108]. The design consists of two diameters of split ring cylinders with porous covering (see Figure 9.39(a)), which act as locally resonant absorbers, and two diameters of rigid cylinders arranged in a fractal (Serpinski triangle) pattern (see the inset of Figure 9.39(b)), which increase the Bragg scattering, i.e. the multiple scattering associated with periodic spacing. In the FEM-simulated insertion loss spectrum shown in Figure 9.39(b), the two sizes of resonant absorber components contribute the peaks labelled 1 and 2, while the two sizes of rigid cylinders are responsible for the peaks labelled 3 and 4. In a laboratory realization of this design [108], 294 component cylinders were suspended from a metal platform. Fourteen larger resonant absorbing elements (external diameter 165 and 4.85 mm wall thickness) and twenty smaller resonant absorbing elements (external diameter 114 and 3 mm wall thickness) were 3 m long hollow iron cylinders with 20 mm wide slots along their lengths, surrounded by a 40 mm thick rock wool layer (density 100 kg/m^3) and an outer perforated metal cover (1 mm thick, 5 mm diameter perforations) to enable them to act as structural elements. Two hundred and fifty-two solid cylinders were used to complete the fractal distribution (126 with external diameter 0.15 m and 126 with external diameter 0.3 m). The results of measurements made in an anechoic chamber are shown in Figure 9.38(c).

Figure 9.38(c) shows that, while the perforated covers have little influence, the absorbent covering has a significant role. Moreover, as is common with SCNB designed to give good insertion loss near and above 1 kHz, the insertion loss below 400 Hz is small or negative. On the other hand, arrays composed from locally resonant elements have been developed that give remarkably high insertion losses for low-frequency tonal noise [109].

The material of conventional noise barrier is supposed to offer enough transmission loss that the barrier efficiency depends only on diffraction over its top. This means that it is less effective when the receiver is more distant or is located outside the geometrical shadow. However, the effectiveness of a SCNB depends mainly on its transmission loss rather than diffraction over

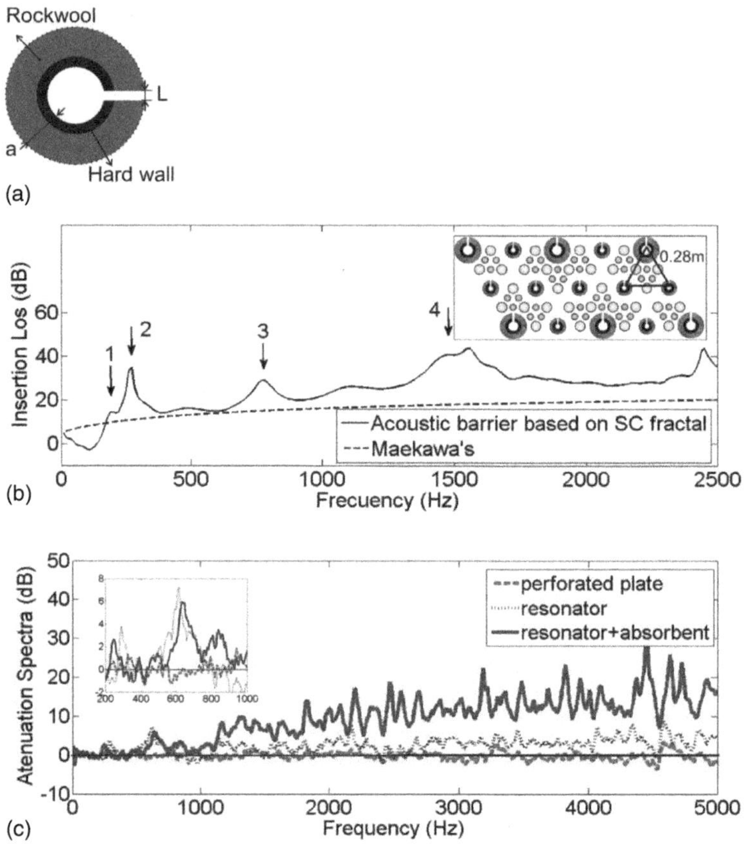

Figure 9.39 (a) Cross section of a resonant absorbing cylinder element (b) FEM-predicted insertion loss of a sonic crystal noise barrier with four types of elements in a fractal distribution (see inset Figure) compared with a Maekawa prediction for a noise barrier with the same external dimensions and (c) measured insertion losses for laboratory realisations of the SCNB and two of its components [108].

the top. Consequently, if it is long enough to avoid significant diffraction from its ends, it will remain effective even for distant receivers.

A conventional noise barrier reduces any existing soft ground effect due to the associated increase in the mean height of propagation. In contrast, an important feature of SCNB is that part of the destructive interference associated with soft ground can be retained and, potentially, exploited. Analytical models have been derived for arrays of cylinders with horizontal axes above a ground plane [110,111]. Numerical simulations and laboratory measurements have been carried out for arrays of vertical cylinders on acoustically soft ground and these are explored in Chapter 10 (Section 10.3) when discussing the acoustical performance of tree belts.

Potential advantages of SCNB are that (a) there is less influence of wind in terms of structural stability and (b) less reduction in their acoustical performance for downwind receivers than encountered with conventional noise barriers (see Section 9.8.2). The stability aspect has been explored in wind tunnel experiments [108,112]. So far, little research has been carried out into the acoustical performance of SCNB in wind. Wind tunnel tests suggest that wind speeds of more than 10 m/s are required to affect the band gap transmission loss. But, at sufficiently high wind speeds, SCNB can themselves become noise sources [113].

9.11 PREDICTED EFFECTS OF SPECTRAL VARIATIONS IN TRAIN NOISE DURING PASS-BY

Models for predicting the effects of noise control treatments rely heavily on the knowledge of the reference noise source spectrum. In road traffic noise studies, a standardized spectrum is used widely to account for the non-linearity in the emitted acoustic spectrum [114]. So far, no similar standard exists to regulate the noise spectrum emitted by high-speed trains. Unlike noise from continuous motorway traffic, which is a relatively stationary process, the noise spectrum of a train can vary significantly during its by-pass. During the pass-by of a high-speed Intercity train, the variation of the sound pressure level (SPL), in the frequency range between 500 and 5000 Hz, can be as high as 20 dB/sec [115]. However, calculations of barrier effectiveness tend to use the average spectrum for the whole by-pass. A likely drawback in adopting an average spectrum is the failure to model the distinctive time-dependent performance of noise control elements (e.g. noise barriers) during the by-pass of a train. Indeed, the engineering measure of the in situ efficiency of a railway noise barrier, the broadband insertion loss, can vary considerably during the by-pass. Subjectively, the noticeable fluctuations in the insertion loss during the train by-pass might be more significant than the average barrier performance and should be considered when the efficiency of expensive railway noise barrier schemes is assessed. This section investigates the way in which variation in the reference noise source spectrum can result in a perceivable fluctuation of the noise efficiency of some noise barrier designs.

Figures 9.40 and 9.41 show typical narrow band spectrograms measured for the passage of the Intercity 125 (diesel) and Intercity 225 (electric) trains, respectively. The distances on the vertical axes in these Figures correspond to the time-dependent separation between the fixed receiver position on the ground and the varying position of the front locomotive in the train.

The measured narrow-band spectrograms have been used to determine the 1/3-octave band reference, time-dependent noise spectra which are required to model the time-dependent insertion loss of three different barrier shapes: a plane screen, a screen with a quadrant of cavities at its upper edge ('spiky top') and a T-shape noise barrier. The 'spiky top' barrier has been designed to achieve its maximum efficiency around 1000–2000 Hz, i.e. in the part of the spectrum where the maximum noise emission occurs (see

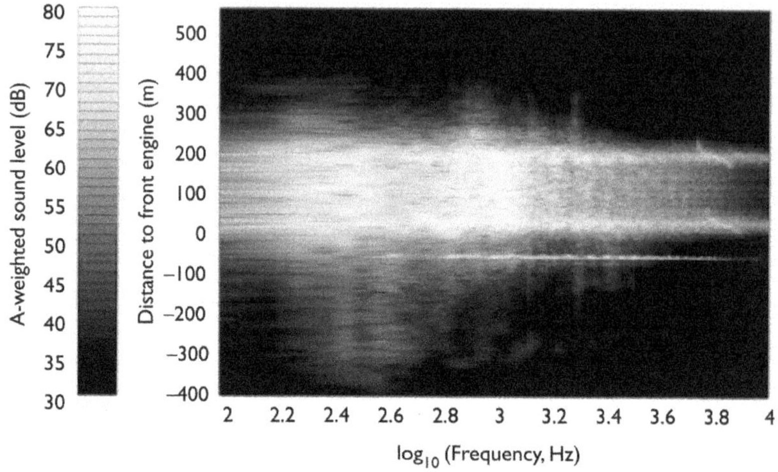

Figure 9.40 The spectrogram of noise emitted by a passing Intercity 125 (diesel) travelling at 184 km/h.

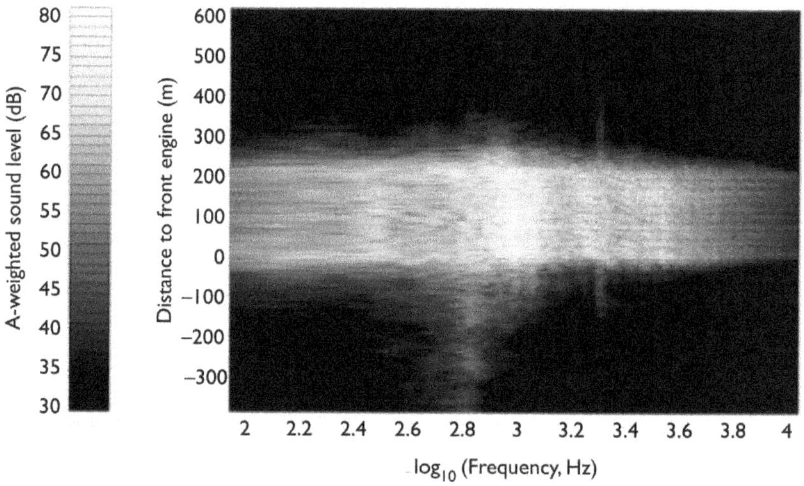

Figure 9.41 The spectrogram of noise emitted by a passing Intercity 225 (electric) travelling at 176 km/h.

Figures 9.40 and 9.41). The supposed noise control configurations are shown in Figures 9.42(a)–(c). Two sources of noise are elevated 0.3 m above the ballast and their positions coincided with the areas of rail-to-wheel interactions (see Figure 9.42). The receiver is assumed to be at ground level 10 m from the barrier.

Predictions have been made using the incoherent line source boundary element model proposed by Duhamel [116]. The model requires the discretization of the surfaces in two dimensions only since it assumes that the

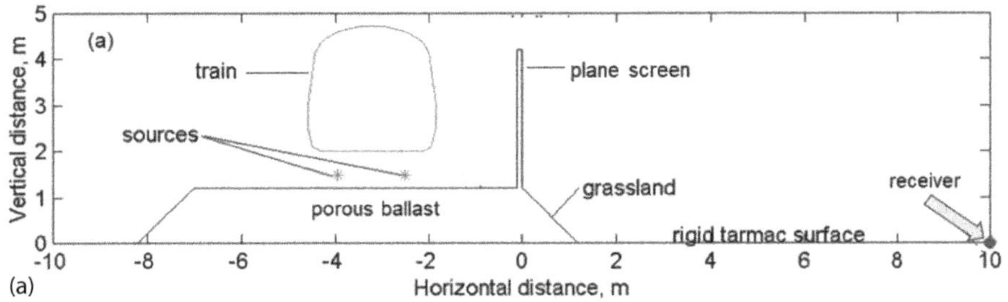

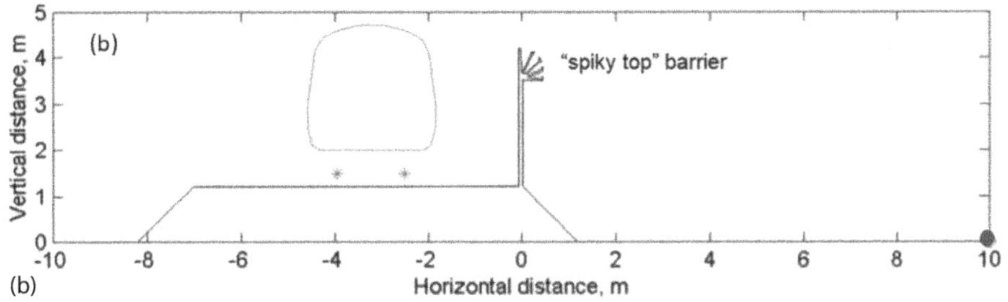

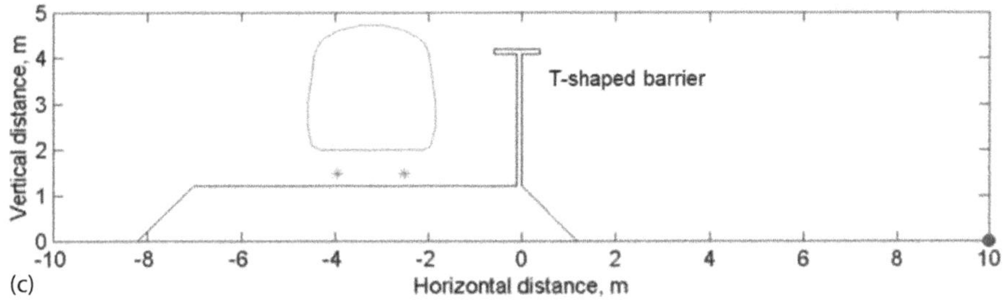

Figure 9.42 Schematic representation of the terrain, train body and the noise barrier profiles assumed for calculations with the BEM model.

barrier and the source are infinitely long. The pseudo-3D pressure field is calculated from

$$p_{3D}(x,y,z,k) = \frac{1}{2\pi}\int_{-\infty}^{+\infty} e^{i\alpha z} p_{2D}\left(x,y,\sqrt{k^2(v)-\alpha^2}\right)d\alpha, \tag{9.56}$$

where $k(v) = \sqrt{k_0^2 + iv}$, $v > 0$ is the attenuation and $p_{2D}(x,y,k)$ is the 2-D, frequency-dependent acoustic field predicted, for example, using the boundary integral equation method detailed in Chapter 8 (Section 8.2.4). In ideal circumstances, a full 3-D model would be required to predict the time-dependent insertion loss of a finite noise barrier length for a moving train of finite dimensions. Practical realization of such a 3-D model is beyond the scope of many numerical methods and likely to be beyond the capability of desk top PCs for the foreseeable future.

Using the measured time-dependent spectral data and predicted values of the insertion loss, the broadband time-dependent noise barrier insertion loss is calculated using

$$IL_B(t) = 10\log_{10}\left(\sum_{n=1}^{18} 10^{\frac{L_0(f_n,t)}{10}}\right) - 10\log_{10}\left(\sum_{n=1}^{18} 10^{\frac{L_B(f_n,t)}{10}}\right), \tag{9.57}$$

where $L_0(f_n,t)$ is experimentally determined time-dependent, 1/3-octave railway noise spectra, $L_B(f_n,t) = L_0(f_n,t) - IL_P(f_n)$ and $IL_P(f_n)$ is the predicted 1/3-octave insertion loss. The values of f_n are the standard 1/3-octave band frequencies defined in the range between 63 and 3150 Hz. The resulting predictions of the time-dependent insertion loss are shown in Figures 9.43(a) and (b) for the diesel and electric trains, respectively.

The noise barrier surface and the body of the train have been assumed to be rigid so that multiple reflections are included. The acoustic surface impedance of the porous ballast and the porous grassland have been modelled using a four-parameter model (see Chapter 5) and the parameter values listed in Table 3.1.

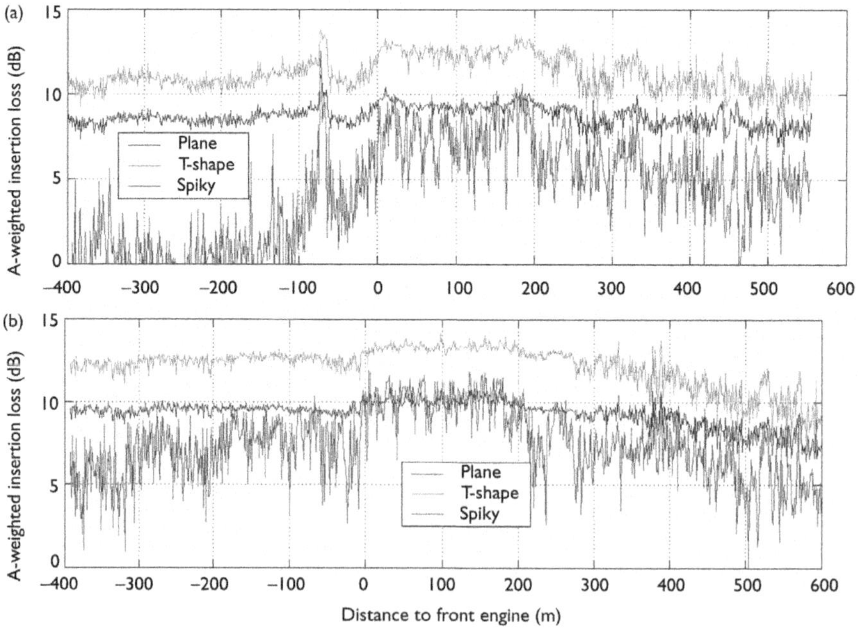

Figure 9.43 Predicted time-dependent, broad-band insertion losses for three barrier designs: (a) Intercity 125 (diesel); (b) Intercity 225 (electric).

The results for the Intercity 125 train demonstrate that the performance of the plane screen barrier is stable throughout the entire passage of the train. The average performance of the T-shape barrier is superior to that of plane screen; however, its dependence upon the instantaneous spectrum of the emitted noise is more pronounced. Although the average performance of the 'spiky top' barrier is similar or lower to that of the plane screen during the by-pass, the variation in the performance of this barrier during the by-pass is considerable. With the Intercity 225 train pass-by spectral variation, the insertion loss for the plane screen is inferior, but remarkably stable in comparison with the more complex barrier shapes. The fluctuation in the insertion loss of the 'spiky top' barrier is between 5 and 7 dB with a maximum fluctuation around the time of the train arrival and departure. Subjectively, this level of fluctuation might be more significant than the predicted gain of between 1 and 2 dB in the average barrier performance. Such effects should be considered when the efficiency of expensive railway noise barriers is assessed.

REFERENCES

[1] N. Sommerfeld, Mathematische Theorie der Diffraktion, *Math. Ann.* **47** 317–374 (1896).

[2] M. Born and E. Wolf, *Principles of Optics, Electromagnetic Theory of Propagation, Interference and Diffraction of Light*, Pergamon Press, Cambridge, UK (1975).

[3] H. S. Carslaw, Some Multiform Solutions of the Partial Differential Equations of Physical Mathematics and their Applications. *Proc. London Math. Soc.* **30** 121–161 (1899).

[4] H. S. Carslaw, Diffraction of waves by a wedge of any angle, *Proc. Lond. Math. Soc.* **18** 1291–1306 (1920).

[5] H. M. MacDonald, A class of diffraction problems, *Proc. Lond. Math. Soc.* **14** 410–427 (1915).

[6] E. T. Copson, On an integral equation arising in the theory of diffraction, *Quart. J. Math.* **17** 19–34 (1946).

[7] H. Levine and J. Schwinger, On the theory of diffraction by an aperture in an infinite plane screen. *I. Phys. Rev.* **74** 958–974 (1948).

[8] H. Levine and J. Schwinger, On the theory of diffraction by an aperture in an infinite plane screen. II., *Phys. Rev.* **75** 1423–1432 (1949).

[9] D. G. Crighton, A. P. Dowling, J. E. Ffowcs Williams, M. Heckl and F. G. Leppington, *Modern Methods in Analytical Acoustics Lecture Notes*, Springer-Verlag, Ch. 5 (1996).

[10] M. C. M. Wright ed. *Lecture Notes on the Mathematics of Acoustics*, Imperial College Press, London (2005).

[11] Tolstoy, Diffraction by a hard truncated wedge and a strip, *IEEE J. Ocean Eng.* **14** 4–16 (1989).

[12] Tolstoy, Exact explicit solutions for diffraction by hard sound barriers and seamounts, *J. Acoust. Soc. Am.* **85** 661–669 (1989).

[13] J. J. Bowman and T. B. A. Senior, *Electromagnetic and Acoustic Scattering by Simple Shapes*, edited by J. J. Bowman, T. B. A. Senior and P. L. E. Uslenghi, North-Holland, Amsterdam (1969) (revised edition Hemisphere Publishing Corp., New York, 1987).

[14] W. J. Hadden and A. D. Pierce, Sound diffraction around screens and wedges for arbitrary point-source locations, *J. Acoust. Soc. Am.* **69**(5) 1266–1276 (1981). See also the Erratum, *J. Acoust. Soc. Am.* 71 1290 (1982).

[15] E. Hecht, *Optics*, Addison-Wesley Publishing Company, 5th edition, Reading MA, USA (1998).

[16] E. J. Skudrzyk, *The foundations of Acoustics*, Springer-Verlag, New York (1971).

[17] T. F. W. Embleton, Line integral theory of barrier attenuation in the presence of the ground, *J. Acoust. Soc. Am.* **67** 42–45 (1980).

[18] G. M. Jebsen and H. Medwin, On the failure of the Kirchhoff assumption in backscatter, *J. Acoust. Soc. Am.* **72** 1607–1611 (1982).

[19] S. W. Redfearn, Some acoustical source-observer problems, *Phil. Mag.* **30** 223–236 (1940).

[20] Z. Maekawa, Noise reduction by screens, *Appl. Acoust.* **1** 157–173 (1968).

[21] E. J. Rathe, Note on two common problems of sound propagation, *J. Sound Vib.* **10** 472–479 (1969).

[22] U. J. Kurze and G. S. Anderson, Sound attenuation by barriers, *Appl. Acoust.* **4** 35–53 (1971).

[23] U. J. Kurze, Noise reduction by barriers, *J. Acoust. Soc. Am.* **55** 504–518 (1974).

[24] J. B. Keller, The geometrical theory of diffraction, *J. Opt. Soc.* **52** 116–130 (1962).

[25] R. B. Tatge, Barrier-wall attenuation with a finite sized source, *J. Acoust. Soc. Am.* **53** 1317–1319 (1973).

[26] K. Yamamoto and K. Takagi, Expressions of Maekawa's chart for computation, *Appl. Acoust.* **37** 75–82 (1992).

[27] P. Menounou, A correction to Maekawa's curve for the insertion loss behind barriers, *J. Acoust. Soc. Am.* **110** 1828–1838 (2001).

[28] T. Isei, T. F. W. Embleton and J. E. Piercy, Noise reduction by barriers on finite impedance ground, *J. Acoust. Soc. Am.* **67** 46–58 (1980).

[29] J. Nicolas, T. F. W. Embleton and J. E. Piercy, Precise model measurements versus theoretical prediction of barrier insertion loss in presence of the ground, *J. Acoust. Soc. Am.* **73** 44–54 (1983).

[30] Y. W. Lam and S. C. Roberts, A simple method for accurate prediction of finite barrier insertion loss, *J. Acoust. Soc. Am.* **93** 1445–1452 (1993).

[31] L'Espérance, The insertion loss of finite length barriers on the ground, *J. Acoust. Soc. Am.* **86** 179–183 (1989).

[32] R. Pirinchieva, Model study of the sound propagation behind barriers of finite length, *J. Acoust. Soc. Am.* **87** 2109–2113 (1990).

[33] H. Medwin, Shadowing by finite noise barriers, *J. Acoust. Soc. Am.* **69** 1060–1064 (1981).

[34] I. Muradai and K. R. Fyfe, A study of 2D and 3D barrier insertion loss using improved diffraction-based methods, *Appl. Acoust.* **53** 49–75 (1998).

[35] G. R. Watts, Effects of sound leakage through noise barriers on screening performance, *6th International Congress on Sound and Vibration*, Copenhagen, Denmark, 2501–2508 (1999).
[36] H. Y. Wong and K. M. Li, Prediction models for sound leakage through noise barriers, *J. Acoust. Soc. Am.* **109** 1011–1022 (2001).
[37] S. I. Thomasson, Diffraction by a screen above an impedance boundary, *J. Acoust. Soc. Am.* **63** 1768–1781 (1978).
[38] K. B. Rasmussen, A note on the calculation of sound propagation over impedance jumps and screens, *J. Sound Vib.* **84** 598–602 (1982).
[39] D. Rawlins, Diffraction of sound by a rigid screen with an absorbent edge, *J. Sound Vib.* **47** 523–541 (1976).
[40] T. Isei, Absorptive noise barrier on finite impedance ground, *J. Acoust. Soc. Jpn. (E)* **1** 3–9 (1980).
[41] K. Fujiwara, Noise control by barriers, *Appl. Acoust.* **10** 147–159 (1977). See also K. Fujiwara, Y. Ando and Z. Maekawa, Noise control by barriers – Part 2: Noise reduction by an absorptive barrier, *Appl. Acoust.* 10: 167–179 (1977).
[42] L. L'Espérance, J. Nicolas and G. A. Daigle, Insertion loss of absorbent barriers on ground, *J. Acoust. Soc. Am.* **86** 1060–1064 (1989).
[43] S. I. Hayek, Mathematical modeling of absorbent highway noise barriers, *Appl. Acoust.* **31** 77–100 (1990).
[44] A. Zwikker and C. Kosten, *Sound Absorbing Materials*, Elsevier, New York (1949).
[45] J. F. Hamet and M. Bérengier, Acoustical Characteristics of Porous Pavements: A New Phenomenological Model, Internoise'93, Leuven, Belgium (1993); see also M. Bérengier, M. Stinson, G. Daigle and J. F. Hamet, Porous road pavements: acoustical characterization and propagation effects, *J. Acoust. Soc. Am.* **101** 155–162 (1997).
[46] T. Van Renterghem, I. Vanginderachter and P. Thomas, Porous stones increase the noise shielding of a gabion, *Appl. Acoust.* **145** 82–88 (2019).
[47] F. Koussa, J. Defrance, P. Jean and P. Blanc-Benon, Acoustic performance of gabions noise barriers: numerical and experimental approaches, *Appl. Acoust.* **74** 189–197 (2013).
[48] M. Garai and P. Guidorzi, European methodology for testing the airborne sound insulation characteristics of noise barriers in situ: experimental verification and comparison with laboratory data, *J Acoust Soc Am.*, **108** 1054–1067 (2000).
[49] EN 1793-3, Road traffic noise reducing devices—Test methods for determining the acoustic performance—Part 3: Normalized traffic noise spectrum, European standards (1997).
[50] BS EN 1793-1, Road traffic noise reducing devices - Test method for determining the acoustic performance - Part 1: Intrinsic characteristics of sound absorption under diffuse sound field conditions (2017).
[51] BS EN 1793-2, Road traffic noise reducing devices - Test method for determining the acoustic performance - Part 2: Intrinsic characteristics of airborne sound insulation under diffuse sound field conditions (2018).
[52] L. S. Wirt, The Control of Diffracted Sound by Means of Thnadners (Shaped Noise Barriers), *Acustica* **42** 73–88 (1979).

[53] S. S. T. Ho, I. J. Busch-Vishniac, and D. T. Blackstock, Noise reduction by a barrier having a random edge profile, *J. Acoust. Soc. Am.* **101** 2669–2676 (1997).
[54] W. Shao, H. P. Lee and S. P. Lim, Performance of noise barriers with random edge profiles, *Appl. Acoust.* **62** 1157–1170 (2001).
[55] N. May and M. M. Osman, The Performance of sound absorptive, reflective, and T-profile noise barriers in Toronto, *J. Sound Vib.* **71** 65–71 (1980).
[56] N. May and M. M. Osman, Highway Noise Barriers – New Shapes, *J. Sound Vib.* **71** 73–101 (1980).
[57] D. A. Hutchins, H. W. Jones and L. T. Russell, Model studies of barrier performance in the presence of ground surfaces. Part I – Thin, perfectly reflecting barriers, *J. Acoust. Soc. Am.* **75** 1807–1816 (1984).
[58] D. A. Hutchins, H. W. Jones and L. T. Russell, Model studies of barrier performance in the presence of ground surfaces. Part II – Different shapes, *J. Acoust. Soc. Am.* **75** 1817–1826 (1984).
[59] D. C. Hothersall, D. H. Crombie and S. N. Chandler-Wilde, The performance of T-Profile and associated noise barriers, *Appl. Acoust.* **32** 269–287 (1991).
[60] J. Jin, H. S. Kim, H. J. Kang and J. S. Kim, Sound diffraction by a partially inclined noise barrier, *Appl. Acoust.* **62** 1107–1121 (2001).
[61] R. G. Kouyoumjian and P. H. Pathak, A uniform geometrical theory of diffraction for an edge in a perfectly conducting surface, *Proc. IEEE* **62** 1448–1461 (1974).
[62] T. Okubo and K. Fujiwara, Efficiency of a noise barrier on the ground with an acoustically soft cylindrical edge, *J. Sound Vib.* **216** 771–790 (1998); See also T. Okubo and K. Fujiwara, Efficiency of a noise barrier with an acoustically soft cylindrical edge for practical use, J. Acoust. Soc. Am. **105** 3326–3335 (1999).
[63] J. Wassilieff, Improving the noise reduction of picket barriers, *J. Acoust. Soc. Am.* **84** 645–650 (1988).
[64] R. Lyons and B. M. Gibbs, Investigation of open screen acoustic performance, *Appl. Acoust.* **49** 263–282 (1996).
[65] B. Viveiros, B. M. Gibbs and S. N. Y. Gerges, Measurement of sound insulation of acoustic louvers by an impulse method, *Appl. Acoust.* **63** 1301–1313 (2002).
[66] R. Watts, D. C. Hothersall and K. V. Horoshenkov, Measured and predicted acoustic performance of vertically louvred noise barriers, *Appl. Acoust.* **63** 1287–1311 (2001).
[67] T. Ishizuka and K. Fujiwara, Performance of noise barriers with variable edge shape and acoustical conditions, *Appl. Acoust.* **65** 125–141 (2004).
[68] S. I. Voropayev, N. C. Ovenden, H. J. S. Fernando and P. R. Donovan, Finding optimal geometries for noise barrier tops using scaled experiments, *J. Acoust. Soc. Am.* **141** 722–736 (2017).
[69] P. A. Morgan, D. C. Hothersall and S. N. Chandler-Wilde, Influence of shape and absorbing surface - A numerical study of railway noise barriers, *J. Sound Vib.* **217** 405–417 (1998).
[70] W. E. Scholes, A. C. Salvidge and J. W. Sargent, Field performance of a noise barrier, *J. Sound Vib.* **16** 627–642 (1971).
[71] M. Salomons, Diffraction by a screen in downwind sound propagation: a parabolic-equation approach, *J. Acoust. Soc. Am.* **95** 3109–3117 (1994).

[72] K. B. Rasmussen, Sound propagation over screened ground under upwind conditions, *J. Acoust. Soc. Am.* **100** 3581–3586 (1996).
[73] S. Muradali and K. R. Fyfe, Accurate barrier modeling in the presence of Atmosphere Effects, *Appl. Acoust.* **56** 157–182 (1999).
[74] K. M. Li and Q. Wang, A BEM approach to assess the acoustic performance of noise barriers in the refracting atmosphere, *J. Sound Vib.* **211** 663–681 (1998).
[75] S. Taherzadeh, K. M. Li and K. Attenborough, A hybrid BIE/FFP scheme for predicting barrier efficiency outdoors, *J. Acoust. Soc. Am.* **110** 918–924 (2001).
[76] G. Daigle, Diffraction of sound by a noise barrier in the presence of atmospheric turbulence, *J. Acoust. Soc. Am.* **71** 847–854 (1982).
[77] J. Forssén and M. Ogren, Thick barrier noise-reduction in the presence of atmospheric turbulence: measurements and numerical modeling, *Appl. Acoust.* **63** 173–187 (2002).
[78] J. Forssén, Calculation of noise barrier performance in a turbulent atmosphere by using substitute-sources above the barrier, *Acta Acust. united Ac.* **86** 269–275 (2000).
[79] J. Forssén, Calculation of noise barrier performance in a three-dimensional turbulent atmosphere using the substitute-sources method, *Acta Acust. united Ac.* **88** 181–189 (2002).
[80] A. Premat and Y. Gabillet, A new boundary-element method for predicting outdoor sound propagation and application to the case of a sound barrier in the presence of downward refraction, *J. Acoust. Soc. Am.* **108** 2775–2783 (2000).
[81] Y. W. Lam, A boundary element method for the calculation of noise barrier insertion loss in the presence of atmospheric turbulence, *Appl. Acoust.* **65** 583–603 (2004).
[82] E. Gilbert, R. Raspet and X. Di, Calculation of turbulence effects in an upward refracting atmosphere, *J. Acoust. Soc. Am.* **87** 2428–2437 (1990).
[83] K. B. Rasmussen and M. G. Arranz, The insertion loss of screens under the influence of wind, *J. Acoust. Soc. Am.* **104** 2692–2698 (1998).
[84] T. Van Renterghem, E. M. Salomons and D. Botteldooren, Efficient FDTD-PE model for sound propagation in situations with complex obstacles and wind profiles, *Acta Acust. united Ac.* **91** 671–679 (2005).
[85] D. Heimann and R. Blumrich, Time-domain simulations of sound propagation through screen-induced turbulence, *Appl. Acoust.* **65** 561–582 (2004).
[86] T. Busch, M. Hodgson and C. Wakefield, Scale-model study of the effectiveness of highway noise barriers, *J. Acoust. Soc. Am.* **114** 1947–1954 (2003).
[87] D. Hutchins, H. Jones and L. Russell, Model studies of barrier performance in the presence of ground surfaces. Part II – Different shapes. *J. Acoust. Soc. Am.* **75** 1817–1826 (1984).
[88] T. Van Renterghem and D. Botteldooren, On the choice between walls and berms for road traffic noise shielding including wind effects, *Landscape and Urban Planning*, **105** 199–210 (2012).
[89] P. Jean, Effectiveness of added roughness on berm surfaces, *Appl. Acoust.* **154** 77–89 (2019).
[90] A. Ekici and H. Bougdah, A review of Research on Environmental Noise Barriers, *Build. Acoust.* **10** 289–323 (2003).
[91] P. Jean, HSNA_WP4_TRP_2011-05-24_Berms-BEM 01.doc unpublished (2011).

[92] T. Van Renterghem and D. Botteldooren, Effect of a row of trees behind noise barriers in wind, *Acta Acust. united Ac.* **88** 869–878 (2002).
[93] J. Hornikx and T. Van Renterghem, Numerical investigation of the effect of crosswind on sound propagation outdoors, *Acta Acust. united Ac.* **102** 558–565 (2016).
[94] J. Arenas, Potential problems with environmental sound barriers when used in mitigating surface transportation noise, *Sci. Total Environ.* **405** 173–179 (2008).
[95] J. Sánchez-Pérez, D. Caballero, R. Mártinez-Sala, C. Rubio, J. Sánchez-Dehesa, F. Meseguer, J. Llinares and F. Gálvez, Sound attenuation by a two-dimensional array of rigid cylinders, *Phys. Rev. Lett.* **80** 5325–5328 (1998).
[96] L. Fredianelli, A. Del Pizzo and G. Licitra, Recent developments in sonic crystals as barriers for road traffic noise mitigation, *Environments* **6** 14–33 (2019).
[97] J. V. Sánchez-Pérez, C. Rubio, R. Martinez-Sala, R. Sanchez-Grandia and V. Gomez, Acoustic barriers based on periodic arrays of scatterers, *Appl. Phys. Lett.* **81** 5240 (2002).
[98] C. Goffaux, F. Maseri, J. O. Vasseur, B. Djafari-Rouhani and P. Lambin, Measurements and calculations of the sound attenuation by a phononic band gap structure suitable for an insulating partition application, *Appl. Phys. Lett.* **83** 281 (2003).
[99] O. Umnova, K. Attenborough and C. Linton, Effects of porous covering on sound attenuation by periodic arrays of cylinders, *J. Acoust. Soc. Am.* **119** 278–284 (2006).
[100] J. Sánchez-Dehesa, V. M. Garcia-Chocano, D. Torrent, F. Cervera, S. Cabrera, and F. Simon, Noise control by sonic crystal barriers made of recycled materials, *J. Acoust. Soc. Am.* **129** 1173–1183 (2011).
[101] K. M. Ho, C. K. Cheng, Z. Yang, X. X. Zhang and P. Sheng, Broadband locally resonant sonic shields, *Appl. Phys. Lett.* **83** 5566 (2003).
[102] X. Hu, C. Chan and J. Zi, Two-dimensional sonic crystals with Helmholtz resonators, *Phys. Rev. E* **71** 055601 (2005).
[103] D. P. Elford, L. Chalmers, F. V. Kusmartsev and G. M. Swallowe, Matryoshka locally resonant sonic crystal, *J. Acoust. Soc. Am.* **130** 2746–2755 (2011).
[104] A. Krynkin, O. Umnova, A. Yung Boon Chong, S. Taherzadeh and K. Attenborough, Predictions and measurements of sound transmission through a periodic array of elastic shells in air, *J. Acoust. Soc. Am.* **128**(6) 3496–3506 (2010).
[105] A. Krynkin, O. Umnova, S. Taherzadeh and K. Attenborough, Analytical approximations for low frequency band gaps in periodic arrays of elastic shells, *J. Acoust. Soc. Am.* **133** 781–791 (2013).
[106] A. Krynkin, O. Umnova, A. Y. B. Chong, S. Taherzadeh and K. Attenborough, Scattering by coupled resonating elements in air, *J. Phys. D. Appl. Phys.* **44** 125501 (2011).
[107] V. Romero-García, J. V. Sánchez-Pérez and L. M. Garcia-Raffi, Tunable wideband bandstop acoustic filter based on two-dimensional multiphysical phenomena periodic systems, *J. Appl. Phys.* **110** 014904 (2011).

[108] S. Castineira-Ibanez, C. Rubio, V. Romero-Garcia, J. V. Sanchez-Perez and L. M. Garcia-Raffi, Design, manufacture and characterization of an acoustic barrier made of multi-phenomena cylindrical scatterers arranged in a fractal-based geometry, *Arch. Acoust.* **37** 455–462 (2012).
[109] https://www.sonobex.com/technologies (last accessed 8/3/20).
[110] A. Krynkin, O. Umnova, J. V. Sánchez-Pérez, A. Y. Boon Chong, S. Taherzadeh and K. Attenborough, Acoustic insertion loss due to two dimensional periodic arrays of circular cylinders parallel to a nearby surface, *J. Acoust. Soc. Am.* **130** 3736–3745 (2011).
[111] V. Romero-Garcıa, J. V. Sanchez-Perez and L. M. Garcia-Raffi, Analytical model to predict the effect of a finite impedance surface on the propagation properties of 2D Sonic Crystals, *J. Phys. D: Appl. Phys.* **44** 265501 (2011).
[112] V. Romero-Garcia, S. Castineira-Ibaneza, J. V. Sanchez-Perez and L. M. Garcia-Raffi, Design of wideband attenuation devices based on Sonic Crystals made of multi-phenomena scatterers, *Proceedings of Acoustics 2012*, Nantes (23–27 April 2012).
[113] T. Elnady, A. Elsabbagh, W. Akl, O. Mohamady, V. M. Garcia-Chocano and J. Sanchez-Dehesa, Quenching of acoustic bandgaps by flow noise, *Appl. Phys. Lett.* **94** 134104 (2009).
[114] BS EN 1793-3: 1998 Road traffic noise reducing devices - Test methods for determining the acoustic performance. Part 3 – Normalized traffic noise spectrum, BSI (1998).
[115] K. V. Horoshenkov, S. Rehman and S. J. Martin, The influence of the noise spectra on the predicted performance of railway noise barriers, *CD-ROM Proceedings of the 9th International Congress on Sound and Vibration*, Orlando, USA (July 2002).
[116] D. Duhamel, Efficient calculation of the three-dimensional sound pressure field around a noise barrier, *J. Sound Vib.* **197**(5) 547–571 (1996).
[117] J. Defrance, K. Attenborough, K. Horoshenkov, F. Koussa, P. Jean, T. Van Renterghem, "Intermediate report of Task 2.2", FP7 HOSANNA Deliverable 2.2, 2011

Chapter 10

Predicting Effects of Vegetation, Trees and Turbulence

10.1 MEASURED EFFECTS OF VEGETATION

10.1.1 Influence of Vegetation on Soil Properties

As plants grow, they create root zones in porous ground and, as mentioned in Chapter 6, in comparison to bare compacted ground with little or no vegetation, the near-surface flow resistivity is lower and the air-filled porosity in the root zone is higher (see Table 6.5). Short-range vertical-level difference measurements using two source–receiver geometries, similar to those used for ground impedance deduction (see Chapter 6), have been made in a field containing 0.45 m–0.55 m high winter wheat crops [1,2]. The soil was a loamy sand, typically composed of 80% sand, 10% silt and 10% clay. After making measurements in the presence of the crops, the stems and leaves of the crops were removed carefully from a 1.88 m × 1.84 m area without altering the ground surface and vertical-level difference measurements were repeated over the cleared area. Both sets of measurements were made on the same day which was dry with little or no wind. Photographs of the measurement area before and after removal of crops are shown as Figures 10.1(a) and (b). The level difference spectra obtained with upper microphone height 0.3 m, lower microphone height 0.15 m, horizontal source–receiver separation 1.0 m and two source heights (geometry E, 0.3 m; geometry F, 0.21 m) before and after stem and leaf clearance are compared in Figures 10.1(c) and (d). For each of the two conditions the level difference spectra with either of the source heights are rather similar. Similar results have been obtained before and after removal of winter wheat stems at a different site and before and after removal of a different crop (oil seed rape plants) [2]. Within the short range (1.0 m) involved in these measurements the presence of above-surface vegetation does not appear to have a significant influence on ground effect.

The measured vertical-level difference spectra in Figures 10.1(c) and (d) can be fitted using a two-parameter slit-pore impedance model (see Chapters 5 and 6 for details of the model and fitting). Figure 10.2 shows an example comparison between the vertical-level difference spectrum measured with (point) source at height of 0.3 m, upper and lower microphones at heights

(a)

(b)

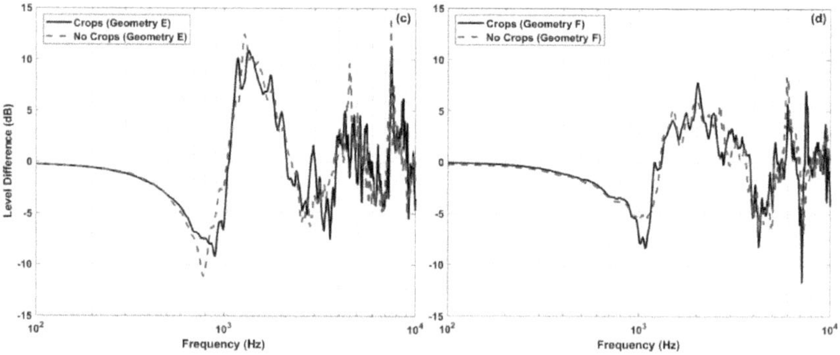

Figure 10.1 (a) and (b) Sites with 0.5 m high crops and after careful removal of crop stems, respectively, used for short range measurements: (c) and (d) level difference spectra between vertically-separated microphones measured at these sites (upper microphone height = 0.3 m, lower microphone height = 0.15 m and source-receiver separation = 1.0 m: Geometry E, Source height = 0.3 m; Geometry F: Source height = 0.21 m) [2].

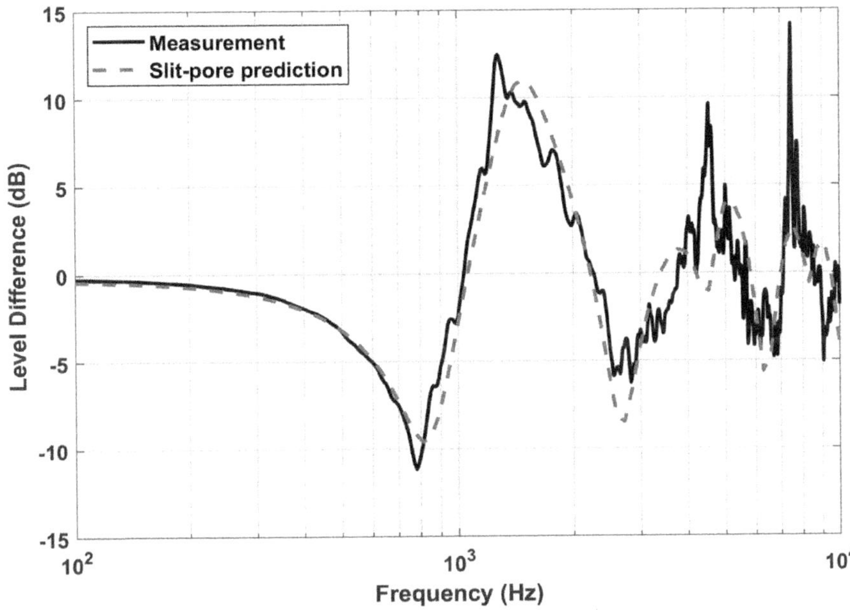

Figure 10.2 Example comparison between level difference spectra (continuous line) measured over an area cleared of winter wheat with source at height of 0.3 m, upper and lower microphones at heights of 0.3 m and 0.15 m respectively and source-receiver separation of 1.0 m with a prediction (broken line) based on the two parameter slit pore model impedance model with effective flow resistivity 300.0 kPasm^{-2} and porosity 0.4 [2].

of 0.3 m and 0.15 m, respectively, and source–upper receiver separation of 1.0 m over the cleared area and predictions using a two-parameter slit-pore model impedance with best-fit flow resistivity of 300 kPa s m^{-2} and porosity 0.4. Data obtained at a second cleared area in the same winter wheat crop were fitted also using the two-parameter slit-pore model with a flow resistivity of 170 kPa s m^{-2} and a porosity of 0.2. The difference in fitted parameter values is within the potential spatial variability in soil properties. A more extensive study of spatial variability is discussed in Chapter 6.

The values of best-fit flow resistivity (i.e. 170 and 300 kPa s m^{-2}) obtained from short-range vertical-level difference measurements over ground containing crops differ significantly from that obtained (2000 kPa s m^{-2}) over nearby bare ground with same type of soil but on which no crops were growing or had been grown [2]. This indicates that the growing of winter wheat plants in this soil reduced its effective flow resistivity by around a factor of 10.

Impedance tube studies [3] have shown that plants have a significant effect on the normal incidence absorption coefficient of soils. Figure 10.3 shows photographs of two soils (high and low porosity) and a winter primula (*primula vulgaris*) plant together with absorption coefficient data obtained without

Figure 10.3 Photographs of (a) a high porosity soil (porosity between 0.7 and 0.85); (b) a low porosity soil (porosity between 0.26 and 0.48); (c) a winter primula vulgaris plant; typical height 0.1 m, number of leaves 22, leaf area per unit volume 143/m; (d) and (e) normal incidence absorption coefficient spectra measured on soil alone (asterisks) and soil plus plant (open circles) with high and low porosity soils respectively [3]. Reproduced from K. Horoshenkov, A. Khan and H. Benkriera, Acoustic properties of low growing plants, *J. Acoust. Soc. Am.*, **133**: 2554–2565 (2013) with the permission of the Acoustical Society of America.

and with this plant on each of the soils. However, in this case, not only may the plant foliage have contributed to the change (see also Section 10.2.2.1), but disturbance of previously compacted soil associated with inserting a plant may have also been partly responsible.

Other studies have investigated the effects of plant leaves placed loosely on porous substrates, i.e. without disturbing the soil [4]. This results in an increase in the low-frequency normal incidence absorption coefficient (below 2 kHz) and a decrease in the absorption at higher frequencies. Laboratory and in situ measurements on the effects of vegetation on the diffuse field absorption of a substrate soil used on green roofs have confirmed this effect [5]. To avoid perturbations of the sound field on green roofs resulting from non-continuous surfaces and architectural features, these measurements were made using a loudspeaker source suspended at 1.75 m above the surface of interest and two microphones 0.8 m and 0.55 m above the surface and vertically below the source. A transfer function method was used to deduce the normal surface impedance (see Chapter 6) which was then used to calculate a diffuse field absorption coefficient.

Beneath mature trees, shrubs and hedges, fallen and decaying leaves have an important influence on the acoustical properties of ground. During measurements of sound transmission through 16 m tall red pine trees with mean trunk diameter 0.23 m and spacing 3.3 m [6,7] it was noted that, over time, leaves falling from the red pine trees had created a 0.025 m deep layer of decaying foliage (litter and humus layers) above the soil. The presence of similar litter and humus layers have been noted during measurements in other types of forest [8,9,70]. Fitted effective flow resistivity values for mature pine forest floors range lie between 10 and 62 kPa s m^{-2} (see Table 6.2). Measured flow resistivity values for litter layers on three forest floors are 9, 22 and 30 kPa s m^{-2} (see Table 6.7). Since forest floor flow resistivity is substantially lower than typical of a grassland, for the same source–receiver geometry, the first destructive interference due to ground effect in a forest is at a lower frequency than over grassland. As a consequence, narrow tree belts alongside roads do not necessarily offer much additional attenuation of road traffic noise compared with the same distances over open grassland (see Section 10.1.3). Nevertheless, additional effects are possible from regular tree planting arrangements (see Section 10.3.1) and if there is dense foliage to ground level (see Sections 10.1.2 and 10.2.2).

10.1.2 Measurements of Sound Transmission through Vegetation

The stems of plants and trunks and branches of trees scatter sound. Foliage converts sound energy into heat through viscous friction and thermal exchanges at the leaf surfaces and, possibly, by induced leaf vibration. Aylor [6,7] measured sound transmission loss through dense corn, hemlock, red pine trees, hardwood brush and densely planted reeds in water and observed that the measured attenuation was greater than that attributable only to viscous and thermal losses for a given vegetation density. He argued that the extra energy loss must be due to multiple scattering effects within the vegetation. He concluded that the overall attenuation is directly proportional to vegetation density, that foliage attenuates the sound at higher frequencies and that to maximize the sound attenuation due to the vegetation it should be planted densely and have a large leaf area per unit volume.

Laboratory measurements of sound propagation through vegetation [10] have confirmed that vegetation behaves as a low pass filter giving attenuation between 2 and 8 kHz. It has been suggested that the vibration of leaves and branches also contributes to transmission loss [11,12]. The vibration of plant leaves in response to incident acoustic energy has been measured using a laser-Doppler-vibrometer [13]. Although the sound energy absorbed by each leaf through vibration is very small, the individual leaf attenuations add together and may give a significant overall effect for a complete plant. It has been found that the size of the leaf is an important parameter in the reflection of sound, i.e. the bigger the leaf size, the larger will be the acoustic reflection [14]. Moreover, it has been confirmed that plants with dense foliage and larger leaf sizes give larger high-frequency sound attenuation during transmission.

Differences in spectra received by horizontally separated microphones have been measured at the same winter wheat site during dry conditions in August 2011, wet conditions in May 2012 and intermediate conditions in June 2012 [1,2]. A reference microphone was placed 1.0 m from the loudspeaker point source (a Tannoy® driver with tube attachment) and further microphones, at the same height as the source, were located at horizontal distances of 2.5 m, 5.0 m, 7.5 m and 10.0 m, respectively, from the source. Measurements were carried out with source and receivers at heights of 0.2 m, 0.3 m and 0.4 m. Since the crop height was between 0.45 m and 0.55 m, the microphones and sound source were always inside the crops. Similar data were obtained at all three heights so only the results of measurements at 0.3 m height are considered further here. Photographs of the different crop conditions during the measurements are shown in Figures 10.1(a) and 10.4. Spectra of differences between signals received at the reference microphone and each further microphone are shown in Figures 10.5(a)–(d).

The spectra have clear differences between 600 Hz and 3 kHz. Incoherent scattering by the largest leaf sizes (in May) causes the first destructive interference of the ground effect to be much reduced at 2.5 m and 5 m range and to be practically non-existent at 10 m range. There is an increase in attenuation above 3 kHz with increase in the path length through the crop and, at each range, the greatest high-frequency attenuation corresponds to the largest leaf sizes.

(a) (b)

Figure 10.4 Photographs of crop conditions during measurements (a) in May 2012 and (b) in June 2012 [2]

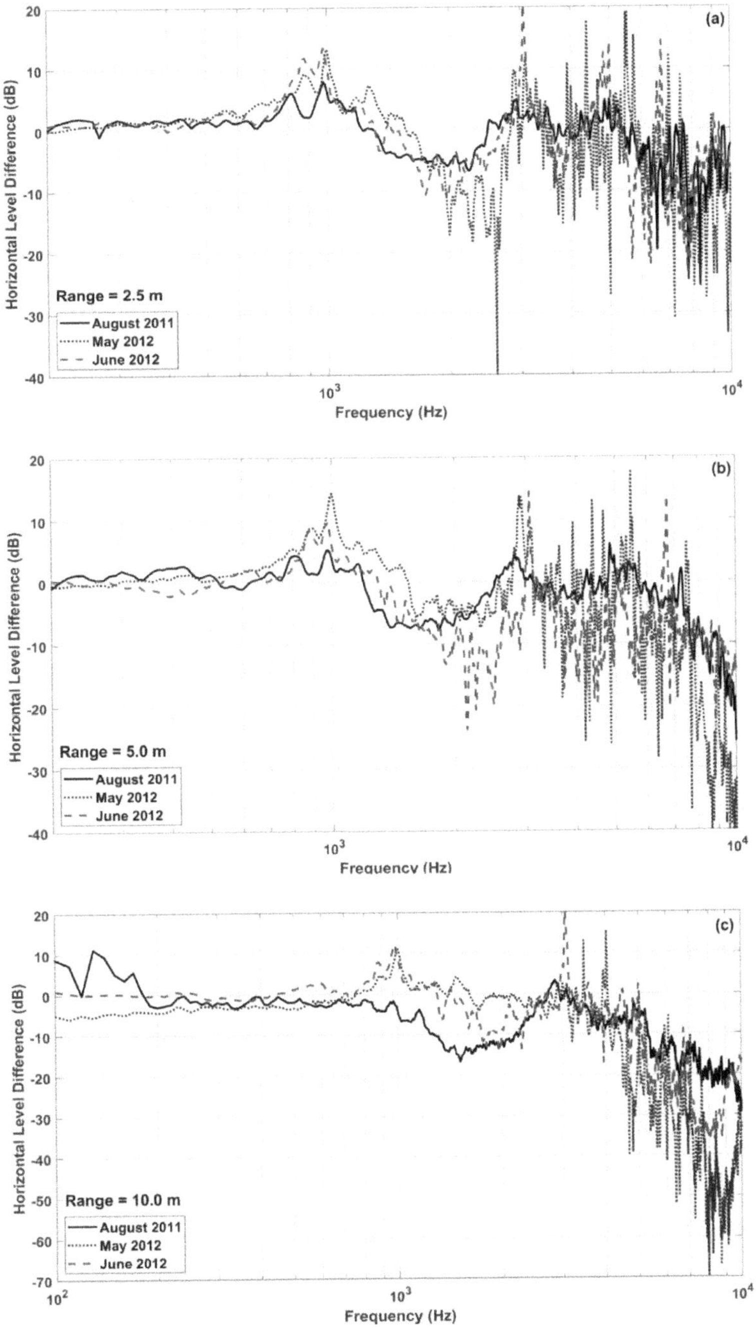

Figure 10.5 Level difference spectra measured with source and receivers 0.3 m above the ground in 0.45 m to 0.5 m high winter wheat crops in August 2011 (see Figure 10.1(a)), May 2012 and June 2012 and at various horizontal ranges (a) 2.5 m, (b) 5.0 m and (c) 10.0 m [2, Figures 10.7–10.10]

10.1.3 Measured Attenuation Due to Trees, Shrubs and Hedges

Measurements made with broadband source 2 m high and receiver height 1.5 m through 500 m of coniferous woodland have shown significant extra attenuation compared with CONCAWE predictions (see Chapters 1 and 12) particularly in 63 Hz (3.2 dB), 125 Hz (9.7 dB), 2 kHz (21.7 dB) and 4 kHz (24.7 dB) octave bands [15]. The United States Department of Agriculture National Agroforestry Center [16] has suggested guidelines for the planting of trees and bushes for noise control based on extensive data collected in the 1970s [17]. The basis for the foliage correction in ISO-9613 part 2 [18] is a Danish study [19] which found that tree belts between 15 and 41 m wide resulted in relative attenuation of only 3 dB in the A-weighted L_{eq} due to traffic noise. On the other hand, 100 m of red pine forest has been found to give 8 dB reduction in the A-weighted L_{eq} due to road traffic compared with open grassland [20]. The edge of the forest was 10 m from the edge of the highway and the trees occupied a gradual downward slope from the roadway extending about 325 m in each direction along the highway from the measurement site. An extra 6 dB reduction in A-weighted L_{10} index due to traffic noise through 30 m wide belt of dense spruce has been found compared with the that over the same width of grassland [21]. A relative reduction of 5 dB in the A-weighted L_{10} index was found after transmission through only 10 m of vegetation. This UK study found also that the effectiveness of the vegetation was greatest closest to the road consistent with findings reported elsewhere [17]. If the belt is close to the road then the point of specular reflection on the ground will be located inside the vegetation belt.

In an investigation of the attenuation of sound by 35 different tree belts [22], a point source playing pre-recorded city traffic noise was placed in front of the tree belts and sound pressure levels were measured at different positions inside the tree belts. The attenuation was found to depend on the width, height, length and density of tree belts. Large shrubs and densely populated tree belts were found to result in 6 dBA more attenuation (normalized to 20-m belt width) than nearby ‘open’ ground. Medium-size shrubs and tree belts attenuated the sound by between 3 and 6 dBA and sparsely distributed tree belts and shrubs attenuated the sound by less than 3 dBA more than ‘open’ ground. The width of vegetation was found to be the most important factor, the longer paths of sound through the vegetation resulting in higher sound absorption and diffusion. In the vegetation belts examined, shrubs were considered the most effective in reducing noise due to scattering from dense foliage and branches at lower source–receiver heights. At higher source–receiver heights, trees were considered to contribute to attenuation due to sound diffusion and absorption processes. It was concluded that tree belts and shrubs should be planted together to provide best attenuation performance. Other measurements of sound transmission through trees and tree belts are discussed in Sections 10.2 and 10.3 which are concerned with predicting these effects.

In a study of the noise attenuation by five different kinds of hedges between 1.2 and 3 m width [23], the source was placed in front of the hedges and receivers were placed at different heights and distances behind the hedges. It was concluded that hedges with denser foliage give better attenuation due to increased scattering and absorption. But, typically, measured road traffic noise reductions were only a few dBA.

Other measurements of the acoustical impact of hedges [24] have involved statistical pass-by measurements of noise levels from road traffic before and after removal of a hedge, controlled pass-bys during which levels due to single vehicles were recorded by a microphone located behind the hedge and a reference microphone and transmission loss measurements using a fixed loudspeaker source and a microphone array. Measured insertion losses ranged from 1.1 dBA to 3.6 dBA, the higher noise reductions being associated with increased ground effect.

Figure 10.6 shows photographs of a hedge next to a car park and the source–receiver locations used for measurements at this site. Figure 10.7

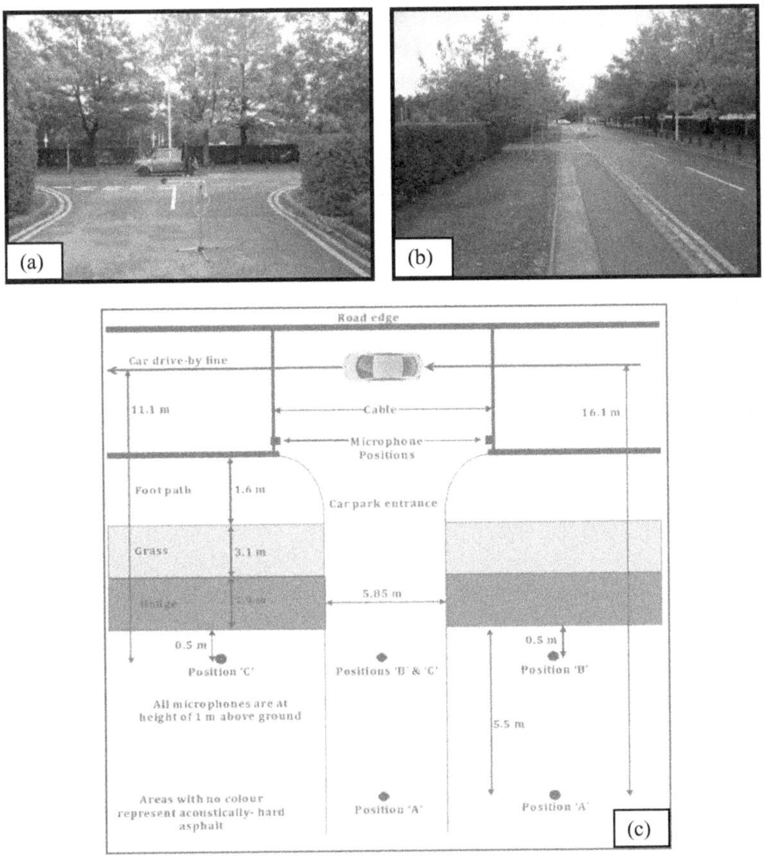

Figure 10.6 Photographs of car-park border hedge and schematic of microphone positions used for controlled pass-by measurements at The Open University, Milton Keynes, UK [2].

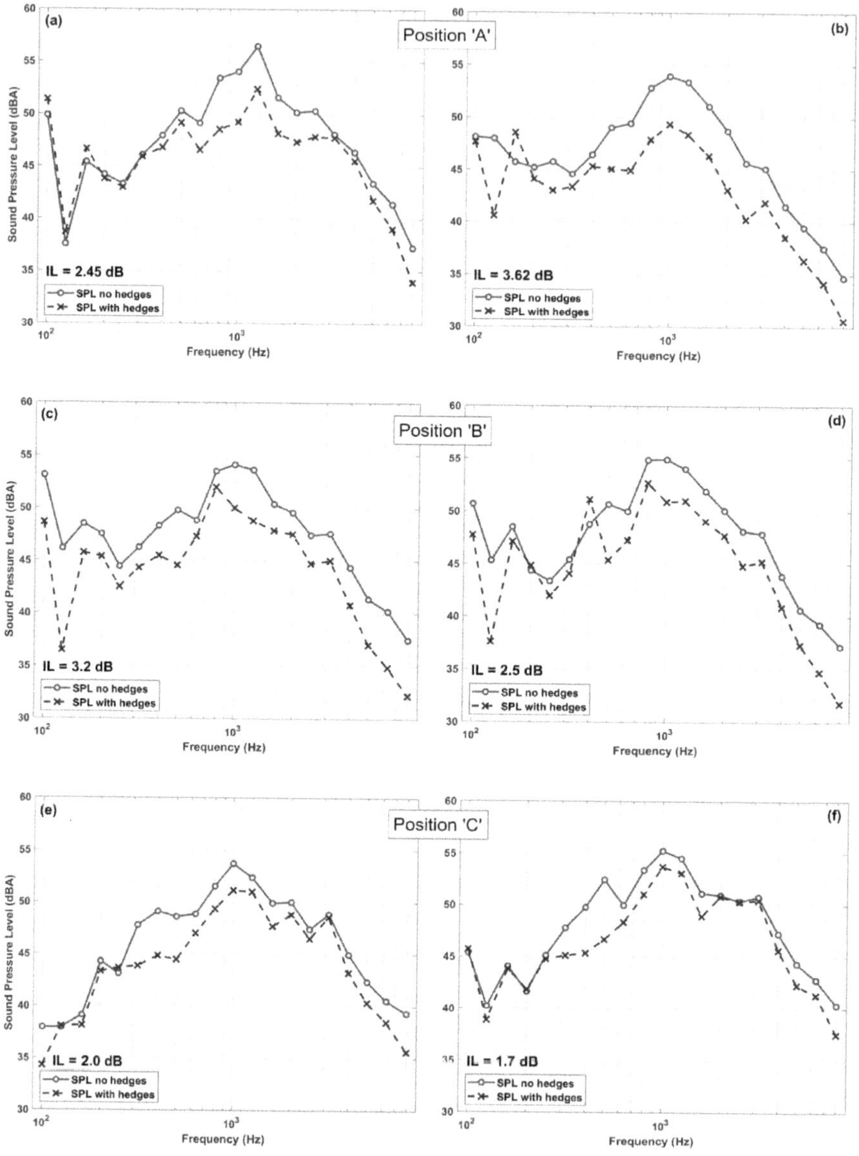

Figure 10.7 One-third-octave band A-weighted Sound Pressure Levels (SPL) during two car pass-bys measured at positions 'A', 'B' and 'C' in the car park entrance and behind the hedge respectively (See Figure 10.6) [2].

shows examples of measured spectra from controlled pass-bys [2]. The insertion loss peaks near 1 kHz are associated with ground effect due to the grass covered border and the 'soft' ground beneath the hedge (visible in Figure 10.6(b)). The insertion loss at higher frequencies (above 3 kHz) is associated with attenuation by the hedge 'superstructure'.

10.2 PREDICTING SOUND TRANSMISSION THROUGH VEGETATION

10.2.1 Ground Effect with Plants and Vegetation

If direct and ground-reflected sound components can combine coherently between source to receiver during near-grazing sound propagation through trees, shrubs, hedges, plants and crops, there will be destructive interference and associated ground effect attenuation. Third octave band measurements have been carried out with a loudspeaker source at 0.8 m height, six 1.2 m high receivers at horizontal source-to-receiver distances of between 2 m and 80 m in forests of poplars (*Populus nigra*), pines (*Pinus pinaste*) and oaks (*Qercus ilex*) on days with very little wind [25]. The forest floors were relatively free from decaying leaf litter. Measured excess attenuation spectra were compared with predictions of the NORD2000 model [26] (also see Chapter 12). Since all measurements were carried out on days with very little wind, a homogeneous atmosphere was assumed in the predictions. It was found that while the measured excess attenuation spectra agree tolerably well with predictions of ground effect alone at a range of 10 m, they departed from the data at longer ranges. Typically, the measured minima associated with the first destructive interference were less pronounced than in the predictions. To obtain the best ground effect fits it was found necessary to vary the effective flow resistivity using the physically inadmissible single parameter model due to Delany and Bazley (see Chapter 5) with range. Comparisons between predictions and these data are considered further in Section 10.2.4.

Spectra of the difference in levels have been measured between a reference microphone at 2 m and second microphones at distances of up to 96 m from a loudspeaker source of broadband random noise in three different woodlands (mixed deciduous, spruce monoculture and mixed coniferous) [9]. The second microphones were 1.2 m high and positioned 12, 24, 48 and 72 m from the source in the mixed deciduous wood; at 12, 24, 48 and 96 m range in the spruce monoculture and at 12, 24, 26 and 40 m in the mixed conifers. The two-way loudspeaker source used for these measurements had a separation of 0.2 m between the cone centres and the centre of the larger low-frequency cone was located at a height of 1.3 m. The data were corrected for wavefront spreading and air absorption. The mixed deciduous woodland contained alternating bands of mature Norway spruce (*Picus abies*) and oak. The oak had dense undergrowth consisting of hawthorn (*Crataegus monogyna*), rose and honeysuckle. The spruce monoculture contained trees with similar heights of between 11 and 13 m, with dead branches only below 4 m. The third woodland consisted of a young mixed coniferous stand containing alternate rows of red cedar, Norway spruce and Corsican pine (*Pinus nigra*) and foliage extended down to ground level. For each measurement, the wind speed and direction and temperature at two heights were recorded. Measurements were carried out only on days with little or

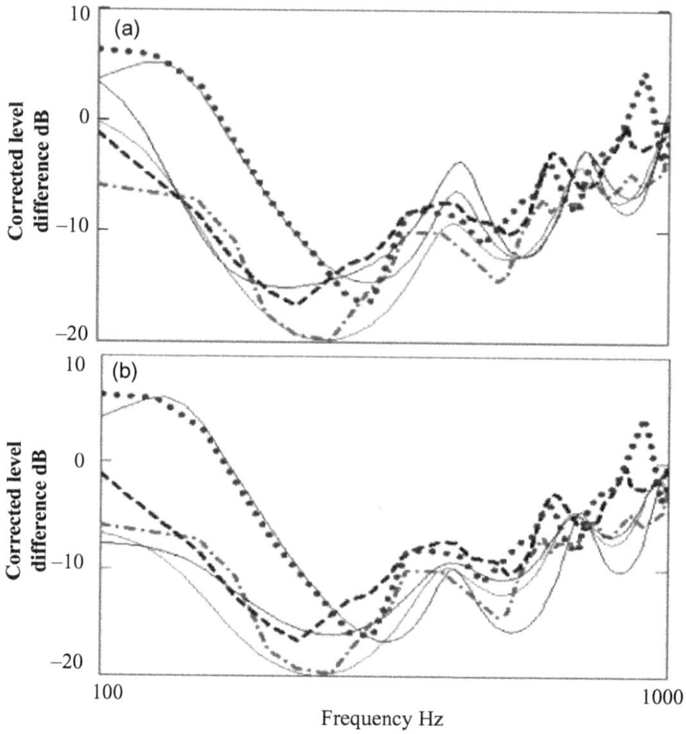

Figure 10.8 Level difference spectra below 1 kHz corrected for distance and air absorption, including maximum (dash-dot line) and minimum (broken line) spectra in summer and the mean level difference spectrum in winter (dotted line), measured with a 1.3 m high loudspeaker source in a mixed deciduous woodland [8] between 1.2 m high receivers at 2 m and 72 m and corresponding ground effect predictions without turbulence (solid lines) using (a) the slit pore layer model and (b) the variable porosity model with the parameter values listed in Table 10.2.

no measurable wind (<1.5 m/s) and small temperature gradients. Data were averaged over three receiver positions at a given range (laterally displaced by ± 0.5 m from a central location) and each measurement was averaged over approximately 90 s.

Figures 10.8 (a) and (b) show maximum and minimum corrected level difference spectra up to 1 kHz measured in summer and the mean corrected level difference spectrum measured in winter (dash-dot, broken and dotted lines, respectively) in a mixed deciduous woodland with loudspeaker source and microphone receiver at heights of 1 m and 1.2 m separated horizontally by 72 m [9]. Also shown (thin solid lines) are predictions of ground effect assuming an effective source height of 1.3 m, an acoustically neutral atmosphere and using (a) slit-pore layer and (b) variable porosity impedance models (see Chapter 5) with the parameter values listed in Table 10.1.

Table 10.1 Parameter values used for predicting ground effect in a mixed deciduous forest during summer and winter (see Figures 10.8(a) and (b)).

	Slit pore layer model				*Variable porosity model*	
Season	*Porosity*	*Flow resistivity kPa s m⁻²*	*Tortuosity*	*Thickness m*	*Effective flow resistivity kPa s m⁻²*	*Porosity rate /m*
Summer minimum	0.6	30	1.7	0.16	10	–25
Summer maximum	0.35	20	2.9	0.18	10	–55
Winter mean	0.35	45	2.9	0.3	45	–20

These data are revisited in Section 10.2.4 where it will be shown that, when incoherence induced by scattering from trunks and branches and foliage attenuation are included, the ground impedance parameters giving best fit over the whole frequency ranges of the data differ from those that enable fitting of ground effect alone up to 1 kHz. Nevertheless, the values in Table 10.1 indicate that the fitted flow resistivity in winter in the mixed deciduous woodland is significantly higher than the values that fit the maximum and minimum summer data in this woodland according to either of impedance models. This is consistent with the woodland floor being wetter in winter. Grassland data showing a similar seasonal change in ground effect are discussed in Chapter 6.

Huisman et al. [10,27] measured the excess attenuation through woodland consisting of a 29-year-old monoculture of Austrian pines (*Pinus nigra subsp. nigra*) located in an area reclaimed from the sea (polder) in the Netherlands. A loudspeaker, with its centre at a height of 1 m, was used as source of swept sine waves and simultaneous measurements were made of wind and temperature gradients. Receivers were at a horizontal distance of 100 m from the source and at three different heights (1 m, 2.5 m and 4.5 m). Best fit to measured ground effect was obtained using the variable porosity impedance model with effective flow resistivity 7.5 kPa s m^{-2} and porosity rate 100 m^{-1}. Below 1 kHz, the calculated ground effects agree with the measured data. At higher frequencies, the data differ significantly from predictions of ground effect alone. Again, this can be attributed (a) to scattering of the sound out of the path between source and receiver by trunks and branches and (b) to attenuation of the sound by scattering and visco-thermal effects in the foliage.

10.2.2 Models for Foliage Effects

10.2.2.1 Empirical Models

Foliage contributes mainly to sound attenuation through trees, bushes, hedges and crops above 1 kHz. Measured spectra of attenuation in dB/m obtained in three different woodlands [9], shown in Figure 10.9(a), suggest that log(frequency) fits provide a close match above 1 kHz:

$$\text{Attenuation in dB/m through mixed conifers} = 0.7(\log f) - 2.03 \tag{10.1}$$

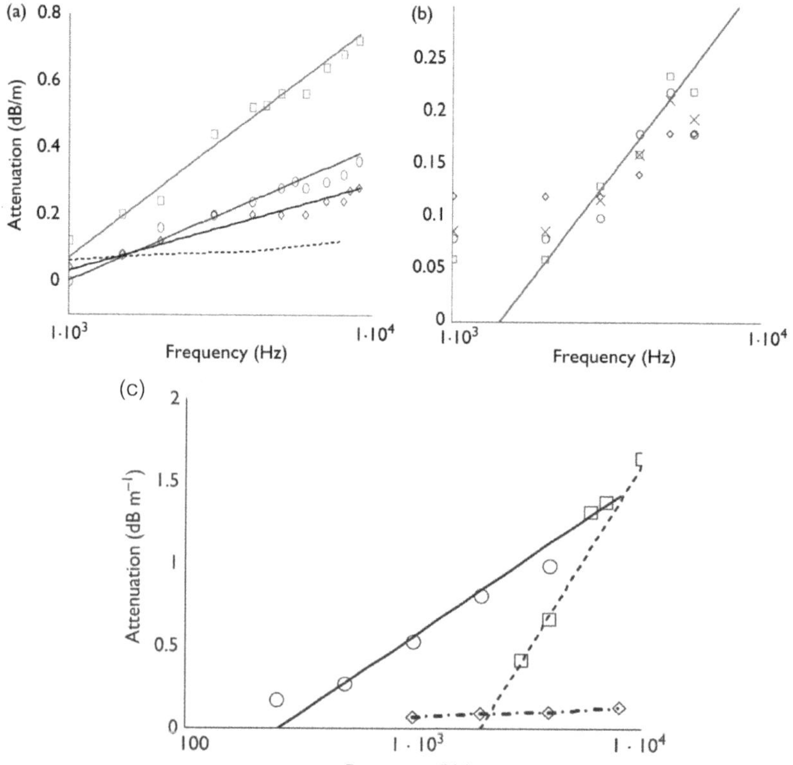

Figure 10.9 (a) Measured variation of corrected level difference with frequency above 1 kHz through three woodlands: mature mixed deciduous forest in summer (circles), Norway spruce monoculture (diamonds) and a mixed conifers (boxes) [6]. Also shown are empirical log(frequency) fits (solid and broken lines) and the foliage attenuation according to ISO9613-2 (broken line). (b) Measured excess attenuation through 100 m of pine forest for three different receiver heights: 4.5 m (boxes), 2.5 m (circles), 1 m (diamonds) and mean values (crosses). The continuous line is a log(frequency) fit to a restricted portion (2 kHz < f < 6 kHz) of these data. (c) Measured variation of attenuation with frequency through corn (circles) and reeds (boxes) [5] and log(frequency) fits (solid and broken lines respectively). Also shown (joined diamonds) is the attenuation due to foliage predicted by ISO 9613 – 2.

$$\text{Attenuation in dB/m through mixed oak and spruce(summer)} = 0.4(\log f) \quad (10.2)$$

$$\text{Attenuation in dB/m through spruce monoculture} = 0.26(\log f) - 0.75 \quad (10.3)$$

The attenuation rates measured at the coniferous sites are rather greater than those measured through the mixed deciduous wood even in summer. Above 1500 Hz, measured attenuation rates are significantly greater than predicted by the foliage correction in ISO9613-2 [18] (see Chapter 12, Section 2), shown as the broken line in Figure 10.9 (a). The attenuation rate measured through the mixed deciduous site for source–receiver separations of up to 96 m is comparable with that obtained over a longer range (300 m) through dense evergreen woods [28]. Figure 10.9 (b) shows results obtained by Huisman [10] through a pine forest. These suggest attenuation in dB/m varies as log(frequency) only in a limited frequency range (2 kHz $< f <$ 6 kHz). Within this range an empirical fit is given by

$$\text{Attenuation in dB/m through pine forest} = 0.39(\log f) - 1.23 \ (f > 2\text{kHz}) \quad (10.4)$$

Figure 10.9(c) shows that a log(frequency) dependence provides good fits also to data for the attenuation of sound above 1 kHz through a dense planting of field corn (*Zea mays*) and reeds (*Phragmites*) [5,6]:

$$\text{Attenuation in dB/m through corn} = 0.95(\log f) - 2.3 \ (f > 500\text{ Hz}) \quad (10.5)$$

$$\text{Attenuation in dB/m through reeds} = 2.3(\log f) - 12.2 \ (f > 2\text{kHz}) \quad (10.6)$$

Again, these data show higher rates of attenuation than predicted by the foliage correction in ISO9613-2 [18] (see also Chapter 12), represented in Figure 10.9(c) by joined diamonds. However, such empirical log(frequency) fits do not relate attenuation to physically measurable parameters such as leaf size and foliage density and may contain contributions from more than one attenuation mechanism.

Aylor [6] has suggested that there is a relationship between the attenuation in excess of that due to ground effect divided by the square root of the product of foliage area per unit volume and the scattering parameter, i.e. the product of wavenumber and a characteristic leaf dimension. Using these suggestions [1], the data in Figure 10.10(a) have been fitted reasonably well by:

$$\frac{EA}{\sqrt{FL}} = A\left[1 - \exp\left(0.3 - 0.5(ka)\right)\right], \quad ka > 0.3, \quad (10.7a)$$

where EA represents the excess attenuation in dB, F/m is the foliage area per unit volume (similar to leaf area density index used in studies of optical and thermal reflectivity of foliage), L m is the length of the propagation path,

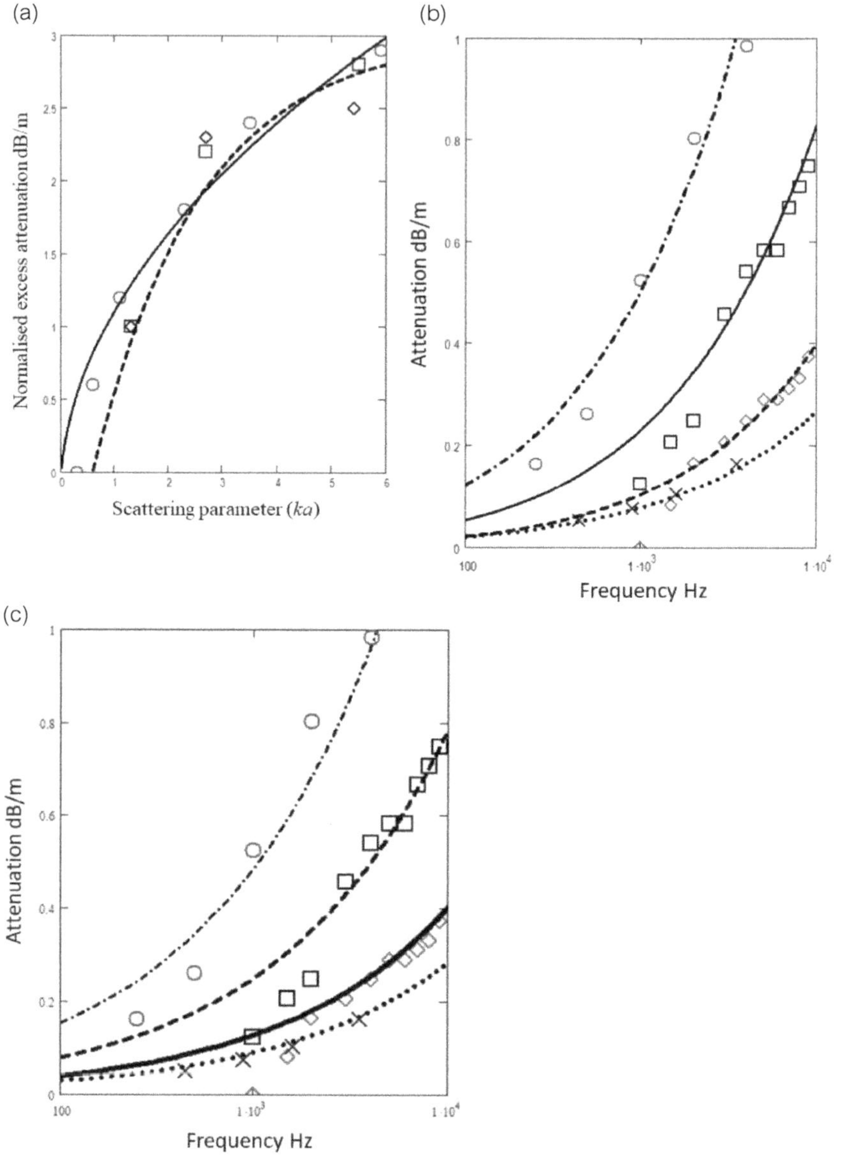

Figure 10.10 (a) Aylor's data for sound attenuation through corn (boxes $F = 3/\text{m}$; diamonds $F = 6.3/\text{m}$) with mean leaf width 0.074 m and reeds (open circles) with mean leaf width 0.032 m normalised by the square root of the product of leaf area density and path length compared with predictions of equations (10.7a) [broken line] and (10.7b) [solid line]. (b) Attenuation in dB per m in dense corn (open circles) [6], a mixed coniferous woodland (open squares) [8], a mixed deciduous woodland in Summer (open diamonds) [8] and an evergreen forest (crosses) [27] and corresponding fits using equation (10.7b) [dash-dot line $F = 4/\text{m}$, $a = 0.074$ m (corn); solid line $F = 0.75/\text{m}$, $a = 0.07$ m (mixed conifers); dotted line $F = 1/\text{m}$, $a = 0.015$ m (mixed deciduous); dotted line $F = 1/\text{m}$, $a = 0.01$ m (evergreen forest)]. (c) Fits to the same data as shown in Figure 10.10(b) using the slit pore model with the parameter values listed in Table 10.3.

$k = 2\pi f/c$ is the wavenumber, c is the adiabatic sound speed in air, a m is the mean leaf width and a value of $A = 3$ fits Aylor's data. The lower limit on ka avoids negative values of EA, implying, for example, a low frequency limit of approximately 1 kHz for a mean leaf width of 0.032 m and a low-frequency limit of about 100 Hz for a mean leaf width of 0.3 m.

While equation (10.7a) has been used to fit measured attenuation through crops [1,2] it has the drawbacks that (i) it requires a low-frequency cut off to avoid predicting negative attenuation and (ii) it predicts that foliage attenuation 'saturates', i.e. reaches a limit with increasing frequency.

An alternative empirical equation without these drawbacks is [1]:

$$\frac{EA}{\sqrt{FL}} = \frac{\sqrt{ka}}{\left(\dfrac{0.146}{\sqrt{ka}} + 0.76\right)}. \tag{10.7b}$$

The fit given by equation (10.7b) is shown also in Figure 10.10(a). Figure 10.10(b) confirms that equation (10.7b) enables fits of data for sound transmission through various types of foliage using plausible values for F and a.

Impedance tube studies [11] have shown that the acoustical properties of 'garden' plants can be modelled as low flow resistivity porous layers with thickness approximately 70% of the plant height. Semi-empirical exponential relationships, depending on leaf orientation, between the effective flow resistivity (R_{eff}) of the equivalent porous layer and the leaf area density (F) are:

$$\log\left(R_{eff}\right) = 0.0083F + 1.413 \text{ for } \theta > 70° \tag{10.8a}$$

$$\log\left(R_{eff}\right) = 0.0067F + 0.746 \text{ for } \theta < 70° \tag{10.8b}$$

The effective tortuosity of the equivalent porous layer is given in terms of leaf orientation by

$$\sqrt{T_{eff}} = \cos\left(\frac{\theta}{2}\right) + 2\sin\left(\frac{\theta}{2}\right). \tag{10.8c}$$

During near-grazing sound propagation over vegetated ground, leaf orientation is likely to have a different influence than that at normal incidence. These studies used the three-parameter Miki model (see Section 5.2.1) which, as discussed in Chapter 6, is physically inadmissible. Moreover, the study did not consider changes in soil effective flow resistivity caused by the root system of the plants or by the act of planting (see Section 10.1.1). Nevertheless, it is interesting to investigate further whether sound transmission through foliage can be modelled as transmission through a rigid porous medium. Figure 10.10(c) shows fits to the same data as in Figure 10.10(b) obtained by using the (physically admissible) slit-pore model with the parameter values listed in Table 10.2. While the fits are similar to those obtained using equation (10.7b), the fitted effective porosity, flow resistivity and tortuosity would require derivation of relationships such as (10.8a–10.8c) to be related to leaf area density and mean leaf size.

Table 10.2 Parameter values used for fitting data for foliage attenuation using the slit pore model for the acoustical properties of a rigid porous medium (see Figure 10.10(c)).

	Slit pore model parameters		
Foliage	*porosity*	*flow resistivity kPa s m^{-2}*	*tortuosity*
Corn	0.6	1.5	1.7
Mixed conifer woodland	0.8	0.3	1.25
Mixed deciduous woodland	0.9	0.07	1.1
Evergreen forest	0.9	0.035	1.1

10.2.2.2 Scattering Models

Several analytical models for propagation through trees and tall vegetation have considered scattering. Bullen and Fricke [29] developed a diffusion model for sound intensity involving adjustable parameters for scattering and absorption cross sections. Embleton [30] was first to use Twersky's multiple scattering theory [31] for propagation through an idealized (random) array of identical parallel impedance-covered cylinders, to explain attenuation through forests.

The propagation constant (k_b) through the array of cylinders may be deduced from

$$k_b^2 = k^2 - 4iNg + \left(g'^2 - g^2\right)\left(2N / k\right)^2, \tag{10.9a}$$

where N is the average number of cylinders per unit area normal to their axes, $k = 2\pi f/c_0$, g and g′ are forward and backward scattering amplitudes given by

$$g = \sum_{n=-\infty}^{\infty} A_n, g' = \sum_{n=-\infty}^{\infty} (-1)^n A_n. \tag{10.9b}$$

The scattering coefficients A_n are given by

$$A_n = \left[iJ_n(ka) + ZJ_n'(ka)\right] / \left[iH_n(ka) + ZH_n'(ka)\right], \tag{10.9c}$$

where Z is the surface normal impedance of the cylinder of radius a, J_n and H_n are Bessel and Hankel functions of the first kind and order n, and primes indicate derivatives.

As well as including ground effect, Price et al. [9] modelled each of the three woodlands that they investigated as an array of large cylinders (representing trunks) with finite impedance plus an array of small rigid cylinders (representing branches and foliage) over an impedance ground. As a first approximation Price et al. [9] simply added the attenuations calculated for the large cylinders, the small cylinders and the ground. Subsequently, through laboratory measurements [32] and numerical simulations [33] this approximation has been shown to be a reasonable for low filling fractions (less than 13%) of scatterers (see also Section 10.3). While their calculations were in qualitative agreement with their measurements, Price et al. [9] found it necessary to adjust several parameters including the assumed finite impedance of the scatterers and the number of scatterers per unit volume to obtain quantitative agreement. Twersky's approach neglects the incoherent scattering by cylinders and thereby introduces an error which is increasingly important at high frequencies.

Laboratory experiments using identical wooden dowel rods as vertical scatterers [9] indicated that a 60% reduction in the number of scatterers per unit area is required to obtain agreement between multiple-scattering predictions using equations (10.9a–10.9d) and data. It was found that, after the empirical correction and parameter adjustment, use of scattering theory for attenuation due to two independent cylinder arrays, one representing trunks and large branches and one representing twigs and foliage, predicts attenuation rates comparable to the measured values in the various woodlands but with a different frequency dependence.

Corrections to equations (10.9a–10.9d) have been reported [34] and found to give better agreement with data without adjusting the number of scatterers per unit area [35]. Equation (10.9d) shows a simplified version of the improved relationship [35]:

$$k_b^2 = k^2 - 4iN\sum_{n=-\infty}^{+\infty} A_n + \frac{8N^2}{k^2}\sum_{n=-\infty}^{+\infty}\sum_{s=-\infty}^{+\infty} A_n A_s (n-s)\,\mathrm{sgn}(n-s). \tag{10.9d}$$

A more extensive 3-D theory for propagation in a forest has derived effective complex sound wavenumbers k_b taking account of scattering by trunks, modelled as finite vertical cylinders, by large branches, modelled as vertical and slanted finite solid cylinders and diffuse scattering in the canopy [36]. For near horizontal propagation below the canopy the theory reduces to that for 2-D multiple scattering by cylinders. However, with this simplification, the theory suggests that, for typical areal densities of trunks, terms in N^2 can be neglected, thereby raising the question of why the corrected forms in equation (10.9d) have been found to lead to any improvement.

Calculations of the attenuation through random arrays of 4 m tall acoustically hard cylinders with axes normal to an acoustically hard plane have been made using a time-domain formulation (based on a numerical solution of Euler's equations) [37]. These have predicted, for example, a mean attenuation of 1.5 dB over the frequency band from 50 Hz to 600 Hz of sound from a 0.5 m high coherent line source 1 m from the edge of an 18 m wide array of 0.2 m diameter cylinders with 0.2 cylinders per m^{-2}. At high frequencies, the simulations showed substantially lower attenuation than predicted by Twersky's theory.

Huisman et al. [8,27] used a stochastic particle-bounce approach to model attenuation and reverberation through an array of parallel vertical cylinders. In conjunction with a ground effect model and an assumption about the dependence of the proportion of incoherent scattering on distance and frequency, good agreement was obtained with data from a pine forest. Another important finding in this work was that, even after propagating 100 m through trees, the sound field is formed mainly by the direct and ground-reflected fields and the contribution of the reverberant field is minimal. This argues against modelling sound propagation through trees as purely diffusive.

The effects of scattering by stems and branches in vegetation have been calculated using a 2-D theory for scattering by arrays of cylinders [2; see also Chapter 5]. Predictions of the reflection of sound by a hard-backed array of acoustically hard cylindrical scatterers (diameters > 0.005 m) have been used to determine an equivalent wave propagation constant for a specific density and mean scatterer size. The propagation constant is obtained by fitting the reflection predictions with those of an equivalent rigid-porous layer described by a three-parameter slit-pore impedance model (see Chapter 5).

The required porosity and layer depth are determined, respectively, from the density or volume fraction (filling fraction) of scatterers and depth of the scatterer-filled area used in the simulation. The equivalent flow resistivity is used as an adjustable parameter.

The complex wave propagation constants for simulated arrays of horizontal cylinders with a log-normal distribution of radii can be calculated from a slit-pore layer model (see Chapter 5) for

- filling fractions in the range 0.001–0.1 and
- scatterer diameters in the range 5–250 mm.

In this range of filling fractions, the porosity may be approximated by the simple linear relationship:

$$Porosity = -0.786 \times FillingFraction + 0.975. \tag{10.10}$$

Knowing the diameter and volume fraction (filling fraction) of scatterers, the ranges of equivalent flow resistivity (R_e) given in Figure 10.11 can be used to calculate the propagation constant. Typically, effective flow resistivity (R_e) increases with scatterer density.

Based on the results plotted in Figure 10.11, a simple empirical formula relating mean scatterer radius (r) and filling fraction ($\mathcal{F}$) to equivalent flow resistivity (R_e) has been proposed [2]:

$$R_e = 0.11\mathcal{F}^{0.95} / r^2. \tag{10.11}$$

The attenuation constant predictions obtained by using the resulting equivalent flow resistivity and porosity in the slit-pore model can be used to determine the additional attenuation due to (reverberant) scattering by branches and twigs as long as the characteristic scattering element sizes are greater than about 5 mm. Section 10.2.4 describes their application to multiple scattering by winter wheat stems [1].

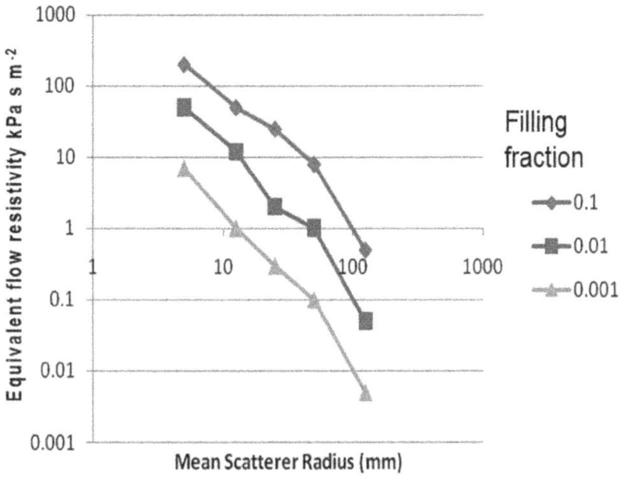

Figure 10.11 Predicted relationships between filling fraction, mean scatterer radius and equivalent flow resistivity deduced from calculations of multiple scattering by acoustically hard cylinder arrays [2].

Full-wave numerical techniques have been used to simulate sound propagation through a forest, where the canopy layer has been represented by a random distribution of small scattering elements. The total volume of the scattering elements is chosen to match the woody biomass present in that zone, and a limited amount of absorption assigned to each scattering cell to model absorption by the canopy [33,38]. These simulations have shown that, when the acoustic source and receiver are both positioned below the canopy, downward scattering of sound reduces transmission loss through trees. Although, typically, this is a small effect when making road traffic noise predictions, an improved spectral fit of measured data has been obtained following this procedure ([38] and Chapter 12, section 12.10).

10.2.3 Reduction of Coherence by Scattering

Scattering by trunks, branches, stems in vegetation reduces the coherence between direct and ground-reflected sound, i.e. it weakens the constructive and destructive interference responsible for the ground effect. This is similar to the effect of turbulence (see also Section 10.7). In the absence of turbulence and refraction, the classical Weyl-Van der Pol formula (see Chapter 2 equation (2.40)), gives the sound pressure field propagating from a point source above an impedance flat boundary as

$$p = \frac{e^{ikR_1}}{R_1} + Q\frac{e^{ikR_2}}{R_2}, \tag{10.12a}$$

where k is the wavenumber, R_1 and R_2 are the direct and image path lengths, respectively, and Q is the (complex) spherical reflection coefficient.

The Sound Pressure Level, L_p, = $10\log\left(\langle p^2\rangle\right)$.

A convenient analytical method for calculating the influence of turbulence on propagation from a point source near the ground is due to Clifford and Lataitis [39]. Their expression for mean square pressure is

$$\langle p^2\rangle = \frac{1}{R_1^2} + \frac{|Q|^2}{R_2^2} + \frac{2|Q|}{R_1R_2}\cos\left[k(R_2 - R_1) + \theta\right]T, \tag{10.12b}$$

where θ is the phase of the spherical wave reflection coefficient, ($Q = |Q|e^{i\theta}$) (a function of source and receiver geometry and ground impedance – see Chapter 2) and, T, the coherence factor associated with the effect of turbulence, which is assumed to have a Gaussian spectrum, is given by

$$T = e^{-\sigma^2(1-\rho)}. \tag{10.12c}$$

In Equation (10.12c), σ^2 is the variance of the phase fluctuation along a path given by

$$\sigma^2 = A\sqrt{\pi}\mu^2k^2RL_0, \tag{10.12d}$$

where L_0 is the outer scale of turbulence, R is the range, $\langle\mu^2\rangle$ is the variance of the index of refraction, k is wavenumber, and the coefficient A is given by

$$A = 0.5, R > kL_0^2 \text{ or } A = 0, R < kL_0^2 \tag{10.12e}$$

ρ is the phase which is a function of L_0 and h the maximum transverse path separation, i.e.

$$\rho = \frac{\sqrt{\pi}}{2}\frac{L_0}{h}\operatorname{erf}\left(\frac{h}{L_0}\right). \tag{10.12f}$$

In the absence of rcfraction, h is given by

$$\frac{1}{h} = \frac{1}{2}\left(\frac{1}{h_s} + \frac{1}{h_r}\right), \tag{10.12g}$$

where h_s and h_r are the source and receiver heights, respectively, and erf(x) is the error function defined by

$$\operatorname{erf}(x) = \frac{2}{\sqrt{\pi}}\int_0^x e^{-t^2}dt. \tag{10.12h}$$

Typically, outdoor propagation studies have assumed that L_0 is the source height and that values of $\langle\mu^2\rangle$ lie between 2×10^{-6} and 10^{-4}.

A heuristic way of modelling loss of coherence due to scattering by vegetation is to regard the scattering components in the vegetation as if they are 'frozen', i.e. stationary turbulence and to use values for $\langle\mu^2\rangle$ and L_0 obtained by fitting data [1,2] (see also Section 10.2.4). In Section 10.7, it is reported that the influence of stem or trunk scattering on coherence between direct and ground-reflected sound can be treated also through a coherence factor, T, as in Equation (10.12b).

The NORD 2000 prediction scheme [25,2] includes the effects of scattering by buildings, trees and other obstacles. The combined excess attenuation (in dB) as a function of frequency due to scattering and reflection from the ground is given by

$$A_{gr+sc} = 10\log\left[\frac{\left(1 - F_{coh}C_{coh}\right)\left|1 + Qe^{ikR_2} / R_2\right|^2 + F_{coh}C_{coh}}{\left(1 + \left|Qe^{ikR_2} / R_2\right|^2\right)}\right] + F_{coh}C_{coh}K_F A_{sc}\left(R_1\right), \quad (10.13a)$$

where F_{coh} is the fraction of coherent sound that is transformed into incoherent sound, which increases with the frequency and trunk diameter, but is zero if there is no scattering (see Table 10.3), $C_{coh} =$ min $[(R_{sc}Nd/K_F)^2, 1]$ is a function intended to provide a smooth transition and avoid overestimation at short distances, N is the number of trees per unit area, d is the average trunk diameter, $K_F = 1.25$ in forests and R_{sc} is the path length inside the forest. Other symbols are as in Equation (10.12a). Values for F_{coh} are listed in Table 10.3.

$$A_{sc}\left(R_1\right) = \Delta L\left(hNd,\ \alpha,\ R_{sc}Nd\right) + 20\log\left(8R_1Nd\right), \quad (10.13b)$$

where h is the mean tree height, α is the mean absorption coefficient of the scatterer surfaces and values of $\Delta L(hNd,\alpha,R_{sc}Nd)$ in dB are listed in Table 10.4.

Comparisons between data and NORD 2000 predictions show that, while the inclusion of loss of coherence in the ground effect calculations improves agreement between predictions and data at the destructive

Table 10.3 Values of F_{coh} in equation (10.13a)

$kd/2$	F_{coh}
<0.7	0.00
1	0.05
1.5	0.20
3	0.70
5	0.82
10	0.95
20	1.00

Table 10.4 Values of $\Delta L(hNd, \alpha, R_{sc}Nd)$ in equation (10.13b)

	$hNd = 0.01$			$hNd = 0.1$			$hNd = 1$		
$R_{sc}Nd$	$\alpha = 0$	$\alpha =$ 0.2	$\alpha =$ 0.4	$\alpha =$ 0	$\alpha =$ 0.2	$\alpha =$ 0.4	$\alpha =$ 0	$\alpha =$ 0.2	$\alpha =$ 0.4
0.0625	6.0	6.0	6.0	6.0	6.0	6.0	6.0	6.0	6.0
0.125	0.0	0.0	0.0	0.0	0.0	0.0	0.0	0.0	0.0
0.25	−7.5	−7.5	−7.5	−6.0	−7.0	−7.5	−6.0	−7.0	−7.5
0.5	−14.0	−14.25	−14.5	−12.5	−13.5	−14.5	−12.5	−13.0	−14.0
0.75	−18.0	−13.8	−19.5	−17.3	−18.0	−19.0	−16.0	−16.8	−17.7
1	−21.5	−22.5	−23.5	−20.5	−21.6	−22.8	19.3	−20.5	−21.3
1.5	−26.3	−27.5	−29.5	−25.5	−27.2	−29.0	−24.0	−25.5	−26.3
2	−31.0	−32.5	−34.5	−30.0	−32.0	−33.3	−27.5	−29.5	−30.8
3	−40.0	−42.5	−45.5	−37.5	−40.5	−42.9	−34.2	−36.0	−37.8
4	−49.5	−52.5	−56.3	−45.5	−49.5	−52.5	−40.4	−42.8	−45.5
6	−67.0	−72.5	−78.0	−62.0	−67.0	−72.0	−52.5	−56.2	−60.0
10	−102.5	−113.0	−122.5	−94.7	−103.7	−112.0	−78.8	−84.0	−89.7

interference frequency range of ground effect, there is consistently poor agreement (i.e. in 10 out of 13 cases) between data and NORD 2000 predictions at high frequencies and longer measurement ranges, even after 'tuning' the effective flow resistivity in the Delany and Bazley impedance model to give the best-fit ground effect at each range [25]. Several factors are suggested as being responsible for poor predictions [25]. These include ground roughness, planting periodicity, upward refraction, an oversimplified impedance model and the fact that the NORD2000 model was not developed exclusively for propagation in forests. As a result of the information in the following sections in this chapter, it can be argued that important contributors to differences between these predictions and data are (a) an over-simplified and physically inadmissible ground impedance model (see Chapter 5), which is undisputedly responsible for poor predictions of ground effect and (b) the omission of an explicit allowance for visco-thermal dissipation in the foliage, which will be a major contribution to the underestimation of measured attenuation higher frequencies at the longer ranges in the pine and oak forests. Furthermore, there might be an error in the reported source–receiver geometry at 10 m range in the pine forest. It can be shown that an alternative method for predicting incoherence due to scattering based on that used for the corresponding influence of turbulence together with the addition of foliage attenuation leads to improved agreement with data.

10.2.4 Predictions of Ground Effect, Scattering and Foliage Attenuation

10.2.4.1 Sound Propagation in Crops

Measurements of sound transmission through winter wheat crops (average stem diameter 2.63 mm) obtained in August 2011 have been compared with predictions including ground effect based on short-range fitting of level difference spectra, multiple scattering by stems (assumed to be cylindrical) with a known distribution of diameters (Equation 10.11), loss of coherence due to scattering (equations 10.12a–10.12h) and foliage attenuation by known leaf size and area density (Equation 10.7a) [2,35]. Figures 10.12(a)–(c) compare spectra of the difference in levels at 0.3 m high receivers 1 m and at (a) 5 m (b) and (b) 10 m from the 0.3 m high source (continuous lines) and calculations of (i) ground effect alone (broken lines), using the two-parameter slit-pore model with parameters deduced from fitting short-range level difference spectra (flow resistivity 100 kPa s m^{-2} and porosity 0.27), (ii) ground effect plus multiple scattering, by an equivalent slit-pore medium with R_e = 29 Pa s m^{-2} and porosity 0.975 and loss of coherence due to scattering using equivalent turbulence (equation 10.12a–10.12h) with $\langle \mu^2 \rangle = 5.0 \times 10^{-4}$ and L_0 = 0.3 m) and (iii) the sum of these effects and visco-thermal attenuation in foliage (equation (10.7a) with F = 20/m and a = 0.008 m). The total of the predicted effects agrees well with the data.

Although the predicted influences of each of the identified physical mechanisms are distinguishable, the major contributors to the overall attenuation are the destructive interference due to ground effect and thermal and viscous losses due to vegetation. This suggests a simplified way of predicting propagation over or through crops which is simply to add only ground and foliage effects with larger *effective* values for foliage per unit area and mean leaf size [2], thus avoiding the need for the detailed calculations of multiple scattering and of the loss of coherence. Predictions of this simple approximation (broken lines) are compared with data (continuous lines) for propagation through winter wheat in Figures 10.13(a)–(f). The data were obtained not only in August 2011 but also with differing foliage and moisture conditions in May and June 2012. Other data on sound transmission through willow crops were obtained using a similar geometry in June 2012 and these together with corresponding predictions are shown in Figures 10.13(g) and (h). Predictions of ground effect alone are represented by broken lines. The parameter values used in the predictions are listed in Table 10.5. The approximate method enables tolerably good fits to the level difference spectra measured in August 2011 in winter wheat and in June 2012 in winter wheat and willow but provides less good agreement with measured spectra in May 2012 winter wheat since it is least successful in capturing the effects of multiple scattering in denser more luxuriant winter wheat foliage conditions (see Figure 10.4 (a)) which more or less removes the destructive interference due to ground effect at 10.0 m range.

Predictions of the simplified method have been compared with data for transmission of sound from different single vehicles passing by a ‘hornbeam' hedge near a tennis court [2]. Figures 10.14(a) and (b) show a photograph of the hedge and a schematic of the measurement positions, respectively.

The hedge was planted on a 0.1 m high kerb and the acoustically hard surface in the gap in the hedge had a 0.1 m high slope. The ground effect including the hard-to-soft profile and the 0.1 m high kerb in front of the

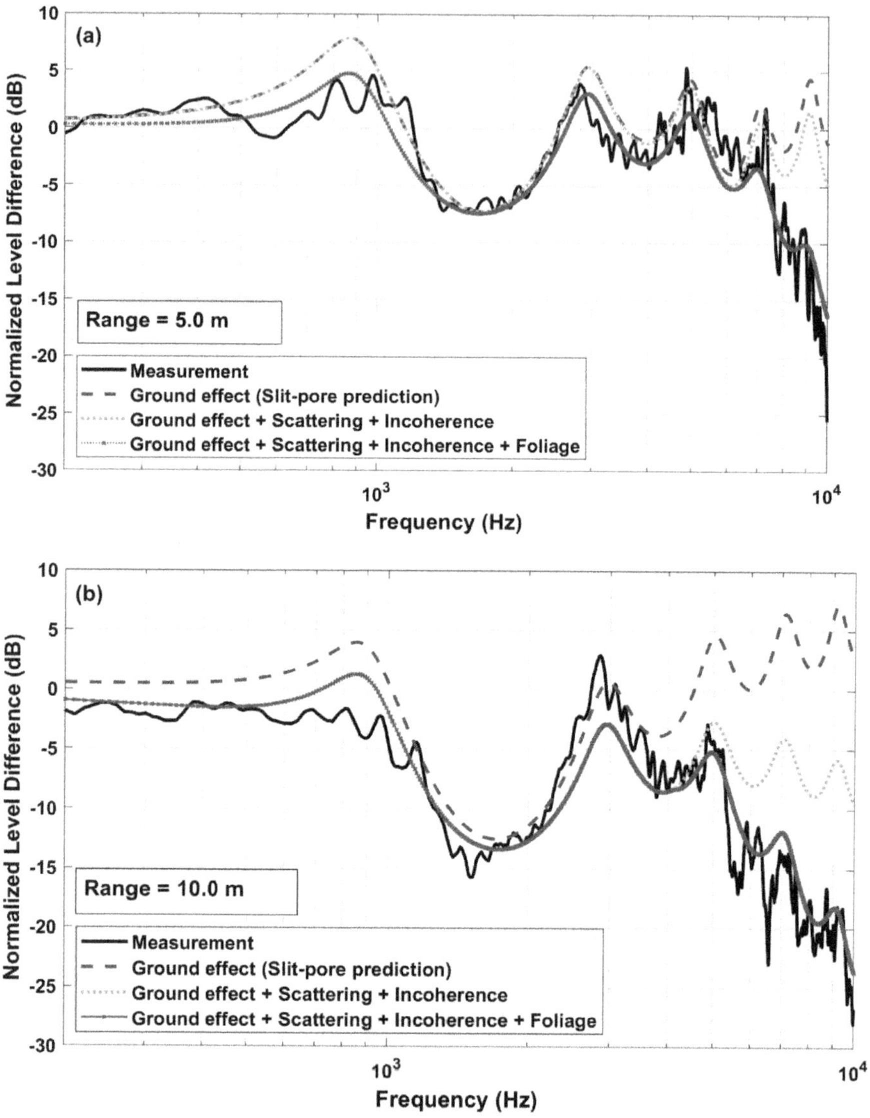

Figure 10.12 Level difference spectra measured through winter wheat crop in August 2011 [2] between a 0.3 m high (reference) receiver at 1 m from a 0.3 m high omnidirectional source and 0.3 m high receivers at distances of (a) 5.0 m and (b) 10.0 m from the source in 0.45 m to 0.5 m high winter wheat (solid lines). Also shown are predictions of ground effect alone (broken lines); ground effect plus incoherence plus multiple scattering by stems (broken lines); ground effect plus incoherence plus multiple scattering by stems plus viscous and thermal attenuation (Equation (10.7a) with F = 20/m and a = 0.008 m) (dash-dot lines) [1,2].

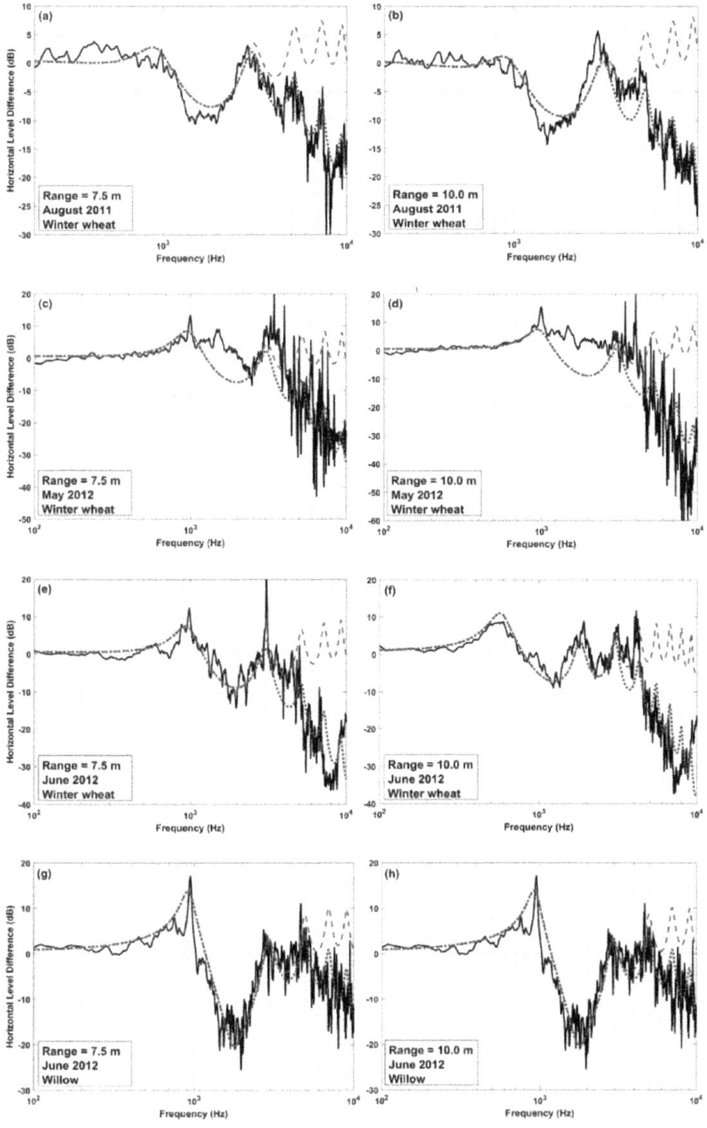

Figure 10.13 Comparison between spectra of the difference in levels measured by receivers at the same height but separated by 7.5 m or 10.0 m in winter wheat and willow (solid lines) and predictions either for ground effect alone (broken lines) or the result of adding ground effect to foliage attenuation predicted by Equation (10.7a) (broken lines). (a) and (b) show data and predictions for August 2011 winter wheat; (c) and (d) show data and predictions for May 2012 winter wheat; (e) and (f) show data and predictions for June 2012 winter wheat and (g) and (h) show data and predictions for June 2012 willow. Source and receivers were at 0.3 m height for 7.5 m range measurements at 0.4 m height for the 10.0 m range measurements [2]. The parameters used in predictions are listed in Table 10.1.

Table 10.5 Parameter values for fitting winter wheat and willow data using an approximate prediction method which sums ground effect and foliage attenuation only.

	Ground (2-parameter slit pore)		Equation (10.7a)	
Crop and month	*Effective flow resistivity kPa s m^{-2}*	*Porosity*	*Leaf area density /m*	*Leaf size m*
Winter wheat				
August 2011	100	0.27	30	0.012
May 2012	200	0.2	50	0.012
June 2012	200	0.2	40	0.012
Willow				
June 2012	100	0.2	10	0.01

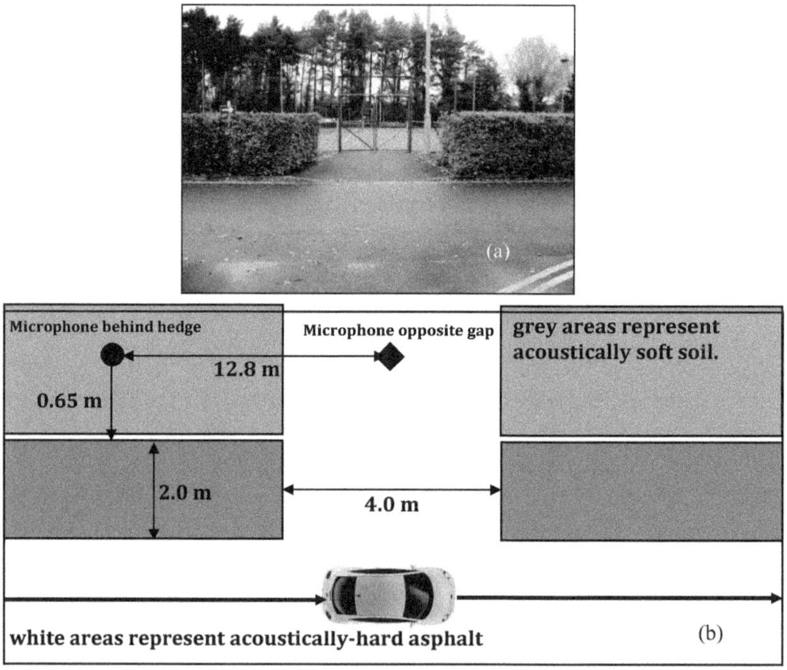

Figure 10.14 Hedge near a tennis court and a schematic of the measurement locations [2].

hedge and the slope near the tennis court has been modelled using a 2-D BEM. The soft ground was modelled using a two-parameter slit-pore impedance model with flow resistivity of 50 kPa s m^{-2} and porosity of 0.5. Since the environment did not permit separate characterization of the 'soft' ground bordering and underneath the hedge these values were estimated from values for similar ground types (see Chapter 6). Excess attenuation has been predicted for a car tyre/road source at height of 0.01 m (a) for a receiver at a height of 1 m above hard ground and (b) for a receiver at a height of 1 m over soft ground, i.e. in the presence of an impedance discontinuity. Figure 10.15 compares predictions with data for four different vehicle pass-bys.

There is reasonable agreement between data and predictions.

10.2.4.2 Sound Propagation in Forests

The solid lines in Figures 10.16 (a)–(c) represent corrected level difference spectra measured between 1.2 m high reference receiver at 2 m and second receivers at 24 m, 72 m and 96 m, respectively, from a 1.3 m high source in a mixed conifer forest, a deciduous woodland and a spruce monoculture [8]. Also shown (broken and dash-dot lines) are predictions of the sum of scattering-modified ground effect (equations 10.12b–10.12h) and parameters listed in Table 10.1) and foliage attenuation calculated according to

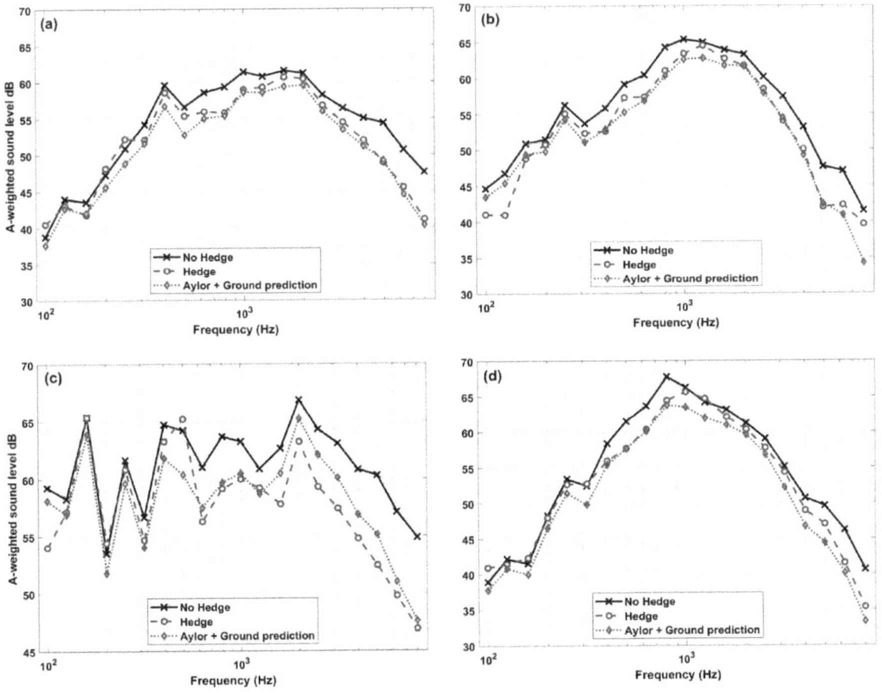

Figure 10.15 Comparison between measured A-weighted SPLs behind the tennis court hedge ('Hedge') and opposite a gap in the hedge ('No hedge) during four vehicle pass-bys with predictions using only attenuation predicted by Eq.10.7(a) (foliage area per unit volume 4.5 /m, mean leaf width 0.03 m, length of propagation path 2.2 m) and plus discontinuous soft ground effect underneath hedge [2].

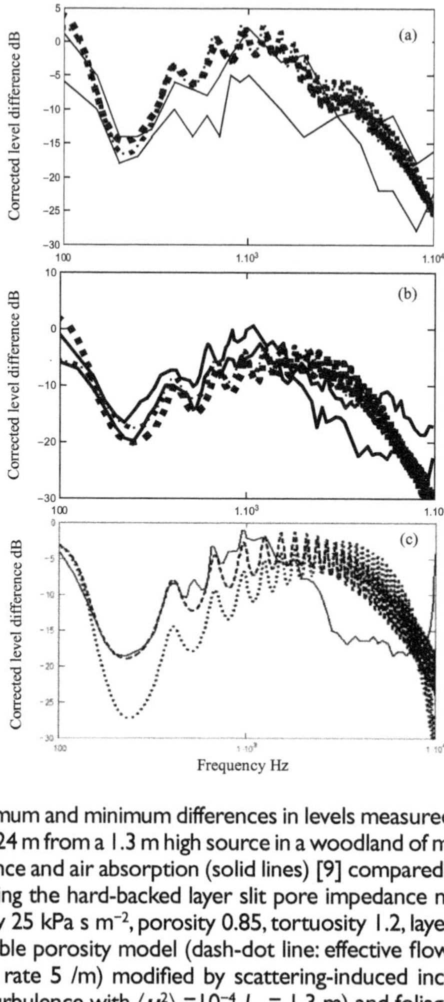

Figure 10.16 (a) Maximum and minimum differences in levels measured at 1.2 m high receivers 2 m and 24 m from a 1.3 m high source in a woodland of mixed conifers corrected for distance and air absorption (solid lines) [9] compared with the sum of ground effect using the hard-backed layer slit pore impedance model (broken line: flow resistivity 25 kPa s m^{-2}, porosity 0.85, tortuosity 1.2, layer thickness 0.12 m) and the variable porosity model (dash-dot line: effective flow resistivity 5 kPa s m^{-2}, porosity rate 5 /m) modified by scattering-induced incoherence (modelled as frozen turbulence with $\langle \mu^2 \rangle = 10^{-4}$, L_0 = 1.3 m) and foliage attenuation according to equation (10.7b) with F = 30/m and a = 0.06 m. (b) Maximum and minimum differences in levels measured at 1.2 m high receivers 2 m and 72 m from a 1.3 m high source in a mixed deciduous woodland in summer corrected for distance and air absorption (solid lines) [9] compared with the sum of ground effect using a hard-backed layer slit pore impedance model (broken line: flow resistivity 30 kPa s m^{-2}, porosity 0.5, tortuosity 2, layer thickness 0.15 m) or a variable porosity model (dash-dot line: effective flow resistivity 15 kPa s m^{-2}, porosity rate -40 /m) modified by scattering-induced incoherence (modelled as frozen turbulence with $\langle \mu^2 \rangle = 10^{-5}$, L_0 = 1.3 m) and foliage attenuation calculated from equation (10.7b) (F = 4.5/m, a = 0.1 m). (c) Mean corrected level difference measured at 1.2 m high receivers 2 m and 96 m from a 1.3 m high source in a spruce monoculture (solid line) [9] compared with the sum of ground effect using the variable porosity impedance model (effective flow resistivity 10 kPa s m^{-2}, porosity rate 0 /m) modified by scattering-induced incoherence (modelled as frozen turbulence with $\langle \mu^2 \rangle = 10^{-4}$, L_0 = 1.3 m) and foliage attenuation calculated from equation (10.7b) (F = 2.5/m, a = 0.08 m) represented by the broken line.Also shown is the prediction without 'frozen' turbulence (dotted line).

equation (10.7b). The fitted leaf area density (F = 20 /m) is highest for the mixed conifer forest and lowest for the spruce monoculture (F = 2.5 /m) which is consistent with the visual observations of the relative foliage density [8]. To indicate the relatively important effect of scattering-induced incoherence on propagation through the spruce monoculture, Figure 10.16(c) shows predictions with and without scattering-induced incoherence modelled as enhanced turbulence. The agreement between predictions and data above 1 kHz is poorest for the spruce monoculture.

Figures 10.17 (a)–(c) compare excess attenuation spectra, obtained by Huisman [7,26] in a pine forest at 100 m range and three receiver heights (4.5 m, 2.5 m and 1 m), with predictions using the variable porosity ground

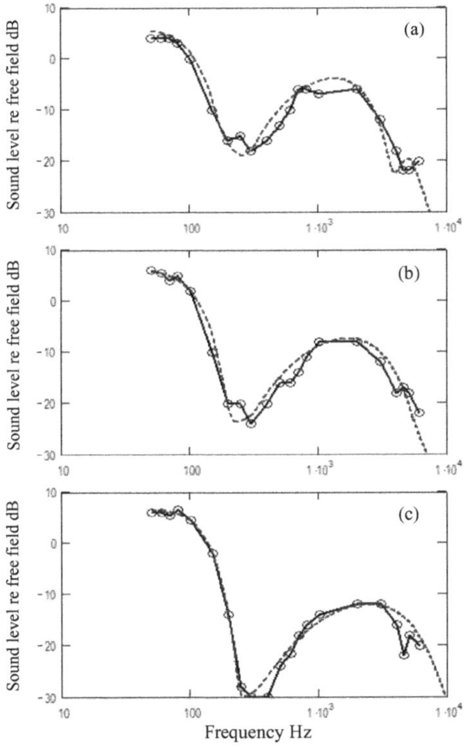

Figure 10.17 Excess attenuation spectra (joined circles) obtained from measurements using swept tones generated by a loudspeaker source at 1 m height and receivers 100 m from the source at heights of (a) 4.5 m (b) 2.5 m and (c) 1 m height [8]. Also shown as broken lines are predictions including ground effect (variable porosity model with effective flow resistivities and porosity rates of (a) 7.5 kPa s m^{-2} and 60/m (b) and (c) 10 kPa s m^{-2} and 50/m respectively, modified by incoherence due to effective turbulence ($\langle\mu^2\rangle = 10^{-5.5}$, L_0 = 1.0 m) added to foliage attenuation (equation (10.7b)) with leaf area densities and mean leaf sizes (a) F = 12 /m and a = 0.05 m (b) F = 7/m and a = 0.07 m and (c) F = 6/m and a = 0.06 m.

impedance model, modified by incoherence using equations (10.12a–10.12f) and augmented by foliage attenuation according to equation (10.7b). Although good agreement is obtained at all receiver heights, the predictions involve adjustments to the various parameter values. The higher foliage density required to fit the data for the 4.5 m high receiver is consistent with the expected variation in foliage density with height. This is expected to be maximum at 0.4 of the mean tree height, which, in this case, is at 0.4×11.2 m = 4.48 m [40]. Also, sound propagation to the highest receiver would involve a distinctly different specular reflection point so the different ground impedance model values required to fit data for the highest receiver height may be consistent with spatial variation in ground properties.

Figures 10.18 (a) and (b) compare third octave band excess attenuation spectra (joined circles) obtained from measurements in a pine forest with an 0.8 m high loudspeaker source and 1.2 m high receivers at 10 m and 80 m from the source in a pine forest [25] with predictions (broken lines) obtained by summing ground effect modified by scattering-induced incoherence and

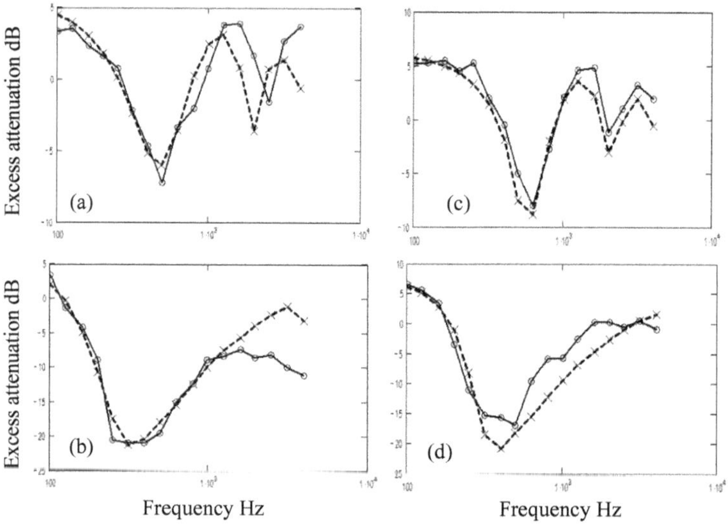

Figure 10.18 Measured excess attenuation spectra (joined circles) at 1.2 m high receivers with a loudspeaker source at 0.8 m height [24] at ranges of 10 m ((a) and (c) respectively) and at 80 m ((b) and (d) respectively) in a pine forest and a poplar forest. Predictions (crosses joined by broken lines) for 10 m range use the variable porosity model (pine forest: σ_e = 40 kPa s m^{-2}, α_e = 10 /m; poplar forest: σ_e= 50 kPa s m^{-2}, α_e = 200 /m) and scattering-induced incoherence modelled as frozen turbulence ($\langle\mu^2\rangle$ = 10^{-6}, L_0 = 0.8 m). At 80 m range, the variable porosity parameters used are σ_e = 20 kPa s m^{-2} and α_e = 15 /m in the pine forest; σ_e = 20 kPa s m^{-2}, α_e = 75 /m in the poplar forest and the scattering-induced incoherence is modelled as frozen turbulence ($\langle\mu^2\rangle$ = 10^{-5}, L_0 = 0.8 m). Foliage attenuation is predicted using equation (10.7b) with F = 0.05/m and a = 0.2 m in the pine forest and F = 0.1/m and a = 0.05 m in the poplar forest.

foliage attenuation. Figures 10.18 (c) and (d) are the corresponding plots for a poplar forest [25].

The broken lines in Figures 10.18(a) and (c) represent ground effect predicted using the variable porosity model (pine forest: R_e = 40 kPa s m^{-2}, α_e = 10 /m; poplar forest: R_e = 50 kPa s m^{-2}, α_e = 200 /m), modified to account for scattering-induced incoherence (eqns. (4) with $\langle\mu^2\rangle = 10^{-6}$ and L_0 = 0.8 m) but assuming no foliage attenuation. There is agreement between prediction and the pine forest data at 10 m range near the first destructive interference, but the fit to the second destructive interference, which is very sensitive to the source–receiver geometry are relatively poor. On the other hand, the fit between prediction and 10 m range data in the poplar forest is reasonable throughout the frequency range. This suggests that, for the measurements at 10 m range in the pine forest, the measurement geometry may have differed from the reported geometry.

The broken lines in Figures 18(b) and (d) represent ground effect predicted using the variable porosity model (pine forest: R_e = 20 kPa s m^{-2}, α_e = 15 /m; poplar forest: R_e = 20 kPa s m^{-2}, α_e = 75 /m), modified to account for scattering-induced incoherence (eqns. (4) with $\langle\mu^2\rangle = 10^{-5}$ and L_0 = 0.8 m) plus foliage attenuation (pine forest: F = 0.05 /m, a = 0.2 m; poplar forest: F = 0.1 /m, a = 0.05 m). With these third octave band data, it is not found possible to fit both the destructive interference due to ground effect and the high frequency in either forest. Also, as when using the Delany and Bazley model [24], the parameter values providing fits to the 80 m range data when using the variable porosity model differ from those that fit the 10 m data in either forest.

10.3 INFLUENCE OF GROUND ON PROPAGATION THROUGH ARRAYS OF VERTICAL CYLINDERS

10.3.1 Laboratory Data Combining 'Sonic Crystal' and Ground Effects

Frequently, in studies of propagation through forests and tree belts, tree trunks are modelled by vertical circular cylinders. There have been many studies of the potential use of regular arrays of cylinders in the form of sonic crystal noise barriers (SCNB), for noise reduction [2, see also Chapter 9, section 9.10]. If cylinders are closely and regularly spaced, such as in the square array depicted by the partial cross section in Figure 10.19, then coherent multiple scattering (sometimes called Bragg scattering) causes constructive and destructive interference effects. In the frequency domain, the results of destructive interference are called band gaps and represent frequency ranges with strongly reduced transmission. If a sound wave is normally incident on a regular scatterer array (sonic crystal) with a scatterer-to-scatterer separation (lattice constant) of d, the frequencies at which the band gaps occur are given by

$$f = \frac{nc}{2d}, \tag{10.14a}$$

where c is the speed of sound in the host medium, e.g. 344 m/s in air.

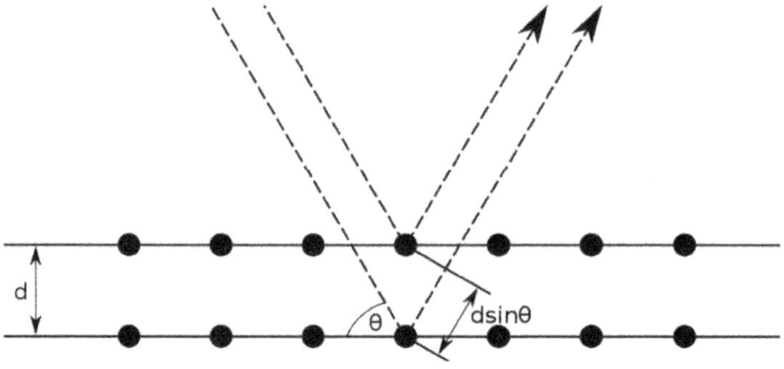

Figure 10.19 Schematic of a plane wave incident at angle θ from the normal on two rows of a square array with scatterer-to-scatterer separation (lattice constant) *d*.

The first band gap corresponds to $n = 1$; the frequencies of subsequent higher order band gaps are obtained by substituting higher values of n. Constructive interferences or pass bands occur between the band gaps and their frequencies are predicted in a similar way.

The magnitude of the insertion loss peaks in the band gaps depends on the filling fraction. In a square array of cylinders of radius R with lattice constant d the filling fraction is given by

$$\mathcal{F} = \frac{\pi R^2}{d^2}. \qquad (10.14b)$$

The scattering elements in SCNB are supported on the ground. The question arises of how ground and scattering effects combine, particularly when the scatterers are in relatively sparse regular arrays. Since they are more convenient than outdoor measurements, laboratory experiments including the presence of ground surfaces have been used to answer this question, albeit at much higher filling fractions than would be practical in forest stands. Results of measurements conducted in an anechoic chamber with a regular 5×10 arrays of PVC pipes, having outer diameter 0.04 m, wall thickness 0.002 m and length 0.5 m, on acoustically hard and acoustically soft surfaces [2] show that the combination of multiple scattering (sonic crystal) and ground effects can be beneficial in the frequency range of the first ground effect destructive interference. Figures 10.20 (a) and (b) show photographs of the pipes in a 5×10 array with lattice constant 0.1 m (corresponding to a filling fraction of 0.13) and axes normal to the MDF and felt layer surfaces, respectively. The distance between the point source and the nearest-to-source row of pipes was 0.51 m, the array was 0.44 m wide and the receiver was only 5 cm from the nearest-to-receiver row of pipes. Figures 10.20 (c) and (d) show the corresponding insertion loss spectra measured with source–receiver separation 1.0 m, receiver height 0.1 m and source heights (c) 0.1 m and (d) 0.02 m.

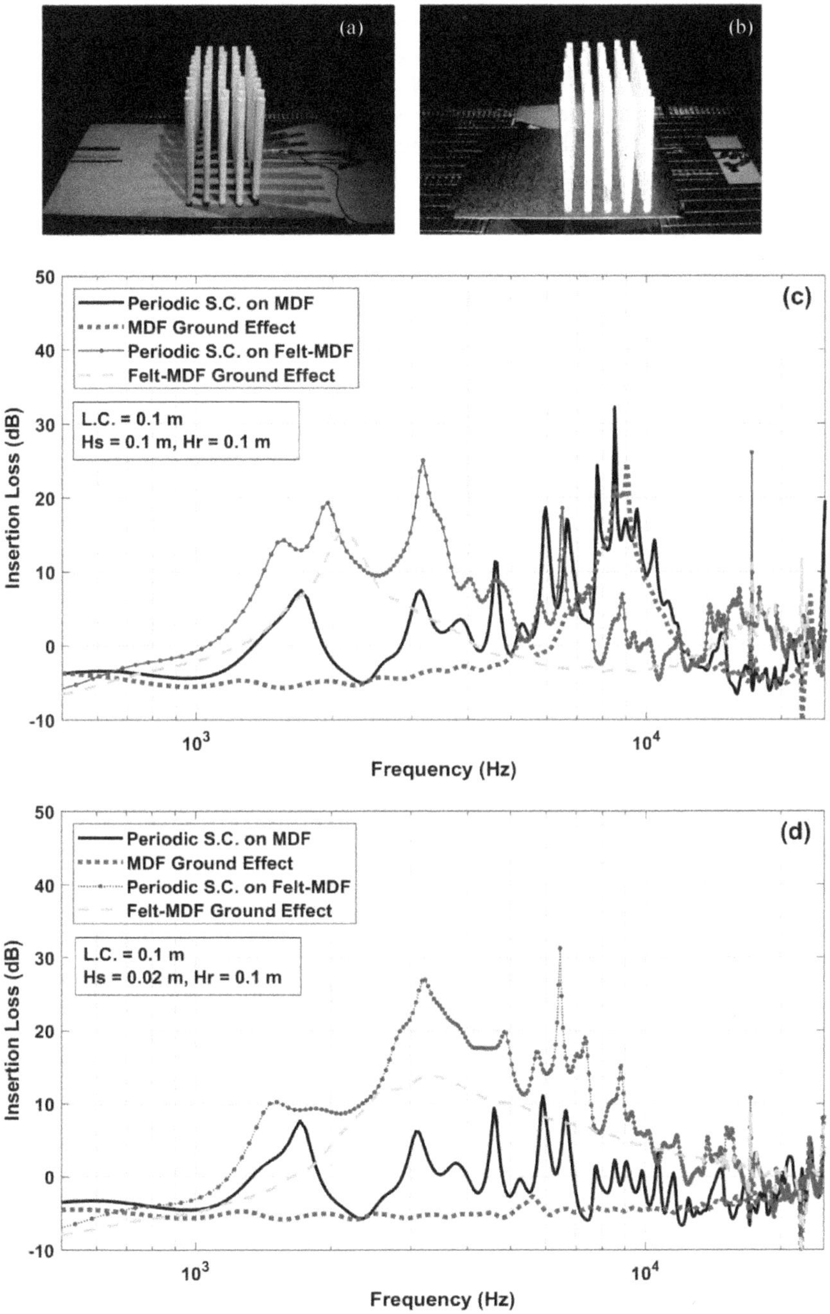

Figure 10.20 A regularly spaced 5 × 10 square array of PVC pipes with lattice constant 0.1 m over (a) MDF board and (b) Felt-over-MDF board [2]. Measured insertion loss spectra with source-receiver separation 1.0 m, receiver height 0.1 m and source heights (c) 0.1 m and (d) 0.02 m. Also shown in (c) and (d) are the attenuation spectra measured over the MDF board and Felt-MDF alone for the same source-receiver geometries [2].

The measured insertion loss peaks near 1.7 kHz, 3.4 kHz and 6.8 kHz correspond to first, second and third multiple scattering (i.e. sonic crystal effect) band gaps, respectively. The insertion loss measurements are with respect to free field. So, the negative insertion loss measured at low frequencies corresponds to pressure doubling due to the introduction of the ground surface. The negative insertion losses in the data between 2 and 3 kHz for both source heights above MDF correspond to the first multiple scattering pass bands. Over MDF, with source heights of 0.1 m and 0.02 m, the lowest frequencies of attenuation peaks due to destructive interference (ground effect alone) are near 8.5 kHz and above 20 kHz, respectively. The corresponding frequencies for the broader destructive interference (ground effect) attenuation peaks over the felt surface are near 2 and 3 kHz, respectively. When the source height is 0.02 m above MDF the peaks associated with multiple scattering band gaps are more less unaltered. When the source height is 0.1 m high over MDF, the destructive interference associated with ground effect near 8.5 kHz enhances the third and higher order band gaps due to the pipe array. When the source height is 0.1 m over the felt layer, the destructive interference near 2 kHz associated with ground effect enhances the first and second-order band gaps due to the pipe array. With the source height at 0.02 m over the felt layer the ground effect near 3 kHz enhances the first, second, third and fourth band gaps due to the pipe array. Consequently, for the lower source height over the soft surface, the overall insertion loss is significantly larger. These data support the idea used earlier in this chapter that, when predicting propagation through vegetation, ground effects and multiple scattering effects are additive even at filling fractions far greater than in forests or tree belts.

10.3.2 Numerical Design of Tree Belts for Traffic Noise Reduction

If tree trunks are arranged regularly, then, in theory, multiple scattering between the trunks may lead to stop- and pass bands similar to those obtained with laboratory arrays of cylinders such as described in Section 10.3.1. The frequencies and magnitudes of such 'sonic crystal' effects are governed by the lattice constant (the spacing between the trunks) and the filling fraction which depends on the trunk diameter and spacing. The stop bands could be associated with a strong reduction of the noise level but in pass bands the sound waves would be transmitted more strongly, and the resulting insertion loss spectrum would show corresponding peaks and troughs. While such 'sonic crystal' effects have indeed been observed in measurements of transmission through arrays of nursery trees in flowerpots placed in regular patterns over hard ground [41] the trees were placed closer together than would be viable in a healthy plantation. The tree filling fractions (0.47) leading to the observation of 'sonic crystal' effects in these measurements largely exceed those commonly used in tree belts and orchards. Other measurements [41] made in pre-existing plantations with much lower filling fractions (0.07–0.1), still denser than encountered typically in forests or tree belts (typically below 0.015), did not show 'sonic crystal' effects in the transmission loss spectra.

Full-wave numerical simulations, inspired by the Harmonoise reference methodology (see Section 12.7), have been carried out to investigate the design of narrow tree belts to reduce road traffic noise [42]. Specifically, the situation of a four-lane road bordered by a tree belt extending for 15 m from the edge of the road (see Figure 10.21(b)) is compared with open grassland (Figure 10.21 (a)).

Each lane is modelled as an incoherent line source. Since the major contributors to the noise shielding in a tree belt, namely the forest floor and the trunks, can be treated separately (see Section 10.3.1) sound propagation is modelled in two orthogonal planes. Interactions between sound waves and the trunks are modelled using a 2-D FDTD technique (see Chapter 4) in a horizontal cross section through the tree belt, parallel to ground, but in absence of ground reflections. The 2-D model assumes that the trunks are infinitely long and that tree crowns are absent. In testing against full 3-D FDTD calculations [33], this assumption has been found to be valid as long as sources and receivers are close to the ground and the trunk heights exceed 1.5–2 m. Absorption by tree bark is included through a frequency-independent purely real impedance of 51 (relative to air) at the trunk surfaces [44]. Also, the simulations rely on the equivalence between a point source and a coherent line source when expressing results relative to free-field sound propagation [43]. Scattering by foliage and branches in the crown is neglected as this is predicted to have only a small effect (less than 0.5 dBA for road traffic spectra). Ground effect, including the impedance discontinuities (see Figure 10.21 (b)) between the acoustically hard road surface, the forest floor zone (assumed to have a flow resistivity of 20 kPas/m^2 and a porosity of 0.5) and between this zone and the grassland behind the tree belt (flow resistivity of 300 kPas/m^2, porosity of 0.75) is calculated in any given vertical plane with the axi-symmetrical Green's Function Parabolic Equation

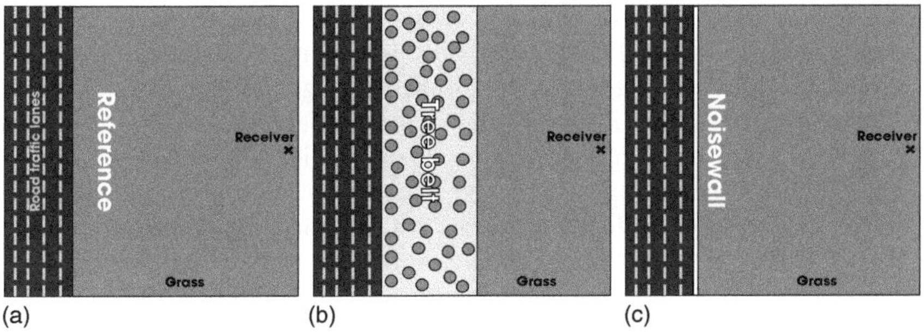

Figure 10.21 Road and receiver geometries assumed for simulating the noise reduction by a narrow tree belt (a) with open grassland only (for reference) (b) with a 15 m wide tree belt over a 'forest floor' and grassland thereafter and (c) noise barrier close to road and grassland thereafter. Reproduced from [42] under the STM agreement.

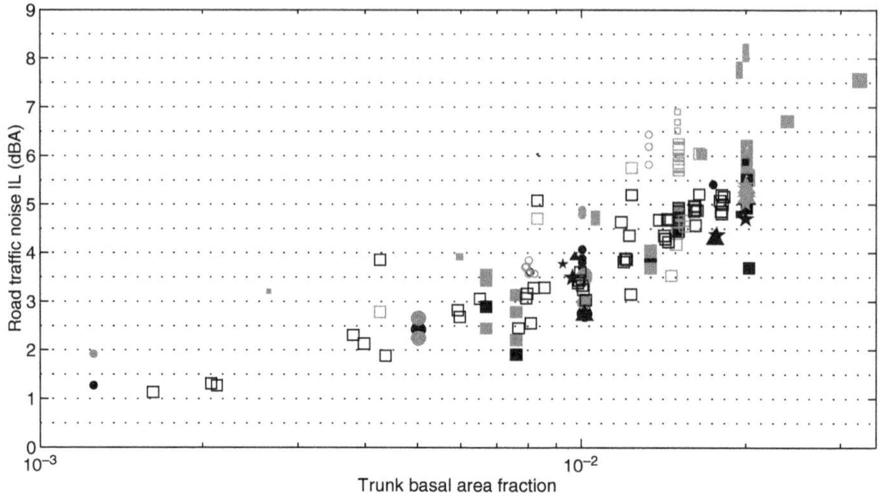

Figure 10.22 Predicted road traffic noise insertion losses (relative to sound propagation over grassland) for a 15-m deep tree belt bordering a 4-lane road (containing 15 % heavy vehicles, all cars driving at 70 km/h and equally distributed over the 4 lanes). The receiver is positioned behind the belt, 30 m from the nearest edge of the road, and at a height of 1.5 m. Trunk basal area and planting scheme pattern have been varied. The size of the symbols is proportional to the diameter of the trunks modelled (ranging from 4 cm to 28 cm). The filled symbols represent fully populated tree belts, the open symbols are used when a fraction of the trees have been omitted. Patterns considered are square grids (circles), rectangular grids (squares), triangular grids (triangles), and face-centred square lattices (pentagrams). Black symbols indicate regularly positioned tree trunks and grey symbols indicate some randomness in tree centre location [42].

(GFPE) method (see Chapter 4). The reference situation is sound propagation from an identical road over grassland without trees (Figure 10.21 (a)).

The simulations show that when the fractional area of ground occupied by trunks (called the stem cover fraction or trunk basal area fraction) increases, the acoustical shielding of the tree belt becomes larger (see Figure 10.22).

Large stem cover fractions correspond to large stem diameters and/or short distances between adjacent tree trunks. Nevertheless, whatever the stem cover fraction, numerical calculations predict that planting schemes that use specific combinations of tree placement, tree spacing and trunk diameter can have a significant influence on the noise reductions that can be obtained. With a trunk basal area fraction of 1%, the predicted extra attenuation due to 15-m wide tree belt is between 2.7 dBA and 5 dBA compared with sound propagation over grassland. When the trunk basal area fraction is 2%, the predicted extra attenuation is between 3.6 dBA and 8.3 dBA.

Such high planting densities are difficult to achieve. However, even at more realistic planting densities, if a tree belt beside a road replaces acoustically hard ground rather than grassland, the predicted insertion losses are

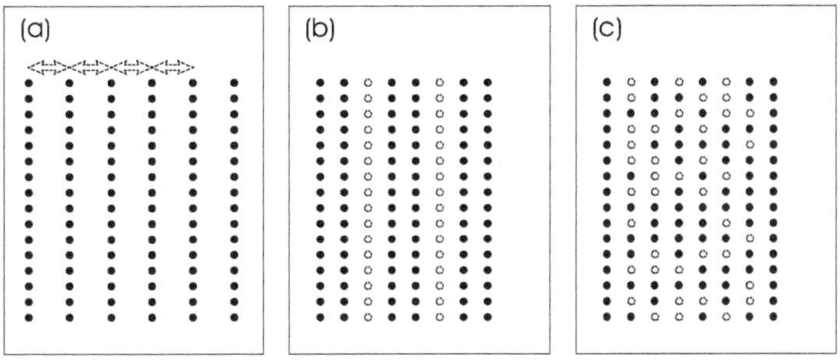

Figure 10.23 Three tree planting patterns that reduce trunk basal area fraction compared with a square array without affecting the predicted traffic noise reduction [33,42].

larger, since a significant part of the noise reduction due to a tree belt can be attributed to the 'soft' ground effect.

Simulations have suggested ways of relaxing the need for high trunk basal area fractions, without affecting noise shielding too much [42]. One possibility is a rectangular planting scheme in which the spacing of rows orthogonal to the road is larger than the spacing between the trees along the road's length axis (see Figure 10.23 (a)). Another is to omit some rows parallel to the road (Figure 10.23 (b)). The resulting closely spaced pairs of rows parallel to the road length axis, separated by open spaces without trees, which might function also as firebreaks, could offer a practical scheme of tree planting for noise reduction. A third possibility is to omit individual trees (thinning) at random inside the belt (but not at its borders) (see Figure 10.23 (c)).

A degree of randomness in trunk positioning and/or a random variation in trunk diameter is predicted to increase the noise shielding. Some deviation from perfect ordering is likely to occur during any tree planting. On the other hand, fully random positioning is predicted to lead to less shielding. For a specific simulated geometry [42], the optimal degree of randomness has been found to be near 20%. The effect of randomness might be greater in denser arrays (more resembling sonic crystals), as discussed in Section 10.3.2.1. A possible physical explanation is that additional periodicity is introduced by the randomness, while focusing effects are reduced.

Numerical calculations for a non-refracting atmosphere predict a linear trend between road traffic noise shielding and tree belt depth (orthogonal to the road) [42]. Also, predictions suggest that the length of the tree belt (along the road axis) should be chosen in a similar manner to that given in guidelines for estimating noise-wall length; longer tree belts might be needed for receivers further away from the road. Conventional noise barrier shielding is governed by diffraction, which means that receivers further away benefit less from the noise barrier. However, if it is long enough, the effect of a tree belt does not depend on receiver distance since sound-level reduction is obtained by transmission loss [42]. At receivers more than 40 m from the edge of the road, the predicted transmission loss due to a 15-m deep optimized tree belt is comparable with that from a 1-m or even 2-m high single thin concrete noise wall positioned where the tree belt would start (see Figure 10.21 c) [45]. The predicted noise shielding of a 30-m deep optimized belt is close to that from a 3-m or even 4-m high noise barrier, depending on the type of ground cover between source and receiver [45].

Clearly the acoustical efficiencies predicted for newly planted tree belts, with juvenile thin trunks, are relatively small. However, a way of improving their shielding is to introduce a judicious distribution of supporting poles around the trees [46]. Such poles are introduced most often to ensure straight growth. By adding poles, the number of scatterers is increased without violating biological limitations associated with close planting. Furthermore, if the poles are made sound absorbing, road traffic noise shielding by a 15-m deep tree belt is predicted to even exceed 10 dBA [46].

The numerical models and the modelling assumptions used for the simulations reported in this section have been validated where possible; moreover, the input parameters used are based on measurements. Nevertheless, validations of the sound reductions predicted for tree belts through comparisons with data from measurements in the field are lacking due to the decade(s) of growth needed to produce a mature tree belt and the fact that most existing tree belts are not optimized to reduce sound pressure levels. According to the simulations discussed, a tree belt will be more useful for road traffic noise abatement if it is designed for the purpose, especially when it is of limited depth.

Only a few studies report basal areas along with measurements of sound transmission, thereby enabling comparisons of predictions with data [47,48]. Basal areas of up to about 70 m^2/ha (0.07) have been found to give attenuations of up to 4 dB for propagation distances between 10 m and 20 m relative to a forest edge. This corresponds with the road traffic noise insertion loss predictions for low stem cover in Figure 10.22.

10.3.3 Measured and Predicted Effects of Irregular Spacing in the Laboratory

Measurements in an anechoic chamber, similar to those described in Section 10.3.1, have compared the insertion loss spectra due to regularly (lattice constant 0.1 m) and irregularly spaced PVC pipes (outer diameter 0.04 m, wall thickness 0.002 m and length 0.5 m) with axes vertical to hard and soft surfaces [2,32]. Figure 10.24 (a) shows a perturbed arrangement (standard deviation 2.0r where r is the cylinder radius) of PVC pipes over MDF. Figure 10.24 (b) and (c) compare insertion loss spectra measured with perturbed and regular pipe arrangements (source height 0.02 m, receiver height 0.1 m separation 1.0 m) over felt and MDF, respectively. The perturbed arrangements result in higher insertion losses above the second band gap frequency. Other calculations show that the insertion losses measured with the perturbed array over either surface are greater than the insertion loss measured for a corresponding but totally random array [32].

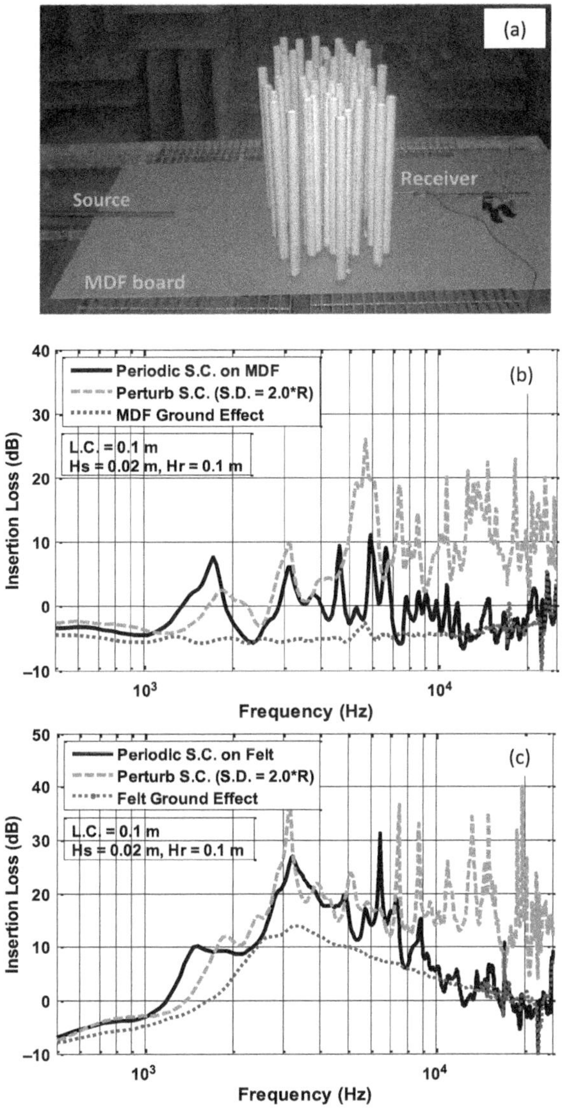

Figure 10.24 (a) Randomly perturbed (standard deviation = diameter) originally square 5×10 array of PVC pipes with original lattice constant 0.1 m. (b) and (c) insertion loss spectra for periodic and perturbed PVC pipe arrays measured with source height 0.02 m receiver height 0.1 m and separation 1.0 m over MDF and a felt layer respectively [2].

Figure 10.25 compares data and predictions for the insertion loss of perturbed arrays over MDF and felt. The predictions are the result of adding the ground effects and the multiple scattering effects. The latter are calculated without the ground plane using a 2-D multiple scattering theory [49]. Excess attenuation spectra measured over the ground surfaces alone are

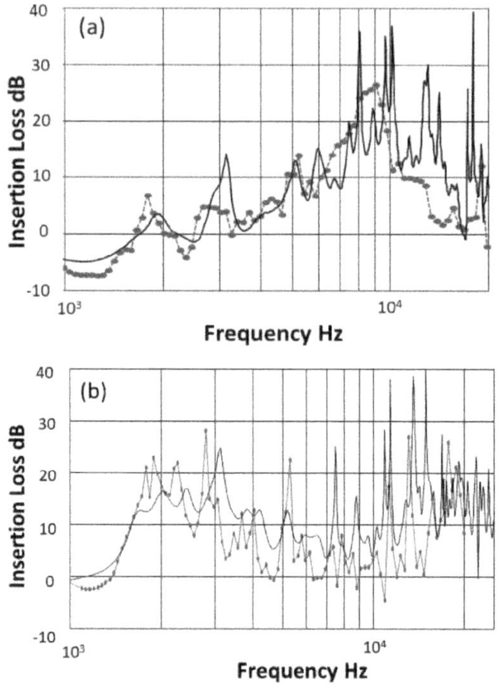

Figure 10.25 Comparison between measured (lines) and predicted (joined dots) insertion loss spectra with source and receiver at height of 0.1 m and separation of 1.0 m due to perturbed 5 × 10 arrays of PVC pipes (a) over MDF and (b) over a felt layer [2].

fitted using the two-parameter variable porosity model (see Chapter 5). Thereby the surface impedance of MDF is characterized by an effective flow resistivity of 5×10^5 kPa s m^{-2}, porosity rate 100 m^{-1} and that of the felt layer by an effective flow resistivity of 20 kPa s m^{-2}, porosity rate 100 m^{-1}. Again, the approximation whereby multiple scattering and ground effects are calculated separately and then added seems to be tolerably successful.

10.4 REFLECTION FROM FOREST EDGES

The edge of a forest or mature tree belt is a transition area between adjacent ecological communities and includes some of the ecological characteristics of both communities. In this region, the structure and quantity of vegetation can change quite dramatically in a relatively short distance. For example, the area separating a forest from an open field/grassland may have a mix of plant species from the adjoining lands, grasses and herbaceous plants from the open area and tree species that are found within the forest along with additional species that depend on the greater daylight that occurs at the edge compared with deep within the forest or tree belt. The trees on the edge often differ in structure from those further inside the forest as greater light exposure allows them to retain more branches and foliage closer to the ground. Together these differences mean more branches and foliage and a structure that is more uniform from top to bottom. The different structure may mean distinct sound reflections and transmission near the edge. Although reflections from the edge of a forest are not considered to be important for road traffic noise, they have been studied in connection with propagation of railway noise and blast noise.

A semi-empirical model developed for predicting railway noise [50,51] assumes that the edge reflection can be modelled as that from arrays of cylinders representing trunks and elevated spheres representing branches. Although incorporated in a railway noise prediction scheme, it was validated by measurements of sound propagation from explosions [51]. The model introduces five parameters: average forest height, average forest density (trees per 100 m^2), average trunk diameter, number of spheres representing scattering from branches and their average diameter. After adjusting the parameters for best fit, the predictions were found to lie between ±4.5 dBA of the majority of equivalent continuous level data at ranges from 20 m to 500 m. Also, it was concluded that the vertical directivity of forest edge reflections is more pronounced at greater distances.

Nevertheless, since the model developed in this work ignores phase-related effects, it is less illuminating about the extent and character of forest edge reflections and the associated transmission loss than the results of a series of measurements using a propane cannon source (with main sound energy content in the 125 Hz third octave band and effective source height 0.62 m) in an open grass-covered field near a pine forest (*pinus resinosa*) and deep within the forest also [52]. The forest, which was on fairly flat terrain, contained 131 trees per hectare with an average height of 11 m planted at approximately 2 m intervals. The average tree diameter at 1.4 m above the ground was 20.8 cm and there was minimal undergrowth apart from areas of 0.3 m high grass. The width of the transition zone from the edge of the field to trees within the forest of average height 7.9 m and uniform canopy with branches down to 1.5 m from the ground was estimated to be 10 m. The grass in the field was growing through a mat of dead grass. The forest floor had a well-developed humus layer of decaying pine needles lying above sandy soil. Acoustical characterization of the ground surfaces was based on waveform fitting to time-domain data from pistol shots (blanks) and revealed that the forest floor had an effective flow resistivity of less than half of that in the adjacent open field resulting in a faster decay inside the forest than outside of the sound energy between 250 Hz and 500 Hz in the propane canon shots spectra.

A particularly interesting finding is that propagation either into or out of a forest is non-reciprocal. Figures 10.26(a) and (b) compare spectra recorded at various distances by 1.4-m high microphones along lines crossing the forest edge with the source outside the forest and inside the forest, respectively. The dips in the spectra between 100 Hz and 1 kHz may be attributed to

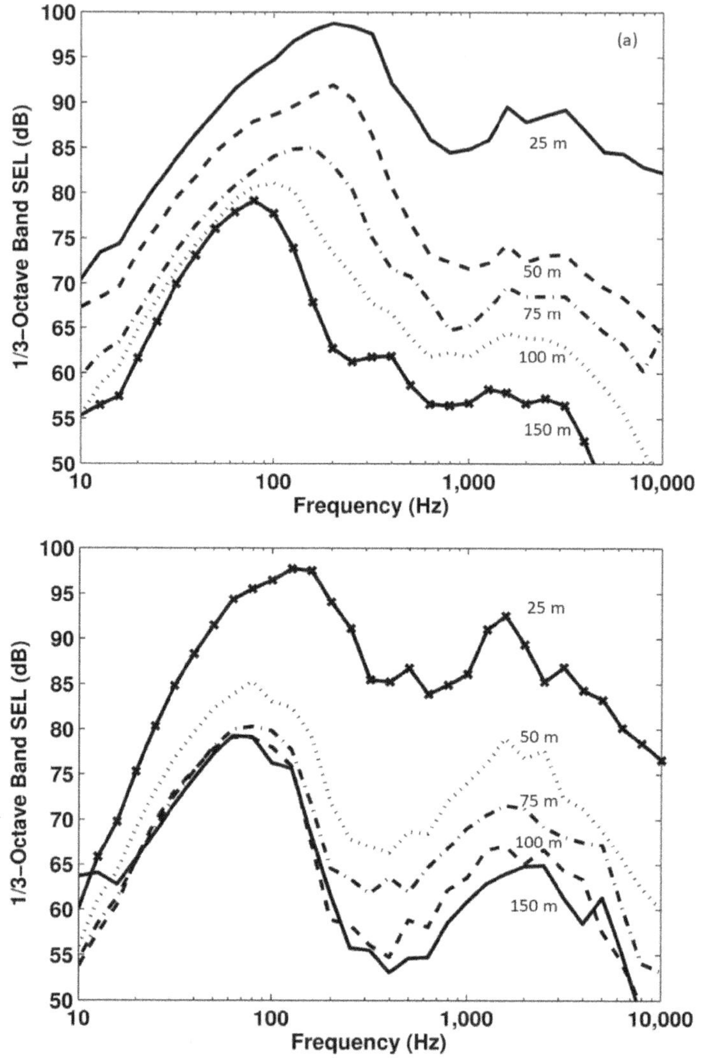

Figure 10.26 Octave band spectra averaged over 90 shots from a propane cannon (effective source height 0.62 m) measured (a) with the source in an open field 50 m from a forest edge and 1.4 m high receivers along a line crossing from the open field into the forest and (b) with the source located 125 m from the forest edge within the forest and 1.4 m high receivers along a line crossing from the forest into the open field. The curves are labeled with distances from the source [52]. Reproduced from M. E. Swearingen, M. J. White, P. J. Guertin, D.G. Albert and A. Tunick, Influence of a forest edge on acoustical propagation: Experimental results, *J. Acoust. Soc. Am.*, **133**(5): 2566–2575 (2013) with permission of the Acoustical Society of America.

ground effect. There is a higher attenuation over the three decades of frequency displayed during propagation from the open field into the forest (the difference between curves labelled 50 m and 75 m in Figure 10.26(a), the forest edge being 50 m from the source) than during propagation from the forest into the open field (the difference between the curves labelled 125 m and 150 m in Figure 10.26(b), the forest edge being 125 m from the source location in the forest). Also, the measured decrease in the frequency of the maximum levels of the propane cannon signal with distance during propagation into the forest is greater than that during propagation from the source within the forest to receivers outside the forest.

The measured difference between the overall attenuations through the forest edge in opposite directions was 4.5 dB. Simulations of likely meteorological effects and effects due to different ground conditions do not account for this difference. As suggested in this study [52], the non-reciprocity of the forest edge propagation effects could be exploited for noise reduction by forests perhaps through the judicious use of firebreaks, i.e. natural or deliberate gaps in otherwise continuous forests that delay the spreading of forest fires.

10.5 METEOROLOGICAL EFFECTS ON SOUND TRANSMISSION THROUGH TREES

Measured wind speed profiles in an open field increase with height. Inside the forest, wind velocities typically decrease from the top of the canopy downwards through the layers of leaves and branches due to increased drag. Also, the forest canopy results in more elevated radiative and moisture exchanges than in an open field. During the daytime, the forest canopy absorbs incoming solar radiation and generates a layer of warmer-than-ambient air that mixes downwards from the top of the canopy through normal mixing processes. As a result, maximum temperatures within the forest interior tend to coincide with the height of maximum leaf density and a local minimum in temperature tends to occur in the trunk volume below the branches and leaf structures. Evapotranspiration processes are also closely linked to the diurnal heating cycle and wind fields within the canopy. As a result, humidity and temperature profiles often behave similarly within and above the forest canopy. In contrast, temperature and moisture profiles in an open field tend to decrease from the ground upwards. Also, measurements show relatively low turbulence underneath the canopy.

After sunset, soil in an open field cools down much faster than the layers of air in the atmospheric boundary layer. This nightly ground-based temperature inversion results in downward bending of sound and larger sound pressure levels compared to sound propagation in an atmosphere having constant air temperature with height. The presence of the forest canopy traps some of the thermal radiation from the forest floor below the canopy, thereby affecting sound propagation. During daytime, soil in an open non-vegetated environment heats up much faster than the air above it. So, the temperature decreases with height and the atmosphere is upward refracting for sound. On the other hand, as a result of the shading provided by the canopy, the forest floor heats up more slowly than the air above the canopy. In this way, a temperature inversion condition is created. Hence, the shadow zone created by upward refraction of sound propagating outside the forest is (partly) counteracted by downward refraction within the forest. Nevertheless, if there is a strong temperature decrease with height outside the forest, a shadow zone will be formed at some distance from the source and the slight increase in sound pressure levels due to the presence of the forest may be of limited importance. Moreover, attenuation associated with forest floor ground effect and scattering and visco-thermal losses in the above-ground biomass will partly counteract this downward-refraction effect.

Compared with open grassland Huisman [8] has predicted an extra 10 dB(A) attenuation of road traffic noise through 100 m of pine forest. He has remarked also that whereas downward-refraction conditions lead to higher sound levels over grassland, the levels in woodland are comparatively unaffected suggesting that extra attenuation through trees should be relatively robust to changing meteorology.

Defrance et al. [53] have compared numerical calculations and outdoor measurements for different meteorological situations. Their numerical method is based on a 2-D GFPE method [54] (see Chapter 4) developed and adapted [55] so that traffic is modelled by a line of equivalent point sources of height 0.5 m. To allow for the non-uniform wind conditions due to the forest, GFPE calculations were made for each equivalent point source by projecting the wind profile on the source–receiver direction (see Figure 10.27).

The measurements were obtained through a pine forest with an average tree height 11.83 m, an average circumference of 53.1 cm and a density of

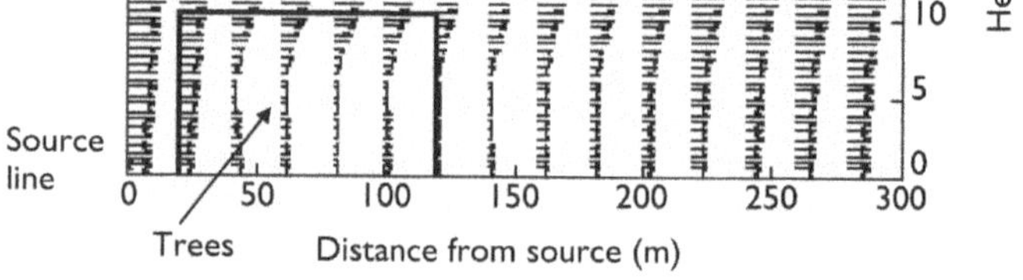

Figure 10.27 Range-dependent sound speed profiles used in simulations by Defrance et al [53]. The rectangle shows the region occupied by trees.

0.1078 trees/m². For different range-dependent sound speed profiles (downward-refraction, homogeneous or upward refraction situations), the acoustic attenuation between the highway source and receivers at 150 m distance was compared with that obtained in the absence of the trees. During the measurement period, the mean flow was of 2200 vehicles per day, for each of the two lanes, with 5% of heavy trucks. The mean speed of light vehicles was 85 km/h. The data showed a reduction in A-weighted L_{eq} due to the trees of 3 dB during downward-refracting conditions, 2 dB during homogeneous conditions and 1 dB during upward-refracting conditions.

The numerical predictions shown in Figure 10.28 suggest that in downward-refracting conditions, the extra attenuation due to the forest is between 2 and 6 dB(A) if the receiver is at least 100 m away from the road.

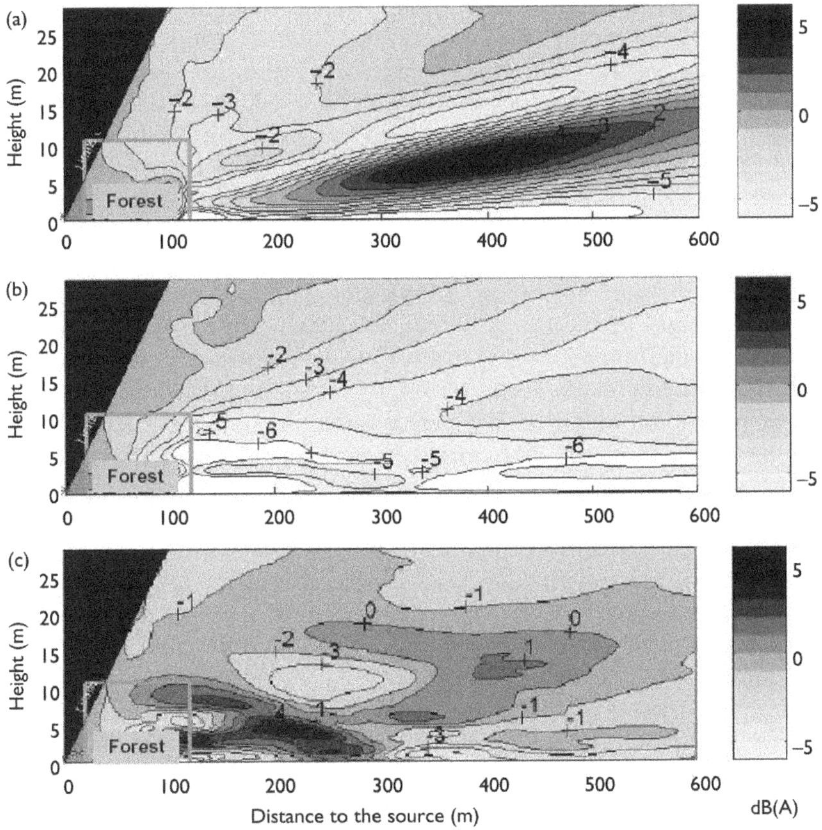

Figure 10.28 Predicted differences in A-weighted traffic noise level due to a 100 m wide and 10.5 m high pine forest relative to the situation without forest in the case of: (upper plot) typical day temperature profile (unfavourable to propagation), (middle plot) typical night temperature profile (favourable) and (lower plot) typical downward-refracting situation with a positive gradient wind speed profile (favourable). Reprinted with permission from ICSV 9 Proc. Orlando, 2002.

This is confirmed by other numerical predictions [56] which show that a sufficiently long 50-m deep forest or mature tree belt may strongly decrease night noise levels in situations with a nocturnal inversion layer. At a 2-m high receiver, centrally located along the belt and at 200 m from the road edge, this benefit could easily exceed 10 dBA compared to sound propagation entirely over grass-covered soil. The forest-induced change in the local microclimate adds to the attenuation provided by the soft forest floor, trunks and canopy. The influence of refraction on the sound pressure levels decreases with receiver height, and consequently the improved attenuation due to the tree belt also decreases with receiver height.

In conditions that are upward-refracting outside the tree covered area the adapted GFPE model predicts that a forest may increase the received sound levels somewhat at large distances [55]. But, as remarked earlier, this prediction is not important in practice since the levels predicted at sufficiently large distances during upward refraction are relatively low anyway.

Other predictions [57] have included an allowance for a large wind speed gradient through the foliage canopy based on a micrometeorological model [58] and assumed effective wavenumbers deduced from multiple scattering theory [59] for scattering effects of trunks, branches and foliage. Figures 10.29 (a) and (b) show predictions in reasonable agreement with third-octave band level data up to 5 kHz obtained at ranges of 174 m and 315 m, respectively, from (5 lb) C4 explosions above ground. It is noticeable, however, that the predicted ground effect 'dips' are greater than measured, particularly at the longer range. Moreover, the predictions significantly underestimate the measured attenuation above 2 kHz. Previous discussions suggest that, respectively, these discrepancies are due to scattering-induced incoherence, which reduces the magnitudes of destructive interferences associated with ground effect, and visco-thermal foliage attenuation at higher frequencies.

Figures 10.29 (c) and (d) show separate predictions of ground effect, ground effect modified by refraction plus trunk scattering and ground effect modified by refraction, trunk scattering and canopy scattering, respectively, at 315 m range for upwind and downwind conditions respectively. Upward refraction due to wind and temperature gradients in the foliage is predicted to have a significant effect. It should be noted also that trunk scattering has an important effect above 500 Hz.

While Defrance et al. [53] have concluded that a forest strip of at least 100 m wide appears to be a useful natural acoustic barrier, it should be noted that both the data and numerical simulations were compared to sound levels without the trees present, i.e. over ground from which the trees had been removed. This means that the ground effects both with and without trees are similar. As discussed previously, this is unlikely to be true of a mature forest and is likely to lead to underestimates of the sound attenuating properties of such forest compared with propagation over grassland.

Further points to be noted about this study are as follows: (i) shielding due to trunk scattering is likely to have been very limited in these measurements,

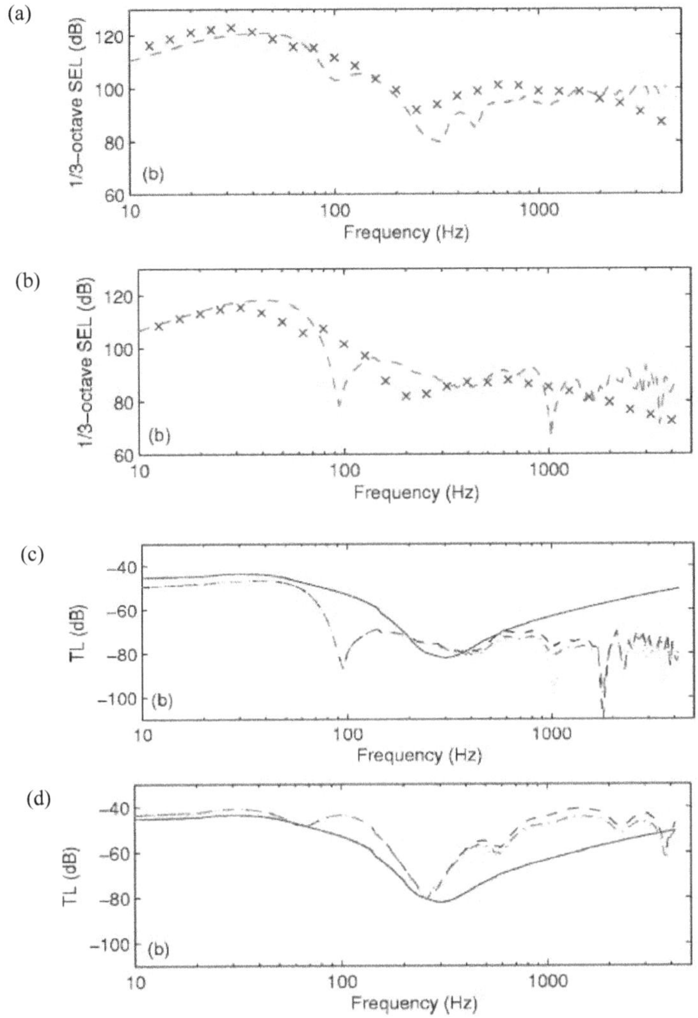

Figure 10.29 Third-octave band level data (crosses) and predictions (broken lines) up to 5 kHz at ranges of (a) 174 m and (b) 315 m from above-ground (5 lb) C4 explosions in a pine forest (0.0124 trunks/m^2, average diameter 0.096 m) with source and receiver heights of 2 m and 1.2 m respectively. The predictions include ground effect (effective flow resistivity 7.5 kPa s m^{-2}, porosity 0.4, tortuosity 4 (see equations (5.28), (5.29) and (5.37)) and the measured layer depth of 0.03 m), an effective sound speed gradient calculated from a micro-meteorological model, and multiple scattering by trunks and canopy. Figures 10.29 (c) and (d) show separate predictions for ground effect before (continuous lines) and after modified by refraction ((c) upward and (d) downward, broken lines) plus trunk scattering (dash-dot lines) plus canopy scattering (dotted lines) [57]. Reproduced from M. E. Swearingen and M. White, Influence of scattering, atmospheric refraction and ground effect on sound propagation through a pine forest, *J. Acoust. Soc. Am.*, **122**: 113–119 (2007) with permission of the Acoustical Society of America.

since the trunk basal area was only 0.24%, (ii) if the further microphone was within the forest during the measurements in trees it would be subject to backscatter from trees around it thereby being subject to a higher level than at a microphone at the same distance but outside the trees and (iii) the specific GFPE models used in Refs. [54,55,57], while including discontinuous impedance and range-dependent atmospheric refraction, do not include turbulence or forest edge effects (see 10.3.4).

10.6 COMBINED EFFECTS OF TREES, BARRIERS AND METEOROLOGY

A significant problem for conventional outdoor noise barriers is their interaction with meteorological conditions. The presence of the barrier disturbs the airflow during wind. Wind speed gradients and intensity of turbulence become larger near the barrier. This results in an increased refraction of sound during downwind sound propagation (sometimes called REfraction of Sound by WINd-induced Gradients (RESWING) or Screen-Induced Refraction Of Sound (SIROS)), which reduces the extent of the shadow region behind a barrier. This has been analysed extensively through wind tunnel experiments, in situ measurements and numerical simulations [60–67]. Results of wind tunnel experiments showing that sound propagation over a thin screen on acoustically soft ground could lead to negative insertion losses in strong wind [66], thereby implying that it would be more advantageous to limit exposure to the noise source without inserting a barrier, have been supported by numerical simulations [67].

Besides the additional refraction caused by the screen, turbulence, including screen-induced turbulence, increases noise levels behind the barrier at higher frequencies [66]. However, the scattering by turbulence will mainly be observed in the deep shadow zone, where sound pressure levels are low. On the other hand, at large distances, the superposition of scattered and diffracted waves will result in fluctuations of phase and amplitude, smoothing out interference patterns. At smaller distances behind thin rectangular barriers, the turbulence effects on broadband traffic noise levels can be small in comparison with the screen-induced refraction which adds to refraction due to gradients in the wind speed profile that would occur above unscreened ground. Usually refraction due to wind speed gradients is important at long range, whereas screen-induced refraction occurs in the zone directly behind the barrier where, in a still atmosphere, the barrier is expected to be most effective.

Since a tree canopy is effective in reducing wind speed, the effect of a row of trees (in leaf) behind a noise barrier (i.e. on the opposite side to the source) in wind has been investigated [68,69]. Measurements near a highway with and without a row of trees behind a noise barrier continuously from the middle of the summer until fall have been compared. For downwind locations and for an orthogonal incident wind, the efficiency of the noise barrier with trees was better than that of the noise barrier without trees. Figure 10.30

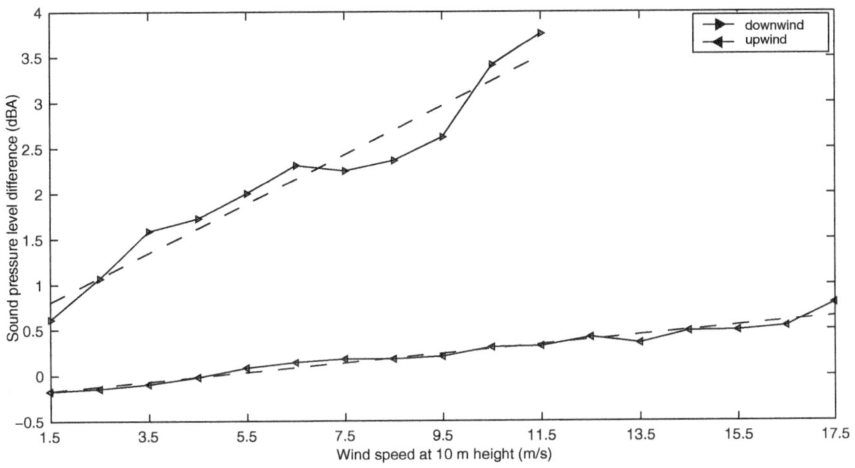

Figure 10.30 Measured average net efficiency (A-weighted level reduction) of a row of trees behind a noise barrier as a function of wind speed. The effect of the wind direction orthogonal to the noise barrier is shown, for both downwind and upwind sound propagation. The best-fit straight lines to these data are shown [68].

shows the difference in A-weighted motorway noise levels measured with and without a row of trees next to a barrier. These data indicate that, downwind, there is a clear improvement due to the trees and it increases with increasing wind speed.

Part of the shielding that would otherwise be lost due to the screen-induced refraction of sound by wind is recovered by the presence of the canopies. Note that the maximum improvement measured is near 4 dBA (for wind speeds exceeding 10 m/s, measured at 10 m height) which represents a significant part of the noise reduction one gets behind a noise wall along a multi-lane highway.

The improvement by the trees is only slightly affected if the wind direction is not perfectly orthogonal to the barrier. Upwind sound propagation is affected only to a small degree by the presence of trees. Since the main sound energy in the frequency spectrum of traffic noise involves wavelengths rather bigger than the leaves and stems in a typical tree canopy, downward scattering by the canopy results in only a slight (no more than 0.7 dBA) increase in total A-weighted sound pressure level [70].

On the other hand, since the presence of trees shifts the downward-refracting gradients near the top of the barrier towards the top of their canopy, dense and tall canopies will have more effect [70,71]. Also, the distribution of biomass with height in the canopy is important. A canopy shape in which most biomass is located near the bottom of the canopy results in a smooth transition with height thereby limiting the generation of wind speed gradients at the top [71]. In this respect, conifers are most suitable not only because their typical canopy shape is that required for the optimum flow effect but also because they have larger needle-area densities, larger drag coefficients and do not drop leaves during winter.

The gap between the top of the barrier and the bottom of the tree canopy is important also. The absence of gaps is predicted mainly to improve shielding at close distances behind the barrier in wind. The fact that the trees shift wind speed gradients upwards has a detrimental effect on shielding from more distant sources or to more distant receivers. Conversely, the presence of a gap is predicted to reduce the benefit from introducing a row of trees for downwind receivers close to the barrier but improves their benefit for more distant receivers and sources [70,71].

Numerical predictions made with a hybrid FDTD-PE method using simulated wind fields (see Chapter 4) indicate the potential benefits from placing a row of trees behind a single noise barrier along a four-lane highway (see Figure 10.32), and how canopy height and gaps influence this effect [70]. The road surface is assumed to be rigid, and typical grassland impedance is assumed between road and receivers at 2 m height (i.e. half screen height) downwind from the road. The results are summarized in Figure 10.31.

The inflow wind speed profile is characterized by a friction velocity u^*=0.4 m/s, assuming a neutral atmosphere, with an aerodynamic roughness length z_0 = 0.01 m (see Chapter 1.5). The air flow through the canopy of the trees results in a pressure drop Δp , quantified by the following expression:

$$\Delta p = k_r \frac{\rho_0 u_x^2}{2}, \tag{10.15a}$$

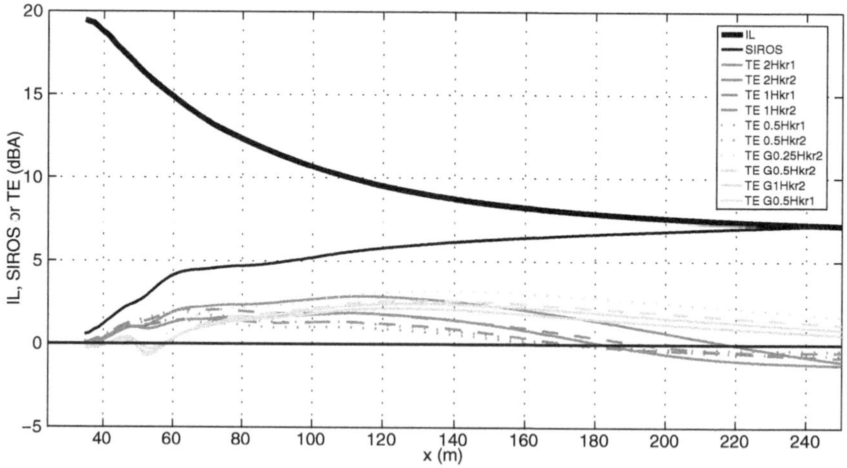

Figure 10.31 Plots of insertion loss *IL*, screen-induced refraction of sound (*SIROS*) and tree effect (*TE*) as a function of receiver distance, for the geometries presented in Figure 10.32 [70].

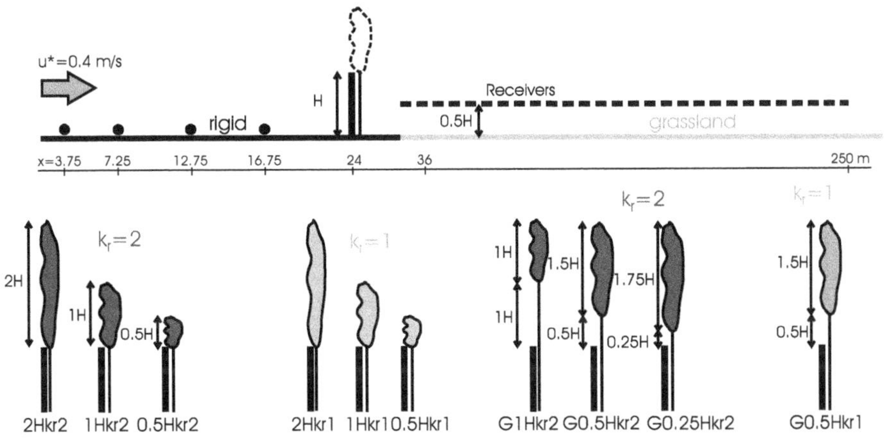

Figure 10.32 Predicted effects of tree canopy configurations in a row of trees behind a noise barrier along 4-lane highway (vehicle speed of 110 km/h), at receivers located at half the screen height above ground in downwind conditions. A uniform pressure resistance coefficient k_r is assumed all over the canopy volume [70].

where k_r is the pressure resistance coefficient, ρ_0 the air mass density and u_x the horizontal component of the wind velocity. The pressure resistance coefficient can be related to physical characteristics of the canopy. If the aerodynamic drag of the canopy is assumed to balance the pressure drop, then [72]:

$$k_r \approx \int_0^D C_d \, LAD \, dx, \tag{10.15b}$$

where D is the width of the canopy layer in the horizontal direction, C_d is the dimensionless drag coefficient of the elements forming the canopy and LAD is the leaf area density which is defined as the total area of leaf per unit volume of the canopy. Assuming further that the LAD and C_d are constant in horizontal direction, Equation (10.15b) can be simplified to

$$k_r \approx C_d \, LAD \, D. \tag{10.15c}$$

For the simulation results shown in Figure 10.31, k_r is assumed to have values of 1 and 2, representing various combinations of the relevant parameters.

The insertion loss, IL, is defined as the sound pressure level in case of unscreened ground, minus the sound pressure level in presence of the barrier, for the same source receiver configuration in the absence of wind. Positive values of IL indicate the predicted effectiveness of the noise barrier in reducing sound pressure levels in absence of wind. The screen-induced refraction of sound by wind, $SIROS$, is the sound pressure level with the noise barrier in the presence of wind, minus the sound pressure level with the noise barrier in absence of wind, for the same source–barrier–receiver configuration. Positive values of *SIROS* indicate the extent to which wind flow is predicted to reduce the barrier efficiency. The tree effect, *TE*, is defined as the sound pressure level in the presence of a noise barrier and wind, minus the sound pressure level in the presence of a noise barrier combined with trees and wind, for the same source–barrier–receiver configuration. Positive values of *TE* indicate the extent to which the presence of the canopies is predicted to increase shielding when there is wind, or in other words, the extent to which it is predicted that part of the reduction in barrier *IL* due to *SIROS* is counteracted.

With increasing distance, *IL* decreases while *SIROS* increases. Both indicators become equal at 230 m, implying that all shielding by the barrier in a still atmosphere would be lost by the action of the wind. Near to the noise barrier in downwind conditions, the *TE* is predicted to be positive for all canopy and gap configurations. However, in this zone, the presence of gaps is predicted to have slightly negative effects. At larger distances downwind, if crowns start at the top of the barrier, they are predicted to have negative effects. Trees planted such that there are gaps between the top of the barrier and bottom of the canopy are predicted to be the global optimal solution for the highway case considered. In general, this suggests that canopies might be optimized depending on where the receiver zone is located.

Wind-induced vegetation noise in trees could increase noise levels behind a barrier. Indeed, some species might be more effective in generating such sounds than others [73–75]. However, this is unlikely to cause a problem in practice. The sound of rustling leaves is a top-rated natural sound [76], whereas many mechanical sounds are disliked. Moreover, wind-induced vegetation noise can mask unwanted sounds. Another benefit is that a noise barrier hidden by trees is less visually intrusive and, moreover, 'greened' noise barriers have been found to be preferred regardless of their noise shielding performance [77].

10.7 TURBULENCE AND ITS EFFECTS

10.7.1 Turbulence Mechanisms

Two types of atmospheric instability, shear and buoyancy, are responsible for the generation of turbulence. Wind-driven flow causes shear instabilities and is associated with mechanical turbulence. High wind conditions and a small temperature difference between the air and ground are the primary causes of mechanical turbulence. Buoyancy or convective turbulence is associated with thermal instabilities. Convective turbulence prevails when the ground is much warmer than the overlying air, for example on a sunny day with little wind. Fluctuating irregularities in the temperature and wind fields cause scattering of sound in the atmosphere and local variations in the effective refractive index for sound waves. Consequently, sound propagating through a turbulent atmosphere fluctuates both in amplitude and phase as well as being scattered out of the line of sight between source and receiver. The amplitude of fluctuations in sound level caused by turbulence initially increases with increasing distance of propagation, sound frequency and strength of turbulence but, fairly quickly, it reaches a limiting value. This means that the fluctuation in overall sound levels from distant sources, such line-of-sight sound levels due to an aircraft a few km away, would have a standard deviation of no more than about 6 dB [78].

Humidity fluctuations can contribute to the variance in the refractive index also and are particularly important for sound propagation over warm wet ground, swamps and warm oceans. Although the effects of humidity fluctuations are not considered in this text they are investigated extensively elsewhere [79].

Turbulence can be visualized as a continuous distribution of eddies (see Chapter 1, Fig.1) in time and space. On a sunny afternoon, the largest eddies can extend to the height of the atmospheric boundary layer, i.e. up to 1–2 km and this represents the outer scale of turbulence. Initially, through inertial effects, the largest eddies break down into smaller eddies without energy loss. The size range of interest for outdoor sound propagation is called the inertial subrange. In this range, the kinetic energy in the larger eddies is transferred continuously to smaller ones. As the eddy size becomes smaller, virtually all their energy is dissipated into heat through viscosity. The length scale, at which viscous dissipation processes begin to dominate in atmospheric turbulence, is between 1 and 10 mm. In describing fluid flow, the ratio of inertial and viscous forces is represented by a dimensionless number called the Reynolds number (uL/ν, where u is velocity, L is a characteristic dimension and ν is the kinematic viscosity of the fluid). Turbulent flow is associated with high Reynolds numbers.

Different statistical characteristics of a sound field are affected by turbulent eddies of different sizes. For example, fluctuations in phase are affected by largest eddies with the size of the outer scale. For a boundary layer depth of 1 km, the eddy size responsible for phase fluctuations is on the order of 230 m. Log-amplitude fluctuations are affected by eddies in the inertial subrange and the coherence between signals received at two microphones is affected by eddy sizes on the order of the distance between them. The scattering of sound by turbulence is affected most by eddies with sizes on the order of the sound wavelength. The size of eddies of most importance to scattering may be estimated by assuming regularly spaced eddies and considering Bragg diffraction. For a sound with wavelength λ being scattered through angle θ (see Figure 10.27), the important scattering structures have a spatial periodicity D satisfying

$$\lambda = 2Dsin(\theta/2). \tag{10.16}$$

This relationship holds also for single scattering by random inhomogeneities [78].

At a frequency of 500 Hz and a scattering angle of 10°, this suggests an eddy size of 4 m.

Typically, during the daytime, the contribution of velocity fluctuations to the effective index of refraction variance always dominates over the contribution from temperature fluctuations. This is true, even on sunny days, when turbulence is generated by buoyancy rather than shear. Strong buoyant instabilities produce vigorous motion of the air. Situations where temperature fluctuations have a more significant effect on acoustic scattering than velocity fluctuations occur most often during clear still nights.

10.7.2 Models for Turbulence Spectra

There are several models for the size distribution (spectrum) of turbulent eddies, but no single model gives an accurate description of the entire turbulence spectrum in the atmospheric boundary layer. The three models of turbulence described here in turn are the von Kármán, Kolmogorov and Gaussian models. Temperature and velocity fluctuations give rise to different spectra. The spectra associated with velocity fluctuations are called energy spectra since they are associated with the kinetic energy of turbulence.

The von Kármán spectrum for velocity fluctuations is known to work reasonably well for high Reynolds number turbulence. The von Kármán spectrum associated with temperature fluctuations and the von Kármán energy spectrum associated with velocity fluctuations are given, respectively, by [79]

$$\Phi_{VK}(K)=\left[\frac{\Gamma(5/6)}{\pi^{3/2}\Gamma(1/3)}\right]\left[\frac{\sigma_T^2L_T^3}{\left(1+(KL_T)^2\right)^{11/6}}\right],$$
$$e_{VK}(K)=\left[\frac{55\Gamma(5/6)}{9\sqrt{\pi}\Gamma(1/3)}\right]\left[\frac{(\sigma_v^2L_v)(KL_v)^4}{\left(1+(KL_V)^2\right)^{17/6}}\right], \tag{10.17a,b}$$

where σ_T^2 and σ_v^2 are the variances of temperature and velocity fluctuations, respectively, Γ is the Gamma function, L_T is the von Kármán length representative of the outer scale of turbulence associated with temperature fluctuations and L_v is the von Kármán length scale associated with velocity fluctuations.

Kolmogorov spectra are considered to offer more realistic representations of the spectra of isotropic turbulence in the inertial subrange. Kolmogorov spectra associated with temperature fluctuations and velocity fluctuations are given, respectively, by [80]:

$$\Phi_K(K)=\frac{5\Gamma(5/6)}{9\pi^{3/2}\Gamma(1/3)}\frac{\sigma_T^2L_T^3}{(KL_T)^{11/3}},\quad e_K(K)=\frac{55\Gamma(5/6)}{9\sqrt{\pi}\Gamma(1/3)}\frac{\sigma_v^2L_v}{(KL_v)^{5/3}} \tag{10.18a,b}$$

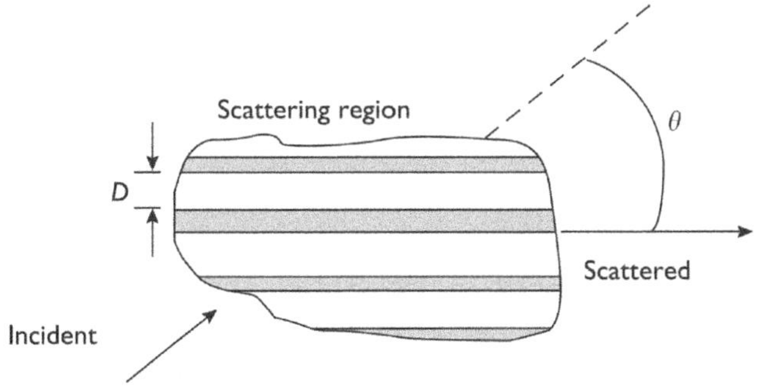

Figure 10.33 Bragg scattering from regularly spaced turbulence.

where the numerical coefficients have been adjusted to use the von Kármán length scales.

In the Gaussian model of turbulence statistics, the spectra associated with temperature fluctuations and velocity fluctuations are given, respectively, by [80]

$$\Phi_G(K) = \frac{\sigma_T^2 L_{GT}^3}{8\sqrt{\pi}} e^{-K^2 L_{GT}^2/4}, e_G(K) = \frac{K^4 L_{Gv}^4 \left(\sigma_v^2 L_{Gv}\right)}{8\sqrt{\pi}} e^{-K^2 L_{Gv}^2/4}, \qquad (10.19\text{a,b})$$

where L_{GT} and L_{G_v} are Gaussian length scales proportional to the inner length scales for temperature and velocity fluctuations, respectively. If it is assumed that the integral length scales of temperature and velocity fluctuations are the same for both the Gaussian and von Kármán models, then it is possible to deduce relationships between L_{GT} and L_T and between L_{Gv} and L_v. These are [80]

$$L_{GT} = \frac{2\Gamma(5/6)}{\Gamma(1/3)} L_T \approx 0.843 L_T,\ L_{Gv} = \frac{2\Gamma(5/6)}{\Gamma(1/3)} L_v \approx 0.843 L_v. \quad (10.20\text{a,b})$$

The temperature fluctuation spectra for the various models are normalized by dividing them by $\sigma_T^2 L_T^3$ and the velocity fluctuation (energy) spectra are normalized by dividing them by $\sigma_T^2 L_v$ [79].

Figures 10.34 (a) and (b) compare the normalized spectra as functions of KL_T and KL_v.

Over the entire range of KL_T, the normalized Kolmogorov spectrum has a power-law dependence of ~−11/3. In the inertial subrange (i.e. for large values of KL_T), the normalized Kolmogorov and von Kármán spectra of temperature fluctuations are practically coincident. For small values of KL_T, the von Kármán spectrum is nearly constant. Over the entire range of KL_v, the Kolmogorov spectrum has a power-law dependence of ~−5/3. The normalized Kolmogorov and von Kármán energy spectra are practically coincident for large KL_v. So, when the high wavenumber part of the spectrum is

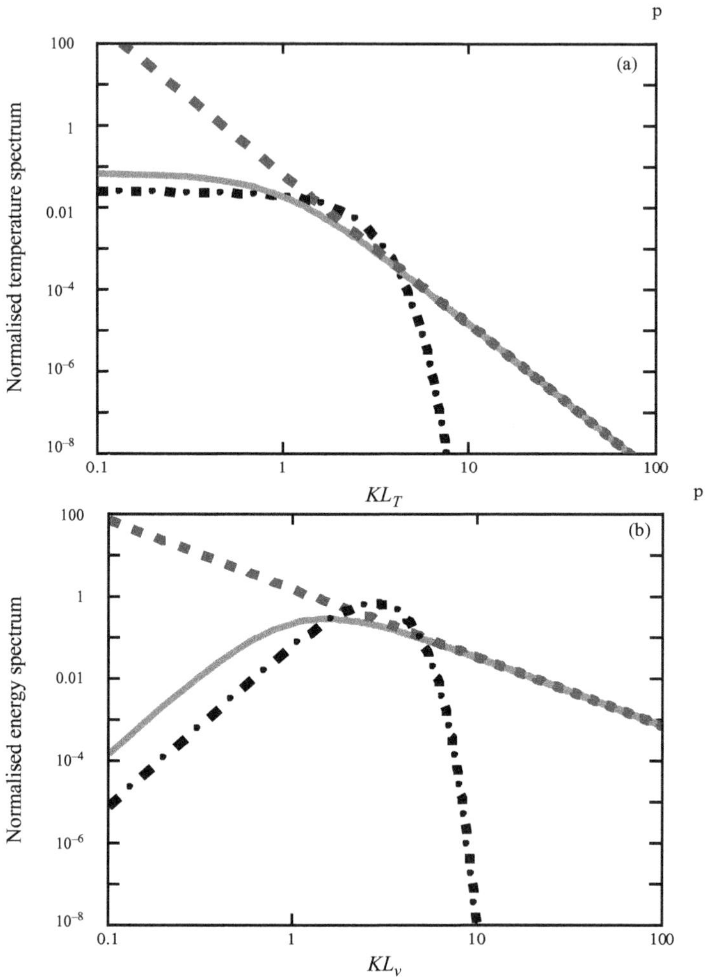

Figure 10.34 Normalised von Karman (solid lines), Kolmogorov (broken lines) and Gaussian (dash-dot lines) turbulence spectra (a) for temperature fluctuations (b) for velocity fluctuations.

mainly responsible for turbulence effects, there is little to choose between the von Kármán and Kolmogorov spectra for modelling outdoor sound propagation.

For small KL_T and KL_v, the normalized Gaussian spectra are similar to the normalized von Kármán spectra, i.e. nearly constant for temperature fluctuations and having the same power-law dependence ($\sim K^4$) for velocity fluctuations. Otherwise, except in limited ranges of KL, the normalized spectra predicted by the Gaussian model are significantly different from the normalized von Kármán and Kolmogorov spectra. On the other hand, the

Gaussian model can be useful since it allows many results to be obtained in simple analytical forms. As well as having been assumed in earlier theories for outdoor sound propagation [80,81], the Gaussian turbulence model is assumed in a stochastic inversion algorithm which calculates correlation functions for travel time fluctuations along sound propagation paths in acoustic tomography of the atmosphere [80].

The plots of normalized Gaussian spectra in Figures 10.34(a) and (b) have been obtained using the length scale relationships (10.20a,b). Although it is not possible for Gaussian and Kolmogorov spectra to be matched, the length scales and numerical coefficients for the normalized Gaussian spectra (10.19a,b) can be adjusted to match the normalized von Kármán spectra fairly closely in any spectral range of interest. If this happens to include the wavenumber range relevant to predicting turbulence effects on outdoor sound propagation, then the Gaussian spectrum may be satisfactory. Moreover, many calculations of turbulence effects on outdoor sound rely on estimated or best-fit values of turbulence parameters rather than measured values. Under these circumstances, there is no reason to assume spectral models other than Gaussian. Typically, the assumption of a Gaussian spectrum results in best-fit parameter turbulence values that are rather less than measured values. When measured values of turbulence parameters are available, it is more realistic to assume Kolmogorov or von Kármán spectra than Gaussian spectra of the turbulence.

10.7.3 Clifford and Lataitis Prediction of Ground Effect in Turbulent Conditions

The approach due to Clifford and Lataitis [34] (see equations 10.12a–10.12h) involves a straightforward extension of the classical theory for the sound field due to a point source over an impedance plane and assumes a Gaussian turbulence spectrum. It has been used extensively for predicting outdoor sound propagation in a non-refracting turbulent atmosphere near the ground [78,80,81]. The Clifford and Lataitis method for calculating the effects of turbulence on ground effect may be summarized as follows.

1. Estimate the outer scale of turbulence, L_0, and the variance of the index of refraction, $\langle \mu^2 \rangle$, or determine from measurements using

$$\left\langle \mu^2 \right\rangle = \frac{\sigma_v^2 \cos^2 \alpha}{c_0^2} + \frac{\sigma_T^2}{4T_0^2},$$

   where σ_v^2 is the variance of the wind velocity, σ_T^2 is the variance of the temperature fluctuations, α is the wind vector direction, c_0 and T_0 are the ambient sound speed and temperature, respectively. Typical values of $\langle \mu^2 \rangle$ are between 2×10^{-6} and 10^{-4} and a typical value of L_0 is the source height.
2. Determine the phase covariance, ρ, from equations (10.12e–10.12g).

3. Evaluate the turbulence coherence factor, T, from equations (10.12b) and (10.12c).
4. Evaluate the mean-squared pressure and the sound pressure level, L_p, from equation (10.12a) and $L_p = 10 \log \langle p^2 \rangle$, respectively.

Figure 10.35 shows example predictions of excess attenuation using this procedure for source and receiver heights of 1.8 m and 1.5m, respectively, separation 600 m, three values of $\langle \mu^2 \rangle$ and L_0 equal to source height. Increasing turbulence reduces the depth of the main ground effect dip.

10.7.4 Ostashev et al. Improvements on the Clifford and Lataitis Approach

The classical Clifford and Lataitis method [34] for incorporating turbulence effects in ground effect calculations, while ostensibly including wind-induced velocity fluctuations, does not allow for the vector nature of wind fluctuations and so, strictly, it is valid only for temperature fluctuations. Wind-induced turbulence is at least as important as temperature-induced turbulence [82]. Furthermore, there may be interaction between wind and temperature fluctuations (see (10.20a,b)). The method assumes a Gaussian turbulence spectrum

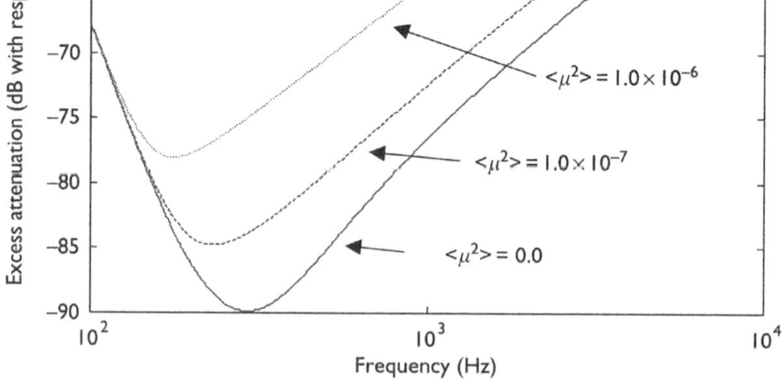

Figure 10.35 Excess Attenuation vs. frequency for a source and receiver above an impedance ground in acoustically neutral atmosphere for three values of $\langle \mu^2 \rangle$. Source and receiver heights are 1.8 m and 1.5 m respectively, and the separation is 600 m. A two-parameter impedance model (equation (5.38)) has been used with effective flow resistivity 30.0 kPa s m^{-2} and rate of change of porosity 0.0 m^{-1}.

and, as pointed out earlier (Section 10.6.2) and elsewhere [79,83], more realistic turbulence spectra are those of von Kármán and Kolmogorov. Also, the method does not allow for ground reflection of waves scattered by the turbulence. A further drawback is that the method assumes homogeneous isotropic turbulence whereas, more realistically, atmospheric turbulence is height dependent (inhomogeneous) and anisotropic.

Ostashev et al. [79,83] have shown that, in general, the coherence factor, T, introduced through equation 10.12a, is the same as the coherence function of a spherical sound wave for line-of-sight propagation. Their result for the Gaussian spectrum turbulence coherence factor, T_G, equivalent to, but more accurate than that given by equations (10.12a–10.12h), is

$$T_G = \exp\left\{\begin{array}{l} -2A\gamma_T^G R\left[1-\dfrac{\sqrt{\pi}}{4D_G}erf\left(D_G\right)\right] \\ -2A\gamma_v^G R\left[1-\dfrac{\sqrt{\pi}}{4D_G}erf\left(D_G\right)-\dfrac{1}{2}e^{-D_G^2}\right]\end{array}\right\}, \qquad (10.21a)$$

where $D_G = h/L_0$, erf(x) is the error function (see equation (10.12g). $\gamma_T^G = \sqrt{\pi}k_0{}^2\sigma_T^2 L_T / 8T_0{}^2$ and $\gamma_v^G = \sqrt{\pi}k_0{}^2\sigma_v^2 L_v / 8c_0{}^2$ are the extinction (attenuation) coefficients of the mean sound field due to temperature fluctuations and wind velocity fluctuations with a Gaussian spectrum, respectively. Expressions (10.21a) and (10.12b–10.12e) are identical if $\sigma_v^2 = 0$.

Figure 10.36 compares predictions of coherence factor using equations (10.12a–10.12h) and (10.21a) for equal source and receiver heights of 1 m and a separation of 100 m assuming a refractive index variance of 10^{-5} and using equations (10.12a–10.12h) and that this total variance includes equal contributions from thermal and velocity fluctuations for predictions using equation (10.21a). A Gaussian scale length of 1 m as in equation (10.12a–10.12h) has been used for thermal turbulence in (10.21a), but four scale lengths (1 m, 3 m, 5 m and 10 m) have been used for the velocity fluctuations. With all scale lengths equal to 1 m, equation (10.20a,b) predicts slightly higher coherence than (10.12a–10.12h). The predictions of (10.21a) and (10.12a–10.12h) are practically coincident when the scale length for velocity fluctuations is increased to 3 m. The other curves show that the equation (10.21a) predictions of coherence factor are significantly influenced by the scale length for velocity fluctuations.

The coherence factor expression for a Kolmogorov turbulence spectrum is [84]

$$T_K = \exp\left(-k^2 h^{\frac{5}{3}} L \frac{3BC_T^2}{8T_0^2} - k^2 h^{\frac{5}{3}} L \frac{11BC_v^2}{4c_0^2}\right), \qquad (10.21b)$$

where $B \approx 0.364$, $C_T^2 = \dfrac{3\Gamma(5/6)}{\pi^{1/2}}\dfrac{\sigma_T^2}{L_T^{2/3}}$, $C_v^2 = \dfrac{3\Gamma(5/6)}{\pi^{1/2}}\dfrac{\sigma_v^2}{L_v^{2/3}}$, L_T and L_v are the length scales of temperature and velocity fluctuations, respectively.

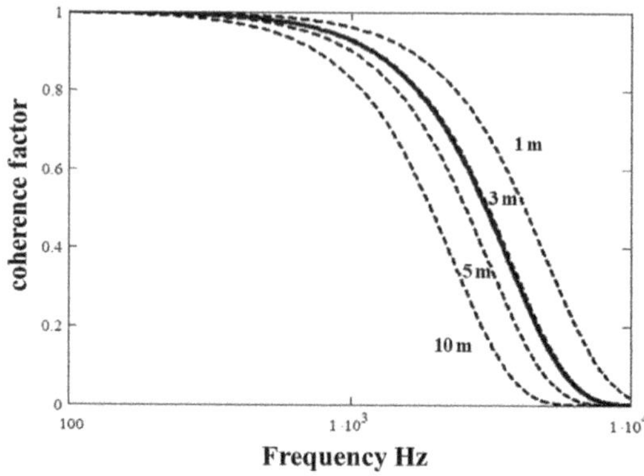

Figure 10.36 Comparison of coherence factor predictions assuming a Gaussian turbulence spectrum for source and receiver heights of 1 m and separation 100 m, Gaussian scale length 1 m: equations (10.12a–10.12h) – solid line; equation (10.18a,b)–broken lines, assuming scale length of 1 m for thermal fluctuations and labelled with the assumed values of the scale length for velocity fluctuations.

The coherence factor expression for a von Kármán turbulence spectrum is [80]:

$$T_{VK} = \exp\left(\begin{array}{l} -\dfrac{2L}{K_0 h}\displaystyle\int_0^{K_0 h} dt \left[\gamma_T^{VK} \left\{ 1 - \dfrac{2^{\frac{1}{6}} t^{\frac{5}{6}}}{\Gamma(5/6)} K_{5/6}(t) \right\} \right] + \\ \gamma_v^{VK} \left\{ 1 - \dfrac{2^{\frac{1}{6}} t^{\frac{5}{6}}}{\Gamma(5/6)} \left(K_{5/6}(t) - \dfrac{t}{2} K_{1/6}(t) \right) \right\} \end{array} \right), \qquad (10.21c)$$

where $K_n(t)$ is the modified Bessel function of order n, $\gamma_T^{VK} = 3\pi^2 A k_0^2 K_0^{-5/3} C_T^2 / 10 T_0^2$ is the extinction (attenuation) coefficient of the mean sound field due to temperature fluctuations with the von Kármán spectrum of turbulence, $\gamma_v^{VK} = 6\pi^2 A k_0^2 K_0^{-5/3} C_v^2 / 5 c_0^2$ is the extinction coefficient of the mean sound field due to wind velocity fluctuations with the von Kármán spectrum of turbulence and $A \approx 0.033$ (see also equations (10.28a) and (10.28b)).

Calculations assuming von Kármán turbulence spectra have been used to distinguish between turbulence effects in sunny and cloudy conditions and show that turbulent sunny conditions lead to the largest decrease in the magnitudes of destructive interferences associated with ground effect [79]. However, the relationships discussed so far assume that turbulence is independent of height above ground and isotropic.

10.7.5 Height Dependence of Turbulence

When sound waves propagate nearly horizontally in the atmosphere, the variance in the effective index of refraction $\langle\mu^2\rangle$ is related approximately to those in velocity and temperature by [65]

$$\langle\mu^2\rangle = \frac{\langle u'^2\rangle}{c_0^2}\cos^2\varphi + \frac{\langle v'^2\rangle}{c_0^2}\sin^2\varphi + \frac{\langle u'T\rangle}{c_0}\cos\varphi + \frac{\langle T^2\rangle}{4T_0^2}, \qquad (10.22a)$$

where T, u' and v' are the fluctuations in temperature, horizontal wind speed parallel to the mean wind and horizontal wind speed perpendicular to the mean wind, respectively. ϕ is the angle between the wind and the wave-front normal.

Use of similarity theory (see Chapter 1) gives [61]

$$\langle\mu^2\rangle = \frac{5u_*^2}{c_0^2} + \frac{2.5u_*T_*}{c_0T_0}\cos\varphi + \frac{T_*^2}{T_0^2}, \qquad (10.22b)$$

where u_* is the friction velocity and $T_* = -Q/(\rho_0 C_p u_*)$ is the scaling temperature (Q being the surface temperature flux, ρ_0 the density of air and C_p the specific heat of air at constant pressure), respectively. Although the third term (the covariance term) in (10.16b) could be at least as important as the temperature term, often it has not been included when predicting acoustic propagation (see equation 10.12b and Section 10.6.3).

In general, turbulence spectra depend on height. The height dependence can be modelled using the Monin–Obukhov similarity theory described in Chapter 1.

Estimates of the fluctuations in terms of u_* and T_* and the Monin–Obukhov length, $L = -\left(u_*^3 T_s \rho_0 C_p\right)/\left(gK_v Q\right)$, where T_s is surface temperature, g is the acceleration due to gravity and K_v is the von Kármán constant (= 0.4), are [84]:

For $L > 0$ (stable conditions, e.g. at night),

$$\sqrt{\langle u'^2\rangle} = \sigma_u = 2.4u_*,\ \sqrt{\langle v'^2\rangle} = \sigma_v = 1.9u_*,\ \sqrt{\langle T^2\rangle} = \sigma_T = 1.5T_*$$

For $L < 0$ (unstable conditions, e.g. daytime),

$$\sigma_u = \left(12 - 0.5\frac{z}{L}\right)^{1/3} u_*,\ \sigma_v = 0.8\left(12 - 0.5\frac{z}{L}\right)^{1/3} u_*,\ \sigma_T = 2\left(1 - 18\frac{z}{L}\right)^{-1/2} T_*$$

More accurate expressions for the height dependencies of the variance and length scale of temperature fluctuations obeying the von Kármán spectrum (10.14a) are [79]:

$$\sigma_T(z) = 2\left[1 - 10(z/L)\right]^{-1/3} T_*,\ L_T(z) = 2z\frac{\left[1 - 7(z/L)\right]}{\left[1 - 10(z/L)\right]} \qquad (10.23a,b)$$

Corresponding height dependencies of the variance and length scale of velocity fluctuations are given by [79]

$$\sigma_v = \sqrt{3}u_*, Lv = 1.8z.$$

Strictly, these relationships are valid only for turbulence that is exclusively shear driven since convective turbulence does not observe Monin–Obukhov similarity. More generally a mixed similarity condition applies in which the Monin–Obukhov length is replaced by the height of the atmospheric boundary layer, z_i, i.e.

$$\sigma_v = \sqrt{0.35}w_*, \ L_v = 0.23z_i,$$

where $w_* = \dfrac{z_i g Q}{\rho_0 C_P T_s}$. This represents the main contribution to sound-level fluctuations unless the surface heat flux is zero or near zero.

A widely used approach adds spectra calculated using both the similarity and mixed similarity relationships [79].

10.7.6 Turbulence-Induced Phase and Log-Amplitude Fluctuations

As well as reducing the coherence between direct and ground-reflected sound as discussed in Sections 10.6.3 and 10.6.4, the presence of turbulence causes fluctuations in the phase and amplitude of sound propagating through it. For line-of-sight propagation in the atmosphere through turbulence with a Gaussian spectrum, the mean squared fluctuations in the phase and log-amplitude of a plane sound wave are [60]

$$\begin{aligned}\langle\phi^2\rangle &= \frac{\sqrt{\pi}k_0^2 x}{8}\left[\left(1+\frac{\arctan D_T}{D_T}\right)\frac{\sigma_T^2 L_{GT}}{T_0^2}+\left(1+\frac{1}{1+D_v^2}\right)\frac{4\sigma_v^2 L_{Gv}}{c_0^2}\right],\\ \langle\chi^2\rangle &= \frac{\sqrt{\pi}k_0^2 x}{8}\left[\left(1-\frac{\arctan D_T}{D_T}\right)\frac{\sigma_T^2 L_{GT}}{T_0^2}+\left(1-\frac{1}{1+D_v^2}\right)\frac{4\sigma_v^2 L_{Gv}}{c_0^2}\right],\end{aligned} \qquad (10.24\text{a,b})$$

where x is the range, $D_T = 4x/k_0 L_{GT}^2$, $D_v = 4x/k_0 L_{Gv}^2$, L_{GT}, L_{Gv} are integral length scales for a Gaussian spectrum (see (10.19a,b)) and σ_T^2, σ_v^2 are variances of temperature and velocity fluctuations, respectively.

If $D_T \gg 1$ and $D_v \gg 1$ (long range and low frequency), these relationships simplify to

$$\langle\phi^2\rangle \approx \langle\chi^2\rangle \approx \frac{\sqrt{\pi}k_0^2 x}{8}\left[\frac{\sigma_T^2 L_{GT}}{T_0^2}+\frac{4\sigma_v^2 L_{Gv}}{c_0^2}\right]. \qquad (10.25)$$

For $D_T \gg 1$ and $D_v \gg 1$, this plane wave result is valid also for spherical wave propagation [79].

10.7.7 Scattering by Turbulence

Apart from the degradation of ground effect and the fluctuation in sound levels caused by turbulence, turbulent eddies scatter sound into the deep shadows that would otherwise form behind barriers or close to the ground during propagation in upward-refracting conditions. The scattering of sound behind a noise barrier can be calculated from the scattering cross section of atmospheric turbulence. Assuming the Kolmogorov turbulence spectrum and ignoring the effects of humidity fluctuations, the scattering cross section σ_s (in reciprocal metres) as a function of scattering angle, θ, measured from the direction of propagation (see Figure 10.33) is [79]

$$\sigma_s = 4.08\times10^{-3}k_0^{1/3}\cos^2\theta\left[\frac{C_T^2}{T_0^2}+\left(\frac{22C_v^2}{3c_0^2}\right)\cos^2\left(\frac{\theta}{2}\right)\right]\cdot\left[2\sin\left(\frac{\theta}{2}\right)\right]^{-11/3} \qquad (10.26)$$

where k_0 is the initial wavenumber, $C_T^2 = \dfrac{3\Gamma(5/6)}{\pi^{1/2}}\dfrac{\sigma_T^2}{L_T^{2/3}}$, $C_v^2 = \dfrac{3\Gamma(5/6)}{\pi^{1/2}}\dfrac{\sigma_v^2}{L_v^{2/3}}$, and, as before, L_T and L_v are the length scales of temperature and velocity fluctuations respectively. This relationship is not valid close to $\theta = 0$ i.e. for forward scattering.

10.7.8 Decrease in Sound Levels Due to Turbulence

Since turbulent eddies scatter sound out of the direct path between source and receiver there is a decrease in mean sound levels with range, i.e. an effective attenuation of line-of-sight propagation due to redirection of sound energy. For a sound wave travelling in the positive x-direction

$$\langle p(x,r) = \exp(-\gamma x)p^0(x,\mathbf{r}) \qquad (10.27)$$

where γ, the extinction coefficient, is the sum of the extinction coefficients of temperature and velocity fluctuations, $\mathbf{r}$ is the vector of the transverse coordinates (y,z) of the receiver location and $p^0(x,\mathbf{r})$ is the sound field in the absence of turbulence. For a von Kármán turbulence spectrum [79]

$$\gamma = \gamma_T + \gamma_v = \frac{\pi^{\frac{1}{2}}\Gamma(5/6)k_0^2}{4\Gamma(1/3)}\left(\frac{\sigma_T^2 L_T}{T_0^2}+\frac{4\sigma_v^2 L_v}{c_0^2}\right), \qquad (10.28a)$$

where Γ is the Gamma function, σ_T^2 and σ_v^2 are the variances of temperature and velocity fluctuations respectively, which depend on height as discussed in 10.6.5, L_T and L_v are height dependent length scales associated with temperature fluctuations and velocity fluctuations. The Kolmogorov turbulence spectrum leads to a prediction of an infinite extinction coefficient and zero mean field [79]. For Gaussian spectra [79]

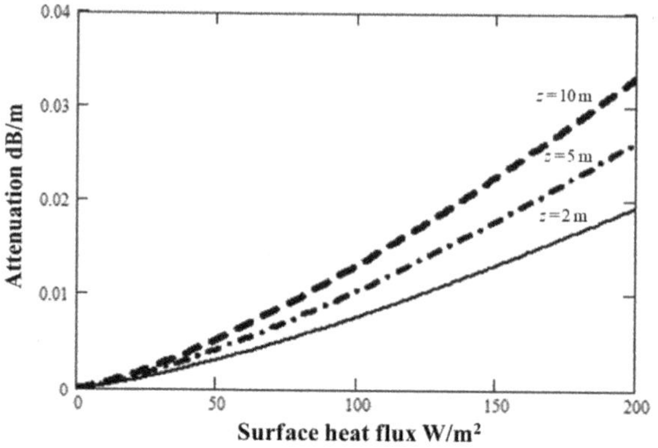

Figure 10.37 Prediction reduction of line-of-sight sound levels (dB/m) at 500 Hz due to scattering by turbulence as a function of surface heat flux and sound path height for a friction velocity of 0.3 m/s and a surface temperature of 20°C.

$$\gamma == \frac{\pi^{\frac{1}{2}} k_0^2}{8}\left(\frac{\sigma_T^2 L_{GT}}{T_0^2} + \frac{4\sigma_v^2 L_{Gv}}{c_0^2}\right), \tag{10.28b}$$

where L_{GT} and L_{Gv} are related to L_T and L_v through equation (10.17a,b). With these substitution (10.26) is identical.

Figure 10.37 shows predictions of attenuation (dB/m) due to von Kármán or Gaussian turbulence spectra as functions of surface heat flux (Q) and sound path height (z) above ground. The predictions assume frequency of 500 Hz, a surface temperature of 20°C and friction velocity $u_* = 0.3$ m/s which is representative of moderate wind. The predicted attenuation increases with path height and with surface heat flux. Low values of heat flux correspond to cloudy conditions and high values to sunny conditions. Although the predicted attenuation increases with friction velocity, it is less sensitive to friction velocity than to path height and surface heat flux.

A relationship between the extinction coefficient (equation (10.25)) and the phase and log amplitude fluctuations, valid for plane and spherical waves, is [79]

$$\gamma = \frac{\langle \varnothing^2 \rangle + \langle \chi^2 \rangle}{2x}. \tag{10.29}$$

10.7.9 Measurement of Turbulence

The measurement of the effective refractive index variance requires measurement of wind velocity and temperature fluctuations. Hot wire and sonic anemometers may be used to measure wind speed fluctuations, giving readings at intervals of between 0.05 s and 0.01 s. Temperature variation is monitored by means of thermometers (thermistors).

The log-amplitude pressure variance of a spherical sound wave is given by

$$\left\langle \chi^2 \right\rangle = 0.031 C_{eff}^2 k_0^{7/6} x^{11/6} \tag{10.30}$$

where x is the source–receiver separation (range) and C_{eff}^2 is the effective structure-function parameter that combines those of velocity (C_v^2) and temperature (C_T^2) fluctuations into a single parameter. Normally the effective structure-function parameter is assumed to have the form

$$C_{eff}^2 = \frac{C_T^2}{T_0^2} + a_v \frac{C_v^2}{c_0^2} \tag{10.31}$$

where T_0 is the average temperature, c_0 is the average sound speed and a_v = 22/3 79]. Typically, the contribution of temperature fluctuations to C_{eff}^2 is small so that

$$C_{eff}^2 \approx a_v \frac{C_v^2}{c_0^2}.$$

C_v^2 can be determined from hot wire or sonic anemometer readings. One method of doing this uses readings from two sensors placed along the mean wind direction [85]. Hence

$$C_v^2 = \frac{\left\langle u(t+\Delta t) - u(t) \right\rangle}{(U\Delta t)^{2/3}}$$

where the sensor spacing $\Delta r = U\Delta t$, t is time and U is the mean wind speed.

10.7.10 Inclusion of Atmospheric Turbulence in the Fast Field Program

The FFP formulation does not include the effect of turbulence. In a turbulent medium the complex sound field fluctuates in amplitude and phase. The time-averaged pressure squared, in terms of the complex pressure and its complex conjugate, is

$$\left\langle \overline{P^2(r,z)} \right\rangle = \frac{1}{2} Re\left(P(r,z)\bar{P}(r,z)\right) \tag{10.32}$$

The bar above a variable or parameter signifies its complex conjugate. The turbulence effect is described by equations (10.12a–10.12h) which assume a Gaussian spectrum of turbulence.

The parameter h represents the maximum transverse separation between a pair of paths and depends on values of the wavenumber K for each wave component. An approximate formula for h between waves corresponding to values of $K_n = n\delta K$ and $K_m = m\delta K$ is given by [86]

$$h = \frac{r}{2}\left[\left(\frac{n'-m}{n'+m}\right)^{1/2} - \left(\frac{n'-n}{n'+n}\right)^{1/2}\right] \tag{10.33}$$

where n' is the integer corresponding to $k_0(=\omega/c) = n'\delta K$.

In integral form the averaged pressure is given by

$$\left\langle \overline{P^2(r,z)} \right\rangle = \frac{1}{2} Re \left[\begin{array}{l} \frac{1}{\pi r} \int_0^\infty \int_0^\infty p(z,K) p(z,K') \\ \exp\left[i(K-K')r \right] T(K,K') \sqrt{KK'} dK dK' \end{array} \right] \tag{10.34}$$

The function $T(K,K')$ is the average effect of the decorrelation in phase and amplitude between the wavenumber components K and K'.

Evaluating this time-averaged pressure is much more costly than the unperturbed pressure P. The computation of this expression becomes somewhat easier if one breaks the integral into its coherent and incoherent components [87]

$$\left\langle \overline{P^2(r,z)} \right\rangle = \frac{1}{2} Re \left[e^{-\sigma^2} \frac{1}{\pi r} \int_0^\infty \int_0^\infty p(z,K) p'(z,K') exp\left[i(K-K')r \right] \sqrt{KK'} dK dK' \right]$$

$$+ \frac{1}{2} \mathrm{Re} \left[\begin{array}{l} \frac{1}{\pi r} \int_0^\infty \int_0^\infty p(z,K) p'(z,K') exp\left[i(K-K')r \right] \\ \left(T(K,K') - e^{-\sigma^2} \right) \sqrt{KK'} dK dK' \end{array} \right] \tag{10.35}$$

The first integral is $\exp(-\sigma^2)$ times the unperturbed mean pressure squared. The unperturbed pressure can be evaluated efficiently by the conventional FFP. This represents the coherent component reduced by the scattered energy.

The second integral is the contribution of the scattered energy from one wavenumber to another. Evaluating this double integral is slightly faster since $(T(K,K') - e^{-\sigma^2})$ is only large for $K = K'$. Moreover, once the integrals are approximated by discrete sums, the summation needs to be performed up to a value of $K = k_0$. Therefore, this component must be evaluated as an $n' \times n'$ double summation where n' corresponds to the value of K nearest to k_0. This is where the main computational effort is spent and the simplifying features of the FFT approximation are no longer available. One further complication is, that to obtain a converging sum, it is necessary to evaluate $p(z,K)$ at many more points than required in the conventional FFP (up to eight times more).

The complex kernel function, $p(z,K)\sqrt{K}$, and its conjugate are evaluated and output from an FFP program. The first integral is evaluated by the FFP program also. The second integral is evaluated by a simple double loop up to the value of n'.

10.7.11 Comparisons with Experimental Data

An example comparison of predictions of the FFP including turbulence with single tone data obtained as a function of range in upward-refracting conditions is shown in Figure 10.38. The classical data shown were obtained upwind i.e. during a strong negative wind speed gradient [27]. The source

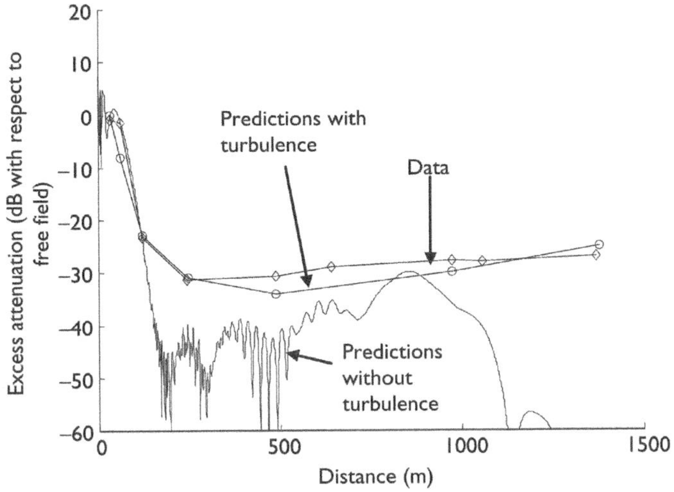

Figure 10.38 A comparison of the FFP modified to include turbulence (joined circles) with data at 424 Hz (joined diamonds) out to 1.5 km range under upwind conditions (data from [88]).

and the receiver heights were 3.66 m and 1.52 m respectively. A logarithmic wind speed profile, deduced by Gilbert *et al* [88] from the meteorological data, is given by

$$C(z) = 340.0 - 2.0\ln\left(z/6\times10^{-3}\right) \tag{10.36}$$

The agreement between the predictions from the FFP including turbulence and the data, while not perfect, represents a large improvement over the results of calculations without turbulence (shown by the solid line) and is similar to that achieved by predictions obtained by a Parabolic Equation method [88,89]. In the FFP computations a total of 400 layers were assumed with a ceiling of 100 m.

A second comparison is with data obtained during tests using a fixed jet engine source (centre of exit nozzle at 2.16 m above ground) at a grass-covered airfield near Hucknall, Notts, UK. During these measurements, there was a negative temperature gradient but no discernible wind speed gradient. Figure 10.39 (a) shows that the temperature profile can be approximated by a two-segment 'linear' profile.

$$T = 22.0 - 0.1z \quad z \le 16.5\text{m} \quad T = 20.55 - 0.0122z \quad z > 16.5\text{m} \tag{10.37}$$

Figure 10.39 (b) shows that the measured wind speed had a large spread of values with height with a mean value ~ 4 ms^{-1}. The lack of a definite wind gradient suggests a strongly turbulent atmosphere. For the corresponding predictions, the wind speed was assumed to be constant with height.

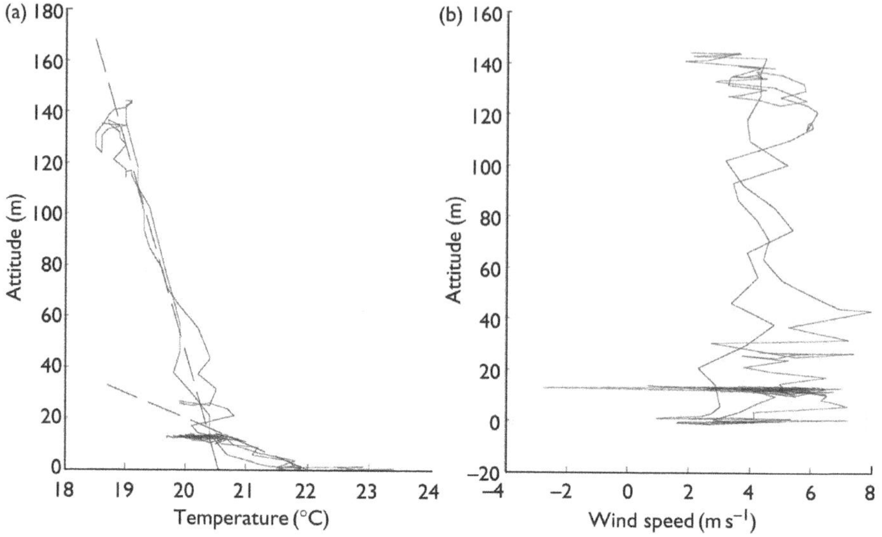

Figure 10.39 (a) Measured and assumed temperature profiles for Hucknall data (b) Measured wind speed profiles in the direction of the line of microphones.

Figure 10.40 compares the measured level difference between 6.4 m high microphones located at 152.4 m and 762 m from the Avon engine source. The lines represent four different 26 second averages of measured data. The diamonds represent predictions of the FFP including effects of turbulence and the crosses represent predictions without turbulence. The agreement is reasonable. Turbulence has a significant influence at frequencies higher than 100 Hz.

10.7.12 Including Turbulence in FDTD Calculations

Although computational fluid dynamics (CFD) simulations produce momentary turbulence fields that can be directly used in a volume discretization technique like the finite-difference time-domain (FDTD) method, the huge computational cost involved is still an important barrier for their wide use. Moreover, the desired spatial resolution in FDTD is often too demanding.

In outdoor sound propagation modelling, it is common to assume ‘frozen turbulence’ whereby the sound propagation medium is perturbed by local variations in the flow velocity and temperature in such a way that the statistics of the turbulent realizations follow specific turbulence models or approach experimental data. By propagating through enough realizations of the turbulence fields, it is possible to predict, for example, the statistics of the variation in (complex) sound pressure level. Not only scattering by atmospheric turbulence but also loss of coherence can be included in this way. An underlying assumption is that the sound speed is much higher than temporal variations in the atmospheric flow and temperature fields.

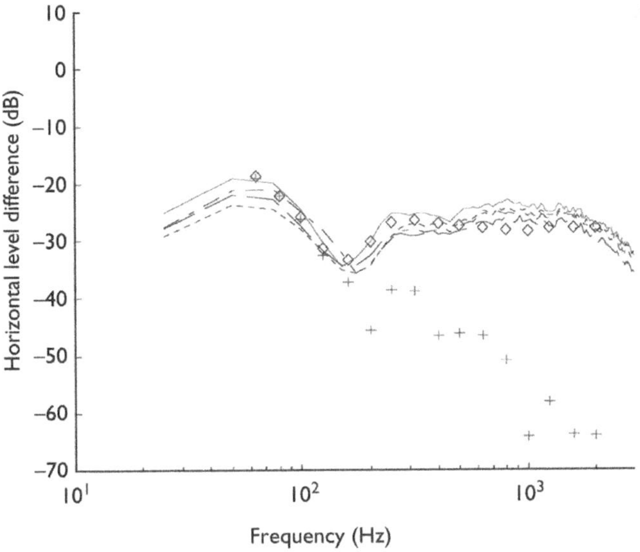

Figure 10.40 Measured and predicted level difference spectra between horizontally separated positions at a height of 6.4 m and distances of 152.4 m and 762 m from a jet engine source at a height of 2.16 m. The diamonds are predictions of the FFP with turbulence at third octave centre frequency values (63Hz–2kHz) and lines represent four different 26 second averages of measured data. A Gaussian spectrum of turbulence was assumed with values of the variance of index of refraction ($\langle\mu^2\rangle$) and the outer scale of turbulence (L_0) equal to 8.0×10^{-6} and 1.1 m respectively. The crosses represent FFP predictions without turbulence.

Although turbulent fields can be realised in various ways, it is relatively easy to use the 'turbule' or quasi-wavelet approach [89,90] in FDTD calculations. The combination of the turbule method and the FDTD technique enables the inclusion of scattering at all propagation angles and multiple scattering by atmospheric turbulence. Also, this approach may be combined with the effect of a moving and inhomogeneous atmosphere on sound propagation (see Chapter 4).

Usually, wind velocity fluctuations dominate over temperature fluctuations in the atmosphere [91] and influence acoustic scattering to a larger extent. So, it is convenient to give results here that allow only for turbulent wind velocity fluctuations. But a similar approach can be used to predict propagation through turbulence caused by temperature fluctuations [90].

A turbulent flow field is constructed by randomly positioning randomly oriented eddies (or 'turbules') of different sizes. Such flow fields might be superimposed on averaged non-turbulent fields. Relations between the parameters at different scales have been derived to obtain either a Kolmogorov spectrum [89] or a von Kármán spectrum [93] (see Section 6.2). Since, in the inertial subrange both spectra are almost identical, only the relations based on the Kolmogorov spectrum are considered here.

A single velocity turbule is characterized by the location of its midpoint b, length scale a_α at scale α and an envelope function describing the decrease of the velocity with distance from the midpoint. A single velocity turbule using a Gaussian envelope function is calculated from [90]:

$$v = \Omega_\alpha \times (r - b) e^{-\frac{|r-b|^2}{a_\alpha^2}}, \tag{10.38}$$

where $\boldsymbol{r}$ indicates a position and $\boldsymbol{\Omega}_\alpha$ is the angular velocity vector at scale α.

The orientation of the angular velocity vector is random, while its amplitude Ω_α depends of the scale of turbulence. Equation (10.38) gives rise to a spherical and symmetrical solenoidal flow. For a homogeneous distribution of the midpoints of the turbules, the Kolmogorov turbulence spectrum in the inertial range will be obtained [90].

The general turbule parameters, the magnitude of the angular velocity Ω_1 and the length scale of the first scale a_1 are linked to the structure velocity parameter C_v^2 of a Kolmogorov spectrum [90]:

$$\Omega_1 = \sqrt{\frac{C_v^2}{0.34 \frac{\varphi}{\mu} a_1^{4/3} \Gamma(17/6)}}, \tag{10.39}$$

where Γ is the gamma function, φ is the packing fraction and μ is a parameter that describes the relation between successive length scales.

The packing fraction is calculated from

$$\varphi = n_\alpha a_\alpha^3, \tag{10.40}$$

where n_α is the number density of turbules with size a_α.

Successively smaller turbule sizes are related to each other by

$$\frac{a_{\alpha+1}}{a_\alpha} = e^{-\mu}. \tag{10.41}$$

The structure velocity parameter C_v is a general characterization of the strength of turbulence. For the Kolmogorov spectrum of turbulence, this parameter is related to the outer length scale L_0 and the variance of velocity fluctuations in one direction σ_v^2 (since turbulence is assumed to be isotropic) [91]:

$$C_v^2 = \frac{2L_0^{\frac{2}{3}}}{3\sigma_v^2}. \tag{10.42}$$

The relation between the angular velocity at different scales is given by the following expression:

$$\Omega_\alpha = \Omega_1 \left(\frac{a_1}{a_\alpha}\right)^{\frac{2}{3}}. \tag{10.43}$$

To obtain a homogenous turbulent field, the part of the atmosphere under consideration is divided at each scale in a number of cubic cells of dimension d_α. In each cell, one turbule is placed at a random location. The size of the cubic cells are related to the length scale:

$$d_\alpha = n_\alpha^{-\frac{1}{3}} \tag{10.44a}$$

or

$$d_\alpha = a_\alpha \varphi^{-\frac{1}{3}} \tag{10.44b}$$

In this approach, the dimensions of the computational domain will limit the largest scale of turbulence that can be taken into account. The smallest scale of turbulence must be greater than the spatial discretization step in order to produce a non-divergent, rotating turbule. This implies that the FDTD grid will limit the range of scales that can be included. However, it is mainly the eddies with sizes comparable to the wavelength of the sound that are important when quantifying scattering. If, as is common in FDTD applications, 10 computational cells per wavelength are used, the relevant scales of turbulence should be present.

As an example, four realizations of such artificially generated turbulent kinetic energy fields are shown in Figure 10.41, for a C_v^2 value equal to 0.01 $m^{4/3}/s^2$. The various length scales can be easily identified.

In the above approach, homogeneous and isotropic turbulence is assumed. Usually, both conditions are not met in the atmospheric boundary layer. Wind speed fluctuations along the wind direction are typically larger than those orthogonal to it. Moreover, the degree of turbulence will increase as the wind speed increases and the outer scale of turbulence is proportional to the height above the ground surface (see Section 10.7). So, it can be concluded that in a refracting atmosphere, a homogenous and isotropic turbulent flow field will only be an approximation. If information is needed on the time-dependency of amplitude and frequency (e.g. to model spectral broadening of acoustic signals), the advection of eddies must be accounted for as well [90].

A possible way to link the turbule model to localized regions with higher turbulence (e.g. near obstacles) is scaling the angular velocities of the turbules based on local values of the turbulent kinetic energy [94,95]. For example, such turbulent kinetic energy fields can be calculated with a standard k-ε turbulence closure CFD model.

10.8 EQUIVALENCE OF TURBULENCE AND SCATTERING INFLUENCES ON COHERENCE

Even though the starting equations for sound propagation through a forest and through a turbulent atmosphere are different, the equations for the correlation function of the sound field have the same form. Consequently, if

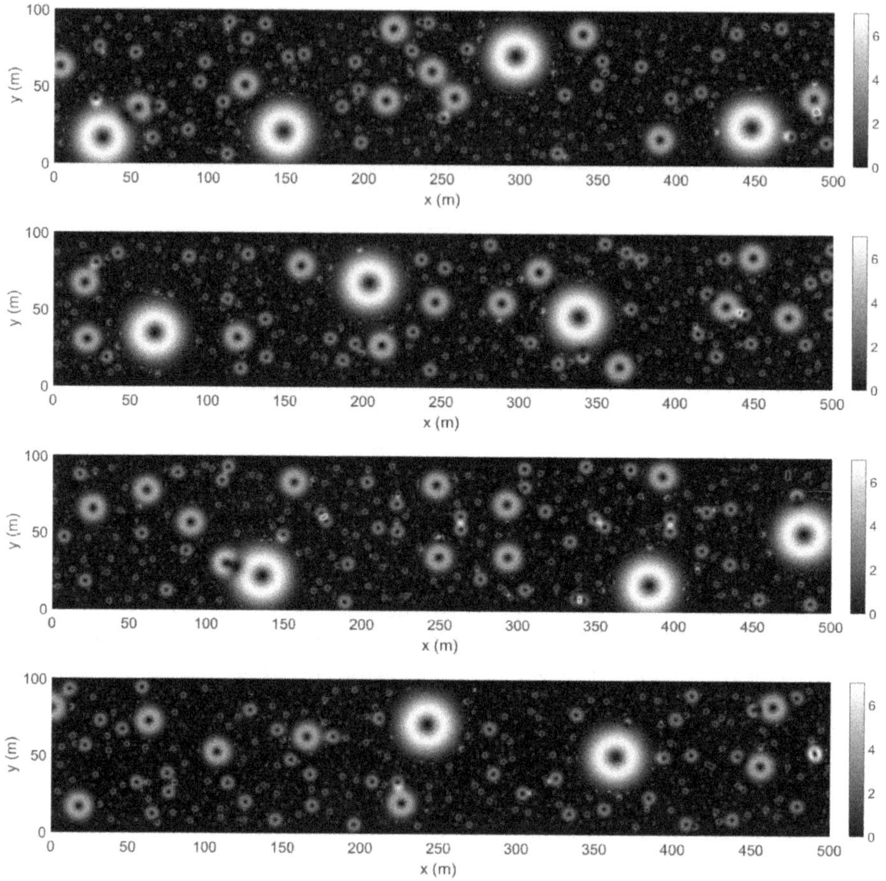

Figure 10.41 Four turbulent kinetic energy field realisations (in m²/s²) employing the turbule theory for flow velocity turbulence (homogeneous and isotropic turbulence; solenoidal eddies, two-dimensional representation). A Kolmogorov spectrum is used, with $C_v^2 = 0.01\,\mathrm{m}^{4/3}\,\mathrm{s}^{-2}$ and eight length scales have been considered.

there is mainly forward scattering, and after employing a full 3-D model of scattering by cylinders of finite length (see Section 10.2.2.2), then the influence of scattering by tree trunks on the sound field can be described by equation (10.12a), with [96]

$$T = \exp\left[-2rNbF\left(h_{SR} / h\right)\right], \tag{10.45}$$

where r is the distance, N is the number of trees per unit area, b is the tree radius, $h_{SR} = 2h_S h_R/(h_S + h_R)$, h_S and h_R being source and receiver heights, respectively, h is the tree height and

$$F\left(h_{SR} / h\right) = \frac{2}{\pi}\int_0^{\infty}\left[1 - \mathrm{sinc}\left(2\eta h_{SR} / h\right)\right]\mathrm{sinc}^2\eta\, d\eta. \tag{10.46}$$

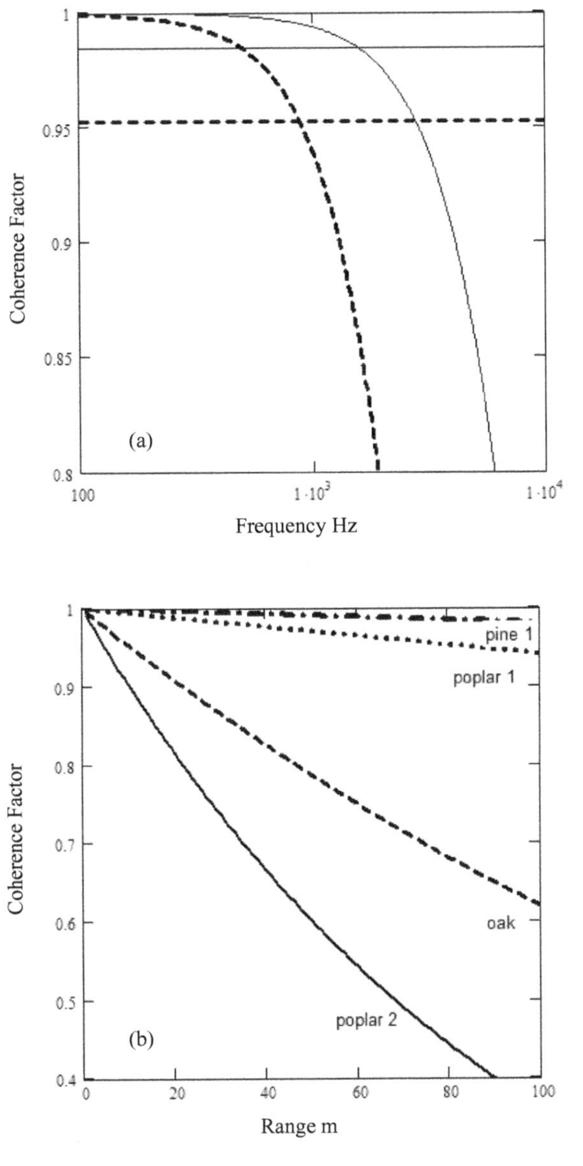

Figure 10.42 (a) Predicted frequency-independent coherence factors (using Equation (10.45)) at a receiver height 1.2 m at a range of 80 m from a source height 0.8 m associated with trunk scattering, with tree height 10 m, number per unit area 0.027 /m^2, radius 0.075 m (solid line) or tree height 9 m, number per unit area 0.042 /m^2, radius 0.135 m (broken line) compared with predicted frequency-dependent values due to turbulence with parameter values based on fitting data (see Figure 10.18) for a pine forest and a poplar forest respectively [25] (b) coherence factors predicted as a function of range for four forest types [25] with the parameters listed in Table 10.4.

Table 10.6 Tree parameter values from [25]

Forest type	*density N trees/m²*	*Height m*	*Radius m*
Pine 1	0.027	10	0.075
Poplar 1	0.042	9	0.135
Oak	0.2	4	0.1
Poplar 2	1.33	3	0.024

If $h_{SR}/h \leq 1$, which requires source and receiver to be sufficiently below the tree tops, then $F(h_{SR}/h) \approx 0.5(h_{SR}/h)$, so (10.43) simplifies to

$$T = \exp\left[-rNb\left(h_{SR} / h\right)\right]. \tag{10.47}$$

A consequence of considering finite length cylinders is that the influence of scattering on the coherence between direct and ground-reflected sound field components depends on the tree height. On the other hand, unlike the coherence factor for turbulence, the high-frequency approximation for trunk scattering given by equation (10.45) is independent of frequency.

Figure 10.42(a) compares the frequency-independent coherence factor for trunk scattering according to (10.45) with the frequency-dependent values predicted for turbulence used to fit the data for propagation in a pine and poplar forest, respectively, in Fig. 10.18. Figure 10.42(b) shows predictions of equation (10.45) for the variation of coherence factor with range in four types of forest having the planting densities, tree heights and diameters listed in Table 10.6 [25].

Although effects of canopy scattering and diffusion above the finite length cylinders representing tree trunks are included in the 3-D model [96], it is likely that, if the canopy extends below the tops of the cylinders, it will alter the contributions from diffraction such that the influence of the finite cylinder lengths is overestimated.

REFERENCES

[1] I. Bashir, S. Taherzadeh, H-C. Shin and K. Attenborough, Sound propagation over soft ground without and with crops and potential for surface transport noise attenuation, *J. Acoust. Soc. Am.*, **137**: 154–164 (2015). Equation (10.7b) was suggested by Kees Harmans (private communication).

[2] I. Bashir, *Acoustical exploitation of rough, mixed impedance and porous surfaces outdoors*, Ph.D. dissertation, The Open University, Milton Keynes, UK, 2013.

[3] K. Horoshenkov, A. Khan and H. Benkriera, Acoustic properties of low growing plants, *J. Acoust. Soc. Am.*, **133**: 2554–2565 (2013).

[4] D. Ling, T. Van Renterghem, D. Botteldooren, K. Horoshenkonv and A Khan, Sound absorption of porous substrates covered by foliage: experimental results and numerical predictions, *J. Acoust. Soc. Am.*, **134**: 4599–4609 (2013).

[5] M. Connelly and M. Hodgson, Experimental investigation of the sound absorption characteristics of vegetated roofs, *Build. Environ.*, **92**: 335–346 (2015).
[6] D. E. Aylor, Noise Reduction by Vegetation and Ground. *J. Acoust. Soc. Am.*, **51**: 197–205 (1972).
[7] D. E. Aylor, Sound transmission through vegetation in relation to leaf area density, leaf width and breadth of canopy, *J. Acoust. Soc. Am.*, **51**: 411–414 (1972).
[8] W. H. T. Huisman, *Sound Propagation over vegetation-covered ground*, Ph.D. Thesis, University of Nijmegen, The Netherlands (1990).
[9] a. M. A. Price, K. Attenborough and N. W. Heap, Sound attenuation through trees: Measurements and Models, *J. Acoust. Soc. Am.*, **84**(5): 1836–1844 (1988) and b. *ibid.* Sound propagation results from three British woodlands, in *Sound propagation in Forested Areas and Shelterbelts* ed. M. Martens, University of Nijmegen, March 1986.
[10] M. J. M Martens, Foliage as a low-pass filter: Experiments with model forests in an anechoic chamber. *J. Acoust. Soc. Am.*, **67**: 66–72 (1980).
[11] S. Tang, P. Ong, H. Woon, Monte-Carlo simulation of sound propagation through leafy foliage using experimentally obtained leaf resonance parameters, *J. Acoust. Soc. Am.*, **80**: 1740–1744 (1986).
[12] T. F. W. Embleton, Sound propagation in homogeneous deciduous and evergreen woods, *J. Acoust. Soc. Am.*, **35**: 1119–1125 (1963).
[13] M. Martens, A. Michelsen, Absorption of acoustic energy by plant-leaves, *J. Acoust. Soc. Am.*, **69**: 303–306 (1981).
[14] M. J. M. Martens, P. P. J. Severens, H. A. W. M. Van Wissen and L. A. M. Van Der Heijden, Acoustic reflection characteristics of deciduous plant leaves. *Environ. Exp. Bot.*, **25**: 285–292 (1985).
[15] G. A. Parry, J. R. Pyke, and C. Robinson, The excess attenuation of environmental noise sources through densely planted forest, Proc. Inst. *Acoust.*, **15**, 1057–1065 (1993).
[16] Overstory #60, Trees as noise buffers http://agroforestry.org/overstory-back-issues/203-overstory-60-trees-as-noise-buffers 1998 (last viewed 6/05/20).
[17] a Cook, D., Van Haverbeke, D., 1971. Trees and Shrubs for Noise Abatement, Research Bulletin 246, Historical materials from University of Nebraska-Lincoln Extension, Paper 1629; b D. I. Cook and D.F. Van Haverbeke, Tree covered land forms for noise control, Univ. Nebr. Agric. Exp. Stat. Res. Bull. Volume no. 263, 47p (1974).
[18] ISO9613 – 2 Propagation of sound outdoors: a general method of calculation
[19] J. Kragh, Road traffic noise attenuation by belts of trees and bushes, *Danish Acoustical Laboratory Report no.*31 1982.
[20] G. M. Heisler, O. H. McDaniel, K. K. Hodgdon, J. J. Portelli and S. B. Glesson, Highway Noise Abatement in Two Forests, Proc. NOISE-CON 87, PSU, USA (1987).
[21] L. R. Huddart, The use of vegetation for traffic noise screening, TRRL Research Report 238, 1990.
[22] C.-F. Fang and D.-L. Ling, Investigation of the noise reduction provided by tree belts, *Landsc. Urban Plan.*, **63**: 187–195 (2003).
[23] C.-F. Fang, The criterion of noise attenuation by hedges, *Design and Nature II*, M. W. Collins & C. A. Brebbia (Editors), WIT Press (2004).

[24] T. Van Renterghem, K. Attenborough, M. Maennel, J. Defrance, K. Horoshenkov, J. Kang, I Bashir, S. Taherzadeh, B. Altreuther, A. Khan, Y. Smyrnova and H-S Yang, Measured light vehicle noise reduction by hedges, *Appl. Acoust.*, **78**: 19–27 (2014).

[25] A. I. Tarrero, M. A. Martín, J. González, M. Machimbarrena, and F. Jacobsen, Sound propagation in forests: A comparison of experimental results and values predicted by the Nord 2000 model, *Appl. Acoust.*, **69**: 662–671 (2008).

[26] J. Kragh and B. Plovsing Nord 2000. Comprehensive Outdoor Sound Propagation Model. Part I-II. DELTA Acoustics & Vibration Report, 1849-1851/00, 2000 (revised 31 December, 2001): http://www.magasbakony.hu/Val/Nord2000_homogeneous_atmosphere_Part_1.pdf, https://assets.madebydelta.com/assets/docs/share/Akustik/Nord2000_Comprehensive_Outdoor_Sound_Propagation_Model-_-Part_2_Propagation_in_an_Atmosphere_with_Refraction.pdf (last viewed 01/06/20).

[27] W. H. T. Huisman and K. Attenborough, Reverberation and attenuation in a pine forest, *J. Acoust. Soc. Am.*, **90**: 2664–2667 (1991).

[28] F. M. Weiner and D.N. Keast, Experimental study of the propagation of sound over ground, *J. Acoust. Soc. Am.*, **31**: 724–733 (1959).

[29] R. Bullen and F. Fricke, Sound Propagation Through Vegetation, *J. Sound Vib.*, **80**: 11–23 (1982).

[30] T. F. W. Embleton, Scattering by an array of cylinders as a function of surface impedance, *J. Acoust. Soc. Am.*, **40**: 667–670 (1966).

[31] V. F. Twersky, On the scattering of waves by random distributions. 1 Free space scatterer formalism, *J. Math. Phys.*, **3**: 700–715 (1962).

[32] S. Taherzadeh, I. Bashir, and K. Attenborough, Aperiodicity effects on sound transmission through arrays of identical cylinders perpendicular to the ground., *J. Acoust. Soc. Am.*, **132**: EL323–EL328 (2012).

[33] T. Van Renterghem, D. Botteldooren, and K. Verheyen, Road traffic noise shielding by vegetation belts of limited depth, *J. Sound Vib.*, **331**: 2404–2425 (2012).

[34] C. M. Linton and P. A. Martin, Multiple scattering by random configurations of circular cylinders: Second-order corrections for the effective wave number, *J. Acoust. Soc. Am.*, **117**: 3413–3423 (2005).

[35] M. E. Swearingen and M. J. White, Influence of scattering, atmospheric refraction, and ground effect on sound propagation through a pine forest, *J. Acoust. Soc. Am.*, **122**: 113–119 (2007).

[36] V. E. Ostashev, D. K. Wilson and M. B. Muhlestein, Effective wave numbers for sound scattering by trunks, branches and canopy in a forest, *J. Acoust. Soc. Am.* **142**: EL177–EL183 (2017).

[37] D. Heimann, Numerical simulation of wind and sound propagation through an idealised stand of trees, *Acta Acust. united Ac.* **89**: 779–788 (2003).

[38] T. Van Renterghem, D. Botteldooren, Landscaping for road traffic noise abatement: model validation, *Environmental Modelling and Software*, **109**: 17–31 (2018).

[39] S. F. Clifford, R. T. Lataitis, Turbulence effects on acoustic wave propagation over a smooth surface, *J. Acoust. Soc. Am.*, **73**: 1545–1550 (1983).

[40] B. Lalic and D. Mihailovic, An Empirical Relation Describing Leaf-Area Density inside the Forest for Environmental Modeling, *J. Appl. Met.* **43**: 641–645 (2004).

[41] R. Martínez-Sala, C. Rubio, L. M. García-Raffi, J. V. Sánchez-Pérez, E. A. Sánchez-Pérez and J. Llinares, Control of noise by trees arranged like sonic crystals, *J. Sound Vib.* **291**: 100–106 (2006).
[42] T. Van Renterghem, Guidelines for optimizing road traffic noise shielding by non-deep tree belts, *Ecol. Eng.* **69**: 276–286 (2014).
[43] T. Van Renterghem, E. Salomons and D. Botteldooren, Efficient FDTD-PE model for sound propagation in situations with complex obstacles and wind profiles, *Acta Acust. united Ac.* **91**: 671–679 (2005).
[44] G. Reethof, L. Frank, O. McDaniel, Sound absorption characteristics of tree bark and forest floor, Proceedings of the Conference on Metropolitan Physical Environment, USDA Forest Service General Technical Report NE-2, 1977.
[45] T. Van Renterghem, B. De Coensel, D. Botteldooren, Loudness evaluation of road traffic noise abatement by tree belts, Proceedings of the 42nd international congress and exposition on noise control engineering (Internoise 2013), Innsbruck, Austria.
[46] T. Van Renterghem, Exploiting Supporting Poles to Increase Road Traffic Noise Shielding of Tree Belts, *Acta Acust. united Ac.* **101**: 1–7 (2015).
[47] K. Tanaka, S. Ikeda, R. Kimura and K. Simazawa, The function of forests in sound-proofing. *Bull. Tottori Univ. For.* **11**: 77–102 (1979).
[48] V. Bucur, *Urban Forest Acoustics*. Springer-Verlag, Berlin (2006).
[49] O. Umnova, K. Attenborough and C. Linton, Effects of porous covering on sound attenuation by periodic arrays of cylinders, *J. Acoust. Soc. Am.*, **119** 278–284 (2006).
[50] J. M. Wunderli and E. Salomons, A model to predict the sound reflection from a forest, *Acta Acust. united Ac. Acust.* **95**: 76–85 (2009).
[51] J. M. Wunderli, An extended model to predict reflections from forests, *Acta Acust. united Ac.* **98**: 263–278 (2012).
[52] M. E. Swearingen, M. J. White, P. J. Guertin, D.G. Albert and A. Tunick, Influence of a forest edge on acoustical propagation: Experimental results, *J. Acoust. Soc. Am.* **133**: 2566–2575 (2013).
[53] J. Defrance, N. Barrière and E. Premat, Forest as a meteorological screen for traffic noise, Proc. 9th ICSV Orlando, International Institute of Acoustics and Vibration, Auburn, USA (2002).
[54] E. M. Salomons, Improved Green's function parabolic equation method for atmospheric sound propagation, *J. Acoust. Soc. Am.* **104**: 100–111 (1998).
[55] N. Barrière, *Theoretical and experimental study of traffic noise propagation through forest*, PhD, Ecole Centrale de Lyon (1999).
[56] T. Van Renterghem, K. Attenborough and P. Jean, Designing vegetation and tree belts along roads, Chapter 5 of *Environmental Methods for Transport Noise Reduction*, Ed. M. Nilsson, M Bengtsson and R. Klaeboe, Taylor and Francis (2015).
[57] M. E. Swearingen and M. White, Influence of scattering, atmospheric refraction and ground effect on sound propagation through a pine forest, *J. Acoust. Soc. Am.*, **122**: 113–119 (2007).
[58] A. Tunick, Calculating the micrometeorological influences on the speed of sound through the atmosphere in forests, *J. Acoust. Soc. Am.*, **114**: 1796–1806 (2003).

[59] C. M. Linton and P. A. Martin, Multiple scattering by random configurations of circular cylinders: Second-order corrections for the effective wave number, *J. Acoust. Soc. Am.*, **117**: 3413–3423 (2005).
[60] R. De Jong and E. Stusnick, Scale model studies of the effect of wind on acoustic barrier performance, *Noise Control Engineering* **6**: 101–109 (1976).
[61] K. Rasmussen and M. Arranz, The insertion loss of screens under the influence of wind, *J. Acoust. Soc. Am.*, **104**: 2692-2698 (1998).
[62] E. M. Salomons, Reduction of the performance of a noise screen due to screen-induced wind-speed gradients. Numerical computations and wind tunnel experiments. *J. Acoust. Soc. Am.*, **105**: 2287–2293 (1999).
[63] E. M. Salomons and K.B. Rasmussen, Numerical computations of sound propagation over a noise screen based on an analytic approximation of the wind speed field, *Appl. Acoust.*, **60**: 327–341 (2000).
[64] N. Barrière and Y. Gabillet, Sound propagation over a barrier with realistic wind gradients. Comparison of wind tunnel experiments with GFPE computations, *Acta Acust. united Ac.* **85**: 325–334 (1999).
[65] J. Forssen, Calculation of sound reduction by a screen in a turbulent atmosphere using the parabolic equation method, *Acta Acust. united Ac.* **84**: 599–606 (1998).
[66] T. Van Renterghem, D. Botteldooren, W. Cornelis, D. Gabriels, Reducing screen-induced refraction of noise barriers in wind with vegetative screens, *Acta Acust. united Ac.* **88**(2): 231–238 (2002).
[67] T. Van Renterghem, D. Botteldooren, On the choice between walls and berms for road traffic noise shielding including wind effects, *Landsc. Urban Plan.*, **105**: 199–210 (2012).
[68] T. Van Renterghem, D. Botteldooren, Effect of a row of trees behind noise barriers in wind, *Acta Acust. united Ac.* **88**: 869–878 (2002).
[69] T. Van Renterghem and D. Botteldooren, Numerical simulation of the effect of trees on downwind noise barrier performance. *Acta Acust. united Ac.* **89**: 764–778 (2003).
[70] T. Van Renterghem and D. Botteldooren, Designing canopies to improve downwind shielding at various barrier configurations at short and long distance, Proceedings of the 21st International Congress on Acoustics (ICA 2013), Montreal, Canada.
[71] T. Van Renterghem and D. Botteldooren, Numerical evaluation of tree canopy shape near noise barriers to improve downwind shielding, *J. Acoust. Soc. Am.*, **123**: 648–657 (2008).
[72] J. Wilson, Numerical studies of flow through a windbreak, *J. Wind Eng. Indust. Aerodyn.*, **21**: 119–154 (1985).
[73] O. Fegeant, Wind-induced vegetation noise. Part I: a prediction model. *Acta Acust. united Ac.* **85**: 228–240 (1999).
[74] O. Fegeant, Wind-induced vegetation noise. Part II: field measurements. *Acta Acust united Ac.* **85**: 241–249 (1999).
[75] K. Bolin, Prediction method for wind-induced vegetation noise. *Acta Acust. united Ac.* **95**: 607–619 (2009).
[76] C. Guastavino, The ideal urban soundscape: Investigating the sound quality of French cities, *Acta Acust. united Ac.* **92**: 945–951 (2006).
[77] J. Hong and J. Jeon, The effects of audio-visual factors on perceptions of environmental noise barrier performance, *Landsc. Urban Plan.* **125**: 28–37 (2014).

[78] L. Sutherland and G. A. Daigle, Atmospheric Sound Propagation, Ch. 32 in *Encyclopedia of Acoustics*, ed. M. J. Crocker, Wiley, New York, 1997.
[79] V. E. Ostachev and D. K. Wilson, *Acoustics in moving inhomogeneous media*, 2nd edition CRC Press, ISBN: 978-0-415-56416-8, 2015.
[80] G.A. Daigle, J.E. Piercy and T.F.W. Embleton, Effects of atmospheric turbulence on the interference of sound waves near a hard boundary. *J. Acoust. Soc. Am.*, **64**: 622–630 (1978).
[81] G.A. Daigle, Effects of atmospheric turbulence on the interference of sound waves above an impedance boundary, *J. Acoust. Soc. Am.*, **65**: 45–49 (1979).
[82] V. E. Ostashev and D. K. Wilson, Relative contributions from temperature and wind velocity fluctuations to the statistical moments of a sound field in a turbulent atmosphere. *Acta Acust. united Ac.* **86**: 260–268 (2000).
[83] V. E. Ostashev, E. Salomons, S. Clifford, R. Lataitis, D. K. Wilson, P. Blanc-Benon and D. Juvé, Sound propagation in a turbulent atmosphere near the ground: A Parabolic equation approach, *J. Acoust. Soc. Am.*, **109**: 1894–1908 (2001).
[84] D. K. Wilson and D. W. Thomson, Acoustic propagation through anisotropic surface layer turbulence, *J. Acoust. Soc. Am.*, **96**: 1080–1095 (1994).
[85] D. K. Wilson, D. I. Havelock, M. Heyd, M. J. Smith, J. M. Noble and H. J. Auvermann, Experimental determination of the effective structure-function parameter for atmospheric turbulence, *J. Acoust. Soc. Am.*, **105**: 912–914 (1999).
[86] P. L'Esperance, P. Herzog, G.A. Daigle and J.R. Nicholas, Heuristic model for outdoor sound propagation based on an extension of the geometrical ray theory in the case of a linear sound speed profile, *Appl. Acoust.*, **37**: 111–139 (1992).
[87] R. Raspet and W. Wu, Calculation of average turbulence effects on sound propagation based on the fast fild program formulation, *J. Acoust. Soc. Am.*, **97**: 147–153, (1995).
[88] K.E. Gilbert, R. Raspet and X. Di, Calculation of turbulence effects in an upward refracting atmosphere, *J. Acoust. Soc. Am.*, **87**: 2428–2437 (1990).
[89] G. Goedecke and H. Auvermann, Acoustic scattering by atmospheric turbules, *J. Acoust. Soc. Am.*, **102**: 759–771 (1997).
[90] G. Goedecke, R. Wood, H. Auvermann, V. Ostashev, D. Havelock, and C. Ting, Spectral broadening of sound scattered by advecting atmospheric turbulence, *J. Acoust. Soc. Am.*, **109**: 1923–1934 (2001).
[91] V. E. Ostashev, *Acoustics in moving inhomogeneous media*, E & FN Spon, London, 1997.
[92] D. K. Wilson, A turbulence spectral model for sound propagation in the atmosphere that incorporates shear and buoyancy forcings, *J. Acoust. Soc. Am.*, **108**: 2021–2038 (2000).
[93] G. H. Goedecke, V. E. Ostashev, D. K. Wilson and H. J. Auvermann, von Kármán spectra of temperature and velocity fluctuations. *Proceedings of 9th International Congress on Sound and Vibration ICSV* (2002), Orlando, USA.
[94] T. Van Renterghem, *The finite-difference time-domain method for sound propagation in a moving medium*, PhD Thesis, Ghent University, 2003.
[95] D. Heimann and R. Blumrich, Time-domain simulations of sound propagation through screen-induced turbulence, *Appl. Acoust.*, **65**: 561–582 (2004).
[96] V.E. Ostashev, D. K. Wilson, M. B. Muhlestein and K. Attenborough, Correspondence between sound propagation in discrete and continuous random media with application to forest acoustics, *J. Acoust. Soc. Am.*, **143**: 1194–1205 (2018).

Chapter 11

Ray Tracing, Analytical and Semi-empirical Approximations for A-Weighted Levels

11.1 RAY TRACING

Most community noise problems require information about the attenuation of A-weighted noise. Full-wave numerical methods, such as those described in Chapter 4, demand considerable computational resources to evaluate A-weighted mean square pressures. Chapter 12 explores some practical engineering alternatives, many of which contain empirical components. Frequently, it is assumed that the atmosphere is vertically stratified, and an effective sound speed gradient is used to account for the influence of wind and temperature gradients. Given the assumption of an effective sound speed profile, a ray tracing approach is both convenient and less computationally demanding. The analytical basis for using ray tracing outdoors is explored in this chapter.

Ray tracing sums the contributions from all rays from a source that are computed to pass through a chosen receiver. These are known as eigenrays. For sources close to the ground, it is necessary to allow for any ground reflections. Rather than using plane wave reflection coefficients to describe these ground reflections, a better approximation is to use spherical wave reflection coefficients (see Chapter 2). This has resulted in a heuristic modification of the Weyl-Van der Pol formula (2.40) [1, 2]. It has been shown [3, 4] that, if the rays have not passed through turning points and there is a single reflection at the ground, which is usually the case at short ranges, the resulting formulation represents the first term of an asymptotic solution of the full-wave Equation. Furthermore, it has been demonstrated numerically, that the ray trace calculations agree reasonably well with the results of other numerical schemes (see also Section 11.4). Nevertheless, it should be noted that a drawback of ray tracing is that it does not predict sound in shadow zones where rays do not penetrate.

The sound field due to a monopole source above a porous ground in the presence of temperature and wind gradients can be written in a form similar to that of the Weyl-Van der Pol formula (2.40) [5]. In a moving stratified medium, the index of refraction $m(z)$ is defined by

$$m(z) = \frac{n(z)}{1 + M(z)\cos(\varepsilon - \psi_w)\sin\mu(z)}, \tag{11.1}$$

where $n(z)$ is the index of refraction in an otherwise homogeneous medium, $M(z) = u/c$ is the Mach number, u and ψ_w are the magnitude and direction of wind, $\mu(z)$ and ε are the polar and azimuthal angles of a wavefront normal and $c(z)$ is the speed of sound. Although the azimuthal angle is constant for a given wavefront normal, the polar angle varies as a function of height in a moving stratified medium. Using the index of refraction in a moving medium, the modified Snell's law for a wavefront normal is derived as

$$m \sin\mu = \sin\mu_0 = \text{constant}, \tag{11.2}$$

where μ_0 is a reference polar angle of a ray at $z = 0$.

In other words, the wavefront normal of any ray can be determined by specifying a polar angle at a given height and its corresponding azimuthal angle. As pointed out elsewhere [6], the trajectory of the wavefront normal should be distinguished from the sound ray that travels from the source to receiver since, in a moving medium, the wavefront normal is not coincident with the sound ray.

The acoustical path length, $R_L(\mu,\varepsilon)$, and the radius of curvature, $R_r(\mu,\varepsilon)$ are defined, respectively, by

$$R_L(\mu,\varepsilon) = \int_{z_<}^{z_>} (r \sin\mu_0 \cos\varepsilon + m cos\mu)\,dz \tag{11.3}$$

and

$$R_r(\mu,\varepsilon) = \frac{D}{D_s}\sqrt{\left\{\frac{\sin^2\mu_0 \left(\dfrac{\rho}{\rho_s}\right)\left(\dfrac{\cos\mu_0}{m\cos\mu}\right)\left(\dfrac{\cos\mu_0}{m_s\cos\mu_s}\right)}{\left|\dfrac{\partial^2 R_L}{\partial\mu_0^2}\dfrac{\partial^2 R_L}{\partial\varepsilon^2} - \left[\dfrac{\partial^2 R_L}{\partial\mu_0 \partial\varepsilon}\right]^2\right|}\right\}}, \tag{11.4}$$

where the subscript s denotes the variables to be evaluated at the source height, z_s.

$z_<$ and $z_>$ are, respectively, the lesser and the larger of the source and receiver heights.

In addition, the Doppler factor $D(z)$ is given by

$$D(z) = \left[1 + M(z)\cos(\varepsilon - \psi_w)\sin\mu\right]^{-1} \tag{11.5}$$

Suppose that the polar and azimuthal angles of the wavefront normal of the direct wave are θ and ψ, respectively, whereas they are $\bar{\theta}$ and $\bar{\psi}$ for the reflected wave. If the receiver position is given in the cylindrical polar coordinate system, with r as the range and z the height of receiver, and, without loss of generality, the azimuthal angle of the receiver is set equal to zero, then, the sound field, above a locally reacting ground surface, can be determined from the classical Weyl-Van der Pol form (see Chapter 2), i.e.

$$p(r,0,z) = \frac{e^{ikR_1'}}{4\pi \bar{R}_1} + \left[R_p + (1-R_p)F(w)\right]\frac{e^{ikR_2'}}{4\pi \bar{R}_2}, \tag{11.6}$$

where k_0 is the reference wave number, R_p is the plane wave reflection coefficient, $F(w)$ is the boundary loss factor and w is the numerical distance.

These parameters can be determined according to

$$R_p = \frac{\cos\bar{\theta}_0 - \beta}{\cos\bar{\theta}_0 + \beta} \tag{11.7a}$$

$$F(w) = 1 + i\sqrt{\pi}w\,exp(-w^2)\,erfc(-iw) \tag{11.7b}$$

$$w^2 = \frac{1}{2}ik_0R_2'\left(\beta + \cos\bar{\theta}_0\right), \tag{11.7c}$$

where, β is the normalized specific admittance.

In Equation (11.6), R_1' $[\equiv R_L(\theta,\psi)]$ and $R_2'\left[\equiv R_L\left(\bar{\theta},\bar{\psi}\right)\right]$ are the path lengths for the direct and reflected sound waves, respectively. $\bar{R}_1$ $[\equiv R_r(\theta,\psi)]$ and $\bar{R}_2$ $\left[\equiv R_r\left(\bar{\theta},\bar{\psi}\right)\right]$ are the effective radii of curvature of the corresponding rays.

The above analysis is based on the method of Fourier transformation and the well-known Wentzel–Kramers–Brillouin (WKB) approximation in which the sound field can be expressed in terms of an integral representation (Chapter 2). By evaluating the integral asymptotically, it is possible to confirm the validity of the heuristic approximation used in previous analysis [7] which includes the ground wave term explicitly (the second term of the curly bracket) in Equation (11.6). This equation is valid for a relatively short separation between the source and receiver. In an upward-refracting medium, the analysis is invalid in the shadow zone and near the shadow boundary. In a downward-refracting medium, the analysis is satisfactory if the reflected wave encounters the ground only once. Also, the receiver must not be close to a caustic (i.e. a ray crossing) where the effective radius of curvature of a ray vanishes.

It is possible to allow for multiple bounces in a temperature-stratified but downward-refracting medium [1]. The asymptotic solution for the total field above an impedance ground is

$$p(r,0,z) = \Sigma\frac{e^{ik\left(R_1'-\chi_1\right)}}{4\pi\bar{R}_1} + \Sigma_j\left[R_p + (1-R_p)F(w)\right]^j\frac{e^{ik\left(R_2'-\chi_2\right)}}{4\pi\bar{R}_2}. \tag{11.8}$$

This solution is simply the sum of contributions from all possible eigenrays linking the source and receiver. When calculating R_1', R_2', $\bar{R}_1$ and $\bar{R}_2$, it is crucial to include all possible branches of the ray traces. The phase shifts, χ_1 and χ_2, result where the direct and reflected rays graze a caustic. There will be a phase reduction of $\pi/2$ each time the ray touches a caustic [8].

Commonly, the method developed by Thompson [9] from the theory derived by Blokhintzev [10] and its derivative [11] is used as the basis for the ray-tracing algorithm in a moving stratified medium. These earlier methods often require the setting up of a pair of first-order differential equations which can be solved either by numerical integration or by the construction of a finite-step wavefront, i.e. by the Euler method.

Usually, the search for an eigenray involves some form of hit-and-miss approach. A ray is launched from the source in a certain direction and then it is determined whether the launched ray hits the 'target' at the receiver location. This is described as 'blind-shooting'. Often, to reduce computational time, the area of the target is restricted. The numerical accuracy is controlled by the size of the target as well as the step size in tracing the ray path: the smaller the step length and the smaller the target, the greater the computational time.

A bracketing scheme has been devised that avoids the need for 'blind-shooting' and offers a better method for finding eigenrays [12]. The stationary points for $R_L(\mu,\varepsilon)$ in Equation (11.3), ((θ,ψ) for the direct wave and $(\bar{\theta},\bar{\psi})$ for the reflected wave), are determined. Then the eigenray is obtained by solving the following pair of non-linear equations for μ and ε:

$$r\sin\varepsilon = \int_{z_<}^{z_>} \frac{M\sin(\varepsilon - \psi_w)}{\cos\mu}\,dz \tag{11.9}$$

and

$$r\cos\varepsilon = \int_{z_<}^{z_>} \frac{\sin\mu + M\cos(\varepsilon - \psi_w)}{\cos\mu}\,dz \tag{11.10}$$

To determine an eigenray, it is sufficient just to specify the polar angle at a given height and the azimuthal angle of a wavefront normal, because the polar angle at other heights can be found by using the modified Snell's law *cf* Equation (11.2). The small ratio between wind and sound speed ($M \approx 0.03$) in a normal atmospheric environment means that the azimuthal angle of the wavefront normal is very close to the azimuthal angle of the receiver. So, $\varepsilon = 0$ may be used as the first approximation and $\mu(z)$ can be determined from (11.2) and (11.10) for a given range r.

A second approximation for ε can be found by using $\mu(z)$ and (11.9) to give

$$\varepsilon = \tan^{-1}\left[\frac{\sin\psi_w \int_{z_<}^{z_>} \frac{M}{\cos\mu}\,dz}{\cos\psi_w \int_{z_<}^{z_>} \frac{M}{\cos\mu}\,dz - r}\right]. \tag{11.11}$$

The integration in (11.11) can be carried out numerically. The process for determining ε and $\mu(z)$ can be repeated iteratively to achieve the required accuracy. In most practical situations, ε is very small and only two or three iterations are necessary.

For convenience, analysis is restricted to a monotonically increasing function of $m(z)$. Although, this implies a monotonically increasing function of $c(z)$, the method should apply equally for other more intricate profiles that may include regions where $dc(z)/dz$ is zero. Also, the approach can be extended to take account of the wind effects.

As mentioned earlier, it is sufficient just to specify $\mu_0(\equiv\mu(0))$, say, as the unknown variable in (11.10). The polar angle at other heights can be determined by using (11.2). Determination of μ_0 involves the evaluation of the integral given in Equation (11.10). There is an integrable singularity at the turning point where the polar angle is $\pi/2$ (or the slope of the wavefront normal is zero). After substituting (11.9) into (11.10),

$$\int_{z_<}^{z_>} \tan\mu \, dz = \frac{r\sin\psi_w}{\sin(\psi_w - \varepsilon)} \tag{11.12}$$

Partial integration removes the integrable singularity of (11.12) thereby making the subsequent numerical analysis somewhat simpler. Hence, (11.12) is recast as

$$I(\mu_0) = \frac{r\sin\psi_w}{\sin(\psi_w - \varepsilon)} - \left[\frac{m\sin\mu\cos\mu}{m'}\right]_{z_<}^{z_>}, \tag{11.13}$$

where

$$I(\mu_0) = \int_{z_<}^{z_>} \sin\mu\cos\mu\left[1 + mm''/(m')^2\right]dz \tag{11.14}$$

and the primes denote derivatives with respect to z.

This analysis is valid if the derivative of the index of refraction, m does not vanish at any point along the ray path. Physically, this situation corresponds to the case of an infinitely narrow sound channel where a ray will be trapped. In general, numerical integration is required to calculate $I(\mu_0)$ in (11.13). On the other hand, the computation of definite integrals may be regarded as an initial value problem. By differentiating both sides of (11.14) with respect to z, it is transformed into a first-order differential equation:

$$\frac{dI}{dz} = \sin\mu\cos\mu\left[1 + mm''/(m')^2\right] \tag{11.15}$$

with the initial condition of $I = 0$ at $z = z_<$.

To compute the function I at $z = z_>$ with due consideration of the need to trace all branches for a complete ray path, it is adequate just to calculate the integral from $z_<$ to the turning point, from $z_>$ to the turning point and from the ground surface to the turning point. An appropriate number of these portions are then added together to trace all rays in all possible ways. A typical approach, sometimes known as the Euler method, involves the use of a finite-step size (Δz). The increment of I can be computed by multiplying the right-hand side of (11.15) by Δz for each step. However, the Euler method is not recommended and more sophisticated numerical methods, such as a fourth order Runge–Kutta or the so-called adaptive Bulirch–Stoer technique, are more accurate and stable [13].

To find all solutions for (11.13), we introduce the eigenray error function,

$$E(\mu_0) = \frac{r \sin \psi_w}{\sin(\psi_w - \varepsilon)} - \left[\frac{m \sin \mu \cos \mu}{m'} \right] - I(\mu_0) \tag{11.16}$$

The eigenrays are then determined by minimizing $|E(\mu_0)|$ for a given source/receiver geometry and wind and temperature profiles. The error function $E(\mu_0)$ is sampled at a range of polar angles specified by the user. This allows the determination of zero crossings which, in turn, provides information about a pair of bracketing polar angles. The spacing of the samples $\Delta\mu_0$ is also set by the user. A smaller $\Delta\mu_0$ and a large range of polar angles increase the probability that all eigenrays will be found at the expense of higher computational time. Brent's root finding algorithm [13] is used to find the eigenray solution if zero crossings are found. Often an initial examination of the wind velocity and sound speed profiles will provide some useful insight in the choice of upper and lower bounds of the polar angles. For example, suppose a ray launches upwards at some angle μ_u that has its turning point at height $z_>$ (see Figure 11.1). A ray launched upwards at a polar

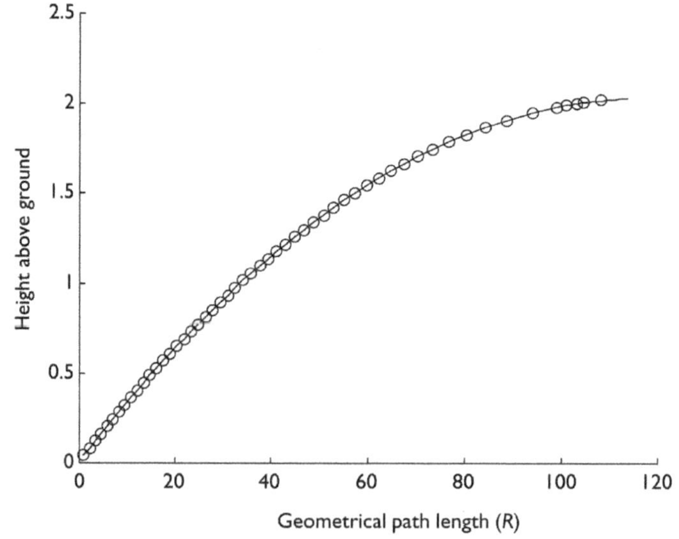

Figure 11.1 Plot of the inverse of the ray amplitude for wavefront positions along a ray trajectory from the ground up to the turning point. The solid line represents the geometrical path length calculated according to (11.4) and the circles *represent values of* the expression $\left(S_i/\sqrt{J_i}\right)^{-1}$ where J is given by (11.18).

angle greater than μ_u cannot reach a receiver situated at a height of $z_<$. Hence, μ_u provides an upper bound for the sampling range for the reflected rays. On the other hand, the choice of the sampling range may also be decided on a trial and error basis.

In a downward-refracting medium, a ray launched at an angle $\pi - \mu_0$ has the same characteristic parameters as the ray launched upwards with the polar angle μ_0 so the down-going rays can be located at the same 'time'. Once the polar angle of the wavefront normal has been determined, its value is substituted into (11.11) thereby allowing ε to be found by the simple iteration procedure detailed earlier. Although the principle described here can be generalized to other more intricate profiles, such developments are beyond the scope of the present text.

On finding the polar and azimuthal angles of all eigenrays, R_1', R_2', $\bar{R}_1$ and $\bar{R}_2$ are computed by direct numerical integration and used, in turn, to compute the total sound field given by (11.8). The determination of $\bar{R}_1$, $\bar{R}_2$ and the reduction in phases (χ_1 and χ_2) can be facilitated by noting

$$\frac{\partial^2 R_L}{\partial \mu^2} = -\cos^2\mu_0 \int_{z_<}^{z_>} \frac{n^2}{m^3\cos^3\mu}\,dz. \tag{11.17a}$$

$$\frac{\partial^2 R_L}{\partial \varepsilon^2} == -\sin^2\mu_0 \int_{z_<}^{z_>} \left[\frac{n}{m\cos\mu} + \frac{M^2\sin^2(\psi_w-\varepsilon)\sin^2\mu_0}{m^3\cos^3\mu}\right]dz \tag{11.17b}$$

and

$$\frac{\partial^2 R_L}{\partial \mu_0\partial\varepsilon} = -\cos\mu_0\sin\mu_0 \int_{z_<}^{z_>} \left[\frac{nM\sin(\psi_w-\varepsilon)\sin\mu_0}{m^3\cos^3\mu}\right]dz \tag{11.17c}$$

The equation of the caustic can be found by setting the Jacobian factor J given by

$$J = \sqrt{\frac{\partial^2 R_L}{\partial \mu_0^2}\frac{\partial^2 R_L}{\partial \varepsilon^2} - \left[\frac{\partial^2 R_L}{\partial \mu_0\partial\varepsilon}\right]^2} \tag{11.18}$$

to zero [1] and it is then possible to determine χ_1 and χ_2 by examining whether the ray grazes the caustic.

Ray tracing may be used to calculate excess attenuation including turbulence (see Chapter 10), defined by 10 log $\langle p^2\rangle$, where

$$p^2 = \sum_i^N \frac{A_i^2\times|Q_i|^2}{R_i} + 2\sum_{i=1}^N\sum_{j=1}^{i-1}\frac{A_i|Q_i|\times A_j|Q_j|}{R_iR_j}\cos\left[k\left(R_j'-R_i'\right)+\arg\left(\frac{Q_j}{Q_i}\right)\right]\times\Gamma \tag{11.19}$$

and R_i represents the Jacobian function divided by the stratification factor, the spherical wave reflection coefficient $Q = R_p + (1 - R_p)F(W)$ and Γ is the turbulence parameter, allowing for the destruction of coherence between the rays. The Clifford and Lataitis method (see Chapter 10) is used whereby

$$\Gamma = \exp\left(-\alpha\sigma^2(1-\rho)\right), \tag{11.20}$$

where $\sigma^2 = \frac{\sqrt{2}}{2} n^2 k^2 R L_0$ is the variance of phase fluctuation along a path, $\rho = \frac{\sqrt{\pi} L_0}{2h} erf\left(\frac{h}{L_0}\right)$ (for equal source and receiver heights) is the covariance between paired rays, $erf(x) = \int_0^x e^{-t^2} dt$, $\langle n^2 \rangle$ is the fluctuation of the refractive index, L_0 is the outer scale of turbulence, h is the maximum separation between paired rays, $\alpha = 1$ if $L_0 \gg \sqrt{R/k}$, $\alpha = 0.5$ otherwise and R is the distance between source and the receiver.

In previous ray tracing methods [2] h may be computed from predictions of the complete ray paths. However, the method of Raspet and Wu [14], defining it as half of the vertical distance between the turning points of the two ray paths, is used here. Since, the locations of the turning points are calculated in the routine anyway no extra computation is necessary.

Figure 11.2 demonstrates the result of including turbulence in this manner for source and receiver at 1.5 and 1.8 m heights, respectively, horizontal

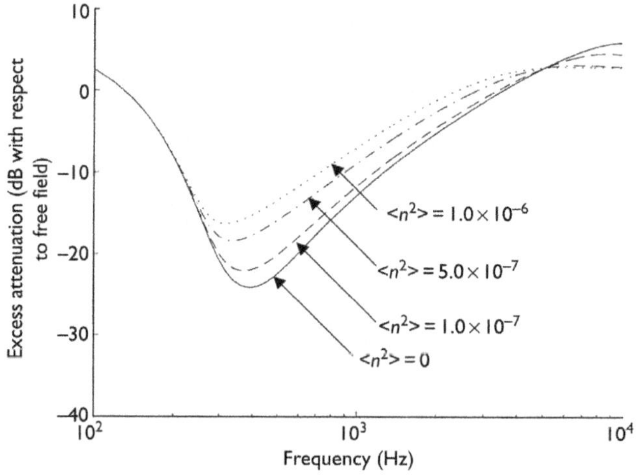

Figure 11.2 Predicted effect of turbulence on excess attenuation spectra under downwind conditions with varying degrees of turbulence. Source and receiver heights are 1.5 and 1.8 m respectively and the horizontal separation is 1000 m. The assumed linear sound speed gradient is 10^{-4} s^{-1} and the ground impedance is calculated using the two-parameter model ((3.48) with σ_e = 300 kPa s m^{-2} and α_e = 20.0 m^{-1}). The solid line indicates $\langle \mu^2 \rangle = 0$, the dashed line corresponds to $\langle \mu^2 \rangle = 1.0 \times 10^{-7}$, the dash-dot line corresponds to $\langle \mu^2 \rangle = 5.0 \times 10^{-7}$ and the dotted line corresponds to $\langle \mu^2 \rangle = 1.0 \times 10^{-6}$.

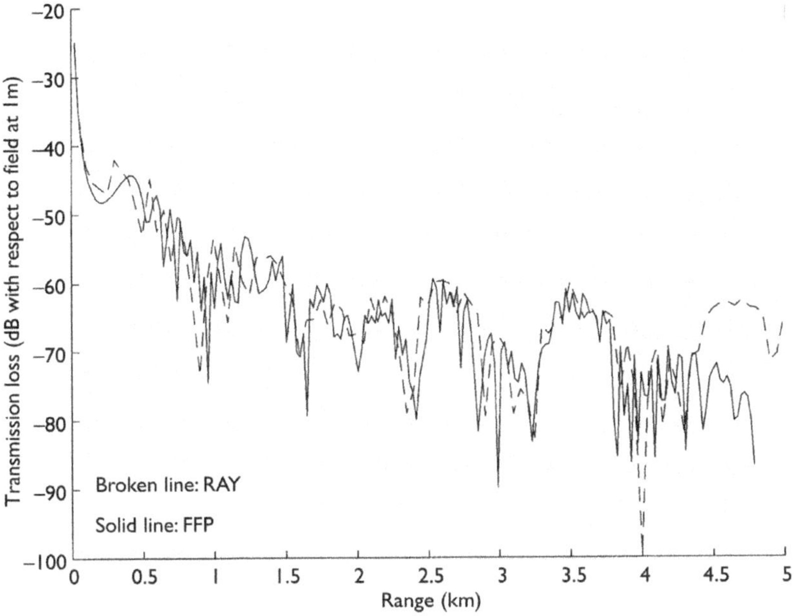

Figure 11.3 Comparison between ray-trace and FFP calculations as a function of range out to 5 km at 1000 Hz under downward refraction conditions. The source and receiver heights and the ground impedance values are assumed to have the same values as for Figure 11.2. The sound speed gradient is assumed to be 0.1 s^{-1}.

separation 1000 m, a linear sound speed gradient of 10^{-1} s^{-1} and $\langle n^2 \rangle$ varying between zero and 10^{-6}.

Figure 11.3 compares predictions of transmission loss at 1 kHz against range obtained from the ray trace procedure with results of FFP calculations in the presence of a downward-refracting linear sound speed gradient of 0.1 s^{-1}.

To obtain the FFP results, 16384 integration points and 1000 layers each 0.5 m thick were used. Typically, this calculation required five hours on a PC with a 3 GHz processor. The FFP calculation results have been plotted at 100-point intervals since finer interference structure than that shown will be destroyed by turbulence and has no practical significance. The ray-trace predictions, allowing for up to four ground reflections, were accomplished in a matter of seconds on the same computer and are in good agreement with the 'smoothed' structure predicted by the FFP out to 4 km range. Beyond 4 km, it appears that there are additional ray arrivals not accounted for in the ray-trace predictions. The remarkable agreement between ray-trace and full-wave calculations at such a long range has been shown to be the result of the finite ground impedance which removes the rays for which the ray-tracing approximations do not hold [15].

11.2 LINEAR SOUND SPEED GRADIENTS AND WEAK REFRACTION

There are distinct advantages in assuming a linear effective sound speed profile in ray tracing since this assumption leads to circular ray paths and analytically tractable solutions. With this assumption, the effective sound speed, c, can be written as

$$c(z) = c_0(1 + \zeta z), \tag{11.21}$$

where ζ is the normalized sound velocity gradient $((dc/dz)/c_0)$ and z is the height above ground. If it is assumed also that the source–receiver distance and the effective sound speed gradient are sufficiently small that there is only a single 'ray bounce', i.e. ground reflection, between source and receiver, it is possible to use a simple adaptation of the Weyl-Van der Pol formula (2.40) replacing the geometrical ray paths defining the direct and reflected path lengths by curved ones.

Consequently, the sound field is approximated by

$$p = \{\exp(ik_0\xi_1) + Q exp(ik_0\xi_2)\} / 4\pi d, \tag{11.22a}$$

where Q is the appropriate spherical wave reflection coefficient, d is the horizontal separation between the source and receiver, ξ_1 and ξ_2 are, respectively, the acoustical path lengths of the direct and reflected waves. These acoustical path lengths can be determined by [16, 17]

$$\xi_1 = \int_{\phi_<}^{\phi_>} \frac{d\phi}{\zeta \sin\phi} = \zeta^{-1}\log_e\left[\tan(\phi_> / 2) / \tan(\phi_< / 2)\right] \tag{11.22b}$$

and

$$\xi_2 = \int_{\theta_<}^{\theta_>} \frac{d\phi}{\zeta \sin\phi} = \zeta^{-1}\log_e\left[\tan(\theta_> / 2)\tan^2(\theta_0 / 2) / \tan(\theta_< / 2)\right], \tag{11.22c}$$

where $\phi(z)$ and $\theta(z)$ are the polar angles (measured from the positive z-axis) of the direct and reflected waves and the path length difference, Δr, is given by

$$\Delta r = \xi_2 - \xi_1. \tag{11.22d}$$

The subscripts > and < denote the corresponding parameters evaluated at $z_>$ and $z_<$, respectively, $z_> \equiv \max(z_s, z_r)$ and $z_< \equiv \min(z_s, z_r)$.

Hidaka et al. [18] have used the travel time along the curved ray to characterize the acoustical path length. Equations (11.22b) and (11.22c) can be reduced to their equations. Additionally, it is possible to show, in the limit of $\zeta \to 0$ (i.e. in a homogeneous medium), the acoustical path lengths can be reduced to $\xi_1 = \sqrt{(z_> - z_<)^2 + d^2}$ and $\xi_2 = \sqrt{(z_> + z_<)^2 + d^2}$ which are the corresponding geometrical path lengths of the direct and reflected waves.

To allow for the computation of $\phi(z)$ and $\theta(z)$, we need to find the corresponding polar angles (ϕ_0 and θ_0) at $z = 0$. Formulae for finding ϕ_0 and θ_0 in terms of sound velocity gradient are given elsewhere [1]. Once the polar angles are determined at $z = 0$, $\phi(z)$ and $\theta(z)$ at other heights can be found by using Snell's law:

$$sin\vartheta = (1+\zeta z)\sin\vartheta_0,$$

where $\vartheta = \phi$ or θ. After substituting these angles into Equations (11.22b) and (11.22c) and, in turn, into Equation (11.22a), the sound field in the presence of a linear sound velocity gradient can be calculated.

In downward refraction, additional rays will cause a discontinuity in the predicted sound level because of the inherent approximation used in the ray-trace model. It is possible to determine the critical range, r_c where there are two additional rays in the ray-trace solution. Although tedious, it is straightforward to show that the critical range, for $\zeta > 0$, is given by

$$r_c = \frac{\left\{ \left[\sqrt{(\zeta z_>)^2 + 2\zeta z_>} + \sqrt{(\zeta z_<)^2 + 2\zeta z_<} \right]^{2/3} + \left[\sqrt{(\zeta z_>)^2 + 2\zeta z_>} - \sqrt{(\zeta z_<)^2 + 2\zeta z_<} \right]^{2/3} \right\}^{3/2}}{\zeta} \tag{11.23a}$$

Figure 11.4 shows that the critical range for the single bounce assumption increases as the source and receiver heights increase and that, if predictions are restricted to a horizontal separation of less than 1 km and a normalized

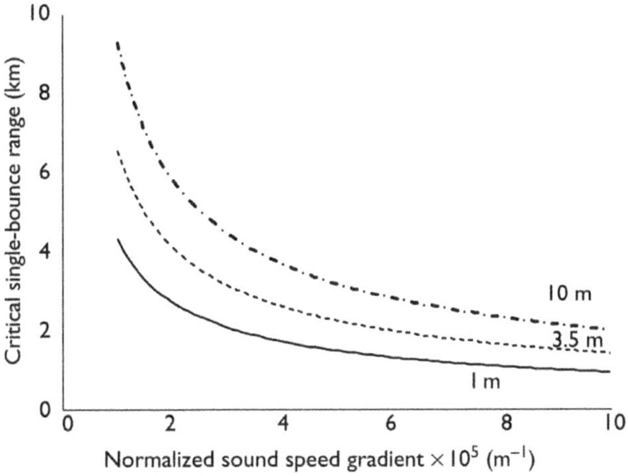

Figure 11.4 Maximum ranges for which single bounce assumption is valid for linear sound speed gradient based on (11.23a) and assuming equal source and receiver heights: 1 m (solid line); 3.5 m (broken line) and 10 m (dot-dash line).

sound speed gradient of less than 0.0001 m^{-1} (corresponding, e.g., to a wind speed gradient < 0.1 s^{-1}), then, for source and receiver at 1 m height, it is reasonable to assume a single-ground bounce in the ray-trace model.

The ray-trace solution for upward-refracting conditions is incorrect when the receiver is in the shadow and penumbra zones. The shadow boundary can be determined from geometrical considerations.

For a given source and receiver heights, the critical range, r_c' is determined from

$$r_c' = \frac{\left\{\left[\sqrt{(\zeta' z_>)^2 + 2\zeta' z_>}\right] + \sqrt{\zeta'^2 z_<(2z_> - z_<) + 2\zeta' z_<}\right\}}{\zeta'}, \tag{11.23b}$$

where

$$\zeta' = \frac{|\zeta|}{1 - |\zeta|\, z>}.$$

Figure 11.5 shows that, for source and receiver heights of 1 m and a normalized sound speed gradient of 0.0001 m^{-1}, the predicted distance to the shadow zone boundary is about 300 m and, as expected, the predicted distance to the shadow zone boundary increases as the source and receiver heights are increased.

Conditions of weak refraction may be said to exist where, under downward-refracting conditions, the ground-reflected ray undergoes only a single bounce, and, under upward-refracting conditions, the receiver is within the illuminated zone. Using Equation (11.19) or alternatives described in Chapter 10, turbulence can be included in (11.2a) to give a heuristic ray-tracing scheme that includes ground effect, refraction and turbulence.

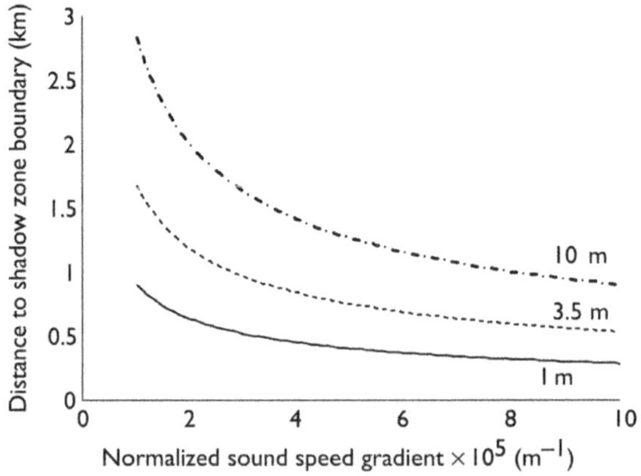

Figure 11.5 Distances to shadow zone boundaries for linear sound speed gradient based on (11.3b) assuming equal source and receiver heights: 1 m (solid line); 3.5 m (broken line) and 10 m (dot-dash line).

11.3 APPROXIMATIONS FOR A-WEIGHTED LEVELS AND GROUND EFFECT OPTIMIZATION IN THE PRESENCE OF WEAK REFRACTION AND TURBULENCE

11.3.1 Ground Effect Optimization

Typically, empirical schemes for predicting assume that all porous ground surfaces may give identical excess attenuation within the accuracy of the scheme. However, as described in Chapter 5, porous ground surfaces display a wide range of porosity and the flow resistivity can vary over nearly three orders of magnitude. The question arises of whether it is possible to optimize the excess attenuation due to ground effect for a given source spectrum and source–receiver geometry. Makarewicz [19] derived an approximation for ground effect at relatively long range which could be used to calculate the effective flow resistivity that maximizes ground effect for a given point source–receiver geometry. He used exponential functions to represent A-weighting and air absorption, and the Delany-Bazley single parameter model of ground impedance which, however, is unphysical (see Chapter 5). Li et al. [1] pointed out sign errors in this work, used a two-parameter impedance model (see Chapter 3) and deduced alternative closed-form results for ground effect and optimum parameters. Makarewicz [20] subsequently extended this model to take account of different source spectra and atmospheric turbulence. Attenborough and Li [21] revised this work to allow for a more general range of ground impedance and the effects of roughness, at scales small compared with the sound wavelengths of interest, on the coherent sound field. For a given source–receiver geometry, this enables prediction of ground types that will achieve maximum excess attenuation and estimation of the magnitudes of the optimized excess attenuation. Also, an approximation for propagation in a linear sound velocity gradient can be used to explore the extent to which the optimized ground effect is affected by atmospheric refraction [21].

In the next sections, the general basis for the approximations including an approximate method for including weak refraction effects is reviewed. Subsequently, expressions for excess attenuation and optimum ground parameters are derived. Numerical explorations of their sensitivities to source–receiver geometry and refraction are made. Finally, conclusions are offered based on these predictions.

11.3.2 Integral Expressions for A-Weighted Mean Square Sound Pressure

Following Makarewicz [19], the total A-weighted power of a (broadband) source is written as

$$\left(P^{(0)} / \mu_A\right)\left[\Gamma(m+1) / \mu_A{}^m\right] \text{ W}, \tag{11.24}$$

where Γ is the gamma function, m, μ_A and $P^{(0)}$ are constants determined by fitting the source power spectrum and $\mu_A = \delta + \mu$.

The A-weighting parameter δ (=6.1413 × 10^{-4}) is chosen such that the spectrum peaks at $f_m = m/\mu_A$ Hz.

Figure 11.6 shows an example simulation of an A-weighted source power spectrum.

The crosses correspond to $m = 2.55$, $P^{(0)} = 10^8$ and $\mu_A = 4.03 \times 10^{-3}$, giving a peak sound power level at 633 Hz and the open boxes represent A-weighted octave band power levels deduced from the measured sound level spectrum at 152.4 m from a fixed Avon engine at a disused airfield during low wind, low turbulence conditions by correcting for spherical spreading, ground effect, air absorption and turbulence [22].

In principle it should be possible to fit m, $P^{(0)}$ and μ_A analytically, given f_m, the peak A-weighted sound power level and P_A. However, in this instance, the values are the result of trial and error fitting.

The basis for this and previous work is the classical Weyl-Van der Pol formula (2.40) for the sound field due to a point source above an impedance plane (see Chapter 2). After further approximation for near-grazing incidence, inclusion of exponential factors to allow for A-weighting, source spectrum, air absorption and turbulence, and integration over frequency (f), an expression for the A-weighted mean square sound pressure may be deduced. In this chapter, the time convention $\exp(i\omega t)$, where $\omega = 2\pi f$, is used rather than the $\exp(-i\omega t)$ convention used in previous chapters.

Following [20, 21],

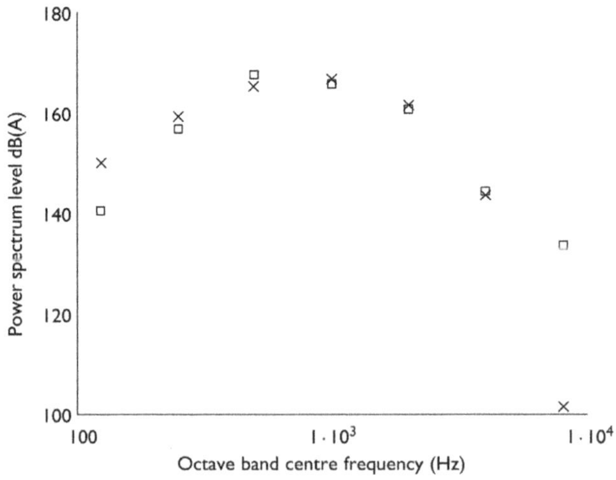

Figure 11.6 Example simulation of an A-weighted octave band power spectrum (×) compared with that deduced from measurements (□) obtained at 1.2 m height with a fixed Rolls Royce Avon jet engine mounted at 2.16 m height over grassland [23]. The simulation uses $m = 2.55$, $P^{(0)} = 10^8$ and $\mu_A = 4.03 \times 10^{-3}$.

$$p_A^2 = \left(\rho_0 c_0 / 4\pi d^2\right)\left[P_A + P_{A1} + P_{A2}\right], \tag{11.25a}$$

where

$$P_A = \int_0^\infty W_0 P^{(0)} f^m \exp\left(-\mu_A f - 2\alpha(f)\right) df, \tag{11.25b}$$

$$P_{A1} = \int_0^\infty W_0 P^{(0)} f^m \left(Q_1^2 + Q_2^2\right)\exp\left(-\mu_A f - 2\alpha(f)\right) df, \tag{11.25c}$$

$$P_{A2} = 2\int_0^\infty W_0 P^{(0)} f^m T\left\{Q_1 \cos(k_0 \Delta r) + Q_2 \sin(k_0 \Delta r)\right\}\exp\left(-\mu_A f - 2\alpha(f)\right) df, \tag{11.25d}$$

where $W_0 = 1.4676 \times 10^{-6}$, $\alpha(f) = a_1 f + a_2 f^2$ is the air absorption parameter, a_1 and a_2 are constants with $a_2 \ll a_1$.

The turbulence parameter, $T = \exp(-\gamma_t f d)$, where $\gamma_t = \varepsilon\left(4\pi^{5/2}/c_0^2\right)n^2 L_0 (1-\rho)$, (see Chapter 10), depends on the mean square refractive index $\langle n^2 \rangle$, the largest turbulence scale (L_0 m), the transverse correlation (ρ), and ε is a constant (= 1/2 or 1).

Δr m is the path length difference between direct and ground-reflected rays, d m is the range and k_0 m^{-1} is the wave number for sound in air.

In the presence of a linear sound speed profile, and subject to certain constraints, Equations (11.25a–11.25d) remain valid as long as Δr in equation (11.25d) and the angle of specular reflection θ at the ground are calculated for curved ray paths (see Section 11.1).

Q_1 and Q_2 are the real and imaginary parts of the spherical wave reflection coefficient.

If it is assumed that (i) A-weighting reduces to insignificance any broadband contributions at frequencies less than 250 Hz, (ii) $d \geq 10$ and (iii) $Z \cos\theta \ll 1$, where $Z = R - iX$ is the normal surface impedance of the ground and θ is the angle of specular reflection, then it is possible to approximate the spherical wave reflection coefficient considerably.

Hence,

$$Q_1 \approx -1 + 2R\cos\theta - \frac{2c_0}{\pi d f} RX - 2\left(R^2 - X^2\right)\cos^2\theta \tag{11.26a}$$

$$Q_2 \approx -2X\cos\theta + 4RX\cos^2\theta - \frac{c_0}{\pi d f}\left(R^2 - X^2\right) \tag{11.26b}$$

and

$$Q_1^2 + Q_2^2 \approx 1 - 4R\cos\theta + \frac{4c_0}{\pi d f} RX + 8R^2\cos^2\theta \tag{11.26c}$$

Note that Equations 11.26a–11.26c correct sign errors in the corresponding equations in Makarewicz [19, 20] and misprints in [21].

If $X > R$, then the terms in $(R^2 - X^2)$ tend to increase Q_1 and Q_2, and, hence, the integrands in Equations (11.25b) and (11.26c).

11.3.3 Approximate Models for Ground Impedance

Expressions for two-parameter impedance models (Chapter 5 Equations (5.37) and (5.38)) are repeated here but with the time convention changed from $\exp(-i\omega t)$ to $\exp(+i\omega t)$. The impedance of a variable porosity medium or a non-hard backed layer can be written as (see (5.38))

$$Z = a(1-i)\sqrt{\frac{\sigma_e}{f}} - \frac{ib\alpha_e}{f}, \tag{11.27}$$

where $a = 1/\sqrt{\pi\rho_0\gamma}$, $b = c_0/8\pi\gamma$, σ_e is an effective flow resistivity and $\alpha e = (n' + 2)\alpha/\Omega$, or $\alpha e = 4/d_e$.

If αe is positive, this impedance expression implies $X > R$.

Based on Equations (5.34) and (5.59) and the altered time-dependence convention, an approximate expression for the impedance of a rough porous surface is given by [22],

$$Z_r \approx \frac{1}{2A\sqrt{f}}(1-i) - \frac{D}{A^2}, \tag{11.28}$$

where $A = \frac{1}{2}\sqrt{\pi\rho_0\gamma / \sigma_e}$, and $c_0D/2\pi$ may be considered as an effective roughness volume per unit area.

11.3.4 Effects of Weak Refraction

Apart from turbulence, Makerewicz [19,20] ignored meteorological effects in deriving a closed-form expression for an A-weighted mean square sound pressure. As discussed in previous chapters, variations in temperature and wind with height cause the speed of sound to vary with height and lead to refraction. If refraction is not included, the applicability of the model is limited. Since the direct and reflected sound waves are not separated out explicitly as two terms, the evaluation of A-weighted mean square pressure by numerical methods demands considerable computational resources. For engineering applications, the ray-trace approach is more convenient.

Equations (11.26a–11.26c) and the methods detailed elsewhere [21] can be used to derive the A-weighted mean square sound pressure in the presence of weak effective sound speed gradients. The expression will be given in the next section where weak refraction and other factors are included in the analysis.

11.3.5 Approximations for Excess Attenuation

11.3.5.1 Variable Porosity or Thin Layer Ground

Substitution of (11.7a–11.7c) into (11.6) leads to integrals of the form

$$G_1(k) = \int_0^\infty x^k e^{-\left(gx+hx^2\right)} \cos(lx)\,dx \text{ and } G_2(k) = \int_0^\infty x^k e^{-\left(gx+hx^2\right)} \sin(lx)\,dx$$

where

$$g = \mu_A + 2a_1 d,\ h = (2a_2 + \gamma_t)d \text{ and } 1 = 0 \text{ or } 2\pi\Delta r/c_0$$

These may be evaluated in closed form in terms of products of gamma and parabolic cylinder functions [23]. If d is sufficiently large and a_2 and γ_t are sufficiently small, then $g^2 + l^2 \gg 6h$ and $l/g \ll 1$, so that

$$\text{For } l = 0,\ G_1(k) \to C(k) \approx \Gamma(k+1)/(\mu_A + 2a_1 d)^{-(k+1)}, \tag{11.29a}$$

$$\text{for } l \neq G_1(k) \approx C(k) - \frac{1}{2}C(k+1)l^2 - (2a_2+\gamma_t)dC(k+2) \tag{11.29b}$$

and

$$G_2(k) \approx C(k+1)l - (2a_2+\gamma_t)dC(k+3)l. \tag{11.29c}$$

Finally, use of these and further approximations leads to

$$p_A{}^2 = \left(P_A / 4\pi d^2\right)\left[S(\sigma_e,\alpha_e,g,m,\gamma_t,l,\theta_0,d)\right], \tag{11.30a}$$

where

$$S = B_0 d + B_1 l^2 - B_2 l cos\theta_0 + B_4 l \tag{11.30b}$$

$$B_0 = \frac{2(2a_2+\gamma_t)}{g^2}\frac{\Gamma(m+3)}{\Gamma(m+1)} \tag{11.30c}$$

$$B_1 = \frac{1}{g^2}\frac{\Gamma(m+3)}{\Gamma(m+1)} \tag{11.30d}$$

$$B_2 = 4\left[\frac{\Gamma\left(m+\frac{3}{2}\right)}{\Gamma(m+1)}\frac{a\sqrt{\sigma_e}}{\sqrt{g}} + b\alpha_e\right] \tag{11.30e}$$

$$B_3 = \frac{4g}{\Gamma(m+1)}\left[2\Gamma(m)a^2\sigma_e + 2\sqrt{g}\Gamma\left(m-\frac{1}{2}\right)ab\sqrt{\sigma_e}\alpha_e + g\Gamma(m-1)b^2\alpha_e{}^2\right] \tag{11.30f}$$

$$B_4 = \frac{2c_0 g}{\pi d\Gamma(m+1)}\left[2\Gamma\left(m-\frac{1}{2}\right)ab\alpha_e\sqrt{\sigma_e g} + g\Gamma(m-1)b^2\alpha_e{}^2\right] \tag{11.30g}$$

The ground parameters that minimize S and hence give maximum excess attenuation, for a given geometry, are given by

$$\sqrt{\sigma_{em}} = \frac{A_2A_4 - 2A_1A_5}{A_4^2 - 4A_3A_5}, \tag{11.31a}$$

and

$$\alpha_{em} = \frac{A_1A_4 - 2A_2A_3}{A_4^2 - 4A_3A_5}, \tag{11.31b}$$

where

$$A_1 = \frac{4a}{\sqrt{g}} \frac{\Gamma\left(m + \frac{3}{2}\right)}{\Gamma(m+1)} l \cos\theta_0, \tag{11.31c}$$

$$A_2 = 4bl\cos\theta_0, \tag{11.31d}$$

$$A_3 = 8a^2 g \frac{\Gamma(m)}{\Gamma(m+1)} \cos^2\theta_0, \tag{11.31e}$$

$$A_4 = 8abg^{\frac{3}{2}} \frac{\Gamma\left(m - \frac{1}{2}\right)}{\Gamma(m+1)} \left(\cos^2\theta_0 + \frac{c_0 l}{2\pi d}\right) \tag{11.31f}$$

and

$$A_4 = 4b^2g^2 \frac{\Gamma(m-1)}{\Gamma(m+1)} \left(\cos^2\theta_0 + \frac{c_0 l}{2\pi d}\right) \tag{11.31g}$$

11.3.5.2 Rough Ground

A similar procedure starting with the approximate rough porous surface impedance given by Equation (11.28) yields Equations (11.30a) and (11.30b) together with

$$B_2 = 4\left[\frac{\Gamma\left(m + \frac{3}{2}\right)}{\Gamma(m+1)} \frac{a\sqrt{\sigma_e}}{\sqrt{g}}\right], \tag{11.32a}$$

$$B_3 = 8a^2\sigma_e\left[g\frac{\Gamma(m)}{\Gamma(m+1)} - 2aD\sqrt{\sigma_e g}\frac{\Gamma\left(m + \frac{1}{2}\right)}{\Gamma(m+1)} + 2a^2\sigma_e D^2\right] \tag{11.32b}$$

$$B_4 = -8a^3\sigma_e D\left[a\sigma_e D - \sqrt{\sigma_e g}\,\frac{\Gamma\left(m+\frac{1}{2}\right)}{\Gamma(m+1)}\right]. \tag{11.32c}$$

The rough ground parameters that minimize S are given by

$$\sqrt{\sigma_{em}} = A_2 / 2\left(A_3 + \frac{A_4^{\;2}}{4A_5}\right) \tag{11.33a}$$

and

$$D_m = \frac{A_4}{2A_5\sqrt{\sigma_{em}}} \tag{11.33b}$$

where

$$A_2 = \frac{4a}{\sqrt{g}}\frac{\Gamma\left(m+\frac{3}{2}\right)}{\Gamma(m+1)} l\cos\theta_0 \tag{11.33c}$$

$$A_3 = 8a^2 g\frac{\Gamma(m)}{\Gamma(m+1)}\cos^2\theta_0 \tag{11.33d}$$

$$A_4 = -16a^3\sqrt{g}\frac{\Gamma\left(m+\frac{1}{2}\right)}{\Gamma(m+1)}\left(\cos^2\theta_0 - \frac{c_0 l}{2\pi d}\right) \tag{11.33e}$$

and

$$A_4 = -16a^4\left(\cos^2\theta_0 - \frac{c_0 l}{2\pi d}\right). \tag{11.33f}$$

If the atmosphere is homogeneous, then $l \approx 4\pi h_r h_s / c_0 d$ and $\cos\,\theta_0 \approx (h_s + h_r)/d$ and

$$\sqrt{\sigma_{em}} = \frac{2\pi}{ac_0 g^{\frac{3}{2}}}\frac{\Gamma\left(m+\frac{3}{2}\right)(h_r h_s)(h_r + h_s)}{2\Gamma(m)(h_r+h_s)^2 - \left(\Gamma\left(m+\frac{1}{2}\right)\right)^2\Gamma(m+1)\left(h_r^{\,2}+h_s^{\,2}\right)} \tag{11.33g}$$

and

$$D_m = \frac{c_0 g^2}{2\pi}\frac{2\Gamma(m)(h_r+h_s)^2 - \left(\Gamma\left(m+\frac{1}{2}\right)\right)^2\Gamma(m+1)\left(h_r^{\,2}+h_s^{\,2}\right)}{(2m+1)\Gamma\left(m+\frac{3}{2}\right)(h_r h_s)(h_r+h_s)} \tag{11.33h}$$

11.3.5.3 Smooth High Flow Resistivity Ground

Completing the picture, the optimum effective flow resistivity for a semi-infinite smooth rigid-porous high flow resistivity surface, such that $Z = a(1-i)\sqrt{\frac{\sigma_e}{f}}$, under an homogeneous atmosphere, is given by

$$\sigma_{em} = \frac{\pi^2}{a^2 c_0{}^2 g^3}\left(\frac{\Gamma\left(m+\frac{3}{2}\right)}{\Gamma(m)}\right)^2 \frac{h_r{}^2 h_s{}^2}{(h_r + h_s)^2}, \qquad (11.33\text{i})$$

where S is given by Equation (11.29b) together with Equations (11.29c)–(11.30g), or (11.29c), with (11.21), the excess attenuation (in dB) may be calculated for optimum or other parameter values from

$$EA = 10\log(S). \qquad (11.34)$$

11.3.6 Numerical Examples and Discussion

11.3.6.1 Comparison with Data: Avon Jet Engine Source

With $\mu_A = 4.03 \times 10^{-3}$, $m = 2.55$, $P^{(0)} = 10^8$, the exponentially modelled A-weighted source spectrum peaks at 633 Hz and simulates noise from a stationary Avon jet engine. In Figure 11.7, the A-weighted sound levels predicted by the approximate theory over a variable porosity ground characterized by $\sigma e = 30$ kPa s m^{-2} and $\alpha e = 16$ m^{-1} are compared with measurements at ranges between 200 and 1200 m obtained under zero wind, low turbulence conditions [22]. Other parameter values used in the predictions are $h_s = 2.16$ m, $h_r = 1.2$ m, $\langle n^2 \rangle = 10^{-8}$, $L_0 = 1\ m$, $\rho = 0$, $\varepsilon = 0.5$, $a_1 = 5 \times 10^{-7}$and $a_2 = 2 \times 10^{-12}$.

11.3.6.2 Sensitivity to Spectrum, Source Height and Distance

Assuming a range of 500 m and using the same air absorption, turbulence and spectrum parameter values as used for Figures 11.6 and 11.7, the optimum parameter values according to the variable porosity impedance model are predicted to be $\sigma_{em} = 58.8$ kPa s m^{-2} and $\alpha_{em} = 7.74$ m^{-1}.

For the assumed A-weighted spectrum, source and receiver height and low turbulence, Figures 11.8(a) and (b) show that the predicted excess attenuation of nearly 16 dB is insensitive both to variation in the effective flow resistivity for $\sigma e < 10^3$ kPa s m^{-2}, which represents acoustically soft ground, and to variation in the porosity gradient parameter (which changes the predicted excess attenuation by only 1 dBA). However, increasing the effective flow resistivity

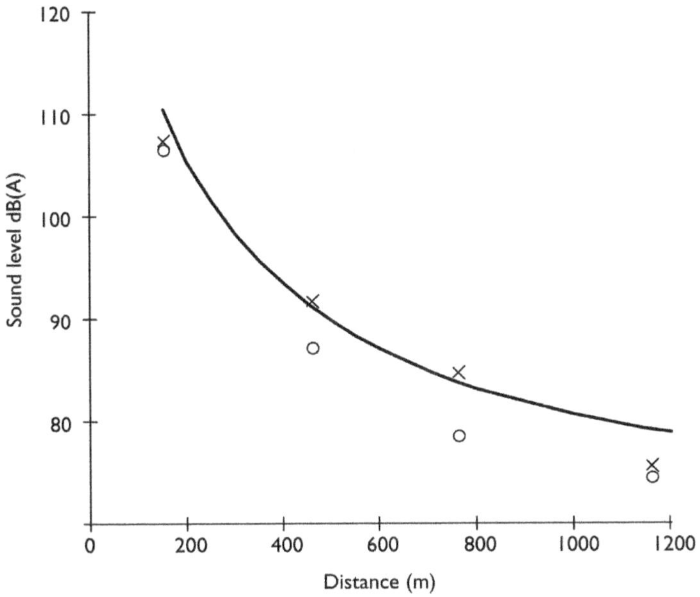

Figure 11.7 Comparison of predicted (solid line) and measured sound level (points) as a function of range from a fixed Avon jet engine (h_s = 2.16 m, h_r = 1.2 m) under zero wind low turbulence conditions [23]. The predictions use equations (11.20) with an A-weighted spectrum simulated by m = 2.55 and $\mu_A = 4.03 \times 10^{-3}$, air absorption given by $a_1 = 5 \times 10^{-7}$, $a_2 = 1 \times 10^{-12}$, turbulence represented by $\langle \mu^2 \rangle = 10^{-8}$, L_0 = 1 m, ρ = 0, and no refraction. The data correspond to averages of two twenty second samples.

above the optimum range ($10^5 < \sigma_e < 10^6$) is predicted to decrease the excess attenuation significantly.

Typical parameter values for relatively 'soft' grassland obtained by fitting short-range level difference spectra [24] are listed in Table 11.1. The fitted parameters for such grassland are predicted to give a ground effect near the optimum. Approximately, the optimum surface is predicted to result in an excess attenuation at 500 m that is more than 20 dBA greater than over an acoustically hard surface (for which the excess attenuation is taken normally to be -3 dBA).

Similar results are predicted when the rough porous surface impedance model (11.28) is used. For example, the optimum rough ground parameters, for the same source–receiver geometry and spectrum as assumed for Figure 11.8, are predicted to be an effective flow resistivity of 107 kPa s m^{-2} and a roughness volume per unit area of 0.0077 m. This represents roughness equivalent to close-packed hemi-spherical bosses of radius 0.08 m. Again, as shown in Figure 11.9, the excess attenuation is predicted to be relatively insensitive to effective flow resistivity below a value of 10^3 kPa s m^{-2}.

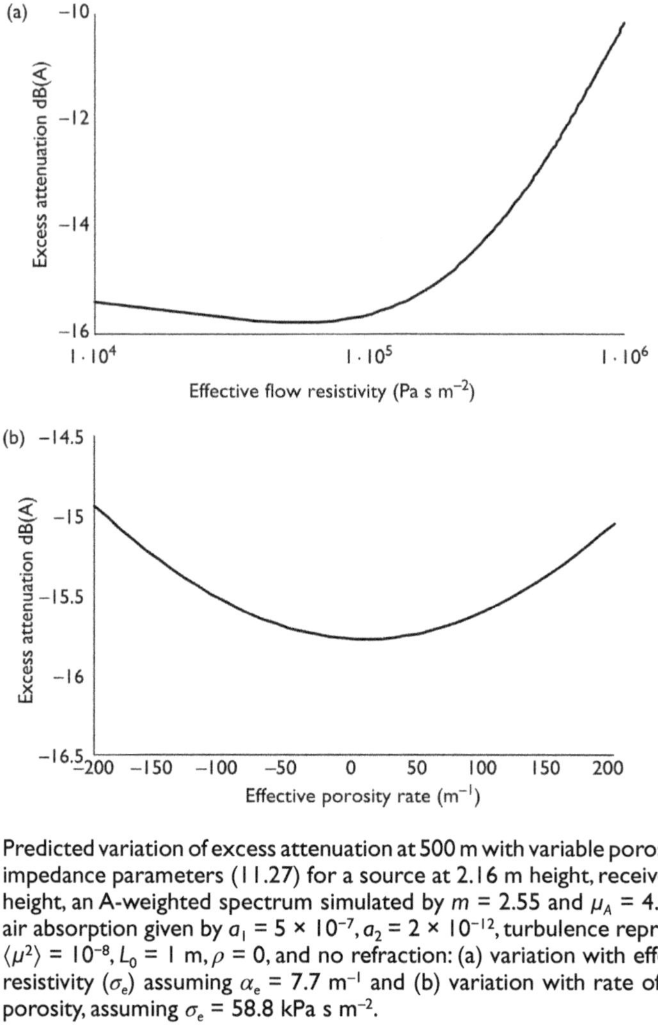

Figure 11.8 Predicted variation of excess attenuation at 500 m with variable porosity ground impedance parameters (11.27) for a source at 2.16 m height, receiver at 1.2 m height, an A-weighted spectrum simulated by $m = 2.55$ and $\mu_A = 4.03 \times 10^{-3}$, air absorption given by $a_1 = 5 \times 10^{-7}$, $a_2 = 2 \times 10^{-12}$, turbulence represented by $\langle \mu^2 \rangle = 10^{-8}$, $L_0 = 1$ m, $\rho = 0$, and no refraction: (a) variation with effective flow resistivity (σ_e) assuming $\alpha_e = 7.7$ m^{-1} and (b) variation with rate of change of porosity, assuming $\sigma_e = 58.8$ kPa s m^{-2}.

Table 11.1 Values of variable porosity impedance model parameters obtained from fitting short-range level difference spectra over relatively 'soft' grassland [24].

Ground description	*Effective flow resistivity kPa s m^{-2}*	*Porosity rate/m*
Heath	114	17
Heath	52	33
Long grass	167	46
Long grass	140	1
Pasture	26	34
Lawn	39	17

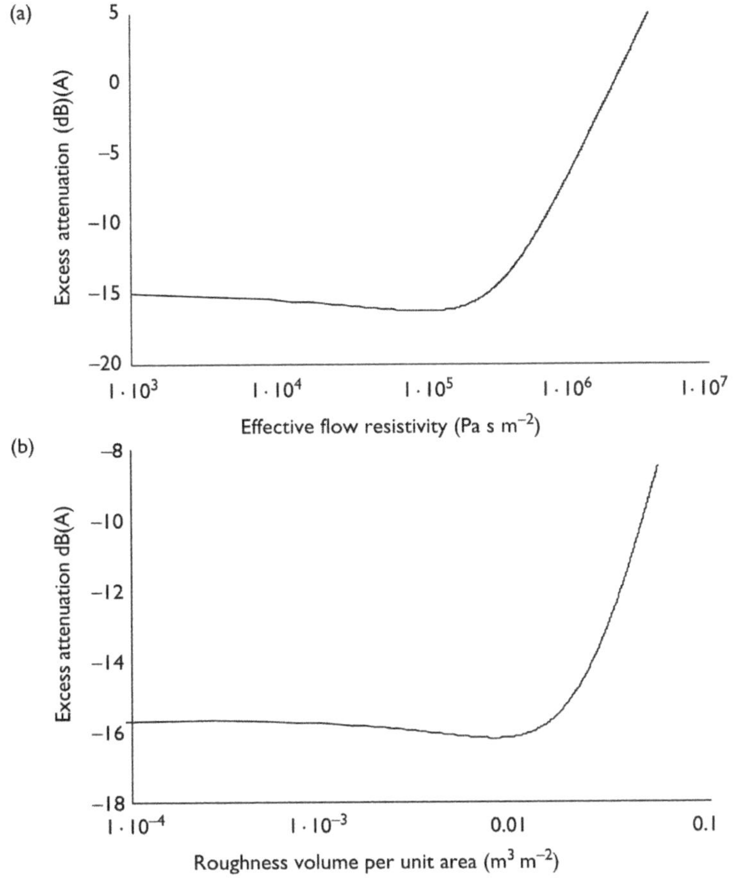

Figure 11.9 Sensitivity of excess attenuation spectrum to ground parameters (rough porous ground impedance approximation, equation (11.28)) for source with A-weighted spectrum peak frequency of 633 Hz at 2.16 m height, receiver at 1.2 m height and range 500 m (a) variation with effective flow resistivity, σ_e, assuming optimum roughness parameter value of 1.41×10^{-4} (b) with roughness, assuming $\sigma_e = 107$ kPa s m^{-2}.

However, increasing roughness beyond the optimum is predicted to have a detrimental effect.

Assuming an Avon engine source spectrum and ground characterized by the variable porosity impedance model, the optimum parameters given by Equation (11.27) are predicted to be very sensitive to source height. This is illustrated in Figure 11.10. As might be expected, with the receiver at 1.2 m height, the higher the source the less important is attenuation due to ground effect and, therefore, the less important the ground impedance. The optimum parameters are predicted to be sensitive also to the peak frequency in the source spectrum. This is illustrated in Figure 11.11 where all parameter values other than the source spectrum peak frequency are as for Figure

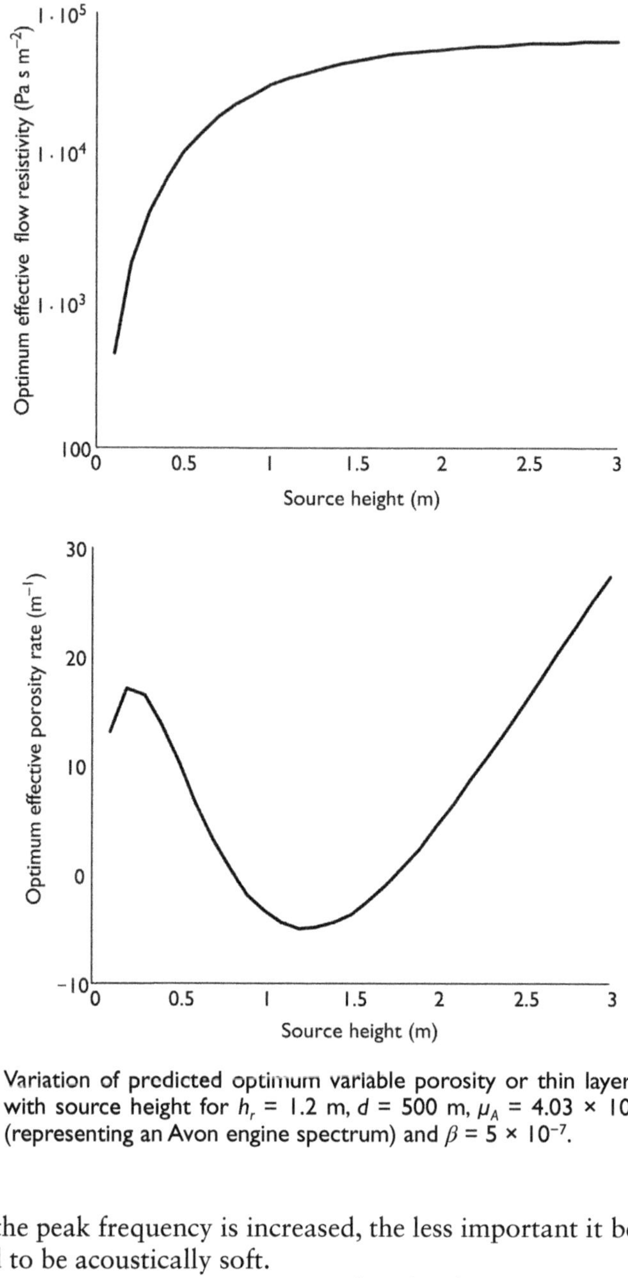

Figure 11.10 Variation of predicted optimum variable porosity or thin layer parameters with source height for h_r = 1.2 m, d = 500 m, $\mu_A = 4.03 \times 10^{-3}$, $m = 2.55$ (representing an Avon engine spectrum) and $\beta = 5 \times 10^{-7}$.

11.10. As the peak frequency is increased, the less important it becomes for the ground to be acoustically soft.

The optimum excess attenuation is predicted to be very sensitive to turbulence. This is demonstrated in Figures 11.12 and 11.13. The predictions for both ground impedance types are similar and suggest, not surprisingly, that ground impedance has little influence when turbulence is strong. Clearly the most useful optimization of the excess attenuation of A-weighted noise due

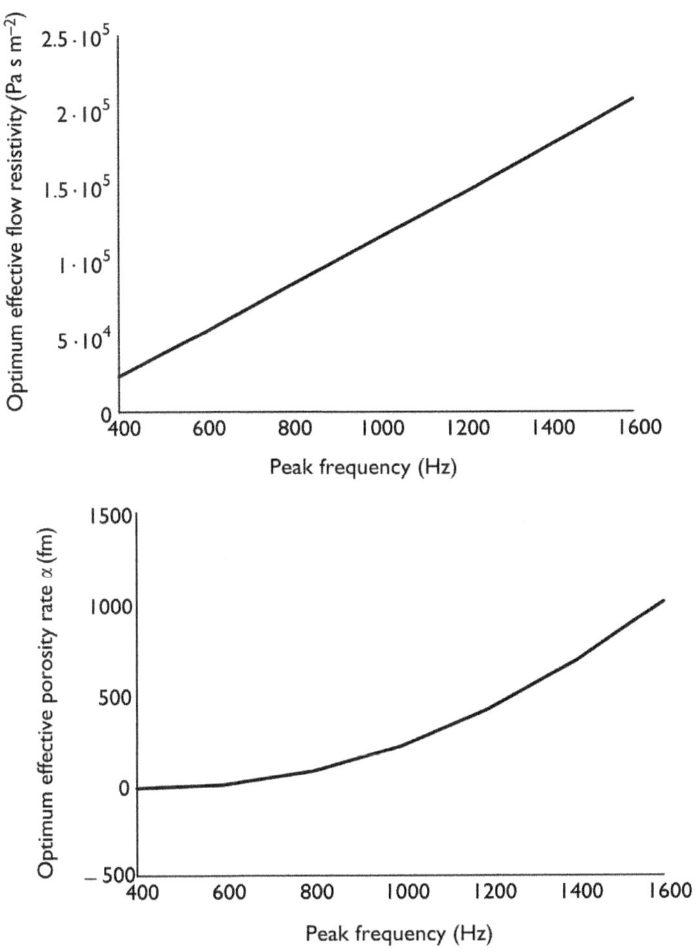

Figure 11.11 Variation of predicted optimum variable porosity or thin layer parameters with A-weighted spectrum peak frequency: $h_s = 2.16$ m, $h_r = 1.2$ m, $\mu_A = 4.03 \times 10^{-3}, \beta = 5 \times 10^{-7}$.

to ground effect will occur for low source heights, low peak frequencies in the A-weighted source spectrum and with little or no turbulence. It should be noted that high impedance and local reaction have been assumed in arriving at these results, so the predictions for the lower range of flow resistivities will be suspected.

11.3.6.3 Variation with Distance

The optimum ground parameters are predicted to depend on distance from the source. For low turbulence and a low height source characterized by an

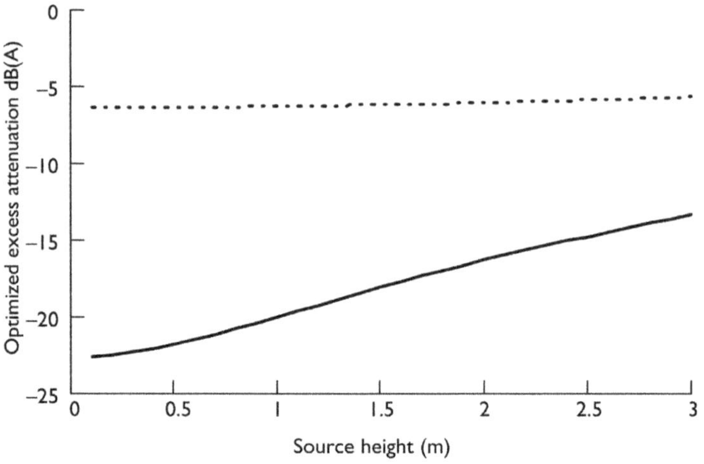

Figure 11.12 Variation of predicted optimum excess attenuation with source height for variable porosity or thin layer ground with source height: h_r = 1.2 m, d = 200 m, m = 2.55, μ_A = 4.03 × 10^{-3}, β = 5 × 10^{-7}. Dotted and solid lines represent predictions with strong and weak turbulence ($<\mu^2>$ = 10^{-6} and 10^{-8} respectively).

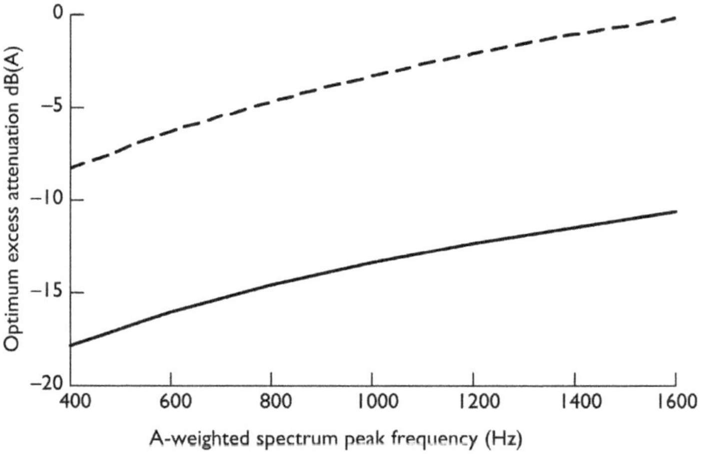

Figure 11.13 Predicted variation of optimum excess attenuation with A-weighted spectrum peak frequency for h_s = 2.16 m, h_r = 1.2 m, d = 500 m, μ_A = 4.03 × 10^{-3} and for two values of turbulence; $\langle\mu^2\rangle$ = 10^{-8} (solid), $\langle\mu^2\rangle$ = 10^{-6} (broken).

A-weighted spectrum peaking at 750 Hz, Figure 11.14 indicates that acoustically softer parameters are predicted to become optimum as the distance from the source increases. However, the dependence is relatively slight and the predicted excess attenuation over ground with constant parameters that have average values within the optimum range is similar to the optimized excess attenuation. This is illustrated in Figure 11.15.

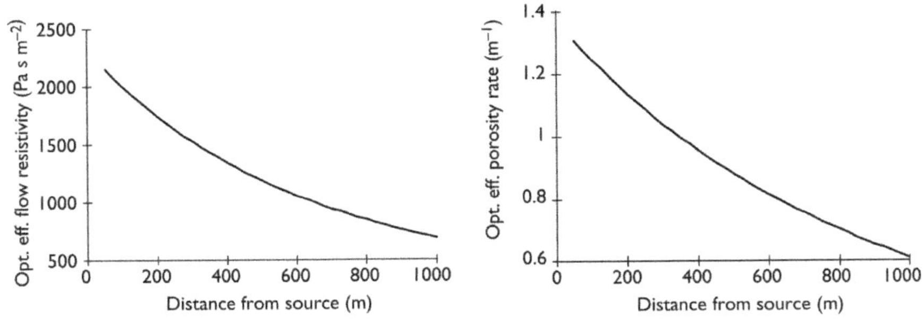

Figure 11.14 Predicted optimum ground parameters as a function of distance for an A-weighted spectrum characterised by $m = 1.5$, $\mu_A = 2 \times 10^{-3}$: $h_s = 0.1$ m, $h_r = 1.2$ m.

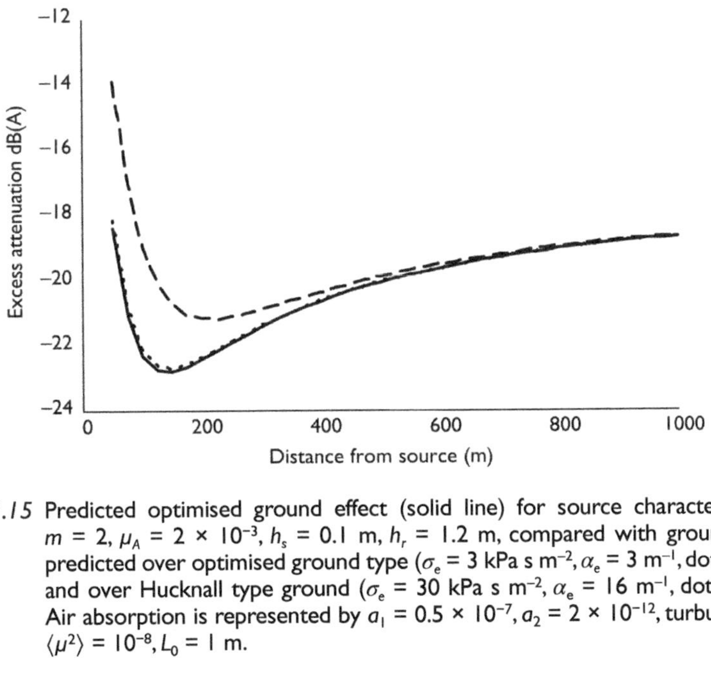

Figure 11.15 Predicted optimised ground effect (solid line) for source characterised by $m = 2$, $\mu_A = 2 \times 10^{-3}$, $h_s = 0.1$ m, $h_r = 1.2$ m, compared with ground effect predicted over optimised ground type ($\sigma_e = 3$ kPa s m^{-2}, $\alpha_e = 3$ m^{-1}, dotted line) and over Hucknall type ground ($\sigma_e = 30$ kPa s m^{-2}, $\alpha_e = 16$ m^{-1}, dotted line). Air absorption is represented by $a_1 = 0.5 \times 10^{-7}$, $a_2 = 2 \times 10^{-12}$, turbulence by $\langle \mu^2 \rangle = 10^{-8}$, $L_0 = 1$ m.

11.3.6.4 Effects of Refraction

For low turbulence, low source and receiver heights (0.1 and 1.2 m, respectively) and a low-frequency (300 Hz) spectrum peak ($m = 1.5$, $\mu_A = 5 \times 10^{-3}$), Figures 11.16–11.19 show that the predicted optimum ground parameters correspond to increasingly hard ground under increasingly downward-refracting conditions and increasingly soft ground under increasingly upward-refracting conditions. The predicted optimum excess attenuation of about 30 dB is little influenced by weak refraction.

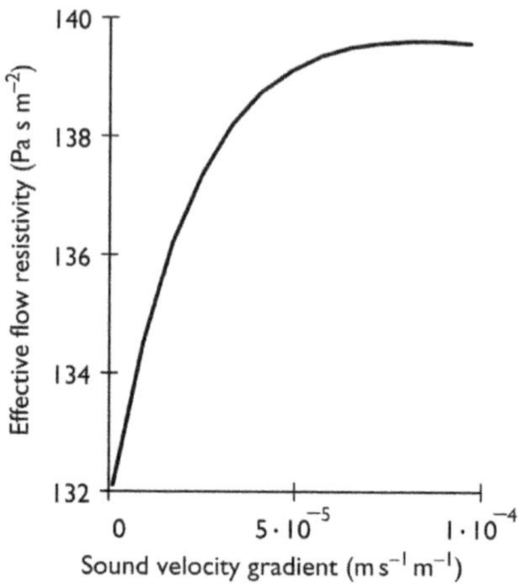

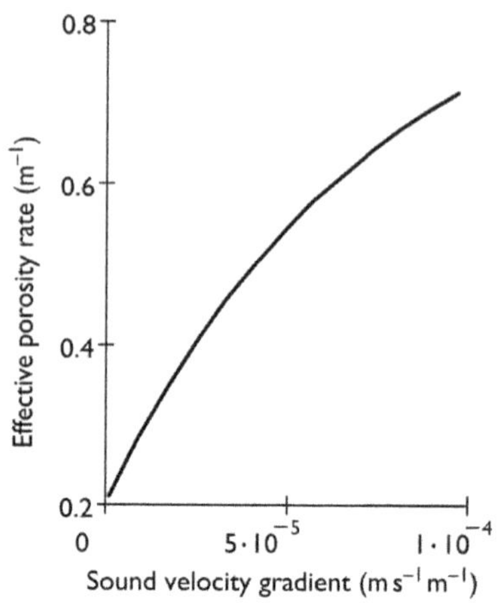

Figure 11.16 Variation of optimum variable porosity ground parameters with downward refracting sound velocity gradient for m = 1.5, μ_A = 5 × 10^{-3}, h_s = 0.1 m, h_r = 1.2 m and range = 200 m.

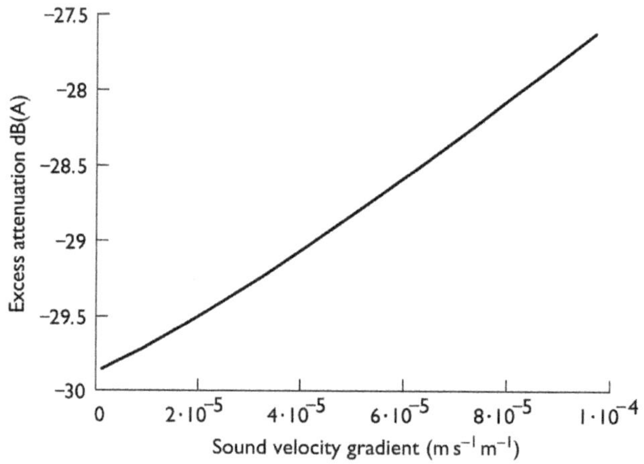

Figure 11.17 Variation of optimum excess attenuations with downward refracting sound velocity gradient for m = 1.5, μ_A = 5 × 10^{-3}; h_s = 0.1 m, h_r = 1.2 m and range = 200 m.

11.3.7 Concluding Remarks

By means of approximations of the classical theory for a point source above an impedance surface and approximate effective impedance models, it is possible to derive closed-form relationships between excess attenuation due to ground effect and parameters relating to the A-weighted broadband source power spectrum, the ground and the source and receiver heights. Although the resulting predictions are approximate, they suggest the possibility for optimizing excess attenuation by controlling the ground characteristics. The greatest opportunities for such optimization are predicted for sources with

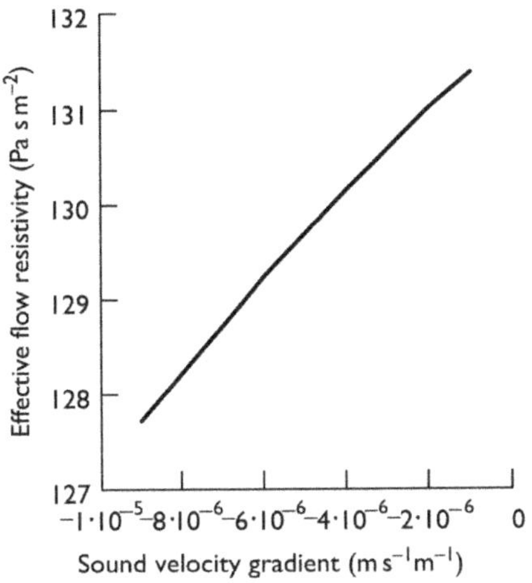

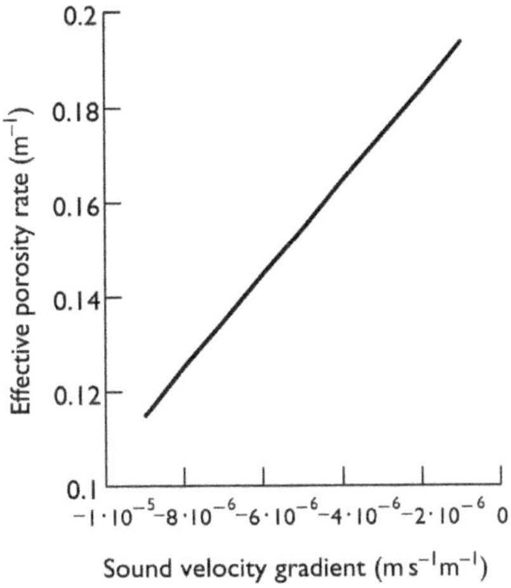

Figure 11.18 Predicted effect of upward refraction on optimum variable porosity ground parameters. Assumed geometry and spectrum are as assumed for Figures 11.16 and 11.17.

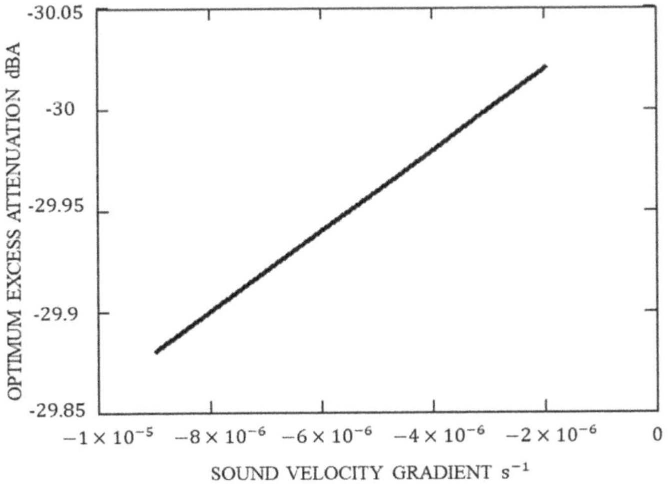

Figure 11.19 Predicted effect of upward refraction on optimum excess attenuation. Assumed geometry and spectrum are as for Figures 11.16 and 11.17.

low-frequency A-weighted spectral peaks, at ranges of less than 200 m in weak turbulence and require rather acoustically soft ground. The resulting predictions are little affected by weak atmospheric refraction. Not surprisingly, the optimum attenuation is predicted to be much higher for sources close to the ground. Since it may not be practicable to lower source heights for many noise sources, an alternative strategy is to raise the ground, perhaps through landscaping. This possibility needs further investigation.

11.4 A SEMI-EMPIRICAL MODEL FOR A-WEIGHTED SOUND LEVELS AT LONG RANGE

Makarewicz and Kokowski [25] have suggested a simple semi-empirical form for carrying out calculations of the variation in A-weighted levels from a stationary source for ranges up to 150 m allowing for wavefront spreading and ground effect. Their result for propagation from a point source is repeated here as Equation (11.35).

If source and receiver heights are denoted by h_s and h_r, respectively, and r is their horizontal separation,

$$L_A = L_{WA} + 10\log\left(\frac{E}{4\pi r^2}\right) - 10\log\left[1 + \gamma_g\left(\frac{r}{h_s + h_r}\right)^2\right]. \tag{11.35}$$

where E is an adjustable parameter intended to include the effect of the presence of the ground on radiation of sound energy from the source ($2 \geq E \geq 1$), L_{WA} is the A-weighted sound power level of the source and γ_g is an adjustable ground parameter.

The lower the impedance of the ground, the larger is the value of γ_g.

Turbulence effects can be included in Equation (11.35) by using $\gamma_g \exp(-\gamma_t)$ instead of γ_g, where γ_t is another adjustable parameter [25]. Air absorption is included [26] by an additional attenuation given by $10 \log (1 + \chi d)$ where χ represents a temperature- and humidity-dependent equivalent absorption coefficient ($= 14\times10^{-4}$ at 10°C and 70% RH). With these modifications, Equation (11.35) may be extended to longer ranges. These modifications have been used for the predictions compared with data in Figure 11.20(b). The value of E has not been included explicitly but assumed to be incorporated in the A-weighted source power deduced from close-range data.

Assuming a linear sound speed gradient, refraction effects may be included [27] by adding a term

$$-10\log\left[1 + \gamma_1\left(\frac{r}{h_s + h_r}\right)^2\right] \tag{11.36}$$

to Equation (11.35), where γ_1 is intended to include effects of both turbulence and refraction. In Figure 11.21, this latter modification has been used to compare with data obtained under downward-refracting conditions at Hucknall [22]. Additional data for A-weighted levels as a function of range have been generated from corrected level difference measurements by Parkin and Scholes under low wind and isothermal conditions at a different airfield [28]. These 'data' and the corresponding predictions are shown in Figure 11.20(a) also. By means of Equations (11.35) and (11.33a–11.33i), differences in the measured attenuation rates at the two airfields under acoustically neutral conditions are predicted to be the result of differences in

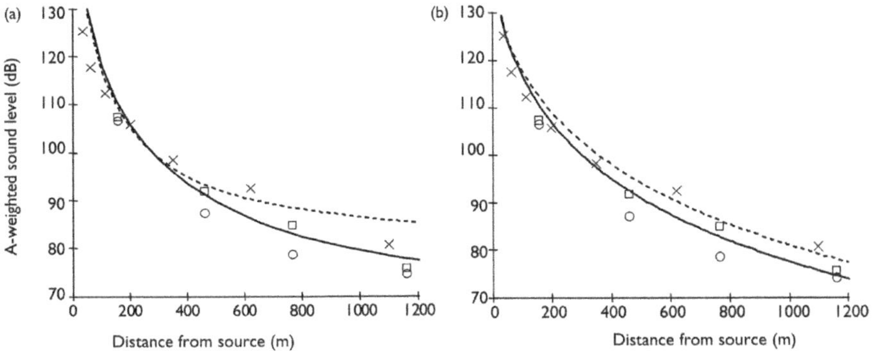

Figure 11.20 A-weighted sound levels during acoustically neutral conditions measured at 1.2 m above grass at Hucknall (open boxes and circles) [23] and deduced from Parkin and Scholes' data at Hatfield (crosses) [29] compared with predictions of (a) the exponential simulation model [10] (solid and broken lines) and (b) Makarewicz *et al* model including turbulence [22] (solid and broken lines using fitted ground parameters $\gamma_g = 7 \times 10^{-4}$ and 3×10^{-4} and fitted turbulence parameters $\gamma_t = 10^{-7}$ and 10^{-6} respectively.

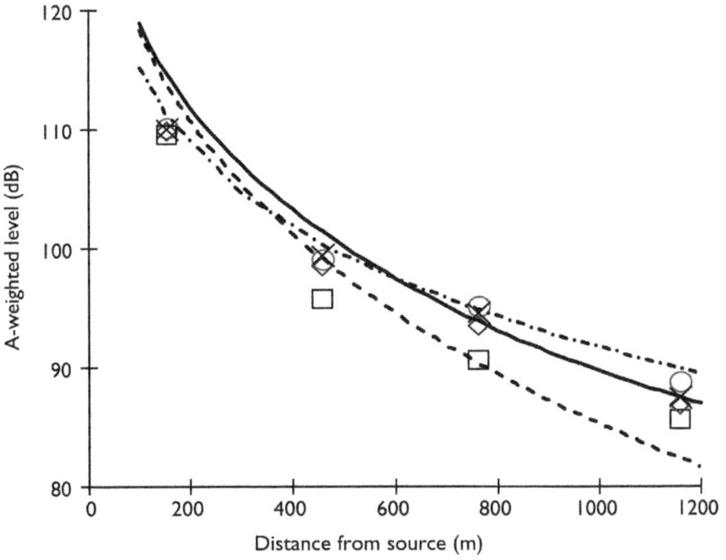

Figure 11.21 Data for A-weighted levels under downwind conditions (wind speed up to 6 m/s at 6.4 m), obtained with receivers 6.4 m above grassland at Hucknall [23], compared with predictions (solid line) using (11.33a–11.33i) augmented by the refraction term (11.36). The values of L_{WA} and γ_g obtained for acoustically-neutral conditions (Figure 11.20(b)) have been used and $\gamma_l = -3.25 \times 10^{-5}$. The broken line represents the corresponding predictions for acoustically neutral conditions. The dash-dot line is the result of the ISO octave band method (see Chapter 12).

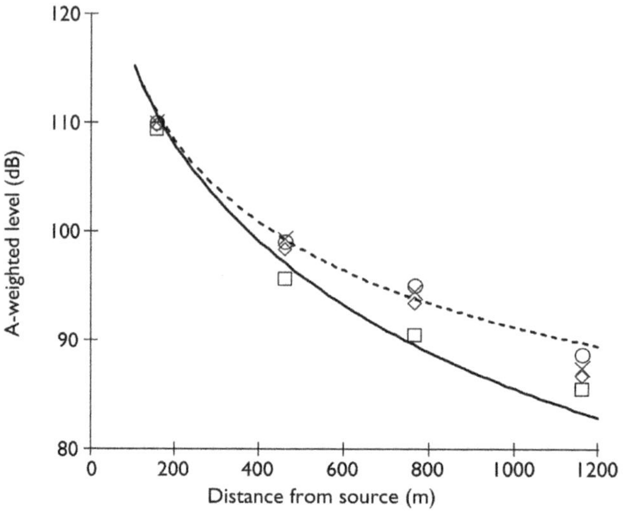

Figure 11.22 Data for A-weighted levels under downwind conditions (wind speed up to 6 m/s at 6.4 m), obtained with receivers 6.4 m above grassland at Hucknall [23], compared with predictions using (11.37) with L_{EWA} = 164 dB, $\gamma_g = 7 \times 10^{-4}$ and $\gamma_t = 8 \times 10^{-3}$ (solid line) or 4×10^{-2} (broken line).

turbulence rather than the result of differences in ground parameters. The ISO octave band method predictions (see Chapter 12) are also in good agreement with these data while slightly underestimating the measured attenuation.

Golębiewski [29] has suggested that the effect of turbulence, and hence γ_t, depends on the sound path height. Consequently, (11.35) and (11.36) may be rewritten in the form

$$L_A = L_{EWA} - 10\log\left[1 + \gamma_g\left(\frac{r}{h_s + h_r}\right)^2 e^{-\gamma_t \frac{d}{h_s + h_r}}\right] - 10\log[1 + \chi d], \quad (11.37)$$

where L_{EWA}, γ_g and γ_t are adjustable parameters and χ depends on temperature and humidity. Figure 11.22 shows that having fixed L_{EWA} and γ_g at values of 164 dB and 7×10^{-4}, respectively, the Hucknall downwind data are bracketed by $8\times10^{-3} \leq \gamma_t \leq 4\times10^{-2}$.

REFERENCES

[1] K. M. Li, K. Attenborough and N. W. Heap, Comment on 'Near-grazing propagation above a soft ground, *J. Acoust. Soc. Am.* 88 1170–1172 (1990).

[2] L'Esperance, P. Herzog, G. A. Daigle and J. R. Nicolas, Heuristic model for outdoor sound propagation based on an extension of the geometrical ray theory in the case of a linear sound speed profile, *Appl. Acoust.* 37 111–139 (1992).

[3] K. M. Li, On the validity of the heuristic ray-trace based modification to the Weyl Van der Pol formula, *J. Acoust. Soc. Am.* **93** 1727–1735 (1993).
[4] K. M. Li, Propagation of sound above an impedance plane in a downward refracting atmosphere, *J. Acoust. Soc. Am.* **99** 746–754 (1996).
[5] K. M. Li, A high-frequency approximation of sound propagation in a stratified moving atmosphere above a porous ground surface, *J. Acoust. Soc. Am.* **95** 1840–1852 (1994).
[6] V. E. Ostashev, Ray acoustics of moving media, *Izv. Atmos. Oceanic Phys.* **25** 661–673 (1989).
[7] M. M. Boone and E. A. Vermaas, A new ray-tracing algorithm for arbitrary inhomogeneous and moving media, *J. Acoust. Soc. Am.* **90** 2109–2117 (1991).
[8] L. M. Brekhovskikh, *Waves in Layered Media*, Academic Press, New York, 2nd Ed., pp. 389–395 (1980).
[9] R. J. Thompson, Ray theory for an inhomogeneous moving medium, *J. Acoust. Soc. Am.* **51** 1675–1682 (1972).
[10] D. I. Blokhintzev, *NACA Tech. Memo.* 1399 (1956).
[11] F. Walkden and M. West, Prediction of enhancement factor for small explosive sources in a stratified moving atmosphere, *J. Acoust. Soc. Am.* **84** 321–326 (1988).
[12] K. M. Li, S. Taherzadeh and K. Attenborough, Some practical considerations for predicting outdoor sound propagation in the presence of wind and temperature gradients, *Appl. Acoust.* **54** 27–44 (1997).
[13] W. H. Press, B. P. Flannery, S. A. Teukolsky and W. T. Vetterling, Numerical Recipes in FORTRAN, Cambridge UP, 2nd Ed. (1992) Ch. 16 for the Bulirsch–Stoer method and Ch. 9 for the Brent's method.
[14] R. Raspet and W. Wu, Calculation of average turbulence effects on sound propagation based on the fast field program formulation, *J. Acoust. Soc. Am.* **97** 147–153 (1995).
[15] R. Raspet, A. L'Espérance and G. A. Daigle, The effect of realistic ground impedance on the accuracy of ray tracing, *J. Acoust. Soc. Am.* **97** 154–158 (1995).
[16] K. M. Li, K. Attenborough and N. W. Heap, Source height determination by ground effect inversion in the presence of a sound velocity gradient, *J. Sound Vib.* **145** 111–128 (1991).
[17] Rudnick, Propagation of sound in the open air, Handbook of Noise Control, Edited by C. M. Harris, McGraw-Hill, New York, Ch. 3 pp. 3:1–3:17 (1957).
[18] T. Hidaka, K. Kageyama and S. Masuda, Sound propagation in the rest atmosphere with linear sound velocity profile, *J. Acoust. Soc. Jpn (E)* **6** 117–125 (1985).
[19] R. Makarewicz, Near-grazing propagation above a soft ground, *J. Acoust. Soc. Am.* **82** 1706–1711 (1987).
[20] R. Makarewicz, Reply to ref.11.1, *J. Acoust. Soc. Am.* **88** 1172–1175 (1990).
[21] K. Attenborough and K. M. Li, Ground effect for A-weighted noise in the presence of turbulence and refraction, *J. Acoust. Soc. Am.* **102** 1013–1022 (1997).
[22] ESDU Data Item No.940355 The correction of measured noise spectra for the effects of ground reflection ESDU International plc, London (1994).
[23] S. Gradshteyn and I. Ryzhik, Tables of Integrals, Series and Products, Academic, New York, 496, 3.953.1 and 2 (1980).
[24] K. Attenborough, I. Bashir and Shahram Taherzadeh, Outdoor ground impedance models, *J. Acoust. Soc. Am.* **129** 2806–2819 (2011).

[25] R. Makarewicz and P. Kokowski, Simplified model of ground effect, *J. Acoust. Soc. Am.* **101** 372–376 (1997).
[26] R. Makarewicz, A simple model of outdoor noise propagation, *Appl. Acoust.* **54** 131–140 (1998).
[27] R. Makarewicz, Industrial noise from a point source, *J. Sound Vib.* **220** 193–201 (1999).
[28] P. H. Parkin and W. E. Scholes, The horizontal propagation of sound from a engine jet close to the ground at Radlett, *J. Sound Vib.* **1** 1–13 (1965).
[29] R. Golębiewski and R. Makarewicz, Engineering formulas for ground effects on broad-band noise, *Appl. Acoust.* **63** 993–1001 (2002).

Chapter 12

Engineering Models

12.1 INTRODUCTION

Outdoor sound propagation involves a complex combination of effects due to source characteristics, source–receiver geometry, intervening terrain and meteorology. Understandably, for engineering applications, empirical and semi-empirical prediction schemes have been preferred to the more complicated analytical and numerical solutions introduced hitherto. Frequently, engineering prediction schemes are 'customized' for a specific type of source and, as a result, there are separate schemes for road traffic, rail traffic, industry and aircraft. Sections 12.3, 12.4 and 12.5 review three of these 'customized' schemes. First, however, in Section 12.2, we review a widely used empirical scheme which is source independent. ISO 9613-2 [1] is used for predicting propagation from a wide range of near ground sound sources. Nordic countries and the EC have funded the production of source-independent schemes NORD2000 [2] and HARMONOISE [3], respectively, which are more sophisticated than ISO9613-2 and are intended to be used together with specific source level predictors. The review of the NORD2000 scheme in Section 12.6 is rather brief since many of its features are repeated in the HARMONOISE scheme, a detailed outline of which is given in Section 12.7. The scheme specified for noise mapping in Europe is considered and criticized briefly in Section 12.8. An investigation of the suitability of engineering schemes for railway noise prediction in a high-rise city is carried out in Section 12.9. The predictions of various schemes are compared with extensive noise data from a complex road traffic source in Section 12.10. Sections 12.11 and 12.12 are concerned with wind turbine noise prediction and the requirements of auralization, respectively.

12.2 ISO 9613–2

12.2.1 Description

The prediction scheme in ISO 9613-2 [1] follows closely the structure of the Nordic scheme for outdoor industrial noise prediction [4]. This section describes essential features of the method, details various criticisms and compares the standard's ground-effect predictions with data from 1992 tests using a fixed jet engine. The aim of the ISO standard is to enable calculation of octave band L_{eq} levels (and hence L_{Aeq}) of environmental noise at distant locations up to about 1 km from various types of ground-based sound sources with known power spectra under 'average' meteorological conditions favourable to propagation. Favourable conditions are defined as those that occur downwind of the source (wind direction within ± 45° of the line between source and receiver and wind speeds up to 5 m s^{-1}) or under an atmospheric temperature inversion (giving rise to a curved ray path of 5 km or less). By restricting attention to moderate downwind conditions or temperature inversions, the ISO working group (WG24) hoped to limit the effects of variations in meteorological conditions as well as to provide a basis for predicting worst-case (i.e. highest) noise exposures. If only the A-weighted sound power of the source is known, rather than the octave band power levels, then the standard recommends use of the 500 Hz attenuation values. However, in addition, it offers an empirical formula for the A-weighted ground effect during propagation from broadband sources (equation 12.8).

The standard is intended to bridge the gap between standards specifying methods for determining the sound power levels emitted by various noise sources, such as machinery and specified equipment (ISO 3740 series), or industrial plants (ISO 8297), and the ISO 1996 series of standards which specify the description of noise outdoors in community environments. ISO 9613-2 allows for source directivity and size, geometrical wavefront spreading, air absorption, ground effect, reflection from surfaces and screening by obstacles. The ground-effect calculation allows only for 'hard' and 'soft' interfaces. Allowances for reflection from obstacles include those from vertical edges of buildings and a detailed correction for façade reflection. Calculations for screening by thin, thick or multiple edge barriers include a correction for the degradation in performance associated with meteorological effects. Various other attenuating environments such as trees (but allowing only for attenuation due to paths through foliage) and arrays of buildings are described in an Annex. The various contributions are assumed to be arithmetically additive except for those due to ground effect and screening by horizontal barriers. A distinction is drawn between short-term predictions, say for a given day, and long-term predictions, corresponding to averages over a month or a year, and a correction for this distinction, thereby accounting for the proportion of non-favourable propagation conditions, is included. Although an outline of the ISO9613-2 scheme is given here, this cannot substitute for reference to the original standard.

12.2.2 Basic Equations

The average octave band *SPL* (L_{eq}) at a receiver location is calculated for each point source and its image sources, and for eight octave bands with nominal mid- or centre frequencies from 63 Hz to 8 kHz. The following equation is used to determine the eight octave band levels

$$L_{fT}(DW) = L_W + D - A, \tag{12.1}$$

where L_W is the octave band sound power level (dB) produced by the point source relative to a reference sound power of 10^{-12} W. D is the directivity correction (dB) that describes the extent to which the L_{eq} from the point source deviates in various directions from the level due to an omni-directional point sound source producing sound power level L_W. For an omni-directional point source radiating into free space $D = 0$.

The total octave band attenuation (A dB) during propagation from the point source to the receiver is the sum of five contributions, i.e.

$$A = A_{div} + A_{atm} + A_{ground} + A_{screen} + A_{misc}, \tag{12.2}$$

where

A_{div} is the attenuation due to geometrical divergence (i.e. wavefront spreading),
A_{atm} is the attenuation due to air absorption,
A_{ground} is the attenuation due to the ground effect,
A_{screen} is the attenuation due to screening and
A_{misc} is the attenuation due to other miscellaneous effects (treated in an Annex).

Methods for calculating these terms for the eight octave bands are detailed in the corresponding sections of the standard.

The total A-weighted downwind sound pressure (dB) for n sources of noise is calculated according to

$$L_{AT}(DW) = 10\log_{10}\left\{\sum_{i=1}^{n}\left(\sum_{j=1}^{8}10^{\frac{L_{fT}(ij)+A_f(j)}{10}}\right)\right\}, \tag{12.3}$$

where $A_f(j)$ is the j-th weight in the A-weighted network.

The long-term average A-weighted sound pressure level (e.g. over several months) is defined by

$$L_{AT}(LT) = L_{AT}(DW) - C_{meteo}, C_{meteo} \geq 0 \tag{12.4}$$

where C_{meteo} is the meteorological correction (see equation (12.13)).

12.2.2.1 Geometrical Divergence

For a point source, and, therefore, for a spherically spreading wave, the geometrical divergence is predicted by

$$A_{div} = 20\log\frac{d}{d_0} + 11, \tag{12.5}$$

where d is the distance to the receiver and $d_0 = 1$ m is used as the reference distance.

12.2.2.2 Atmospheric Absorption

The attenuation (dB) due to atmospheric absorption during propagation through a distance d (m) is given by

$$A_{atm} = \alpha d / 1000, \tag{12.6}$$

where the frequency-dependent atmospheric attenuation coefficient α (dB/km) is predicted using expressions detailed in ISO 9613-1.

12.2.2.3 Ground Effect

ISO 9613-2 specifies three distinct regions for ground attenuation, resulting from the interference between sound reflected from the ground surface and the sound propagating directly between the source and receiver (see Figure 12.1). The source region occupies a distance $30h_s < d_p$ from the source towards the receiver. The receiver region occupies a distance $30h_r < d_p$ from the receiver towards the source. The middle region occupies the distance between the source and receiver regions. If $d_p < (30h_r + 30h_s)$, the source and receiver regions will overlap, and there is no middle region.

The acoustic properties of each ground region are specified by a ground factor, G. Three categories of reflecting surface are specified:

- hard ground ($G = 0$): paving, water, ice, concrete and other surfaces with low porosity,

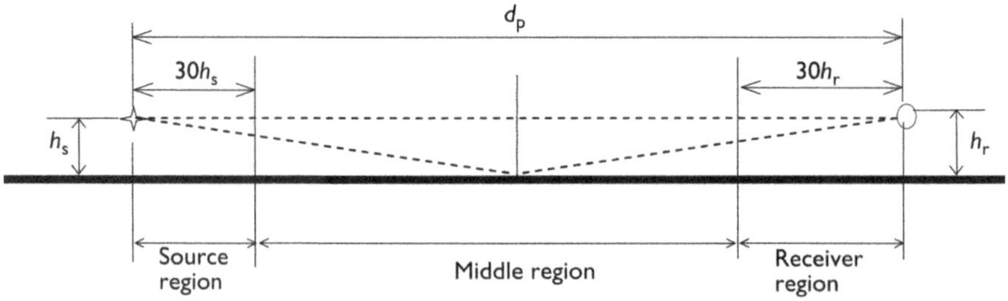

Figure 12.1 Regions identified for the calculation of ground effect in ISO 9613-2.

- porous ground ($G = 1$): grassland, trees, vegetation, farmland,
- mixed ground ($0 < G < 1$): mixture of hard and porous ground.

The combined ground effect is calculated from

$$A_{ground} = A_s + A_r + A_m, \tag{12.7}$$

where A_s, A_r and A_m are the source, receiver and middle region components of the attenuation with corresponding ground factors G_s, G_r and G_m. These values are defined using the expressions provided in Table 2 and Figure 2 of ISO 9613-2.

Alternatively, for some specific conditions when much of the ground is porous and the sound is not a pure tone, the ground attenuation (dB) can be calculated by the formula:

$$A_{ground} = 4.8 - (2h_m / d)(17 + 300 / d), \tag{12.8}$$

where h_m is the mean height of the propagation path above the ground (see Figure 3 in ISO 9613-2). In this case, the directivity correction is given by

$$D = 10\log_{10}\left\{1 + \left(d_p^2 + (h_s - h_r)^2\right) / \left(d_p^2 + (h_s + h_r)^2\right)\right\} \tag{12.9}$$

which accounts for the increase in sound power level of the source due to reflections from the ground near the source.

12.2.2.4 Screening

According to the standard, the effect of an object which obstructs propagation of sound along the direct line-of-sight between source and receiver should be included if

- the surface density is greater than 10 kg/m^2
- the object has a solid surface without large cracks or gaps
- the horizontal dimension of the object normal to the source–receiver line is larger than the acoustic wavelength λ at the nominal mid-band frequency for the octave band of interest.

Not only diffraction over the top edge of a screen but also that around the nearest of its vertical edges can be important. In this case, the screening correction (dB) for the top edge is calculated by

$$A_{screen} = D_z - A_{ground} > 0 \tag{12.10}$$

and for the vertical edge by

$$A_{screen} = D_z, \tag{12.11}$$

where the screening attenuation in the octave band is

$$D_z = 10\log_{10}\left(3 + \left(C_2 / \lambda\right) C_3 z K_w\right). \tag{12.12}$$

The symbols in expression (12.12) are defined by

- $C_2 = 20$ (including the effect of ground reflections)
- $C_3 = 1$ for single diffraction and
- $C_3 = \{1 + (5\lambda/e)^2\}/\{1/3 + (5\lambda/e)^2\}$ for double diffraction, where e is the width of the obstacle,
- z is the difference between the path lengths of the direct and diffracted sound (m)
- $K_W = e^{-1/2000\sqrt{d_{ss}d_{sr}/2z}}$ for $z > 0$ and $K_W = 1$ for $z \leq 0$, is the correction factor for meteorological effects

To calculate the screening attenuation, it is assumed that only one significant sound propagation path exists from the sound source to the receiver. The calculated screening attenuation should be limited to a maximum of 20 dB for single diffraction and to 25 dB for double diffraction.

12.2.2.5 Meteorological Correction

Calculation of a long-term average A-weighted sound pressure level requires allowance for a variety of meteorological conditions, i.e. not only those that are favourable for sound propagation, whereas the standard predicts levels for moderate downwind or temperature inversion to limit the meteorological variability. However, the standard suggests a meteorological correction term (see equation (12.4)) through which short-term predictions may be corrected to long-term predictions by means of a reduction (of less than 5 dB) to account for the fact that conditions favourable for sound propagation constitute only a fraction of the total period calculated on a total energy basis.

This correction is given by

$$\begin{cases} C_{meteo} = 0, \text{if } d_p \leq 10\left(h_s + h_r\right), \text{ i.e. for a short range of propagation} \\ C_{meteo} = C_0\left\{1 - 10\left(h_s + h_r\right)/d_p, \text{ if } d_p > 10\left(h_s + h_r\right), \text{ i.e. long range propagation}\right\} \end{cases}, \tag{12.13}$$

where C_0 (dB) is a constant which depends on the local meteorological statistics.

Typical values are $0 \leq C_0 \leq 5$.

The ISO9613-2 method claims ±3 dB accuracy at ranges up to 1 km for average sound propagation heights of less than 5 m. Even greater accuracy is claimed for higher source heights and ranges of less than 100 m. However, it is important to note that this accuracy is claimed for the prediction of overall A-weighted levels from broadband sources. It is accepted that errors in individual octave bands may be larger. Nevertheless, the claimed accuracy is comparable to that validated for road traffic noise prediction schemes and greater than that validated for comparable existing industrial noise schemes such as the CONCAWE method [5]. The draft of the standard referred to an Appendix intended to demonstrate the existence of a substantial validating database. However, the published version does not contain this Appendix.

12.2.3 General Critique

The scheme does not offer much improvement in scope or accuracy over existing schemes for predicting noise from roads and railways in the UK [5,6] except in its explicit account of meteorological effects. However, it has filled an unoccupied niche as an international standard method for the prediction of outdoor industrial noise and it represents a considerable advance on methods such as that proposed in the British Standard for predicting construction noise (BS5228). In assessing sound levels likely to be caused by industrial and commercial sources before they are in operation, the relevant British Standard (BS4142:2014) notes that '*Using a suitable method for the sound propagation from source to receiver, the sound pressure level at the assessment point can be calculated. It is necessary to relate the sound propagation to well-defined meteorological and ground conditions. Most calculation models refer to neutral or favourable sound propagation conditions, as other propagation conditions are much more difficult to predict*'. Clearly a scheme such as that offered by ISO 9613-2 is appropriate and is likely to be invaluable when predicting noise for planning purposes.

When the first draft of ISO9613-2 was circulated for comment in the UK, it was accompanied by the statement that '*This draft standard is unlikely to be implemented as a British Standard because the relevant UK committee does not consider that there is a need for it in the UK*'. This stance arose from the declared first mission for the relevant ISO working group (WG 24) which was to develop an internationally accepted method for predicting atmospheric absorption. The resulting atmospheric absorption calculation method, rather similar to the American National Standards Institute (ANSI) method, was published as ISO 9613-1.

A general criticism is that it remains an empirical method at a time when there are an increasing number of validated theoretical models for outdoor sound propagation that could be used [7]. A problem shared by all empirical schemes is they are valid only for the data set on which they are based. The ISO standard recognizes this limitation explicitly and states that its use should be confined to '*situations where there exists a substantial data-base of measurements for verification*'. The ISO standard claims also that its measurement database is extensive. However, as discussed below, at least in one respect this statement is a controversial one. During the 'draft for comment' stage of the standard, two countries noted that, in their opinion, the method proposed in the standard is worse than others.

Another general criticism relates to the inconsistent complexity of the standard. For example, the proposed frequency-dependent ground-effect correction is rather more complicated than that proposed for some of the other types of attenuation considered within the standard, for example the attenuation through housing.

The ISO standard claims to be '*applicable in practice to a great variety of noise sources and environments*',but its scope is limited to a variety of fixed industrial sources or to relatively slow-moving (i.e. negligible Doppler effect) sources (see Chapter 7), including road and rail traffic and construction equipment. Specific exclusions are aircraft in flight and blast noise from mining, military or similar operations. Moreover, the Nordic scheme on which the standard is based has been validated only against data collected around fixed industrial premises (an asphalt mixing plant, a plant for feedstuff and an oil refinery) [8]. Transportation or construction noise sources were not included in the validation exercises. Indeed, validated schemes are in use already for road and rail traffic [6,9]. Its potential ambiguity and inaccuracy in predicting ground effect for elevated sources has been criticized [10,11]. An extensive critique [12] has indicated several aspects of the standard that make it unsuitable for railway noise. For example, the standard assumes that any source may be described by an array of directional point sources whereas a finite line of dipole sources has been found more appropriate for noise from trains (but see also Chapter 7).

Several potential sources of outdoor noise nuisance such as open-air music festivals, Theme Parks, sporting stadiums and clay pigeon shooting are not excluded specifically, yet, clearly, are not covered by the methods proposed within ISO 9613–2. An ISO working group (WG61) has considered updating the standard to make it more applicable to the prediction of wind turbine noise [13]. A study has found that at distances above 2 km from wind turbines, low-frequency noise below 200 Hz is dominant [14]. Also, while there is some evidence that low-frequency noise attenuates less rapidly than 6 dB per distance doubling downwind, at ranges where multiple ray arrivals can be expected, this is not consistently supported by data (see also Section 12.11).

12.2.4 Accuracy of ISO 9613-2 Ground Effect

The ISO method gives empirical formulae for calculating ground effect in each of the octave bands from 63 to 8000 Hz and for each of the regions near source, receiver and in the middle of the propagation path. A consequence of these formulae is that the presence of ground adds 4.5 dB to the level in the 63 Hz octave band irrespective of source and receiver heights, range and ground cover. The predictions for the 2000 Hz octave band and above are zero where the ground is completely soft. As in other schemes [6,9], distinction is made only between acoustically hard and acoustically soft ground. According to ISO 9613-2, any ground of low porosity is to be considered acoustically hard and any grass-, tree- or potentially vegetation-covered ground is to be considered acoustically soft.

Data from trials using a fixed jet engine source [15] have been used to compare with ISO 9613-2 predictions from a source at 2.16 m height to a receiver at 1.2 m height above continuous soft ground. The data used for comparison with these prediction formulae were acquired using a Rolls-Royce Avon single-stream jet engine mounted on a stand such that the centre of the exit nozzle was 2.16 m above the ground. Microphone arrays were deployed over grass along a line at 22.5° to the engine exhaust centre line and at 7.5° to the peak jet noise direction. The source-to-receiver direction was 57° West of South. Each microphone array consisted of microphones at 1.2 m and 6.4 m above the ground and arrays were positioned at 152.4 m, 457.2 m, 762 m and 1158 m from the source. Temperature, wind speed and direction were measured at 0.025 m and 6.4 m heights at a weather station approximately 500 m from the source. Within each trial run, the data were averaged over 30 s. From comparison of ISO octave band predictions, assuming continuous soft ground cover, air absorption at 10° C and 70% relative humidity, with data for the horizontal level difference between receivers at 152.4 and 1158 m range for the zero wind and downwind conditions, it is clear that the ISO scheme predicts maximum level differences in a lower octave band than found in the measured data, even under downwind conditions. The poor comparison with zero wind octave band data has been noted previously by Lam [16]. But, as mentioned earlier, the standard itself points out that errors in individual octave bands may be larger than the ±3 dB accuracy claimed for overall A-weighted level predictions.

To obtain A-weighted level predictions from the ISO scheme for comparison with values obtained from the fixed jet engine data at Hucknall, a notional source power spectrum level is needed. This has been obtained from measured no wind/low turbulence data measured at 152.4 m at Hucknall (Table 12.1) after correcting for theoretically predicted effects of ground effect and turbulence [15]. The resulting A-weighted predictions for receivers at 1.2 m height are shown as a function of range in Figure 12.2.

If only octave band calculations are made, the ISO scheme predicts an A-weighted attenuation comparable to that measured for downwind conditions at Hucknall. But the ISO equation for the A-weighted ground effect for broadband sources (equation (12.8)) predicts much less attenuation than the octave band method (see also 12.4.2) for the jet engine source. Also shown in Figure 12.2 are fits using relatively simple functions.

ISO 9613-2 was formulated in response to a need for an internationally recognized standard method of predicting noise from fixed noise sources outdoors and, clearly, it represents an important step towards such a standard. Nevertheless, there are several deficiencies and controversial aspects in the ISO scheme. These concern its empirical basis, its scope of applicability and the chosen reference meteorological condition.

Table 12.1 Meteorological data corresponding to sound level data in Figure 12.2

Run 453 Block No.	Wind speed at ground (m s^{-1})	Wind speed at 6.4 m height (m s^{-1})	Direction relative to line of mics. (°)	Temperature at ground (°C)	Temperature at 6.4 m (°C)	Turbulence variable (squared wind speed ratio)
3	4.09	6.44	10.0	15.0	15.0	0.1202
4	4.09	6.11	20.5	14.9	15.0	0.1606
5	4.33	5.93	17.8	14.9	15.0	0.1729
6	3.89	6.07	10.8	14.9	15.0	0.1028

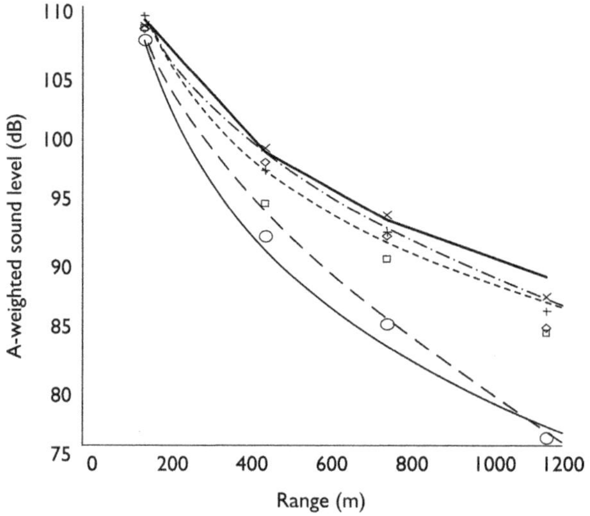

Figure 12.2 Comparison of measured downwind levels at Hucknall (run 453, blocks 3-6 (see table 12.1): ×, +, ◊, □) with. (a) ISO 9613 – 2, normalised at the 152.4 m downwind level (thick solid line). (b) neutral $L_A(152.4) + 2 - 26\log(r/152.4)$ (dotted). (c) spherical spreading (ss) plus 0.5 dB per 100m from 152.4 m downwind L_A (dash-dot) and measured no wind, low turbulence data (run 454, block 20: o) with. (d) ss plus 1.35 dB per 100 m from 152.4 m neutral L_A (dash). (e) L_A (152.4 neutral) – $35\log(r/152.4)$ (thin solid)

The comparison made earlier in this section with data obtained using a fixed jet engine source suggests that while the scheme gives incorrect predictions of ground effect in individual octave bands, it enables reasonable estimates of A-weighted downwind levels over the airfield, as long as octave band calculations are used as the basis. The Environmental Sciences Data Unit (ESDU) scheme [17] and other schemes discussed later in this chapter, use a full-wave calculation for ground effect (see Chapter 2) and enable predictions of narrow band levels.

As was discussed in Chapter 11, if only A-weighted levels from broadband sources are of interest, then it is possible to make use of several approximations and to use the resulting expressions to investigate possibilities for optimizing ground effect [18].

12.3 CONCAWE

12.3.1 Introduction

Like the ISO scheme, the CONCAWE scheme [19] (see also Chapter 1) is essentially empirical and was derived for use by the petrochemical industry. Although it was derived for a specific application, it contains generic material which could permit its wider use for predicting outdoor noise from fixed sources including industrial and construction plant. In some ways the CONCAWE scheme does not contain as many provisions as the ISO scheme. For example, the CONCAWE scheme does not provide for screening by vertical edges, and, while acknowledging the possibility, does not contain specific provision for attenuation during propagation through buildings. In two ways, it is more sophisticated than the ISO scheme. First, the CONCAWE scheme permits predictions for a wide range of meteorological conditions rather than being confined only to those that are favourable for sound propagation. Second, it enables calculation of the reduction of ground effect in octave bands due to the presence of a (horizontal edge) barrier, rather than assuming that the ground effect is lost if a barrier is present.

Here we provide a survey of important elements of the scheme and comment on its applicability in the light of current knowledge. As a source of information about the CONCAWE scheme, this contribution is no substitute for a thorough reading of the source report and the details in the background references that it cites.

12.3.2 Basis and Provisions of Scheme

The scheme is based on a survey of literature that was available prior to 1980, field data collected at ranges up to 1300 m around three complex sources ranging from a small process site to a major petroleum and petro-chemical industry complex and a series of measurements using octave band filtered white noise from source heights between 3 m and 9 m and ranges between 100 m and 1000 m over flat grassland. The component sources for the three sites were between 0 m and 25 m above ground level. The scheme allows for spherical spreading and corrections for atmospheric absorption, ground effect (including propagation path height influence), barrier effect and meteorological (refraction) effects. The corresponding calculations are in octave bands and require an octave band source power spectrum. Meteorological conditions are described using the the Pasquill classification system for predicting air quality and the vector wind speed between source and receiver. The six Pasquill stability classes (A–F) are based on incoming solar radiation, time of day and wind speed. Class A represents a very unstable atmosphere with strong vertical air transport and mixing. Class F represents a very stable atmosphere, with weak vertical air transport. Class D represents a meteorologically neutral atmosphere with a logarithmic wind speed profile and a temperature gradient corresponding to the normal decrease with height (adiabatic lapse rate). It is important to note that this is not the same as an acoustically neutral atmosphere since it includes both wind velocity and temperature gradients. In a very stable atmosphere, the temperature increases with height and the wind speed gradients are larger than is usual for a meteorologically neutral atmosphere. In an unstable atmosphere, the temperature always decreases with height and wind speed gradients are smaller than usual for a meteorologically neutral atmosphere. Usually, the atmosphere is unstable (classes A, B, C and D) by day and stable (classes D, E and F) by night. Acoustically neutral conditions are assumed to correspond to CONCAWE meteorological category 4 and strong downwind conditions to meteorological category 6. The meteorological corrections take the form of polynomials of order 3 in terms of the logarithm of the horizontal separation of source and receiver, i.e.

$$a_1 + a_2 \log(d) + a_3 (\log(d))^2 + a_4 (\log(d))^3, \tag{12.14}$$

where a_i, $i = 1 - 4$ are positive or negative constants having different values for each octave band centre frequency and each meteorological category.

The basic octave band ground-effect corrections for acoustically neutral and various vector wind conditions are derived from the classical data obtained by Parkin and Scholes [20, 21] for a fixed jet engine source at a height of 1.85 m and for ranges to 1097 m over a grass covered airfield. To extrapolate from these data involving a single source height to a scheme applicable to any source height, the CONCAWE scheme assumes that the ground effect diminishes non-linearly as the grazing angle increases until it is zero at and above a grazing angle of 5°. The validation experiments with filtered octave band white noise were restricted to grazing angles up to 2° and so were unable to verify this assumption about the variation of ground effect with source height. As in the ISO scheme, CONCAWE allows for the simultaneous presence of acoustically hard and acoustically soft ground along the propagation paths. However, it suggests that only the distance travelled over the soft ground should be used when calculating the ground-effect correction in octave bands, rather than defining source, middle and receiver ground regions as in the ISO scheme.

Since separate corrections for wind and ground are given in the CONCAWE scheme, it is tempting to think of them as separate effects. However, the presence of wind and temperature gradients influence propagation by altering the ground effect, so it is more accurate to regard the *combination* of the ground and wind corrections as the excess attenuation due to ground in the presence of wind.

In the CONCAWE scheme, the calculation of barrier effects is based on the widely accepted and extensively used formula due to Maekawa [22] (see Equation (9.28)). For a barrier of negligible thickness, the formula gives a relationship between the barrier-induced attenuation, the wavelength of the sound and the difference between the length of the path from source to receiver diffracted around the top of the barrier and that travelling directly from source to receiver. To predict propagation in the presence of a barrier close to the source, the CONCAWE scheme applies a correction (reduction) to the ground-effect term based on a change in the effective source height to coincide with the top of the barrier. Although the scheme does not give an explicit method, it recommends use of a ray tracing method [23], such as discussed in Chapter 11, for calculating meteorological influences on barrier performance particularly for downwind receivers. This method involves calculation of the ray path curvature (assumed circular) from the sound speed gradient. Subsequently, the barrier effect is calculated by using the curvature to define the locations of a virtual source and receiver corresponding to positions where an equal barrier effect would be produced under no-wind conditions. Barrier-induced turbulence (see Chapter 9) is not accounted for.

The 95% confidence limits for predictions using the scheme based on comparison with validation measurements vary between 6.9 dBA for meteorological category 3 (slight upwind) and 4.5 dBA for meteorological category 6 (strong downwind). The limits in individual octave bands are consistently higher for the 250 Hz octave band where ground effects may be expected to be important and in the 4 kHz octave band where signal-to-noise ratio is often a problem during measurements.

Measurements of sound propagation from power plants at distances between 400 m and 1800 m [24] have shown that, while reasonable predictions could be obtained using CONCAWE and ISO9613-2 for propagation from sources low to the ground, i.e. transformers and cooler units, both schemes resulted in poor predictions at both 400 m and 1800 m from the centre of a power plant containing 35-m high exhaust stacks.

12.3.3 Criticisms of CONCAWE

Since some of the largest discrepancies between measurement and prediction and variations between measurements are to be found in octave bands where ground effects are dominant, it seems likely that the CONCAWE scheme could be improved by allowing for more classes of the acoustical properties of ground than simply 'hard' or 'soft'. This would mirror the number of classes used to describe meteorological conditions in CONCAWE and accord with the eight different ground types specified in more recent schemes such as HARMONOISE described later.

The meteorological categories in CONCAWE are based, in part, on the observed wind speed. However, there is no indication, within the part of the scheme in which the meteorological categories are defined, of the height at which this wind speed is to be measured. This is a vital omission since the wind's influence on propagation depends on wind speed *gradient* rather than wind speed alone. It is reported elsewhere in the report proposing the scheme [19] that wind speeds for the Parkin and Scholes measurements and the measurements carried out to validate the CONCAWE scheme were obtained at 11 m (33 feet) height. If a linear wind speed profile is assumed, this implies a positive gradient of greater than 0.27 s^{-1} for meteorological category 6. Use of the recommended method for calculating downwind effects on barrier performance requires knowledge of the wind speed *gradient.* If the input gradient exceeds the implied value for meteorological category 6 to a significant extent, then calculations may indicate that the total attenuation during downwind conditions with a barrier present is less than that without a barrier. This somewhat counter-intuitive result would require the rather special circumstances of very acoustically soft ground, enhanced soft ground attenuation due to downward refraction in the absence of the barrier and that most of the source output power spectrum is in the frequency range corresponding to destructive interference due to ground effect.

A high degree of turbulence can have an important influence on ground effect at long ranges, even under otherwise acoustically neutral conditions, since it destroys the coherence and hence the interference between direct and ground-reflected sound. Also, destruction of coherence between various sound paths by turbulence together with scattering into the shadow zone will influence barrier performance. Consequently, it is important to define the turbulence values implied by each of the meteorological categories. Probably because the effect of turbulence was not realized at the time that Parkin and Scholes took their data, it was not monitored by them. Neither was it deduced, in terms of the variation statistics, from the continuous monitoring of wind speed fluctuations during the validating measurements for the CONCAWE scheme.

Nevertheless, the CONCAWE scheme is an important background to any study of outdoor sound prediction. Its provisions for differing meteorological conditions and for interactions between barrier, ground and meteorological effects represent more sophisticated treatments than those in ISO9613-2. The CONCAWE report acknowledges that, for its intended use by the petrochemical industry, improvements could follow from better allowance for the variation in ground effect with source height, consideration of partial barrier effects and a more definitive allowance for in-plant screening. Ways in which the allowances for meteorological and ground effects could be improved have been pointed out here.

12.4 CALCULATION OF ROAD TRAFFIC NOISE (CRTN)

12.4.1 Introduction

CRTN offers a practical tool for the prediction of *broadband* road *traffic noise* levels in urban environments. It is extensively used by acoustical consultants, land planners and highways engineers. It is semi-empirical in that it is largely based on data from measurements to evaluate specific parametric variations but, nevertheless, it is supported by theoretical analysis. One of the original motivations for the derivation of semi-empirical modelling tools [25] for the UK's Department of the Environment was to enable calculation of entitlement to sound insulation treatment for residential properties. The method also provides guidance on the prediction of noise in connection with the design and location of highways and other aspects of environmental planning. A flow chart was provided for predicting noise from single roads (see Figure 12.3).

This algorithm, typical of semi-empirical methods for traffic noise level assessment, depends upon the flow rate, the speed of traffic, the composition of traffic, the gradient of the road and parameters related to the propagation from the source to the receiver. In developing the method, data from over 2000 field measurements were collected to form a databank and used to derive the equations in Calculation of Road Traffic Noise (CRTN) [6]. In CRTN all the levels are expressed in terms of the A-weighted sound level exceeded for 10% of the time, i.e. the hourly or 18-hour L_{10} index. The value of hourly L_{10} index is the noise level exceeded for just 10% of the time over a period of one hour. The 10-hour L_{10} index is the arithmetic average of the values of hourly L_{10} for each of the eighteen one-hour periods between 0600 to 2400 hours.

The source of traffic noise (the source line) is taken to be a line 0.5 m above the carriageway level and 3.5 m from the nearside carriageway edge (see Figure 12.4).

The method of predicting traffic noise at a reception point consists of five main parts:

- divide the road scheme into one or more segments such that the variation of noise within the segment is small,
- calculate the basic level at a reference distance of 10 m away from the nearside carriageway edge for each segment,
- assess for each segment the noise level at the reception point taking into account distance attenuation and screening of the source line,
- correct the noise level at the reception point to account for site layout features, including reflection from buildings and façades, and the size of the source segment,
- combine the contributions from all segments to give the predicted noise level at the reception point for the whole road scheme.

If the noise levels vary significantly along the length of the road, then the road is divided into a small number of separate segments so that within any one segment the noise level variation is less than 2 dBA. Each segment is treated as a separate noise source and its contribution is determined accordingly. The final stage of the calculation process, to arrive at the predicted noise level, requires the combination of noise levels contributions from all the source segments, which comprise the total road scheme. For a single road segment scheme there is no adjustment to be made.

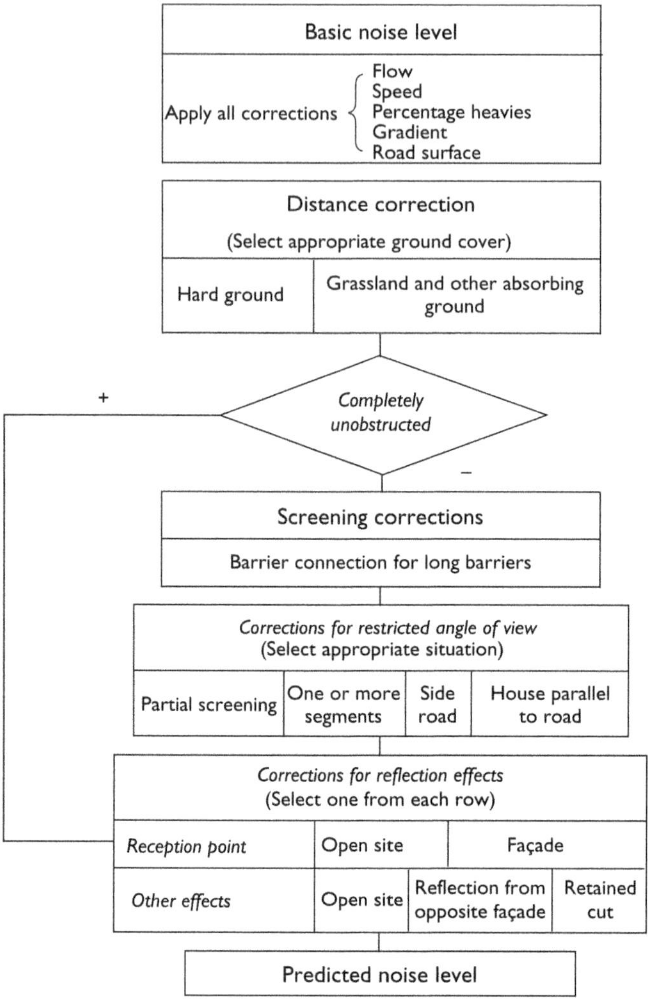

Figure 12.3. Delany's flow-chart for prediction of traffic noise level from a single road.

12.4.2 Basic Equations

12.4.2.1 L_{10} Levels

For road schemes consisting of more than one segment, the predicted level at the reception point is to be calculated by combining the basic hourly or 18-hour levels predicted for N segments, $L_{10,i}$, using

$$L_{10}^{tot} = 10\log_{10}\left(\sum_{i=1}^{N} 10^{\frac{L_{10,i}}{10}}\right), \tag{12.15}$$

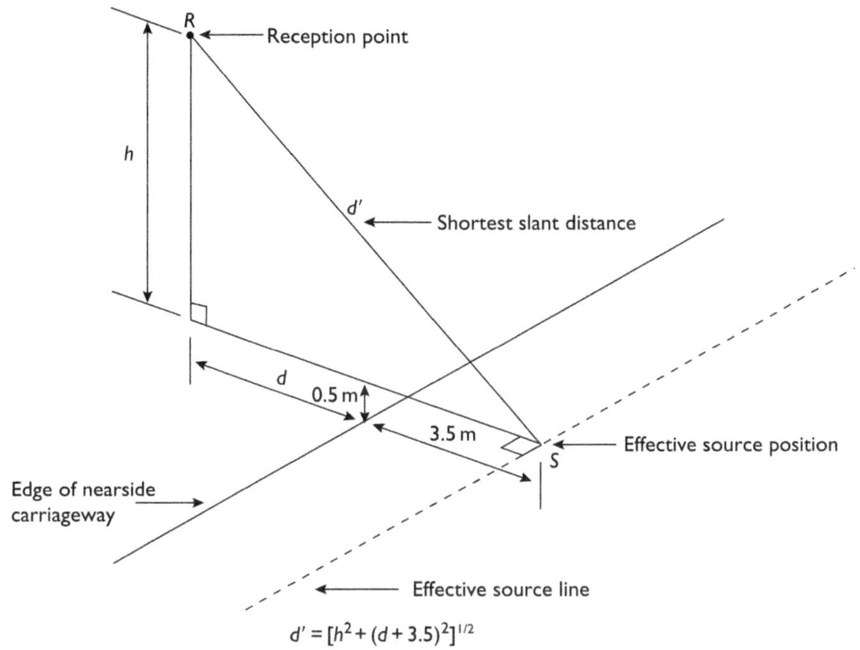

Figure 12.4. An illustration of shortest slant distance between a reception point and an effective source line representing a flow of traffic (Figure 1 from CRTN).

where

$$L_{10,i} = L_i + \Delta_{pV,i} + \Delta_{G,i} + \Delta_{TD,i} + \Delta_{D,i} + \Delta_{GC,i} + A_i + \Delta_{OF,i} + \Delta_{S,i}. \quad (12.16)$$

The following corrections are required for each segment:

- $\Delta_{pV,i}$ correction for mean traffic speed
- $\Delta_{G,i}$ correction for road gradient
- $\Delta_{TD,i}$ correction for road surface texture
- $\Delta_{D,i}$ correction for distance
- $\Delta_{GC,i}$ correction for ground cover
- A_i barrier correction for
- $\Delta_{OF,i}$ correction for opposite façade
- $\Delta_{S,i}$ correction for size of segment.

The basic A-weighted hourly noise level hourly is predicted at 10 m from the nearside carriageway according to the following equation

$$L_i(\text{hourly}) = 42.2 + 10\log q, \quad (12.17)$$

where q is the hourly traffic flow (in vehicles/hour).

Alternatively, the basic A-weighted noise level in terms of total 18-hour flow is

$$L_{Ai}(18-\text{hour}) = 29.1 + 10\log Q, \tag{12.18}$$

where Q is the 18-hour flow.

Here, it is assumed that the basic velocity V is 75 km/h, the percentage of heavy vehicles p is 0 and the gradient G is 0%. It is also assumed that the source line is 3.5 m from the nearside edge of the road for carriageways separated by less than 5.0 m. If the two carriageways are separated by more than 5.0 m, or where the heights of the outer edges of the two carriageways differ by more than 1.0 m, the noise level produced by each of the two carriageways must be evaluated separately and then combined using expression (12.15). For the far carriageway, the source line is assumed to be 3.5 m in from the far kerb and the effective edge of the carriageway used in the distance correction is 3.5 m nearer than this, i.e. 7.0 m in from the edge of the far-side carriageways.

12.4.2.2 Corrections for Mean Traffic Speed, Percentage of Heavy Vehicles and Gradient

The correction for percentage of heavy vehicles and traffic speed is determined using the following expressions:

$$\Delta_{pV} = 33\log\left(V + 40 + \frac{500}{V}\right) + 10\log\left(1 + \frac{5p}{V}\right) - 68.8, \text{dB(A)}. \tag{12.19}$$

In this expression, the percentage of heavy vehicles is $p = \dfrac{100f}{q} = \dfrac{100F}{Q}$, where f and F are the hourly and 18-hour flows of heavy vehicles, respectively. Equation (12.19) is applied to the basic hourly or 18-hour levels. The value of V to be used in (12.19) depends upon whether the road is level or on a gradient.

For level roads, the traffic speed, V, to be used in the calculation (12.19) is the value for the appropriate class of road. For roads with a gradient, traffic speeds must be reduced by ΔV km/h evaluated according to the gradient and the percentage of heavy vehicles from

$$\Delta V = \left[0.73 + \left(2.3 - \frac{1.15p}{100}\right)\frac{p}{100}\right]G, \tag{12.20}$$

where G is the gradient expressed as a percentage. Once the speed of traffic is known, then the adjustment to the A-weighted level for the extra noise from traffic on a gradient is calculated from

$$\Delta_G = 0.3G. \tag{12.21}$$

12.4.2.3 Correction for Type of Road Surface

The A-weighted noise level depends upon the amount of texture on the road surface. For roads which are impervious to surface water and where, in expression (12.19) for $V \geq 75$ km/h, a correction to the basic noise level is applied, the corrections for concrete surfaces and bituminous surfaces are given, respectively, by

$$\Delta_{TD} = 10\log_{10}(90TD + 30) - 20, \tag{12.22}$$

and

$$\Delta_{TD} = 10\log_{10}(20TD + 60) - 20, \tag{12.23}$$

where TD is the texture depth measured by the sand-patch test.

If $V < 75$ km/h, then for impervious bituminous road surfaces, $\Delta_{TD} = -1$ dBA and, for pervious road surfaces, $\Delta_{TD} = -3.5$.

12.4.2.4 Distance Correction

For the reception points located at distances $d \geq 4.0$ m from the edge of the nearside carriageway, the distance correction is given by

$$\Delta_d = -10\log_{10}(d'/13.5), \tag{12.24}$$

where d' is the shortest slant distance between the effective source and the receiver.

Now it is necessary to decide whether the source line of the road segment is obstructed or unobstructed. In some cases, the source line can be obscured by intervening obstacles. For these cases, it is necessary to calculate the noise levels assuming both obstructed and unobstructed propagation taking the lower of the two resulting levels.

12.4.2.5 Ground Cover Correction

If the ground surface between the edge of the nearside carriageway of the road or road segment and the reception point is totally or partially of an absorbing nature (e.g. grass land, cultivated fields or plantations), an additional correction to the A-weighted level for ground cover is required. This correction is progressive with distance and particularly affects the reception points close to the ground. The correction for ground-effect attenuation as a result of destructive interference as a function of horizontal distance from edge of nearside carriageway dm, the average height of propagation, $\bar{h} = 0.5(h+1)$ m and the proportion of absorbent ground, I, is given by

$$\Delta_{GC} = 5.2I\log_{10}\left(\frac{6\bar{h} - 1.5}{d + 3.5}\right) \text{if } 0.75 \leq \bar{h} < \frac{d+5}{6} \tag{12.25}$$

$$\Delta_{GC} = 5.2I \log_{10}\left(\frac{3}{d+3.5}\right) \text{if } \bar{h} < 0.75 \tag{12.26}$$

$$\Delta_{GC} = 0 \text{ if } \bar{h} \geq \frac{d+5}{6}. \tag{12.27}$$

Expressions (12.25), (12.26) and (12.27) are valid for $d \geq 4$ m. In these expressions, the value of $\bar{h}$ is taken to be the average height above the intervening ground of the propagation paths between the segment source line and the reception point, and it is suggested that the intervening ground is assumed to be primarily flat.

If the ground cover is perfectly reflecting, then $I = 0$, so no ground cover correction is applied. If the ground can be split into clear areas of reflecting and absorbing cover, then segmentation is required to separate areas where the ground cover can be defined as either absorbing or non-absorbing, so that the 2 dB(A) variation within a segment is not reached. In certain cases, where the intervening ground cover is a mixture of absorbing and non-absorbing areas which cannot be separated, then the ground cover correction should be calculated in accordance with expressions (12.25), (12.26) and (12.27), but with the value of I as shown in Table 12.2.

12.4.2.6 Screening Correction

The screening effect of intervening obstructions such as buildings, walls, purpose-built noise barriers, etc. is taken into account. The degree of screening depends on the relative positions of the effective source position S, the reception point R and the point B where the diffracting edge along the top of the obstruction cuts the vertical plane (see Figure 12.5), i.e. normal to the road surface, containing both S and R. The region between the obstruction and the reception point is divided into the illuminated zone and shadow zone by the extended line SB as shown in Figure 12.5.

The degree of screening is calculated from the path-length difference, δ m, of the diffracted ray path SBR and the direct ray path SR:

$$\delta = \text{SB} + \text{BR} - \text{SR} = \text{SB} + \text{BR} - \text{d'} \tag{12.28}$$

Table 12.2 Values for ***I*** (CRTN)

Percentage of absorbent ground cover within the segment	*I*
<10	0
10–39	0.25
40–59	0.5
60–89	0.75
≥90	1.0

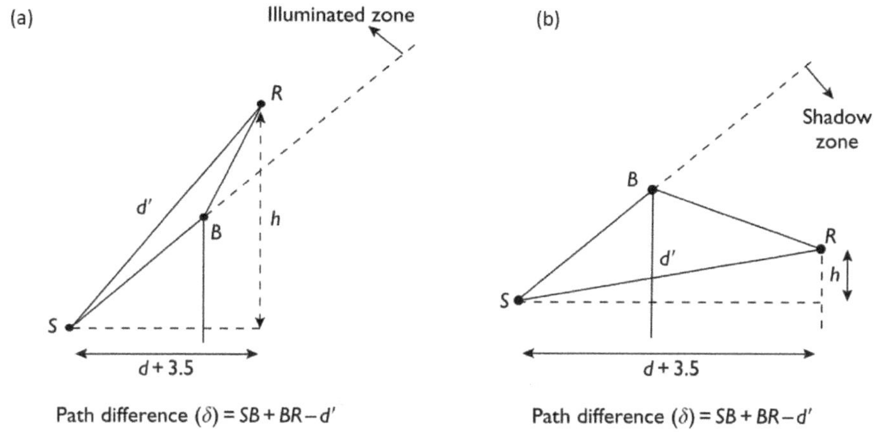

Figure 12.5. On the calculation of the barrier correction (from CRTN).

Table 12.3 Values of the coefficients in expression (12.29) (from CRTN)

Coefficients	*Shadow zone* $(-3 \leq x \leq +1.2)$	*Illuminated zone* $(-4 \leq x < 0)$
A_0	−15.4	0
A_1	−8.26	0.109
A_2	−2.787	−0.815
A_3	−0.831	0.479
A_4	−0.198	0.3284
A_5	0.1539	0.04385
A_6	0.12248	
A_7	0.02175	

The path difference is used to calculate the potential barrier correction to A-weighted SPL

$$A = \sum_{i=0}^{n} A_i x^i, \tag{12.29}$$

where $x = \log \delta$, the coefficients A_i and the value n for the shadow and for the illuminated zones are given in Table 12.3 together with the ranges of validity.

The method suggests that, outside these ranges of validity, the barrier correction is $A = -5$ dB for $x < -3$ and 0 dB for $x > 1.2$ if the receiver is in the shadow zone. If the receiver is outside the shadow zone, then $A = -5$ dB for $x < -4$ and $A = 0$ dB for $x > 0$. Predictions of $A > 20$ dB(A) should be considered unrealistic.

If the barrier is not parallel to the source line, then the potential barrier correction will vary along the barrier length. In this case, it may be necessary to divide the barrier into a number of smaller segments so that, within each, the variation in the barrier correction is less than 2 dB(A). If barriers are installed, then the ground cover correction is ignored assuming that the ground-reflected rays are obstructed.

12.4.2.7 Site Layout

The effects of reflections from buildings and other rigid surfaces result in the increase in the noise level and need to be considered. If the receiver is 1 m in front of a façade, then a correction of 2.5 dB(A) is added to the basic noise level. Calculations of noise levels alongside roads lined with houses but away from the façades also require the same addition of the 2.5 dB(A).

If there is a continuous line of houses along the opposite side of the road, then a correction for the reflections is required. The correction only applies if the height of the reflecting surface is at least 1.5 m above the road surface. The correction to the A-weighted level for the reflection from the opposite façade is

$$\Delta_{OF} = 1.5\frac{\theta'}{\theta}, \tag{12.30}$$

where θ' is the sum of the angles subtended by the reflecting façades on the opposite side of the road facing the reception point, and θ' is the total angle subtended by the source line at the reception point (see Figure 12.6). This correction is required in addition to the 2.5 dB(A) façade correction detailed above.

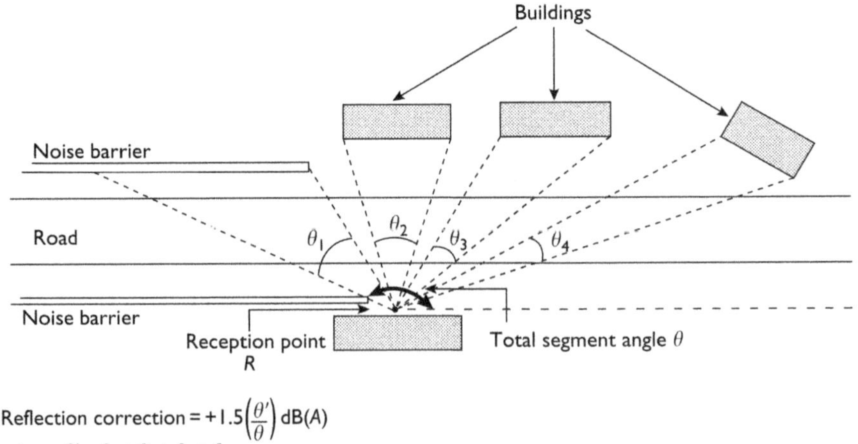

Reflection correction $= +1.5\left(\frac{\theta'}{\theta}\right)$ dB(A)
where $\theta' = \theta_1 + \theta_2 + \theta_3 + \theta_4$
and $\theta =$ total segment angle.

Figure 12.6 On the calculation of the reflection correction from the opposite façades and size of segment correction (from CRTN).

12.4.2.8 Segments and Road Junctions

The noise level at the reception point from the segment of the road scheme depends upon the angle θ (degrees) subtended by the segment boundaries at the reception point (see Figure 12.6). This angle is often referred to as the angle of view. The correction for angle of view is obtained using the expression

$$\Delta_S = 10\log\left(\frac{\theta}{180}\right). \tag{12.31}$$

Calculations of noise levels from multiple roads are achieved by combining the contributions from each individual length of road using the appropriate mean speed and ignoring any speed change at the junction. The overall predicted noise level is then obtained using expression (12.1).

12.5 CALCULATION OF RAILWAY NOISE (CRN)

An empirical scheme (CRN) is associated with the *Noise Insulation (Railways and Other Guided Transport Systems) Regulations 1993*. It is similar to CRTN, but there are several differences. First, it aims to predict either $L_{\text{Aeq, 18h}}$ (daytime) or $L_{\text{Aeq, 6h}}$ (nighttime) rather than the L_{10} index. Second, it works mainly with the sound exposure level, L_{AE}, for each train which is converted, after allowances for distance, ground effect, reflections, gradient, source enhancements (e.g. by bridges), angle of view, screening and the number of trains into the required L_{Aeq}. Apart from diesel locomotives, which are assumed to have a higher source height, the railway vehicle source is assumed to be at the *rail head*, i.e. the top of the rail which is taken as zero height.

The reference sound exposure level for a train depends on its speed and is predicted by

$$L_{\text{AE}} = 31.2 + 20\log(V) + 10\log(N), \tag{12.32a}$$

where V is the speed and N is the number of vehicles in the train.

The distance correction is given by

$$\Delta_d = -10\log(d_s / 25). \tag{12.32b}$$

The railway noise scheme also has an explicit allowance for air absorption given by

$$\Delta_A = -0.008d - 0.2. \tag{12.32c}$$

The ground-effect correction for propagation over acoustically soft ground is given by

$$\Delta_{GC} = -31\log\left(\frac{d}{25}\right) \quad \bar{h} \le 1$$

$$\Delta_{GC} = -0.61\left(6-\bar{h}\right)\log\left(d/25\right) 1 \le \bar{h} \le 6, \tag{12.32d}$$

$$\Delta_{GC} = 0 \quad 6 < \bar{h}.$$

where d is the horizontal distance between near side rail and the reception point, d_s is the slant distance from source to reception point (which must be at least 10 m) and $\bar{h}$ is the mean propagation height.

Figure 12.7 compares predictions of attenuation rates with (horizontal) distance between source and receiver at 1.2 m height, excluding wavefront spreading which is different for the different sources) with reference to the values at 25 m (this is the reference distance for CRN [9]).

The CRTN and CRN curves have been calculated assuming source heights of 0.5 m and 0 m, respectively, and give roughly comparable results. The air absorption term ($0.002d + 0.2$) causes the linear behaviour of the CRN predictions at longer ranges. The ISO prediction curves are based on use of

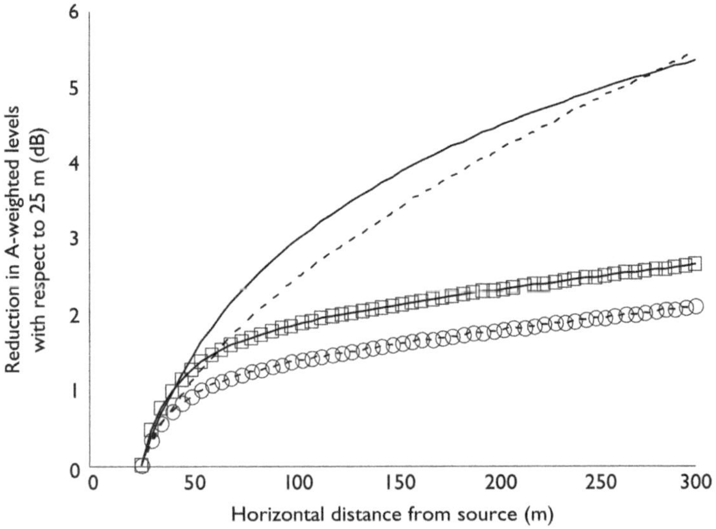

Figure 12.7. Attenuation rates (re level at 25 m) due to ground effect and air absorption predicted by CRTN (solid line), CRN (broken line) and ISO 9613-2 for roads (joined squares) and rail (joined circles).

the formula for attenuation due to ground effect of A-weighted levels from broadband sources (see (12.8)) and the air absorption value for 500 Hz assuming 20°C and 70 % RH (α = 2.8 dB/km)). It is noticeable that the ISO scheme predicts rather lower attenuation rates than the other two schemes beyond 50 m.

12.6 NORD2000

The development of the NORD2000 scheme [2] was financed by the Environmental Protection Agencies in the 5 Nordic countries. It was started in 1996 and finished in 2000. It is intended to predict noise from all types of sources in 1/3 octave bands with centre frequencies from 25 to 10000 Hz, including any terrain profile, any combination of ground surfaces and any 'regular' weather condition. The goal was to achieve acceptable accuracy within 3000 m. The scheme includes allowances for air absorption, based on ISO 9613 – 1, ground effect, evaluated from geometrical ray theory and spherical wave reflection coefficient (see Chapter 2), screening evaluated from diffraction theory combined with geometrical theory and reflections from obstacles by adding a mirror source and using a Fresnel-zone approach (see Chapter 8). The effect of atmospheric refraction is calculated using a heuristic model based on geometrical (curved) ray theory (see Chapter 11). The model allows for six categories of ground (Table 12.4).

12.7 HARMONOISE

12.7.1 Introduction and Background

Methods for predicting outdoor noise have undergone considerable assessment and change in Europe following the EC Noise Directive [26] and the associated requirements for noise mapping. A European project

Table 12.4 Categories of ground included in the NORD2000 model

Categories	*Examples*
A. Very soft	Snow or moss like
B. Soft forest floor	Short, dense heather-like or thick moss
C. Uncompacted, loose ground	Turf, grass, loose soil
D. Normal uncompacted ground	Forest floor, pasture field
E. Compacted field and gravel	Compacted lawns, park area
F. Compacted dense ground	Gravel road, parking lot
G. Hard surfaces	Dense asphalt, concrete, water

HARMONOISE [3] developed comprehensive source-independent schemes for outdoor sound prediction. The approach was to develop a state-of-the-art numerical reference model and to use it to validate a simpler engineering model which could be implemented in commercial software. As in NORD2000, various relatively simple formulae, predicting the effect of topography for example, have been derived and tested against the more sophisticated numerical predictions of the reference model.

12.7.2 General Methodology

The main objective of the reference model is the calculation of long-term equivalent sound pressure levels, with a focus on the L_{den} and L_{night} noise indicators that must be used in the noise maps reported to the EC [26]. Given that road traffic noise is the dominant source of the population's exposure to environmental noise in Europe, this method focuses on predicting propagation from incoherent line sources [27]. However, sound propagation from a point source is covered as well, giving the HARMONOISE methodology a wider applicability.

12.7.2.1 Basic Equations

For the situation portrayed in Figure 12.8 where a flow of vehicles (where Q is the number of vehicles per unit time) is moving along the y-axis at

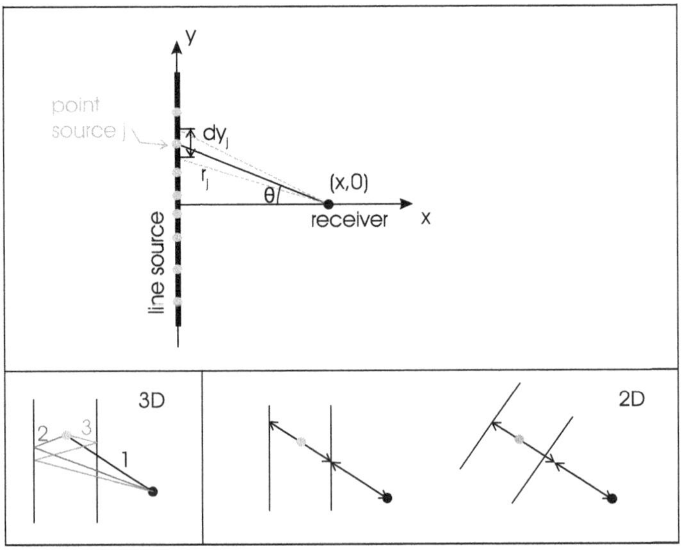

Figure 12.8. Schematic representation (top view) of the equiangular incoherent line source segmentation and propagation plane(s) identification (for a point source in between parallel reflecting planes) in the HARMONOISE reference model.

constant speed ($v_{vehicle}$), past a receiver at $(x, 0)$, the equivalent sound pressure level at a specific frequency band $L_{eq,f}$ can be calculated as follows [27]:

$$L_{eq,f} = L_{W,0,f} + 10\log\left(\frac{Q}{v_{vehicle}}\right) - A_{line,f}, \tag{12.33a}$$

$$L_{W,f}(\theta) = L_{W,0,f} + \Delta L_{W,f}(\theta), \tag{12.33b}$$

$$A_{line,f} = -10\log\left(\sum_j 10^{\frac{\left[\Delta L_{w,j,f} - 10\log_{10}\left(4\pi r_j^2\right) - A_{excess,j,f}\right]}{10}} dy_j\right). \tag{12.33c}$$

In these expressions, $A_{line,f}$ is the total attenuation during propagation from the incoherent line source to the receiver for a specific frequency band. $L_{W,f}(\theta)$ is the (momentary) source power level emitted towards the receiver from a specific position along the line source at a specific frequency band, dependent on its angle of incidence θ. $\Delta L_{W,f}(\theta)$ is the directivity correction (see Chapter 3, section 3.5) relative to the omni-directional emission term $L_{W,0,f}$. The term $L_{W,0,f} + 10\log\left(\frac{Q}{v_{vehicle}}\right)$ can be considered as the sound power level per unit length of the line source in a specific frequency band.

$A_{excess,j,f}$ is the point source excess attenuation for propagation from point j to the receiver in a specific frequency band. The excess attenuation is the free field sound pressure level minus the actual sound pressure level and should cover all relevant propagation effects (due to ground reflection(s), diffraction by terrain and barriers, air absorption and refraction). In the free field sound propagation term, $10\log\left(4\pi r_j^2\right)$, r_j is the length of the direct path between source point j and the receiver.

This approach allows the inclusion of results from most numerical techniques. A simulation including all propagation effects is needed for the actual case and a separate simulation under free field conditions for the same source characteristics and source–receiver separation. Note that, for the case of a thin noise barrier, even for two-dimensional numerical simulations which assume a coherent line source rather than the desired point source, expressing levels relative to free field sound propagation gives results very close to true point source calculations [28].

Air absorption, when not explicitly included in the numerical technique used to calculate the (overall) excess attenuation, can be added afterwards to $A_{excess,j,f}$, i.e.

$$A_{air,j,f} = \alpha_f r_j, \tag{12.34}$$

where α_f, the frequency-dependent absorption coefficient, depends on air temperature, relative humidity and ambient atmospheric pressure.

Absorption coefficients for pure tones are provided in ISO9613-1 [29, see also Chapter 1, section 1.8].

Use of Eq. (12.34) might cause some loss in accuracy, if sound paths are effectively curved due to refraction, thereby making the actual paths longer. For such a case, including air absorption in the numerical model for the excess attenuation is more correct. However, a drawback is that the simulations have to be repeated for each combination of the ambient parameters governing the atmospheric absorption processes. So, decoupling atmospheric absorption from the other propagation effects limits calculation times.

Equiangular discretization of a line source is advised in the HARMONOISE reference model. However, incoherent line source discretization can be further optimized by combining both linear and angular segmentation [30]. If receivers are positioned close to the line source or on the line source itself (which will occur in a noise map), optimized sampling strategies might be used. At longer distances from the line source, such choices are less important. Overall, an angular discretization of 5 to maximum 20 degrees is advised. If a finer angular step is used, the variation in the attenuation of the sound along the line source is predicted more accurately but at increased computational cost.

12.7.2.2 Identification of Propagation Planes

The HARMONOISE reference model uses the idea of propagation planes. A propagation plane is a vertical plane through a single point source and the receiver. In case of multiple reflections in between opposing reflecting planes, some simplifications are needed. If there are parallel walls at either side of the source, and a receiver outside these walls, many relevant propagation planes might be identified. HARMONOISE advises use of the three propagation planes shown in Fig. 12.8 (bottom left). Note that such an approach violates Keller's diffraction law [31], since the true sound paths are not exactly confined to these propagation planes.

When using 2-D numerical techniques to calculate the excess attenuation in such a setup, the barrier surfaces should be rotated so they become orthogonal to the plane containing source and receiver (see Fig. 12.8, bottom right). While this is an approximation for the many zig-zag reflections in between the walls, it has been found to be reasonably accurate [32], even when predicting sound propagation between adjacent city canyons, which requires that this approach is used twice [33].

12.7.2.3 Recommended Numerical Techniques

The HARMONOISE model explicitly mentions a number of numerical techniques able to model sound propagation in a refracting atmosphere like the finite-difference time-domain method (FDTD), the boundary element method (BEM), possibly including simplified refraction ('meteo-BEM' [34]), the PE method, possibly including approaches for multiple reflections [35], the Fast-Field Program (FFP, see Chapter 10) and the curved-ray model [36 and Chapter 11]. Calculations of excess attenuation are needed under specific meteorological conditions.

Clearly, this list is not exhaustive. Any numerical technique can be used, depending on the complexity of the sound propagation situation and the available computational resources. Of special interest are hybrid models that have been shown to be efficient when a clear source region can be identified. Examples of hybrid models for outdoor sound propagation are BEM-PE [37], FDTD-PE [28] or BEM combined with ray tracing [38]. However, note that only FDTD-PE (see Chapter 4, section 4.3.2 for a detailed description) allows refraction in both the source and receiver region to be included to a large extent.

12.7.2.4 Meteorological Conditions

The HARMONOISE reference model recommends the use of linear-logarithmic effective sound speed profiles, discussed in detail in Chapter 1, section 1.7. For a complete assessment, 25 meteorological classes should be considered, for which separate simulations are needed. The acoustic energy resulting from each of these contributions should be incoherently added, weighted by the frequency of occurrence, based on local (yearly) meteorological statistics. The meteorologically weighted excess attenuation for point j, at a specific frequency band, then gives

$$A_{excess,j,f} = -10\log\left(\sum_m w_m 10^{\frac{-A_{excess,j,f,m}}{10}}\right), \sum_m w_m = 1, \qquad (12.35)$$

12.7.2.5 Frequency Resolution

The HARMONOISE model is intended to give predictions in 1/3 octave bands thereby allowing a sufficiently detailed analysis for most applications in outdoor sound propagation. The number of frequencies used to constitute each fractional octave band should be chosen so that convergence is reached in the excess attenuation. When using frequency domain numerical techniques, a new calculation is needed for each sound frequency of interest. A minimum number of 4 or 5 is recommended, although this depends strongly on the case under study. When making time-domain predictions with a sound pulse at the source position, a large range of sound frequencies is included in a single simulation run.

12.7.2.6 Long-Term Integrated Levels

The most common assumption when using a two-dimensional numerical technique to calculate the excess attenuation is that there is only a single propagation plane for the point source under consideration. If multiple propagation planes are considered for a single point source and receiver combination, such contributions should be added incoherently.

Depending on the source model employed, including various source heights for engine and rolling noise adds another dimension to this summation. It is expected that in many practical cases, the rather small difference between rolling and engine noise height for light vehicles does not lead to significantly different results when assigning the engine and rolling source powers to a single source height. This has been illustrated by numerical examples when predicting sound propagation through vegetation belts [39]. Also, in the CNOSSOS (END) model [40] (see also Section 12.8) which has been introduced since the HARMONOISE model, a single road traffic source point has been chosen. Finally, L_{den} values are obtained by adding the contributions from all frequency bands considered, applying A-weighting and incorporating penalties for evening and night equivalent sound pressure levels [26].

Some care is needed when averaging meteorological conditions and traffic data over time, as there might be interactions between them [41]. Morning rush hours, with higher vehicle intensities, often coincide with favourable sound propagation conditions, such as an air temperature inversion that has built up during the night under clear sky conditions, and, therefore, could give important contributions to long-term equivalent sound pressure level indicators like L_{den}.

12.7.2.7 Validation

During its development, the HARMONOISE reference model was validated extensively by measurement campaigns. Two campaigns were undertaken near motorways [27]. In the first campaign, the highway was in a flat environment. In the other, the highway was on an embankment with noise walls, while the surroundings were flat. Excess attenuation was simulated with the PE model, and single-height meteorological data was available. The accuracy of long-term simulated levels was between -1.6 and +0.5 dBA L_{den} at 4 m and 6 m high receivers, located between 25 m and 1115 m normal to the highway.

The overall accuracy goals of the HARMONOISE reference propagation model are summarized in Table 12.5. It is claimed that these goals were fulfilled in 47 (96 %) out of 49 measurements [27,42]. Most (14) of the measurement sites were comparatively flat. The sources included road traffic and railways. Some experiments were carried out with loudspeakers. Most of the microphones were at 4 m height, but some lower and higher ones were included as well. The maximum distance relative to the source was 1.2 km. The measurement durations ranged from one week to a few weeks.

Table 12.5 Accuracy goals of the Harmonoise reference model for predicted L_{den} values.

	95% confidence intervals
up to 100 m (flat terrain)	± 1 dBA
up to 2000 m (flat terrain)	± 2 dBA
up to 2000 m (hilly terrain)	± 5 dBA

12.7.3 Analytical Point-to-Point Model

12.7.3.1 Introduction

The HARMONOISE project provided an analytical model to calculate excess attenuation by the presence of ground and barriers between source and receiver, thus avoiding computationally costly numerical techniques (see Sections 12.7.2) which demand a high level of expertise for their correct use, and, for which, commercially or freely available codes are scarce. The heuristic and analytical HARMONOISE point-to-point model (HP2P) [43,44] was largely inspired by the NORD2000 model [2] for predicting ground reflections and terrain/barrier diffraction. In contrast to ISO9613-2 [1] and CONCAWE [5], ground reflection(s) and diffractions are not treated independently. Accounting for the interactions between ground effect and diffraction captures more of the propagation physics. A main achievement of the HP2P model is that it enables logical and consistent combinations of effects while avoiding abrupt changes in excess attenuation predictions that might result, otherwise, from minimal changes in the geometry of the sound propagation problem, through the so-called 'transition procedures'.

This section outlines the main ideas and basic formulae on which the analytical model relies. The description of workable numerical codes for all possible cases that might be encountered outdoors is beyond the scope of this chapter.

12.7.3.2 Methodology for Combining Ground and Barrier Effect

The general formula to calculate excess attenuation (see Eq. (12.36)) adds ground reflections and diffractions between the source and receiver in the propagation plane considered (see Section 12.7.2.2). The HP2P model does not discriminate between diffraction by terrain and by artificial barriers, thereby presenting a general methodology for diffraction. The terrain and barriers must be represented by a sequence of connected linear segments, with points $P_i\ (x_i, y_i)$, with $i = 0–N$ (see Fig. 12.9). The source is located at $S(x_0, y_0+h_S)$ and the receiver at $R(x_N, y_N+h_R)$. Each segment $[P_i, P_{i+1}]$ can be assigned its own

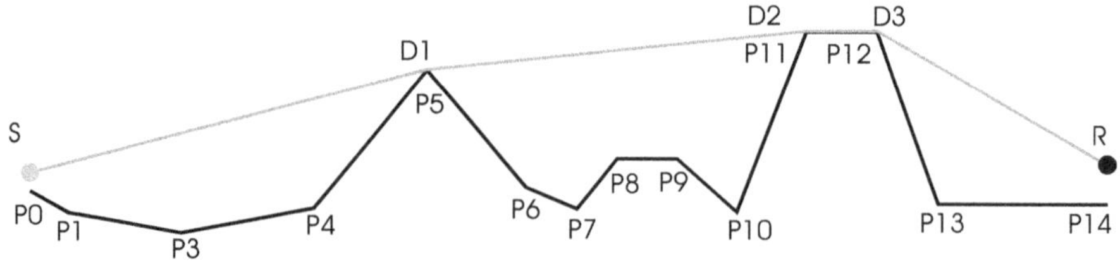

Figure 12.9 The convex hull for the example ground profile P_i between source S and receiver R is indicated by the gray line, connecting S-P5-P11-P12-R. There are three relevant diffracting edges D1, D2 and D3, and four zones where ground reflections will appear, namely in between S-D1, D1-D2, D2-D3 and D3-R. The segment between D2-D3 is part of the convex hull and will need dedicated treatment for the ground reflection. The zone in between S and D1 is a concave ground profile ("valley-type"). The zone in between D1 and D2 has convex-shaped parts and will need the "transition model" (see Section 12.7.3.5). The part between D3 and R can be considered as flat ground.

impedance. Even if there are no changes in the ground slope, a different segment should be defined when changes in ground properties are encountered.

The excess attenuation is calculated by the following expression:

$$A_{excess,G+D} = -20\log\left(\left|\frac{p}{p_{ff}}\right|\right) = -\left(\sum_{i=1}^{M}\Delta L_{D,i} + \sum_{i=0}^{M}\Delta L_{G,i}\right), \tag{12.36}$$

where p is the (total) sound pressure at the receiver due to diffraction and ground reflections, and p_{ff} is the free field sound pressure. The terms $\Delta L_{D,i}$ represent the attenuation due to diffraction over all segment points defining the convex 'hull' which connects the peaks in the ground topology sequence. The terms $\Delta L_{G,i}$ represent the level change due to ground reflections; the sum is taken over sequences of segments between consecutive convex 'hull' diffraction points. An example is given in Fig. 12.9, with three diffracting edges and four zones that include ground reflections.

12.7.3.3 Ground Reflection Model

For uniform flat ground, the two-ray approach (see Chapter 10) can be used. The contribution of the ground reflection, relative to free field sound propagation, is given by

$$\Delta L_{G,flat,uniform} = 10\log\left(\begin{array}{c}\left|1 + C_{coh}Q\dfrac{R_d}{R_g}e^{jk(R_g-R_d)}\right|^2 \\ +\left(1-C_{coh}^2\right)\left|Q\dfrac{R_d}{R_g}e^{jk(R_g-R_d)}\right|^2\end{array}\right), \tag{12.37}$$

where R_d is the direct path length between source and receiver, R_g is the ground-reflected path length, k is the wave number and Q is the spherical ground reflection coefficient (see Chapter 2) and C_{coh} is a coherence factor. For fully coherent summation, $C_{coh} = 1$, while, for completely incoherent interactions between both paths, $C_{coh} = 0$ (see Section 12.7.3.7 for more information on quantification of this coherence factor).

In the presence of ground impedance discontinuities, the H2P2 model uses (linear) Fresnel weighting:

$$\Delta L_{G,flat,mixed} = \sum_{i=1}^{N_m} w_i \Delta L_{G,flat,uniform,i}, \tag{12.38}$$

where w_i is the Fresnel weight of ground in zone i, N_m is the number of different ground types encountered in between source and receiver and $\Delta L_{G,\,flat,\,uniform,\,i}$ the partial ground reflection contribution as if ground type i would appear everywhere between source and receiver.

The Fresnel zone concept is used to account for the fact that the reflection at the ground is not limited to the exact specular reflection point but involves a wider zone, and thus potentially interacts with different ground types in case of impedance discontinuities. The latter becomes more pronounced with increasing wavelength. An ellipsoid is constructed with foci at the image source and receiver, where each point on the Fresnel ellipsoid is defined by the following distances relationship:

$$\left|S_{img}P\right| + \left|PR\right| = \left|S_{img}R\right| + \frac{\lambda}{n_F}, \tag{12.39}$$

where S_{img} is the location of the image source, R is the position of the receiver and n_F is the Fresnel parameter, defining the fraction of the wavelength involved, which is a parameter to be tuned (a value of 8 might be appropriate [45]). The extent of the intersection line between the Fresnel ellipsoid and (each) ground segment is calculated to find the appropriate weights w_i. If this intersection line fully falls within a particular segment, w_i becomes 1 (and consequently zero for all other segments). Mathematical expressions for the geometrical parameters related to Fresnel ellipses and their intersections with the ground can be found in Chapter 8, section 8.3.4.

Improvements have been proposed (the so-called modified Fresnel weighting) to increase accuracy in the higher frequency range, including a frequency-dependent transition function. A low-pass filter approach was considered appropriate, which needs the definition of a cut-off frequency [43,44]. Basically, this modified approach makes the Fresnel parameter, n_F, frequency dependent.

If relevant ground reflections in a concave sequence of segments (such as would be the case for 'valley-shaped' terrain) can be identified, the procedure to be used is based on Fresnel weighting. A transition function F_g is applied for the final ground effect, interpolating between the one for a flat ground and valley-sloped terrain:

$$\Delta L_G = \Delta L_{G,flat} + \left(1 - F_g\right)\Delta L_{G,valley}. \tag{12.40}$$

If the direct path between source/receiver or intermediate diffracting edges is not blocked by a sequence of convex-shaped ground segments, a transition model procedure is proposed which balances ground reflections and diffractions (see Section 12.7.3.4).

12.7.3.4 Sound Diffraction Model

12.7.3.4.1 Single Diffraction

Sound diffraction is based on a 'knife edge' (thin screen) approximation, which allows strong simplifications from more advanced diffraction theories, on condition that source and receiver are at sufficient distance from the diffracting edge, relative to the wavelength of the sound, i.e. for $kr_{S,i} \gg 1$, $kr_{R,i} \gg 1$ (see Figure 12.10).

HP2P's diffraction formula relies on Deygout's approach [46], which calculates the sound pressure level relative to free field propagation. The expressions to be used, which impose a limited computational burden, are

$$\Delta L_{D,i} = \begin{cases} 0 & N_{F,i} < -0.25 \\ -6 + 12\sqrt{-N_{F,i}} & -0.25 \le N_{F,i} < 0 \\ -6 - 12\sqrt{N_{F,i}} & 0 \le N_{F,i} < 0.25 \\ -8 - 8\sqrt{N_{F,i}} & 0.25 \le N_{F,i} < 1 \\ -16 - 10\log_{10}\left(N_{F,i}\right) & N_{F,i} \ge 1 \end{cases}, \tag{12.41}$$

and

$$N_{F,i} = \pm\left(\frac{2}{\lambda}\right)\left(r_{S,i} + r_{R,i} - d_{SR,i}\right). \tag{12.42}$$

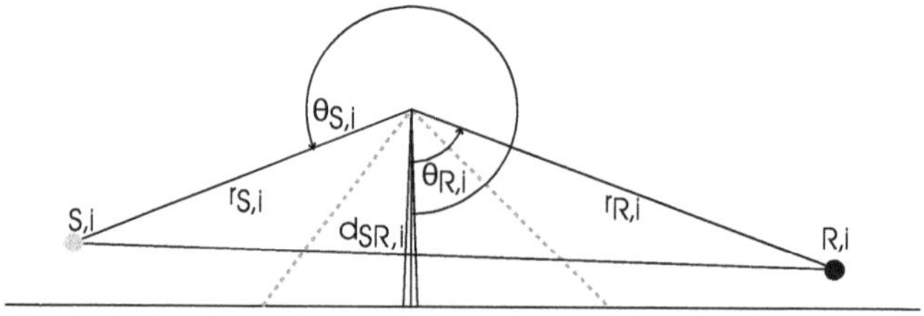

Figure 12.10 Geometrical parameters to evaluate the diffraction function for the simplification of a knife-edge profile

$N_{F,i}$ is the Fresnel number, $r_{S,i}$ is the distance between the source and the diffracting edge, $r_{R,i}$ is the distance between the diffracting edge and the receiver and $d_{SR,i}$ is the direct distance between the source and the receiver (neglecting the presence of the barrier). When a diffracting edge blocks the line of sight between the source and the receiver thereby creating an acoustic shadow zone, so that $\Delta\theta_i = \theta_{S,i} - \theta_{R,i} < \pi$, where $\theta_{S,i}$ is the angle between the right face of the barrier and the source-diffraction edge line and $\theta_{R,i}$ is the angle between the right face of the barrier and the diffraction edge-receiver line, a positive sign is needed in Eq. (12.42). A negative sign should be used if there is a direct line of sight between source and receiver in the so-called 'illuminated' zone, $\Delta\theta_i > \pi$.

In the deep shadow zone, the simple approach presented by Eqs. (12.41) and (12.42) becomes inaccurate. Therefore, a modified Fresnel number approach should be used [43,45]:

$$N_{F_i'} = \left(\frac{2}{\lambda}\right)\left(r_{S,i} + r_{R,i}\right)\left[\frac{\varepsilon^2}{2} + \frac{\varepsilon^4}{3}\right] \tag{12.43}$$

$$\varepsilon = \frac{\sqrt{r_{S,i} r_{R,i}}}{r_{S,i} + r_{R,i}}\left(\Delta\theta_i - \pi\right). \tag{12.44}$$

Note that if the diffracted sound paths reflecting on the ground on both the source and receiver sides are neglected when considering sound propagation over a thin screen (see Chapter 9), it is possible to separate the ground reflection term and diffraction term [43,47], thereby justifying Eq. (12.37).

The simplified diffraction formula was found to give reasonably good results. Nevertheless, the HP2P model architecture is sufficiently flexible to allow use of more advanced diffraction models such as that by Hadden and Pierce [48].

12.7.3.4.2 Multiple Diffractions

If there are multiple diffracting edges along the convex hull between source and receiver, first, the most efficient barrier (i.e. the one leading to the highest insertion loss) is sought, by considering each edge along the convex hull separately using the actual source and receiver position. Second, the most efficient diffracting edge becomes a virtual source for diffraction over the second most efficient diffraction edge. Third, the insertion losses of all relevant diffractions are added linearly as in Eq. (12.37).

The (normalized) pressure at an (intermediate) diffraction point can be calculated as follows:

$$p_{D,i} = 10^{\frac{\Delta L_{D,i}}{20}} \frac{e^{jkd_{SR,i}}}{d_{SR,i}}. \tag{12.45}$$

Given that the diffraction formula is basically governed by the Fresnel number, the ranking of the diffracting edges can be based purely on geometrical considerations and does not require evaluations of diffraction formulae.

When the diffracted ray path coincides with the ground segment, so making a 'thick barrier' (also called a 'hull' ground segment, see Figure 12.9 between P11 and P12), some care is needed if the barrier top is acoustically soft, for which dedicated approximations are provided [43]. The latter include a transition function for consistency when source and receiver are moving away from such shearing propagation. Note that in case of a rigid thick barrier, applying Eq. (12.38) leads to a +6 dB contribution relative to free field sound propagation, corresponding to the pressure doubling that will be observed during diffraction.

12.7.3.5 Transition Model

If there are convex-shaped sequences in zones where a ground reflection is expected (e.g. in between P5 and P11 in Fig. 12.9), it is necessary to balance between accounting for ground reflections only $A_{excess,\ G}$, and combining ground reflections with diffractions $A_{excess,\ G\ +\ D}$, so a transition function, X, is used in weighting these contributions

$$\begin{cases} A_{excess,G} = -\Delta L_G \\ A_{excess,G+D} = -\left(\Delta L_G + \Delta L_D\right) \end{cases} \tag{12.46}$$

$$A'_{excess,G+D} = XA_{excess,G} + \left(1 - X\right) A_{excess,G+D}. \tag{12.47}$$

Conceptually, X takes the following form:

$$X = X_3 + X_1\left(1 - X_2\right)\left(1 - X_3\right), \tag{12.48}$$

expressing the fact that the diffraction model is to be preferred if the diffraction occurs within the Fresnel zone associated with the specular reflection point (X_1) and if the screening obstacle is high enough not to be seen as surface roughness (X_2), or if the receiver is located in the transition region close to the line of sight (X_3). Specifications of the functions X_1, X_2 and X_3 are available [43,44].

12.7.3.6 Refraction

Two approaches are commonly followed when modelling atmospheric refraction in more advanced engineering methods. The first uses curved rays as in the NORD2000 method. The second approach, followed in the HP2P model, uses the analogy between sound propagating under the influence of specific sound speed profiles and sound propagation over curved grounds with a constant sound speed [49–51]. This allows an easier geometrical treatment in case of multiple diffractions since straight rays can still be used after artificially curving the ground surface. Another advantage is that angles of incidence on the ground are preserved during a conformal mapping coordinate transformation which means that ground impedances do not have to be modified to give correct reflection coefficients.

In both approaches, the radius of curvature R_c, either for the equivalent ground profile, or for the curved ray, must be quantified. This radius is inversely proportional to the gradient of the effective sound speed with height. A linear sound speed profile corresponds to a circular ground surface. The logarithmic-linear sound speed profile (see Chapter 1, section 1.7) can be linked to R_c in the following way:

$$\frac{1}{R_c} = \frac{1}{R_{lin}} + \frac{1}{R_{\log}}. \tag{12.49}$$

The radii of curvature R_{lin} and R_{log} can be approximated as follows:

$$\frac{1}{R_{lin}} = \frac{1}{\sqrt{\left(\frac{c_0}{a_{lin}}\right)^2 + \left(\frac{d}{2}\right)^2}}, \tag{12.50}$$

$$\frac{1}{R_{\log}} = \frac{8}{d}\sqrt{\frac{a_{\log}}{2\pi c_0}}, \tag{12.51}$$

where d is the ground projected distance between source and receiver, and c_0 the reference sound speed at ground level. Note that these quantifications include some simplifications, e.g. it is assumed that source–receiver separations are large.

Positive values for R_c mean that the ground profile will be concave, equivalent to a downward-refracting atmosphere. If the centre of the circle whose arc defines the adapted ground profile between source and receiver is below the ground, then R_c is negative and corresponds to convex ground, representing upward refraction.

As discussed in Chapter 1, section 1.7, a_{log} and a_{lin} can be derived either from fitting meteorological tower data, needing air temperature and wind speed at various heights, and the mean wind direction. Alternatively, these parameters might be quantified upon knowledge of the friction velocity u^*, the temperature scale T^* and the Monin–Obukhov length L, and time of the day. The meteorological classification proposed by HARMONOISE provides averaged values for these parameters based on more commonly accessible meteorological observations, i.e. more precisely, the wind speed at

10 m height, the cloud cover fraction and the time of the day (see Chapter 1, section 1.6).

The conformal mapping can be efficiently performed using a complex representation of the coordinate system ($w = x + jz$), where w is the original coordinate system in the x-z space, which is transformed to the ζ domain, dependent on the radius of curvature, R_c, in the following manner:

$$\zeta = u + jv = \frac{j2wR_c}{w + j2R_c}, \tag{12.52}$$

assuming the ground segment below the source is positioned at $(x,z) = (0,0)$. Examples of ground profiles modified in this manner are shown in Figure 12.11.

The opposite way of thinking has been used as well when simulating outdoor sound propagation. In wave-based numerical models like the PE method, ground curvature is more difficult to implement than the effect of a

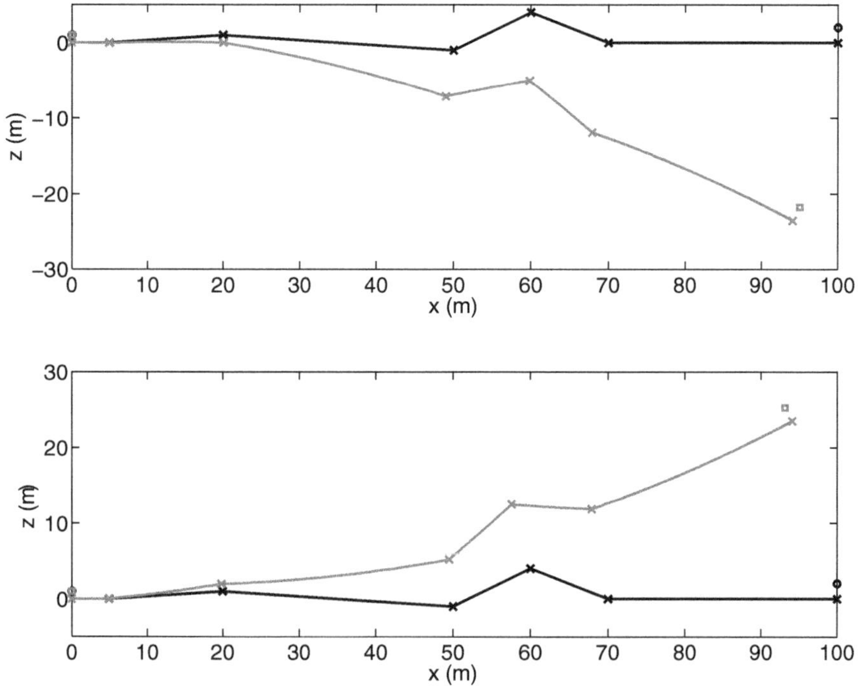

Figure 12.11 A sequence of linear ground segments simulating the influence of a refracting atmosphere. The gray line in the upper figure represents a sequence simulating the influence of an upward refracting atmosphere (R_c = −200 m) and that in the figure below a sequence simulating the influence of a downward refracting atmosphere (R_c = 200 m). The black lines show the reference ground profiles without curvature. The open circles and squares indicate the source and receiver positions in the non-refracting and refracting atmosphere, respectively.

refracting atmosphere. So, circularly curved ground is replaced by an equivalent refracting atmosphere and, subsequently, sound propagation over flat ground is assumed [36,52].

12.7.3.7 Coherence Losses

When sound propagates through the atmosphere over ground and barriers, there are several causes of loss of coherence between sound paths arriving at a single receiver, which leads to interferences becoming less pronounced.

A Gaussian distributed coherence function has been proposed as a general function to express the coherence factor C:

$$C_{misc} = e^{-\frac{\sigma_\phi^2}{2}}, \tag{12.53}$$

where σ_ϕ is the standard deviation of the random variation in phase.

Sources of incoherence include fluctuations in the refractive state of the atmosphere (see Chapter 10) and scattering by surface roughness (see Chapters 5 and 6).

Effectively, the grouping of acoustical energy in 1/3 octave bands can be considered as a coherence loss factor as well treated in the same way as for the other causes. The standard deviation normalized by the centre frequency can be expressed by [43,44]:

$$\frac{\sigma_f}{f} = \frac{1}{3}\left(2^{\frac{1}{6}} - 2^{-\frac{1}{6}}\right). \tag{12.54}$$

If the squares of the (normalized) standard deviations σ_l due to the various factors are assumed to be independent, they may be added:

$$\sigma_\phi = k\Delta d\sqrt{\sum_l \sigma_l^2}, \tag{12.55}$$

where k is the wave number, and Δd is the path-length difference between the sound paths considered.

The major loss of coherence due to turbulence is based on a more physical description by using the following equation [53, see also Chapter 10]:

$$C_{turbulence} = e^{-\frac{3}{8}D\gamma_T k^2 \rho^{\frac{5}{3}} d}, \tag{12.56}$$

where D is a constant (= 0.364) and d is the length of the sound path,

$$\rho = \frac{h_s h_R}{h_S + h_R}, \tag{12.57}$$

and

$$\gamma_T = \left(\frac{C_T}{T_0}\right)^2 + \frac{22}{3}\left(\frac{C_W}{c_0}\right)^2, \tag{12.58}$$

where C_T and C_W are the turbulent structure parameters for temperature and wind speed, T_0 is a reference temperature and c_0 a reference sound speed. A typical value for moderately strong turbulence in the atmospheric boundary layer is $\gamma_T = 5 \cdot 10^{-6}$.

The total coherence factor is obtained by multiplying the component factors [44]:

$$C_{coh} = C_{misc} C_{turbulence}. \tag{12.59}$$

12.7.3.8 Scattering by Turbulence

A consequence of atmospheric turbulence for sound propagation outdoors viz. phase fluctuations is represented by the coherence loss factor (Eq. (12.56)). But scattering by turbulence leads also to increases in amplitude and level (see also Chapter 10), which can be estimated from the following expression [54]:

$$A_{excess,sc} = -\left(25 + 10\log\gamma_T + 3\log\frac{f}{1000} + 10\log\frac{d}{100}\right). \tag{12.60}$$

With realistic input parameters $A_{excess,\,sc}$ is positive. This effect of turbulence is added, in energy terms, to the combined effect of barrier and ground on sound propagation:

$$A_{excess,G+D+sc} = -10\log\left(10^{\frac{-A_{excess,G+D}}{10}} + 10^{\frac{-A_{excess,sc}}{10}}\right). \tag{12.61}$$

Scattered contributions will be important in zones where large attenuations are predicted due to the presence of barriers and ground.

12.8 THE ENVIRONMENTAL NOISE DIRECTIVE (END) SCHEME (CNOSSOS-EU)

12.8.1 Ground Effect

According to the common methodological framework to be used for strategic noise mapping under the Environmental Noise Directive (2002/49/EC) [26,54,55], 'the acoustic absorption properties of ground are mainly linked to its porosity'. While the ground effect depends on surface porosity, there can be 'soft' ground effect related to surface roughness (see Chapters 5 and 6) also; but this is not mentioned. Also, as discussed in Chapters 5 and 6, the acoustical properties of porous ground surfaces are affected more by their *flow resistivity* than by their porosity. The porosity of naturally occurring ground surfaces does not vary as much as their flow resistivity (see Chapter 6).

The END scheme [55], is similar to ISO9613-2 [1] in that it allows for frequency-dependent ground effect over non-flat ground by defining equivalent heights (see Section 12.10) and using a dimensionless frequency-independent coefficient, ground factor, G, with values between 0 (acoustically hard) and 1 (acoustically soft) according to the type of ground [55]. As in ISO9613-2, the END scheme introduces a factor, G_{path}, which is the fraction of the path occupied by porous ground but with a correction if the receiver is sufficiently close to the source.

$$G'_{path} = G_{path}\left[\frac{d_p}{30(z_S + z_R)}\right] + G_S\left[1 - \frac{d_p}{30(z_S + z_R)}\right], \quad \text{if } d_p \le 30(z_S + z_R) = G_{path} \quad \text{else,} \tag{12.62}$$

where d_{p} is the source to receiver distance, z_S, z_R are the source and receiver heights above the potentially slanted mean plane between source and receiver and G_S is the ground factor near the source. $G_S = 0$ for road surfaces and railways on slab tracks and $G_S = 1$ for railways on ballast.

The END scheme lists ground factor values corresponding to eight effective flow resistivity values (see Table 12.6) based on the Delany and Bazley impedance model (see Chapter 5), used also by NORD2000 and HARMONOISE [2,3]. However, the eight flow resistivity classes are associated with only four values of the G factor.

The END scheme includes equations (12.63) for attenuation due to ground effect given homogeneous atmospheric conditions (no refraction) and no impedance discontinuities or diffracting edges.

$$A_{ground,H} = \max\left\{\left[-10\log\left(4\frac{k^2}{d_p^2}\left\{z_S^2 - \sqrt{\frac{2C_f}{k}}z_S + \frac{C_f}{k}\right\}\left\{z_R^2 - \sqrt{\frac{2C_f}{k}}z_R\frac{C_f}{k}\right\}\right)\right], A_{ground,H,\min}\right\},$$

$$C_f = \frac{1 + 3wd_p e^{-\sqrt{wd_p}}}{1 + wd_p}$$

$$w = 0.0185 \frac{f_m^{2.5} G'_{path}{}^{2.6}}{f_m^{1.5} G'_{path}{}^{2.6} + 1300 f_m^{0.75} G'_{path}{}^{1.3} + 1160 \times 10^3} A_{ground,H,\min}$$
$$= -3\left(1 - G'_{path}\right), \qquad (12.63a, b, c, d)$$

where $k = 2\pi f_m/c$, f_m (Hz) being the band centre frequency considered.

Since the ground near the source is relevant until thirty times the sum of source and receiver heights, these equations imply that source and receiver are not interchangeable, as should be required by reciprocity [56].

Table 12.6 Ground types, flow resistivity classes and corresponding ground factor (**G**) values in the END prediction scheme [55].

Description	*Type*	*Flow resistivity kPa s m⁻²*	*G*
Very soft (snow or moss like)	A	12.5	1
Soft forest floor (short, dense heather-like or thick moss)	B	31.5	1
Uncompacted, loose ground (turf, grass, loose soil)	C	80.0	1
Normal uncompacted ground (forest floor, pasture field)	D	200	1
Compacted field and gravel (compacted lawns, park area)	E	500	0.7
Compacted dense ground (gravel road, parking lot)	F	2000	0.3
Hard surfaces (most normal asphalt, concrete)	G	20000	0
Very hard and dense surfaces (dense asphalt, concrete, water)	H	200000	0

For favourable propagation conditions, z_S and z_R are modified to account for refracted ray paths and turbulence, according to equations (12.64a–12.64d).

$$z_S' = z_S + \delta z_S + \delta z_T \tag{12.64a}$$

$$z_R' = z_R + \delta z_{R=} + \delta z_T \tag{12.64b}$$

$$\delta z_S = a_0\left(\frac{z_S}{z_S + z_R}\right)\frac{d_p^2}{2} \tag{12.64c}$$

$$\delta z_R = a_0\left(\frac{z_R}{z_S + z_R}\right)\frac{d_p^2}{2} \tag{12.64c}$$

$$\delta z_T = 6\times10^{-3}\left(\frac{d_p}{z_S + z_R}\right) \tag{12.64d}$$

$a_0 = 2 \times 10^{-4}$ is the inverse of the radius of curvature of the assumed circular downward-refracted ray paths.

Figure 12.12 shows that, while the spectrum of the attenuation over 'soft' ground predicted by the END scheme for acoustically neutral conditions with source height 0.5 m, receiver height 4 m and separation of 200 m is similar to that predicted for the same geometry by ISO9613-2 for 'moderate' downwind conditions, albeit with an octave higher maximum, the attenuation predicted by the END scheme for downwind conditions is substantially less than that predicted by ISO9613-2. This has been highlighted for several source–receiver geometries elsewhere [56].

12.8.2 Criticisms

Various clarifications of the END scheme have been suggested when accounting for points below the mean ground plane and for diffraction by low barriers [56]. The scheme compares the path-length difference introduced by the diffracting edge (e.g. top of the barrier – see Chapter 9) in each octave band with either wavelength/20 or with the Rayleigh criterion without specifying how the latter should be implemented. The Rayleigh criterion distinguishes between specular reflection and diffraction. If the Rayleigh criterion is used by comparing the sum of the path-length difference introduced by the diffracting edge and that corresponding to the mirror image source and receiver with wavelength/4, then the results can be quite different than obtained when using the wavelength/20 method [56].

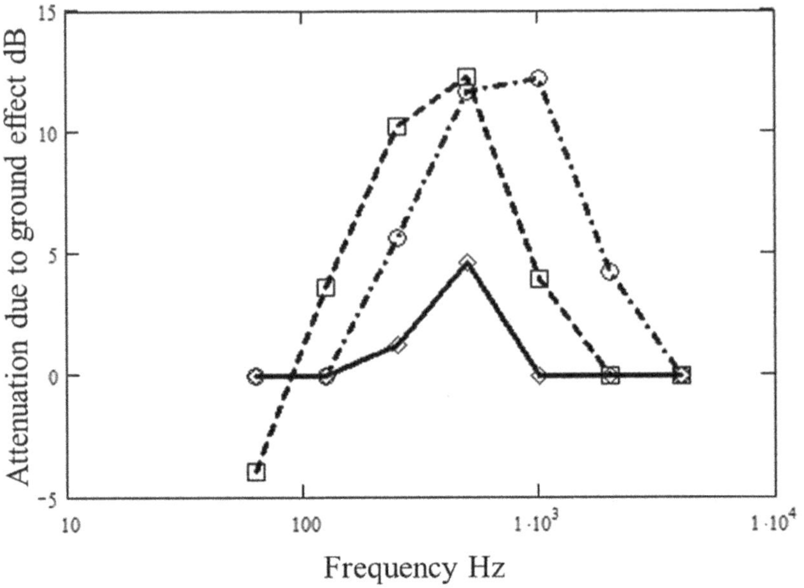

Figure 12.12 Spectra of attenuation due to soft ground effect, with source height 0.5 m receiver height 4 m and 200 m separation, predicted by ISO9613-2 (squares joined by broken line), the END scheme for an acoustically neutral atmospheric condition (circles joined by dot dash line) and the END scheme for downwind condition (diamonds joined by continuous line).

Apart from the problems with reciprocity and soft ground effect already mentioned, there is an unresolved problem with the way diffraction is included the END scheme in downward-refracting conditions. According to the scheme, the presence of more than one diffracting edge in downwind propagation conditions can lead, counter-intuitively, to less attenuation than when there is only a single diffracting edge [56].

12.9 PERFORMANCE OF RAILWAY NOISE PREDICTION SCHEMES IN HIGH-RISE CITIES

The lack of available land means that many cities contain high-rise residential and commercial tower blocks with heights of over 100 m (over 40 stories). Railway noise can be a particularly acute problem near such buildings which can be located very close to railway viaducts so the resulting noise levels in the adjacent residential areas are significant. Prediction schemes for railway noise [9,57,58] do not offer a way of predicting railway noise in high-rise cities. Neither do they offer a way of predicting railway noise from an elevated viaduct. Van Leeuwen [57] has identified the inherent differences between the various models including the assumption of source positions, the levels of noise emitted by pass-by trains, the characteristics of sound radiation and the correction factors adopted to account for the effect of reflection. Consequently, the noise levels predicted by the models for any given situation are different.

Field measurements have been carried out on a residential building to assess the noise levels caused by trains passing on a nearby viaduct and the experimental results have been compared with different schemes for predicting noise from trains [59]. The schemes included the *Calculation of Railway Noise* (CRN) (see 12.4.2), the *Nordic Prediction Method for Train Noise* (NMT) [60], MITHRA (a commercial software package developed by CSTB in France) [61] and ISO 9613-2 (see 12.2).

MITHRA has been developed to model the propagation of sound in outdoor environments and is based on a fast algorithm identifying all possible paths that link noise sources with receivers in a complex urban site. Not only are the direct, diffracted and reflected ray paths included in the predictions, but the combined effects of diffraction and reflection are included also. The software allows use of any of three methods for predicting the propagation of sound outdoors: CSTB 92, ISO 9613-2 and NMPB 95. However, only the first two methods have been used in the comparisons reported here.

The CRN methodology divides the railway line into one or more segments such that the variation of noise within the track segment is less than 2 dB(A). For each segment, the reference sound exposure level A-weighted SEL_{ref} in dB is calculated for each type of train on each track. On the other hand, the NMT calculations are carried out in octave bands from 63 to 4000 Hz. Based on measurements of the noise of octave band emissions for different types of trains, the sound power generated per metre of track can be determined if the speed of the train and the volume of traffic are known. The source line is then divided into elements in the direction along the track. Each element has a length of no more than 50% of the distance between the track and the reception point. The CSTB (MITHRA) software does not allow specifically for trains on an elevated railway viaduct. For the calculations based on MITHRA, the railway track is assumed to be on a reflective embankment instead of a viaduct.

The measurement locations on the selected residential building close to a viaduct are shown in Figure 12.13.

The viaduct is flanked on both sides by residential buildings and built with dual tracks in the vicinity of a road flyover with a single traffic lane in each direction. The viaduct and the flyover are above a busy trunk road with five lanes in both directions. The measurements were made with trains running on the far-side track located at a nominal distance of about 26 m from the selected residential building. Simultaneous measurements were made with 11 microphones protruding 1 m from the façade of the building. Four additional microphones were mounted on an extension pole erected on the floor of the rooftop as indicated in Figure 12.13. Some parts of the viaduct

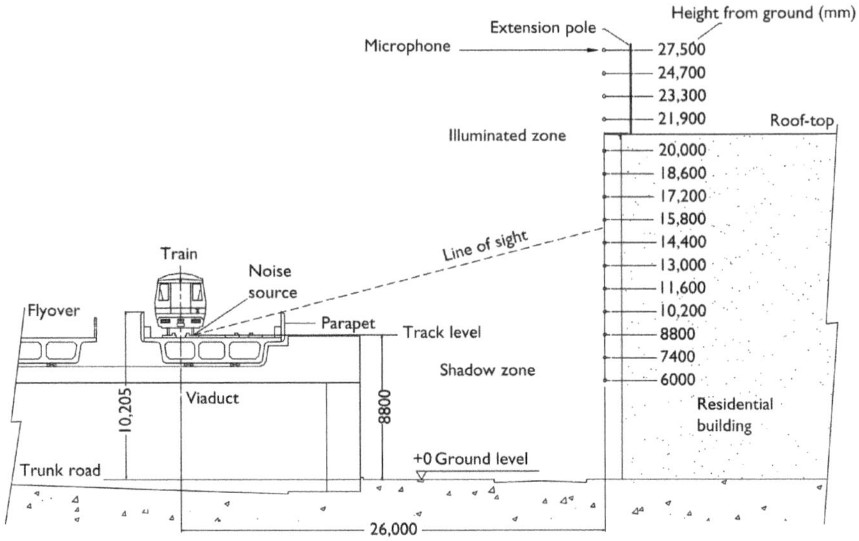

Figure 12.13 Schematic of measurement locations for railway noise study.

at the measurement site are shielded by nearby buildings. To simplify the experimental process and the subsequent numerical analysis, each measurement was started when the train came into sight and stopped when it was out of sight of the selected residential building. The nominal speed of the train as it travelled through the section was about 50 km/h, and the duration of the measurement for each pass-by train was about 17 seconds. The noise data were recorded through a 16-channel digital data recorder and subsequently analysed in the laboratory with the aid of post-processing software.

Since the section of the viaduct selected for this study was in the vicinity of a busy trunk road, the measurements were influenced by traffic noise, even though the measurements were made after midnight. In the measurement periods, the noise from the traffic on the road was monitored closely before and after the noise from a passing train was measured. Motor vehicles on the flyover and the trunk road below were counted during the measurement period. The vehicles were classified into two types according to the CRTN procedure (see 12.4), i.e. heavy vehicles weighing more than 2.8 tonne and light vehicles weighing less than 2.8 tonne. Control measurements were carried out to record the traffic noise under similar traffic conditions to those used for the railway noise data when there were no passing trains. The class and number of motor vehicles on each road were recorded during the 'standard' 17-second measurement period. By assuming that contributions from road traffic noise at the receiver points were unchanged under similar road traffic conditions, the net noise from a passing train was

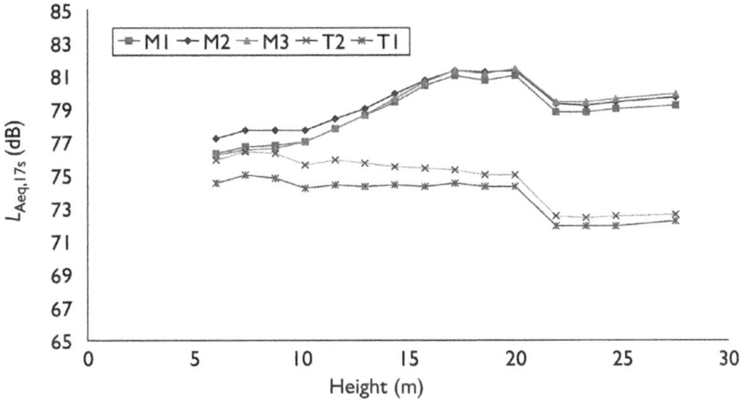

Figure 12.14 Results of measurements of train noise mixed with road traffic noise and control measurements of road traffic noise. M1, M2 and M3 represent mixed noise measurements. T1 and T2 represent two control measurements of road traffic noise

determined by logarithmically subtracting the traffic noise levels obtained in the control measurements from the overall noise levels when the train passed the observation station. Hence, the net levels of noise from a passing train were given by

$$L_{\mathrm{Aeq}} = 10\log\left(10^{\frac{L_{\mathrm{Aeq},m}}{10}} - 10^{\frac{L_{\mathrm{Aeq},t}}{10}}\right), \tag{12.65}$$

where $L_{\mathrm{Aeq},\,m}$ is the measured overall noise level and $L_{\mathrm{Aeq},\,t}$ is the measured level of traffic noise in the control measurements. All noise levels were expressed in terms of A-weighted sound pressure level. As remarked earlier, the measurements took place after midnight in the early morning hours. At that time there were reduced volumes of traffic and the noise from road traffic was greatly diminished. Consequently, the measurements were taken during the periods when they were least affected by noise from road traffic. Moreover, the activity levels and hence interfering noises from occupants of the building were minimized.

Three typical sets of A-weighted railway noise levels are shown in Figure 12.14.

Also Figure 12.14 shows the results of two typical control measurements of road traffic noise that were obtained for representative traffic flow conditions. The symbol M represents the measurement for train noise mixed with road traffic noise and the symbol T denotes the control noise measurement in the absence of the train noise where the volume of road traffic flows was comparable to that observed during the period of the measurement of train noise. Different sets of noise measurements were conducted in the absence of passing trains but only two typical sets, which are marked as *T1* and *T2*, are shown in Figure 12.9. On the other hand, only three typical measurement results with a pass-by train (marked as *M1*, *M2* and *M3*) are shown for comparison of the respective A-weighted equivalent noise levels, $L_{\mathrm{Aeq},\,17s}$. The difference between the L_{Aeq} due to a passing train and road vehicles is about 7 dB for the measurement points above the rooftop of the building (see Figure 12.13). However, the L_{Aeq} due to the passing trains and the road traffic below the track level of the viaduct are very similar. The differences vary from 0 to 2 dB because these receiver locations were exposed directly to the road traffic noise during the measurements.

It was found that the influence of traffic noise is not significant for receivers located in the illuminated zone. However, the measured noise levels at the lower heights were affected adversely by the road traffic noise because the receiver was in the shadow zone. The A-weighted equivalent noise levels due to a passing train have been obtained from the data by logarithmically subtracting the road traffic noise from the overall noise levels using Eqs. (12.33a–12.33c). The L_{Aeq} at different receiver positions due to a passing train after the adjustment are shown in Figure 12.15.

For comparison with predictions, the octave band sound power level per unit length is required for the calculation of the overall L_{eq}. Although the NMT and MITHRA software provide typical sound power levels in octave bands for different trains, neither scheme offers an appropriate source model for the railway stock that were used in the measurements. As a result, the standard procedures proposed by the NMT method have been used to obtain the octave band sound power levels from the measured data. First, the train (source line) was divided into several smaller elements in the direction along the track. Each smaller element could then be treated as a line of infinitesimal point sources. The sound pressure levels generated by all point sources were replaced by an equivalent point source located at the geometric

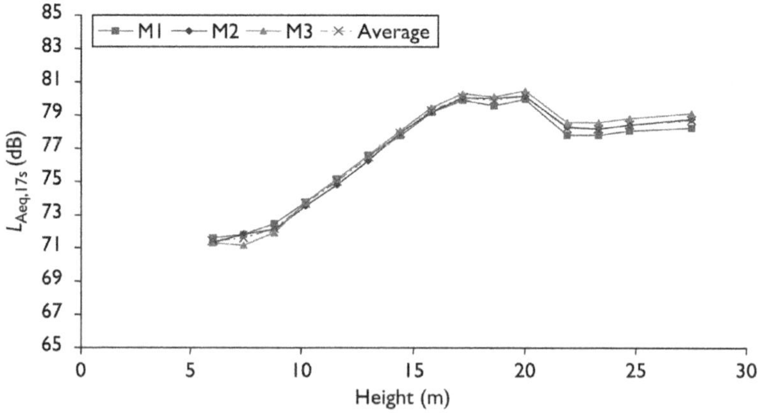

Figure 12.15 A-weighted train noise levels after the logarithmic subtraction of measurements M1, M2 and M3 from the measurements T1 and T2.

centers of the respective elements. The measured noise levels at a representative location were used to deduce the octave band sound power levels per unit length. The highest measurement point (27.5 m above the ground level) was chosen as the representative location for the calculation since it was (i) directly exposed to the train noise without the possible screening effect due to the parapet of the viaduct and (ii) it was relatively free from any reflecting surfaces. Air absorption was not considered due to the relatively short distance, 30 m or less, between the equivalent sources and receiver. The deduced octave band sound power levels were used as an input for the NMT, CSTB 92 and ISO 9613-2 calculation methods.

Although ISO 9613-2 provides an algorithm to account for the long-term meteorological effect on sound propagation outdoors, meteorological effects were considered unimportant in a confined urban space.

The reference sound exposure level (L_{AE}) is required as an input data for the calculation of noise levels of a passing train according to the CRN method. To ensure a comparable sound power level of the source between different prediction schemes, the L_{AE} of the passing trains was determined from the measured data rather than by using the standard figures provided by the CRN method. Again, the highest measurement point was selected as the reference position for the determination of the reference L_{AE}. The correction factors for the distance attenuation and the angle of view, determined according to the CRN method, were used. Only a single segment of track was needed.

The A-weighted equivalent noise levels of a passing train deduced from measurements at different heights above the ground level are shown in Figure 12.16.

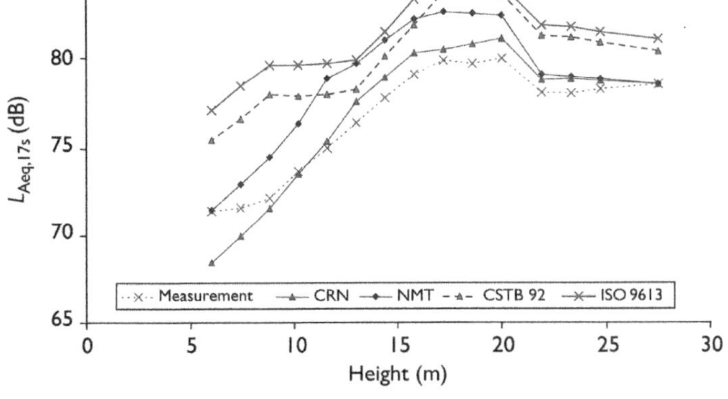

Figure 12.16 Comparisons between the A-weighted equivalent train noise levels deduced from measurements and predictions using the CRN, NMT, CSTB 92 and ISO9613-2 schemes.

The noise level value at each receiver location was obtained by taking an arithmetic average of three sets of A-weighted equivalent noise levels measured at the same position but at different times. The measured noise levels increase progressively from the minimum level at the lowest point located at about 8.8 m above the ground level. Since the average speed of the passing trains is about 50 km/h, the noise radiating from pass-by trains is mainly dominated by the noise from the wheels/rail interaction. It should be noted that the direct ray path is shielded from the parapet of the viaduct for receivers located below 15.8 m. Indeed, the measured noise level increases steadily between the heights of 8.8 m and 15.8 m above the ground level. The noise levels increase because the screening due the parapet of the viaduct becomes less effective when the receiver approaches the boundary of the shadow zone. After the shadow boundary, the measured noise levels remain relatively unchanged at heights between 17.2 m and 20 m above the ground level. At these heights, the measurement points were exposed directly to the source and the change in the propagation distances from the source to the measurement points was relatively small, given that the horizontal distance between the viaduct and the building was about 26 m. Between 20 m and 22 m above the ground level, the measured noise level drops abruptly as the corresponding receiver points were located above the roof top of the building. The noticeable drop in the noise level at these heights results largely from the fact that the measurement points were above the reflecting elements of the building façade. Above a height of 22 m, the measured noise levels on the roof top increase marginally by up to 1 dB(A) as the heights of the receiver points increase. The data obtained above the roof top of the building suggest that the source exhibits a directivity pattern in the vertical plane. Chew [62–64] measured A-weighted equivalent noise levels due to a train running at grade on a viaduct from a building and found similar results, i.e. an increase in noise levels of up to 7 dBA as the polar angle of the receiver in the vertical plane increased from 8° to 30°.

Figure 12.16 also shows the levels predicted by the various schemes. The numerical results predicted by the CRN method give the best quantitative agreement with the measurements as a function of the height above ground in an urban environment. The discrepancies between the predicted noise levels and measured data in the shadow zone are probably due to the measured data being mainly dominated by the road traffic noise. According to the CRN calculations, the noise levels over the roof top are predicted to decrease with increased height above the ground. This differs from the measurement results because the CRN method does not take the directivity of train noise into account.

The predictions by the CRN and NMT methods are in reasonable agreement with the general trend of the noise levels. However, the results predicted by the NMT method are about 2–3 dB higher than those predicted by the CRN method for the receiver positions that are 1 m away from the façades of buildings and there is almost no difference for receiver positions located above the roof. The predictions from the NMT and CRN methods at the highest receiver position are identical with the data because this point was selected as the reference for calculating the octave band sound power levels and the SEL of the passing trains. The relatively large discrepancies between the predictions in front of the building's façade are the consequence of two contributing factors. First, there is a difference between the assumed source heights above the railhead in the two prediction models which result in different predictions of screening effect at the same receiver position. The CRN method assumes that the source is located at the railhead while the NMT method assumes the source to be between 0.3 and 2 m above the railhead depending on the frequency. Second, there are differences in the assumed façade correction factors (2.5 dB in CRN but 3 dB in NMT).

The predictions given by the methods of CSTB 92 and ISO 9613-2 overestimate the A-weighted equivalent noise levels by up to 8 dB for measurement points below the track level. Moreover, none of the methods predict the increase in the measured noise levels in the shadow zone. Although MITHRA provides an optional algorithm for specifying the vertical directivity of a passing train, the noise source was assumed omni-directional in the calculations to enable the comparisons with the other prediction methods.

In summary, the CRN method was found to give the best agreement with the measured results. The NMT scheme was found to predict the general trend of the data, except in the shadow zone, but it overestimated the measured noise levels. The CSTB 92 and ISO 9613-2 prediction schemes were found to be comparatively less satisfactory.

12.10 PERFORMANCE OF ENGINEERING MODELS IN A COMPLEX ROAD TRAFFIC NOISE EXAMPLE

12.10.1 Site and Models

In this section, the respective prediction accuracies of ISO9613-2, CNOSSOS-EU (END) and the HARMONOISE point-to-point engineering model (HP2P) are assessed for the complicated and busy road configuration shown in Figure 12.17. The site involves terrain shielding, the presence of mixed natural grounds, a multi-lane highway (8–9 lanes) with a raised central reservation (0.7 m high) and vegetation near the top of the slope [65].

Sound pressure level differences at two microphones were measured continuously during almost a full month; the first microphone (MP1) was positioned directly bordering the road and the second one (MP2) at 80 m from the centre of the road, on a 6.3-m high embankment. The measurements were found to be stable over time, both as regards total A-weighted road traffic noise levels and in 1/3 octave bands and will be considered here as the ground truth.

Figure 12.17 Part of the highway under study, showing microphone positions MP1 and MP2 [65].

Detailed data on the traffic flow (per driving lane) were obtained near to the cross section under study. There was also access to highly detailed digital terrain elevation data. Also, as a reference, a full-wave numerical model (FDTD) was used to predict the level difference between the measurement points.

Although such a large amount of detailed data is not commonly available in most situations where engineering methods are to be applied, it allows in-depth assessment of the prediction methods. Given that sound propagation outdoors is strongly frequency dependent, the only true validation should involve spectral analysis to assess the degree to which the propagation physics are captured. Also, this would expose whether over-predictions at some frequency bands compensate under-predictions at others, thereby leading, accidentally, to fair total level predictions in particular cases which cannot then be generalized. Including a full-wave numerical model as a reference is helpful to show the maximum accuracy (albeit, of course, requiring maximum computational effort) that could possibly be achieved and the extent to which the engineering models fail to give good predictions.

12.10.2 Approximating the Berm Slope

Although primarily developed to model sound propagation over flat ground, ISO9613-2 allows for undulating terrain by the best-fit hypothetical ground surface line between source and receiver (see Figure 12.18). Source and receiver heights referenced to this best-fit linear profile are called 'effective heights'. The END model (see Section 12.8) uses a similar approach.

In the HP2P model, the terrain cross section is approximated by connected linear segments. The iterative end-point fit algorithm is used here to minimize the number of segments [66] while keeping the deviation from the actual profile as limited as possible. A maximum deviation relative to the actual profile was set at 0.1 m.

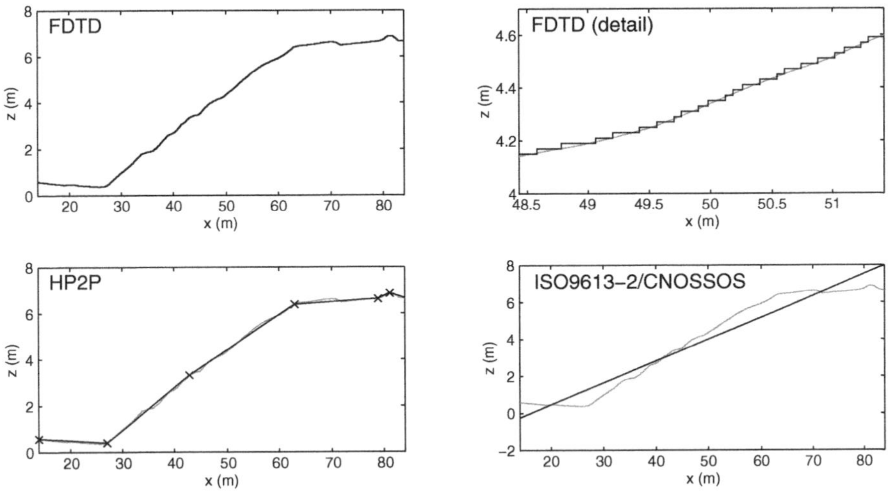

Figure 12.18 Approximations to the berm slope (see Fig. 12.17) used for predictions with the FDTD method, the Harmonoise point-to-point (HP2P) model and ISO9613-2/CNOSSOS. The top left plot and gray lines in the other plots are of the actual profile, the black lines represent the approximations used in the models (axes not to scale). The upper right plot indicates that a finely stepped profile is used for the FDTD calculations [65].

In the FDTD reference model, the slope was approached by means of staircase fitting. The spatial resolution of 0.02 m was sufficiently small to achieve a very close resemblance with the actual slope, including almost any detail present in the digital elevation map.

The first part of the slope can be categorized as rough grassland, while, near the top, the presence of sparsely positioned tall trees resulted in leaf litter such as frequently found on a forest floor (see Chapter 5). The parameter values for the ground and vegetation used in the models are in Table 12.7.

12.10.3 Road Traffic Source Power Modelling

The HARMONOISE/Imagine source power model (see Chapter3, section 3.5) was used for all of the prediction methods, not only for consistency but also because this model provides predictions at the detail of 1/3 octave bands. The use of a single-source power model allows pursuit of the main goal of the current analysis which is to quantify differences in propagation model performance. Detailed data on a one-minute basis during the acoustic monitoring period are available for the vehicle speed and vehicle flow rate in each lane, with discrimination between three categories (light, medium-heavy and heavy vehicles). Each lane has a different source power and a different distribution of the energy over the frequency bands, dependent on the averaged vehicle speed per lane and proportions of light/heavy traffic.

Although the main interest in this analysis is predicting level differences between two points, the acoustic energy distribution over the lanes and over the frequency bands is needed for accurate predictions.

12.10.4 Daytime vs Nighttime Measurements and Predictions

The sound emission of the highway during daytime (0700 – 2259) and nighttime (2300 – 0659) is different due to changes in the share of heavy relative to light traffic, changes in the average vehicle speed per lane and due to different lane use. Therefore, measurements were analysed separately for the day and night period, and dedicated simulations are performed for both periods. Also, atmospheric absorption might be slightly different between day and night. Averaged values of the meteorological parameters relevant for atmospheric absorption during daytime were air temperature (15.2°C), relative humidity (73.7%) and atmospheric pressure (101480 Pa); during the night period, these values were 11.9°C, 84.9% and 101440 Pa, respectively. Given the focus on level differences and short propagation distances, both microphones are expected to be influenced by atmospheric refraction only to a limited extent. In any case, observations for wind speeds above 5 m/s at a reference height of 10 m, based on meteorological data from a nearby meteorological observation station, were removed already to prevent excessive wind-induced microphone noise.

12.10.5 Model Performance

Figures 12.19 (a) and Figure 12.19 (b) compare model predictions with the measurements of the spectral difference and the total difference in sound pressure levels between the microphone directly bordering the road (MP1) and that on the embankment (MP2), respectively.

The ISO9613-2 predictions of both spectral and total A-weighted sound pressure level differences are poor. While the END method gives somewhat improved predictions of the ground effect and captures more of the physics relative to ISO9613-2, it still underpredicts the transmission loss by several dB. If these models are used to assess different noise abatement solutions, then the acoustical effectiveness of a natural berm near a highway, such as in this example, could be underestimated and thereby wrongly discounted during planning.

The HP2P model predictions, which account for terrain diffraction, give a good spectral fit. The HP2P scheme enables accurate modelling of the slope's ground effect, which is the main advantage of a berm relative to a noise wall (see Chapter 9). On the other hand, although in this example it is less than 1 dB(A) in most of the fractional octave bands, the HP2P model could not predict the difference in levels between daytime and nighttime.

Despite the maximum frequency of prediction being limited to the upper frequency of the 1/3 octave band with centre frequency 1.6 kHz to reduce the computational cost, the full-wave FDTD method yields predictions that have the closest spectral resemblance with the measured level difference data and the measured total A-weighted sound pressure level differences. Moreover, in contrast to the engineering methods, the FDTD simulation accords with the measurements when comparing daytime to nighttime transmission losses. The ability to correctly predict such effects, although small, shows the accuracy and high sensitivity of the FDTD method to the details in such a complex example. However, calculation times with FDTD are a few orders of magnitude larger than when using the engineering methods.

The FDTD model used scattering elements to represent the tree canopies (see Table 12.7). This resulted in an improved spectral fit with the measurement data above 1 kHz. It should be noted that MP2 is surrounded by canopies and so downward scattering will lead to slightly increased high-frequency sound pressure levels. In contrast, the levels at MP1 are not expected to be affected by the vegetation. Consequently, the level difference between MP1 and MP2 will drop as shown in Figure 12.19 (a). Additional analysis with the FDTD model has revealed the importance of the vegetated zone near the top of the slope [65]. The level increase due to downward scattering by the canopies (i.e. a negative effect by the trees) is subordinate to the influence of

Table 12.7 Ground and vegetation parameters used for the complex road example.

	ISO9613-2/ CNOSSOS	*HP2P*	*FDTD*
Road surface	G=0	rigid	Rigid
Grass-covered slope	G=1	flow resistivity= 200 kPa s/m² [Delany and Bazley model (see Chapter 5), tuned parameter]	flow resistivity=300 kPas/m², porosity=0.75 (Zwikker and Kosten model [see Chapter 5])
Forest floor	G=1	flow resistivity= 30 kPas/m² (Delany and Bazley model, tuned parameter)	flow resistivity = 20 kPas/m², porosity= 0.50 (Zwikker and Kosten model)
Scattering by vegetation	Not included	Not included	Random distribution of small scattering elements occupying 0.1% of the canopy volume.

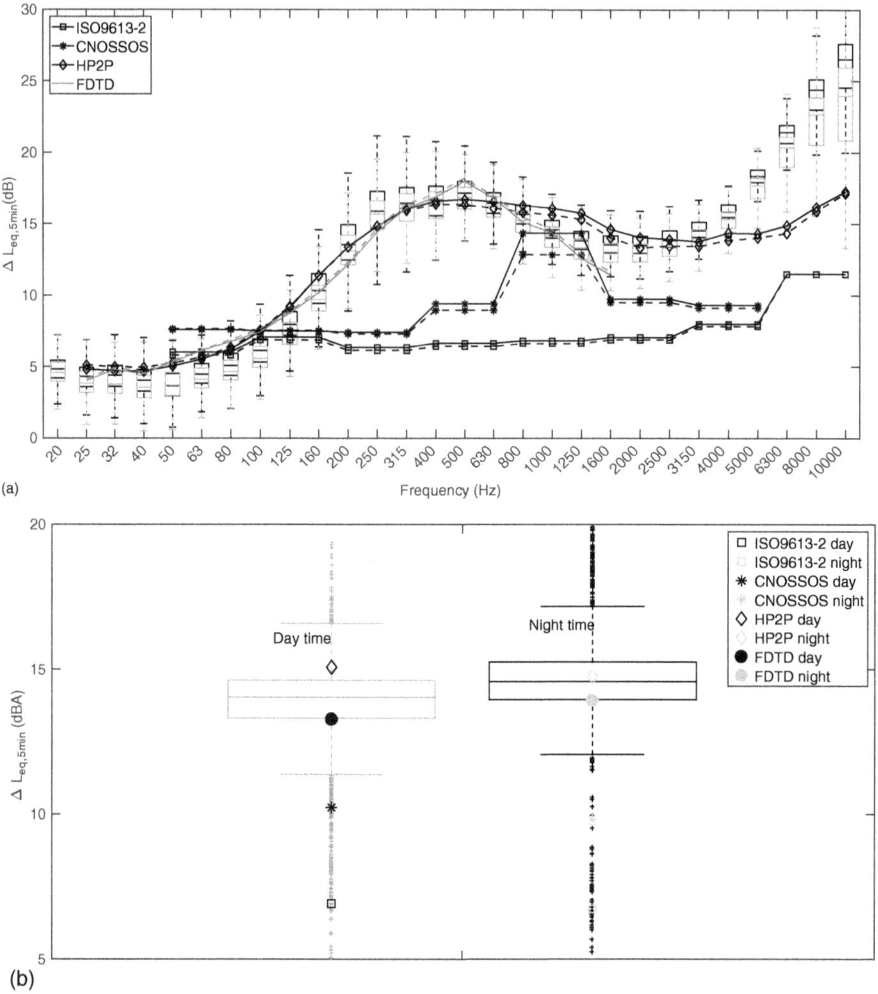

Figure 12.19 (a) Spectra of the measured level differences between MP1 and MP2. The boxplots indicate the range of measurements (gray boxes are for daytime measurements, black boxes for nighttime measurements). The (middle) horizontal line in each boxplot indicates the median of the data. The boxes are closed by the first and third quartile and whiskers extend to 1.5 times the interquartile distance above the maximum value inside each box, and to 1.5 times the interquartile distance below the minimum value inside each box. Outliers are not shown. The predictions of the 3 engineering methods and the reference model are plotted as lines joining the symbols identified in the key. The broken lines indicate the nighttime predictions and the continuous lines the daytime predictions. (b) Measured (boxplots) and predicted total A-weighted road traffic noise level differences during daytime and nighttime periods [65].

the pronounced ground dip due to the presence of the forest floor (i.e. a positive indirect effect by the trees). When assuming the embankment to be completely grass-covered, the predicted level difference between the two assessment points is about 3 dBA lower. This suggests that, in situations such as in this example, detailed ground impedance modelling has a significant influence on the accuracy of predictions.

12.11 PREDICTING WIND TURBINE NOISE

12.11.1 An Untypical Industrial Source

Although wind turbines are considered as industrial infrastructure, the behaviour of their sound emissions is quite different from those of typical industrial sound sources. The fact that the sound source is relatively high means that knowledge about sound propagation from low height sound sources is not necessarily applicable. In addition, sound propagation from a wind turbine is influenced by the complex flow fields which appear upwind and downwind from the turbine. While direct sound contributions interacting with ground reflections might be the dominant propagation effects close to a turbine, at longer ranges accurate predictions require advanced sound propagation models that account for height-dependent refraction.

Specific characteristics of the sound fields produced by wind turbines, including amplitude modulation (see Chapter 3, section 3.6), have led to stringent legal limits to prevent noise annoyance reactions at nearby dwellings. This implies that the propagation distances of concern might exceed 1 km from the infrastructure and care is required in predicting the resulting low levels. In many countries, ISO9613-2 is used for impact assessment during the planning of wind turbine projects. But some care is needed.

12.11.2 Complex Meteorologically Induced Propagation Effects

In contrast to ISO9613-2, full-wave numerical simulations allow in-depth investigations of sound propagation from wind turbines. Such research has led to some counter-intuitive predictions. For example, up to a few hundred metres from the mast, upwind sound pressure levels are predicted to be higher at all frequencies than at the same distance downwind [67], whereas, generally, for low-height sound sources, the opposite is true.

Up to roughly 1 km downwind from the turbine, higher levels are observed as expected. Beyond that distance, mainly wake-induced sound pressure level amplification is observed [67,68]. Especially in very stable atmospheric conditions, long and persistent wakes appear, leading to more pronounced wake amplification effects. In contrast, if the incoming turbulent intensity is strong, wake amplification is poorer. The presence of the turbulent wake has been shown to increase amplitude modulation significantly downwind, and its magnitude increases with distance [69].

There may be more intense refraction effects around noise sources present in the upper emission quadrant of the rotor plane. Wind velocity depletion near hub height means that higher sources are more subject to downward refraction than lower sources which are influenced, initially, by an upward-refracting zone in the atmosphere. Relatively near to the turbine, directivities of the aero-acoustic sources present at the blades are thought to be the main drivers for amplitude modulation, while at larger distances, refraction effects become dominant, depending on source position, as a result of the height-dependent refraction [70].

12.11.3 Ground Effect for Wind Turbine Sound Propagation

Many detailed outdoor sound propagation models and engineering models assume point source radiation. Accordingly, wind turbine sound emission is represented by an 'effective' point source placed at hub height [71]. The source power spectrum can then be estimated by calculating back from a sound pressure level spectrum measured close to this virtual hub height position. In this way, the energy from the noise sources distributed all over the wind turbine rotor plane is assumed to be concentrated at the effective point source position.

A second approach is to model sound propagation from a wind turbine as though it is from a series of (incoherent) noise sources in the rotor plane, positioned at the blade tips or very close to it. Simple analytical considerations in a non-refracting atmosphere indicated that the difference in sound pressure level between a single hub source and distributed tip noise sources is less than 1 dB at a distance exceeding the rotor diameter [72]. Some doubts have been cast on the validity of this approach (i) to predict ground effects accurately [73], (ii) to capture amplitude modulation due to blade passages [69] and (iii) to study the different atmospheric influences on sound propagation depending on the blade tip position [70]. Acoustic camera visualizations [74] have shown that most noise is produced during the downward stroke of a blade, and so a uniform distribution of point sources would not be accurate. A variation in level, but also a dedicated directivity, can be imposed at each distributed source position in the rotor plane [75].

In this section, a comparison is made between assuming source power concentration at hub height and distributing this same amount of acoustic energy over 100 blade tip positions such that the energy contributed by each source is added at the receiver using both the ISO9613-2 model and the HARMONOISE point-to-point model (HP2P). For simplicity, refraction by the atmosphere is neglected. This is justified by the relatively short propagation distance (see Figure 12.20).

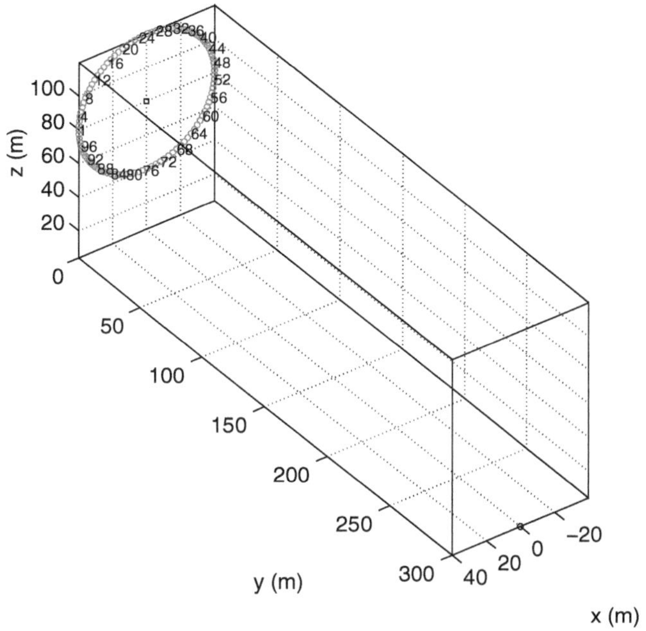

Figure 12.20 Geometry used for the example calculations of wind turbine sound propagation with engineering methods. The numbers indicate the blade tip point source positions in the rotor plane.

The wind turbine source spectrum described in Chapter 3, section 3.6 is used. The ground near the wind turbine is assumed to be 'grassland'. Predicted spectra at the receiver position (4 m high and at a horizontal distance of 400 m from the perpendicular from the centre of the rotor plane to the ground) are depicted in Figure 12.21.

Although single point source modelling leads to pronounced spectral dips in fractional octave bands, they disappear when integrating the contributions from multiple source points. In Figure 12.22, a detailed analysis is made of the ground effect for each source position along the circumference of the rotor plane. The shift of the first destructive interference in function of frequency at the (fixed) receiver due to change in the source height is clearly visible. When using the HP2P engineering method, distributed sources result in a total sound pressure level that is 0.7 dBA lower than that obtained by concentrating the sound energy at hub height. A single hub height source modelled by ISO9613-2 gives a total exposure level that is 2.2 dBA lower. A modification to ISO9613-2 has been proposed specifically for wind turbine sound pressure level calculations [77]. The ground factor G is limited to 0.5, even in case of a natural soil, to account for the loss of ground effect for highly elevated sound sources. This modification seems to work quite well, leading to a difference of only 0.5 dBA in total sound pressure level relative to using the more advanced HP2P model (single hub source).

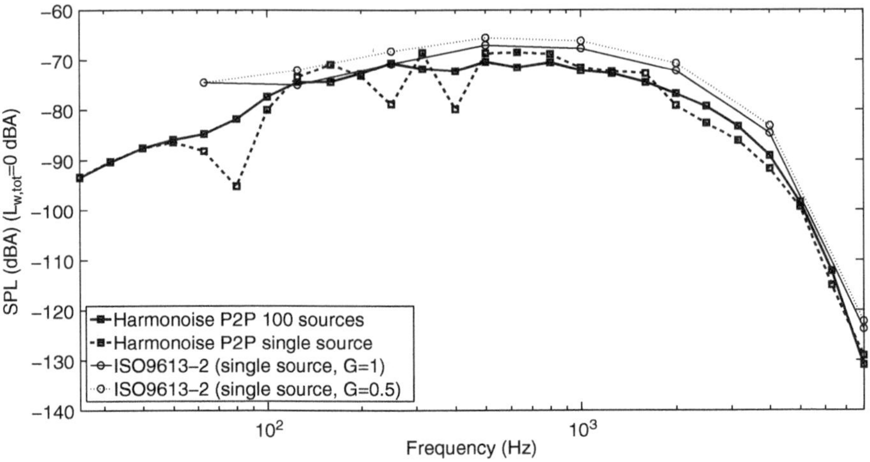

Figure 12.21 Predicted sound pressure level spectra for the wind turbine setup depicted in Fig. 12.20. Note that the ISO9613-2 model gives results in full octave bands, whereas the HARMONOISE point-to-point model outputs in 1/3-octave bands.

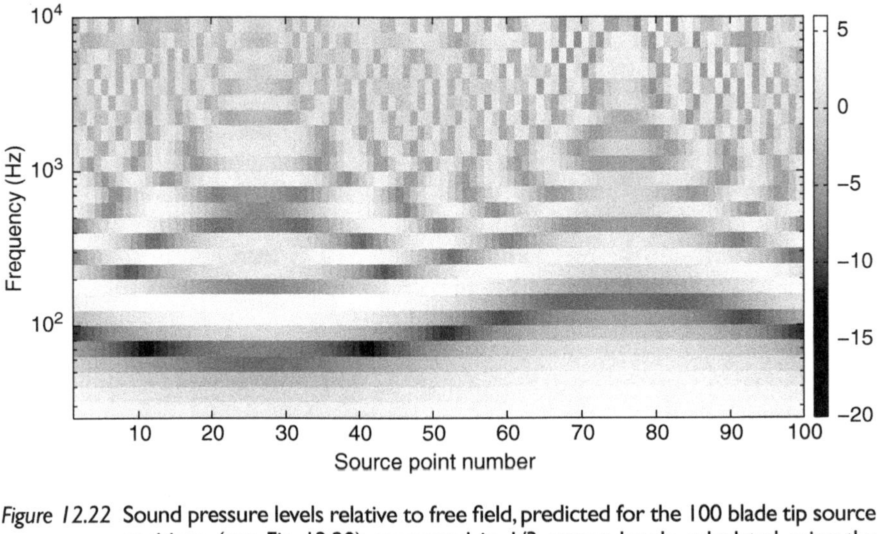

Figure 12.22 Sound pressure levels relative to free field, predicted for the 100 blade tip source positions (see Fig. 12.20), expressed in 1/3 octave bands, calculated using the HARMONOISE point-to-point model, for sound propagation over grassland.

12.11.4 Propagation over Non-Flat Terrain

Often wind turbines are installed in undulating terrain, for example on ridges, to increase their energy capture potential through the wind speed-up effect. However, this further complicates sound propagation predictions. For example, the ISO9613-2 engineering model assumes a fully flat ground. Although the idea of a 'mean ground plane' is mentioned as an approximation for non-flat terrain, this approach should be used with care and the resulting predictions might be highly inaccurate (see Section 12.10). Moreover, measurement campaigns near wind turbines have shown that topography could impact exposure levels significantly [76]. If ISO9613-2 is used for predicting propagation across a concave ground profile, the addition of 3 dB to the calculated overall A-weighted noise level has been proposed [77].

Detailed predictions of sound propagating across a valley using a hybrid, analytical, conformal mapping Green's Function Parabolic Equation approach have shown that exposure levels strongly depend on the position in the valley and on the ground type [78]. Noise shielding is predicted on the source side of the valley. Starting from the centre of the valley, near-free-field sound propagation is predicted in all octave bands when the ground is rigid, while a strong level reduction is predicted along a grass-covered valley.

12.12 PREDICTION REQUIREMENTS FOR OUTDOOR SOUND AURALIZATION

12.12.1 Introduction

Potentially, auralization is a powerful aid to designing a noise abatement solution. In principle, it allows residents to experience the sounds or noise they will be faced with after deployment [79,80] and allows comparison of alternatives at the planning stage by means of 'auditory previews'. Values of long-term averaged sound pressure levels presented in noise maps are not always easy to disseminate to the public at large, or even to planners and architects, since, often, the acoustical knowledge needed to interpret such data is lacking. Moreover, many noise abatement solutions change the spectral balance of the environmental sound and might consequently change the psychoacoustics. Clearly, since sound perception is difficult to predict listening tests could be very useful.

Good auralizations are a logical response to the emergence of virtual reality (VR) techniques which focus strongly on visual representation at the expense of aural representation. On the other hand, the human perception of outdoor environments is strongly governed by audio–visual interactions [81]. Human hearing is incredibly sensitive to small changes in auditory exposure. The hearing threshold, i.e. the smallest amplitude a normal hearing young adult can perceive, is very low. Interaural level differences and interaural time differences, often in the order of milliseconds, allow efficient source localization. Specific sources have distinct and characteristic acoustical features including Doppler shifts in emissions from moving sources, amplitude modulations in the noise from wind turbines or rattling sounds from road vehicles. The need for synthesized signals to be highly accurate makes auralization of outdoor sounds challenging.

Auralization has three main aspects. First, there is a need for audio samples of the sources. This can be achieved either by emission synthesis starting from a collection of sinusoids plus filtered noise components [82], or by using data sets of pre-recorded (sub)sources that should then be combined [83,84]. Specific types of sources might need specific procedures to capture the fine spectro-temporal detail that is often important for the listening experience. Directivity patterns can be complex and strongly dependent on the type of source (see Chapter 3, sections 3.5 and 3.6).

Second, during propagation outdoors, sound pressure levels and the spectral balance will change. This means that propagation models should be sufficiently detailed to provide realistic acoustic rendering of environmental noise sources.

Third, a reproduction system is needed. The playback equipment plays a relevant role too, and ranges from the use of simple headphones (with or without head-tracking) to full blown VR theatres. Typically, an outdoor environment involves a multitude of sources. Since it might be impractical or too computationally demanding to consider every source, adding general background noise might be a solution [85]. This section outlines a few aspects of auralization of the outdoor environment and a procedure for noise abatements is discussed.

12.12.2 Simulating Outdoor Attenuation by Filters

Filters offer a straightforward way of allowing for effects of distance and atmospheric absorption. Geometrical divergence, for example when assuming a point source, is modelled by an all-pass filter having a specific frequency-independent attenuation based on the travelled distance between source and receiver. Atmospheric absorption can be incorporated by a low-pass filter adapted to account for air temperature and relative humidity (see Chapter 1). Low-order Infinite Impulse Response (IIR) filters might be suitable [86].

Accounting for the ground reflection is important for a good auralization. However, a filter based on the theoretically correct and distinct interferences predicted over perfectly flat uniform ground in a non-refracting atmosphere will result in the production of unrealistic sound due to 'flanging' [87,88] and give the impression of 'rasping' sound [87]. In practice, the ground-effect destructive inferences are not as pronounced as predicted over flat ground beneath a stationary homogeneous atmosphere since additional effects such as small changes in ground impedance along the path of propagation (Chapter 6), externally reacting areas of ground (Chapter 2), small scale unevenness of the ground (Chapters 5 and 6) and atmospheric turbulence (Chapter 10), combine to reduce the depth of destructive interferences and, indeed, at high frequencies, remove them altogether.

In an engineering model like ISO9613-2, smoothed ground-effect functions (Section 12.2) are used for simplicity, an advantage of which is that the aforementioned problem will be avoided. Also, the coherence loss correction

(Section 12.7) in the HARMONOISE analytical point-to-point model limits the depth of destructive interferences.

This does not mean that the more theoretical models should not be used for auralization applications, but, if they are, it is necessary to include the 'additional' propagation effects [89]. If this is done, their advantage over engineering type models is that it is possible to control the additional effects, e.g. the degree of the turbulence, and this will further increase the realism of the auralization. Also, temporal variations due to turbulence might be relevant.

Figure 12.23 shows the performance of Finite Impulse Response (FIR) filters of different orders (40 and 400 taps), tailored to sound propagation over either a rigid ground or grass-covered ground [88]. The simulations are (a) for a 0.3 m high source and 1.2 m high receiver horizontally separated by 75 m over hard ground (flow resistivity 20 MPa s m^{-2}) and (b) for a 0.3 m high source and 2 m high receiver horizontally separated by 100 m over soft ground (described using a single-parameter impedance model with effective flow resistivity 200 kPa s m^{-2}). Depending on the type of ground considered, the order of the filter may have to be adjusted for adequate spectral accuracy. In this respect, modelling propagation over rigid ground seems to be less demanding than over acoustically soft soil for the purposes of auralization.

12.12.3 Auralization of a Noise Abatement Based on a Priori Recordings

To avoid the auralization of the sources themselves, which might be especially challenging [88], a methodology has been investigated based on recordings before a noise abatement solution was put in place [90]. It concerned exposure to highway noise, for a receiver at rather close distance. A 32-channel spherical microphone array was used to record the angle-dependent contributions near the receiver. The noise reducing solution was a complex 6 m high L-shaped earth berm, showing angle-dependent insertion losses. The a priori recordings (representing the reference situation) were then filtered in accordance with the predicted angular attenuations. Among others, the HARMONOISE model and ISO9613-2 were used to make appropriate predictions. Comparison of the predicted insertion losses, combining all the angle-dependent contributions, with the measured ones after the berm was constructed, showed that while the HARMONOISE model predictions resulted in an adequate spectral agreement with data, the spectral predictions using ISO9613-2 were relatively poor [90].

The auralized highway sounds were then evaluated by a test panel. A comparison was made relative to the sound field recorded after the berm was constructed (using headphones) [90]. Despite the differences in the predicted insertion losses of the berm, both propagation models resulted in a valid impression of the future sound environment, i.e. the recorded auralized sound fields could not be discriminated (on average) by the test panel. On the other hand, when the number of channels was reduced to 2 (at the

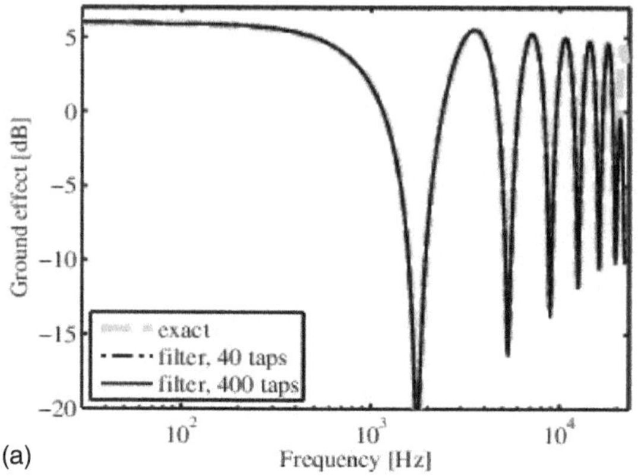

(a)

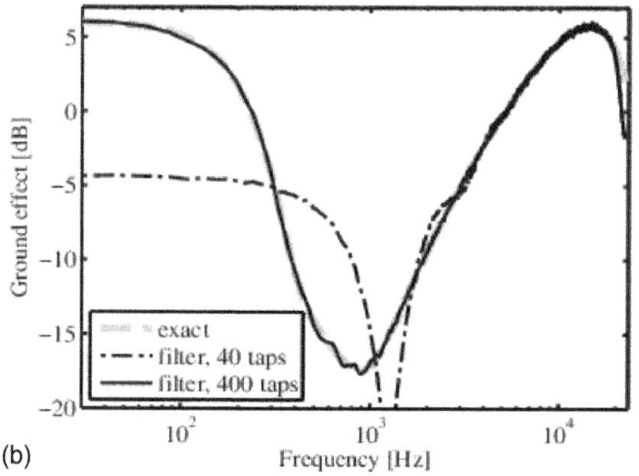

(b)

Figure 12.23 Predicted ground effect (i.e. sound level relative to free field) for a source height 0.3 m and receiver height 1.2 m (a) at a distance of 7.5 m over a rigid ground, and (b) at a distance of 100 m over a grassy ground (characterised by the Delany and Bazley model, with an effective flow resistivity of 200 kPas/m^2) assuming perfectly flat and homogeneous ground and a non-refracting, non-turbulent atmosphere. The best fitted FIR filter amplitude responses are shown as well, using a different number of orders [88]. Reproduced under open access Creative Common CC.

position of the ear), such a setup is more likely to be widely available than multi-channel recording equipment, the ISO9613-2 filtered sound fields resulted in a worse auralization performance [90].

REFERENCES

[1] ISO9613-2 Acoustics - Attenuation of sound during propagation outdoors - Part 2: A general method of calculation, (1996).

[2] J. Kragh and B. Plovsing, Nord2000. Comprehensive Outdoor Sound Propagation Model. Part I-II. DELTA Acoustics & Vibration Report, 1849-1851/00, 2000 (revised 31 December, 2001).

[3] HARMONOISE contract funded by the European Commission IST-2000-28419, https://cordis.europa.eu/project/id/IST-2000-28419 (last viewed 16/06/06), (2000).
[4] J. Kragh, B. Anderson, and J. Jakobsen, *Environmental Noise from Industrial Plants, General prediction Method* Report No. 32, Danish Acoustical Institute, Lyngby, Denmark, (1982).
[5] K. J. Marsh, The CONCAWE model for calculating the propagation of noise from open-air industrial plants, *Appl. Acoust.*, **15**: 411–428 (1982).
[6] Calculation of Road Traffic Noise, Department of Transport and Welsh Office, *HMSO* (1988).
[7] K. Attenborough et al "Benchmark cases for outdoor sound propagation models", *J. Acoust. Soc. Am.*, **97**(1): 173–191 (1995).
[8] J. Jakobsen, Testing the joint Nordic system for prediction and measurement of environmental noise from industry, *Proc. Internoise* **83** 1–4 (1983).
[9] Calculation of Railway Noise, Department of Transport, HMSO (1995).
[10] F. Brittain and P. Charalampous, Assessing Accuracy of ISO 9613-2 for Calculating Ground Effects of Stack Height and Distance Using Olive Tree Labs OTL Suite, Proc. Noise-Con **2016**, Providence, *RI* (2016).
[11] D. Notkovic, M. Stoviljkovic and S. Lloyd, Predictions of Noise Generated by Elevated Sources in Industrial Environments, *Proc. 4th International Conference on Electrical, Electronics and Computing Engineering, IcETRAN 2017*, Kladovo, Serbia, June 05-08, ISBN 978-86-7466-692-0 (2017).
[12] Ashdown Environmental Limited, Task Instruction 511/106 Working Notes for Union Railways 1995 (unpublished).
[13] W. Probst, Proposals for a revision of the International Standard ISO 9613-2, translation of a paper published in the German journal *Zeitschrift für Lärmbekämpfung Nr. 6; 2017-11* (2019).
[14] K. Hansen, G. Hessler, C. H. Hansen, B. Jazamsek, Prediction of infrasound and low frequency noise propagation for modern wind turbines – a proposed supplement to ISO 9613-2, Proc. Sixth international meeting on wind turbine noise, *Glasgow* (2015).
[15] K. Attenborough, K. M. Li and S. Taherzadeh, Propagation from a broad-band source over grassland: Comparison of data and models, Proc. Internoise 95 319–321, INCE (1995).
[16] Y. W. Lam, On the modelling of the effect of ground terrain profile in environmental noise calculations, *Appl. Acoust.*, **42**: 99–123 (1994).
[17] ESDU Data Item No. 940355, The correction of measured noise spectra for the effects of ground reflection, ESDU International plc, London (1994).
[18] K. Attenborough and K.M. Li, Ground effect for A-weighted noise in the presence of turbulence and refraction, *J. Acoust. Soc. Am.*, **102**(2): 1013–1022 (1997).
[19] The propagation of noise from petroleum and petrochemical complexes to neighbouring communities, CONCAWE Report no. 4/81, Den Haag, (1981).
[20] P. Parkin and W. E. Scholes, The horizontal propagation of sound from a jet engine close to the ground at Radlett, *J. Sound Vib.*, **1**: 1–13 (1965).
[21] P. Parkin and W. E. Scholes, The horizontal propagation of sound from a jet engine close to the ground at Hatfield, *J. Sound Vib.*, **2**: 353–374 (1965).
[22] Z. Maekawa, Noise Reduction by Screens, *Mem. Faculty of Eng., Kobe University* **11**, 29.53, (1965).

[23] R. De Jong and E. Stusnik, Scale model studies of the effects of wind on acoustic barrier performance, *Noise Control Eng.*, **6**: 101–109 (1976).
[24] R. Hetzel and J. Halbritter, Influence of temperature inversion on outdoor sound propagation – a case study, Proc. InterNoise 2013, *Innsbruck* (2013).
[25] M. E. Delany, D. G. Harland, R. A. Wood and W. E. Scholes, Prediction of noise levels L_{10} due to road traffic, *J. Sound. Vib.* **48**: 305–325 (1976).
[26] Directive of the European parliament and of the council relating to the assessment and management of noise 2002/EC/49, 25 June (2002).
[27] J. Defrance, E. Salomons, I. Noordhoek, D. Heimann, B. Plovsing, G. Watts, H. Jonasson, X. Zhang, E. Premat, I. Schmich, F. Aballea, M. Baulac, F. de Roo: Outdoor sound propagation reference model developed in the European Harmonoise project. *Acta Acust. united Ac.* **93**: 213–227 (2007).
[28] T. Van Renterghem, E. Salomons and D. Botteldooren, Efficient FDTD-PE model for sound propagation in situations with complex obstacles and wind profiles, *Acta Acust. united Ac.* **91**: 671–679 (2005).
[29] ISO 9613-1 Acoustics – attenuation of sound during propagation outdoors – Part 1. International Organisation for Standardisation, Geneva, Switzerland, (1996).
[30] E. M. Salomons, H. Zhou and W. J. A. Lohman, Efficient numerical modeling of traffic noise, *J. Acoust. Soc. Am.*, **127**: 796–803 (2010).
[31] J. B. Keller, Geometrical theory of diffraction, *J. Opt. Soc. Am.*, **52**: 116–130 (1962).
[32] Harmonoise WP2 team: Task 2.5: Modeling solutions and algorithms. TNO Report HAR25TR-021104-TNO06, November (2003).
[33] T. Van Renterghem and D. Botteldooren, Numerical evaluation of sound propagating over green roofs, *J. Sound Vib.*, **317**: 781–799 (2008).
[34] E. Premat and Y. Gabillet, A new boundary-element method for predicting outdoor sound propagation and application to the case of sound barrier in the presence of downward refraction, *J. Acoust. Soc. Am.*, **108**: 2775–2783 (2000).
[35] F. Aballéa and J. Defrance, Multiple reflection phenomena under meteorological conditions using the PE method with a complementary Kirchhoff approach, *Acta Acust. united Ac.* **93**: 22–30 (2007).
[36] E. Salomons, *Computational Atmospheric Acoustics*, Kluwer, The Netherlands, (2001).
[37] E. Premat, J. Defrance, M. Priour and F. Aballea, Coupling BEM and GFPE for complex outdoor sound propagation, *Proceedings of Euronoise* **2003**, Naples, Italy (2003).
[38] S. Hampel, S. Langer and A. Cisilino, Coupling boundary elements to a ray tracing procedure, *International Journal for numerical methods in engineering*, 73: 427-445 (2008).
[39] T. Van Renterghem, D. Botteldooren and K. Verheyen Road traffic noise shielding by vegetation belts of limited depth, *J. Sound Vib.*, **331**: 2404–2425 (2012).
[40] S. Kephalopoulos, M. Paviotti and F. Anfosso-Lédée. Master report on Common Noise Assessments Methods in Europe (CNOSSOS-EU) Outcome and Resolutions of the CNOSSOS-EU Technical Committee & Working Groups (2012).
[41] T. Van Renterghem and D. Botteldooren, Influence of temporal resolution of meteorological and traffic data on long-term average sound levels, *Acta Acust. united Ac.* **93**: 976–990 (2007).

[42] Harmonoise WP2-WP3-WP4 teams: Task 2.8, deliverable 21: Validation of the harmonoise models. *TNO Report HAR28TR-041109-TNO10.doc*, December (2004).

[43] D. van Maercke and J. Defrance, Development of an analytical model for outdoor sound propagation within the harmonoise project, *Acta Acust. united Ac.* **93**: 201–212 (2007).

[44] R. Nota, R. Barelds and D. Van Maercke, Technical Report HAR32TR-040922-DGMR20 Harmonoise WP 3 Engineering method for road traffic and railway noise after validation and fine-tuning Type of document: Technical Report - Deliverable 18 (2005).

[45] Technical advice on the preparation of the common European assessment methods to be used by the EU Member States for strategic noise mapping after adoption as specified in the Directive 2002/49/EC.

[46] J. Deygout, Multiple knife-edge diffraction by microwaves, *IEEE Trans. Antennas and Propagation*, **14**: 480–489 (1966).

[47] L. K. Leang, Y. Yamashita, M. Matsui: Simplified calculation method for noise reduction by barriers on the ground, *J. Acoust. Soc. Jpn.*, **11** 199–209 (1990).

[48] J. W. Hadden and A. D. Pierce, Sound diffraction around screens and wedges for arbitrary point source locations. *J. Acoust. Soc. Am.*, **69**: 630–637 (1981).

[49] M. Almgren, Simulation by using a curved ground scale model of outdoor sound propagation under the influence of a constant sound speed gradient, *J. Sound Vib.*, **118**: 353–370 (1987).

[50] T. F. W. Embleton, Analogies between nonflat ground and nonuniform meteorological profiles in outdoor sound propagation, *J. Acoust. Soc. Am.*, Suppl. 1 78, S86 (1985).

[51] A. Berry and G. A. Daigle, Controlled experiments of the diffraction of sound by a curved surface, *J. Acoust. Soc. Am.*, **83**: 2047–2058 (1988).

[52] G. R. Watts, Harmonoise models for predicting road traffic noise, Acoustics Bulletin, Institute of Acoustics (UK), **30**: 19–25 (2005).

[53] V. E. Ostashev, *Acoustics in moving inhomogeneous media*, E & FN Spon (an imprint of Thomson Professional), London, (1997).

[54] J. Forssén, Influence of atmospheric turbulence on sound reduction by a thin, hard screen: A parameter study using the sound scattering cross-section. *Proc. 8th Int. Symp. on Long-Range Sound Propagation*, The Pennsylvania State University, USA, 352–364 (1998).

[55] Annex to Commission Directive 2015/996 in Official Journal of the European Union L168 (2015).

[56] A. Kok, Refining the CNOSSOS-EU calculation method for environmental noise, *Proc. Inter Noise* 2019, *Madrid* (2019).

[57] H. J. A. van Leeuwen, Railway noise prediction models: A comparison, *J. Sound Vib.*, **231**: 975–987 (2000).

[58] Victor V. Krylov, *Noise and vibration from high-speed trains* (London: Thomas Telford).

[59] W. K. Lui, K. M. Li and P. L. Ng, A comparative study of different numerical models for predicting train noise in high-rise cities, *Appl. Acoust.*, **67**: 432–449 (2006).

[60] Copenhagen [Denmark]: Nordic Council of Ministers, *Railway traffic noise: the Nordic prediction method*, (1996).

[61] *Mithra Environmental Prediction Software version 5.0: User Manual*, (1998).

[62] C. H. Chew, Vertical directivity of train noise, *Appl. Acoust.*, **51**: 157–168 (1997).
[63] C. H. Chew, Vertical directivity pattern of train noise, *Appl. Acoust.*, **55**: 243–250 (1998).
[64] C. H. Chew, Directivity of train noise in the vertical plane, *Bldg. Acoust.* **16**: 41–56 (2000).
[65] T. Van Renterghem and D. Botteldooren, Landscaping for road traffic noise abatement: model validation, *Environ. Model. Softw.*, **109**: 17–31 (2018).
[66] U. Ramer, An iterative procedure for the polygonal approximation of plane curves. *Comput. Vision Graph.*, **1**: 244–256 (1972).
[67] E. Barlas, W. J. Zhu, W. Z. Shen, K. O. Dag and P. Moriarty, Consistent modelling of wind turbine noise propagation from source to receiver, *J. Acoust. Soc. Am.*, **142**: 3297–3310 (2017).
[68] D. Heimann, Y. Kaesler and G. Gross, The wake of a wind turbine and its influence on sound propagation, *Meteorologische Zeitschrift*, **20**: 449–460 (2011).
[69] E. Barlas, W. J. Zhu, W. Z. Shen, M. Kelly and S. J. Andersen, Effects of wind turbine wake on atmospheric sound propagation, *Appl. Acoust.*, **122**: 51–61 (2017).
[70] D. Heimann, A. Englberger and A. Schady, Sound propagation through the wake flow of a hilltop wind turbine - A numerical study, *Wind Energy*, **21**: 650–662 (2018).
[71] H. Møller and C. S. Pedersen, Low-frequency noise from large wind turbines, *J. Acoust. Soc. Am.*, **129**: 3727–3744 (2011).
[72] R. Makarewicz, Is a wind turbine a point source? *J. Acoust. Soc. Am.*, **129**: 579–581 (2011).
[73] K. Heutschi, R. Pieren, M. Müller, M. Manyoky, U. W. Hayek and K. Eggenschwiler, Auralization of wind turbine noise: propagation filtering and vegetation noise synthesis, *Acta Acust. united Ac.,* **100**: 13–24 (2014).
[74] S. Oerlemans, P. Sijtsma and B. Mendez Lopez, Location and quantification of noise sources on a wind turbine, *J. Sound Vib.*, **299**: 869–883 (2007).
[75] J. E. Ffowcs-Williams and L. H. Hall, Aerodynamic sound generation by turbulent flow in the vicinity of a scattering half plane, *J. Fluid Mech.*, **40**: 657–670 (1970).
[76] T. Evans and J. Cooper, Comparison of predicted and measured wind farm noise levels and implications for assessments of new wind farms. In *Proc. of ACOUSTICS* 2011, Gold Coast, Australia (2011).
[77] M. Cand, R. Davis, C. Jordan, M. Hayes and R. Perkins, A good practice guide to the application of etsu-r-97 for the assessment and rating of wind turbine noise, Institute of Acoustics, (2013).
[78] T. Van Renterghem, Sound propagation from a ridge wind turbine across a valley, *Phil. Trans. Roy. Soc. A: Mathematical, Physical and Engineering sciences*, **375**: 20160105 (2017).
[79] D. Jimenez, M. Allen and C. Nugroho, Ambisonic auralisations for community consultation of traffic noise impacts and mitigation measures, *Internoise* (2018).
[80] P. Finne and J. Fryd, Road noise auralisation for planning new roads, *Internoise* (2016).
[81] M. Southworth, The sonic environment of cities, *Environ. Behav.* **1**: 49–70 (1969).

[82] Y. Fu and D. Murphy, Spectral modelling synthesis of vehicle pass-by noise. *Internoise* (2017).
[83] J. Jagla, J. Maillard and N. Martin, Sample-based engine noise synthesis using an enhanced pitch-synchronous overlap-and-add method, *J. Acoust. Soc. Am.*, **132**: 3098–3108 (2012).
[84] A. Southern and D. Murphy, *A framework for road traffic noise auralisation.* Euronoise (2015).
[85] G. Zachos, J. Forssen, W. Kropp and L. Estevez-Mauriz, Background traffic noise synthesis, *Internoise* (2016).
[86] J. Huopaniemi, L. Savioja and M. Karjalainen, Modeling of reflections and air absorption in acoustical spaces - a digital filter design approach, *Proceedings of 1997 Workshop on Applications of Signal Processing to Audio and Acoustics.*
[87] M. Arntzen and D. Simons, Modeling and synthesis of aircraft flyover noise, *Appl. Acoust.*, **84**: 99–106 (2014).
[88] R. Pieren, T. Bütler and K. Heutschi, Auralization of Accelerating Passenger Cars Using Spectral Modeling Synthesis, *Appl. Sciences* **6**: 5 (2016).
[89] M. Arntzen and D. Simons, Ground reflection with turbulence induced coherence loss in flyover auralization, *Aeroacoust.*, **13**: 449–462 (2014).
[90] P. Thomas, W. Wei, T. Van Renterghem, D. Botteldooren, Measurement-based auralization methodology for the assessment of noise mitigation measures, *J. Sound Vib.*, **379**: 232–244 (2016).

Index

Note: Numerals in *italics* indicate figures.

G

H

I

J

P

Q

R

S